ESSENTIALS OF PHYSICAL GEOGRAPHY

SEVENTH EDITION

Robert E. Gabler
Emeritus, Western Illinois University

James F. Petersen
Southwest Texas State University

L. Michael Trapasso
Western Kentucky University

Contributing Authors

Robert J. Sager
Pierce College, Washington

Daniel L. Wise
Western Illinois University

THOMSON
™
BROOKS/COLE

Australia • Canada • Mexico • Singapore • Spain • United Kingdom • United States

THOMSON
™
BROOKS/COLE

Earth Sciences Editor: Keith Dodson
Development Editor: Marie Carigma-Sambilay
Assistant Editor: Carol Benedict
Editorial Assistant: Melissa Newt
Technology Project Manager: Samuel Subity
Marketing Manager: Kelley McAllister
Marketing Assistant: Sandra Perin
Advertising Project Manager: Nathaniel Bergson-Michelson
Project Manager, Editorial Production: Tom Novak
Print/Media Buyer: Kristine Waller
Permissions Editor: Joohee Lee
Production Service: Ben Kolstad, The Cooper Company
Text Designer: Patrick Devine Design
Photo Researcher: Terri Wright

Copy Editor: Margaret C. Tropp
Illustrators: Accurate Art, Rolin Graphics
Cover Designer: Irene Morris
Cover Image: Yosemite National Park by Kerrick James, Getty Images
Compositor: Thompson Type
Printer: Transcontinental Printing/Interglobe
Title Page Image: Getty Images
Image on p. 3 by R.B. Husar, Washington University; the land layer from the Sea WiFs Project; fire maps from the European Space Agency; the sea surface temperature from the Naval Oceanographic Office's Visualization laboratory; and cloud layer from SSEC, University of Wisconsin

Printed in Canada
1 2 3 4 5 6 7 07 06 05 04 03

For more information about our products, contact us at:
Thomson Learning Academic Resource Center
1-800-423-0563
For permission to use material from this text, contact us by:
Phone: 1-800-730-2214
Fax: 1-800-730-2215
Web: http://www.thomsonrights.com

Library of Congress Control Number: 2002116966

Student Edition: ISBN 0-03-033822-0
Instructor's Edition: ISBN 0-534-40783-8

Brooks/Cole–Thomson Learning
10 Davis Drive
Belmont, CA 94002
USA

Asia
Thomson Learning
5 Shenton Way #01-01
UIC Building
Singapore 068808

Australia/New Zealand
Thomson Learning
102 Dodds Street
Southbank, Victoria 3006
Australia

Canada
Nelson
1120 Birchmount Road
Toronto, Ontario M1K 5G4
Canada

Europe/Middle East/Africa
Thomson Learning
High Holborn House
50/51 Bedford Row
London WC1R 4LR
United Kingdom

Latin America
Thomson Learning
Seneca, 53
Colonia Polanco
11560 Mexico D.F.
Mexico

Spain/Portugal
Paraninfo
Calle/Magallanes, 25
28015 Madrid, Spain

Earth, the life-support system for humankind, is a wondrous and complex place, yet it is also enigmatic. Our planet is in many ways robust and adaptable to environmental change, but in other ways it is fragile and threatened. In our modern day-to-day life, our use of technology has a tendency to insulate us from fully experiencing our environment, and we have become lulled into forgetting about our direct reliance on Earth's natural resources. Sometimes it is hard to imagine that as you read this you are also moving through space on a life-giving planetary oasis surrounded by the vastness of space—empty and, as far as we know, devoid of life. Yet a knowledge of our planet, its environments, and how they operate is as critical as it has ever been for human society.

For as long as humans have existed on Earth, the resources provided by their physical environment have been the key to their survival. Primitive, preindustrial societies, such as those dependent on hunting and gathering or small-scale agriculture, tended to have small populations and exerted relatively little impact on their natural surroundings. In contrast, today's industrialized societies have large populations, demand huge quantities of natural resources, and can influence or cause environmental change, not only locally, but also globally. As the human population has increased, so have the scale, degree, and cumulative effects of its impacts on the environment. Unfortunately, many of these impacts are detrimental. We have polluted the air and water. We have used up tremendous amounts of nonrenewable resources and have altered the natural landscape without fully assessing the consequences. Too often, we have failed to respect the power of Earth's natural forces when constructing our homes and cities or while pursuing our economic activities. In the twenty-first century, it is evident that if we fail to comprehend Earth's potential and respect its limitations as a human habitat, we may be putting ourselves and future generations at risk. Despite the many differences between our current life-styles and those of early humans, one aspect has remained the same: Our use of our physical environment in all its diversity is still the key to our survival.

Today, this important message is getting through. We understand that Earth has limits and does not offer boundless natural resources. Broadcast and print media have expanded their coverage of the environment and human/environment relationships. Politicians are drafting legislation designed to address environmental problems. Scientists and statesmen from around the world meet to discuss environmental issues that increasingly cross international boundaries. Humanitarian organizations, funded by governments as well as private citizens, struggle to alleviate the suffering that results when human beings cannot cope with the limitations of their physical surroundings or when they wittingly or unwittingly degrade the environments in which they live.

The study of physical geography is essential to developing a thorough knowledge, understanding, and appreciation of Earth. This study renders invaluable aid to the people who occupy, alter, and rely upon our planet for their continued existence. In this textbook the study is recognized as a three-dimensional examination of planet Earth and its varied physical environments. As a spatial science, physical geography focuses on the location, distribution, and spatial interactions of all natural phenomena on Earth. As a physical science, the subject incorporates the techniques of models and systems research, as well as the accumulated knowledge of the companion natural and physical sciences as Earth processes and patterns are examined. As an environmental science, physical geography plays a special role because all geographers have a mandate to give equal consideration to both parties in any human-environment relationship.

Long a highly regarded subject in most nations of the world, geography in the last 25 years has undergone a renaissance in the United States. More and more people are realizing the importance of knowing and respecting the people, cultural contributions, and environmental resources of other nations in addition to those of the United States. National standards that include physical geography have been written to ensure that a high quality geography curriculum is offered in United States elementary, secondary, and postsecondary schools. Employers across the nation are recognizing the value and importance of geographic knowledge, skills, and techniques in the workplace. Physical geography as an applied field makes use of computer-assisted and space-age technologies, such as interpretation of satellite images, geographic information systems (GIS), computer cartography, and the global positioning system. At the collegiate level, physical geography offers an introduction to the concerns, ideas, knowledge, and tools that are necessary for further study of our planet. More than ever before, physical geography is being recognized as an ideal science course for general-education students—students who will make decisions that weigh human needs and desires against environmental limits and possibilities. It is for these students that *Essentials of Physical Geography* has been written.

Features

Comprehensive View of the Earth System *Essentials of Physical Geography* introduces all major aspects of the Earth system, identifying physical phenomena and stressing their distribution and relationships. It covers a wide range of topics, including the atmosphere, oceans and other water bodies, the solid Earth, and the patterns of life on our planet.

Clear Explanation The text uses an understandable and often narrative style to explain the origins, development, significance, and distribution of processes, physical features, and events that occur within, on, or above the surface of Earth. The author's writing style is targeted toward rapid reader comprehension and toward making the study of physical geography an enjoyable undertaking.

Introduction to the Geographer's Tools   The space age and the computer have revolutionized the ways that we can study the environments of our planet. The text provides a full chapter devoted to the tools used by geographers, and illustrations throughout the book include many examples of images gathered from space, accompanied by interpretations of those aspects of the environment that the scenes illustrate. There are also introductory discussions of techniques used by geographers in analyzing or displaying locational and environmental aspects of Earth, including remote sensing, geographic information systems, computer-assisted cartography, and the global positioning system.

Focus on Student Interaction   The text provides numerous ways to promote continuous interaction between students and the subject matter of physical geography. The Consider and Respond activities at the end of each chapter are designed to encourage students to apply their newly acquired knowledge in different and, when possible, problem-solving situations. Questions following the captions of most illustrations prompt students to think beyond the map, graph, diagram, or photograph and further consider the subject under consideration.

Map Interpretation Series   Learning map-interpretation skills is a major priority in a physical geography course. To meet the needs of students who do not have the opportunity to interact with physical geography in a laboratory setting, the text includes an organized program of map exercises withaccompanying explanations, illustrations, and interpretation questions. This popular and dynamic feature gives all students the opportunity to learn and practice valuable map skills.

Objectives

Since the first edition, the authors of this book have sought to accomplish several major objectives:

To Address the Academic Needs of the Student   Those instructors who are familiar with the style and content of *Essentials of Physical Geography* know that this is one textbook written specifically for the student. It has been designed to help meet the major purposes of a liberal education by providing students with sufficient knowledge and understanding to make informed decisions involving the physical environments with which they will interact throughout their lives. The text assumes little or no background in physical geography or other Earth sciences on the part of introductory-level students. The authors include numerous examples from

throughout the world to illustrate difficult concepts and help students bridge the gap between scientific theory and its practical application.

To Integrate the Illustration Program with the Text   Numerous photographs, block diagrams, and other line drawings have been carefully chosen to illustrate each of the complex concepts in physical geography the text addresses. Text discussions of these concepts often contain repeated references to the illustrations, and the student is able to examine in model form, as well as mentally reconstruct, the physical processes and phenomena involved. Particularly good examples of the integration of text and illustration include the sections on seasons (Chapter 3), heat energy budget (Chapter 4), surface wind systems (Chapter 5), atmospheric disturbances (Chapter 7), plate tectonics (Chapter 13), and ice sheets (Chapter 19).

To Communicate the Nature of Geography   The nature of geography and three roles of geography (as the spatial science, as a physical science, and as environmental science) are discussed at length in Chapter 1. In subsequent chapters, important topics of geography as the spatial science are used as the organizational bases for chapter content. As examples, location is the major topic of Chapter 2 and remains an important ingredient throughout the text; spatial distributions are emphasized as each of the climatic elements are discussed in Chapters 4 to 6; a changing Earth system is the central consideration in Chapter 8; characteristics of places constitute the content of Chapters 9 and 10; spatial interactions are frequently demonstrated in discussions of weather systems (Chapter 7) and tectonic activity (Chapters 13 and 14). As a part of the course summary, all of the topics are revisited in Chapter 22.

To Fulfill the Major Requirements of Introductory Physical Science College Courses   *Essentials of Physical Geography* offers a full chapter on the important tools and the methodologies of physical geography. Throughout the book, the physical processes that are responsible for the location, distribution, and spatial relationships of physical phenomena beneath, at, and above Earth's surface are examined in detail. Scientific method, hypothesis, theory, and explanation are continually stressed. Models and systems are frequently cited in the discussion of important concepts, and scientific classification is presented in several chapters—specifically, air masses (Chapter 7), climates (Chapters 8 and 9), biomes (Chapter 11), soils (Chapter 12), and coasts (Chapter 21).

Seventh Edition Revision

The challenge of revising *Essentials of Physical Geography* for a seventh edition turned into a delicate balancing act. Several reviewers recommended a slight reduction in chapter length and number, with chapters focused more directly on specific topics. Others urged expanded coverage in critical areas of physical geography that have recently sparked increased interest on the part of students. The authors had already agreed that the time had come to clearly demonstrate that physical geography could open the door to an interesting, exciting, and highly rewarding future for those individuals who decide to take advantage of its career opportunities. The challenge was to accomplish all of this in addition to thoroughly revising the text; preparing new graphs, maps, and diagrams; selecting dozens of new photographs; and updating the numerous examples of worldwide environmental events that keep physical geography current and topical. What follows is a brief review of major changes incorporated in the seventh edition in an attempt to meet the challenge.

New Coauthor   The seventh edition of *Essentials of Physical Geography* is privileged to have L. Michael Trapasso of Western Kentucky University as a new coauthor. Michael's expertise in geographic education, climatology, and applied physical geography has been an invaluable asset to this project. His work throughout the book has allowed us to expand in a number of areas, which is sure to benefit students and teachers alike.

Chapter Reorganization   The number of chapters has been adjusted to 22. The fundamentals of Earth-sun relationships are covered in a separate chapter preceding the discussion of Earth's energy budget. Atmospheric pressure, wind systems, and ocean currents have been combined and reorganized in a chapter on global circulation. The reorganization allows the concept of circulation to progress from smaller to large scales. Climate classification and climate change are treated together in the same chapter. Groundwater is covered in a separate chapter, and the erosional processes associated with streams in both humid and arid regions are now discussed together in a chapter entirely devoted to fluvial geomorphology. Arid landforms and eolian processes are treated in a single chapter which now precedes the chapter on glacial systems. The summary chapter at the end of the textbook should help students reflect on their experiences with a course in physical geography. The authors believe these changes will provide increased course flexibility and that they have been accomplished without altering significantly the original sequence of topics or forcing instructors to make major changes in syllabi.

New and Revised Text   In addition to the text revision and chapter reorganization, major new material has been added on a variety of topics. These include geography as spatial, physical, and environmental science (Chapter 1), GPS, GIS, and remote sensing (Chapter 2), Earth in space (Chapter 3), upper air circulation (Chapter 5), weather systems (Chapter 7), soil-forming processes (Chapter 12), tectonic landforms (Chapter 14), fluvial systems (Chapter 17), and glacial systems (Chapter 19).

Enhanced Illustrations   The illustration program has undergone thorough revision for the seventh edition. The new text required new figures, which resulted in the drafting of many graphs and line drawings, the addition of numerous impressive photographs and satellite images, and the development of new maps treating subjects such as climate, soils, and landforms. In addition, hundreds of figures have been replaced by new photographs and new or revised line drawings. It is anticipated that instructors and students alike will be impressed by the increased number, the selection, and the quality of reproduction of the illustrations in the seventh edition.

Increased Focus on the Discipline   Several changes in the text have been introduced in the seventh edition to provide students with a better understanding and appreciation of geography as a discipline worthy of continued study and meriting serious consideration as a career choice. The definition of geography, the discipline's tools and methodologies, selected topics to illustrate the role of geography as the spatial science, and the practical applications of the discipline are all introduced in Chapter 1. In addition, the authors believe there are few better ways to understand an academic discipline than to learn firsthand from its practitioners. To that end, the Career Vision series has been continued in the seventh edition to provide a close look at geographers in the workplace, how they earn their living, and what educational programs prepared them for their occupations. These brief interviews serve to inform students of the essential knowledge and skills associated with the exciting new career opportunities offered those who choose physical geography as their preferred field of study.

The seventh edition also introduces a new feature in support of the geography discipline in its three-fold roles among the sciences. A series of boxed articles illustrates each role with an average of two articles for each chapter. The articles titled "Geography as the Spatial Science" demonstrate where geography differs from the other sciences. The nature of geography mandates a study of how phenomena are distributed in space. To this end, the spatial point of view becomes that of the geographer. Articles in the "Geography as a Physical Science" series emphasize geography as a synthesis and applied representative of all the major sciences. In these articles, physics, chemistry, biology, geology, and mathematics are used to highlight selected aspects or questions derived from associated chapter text. "Geography as Environmental Science" reminds the students that our physical environment comprises Earth, water, vegetation, and the atmosphere, all of which are essential components of physical geography. Thus, geography plays a central role in an understanding of environmental issues and in the solution of environmental problems.

A Look at the Future   The inclusion of Chapter 22 in the seventh edition has provided authors with more than just a chance to summarize the important knowledge and valuable perspectives about our planet gained from the course. It also provides students with an opportunity to imagine the future role of physical geography in their lives. No student should put *Essentials of Physical Geography* aside without reading the last chapter.

Ancillaries

Instructors and students alike will greatly benefit from the comprehensive ancillary package that accompanies this text.

Available Free with Each Copy of the Text

Blue Skies™ College Edition CD-ROM   This engaging and Internet-integrated software is an interactive exploration of the dynamic world of weather and climate. *Blue Skies* allows students to fully explore and appreciate the diverse phenomena of the dynamic world of atmospheric science.

Earth Systems Today CD-ROM   This innovative and interactive software provides a solid introduction to the physical processes that shape Earth. Students can grasp of the relevance of concepts by manipulating variables and watching the results. *Earth Systems Today* offers hands-on activities with simulations, animations, and live data.

InfoTrac® College Edition   The Seventh Edition is packaged with four months of FREE access to this extensive online library, offering the latest news and research articles online. A dynamic feature has been added to *InfoTrac College Edition: NewsEdge,* an authoritative source of late-breaking local, national, and international news of interest to you and your students.

The Brooks/Cole Earth Science Resource Center
http://earthscience.brookscole.com
This outstanding site features student and instructor resources, including links to other relevant sites.

Book Companion Web Site
http://info.brookscole.com/gabler
For the Seventh Edition, you'll find a detailed and student-focused Web site that includes quizzes, links to other sites, a glossary, *InfoTrac College Edition* activities, Internet exercises, learning objectives, chapter outlines, and more!

For the Instructor

Instructor's Manual with Test Bank and Lab Pack   This comprehensive manual contains teaching suggestions as well as evaluation resources such as course syllabi, lists of key concepts, chapter outlines, chapter notes, answers to review questions, and recommended readings. The *Lab Pack* is included for instructors' use.

Transparency Acetates   An updated collection of 100 acetates that include full-color images from the text.

ExamView® Computerized Testing   Create, deliver, and customize tests and study guides (both print and online) in minutes with this easy-to-use assessment and tutorial system. *ExamView* offers both a *Quick Test Wizard* and an *Online Test Wizard*. You can build tests of up to 250 questions using up to 12 question types. Using *ExamView's* complete word processing capabilities, you can enter an unlimited number of new questions or edit existing questions.

WebTutor™ ToolBox for Web CT and Blackboard   Preloaded with content and available **free** via PIN code when packaged with this text, *WebTutor Toolbox* pairs all the content of the Companion Web Site with all the sophisticated functionality of a WebCT or Blackboard product. Toolbox comes ready to use or you can customize its preloaded content. Students only have access to student resources on the Web site. Contact your Thomson–Brooks/Cole representative for information on packaging *WebTutor ToolBox* with this text.

Multimedia Manager: A Microsoft® PowerPoint® Link Tool   Multimedia lecture tool allows you to quickly assemble art and database files with notes to create fluid lectures. Available on CD, the *Multimedia Manager* includes a database of animations and images from Brooks/Cole Earth Sciences titles. The simple interface makes it easy for you to incorporate graphics, digital video, animations, and audio clips into your lectures.

MyCourse 2.1
http://mycourse.thomsonlearning.com   A **FREE** online course builder, with easy-to-use tools to build it and content to furnish it. A simple solution for a custom course Web site. MyCourse 2.1 allows you to assign, track, and report on student progress; load your syllabus; and more.

For the Student

Study Mate   ISBN: 0-03-033846-8
An essential tool for any student, the *Study Guide* offers a *Chapter Summary* and a *Chapter Guide* for each of the 22 chapters of *Essentials of Physical Geography, Seventh Edition*. Both sections provide a broad overview of chapter content and help the student identify the most important facts, concepts, and underlying explanations. Supplementary reading lists, Web sites, and practice exam questions are included for each chapter.

Lab Pack   ISBN: 0-03-033868-9
The perfect lab complement to the text, this *Lab Pack* contains over 50 exercises, varying in length and difficulty, designed to help students achieve a greater understanding and appreciation of physical geography.

GIS Investigations   Michelle K. Hall-Wallace, C. Scott Walker, Larry P. Kendall, Christian J. Schaller, and Robert F. Butler of the University of Arizona, Tucson.

Exploring the Dynamic Earth:
GIS Investigations for the Earth Sciences
ISBN: 0-534-39138-9

Exploring Tropical Cyclones:
GIS Investigations for the Earth Sciences
ISBN: 0-534-39147-8

Exploring Water Resources:
GIS Investigations for the Earth Sciences
ISBN: 0-534-39156-7

The perfect accompaniment to any physical geography course, these three groundbreaking guides tap the power of *ArcView*® GIS to explore, manipulate, and analyze large data sets. The guides emphasize the visualization, analysis, and multimedia integration capabilities inherent to GIS and enable students to "learn by doing" with a full complement of GIS capabilities. The guides contain all the software and data sets needed to complete the exercises.

InfoTrac® College Edition Student Guide for Earth Science   ISBN: 0-534-24742-3
This 24-page booklet provides detailed user guides for students illustrating how to use the *InfoTrac College Edition* database. Special features include log-in help, a complete search tips cheat sheet, and a topic list of suggested keywords for Earth science.

Opposing Viewpoints Resource Center
http://infotrac.galegroup.com/itweb/wadsworth
The *Opposing Viewpoints Resource Center* draws on Greenhaven Press's acclaimed social issues series, as well as core reference content from other Gale and Macmillan Reference USA sources. A dynamic online library of current event topics—the facts as well as the arguments of each topic's proponents and detractors. For more information, contact your Thomson–Brooks/Cole representative or visit http://www.gale.com/opposingviewpoints/ and take a guided tour!

Acknowledgments

This edition of *Essentials of Physical Geography* would not have been possible without the encouragement and assistance of editors, friends, and colleagues from throughout the country. Great appreciation is extended to Sarah Gabler, and Martha, Emily, and Hannah Petersen for their patience, support, and understanding.

Special thanks go to the splendid staff members of Brooks/Cole Thomson Learning. These include Keith Dodson, Earth Sciences Editor; Marie Carigma-Sambilay, Development Editor; Carol Benedict, Assistant Editor; Samuel Subity, Technology Project Manager; Kelley McAllister, Marketing Manager; Tom Novak and Hal Humphrey, Project Managers; Patrick Devine, Designer; Terri Wright, Photo Researcher; Accurate Art, Illustrator; Ben Kolstad, The Cooper Company, Production Service; and Melissa Newt, Editorial Assistant.

Colleagues who reviewed the plans and manuscript for this and previous editions of this text include: Brock Brown, Southwest Texas State University; Leland R. Dexter, Northern Arizona University; Roberto Garza, San Antonio College; Perry Hardin, Brigham Young University; David Helgren, San Jose State University; Jeffrey Lee, Texas Tech University; John Lyman, Bakersfield College; Fritz C. Kessler, Frostburg State University; Michael E. Lewis, Univerity of North Carolina, Greensboro; Charles Martin, Kansas State University; James R. Powers, Pasadena City College; Joyce Quinn, California State University, Fresno; Colin Thorn, University of Illinois at Urbana-Champaign; Dorothy Sack, Ohio University; George A. Schnell, State University of New York, New Paltz; David L. Weide, University of Nevada, Las Vegas; Thomas Wikle, Oklahoma State University, Stillwater; Amy Wyman, University of Nevada, Las Vegas; Craig ZumBrunnen, University of Washington, Redmond. Reviewers of the fluvial chapter: Joanna Crowe, Richard Earl, and Mark Fonstad, all of Southwest Texas State University.

Photos courtesy of: Center for Cave and Karst Studies, Western Kentucky University; Richard Hackney, Western Kentucky University; Susan Jones, Nashville, Tennessee; Bob Jorstad, Eastern Illinois University; Carter Keairns, Southwest Texas State University; Parris Lyew-Ayee, Oxford University, UK; Kamlesh Lulla, Chief, Earth Sciences, NASA Johnson Space Center; L. Elliot Jones, U.S. Geological Survey; Dorothy Sack, Ohio University; Justin Wilkinson, Earth Sciences, NASA Johnson Space Center.

The detailed comments and suggestions of all of the above individuals have been instrumental in bringing about the many changes and improvements incorporated in this latest revision of the text. Countless others, both known and unknown, deserve heartfelt thanks for their interest and support over the years.

Despite the painstaking efforts of all reviewers, there will always be questions of content, approach, and opinion associated with the text. The authors wish to make it clear that they accept full responsibility for all that is included in the seventh edition of *Essentials*.

Robert E. Gabler
James F. Petersen
L. Michael Trapasso

Why Study Geography?

In this global age, the study of geography is absolutely essential to an educated citizenry of a nation whose influence extends throughout the world. Geography deals with location, and a good sense of where things are, especially in relation to other things in the world, is an invaluable asset whether you are traveling, conducting international business, or sitting at home reading the newspaper.

Geography examines the characteristics of all the various places on Earth and their relationships. Most important in this regard, geography provides special insights into the relationships between humans and their environments. If all the world's people had one goal in common, it should be to better understand the physical environment and protect it for the generations to come.

Geography provides essential information about the distribution of things and the interconnections of places. The distribution pattern of Earth's volcanoes, for example, provides an excellent indication of where Earth's great crustal plates come in contact with one another; and the violent thunderstorms that plague Illinois on a given day may be directly associated with the low pressure system spawned in Texas two days before. Geography, through a study of regions, provides a focus and a level of generalization that allows people to examine and understand the immensely varied characteristics of Earth.

As you will note when reading Chapter 1, there are many approaches to the study of geography. Some courses are regional in nature; they may include an examination of one or all of the world's political, cultural, economic, or physical regions. Some courses are topical or systematic in nature, dealing with human geography, physical geography, or one of the major subfields of the two.

The great advantage to the study of a general course in physical geography is the permanence of the knowledge learned. Although change is constant and is often sudden and dramatic in the human aspects of geography, alterations of the physical environment on a global scale are exceedingly slow when not influenced by human intervention. Theories and explanations may differ, but the broad patterns of atmospheric and oceanic circulation and of world climates, landforms, soils, natural vegetation, and physical landscapes will be the same tomorrow as they are today.

Keys to Successful Study

Good study habits are essential if you are to master science courses such as physical geography, where the topics, explanations, and terminology are often complex and unfamiliar. To help you succeed in the course in which you are currently enrolled, we offer the following suggestions.

Reading Assignments

- Read the assignments before the material contained therein is covered in class by the instructor.
- Compare what you have read with the instructor's presentation in class. Pay particular attention if the instructor introduces new examples or course content not included in the reading assignment.
- Do not be afraid to ask questions in class and seek a full understanding of material that may have been a problem during your first reading of the assignment.
- Reread the assignment as soon after class as possible, concentrating on those areas that were emphasized in class. Highlight only those items or phrases that you now consider to be important, and pass over lightly those sections that are already mastered.
- Add to your class notes important terms, your own comments, and summarized information from each reading assignment.

Understanding Vocabulary

Mastery of the basic vocabulary often becomes the critical issue in the success or failure of the student in the beginning science course.

- Focus on the terms that appear in boldface type in your reading assignments. Do not overlook any additional terms that the instructor may introduce in class.
- Develop your own definition of each term or phrase and associate it with other terms in physical geography.
- Identify any physical processes associated with the term. Knowing the process helps to define the term.
- Whenever possible, associate terms with location.
- Consider the significance to humans of terms you are defining. Recognizing the significance of terms and phrases can make them relevant and easier to recall.

Learning Earth Locations

A good knowledge of place names and of the relative locations of physical and cultural phenomena on Earth is fundamental to the study of geography.

- Take personal responsibility for learning locations on Earth. Your instructor may identify important physical features and place names, but you must learn their locations for yourself.
- Thoroughly understand latitude, longitude, and the Earth grid. They are fundamental to location on maps as well as on a globe. Practice locating features by their latitude and longitude until you are entirely comfortable using the system.

- Develop a general knowledge of the world political map. The most common way of expressing the location of physical features is by identifying the political unit (state, country, or region) in which it can be found.
- Make liberal use of outline maps. They are the key to learning the names of states and countries and they can be used to learn the locations of specific physical features. Personally placing features correctly on an outline map is still the best way to learn location.
- Cultivate the atlas habit. The atlas does for the individual who encounters place names or the features they represent what the dictionary does for the individual who encounters a new vocabulary word.

Utilizing Textbook Illustrations

The secret to making good use of maps, diagrams, and photographs lies in understanding why the illustration has been included in the text or incorporated as part of your instructor's presentation.

- Concentrate on the instructor's discussion. Taking notes on slides, overhead transparencies, and illustrations will allow you to follow the same line of thought at a later date.
- Study all textbook illustrations on your own. Be sure to note which were the focus of considerable classroom attention. Do not quit your examination of an illustration until it makes sense to you, until you can read the map or graph, or until you can recognize what a diagram or photograph has been selected to explain.
- Hand-copy important diagrams and graphs. Few of us are graphic artists, but you might be surprised at how much better you understand a graph or line drawing after you reproduce it yourself.
- Read the captions of photos and illustrations thoroughly and thoughtfully. If the information is included, be certain to note where a photograph was taken and in what way it is representative. What does it tell you about the region or site being illustrated?
- Attempt to place the principle being illustrated in new situations. Seek other opportunities to test your skills at interpreting similar maps, graphs, and photographs and think of other examples that support the text being illustrated.
- Remember that all illustrations are reference tools, particularly tables, graphs, and diagrams. Refer to them as often as you need to.

Taking Class Notes

The password to a good set of class notes is selectivity. You simply cannot and, indeed, you should not try to write down every word uttered by your classroom instructor.

- Learn to paraphrase. With the exception of specific quotations or definitions, put the instructor's ideas, explanations, and comments into your own words. You will understand them better when you read them over at a later time.
- Be succinct. Never use a sentence when a phrase will do, and never use a phrase when a word will do. Start your recall process with your note-taking by forcing yourself to rebuild an image, an explanation, or a concept from a few words.
- Outline where possible. Preparing an outline helps you to discern the logical organization of information. As you take notes, organize them under main headings and subheadings.
- Take the instructor at his or her word. If the instructor takes the time to make a list, then you should do so too. If he or she writes something on the board, it should be in your notes. If the instructor's voice indicates special concern, take special notes.
- Come to class and take your own notes. Notes trigger the memory, but only if they are your notes.

Doing Well on Tests

Follow these important study techniques to make the most of your time and effort preparing for tests.

- Practice distillation. Do not try to reread but skim the assignments carefully, taking notes in your own words that record as economically as possible the important definitions, descriptions, and explanations. Do the same with any supplementary readings, handouts, and laboratory exercises. It takes practice to use this technique, but it is a lot easier to remember a few key phrases that lead to ever increasing amounts of organized information than it is to memorize all of your notes. And the act of distillation in itself is a splendid memory device.
- Combine and reorganize. Merge all your notes into a coherent study outline.
- Become familiar with the type of questions that will be asked. Knowing whether the questions will be objective, short-answer, essay, or related to diagrams and other illustrations can help in your preparation. Some instructors place old tests on file where you can examine them or will forewarn you of their evaluation styles if you inquire. If not, then turn to former students; there are usually some around the department or residence halls who have already experienced the instructor's tests.
- Anticipate the actual question that will likely be on the test. The really successful students almost seem to be able to predict the test items before they appear. Take your educated guesses and turn them into real questions.

- Try cooperative study. This can best be described as role playing and consists very simply of serving temporarily as the instructor. So go ahead and teach. If you can demonstrate a technique, illustrate an idea, or explain a process or theory to another student so that he or she can understand it, there is little doubt that you can answer test questions over the same material.

- Avoid the "all-nighter." Use the early evening hours the night before the test for a final unhurried review of your study outline. Then get a good night's sleep.

The Importance of Maps

Like graphs, tables, and diagrams, maps are an excellent reference tool. Familiarize yourself with the maps in your textbook in order to better judge when it is appropriate to seek information from these important sources.

Maps are especially useful for comparison purposes and to illustrate relationships or possible associations of things. But the map reader must beware. Only a small portion of the apparent associations of phenomena in space (areal associations) are actually cause-and-effect relationships. In some instances the similarities in distribution are a result of a third factor that has not been mapped. For instance, a map of world-wide volcano distribution is almost exactly congruent with one of incidence of earthquakes, yet volcanoes are not the cause of earthquakes, nor is the obverse true. A third factor, the location of tectonic plate boundaries, explains the first two phenomena.

Finally, remember that the map is the most important statement of the professional geographer. It is useful to all natural and social scientists, engineers, politicians, military planners, road builders, farmers, and countless others, but it is the essential expression of the geographer's primary concern with location, distribution, and spatial interaction.

About Your Textbook

This textbook has been written for you, the student. It has been written so that the text can be read and understood easily. Explanations are as clear, concise, and uncomplicated as possible. Illustrations have been designed to complement the text and to help you visualize the processes, places, and phenomena being discussed. In addition, the authors do not believe it is sufficient to offer you a textbook that simply provides information to pass a course. We urge you to think critically about what you read in the textbook and hear in class.

As you learn about the physical aspects of Earth environments, ask yourself what they mean to you and to your fellow human beings throughout the world. Make an honest attempt to consider how what you are learning in your course relates to the problems and issues of today and tomorrow. Practice using your geographic skills and knowledge in new situations so that you will continue to use them in the years ahead. Your textbook includes several special features that will encourage you to go beyond memorization and reason geographically.

Consider and Respond   At the end of each chapter, Consider and Respond questions require you to go well beyond routine chapter review. The questions are designed specifically so that you may apply your knowledge of physical geography and on occasion personally respond to critical issues in society today. Check with your instructor for answers to the problems.

Caption Questions   With almost every illustration and photo in your textbook a caption links the image with the chapter text it supports. Read each caption carefully because it explains the illustration and may also contain new information. Wherever appropriate, questions at the ends of captions have been designed to help you seize the opportunity to consider your own personal reaction to the subject under consideration.

Map Interpretation Series   It is a major goal of your textbook to help you become an adept map reader, and the Map Interpretation Series in your text has been designed to help you reach that goal.

Environmental Systems Diagrams   Viewing Earth as a system comprising many subsystems is a fundamental concept in physical geography for researchers and instructors alike. The concept is introduced in Chapter 1 and reappears frequently throughout your textbook. The interrelationships and dependencies among the variables or components of Earth systems are so important that a series of special diagrams (see, for example, Figure 1.7) have been included with the text to help you visualize how the systems work. Each diagram depicts the system and its variables and also demonstrates their interdependence and the movement or exchanges that occur within each system. The diagrams are designed to help you understand how human activity can affect the delicate balance that exists within many Earth systems.

As authors of your textbook, we wish you well in your studies. It is our fond hope that you will become better informed about Earth and its varied environments and that you will enjoy the study of physical geography.

Brief Contents

Contents

List of Major Maps

Robert E. Gabler During his nearly five decades of professional experience, Professor Gabler has taught geography at Hunter College, City of New York, Columbia University, and Western Illinois University, in addition to five years in public elementary and secondary schools. At times in his career at Western he served as Chairperson of the Geography and Geology Department, Chairperson of the Geography Department, and University Director of International Programs. He received three University Presidential Citations for Teaching Excellence and University Service, served two terms as Chairperson of the Faculty Senate, edited the *Bulletin of the Illinois Geographical Society,* and authored numerous articles in state and national periodicals. He is a Past President of the Illinois Geographical Society, former Director of Coordinators and Past President of the National Council for Geographic Education, and the recipient of the NCGE George J. Miller Distinguished Service Award.

James F. Petersen James F. Petersen is Professor of Geography at Southwest Texas State University, in San Marcos, Texas. He is a broadly trained physical geographer with strong interests in geomorphology and Earth Science education. He enjoys writing about topics relating to physical geography for the public, particularly environmental interpretation, and has written a landform guidebook for Enchanted Rock State Natural Area in Central Texas and a number of field guides. He is a strong supporter of geographic education, having served as President of the NCGE in 2000 after more than 15 years of service to that organization. He has also written or served as a senior consultant for nationally published educational materials at levels from middle school through university, and has done many workshops for geography teachers. Most recently, he co-edited a volume of the journal *Geomorphology,* with a collection of papers dealing with education in geomorphology that represented the proceedings from the Binghamton Geomorphology Symposium.

L. Michael Trapasso L. Michael Trapasso is Professor of Geography at Western Kentucky University, as well as Director of the College Heights Weather Station and a Research Associate with the Kentucky Climate Center. His research interests include human biometeorology, forensic meteorology, and environmental perception. In 1986 he received the Ogden College Faculty Excellence Award, and has also received Fulbright and Malone Fellowships to conduct research in various countries through time. His explorations have extended to all seven continents of the globe and he has subsequently written and lectured extensively on these travels. He has also contributed to a lab manual, a climatology text, and topical encyclopedias; and he has written and narrated educational television programming concerning weather and climate for the Kentucky Educational Television (KET) Network and WKYU-TV 24, Western Kentucky University Television.

ESSENTIALS OF PHYSICAL GEOGRAPHY

SEVENTH EDITION

PHYSICAL GEOGRAPHY AMONG THE SCIENCES

Understanding and living in harmony with the physical geography of planet Earth have been major keys to their survival since humans first appeared.

What areas of interest do physical geographers have in common with early humans?

What activity of early humans might be considered a contribution to the development of the physical sciences?

What did individuals in early civilizations do in order to be credited with originating science as a field of study?

What areas of interest do scientists educated as physical geographers have in common with early humans?

Although the science of geography is closely related to many other sciences, it is unique both in its focus and in its subjects of study.

How do physical geographers differ from other physical scientists when they select topics or phenomena to study?

Why is geography often called the spatial science?

What distinguishes the subfield of physical geography from human or regional geography?

A major role of physical geography among the sciences is to investigate and attempt to explain the spatial aspects of Earth's physical phenomena.

In what ways are location and places fundamental in the science of geography?

Why are the topics of spatial interaction and change so important in physical geography?

The use of models and the analysis of various Earth systems are important re-

search and educational techniques used by scientists who emphasize geography as a physical science.

What is a system?

Why is Earth described as a life-support system?

What kinds of models may be used to portray Earth and its physical processes?

In what ways can systems analysis lead to a more quantitative approach to understanding complex research problems?

What precautions should be exercised when applying models and systems in the examination of a problem?

Unlike some other physical sciences, physical geography places a special emphasis on human-environment relationships.

How does the nature of geography explain the role of geographers as environmental scientists?

Why is this role of geography so important in the study of the physical sciences today?

Why does the study of ecosystems provide such an excellent opportunity for physical geographers to observe the interactions between humans and the natural environment?

Every physical environment offers an array of advantages as well as challenges and potential problems or hazards to the human residents of that location.

What adaptations are necessary for humans to live in your area?

What environmental advantages do you enjoy where you live?

What impacts do humans have on the environment where you live?

Bountiful Earth: Home to humans and laboratory of the sciences since the dawn of history

In some ways, early humans exhibited characteristics similar to those associated with the physical geographers of today. They closely and carefully examined the details of the world around them in order to build a store of environmental information that would improve life for themselves and those around them. In doing so, they inaugurated the accumulation of a body of knowledge that we now call the physical sciences. Although humans may have an innate desire to understand the elements and environmental settings that affect their existence, this knowledge was also a practical necessity. In order to survive, early humans had to be keen observers of their surroundings, and know how to apply this knowledge of their physical environment. They examined the plant life by trial and error to expand their food supply. They studied the habits of other animals to judge which were prey and which were predators and noted their patterns of movement in order to intercept or avoid them.

Long before recorded history, humans learned how to use the natural resources of the physical environment as sources of food, clothing, and shelter. They learned ways to survive the hazards of Earth movements, fires, floods, and storms. As they traveled, our inquisitive ancestors attempted to understand the nature of the world they encountered. They imagined explanations for the environmental characteristics and events that they observed, and they frequently patterned their behavior based on information gained from their observations. As they shared their knowledge with each other and passed on to successive generations what they had learned about that portion of Earth that they occupied, they were developing information that would become a foundation for the organized study of the physical sciences.

As ancestral humans accumulated knowledge, pondered the nature of their physical surroundings, and tested their knowledge against their experiences, they shared some methods of seeking understanding with scientists today. Science as a field of study, however, did not begin until gifted individuals in early civilizations systematically arranged facts and information that they observed to develop a body of physical laws. These laws could then be applied to explain the process or feature that had been observed. Since those beginnings, the goal

of scientific study has been the systematic knowledge of the physical world gained through the processes of observation, hypothesis development, experimentation, and the proposing of theories. Theories may become widely accepted scientific laws after being repeatedly and successfully tested.

Of course, science has changed considerably from its early origins. Modern scientists use techniques and technologies that could not have been conceived a century ago (Fig. 1.1). Yet the methods and the mission of scientists today are not greatly unlike those of our ancestors. The methods of modern scientists include careful observation, proposed explanations or hypotheses about patterns they observe, theories that might explain the patterns, and repeated experimentation to test the validity of their findings before they can be accepted as legitimate additions to our ever-expanding body of scientific knowledge. Scientists today have something critically important in common with the earliest humans: The knowledge gained by each group in its turn held, and will hold, the key to human survival.

The Study of Geography

As early humans were studying their surroundings and gaining knowledge that eventually became the subject of scientific study, why do we suggest that they had something in common with those scientists recognized today as physical geographers? The answer lies in the nature of geography as a science and as a field of study. Geography is unique among the sciences in several ways. Like early humans who studied aspects of the environment that

Figure 1.1 Computer technology can be used to develop accurate scenes for studying the geographic elements of a landscape. Satellite imagery and elevation data were merged to create this digital view of Salt Lake City and the Wasatch Front, the western edge of the Rocky Mountains. ■ **What aspects of this landscape are particularly striking to you?**

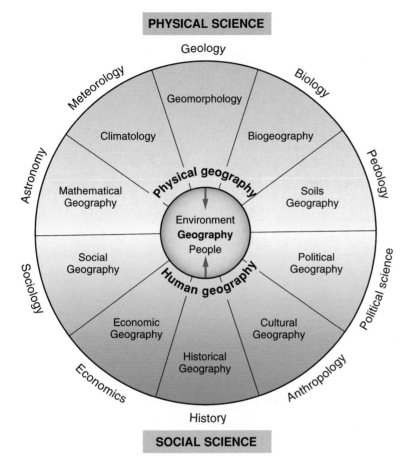

PHYSICAL SCIENCE

Geology

Meteorology

Biology

Astronomy

Pedology

Physical geography

Geomorphology

Climatology

Biogeography

Mathematical Geography

Soils Geography

Environment
Geography
People

Social Geography

Political Geography

Human geography

Sociology

Political science

Economic Geography

Cultural Geography

Historical Geography

Economics

Anthropology

History

SOCIAL SCIENCE

Figure 1.2   When conducting research or examining one of society's many problems, geographers are prepared to consider any information or aspect of a topic that relates to their study.   ■   **What advantage might the geographer have when working together with other physical scientists seeking a solution to a problem?**

surrounded, supported, and sometimes threatened them, geographers consider the characteristics of Earth and its inhabitants when conducting their scientific study. Unlike most scientists in related disciplines (e.g., biologists, geologists, chemists, economists), who are bounded by the phenomena they study, geographers may focus their research on nearly any topic or subject related to the scientific analysis of human or natural processes on Earth (Fig. 1.2). Geographers generally consider and examine *all* phenomena that are relevant to a given problem or issue; in other words, they often take a *holistic* approach to research.

Geography is distinctive among the sciences by virtue of its definition and central purpose. *Geography* is a word that comes from two Greek roots: *Geo-* refers to Earth, and *-graphy* means picture or writing. The primary objective of geography is the examination, description, and explanation of Earth—its variation from place to place, and how places and features change over time. Geography is often called "the spatial science" because it includes recognizing, analyzing, and explaining the variations, similarities, or differences in phenomena located (or distributed) on Earth's surface (through Earth space).

In 1994, a set of national standards for geography education in the schools was published by the major professional geography organizations in the United States in

a book titled *Geography for Life*. The question "What is geography?" is answered in the book as follows:

> *Where is something located? Why is it there? How did it get there? How does it interact with other things?*
>
> *Geography is not a collection of arcane information. Rather it is the study of spatial aspects of human existence. People everywhere need to know about the nature of their world and their place in it. Geography has much more to do with asking questions and solving problems than it does with rote memorization of facts.*
>
> *So what exactly is geography? It is an integrative discipline that brings together the physical and human dimensions of the world in the study of people, places, and environments. Its subject matter is the Earth's surface and the processes that shape it, the relationships between people and environments, and the connections between people and places.*

Geography Education Standards Project, 1994
Geography for Life

In order to answer questions like the ones posed in *Geography for Life*, geographers gather, organize, and analyze geographic information. They also apply many skills, techniques, and tools to the task of answering these geographic—or spatial—questions that have infinite variations in the nature of the phenomena being studied. So it is not surprising that the discipline of

©Dilip Mehta/ Contact Press Images/PictureQuest

Figure 1.3 Vancouver, British Columbia's magnificent physical setting strongly influences the human geography of this urban area. Settlement patterns, economic activities, recreational opportunities, and many other aspects of human activities are a function of interactions among geographic factors, both human (cultural) and physical. ■ **What human geographic characteristics can you interpret from this scene?**

geography has many subdivisions, yet a unifying factor among them is their focus on understanding and explaining spatial locations, distributions, and relationships. Each of these three topics is representative of geography as a spatial science—a study of where certain features are located on the surface of the Earth, how they are distributed, and how features in one place on Earth may be related to one another and to similar features in other places. Geographers study the human and/or physical characteristics of places, attempting to identify and explain the aspects that two or more locations may have in common as well as why places vary in their geographic characteristics. Geographers are interested in how to divide and synthesize areas into meaningful divisions called regions. Geographers also study the processes that influenced the landscapes of the Earth in the past, how they continue to affect them today, how the landscape may change in the future, and the significance or impact of these changes.

Because geography embraces the study of such a broad spectrum of global phenomena, it is common for geographers to specialize in one or more subfields of the discipline. The discipline is also subdivided along academic lines; some geographers are social scientists, and some are natural scientists, but most are involved in studying the

human or natural processes that affect our planet, and the interactions among these processes. The main subdivision that deals with human activities and the results of these activities is called cultural or **human geography.** Human geographers are concerned with such subjects as population distributions, cultural patterns, cities and urbanization, natural resource utilization, industrial location, and transportation networks (Fig. 1.3). When geographic study concentrates on the general physical and human characteristics of a specific region, such as Canada, the Great Plains, the Caribbean, or the Middle East, the subdivision is termed **regional geography.**

Physical Geography

Physical geography, the focus of this textbook, encompasses the study of the processes and features that make up Earth, including human activities where they interface with the physical environment. In fact, physical geographers are concerned with nearly all aspects of Earth and can be considered generalists because they are trained to view a natural environment in its entirety, functioning as a unit (Fig. 1.4). However, after a thorough and broad education in basic physical geography, most physical geographers focus their expertise on advanced

Bruce Heinemann/ Getty Images

Figure 1.4   The Henry Mountains form a backdrop for the physical landscape of Canyonlands National Park, Utah. Physical geographers study the elements and processes that affect natural environments. These include rock structures, landforms, soils, vegetation, climate, weather, and human impacts.   ■   **What physical geography characteristics can you interpret from this scene?**

study in one or two specialties. For example, *meteorologists* and *climatologists* consider the atmospheric components that affect Earth's surface and that interact to influence weather and climate. Meteorologists are interested in the atmospheric processes that affect daily weather, and they use current weather data to forecast weather conditions. Climatologists are interested in the averages and extremes of long-term weather data, the classification of climates into meaningful regions, monitoring and understanding climatic change and climatic hazards, and the long-range impact of atmospheric conditions on human activity and the environment.

The study of the nature, development, and modification of landforms is a specialty called *geomorphology*, a major subfield of physical geography. Geomorphologists are interested in understanding and explaining variation in landforms, the processes that produce physical landscapes, and the nature and geometry of Earth's surface. The factors involved in landform development are as varied as the environments on Earth and include, for example, gravity, running water in streams, stresses in the Earth's crust, the flow of ice in glaciers, volcanic activity, and the erosion or deposition of Earth materials. *Biogeographers* examine natural and human-modified environments and the ecological processes that influence their

nature and distribution, including vegetation change over time. They also study the ranges and patterns of vegetation and animal species, seeking to discover the environmental factors that limit or facilitate their distributions. Many *soil scientists* are geographers, and these individuals are often involved in the mapping and analysis of soil types, determining the suitability of a soil for certain uses, such as agriculture, and working to conserve soil as a natural resource.

Finally, because of the critical importance of water to the existence of life on Earth, geographers are widely involved in the study of water bodies and their processes, movements, impact, quality, and other characteristics. They may serve as *hydrologists, oceanographers*, or *glaciologists*. Many geographers involved with water studies also function as *water resource managers* to ensure that lakes, watersheds, springs, and groundwater sources are suitable for human use, provide an adequate supply of water, and are as free of pollution as possible.

Technology, Tools, and Methods

The technologies being applied by physical geographers and other scientists in their efforts to learn more about Earth are rapidly changing. The abilities of computer systems to capture, process, model, and display spatial data—all functions that today are possible on a desktop computer—were only a dream a quarter of a century ago. Computer access to the Internet provides instant connections to information sources on virtually any topic. The amounts of data, information, and imagery available for studying the environments of Earth have exploded. Graphic displays of these sorts of data and information are becoming more vivid and striking as a result of better methods of data processing and visual representation. Increased computer power allows the presentation of clearer images, three-dimensional images, and even animated images of Earth features, changes, and processes.

Continuous satellite imaging of Earth's environments has been ongoing for more than 30 years, giving us a better perspective on changes as they occur. It is possible to track change in a single place over that time period or to compare two different places at the same point in time. Using various energy sources to produce an image from space, we are now able to see, measure, monitor, and map processes and the effects of these processes that would be invisible to the naked eye. Satellite technology can be used to determine the precise location of a positioning receiver on Earth's surface, a capability that has many useful applications for geography and mapping.

Today, mapmaking (*cartography*) and the analysis of maps are becoming digital, computer-assisted operations. Making direct observations in the field and conducting fieldwork to gather data remain important to most physical geographers, but they must also keep up with new technologies in order to make use of them. Many of these technologies directly support and facilitate traditional methods of fieldwork. In fact, the development of new technology exceeds our ability to understand or apply the full range of potential uses for these new tools. Most advancements in technology are later discovered to have new applications when people begin to test and apply them in a problem-solving environment. This situation of a technology-application lag creates many opportunities for geographers in the applied workplace. Many geographers are gainfully employed in positions that apply technology to the problems of understanding our planet and its environments, and their numbers are certain to increase in the future.

Studying Physical Geography

Your textbook has been designed to demonstrate three major roles that physical geography plays among the sciences today: geography as the spatial science, geography as a physical science, and geography as environmental science. Although the emphasis on each role may vary from chapter to chapter, the significant contributions of physical geography to all three approaches to scientific study will be apparent throughout the book. As you read through the remainder of this chapter, take note of how directly each role of geography in the sciences relates to the unique perspective of the geographic discipline.

Geography as the Spatial Science

The central role of geography among the sciences is best illustrated by its definition as the **spatial science** (the science of Earth space). No other science has the specific responsibility for investigating and attempting to explain the spatial aspects of Earth phenomena. Even though physical geographers may be a varied lot, with many divergent interests, they share the common goal of understanding and explaining the spatial variation displayed by Earth's environments. As a part of this common ground, it is useful to consider some of the spatial questions that physical geographers raise and some of the problems they seek to understand and solve.

How do physical geographers examine Earth from a spatial point of view? On what topics of particular interest do they focus their study? From among the almost unlimited number of topics available to physical geographers, we have chosen five that clearly illustrate the role of geography as the spatial science. In keeping with the quote from *Geography for Life* suggesting that geography is about asking questions and solving problems, frequently pursued directions of study have been noted for each topic.

Location

As a spatial science, geography often begins with location. The location of a feature is usually expressed by one of two methods: **absolute location,** expressed by a coordinate system (or address), and **relative location,** which identifies where something is in relation to something else, usually a fairly well-known location. For example, Pikes Peak, in the Rocky Mountains of Colorado, with an elevation of 4301 meters (14,110 ft), has a location of latitude 38°51' North and longitude 105°03' West. A global address like this is an absolute location. However, another way to report its location would be to state that it is 36 kilometers (22 mi) west of Colorado Springs. This is an example of relative location (relative to Colorado Springs). Typical spatial questions involving location include the following: *Where is a certain type of Earth feature found, and where is it not found? Why is a certain feature located where it is? What methods can we use to locate a feature on Earth? How can we describe its location? What is the most likely or least likely location for a certain Earth feature? Does its location seem to fit with its surrounding environment?*

Characteristics of Places

Physical geographers are interested in the environmental features and processes that combine to make a place unique, and they are also interested in shared character-

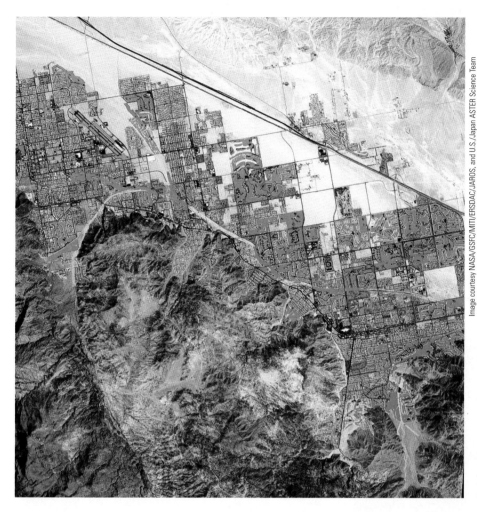

Image courtesy NASA/GSFC/MITI/ERSDAC/JAROS, and U.S./Japan ASTER Science Team

Figure 1.5 Palm Springs, California provides several good illustrations of distribution and pattern. Spatial distribution is where features are located (or, perhaps, where they are absent), and spatial pattern refers to their arrangement. Geographers also attempt to explain these spatial relationships.

■ **Can you locate and propose possible explanations for two patterns and two distributions in this scene?**

istics between places. For example, what physical geographic features make the Rocky Mountains appear as they do? As another example, how are the Andes Mountains different from the Rockies, and what characteristics are common to both of these mountain ranges? Other examples might include: *How does an Australian desert compare to the Sonoran desert of the southwestern United States? How do the grasslands of the Great Plains of the United States compare to the grasslands of Argentina? What are the environmental conditions at a particular site? How do places on Earth vary in their environments, and why? In what ways are places unique, and in what ways do they share similar characteristics with other places?*

Spatial Distributions and Spatial Patterns

When studying how features are arranged in space, geographers are usually interested in two spatial factors (Fig. 1.5). **Spatial distribution** means the extent of the area or areas where a feature exists. For example, where on Earth do we find the tropical rainforests? What is the distribution of rainfall in the United States on a particular day? **Spatial pattern** refers to the arrangement of features in space—are they regular or random, clustered or widely

spaced? From the window seat of an airplane on a clear day, it soon becomes obvious that population distribution on Earth can be either dense or sparse. The spatial pattern of earthquakes may be linear on a map, suggesting that a fault is located there, because faults display similar linear patterns. *Where are certain features abundant, and where are they rare? How are particular factors or elements of physical geography arranged in space, and what spatial patterns exist, if any? What processes are responsible for these patterns? If a pattern exists, what does it signify?*

Spatial Interaction

Few processes on Earth operate in isolation; areas on our planet are interconnected, or linked to conditions elsewhere on Earth. A condition, an occurrence, or a process in one place generally has an impact on other places. Unfortunately, the exact nature of this **spatial interaction** is often difficult to establish with certainty except after years of intense study. A cause-effect relationship can often only be suspected, as a direct relationship is often difficult to prove. It is much easier to observe that changes seem to be associated with each other, without knowing if one event causes the other or if this result is coincidental.

The Regional Concept: Natural and Environmental Regions

The term *region* is familiar to us all, but it has a precise meaning and special significance to geographers. Simply stated, a region is an area that shares a certain characteristic (or a set of characteristics) within its boundaries. Regions are spatial models, just as systems are operational models. Systems help us to understand how things work, and regions help us to make spatial sense of our world. The concept of a region is a tool for thinking about and analyzing logical divisions based on their areal extent. Just as it helps us to understand Earth by considering smaller parts of its overall system, dividing space into coherent segments helps us understand the arrangement and nature of areas on our planet.

Regions can be described based on factors related to either human or natural characteristics, or a combination of the two. Regions can also be divided into subregions. For example, North America is a region, but it can be subdivided into many subregions. Examples of subregions based on natural characteristics include the Atlantic Coastal Plain (similarity of landforms, geology, and locality), the Prairies (ecological type), the Sonoran Desert (climate type and locality), the Pacific Northwest (general locality), and Tornado Alley (region of high potential for these storms).

Physical geographers are mainly interested in natural regions and human-environmental regions. The term *natural,* as used here, means primarily related to natural processes and features involved in the four major spheres of the Earth system. However, we recognize that today human activities have an impact on virtually every natural process, and human-environmental regions offer significant opportunities for geographic analysis. Geographers not only

Ice shelves are thick plates of ice, fed by glaciers, that float on the ocean around much of Antarctica. The Larsen ice shelf was about 220 meters thick. About 3,250 square kilometers of ice shelf disintegrated in a 35-day period beginning on January 31, 2002 (the second image is from March 5th, 2002). The area lost in this breakup is 20% larger than the state of Rhode Island (from NASA Earth Observatory).

For example, the presence of abnormally warm ocean waters off the west coast of South America, a condition called El Niño, seems to occur at times of unusual weather in other parts of the world. There is also a concern that clearing of the tropical rainforest may have an impact on world climates. Interconnections are one reason for considering Earth as a whole. *What are the relationships among places and features on Earth? How do they affect one another? What important interconnections link* the oceans to the atmosphere, and the atmosphere to the solid land below? If two events occur at the same time, why is it so difficult to prove which one caused the other to happen?

Ever-Changing Earth

As you pursue your study of phenomena on Earth, it will become apparent that spatial phenomena are dynamic and that environmental conditions are continuously study and explain regions, their locations and characteristics, but they also strive to delimit them—to outline their boundaries on a map. An unlimited number of regions can be derived for each of the four major Earth subsystems.

There are three important points to remember about natural and environmental regions. Each of these points has endless applications and adds considerably to the questions that the process of defining regions based on spatial characteristics seeks to answer.

Natural regions can change in size and shape over time in response to environmental changes. These changes can be fast enough to observe as they occur, or so gradual that they require intensive study to detect. An example is the changes in the Antarctic ice sheet, the glacier-covered part of Antarctica, and adjacent ice shelves that extend out and over the ocean along parts of the coast. In recent years, the region of Antarctica that is covered with glacial ice appears to be shrinking in response to global warming. This means that the boundaries of that region are shifting their position. The ice-covered region is apparently getting smaller. We can see and monitor the glacial cover from space. Changes in other environmental and climatic regions are very likely associated with this change, but these changes are much more difficult to monitor because they may be more subtle and gradual.

Boundaries separating different natural or environmental regions tend to be indistinct or transitional, rather than sharp. For example, on a climate map, lines separating desert from nondesert regions do not imply that extremely arid conditions instantly appear when the line is crossed rather, if we travel to a desert, it is likely to get progressively more arid as we approach our destination. This human-environmental map divides the United States into three regions based on annual heating and cooling needs. Blue means that heating is required more often than cooling.

The yellow region represents roughly equal heating and cooling needs. Red represents a stronger or greater demand for cooling than heating. The map is clearly related to climate regions. Do you think that the boundaries between these regions are as sharply defined in reality as they are on this map? Can you recognize the spatial patterns that you see? Do the shapes of these regions, and how they are related to each other, seem spatially logical?

Regions are spatial models, devised by humans, for geographic analysis, study, and understanding. Natural or environmental regions, like all regions, are conceptual models that are specifically designed to help us comprehend and organize spatial relationships and geographic distributions. Learning geography is an invitation to think spatially, and regions provide an essential, extremely useful, conceptual framework in that process.

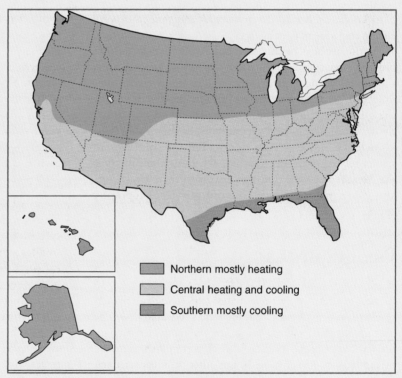

Regions of home heating and cooling requirements in the United States

Legend:
- Northern mostly heating
- Central heating and cooling
- Southern mostly cooling

changing. Weather and weather elements change from day to day, over the seasons, and from year to year. A future Hawaiian island is now forming beneath the waters of the Pacific Ocean. Storms, earthquakes, landslides, and stream processes modify the landscape (Fig. 1.6). Today, vegetation and wildlife are becoming reestablished in areas that were devastated by recent volcanic eruptions or wildfires. World climates have changed throughout Earth's history, with attendant shifts in the nature and distribution of plant and animal life. Change is always occurring, but not all of this change can be directly monitored. *How are Earth features changing? What processes contribute to the change, and what is the nature and impact of this change? What is the rate of change? Does change occur in a cycle? Can humans witness this change as it is taking place, or is a long-term study required to recognize the change? Do all places on Earth experience the same levels of change, or is there spatial variation? Within one*

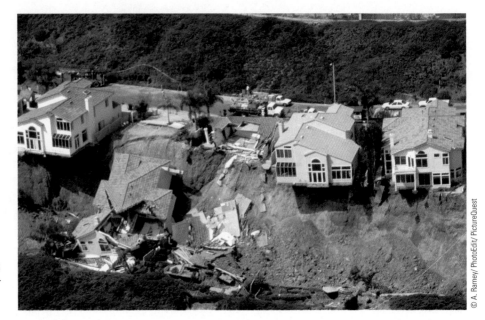

Figure 1.6 Natural changes in a landscape can be either beneficial or detrimental to human existence. Heavy rains triggered a landslide that undermined or destroyed homes that were built on steep slopes made of weak earth materials. The better we understand an environment's characteristics and the processes that affect it, the more able we are to avoid situations such as this scene in Southern California.

© A. Ramey/ PhotoEdit/ PictureQuest

place on Earth, are all environmental components experiencing the same rates of change?

As the authors of this textbook, we have selected the previous five topics to illustrate the role of geography as the spatial science. Other geographers might favor other topics or demonstrate these topics in another way. But the questions that accompany the topics will be much the same for all physical geographers, as they apply their knowledge and expertise in a search for solutions to spatial-related problems. Learning what questions to ask is the first step toward finding answers to these problems, and it is a major objective of your physical geography course.

Geography as a Physical Science

As geographers apply their spatial expertise to the study of Earth, it is axiomatic that physical geographers will be examining phenomena, compiling data, and seeking solutions to problems that are also of particular interest to researchers in one or more of the other physical sciences. Physical geographers, as climatologists, share ideas and information with atmospheric physicists. Soil geographers study the same elements and compounds analyzed by chemists. Biogeographers are concerned about the environments of the same plants and animals that are classified by biologists. However, to whatever question is raised and whichever problem requires a solution, the physical geographer brings a unique point of view—a spatial perspective and a mandate to carefully consider all Earth phenomena that may be involved. This is why an introductory course in physical geography, and your textbook as well, are so often focused on a global scale, although local and regional processes are important as

well. Physical geographers are also concerned with the processes that affect interactions and changes in the Earth's physical environment. By examining the factors, features, and processes that interact to influence the environment and learning how these elements work together to create the ever-changing physical geography of our planet, you can understand the importance of viewing Earth in its entirety as a constantly functioning system—as a whole greater than, but dependent upon, each and all of its parts.

The Earth System

The examination by physical geographers of Earth as a system made up of an almost unlimited number of smaller subsystems constitutes a major contribution to the physical sciences. Because a **system** may be thought of as a set of parts or components that are interrelated through space, the analysis of systems provides geographers with ideal opportunities to study these relationships. The components of a system, termed **variables,** are studied or grouped together because the variables interact with one another as a functioning unit. Earth certainly fits this definition, because many continuously changing variables combine to make our home, the **Earth system,** function the way that it does. Virtually no part of a system operates in isolation from another. A change in one part of the Earth system affects other parts, and the impact of these changes can be significant enough to appear in worldwide patterns, clearly demonstrating the interconnections among these variables. For example, the location and position of mountains influence the global distribution of rainfall, and variation in rainfall affects the density, type, and variety of vegetation. Plants, moisture, and the underlying rock are factors that affect the kind of soil that forms in an area.

Atmosphere

Biosphere

Earth System

Hydrosphere

Lithosphere

Gary Braasch/ Getty Images; NASA; NASA; Olaf Soot/ Getty Images; Schafer and Hill/ Getty Images

Figure 1.7   Environmental Systems: Earth's Major Subsystems. The study of Earth as a system is central to understanding changes in our planet's environments and adjusting to or dealing with these changes. A system as large and complex as Earth consists of many interconnected subsystems that lend themselves more readily to study through the use of computers, theoretical models, and mathematical computation. ■ **How do these systems overlap? For example, how does the atmosphere overlap with the hydrosphere, or the biosphere?**

The characteristics of vegetation and soils influence the runoff of water from the land, leading to completion of the circle, since the amount of runoff is a major factor in stream erosion, which eventually can reduce the height of mountains. Many cycles such as this operate to change our planet daily, but the environment of Earth is complex, and these cycles and processes operate at widely varying rates and over widely varying time spans.

We do know that the Earth system is *dynamic* (ever-changing), and that we can directly observe some of these changes—the seasons, the ocean tides, earthquakes, floods, volcanic eruptions. Other aspects of our planet may take years, or even more than a lifetime, to accumulate enough total change to be directly noticed by humans. Such long-term changes in our planet are often difficult to understand or predict with certainty, so they must be carefully and scientifically studied in order to determine what is really happening and what the consequences might be. Changes of this type include shifts in world climates, drought cycles, the spread of deserts, worldwide rise or fall in sea level, erosion of coastlines, and major changes in river systems. Yet understanding changes in our planet is critical to human existence. We are, after all, a part of the Earth system. Changes in the

system may be naturally caused or human induced, or they may result from a combination of these factors. To understand our planet, therefore, we must learn about its components and the processes that operate to change or regulate the Earth system. Such knowledge is in the best interest not only of humankind but also of Earth, as a habitat for all living things.

Major Divisions

Four major divisions of our planet comprise the Earth system. The **atmosphere** is the gaseous blanket of air that envelops, shields, and insulates Earth. The movements and processes of the atmosphere create the changing conditions that we know as weather and climate. The solid Earth—landforms, rocks, soils, and minerals—makes up the **lithosphere.** The waters of the Earth system—oceans, lakes, rivers, and glaciers—constitute the **hydrosphere.** The fourth major division, the **biosphere,** is composed of all living things: people, other animals, and plants. It is the nature of these four major subsystems and the interactions among them that create and nurture the conditions necessary for life on Earth (Fig. 1.7).

For example, the hydrosphere serves as the water supply for all life, including humans, and provides a home environment for many types of aquatic plants and animals. The hydrosphere directly affects the lithosphere as the moving water in streams, waves, and currents shapes landforms. It also influences the atmosphere through evaporation, condensation, and the effects of ocean temperatures on climate. The impact or intensity of interactions among Earth's subsystems is not identical everywhere on the planet, and it is this variation that leads to the patterns of environmental diversity that are so apparent to orbiting astronauts.

Many other examples of overlap exist among the four divisions. Soil can be examined as part of the biosphere, the hydrosphere, or the lithosphere. The water stored in plants and animals is part of both the biosphere and the hydrosphere, and the water in clouds is a component of the atmosphere as well as the hydrosphere. The fact that we cannot draw sharp boundaries between these divisions underscores the interrelatedness of the various parts of the Earth system. However, like a machine, a computer, or the human body, planet Earth is a system that functions well only when all of its parts (and its subsystems) work together harmoniously.

A Life-Support System

Certainly the most important attribute of Earth is that it is a **life-support system.** Like space vehicles that support astronauts in orbit, the Earth system provides the necessary environmental constituents and conditions to permit life as we know it to exist. If a critical part of a life-support system is significantly changed or fails to operate properly, living organisms may no longer be able to survive. For instance, if all the oxygen in a spacecraft is used up, the crew inside will die. If there is no way to keep the spacecraft within the proper temperature range, its occupants may burn or freeze. If food supplies run out, the astronauts will starve. Similarly, on Earth, natural processes must provide an adequate supply of oxygen; the sun must interact with the atmosphere, oceans, and land to maintain tolerable temperatures; and photosynthesis or other continuous cycles of creation must provide new food supplies for living things.

Earth, then, is made up of a set of interrelated components that are vital and necessary for the existence of all living creatures. About 40 years ago, Buckminster Fuller, a distinguished scientist, philosopher, and inventor, coined the notion of Spaceship Earth—the idea that our planet is a life-support system, transporting us through space. Fuller also thought that knowing how Earth works is important—indeed this knowledge may be required for human survival—but that humans are only slowly learning the processes involved. He compared this information to an operating manual, like the owner's manual for an automobile.

One of the most interesting things to me about our spaceship is that it is a mechanical vehicle, just as is an automobile. If you own an automobile, you realize that you must put oil and gas into it, and you must put water in the radiator and take care of the car as a whole. You know that you are going to have to keep the machine in good order or it's going to be in trouble and fail to function.

Figure 1.8 Toxic chemicals, such as the ones discovered in this solid-waste dump, pose a serious health hazard and threaten local water supplies. ■ **What pollutants form the major threat to the air and water supply in your community?**

Lee Malis

(a)

(b)

Figure 1.9   (a) On a clear day, it is not difficult to understand why Los Angeles, California, was named "The City of Angels." (b) But far too often, this great metropolis justifies its designation as "Smog Capital" of the United States.   ■   **If you were choosing whether to live in a small town or a major city, would pollution affect your decision?**

We have not been seeing our Spaceship Earth as an integrally-designed machine which to be persistently successful must be comprehended and serviced in total . . . . there is one outstandingly important fact regarding Spaceship Earth, and that is that no instruction book came with it.

R. Buckminster Fuller
Operating Manual for Spaceship Earth

Today, as we embark on a new century, we realize that critical parts of our life-support system, the **natural resources,** can be abused, wasted, or exhausted, creating a potential threat to the functioning of planet Earth as a human life-support system. One such abuse is **pollution,** an undesirable or unhealthy concentration of a contaminant in an environment as a result of human activities (Fig. 1.8). We are aware that some of Earth's resources, such as air and water, can be polluted to the point where they become unusable or even lethal to some life forms. By polluting the oceans, we may be killing off important fish species, allowing less desirable species to increase in number. Acid rain, caused by the release into the atmosphere of pollutants from industries and power plants, is damaging forests and killing fish in freshwater lakes. Air pollution has become a serious environmental problem for urban centers throughout the world (Fig. 1.9).

Because pollution is associated with human activity, it is not surprising that it represents a significant problem in places that have high population densities. What some people do not realize, however, is that pollutants are often transported by winds and waterways hundreds or even thousands of kilometers from their source. Lead from automobile exhausts has been found in the snow of Antarctica, as has the insecticide DDT. Pollution is a worldwide problem that does not stop at political boundaries.

Another concern is that humans may be rapidly depleting critical natural resources, especially those needed for fuel. Many natural resources on our planet are non-renewable, meaning that nature will not replace them once they are exhausted. Coal and oil are nonrenewable resources. Although we probably have enough coal to last several hundred years, we have often been warned about future shortages in our petroleum supplies. When nonrenewable resources such as these mineral fuels are gone, the alternative resources may be less desirable or more expensive.

We are learning that, much like living on a spaceship, there are limits to the amount of suitable living space on Earth, and we must use it wisely. In our search for livable space, we occasionally construct buildings in locations that are not environmentally safe or that are already overcrowded. Also, we sometimes plant crops in areas that are ill suited to agriculture while at the same time paving over prime farmland for other uses.

Our space exploration programs are helping us learn more and more about the world in which we live. With the use of spacecraft-mounted cameras and sophisticated imaging systems, we can observe, and monitor over time, the changes on Earth's surface that are the result of human activity. All citizens of Earth must understand the impact of their actions on the complex Earth system. It is to the science of physical geography that we can turn to learn the consequences of those actions.

Models and Systems

As physical geographers work to describe, understand, and explain the often complex features of planet Earth, they support these efforts, as other scientists do, by developing representations of the real world called **models.** A model is a useful simplification of a more complex reality that permits prediction, and each model is designed with a specific purpose in mind. As examples, maps and globes are models—representations of Earth that provide us with useful information required to meet specific needs. But models are greatly simplified versions of what

The Scientific Method

Science . . . is the systematic and organized inquiry into the natural world and its phenomena. Science is about gaining a deeper and often useful understanding of the world.
Multicultural History of Science
Web page, Vanderbilt University

The real purpose of the scientific method is to make sure nature hasn't misled you into thinking you know something you don't actually know.
Robert M. Pirsig,
Zen and the Art of Motorcycle Maintenance

Physical geography is a science that focuses on the Earth system, how its components and processes interact, and how and why aspects that affect Earth's surface are spatially arranged, as well as how humans and their environments are interrelated.

To wonder about your environment, and attempt to understand it, is a fundamental basis of human life. Increasing our awareness, satisfying our curiosity, learning how our world works, and determining how we can best function within it are all parts of a satisfying but never-ending quest. Without curiosity about the world, supported by making observations, noting relationships and patterns, and applying the knowledge discovered, humans would have not survived beyond their earliest beginnings. Science gives us a method for answering questions and testing ideas by examining evidence, drawing conclusions, and making new discoveries.

The sciences search for new knowledge using a strategy that minimizes the possibility of erroneous conclusions. This highly adaptable process is called the *scientific method.* The scientific method is a general framework for research, but it can accommodate an infinite number of topics and strategies for deriving conclusions.

Scientific method generally involves the following steps:

Make an observation that requires an explanation. We may wonder if the observation represents a general pattern or is a "fluke" occurrence. For example, on a trip to the mountains, you notice that it gets colder as you go up in elevation. Is that just a result of conditions the day you were there, or just the conditions at the location where you were, or is it a relationship that occurs everywhere?

Restate the observation as a hypothesis. Here is an example: As we go higher in elevation, the temperature gets cooler (or, as a question, Does it get cooler as we go up in elevation?). The answer may seem obvious, yet it is generally but not always true, depending on environmental conditions that will be discussed in later chapters. Many scientists recommend a strategy called *multiple working hypotheses,* which means that we consider and test many possible hypotheses to discover which one best answers the questions while eliminating other possibilities.

Determine a technique for testing the hypothesis and collect necessary data. The next step is finding a technique for evaluating data (numerical information) and or facts that concern that hypothesis. In our example, we would gather temperature and elevation data (taken at about the same time for all data points) for the area we are studying.

Apply the technique or strategy to test the validity of the hypothesis. Here we discover if the hypothesis is supported by adequate evidence, collected under similar conditions to minimize bias. The technique will recommend either acceptance or rejection of the hypothesis.

If the hypothesis is rejected, we can test an alternate hypothesis or modify our existing one and try again, until all of the conceivable hypotheses are rejected or we discover one that is supported by the data. If the test supports the hypothesis, we have confirmed our observation, at least for

they depict, conveying the most important information about a feature or process without too much (often confusing) detail. Models are essential to understanding and predicting the way that nature operates, and they vary greatly in their levels of complexity (Fig. 1.10). Today, many models are generated by computers, because computers can handle great amounts of data and perform the mathematical calculations that are often necessary to construct and display certain types of models.

Models are of many kinds. **Physical models** are three-dimensional representations, such as globes or a representation of a mountain. **Pictorial/graphic models** include photographs, maps, images, graphs, diagrams, and drawings. **Mathematical or statistical models** are used to predict possibilities, such as the future flooding of major rivers or changes in weather conditions that may result from a rise in atmospheric temperatures. Words, language, and the definitions of terms or ideas can also serve as models. Another important type of model is a **conceptual model**—the mind imagery that we use for understanding our surroundings and experiences.

Imagine for a minute (perhaps with your eyes closed) the image that the word *mountain* (or *waterfall, cloud, tornado, beach, forest, desert*) generates in your mind. Can you describe this feature and its characteristics in detail? Most likely what you "see" (conceptualize) in your mind the location and environmental conditions in which our data and information were gathered.

After similar tests are conducted and the hypothesis is supported in many places and under other conditions, then the hypothesis may become a theory. Theories are well-tested concepts or relationships that, given specified circumstances, can be used to explain and predict outcomes.

The processes of asking questions, seeking answers, and finding solutions through the scientific method have contributed greatly to human existence, our technologies, and our quality of life. Obviously, there are many more questions to be answered and problems yet to be solved. In fact, new findings typically yield new questions.

Human curiosity, along with an intrinsic need for knowledge through observation and experience, has formed the basis for scientific method, an objective, structured approach that leads us toward the primary goal of physical geography—understanding how our world works.

Steps in the scientific method

is sketchy rather than detailed, but enough information is there to convey a mental idea of a mountain. This image is a conceptual model.

For geographers, a particularly important type of conceptual model is the **mental map,** which we use to think about places, travel routes, and the distribution of features in space. Psychologists have shown in many studies that such maps are very efficient media for conveying a great amount of spatial information that the brain can recognize, store, and access in an economical fashion. Try to think of other varieties of conceptual models that represent our planet's environment. How could we even begin to understand our world without conceptual models?

Systems Theory If you try to think about Earth in its entirety, or to understand how even a part of the Earth system works, there are just too many factors to envision without resorting to excessive oversimplification. Our planet is too complex to permit a single model to represent its entire workings or to explain all of its environmental components and how they affect one another. Yet it is often said that in order to be responsible citizens of Earth, we should "think globally, but act locally." In order to begin to comprehend Earth as a whole, or to understand most of its environmental components, physical geographers use a powerful strategy for analysis called **systems theory.** Systems theory suggests that the

Cartesia/Getty Images

U.S. Park Service

NASA

EPA, South Florida Water Management Division

Figure 1.10 Examples of models used to study Earth. Geographers use many kinds of models to understand the Earth system, its subsystems, and how they operate, by simplifying the presentation of a complex reality. Models help us to focus attention on major features or processes, without unnecessary and distracting detail. **Top:** Globes are models, or simplified presentations of Earth, that demonstrate many terrestrial characteristics—planetary shape, configuration of landmasses and oceans, their distributions, and spatial relationships. **Center:** Two digital landscape models. A digital model of the big island of Hawaii made to show the environment of Hawaii Volcanoes National Park. Computer-generated clouds, shadows, and reflections were added to add "realism" to the scene and to produce and aesthetically pleasing image. The terrain is faithfully rendered. The model to the right shows a ridge of active volcanoes on the Kamchatka Peninsula of Russia. The scene was generated using a radar system designed to map topography from the space shuttle and combined with a satellite image. The vertical expression of the terrain has been stretched three times to enchance detail. **Bottom:** This working model of the Kissimmee River in Florida was constructed to investigate ways to restore the river. Proposed modifications could be analyzed on this model before work was done on the actual river (see the discussion in Figure 1.16). Similar models exist of the Mississippi River and San Francisco Bay.

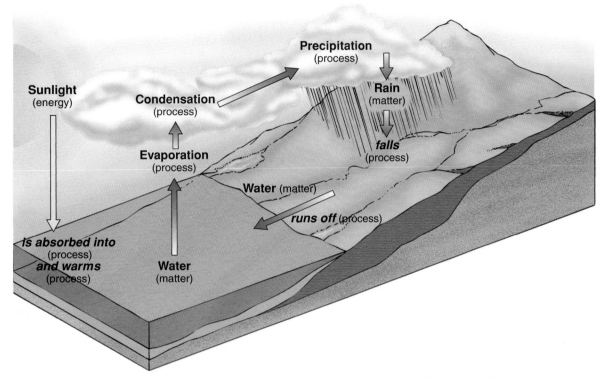

Figure 1.11 Simplified example of environmental interactions: Energy, matter, process. Being aware of energy and matter, and the interactive processes that link them, is an important part of understanding how environmental systems operate. ■ **Can you think of another environmental system and break it down into its components of energy, matter, and process?**

way to understand how anything works is to use the following strategy:

1. **Clearly define the system that you are studying.**
   What are the boundaries (limits) of the system?

2. **Break the defined system down into its component parts (variables). The variables in a system are either matter or energy.**
   What important parts and processes are involved in this system?

3. **Attempt to understand how these variables are related to (or affect, react with, or impact) one another.**
   How do the parts interact with one another to make the system work? What will happen in the system if a part changes?

With environmental systems, an important question that we continually try to answer is how much change a system can tolerate without becoming drastically or irreversibly altered, particularly if the change is for the worse.

The systems approach is a beneficial tool for studying any level of environmental condition on Earth, from global to microscopic. Systems can be divided into **subsystems,** or units that demonstrate strong internal connections. For example, the Earth system consists of the atmosphere, hydrosphere, lithosphere, and biosphere,

each a subsystem of the whole. The human body is a system that is composed of many subsystems (for example, the respiratory system, circulatory system, and digestive system). Subsystems can also be divided into subsystems, and so on.

Geographers often divide the Earth system into smaller subsystems in order to focus their attention on a particular part of the whole. Examples of subsystems examined by physical geographers include the atmospheric water cycle, climatic systems, storm systems, stream systems, the systematic heating of the atmosphere, and ecosystems. A great advantage of systems analysis is that it can be applied to environments at virtually any scale.

How Systems Work   Basically, the world "works" by the movement (or transfer) of matter and energy and the processes attending these transfers. For example, as shown in Figure 1.11, sunlight (*energy*) warms (*process) a* body of water (*matter*), and the water evaporates (*process*) into the atmosphere. Later, the water condenses (*process*) back into a liquid, and the rain (*matter*) falls (*process*) on the land and runs off (*process*) downslope back to the sea. The processes mentioned here each involve the movement of both matter and energy.

In a systems model, geographers can trace the movement of energy or matter into the system (**inputs**) and out of the system (**outputs**), as well as the interactions between energy and matter within the system (Fig. 1.12).

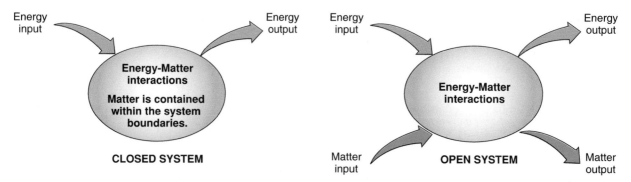

Figure 1.12 Diagram of closed versus open systems. This diagram illustrates the difference between closed systems, which allow only the passage of energy in and out of the system, and open systems that involve the inputs and outputs of both energy and matter. Earth is a good approximation of a closed system: Solar energy (input) enters the Earth system, and that energy is dissipated to space (output) mainly as heat. Energy interacts with matter in the Earth system, but external inputs of matter are virtually nil, consisting mainly of meteorites and Moon rock samples. Except for outgoing space vehicles, equipment, or space "junk," virtually no matter is output from the Earth system. Because Earth is a closed system, humans and other living things on the planet face limits to their available natural resources. Although the Earth system as a whole operates as a (virtually) closed system, most subsystems on the planet are best described as open systems, involving the flows (exchanges) of matter and energy into and out of the subsystems.
■ **The environment of a lake provides a good example of an open system. Can you outline some of the matter-energy inputs and outputs involved in such a system?**

The subsystem involved in the heating of the atmosphere is an example of an **open system,** because energy and matter are not confined to the boundaries of the system, but instead are constantly entering and leaving.

A **closed system** is one in which no substantial amount of matter crosses its boundaries, although energy can go in and out of a closed system. Planet Earth, or the Earth system as a whole, is essentially a closed system, much like the spacecraft that provides the home for our orbiting astronauts. This shared characteristic is another reason why our planet has been referred to as Spaceship Earth. Except for meteorites that reach Earth's surface, the escape of gas molecules or spacecraft from the atmosphere, and a few Moon rocks brought back by astronauts, the Earth system is essentially closed to the movement of material. The subsystem that involves the movement of water from the ocean to the atmosphere to the land and back to the ocean again is another good example of a closed system. Water may exist in the system in all three of its states—as liquid, gas, or solid ice—and may be transformed from one state to another many times, but there is virtually no gain or loss of water (no output of matter) in the system.

Most Earth subsystems, however, are open systems, as both energy and material move freely across subsystem boundaries. A stream is an excellent illustration of an open subsystem, in which matter and energy in the form of soil, rock fragments, solar energy, and precipitation enter the stream while heat energy dissipates into the atmosphere and the stream bed. Water and sediments leave the stream where it empties into the ocean or some other standing body of water, and precipitation provides the input of water to the stream system.

When we describe Earth as a system, or a complex set of interrelated systems, we are using models to help us organize what we are observing. Models also assist us in explaining the processes involved in changing, maintaining, or regulating our planet's life-support systems. Throughout the chapters that follow, we will use the concept of the Earth system as a model, as well as many kinds of other models, to help us simplify a complex planet so that we can focus on the most important elements in the physical environment.

Equilibrium in Earth Systems The parts, or variables, of a system have a tendency to reach a balance with one another and with the factors that influence the system from outside its boundaries. If the inputs entering the system are balanced by its outputs, the system is said to have reached a state of **equilibrium.** Most natural systems have a tendency toward stability (equilibrium); regarding environmental systems, we often hear this called the "balance of nature." What this means is that natural systems have built-in mechanisms that tend to counterbalance, or accommodate, change without changing the system dramatically. Animal populations—deer, for example—will adjust naturally to the food supply of their habitats. If the vegetation on which they browse is sparse in one particular year from drought, fire, overpopulation, or human impact, deer may starve, reducing the population. The smaller deer population may enable the vegetation to recover, and in the next season the deer may increase in numbers. Most systems are continually shifting slightly one way or another as a reaction to external conditions. This change within a range of tolerance is called **dynamic equilibrium;** that is, a balance exists, but maintaining it requires adjustment of the parts to changing conditions, much as tightrope walkers sway back and forth and move their hands up and down to keep their balance. Dynamic equilibrium means that the balance is not static.

The interactions that cause change or adjustment between parts of a system are called **feedback.** Two kinds of feedback are possible in a system. **Negative feedback,** where one change tends to offset another, creates a natural counteracting effect that is generally beneficial because it tends to help the system maintain equilibrium. Earth subsystems can also exhibit **positive feedback** sequences for a while—that is, changes that reinforce the direction of initial change. For example, several times in the past 2 million years, Earth has experienced significant decreases in global temperatures. This cooling of the atmospheric system led to the growth of great ice sheets, which covered large portions of Earth's surface. The massive ice sheets increased the amount of solar energy that was reflected back to space from Earth's surface, thus increasing the cooling trend in the atmosphere and the further growth of the ice sheets. The result over a considerable period of time was positive feedback. But ultimately the climate got so cold that evaporation from the oceans decreased, cutting off the supply of moisture to storms that fed snow to the glaciers. The reduction of moisture is an example of what is called a **threshold,** a condition that causes a system to change dramatically, in this case bringing the positive feedback to a halt. The decrease in snowfall caused the glaciers to shrink and the climate to warm, thus beginning another cycle.

Thresholds are trigger mechanisms—conditions that, if met or exceeded, can cause a fundamental change in the system and the way that it behaves. For example, stresses may build along an earthquake fault, but an earthquake will not occur until the strength of the rocks to resist that stress is exceeded. Thresholds are common regulators of any systems process. As another example, putting fertilizer on a plant will help it to grow larger and faster. But will more and more fertilizer continue this positive feedback relationship forever? Too much fertilizer may actually poison the plant and cause it to die. Either exceeding or not meeting certain critical conditions (thresholds) changes the system dramatically.

To further illustrate how feedback works, let us consider a simplified example—a hypothetical scenario of what might happen if human-caused damage to the atmosphere's ozone layer continues unimpeded by human counteraction (Fig. 1.13). Figure 1.14 shows what is called a **feedback loop**—a circular set of feedback operations that can be repeated as a cycle. Generally in natural systems, the overall result of a feedback loop is negative feedback, because the sequence of changes serves to counteract the direction of change in the initial element. The example is intended to show you how to think about Earth processes as a system. After you fol-

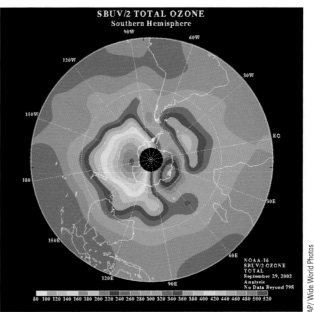

Figure 1.13 Satellite image of the ozone "hole" over Antarctica. Concern about changes in the ozone layer led to a cooperative agreement between the United States and the former Soviet Union to launch a satellite with a special imaging apparatus especially designed to send back images of ozone levels in the upper atmosphere. The system, called TOMS (Total Ozone Mapping Spectrometer), beams data back to Earth, from which this kind of image is constructed through computer analysis and display. The ozone hole is shown in shades of purple. Its pattern changes in both size and position in response to wind patterns that circle the South Polar region, as well as other meteorological factors.

■ **Is it important for scientists to study and monitor changes in the ozone layer? Why or why not?**

low the steps (feedback relationships) in Figure 1.14, try to think of some other examples of feedback operations in natural systems.

Let us look at our example of a feedback loop and examine how the factors are related. First we must start with some facts.

1. We know that the ozone layer in the upper atmosphere protects us by blocking harmful ultraviolet (UV) radiation from space, radiation that could otherwise cause harmful skin cancers and cell mutations.
2. We also know that chlorofluorocarbons (CFCs), chemicals widely used in air conditioners (as freon), can migrate to the upper atmosphere and cause chemical reactions that destroy ozone.

Knowing these facts, keep in mind that the following systems example is simplified and, like all models, is based on assumptions that may or may not be scientifically verified. In fact, efforts have been undertaken in the last 20 years or so to minimize the use of CFCs in the United States. Today, new automobiles and trucks are sold with air conditioners that use an "ozone-friendly," non-CFC unit to cool the vehicles' interiors.

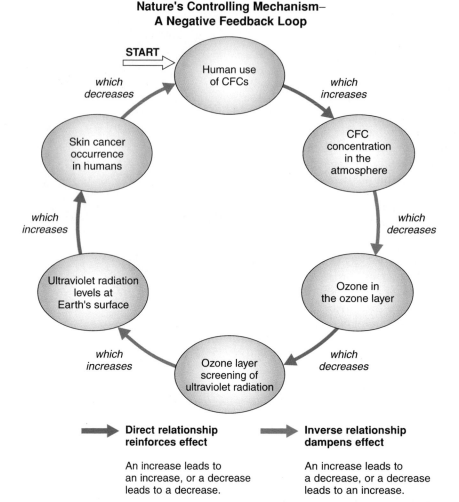

**Nature's Controlling Mechanism—
A Negative Feedback Loop**

Figure 1.14   Nature's controlling mechanism: A negative feedback loop. This diagram of a feedback loop illustrates how negative feedback acts to moderate change and add stability to a system. Relationships between two interacting variables (one link to the next in the loop) can be either direct or inverse. A direct relationship means that either an increase or a decrease in the first variable will lead to the same effect on the next. For example, a decrease in ozone leads to a decrease in ozone-layer screening of ultraviolet radiation. An inverse relationship means that the change in the first variable will result in an opposite change in the next. For example, an increase in CFCs leads to a decrease in ozone in the ozone layer. After completing one pass through the negative feedback loop, an important shift will occur in how the first variable is affected, and this will reverse all of the subsequent changes in the next cycle (although variables will maintain the same relationship to each other, either direct or inverse). You can follow a second pass through the feedback loop (reversing the increase/decrease interactions) to understand how this works. This feedback loop shows that human decision making can play a role in environmental systems. The last link in the loop between skin cancer and human use of CFCs would likely be a result of people acting on their own to reduce the problem.   ■   **What might be the potential (and extreme) alternative resulting from a lack of corrective action by humans?**

The diagram in Figure 1.14 shows six of the most important factors related to ozone-layer damage by CFCs. Each of these factors is linked by an interaction to the next variable in the loop. Systems analysis allows us to see how these processes will affect the variables and helps us to answer "What if?" questions. For example, if CFCs continue to erode the ozone layer, what will happen?

Follow Figure 1.14, starting with the human use of CFCs in the upper left and trace the feedback links outlined below.

1. If the amount of CFCs used by humans *increases,* the amount of CFCs in the atmosphere will also *increase.* An increase leads to an increase in the next factor, so this is a *direct* (positive) relationship.

2. *Increasing* the CFCs in the atmosphere will lead to a *decrease* of ozone in the ozone layer. Here an increase leads to a decrease in the next factor, so this is an *inverse* (negative) relationship between atmospheric CFCs and ozone.

3. *Decreasing* the ozone in the upper atmosphere will *decrease* the amount of harmful ultraviolet (UV) radiation that is blocked by the ozone layer. Here a decrease leads to a decrease; this is a *direct* relationship because the decreasing effect is reinforced.

4. *Decreasing* the blocking of harmful UV radiation will cause an *increased* amount of harmful ultraviolet radiation at Earth's surface. A decrease; leads to an increase, so this is an *inverse* relationship.

5. *Increasing* the level of UV radiation at Earth's surface will cause an *increased* amount of skin cancer in humans, which can be fatal. An increase leads to an increase, so this is a *direct* relationship.

6. *Increasing* skin cancer in humans could lead to policy changes that *decrease* the release of CFCs into the atmosphere, producing negative feedback relative to the initial variable (#1 above) in the feedback loop.

Finally, there remains an important question: What is likely to happen to the human use of CFCs if the occurrence of skin cancer continues to increase? Will humans act to correct the problem, or will they be likely to do nothing? What would be the potential consequences in either case? Ironically, negative feedback loop operations are good because they regulate a system through a strong tendency toward balance. As we have seen in Figure 1.14, feedback loops in nature normally do not operate for extended periods on positive feedback because environmental limiting factors act to return the process to a state of equilibrium.

There are both advantages and disadvantages in viewing Earth as a vast system made up of countless subsystems. As we have mentioned, systems are models, and like all models they are not the same as reality. They are products of the human mind and are only a way of looking at the real world. The examination of various Earth subsystems helps us to understand the natural processes involved in the development of the atmosphere, lithosphere, hydrosphere, and biosphere, as we know them today. Models may even help us to simulate past events or predict future change. But we must be careful not to confuse simplified models with the complexities of the real world.

Geography as Environmental Science

One of the great advantages of considering systems when studying physical geography is that systems serve to illustrate relationships among the elements or variables

Figure 1.15 The physical and cultural attributes of a site combine to form a unique environment. The humans who occupy the site, such as these boat dwellers of Aberdeen, Hong Kong, are both influenced by the environment and an integral part of it. ■ **What can you learn about human–environment relationships by studying this photograph?**

Donald Marshall Collection

Human—Environment Relationships

Throughout much of history, little thought was given to the human-environment relationship, but two negative aspects of that relationship have gained serious attention in recent years. Certain environmental processes, with little or no warning, can become dangerous to human life and property, and certain human activity threatens to cause major, and possibly irrevocable, damage to Earth environments.

Although all four of the major Earth subsystems inevitably become involved, the environment becomes a hazard to humans and other life forms because of relatively uncommon and extraordinary events associated most directly with the atmosphere and the lithosphere. The routine processes of these two subsystems become a problem and spawn environmental hazards for two reasons. First, on occasion and often unpredictably, they operate in excessive or violent fashion. Summer showers become torrential rains that occur repeatedly for days or even weeks. Ordinary tropical storms gain momentum as they travel over warm ocean waters, and they reach coastlines as full-blown hurricanes. Molten rock and associated gases from deep beneath Earth move slowly toward the surface and suddenly trigger massive eruptions that literally blow apart volcanic mountains. Each of these examples of Earth systems operating in sudden or excessive fashion is a noteworthy environmental event, but it does not become an environmental hazard unless people or their property is affected. Thus, the second reason environmental hazards exist is because people live where catastrophic environmental events occur. The greater the number of people and the greater the value of the property involved, the greater the catastrophe.

Why do people live where environmental hazards pose a major threat? Actually, there are many reasons. Some people have no choice. The land they live on is their land by birthright; it was the land of their fathers and grandfathers. Especially in densely populated developing nations, there is often no other place to go, except to overcrowded cities where even greater dangers await. Other people choose to live in hazardous areas because they believe the advantages outweigh the potential for natural disaster. They are attracted by the productivity of the farmland, the natural beauty of a building site, or the economic possibilities associated with a particular location. In addition, there are few populated areas of the world that are not associated with one or another environmental hazard. Forested regions are subject to fire; earthquake, landslide, and volcanic activities plague mountain regions; violent storms threaten interior plains; and many coastal regions experience periodic hurricanes or typhoons (the Pacific Ocean equivalent).

Just as the environment can pose an ever-present danger to humans, humans, in return, can constitute a serious threat to the environment. Issues such as global warming, acid precipitation, deforestation and the extinction of biological species in tropical areas, damage to the ozone layer of the atmosphere, and desertification have risen to the top of agendas when world leaders meet and international conferences are held. They are recurring subjects of magazine and newspaper articles, books, and television programs.

Much of the damage to the environment is a result of atmospheric pollution associated with industrialization in involved in the study. As we have noted, physical geographers are keenly interested in relationships between processes and interactions in the Earth system, and they pay special attention to the relationships between humans and the physical environment. Some geographers even consider the relationships or interactions between humans and the environments they occupy to be the central theme of geography. Whether or not this is the case, much of human civilization throughout time has been a product of the adaptations various cultures have made to, and the modifications they have imposed on, their natural surroundings. Primitive skills and technology generally require people to make greater adjustments in adapting to their environment. The more sophisticated the culture's technology, the greater the amount of environmental modification. Thus, human-environment interaction always was, and always will be, a two-way relationship, with the environment influencing human behavior and humans impacting the environment. This two-way relationship that is so important to the physical geographer is a central concern in environmental science.

The human-environment relationship is well illustrated through the examination by physical geographers of human impacts on ecosystems. During the second half of the twentieth century, people became increasingly conscious of their environment. Today, we regularly hear talk about the environment and ecology and worry about damage to ecosystems caused by human activity. Newspapers and newsmagazines often devote entire sections to discussions of environmental issues. But what are we really talking about when we use words like *environment, ecology,* or *ecosystem?*

In the broadest sense, our **environment** can be defined as our surroundings; it is made up of all the physi-

The Drax coal-fired power station near Selby, North Yorkshire, England: The steam, laden with CO_2 and other harmful gases, provides graphic evidence of atmospheric pollution.

Destruction of property by a tornado that struck Andover, Kansas.

the wealthy, developed nations of the world. But as population pressures mount and the less wealthy nations struggle to industrialize, human activity is exacting an increasing toll on the soils, forests, air, and water of the developing world as well. Issues of environmental deterioration are worldwide, and solutions must involve all of the world's nations to be successful. As citizens of the wealthiest nation in the world, Americans must seriously consider what steps can be taken to counter the major threats to the environment caused by human activity. What are the causes of these threats? Are the threats real and well documented? What can I personally do to help solve environmental problems?

Examining environmental issues from the physical geographer's perspective requires that the characteristics of both the environment and the humans involved in those issues be given equal consideration. As will become apparent in this study of geography, physical environments are changing constantly, and all too frequently, negative changes are a result of human activity. In addition, throughout Earth, humans live in constant threat from environmental hazards such as earthquake, fire, flood, and storm. The issues are directly related to the physical environment, but answers to questions regarding causes and solutions are deeply imbedded in the economic, political, and social characteristics of the cultures involved. The recognition that geography is a holistic discipline—that it includes the study of all phenomena on Earth—requires that physical geographers play a major role in the environmental sciences.

cal, social, and cultural aspects of our world that affect our growth, our health, and our way of living (Fig. 1.15). Humans also share environments with plants and animals. We can speak of a plant's environment and include in our discussion the soil in which the plant grows, the amount of sunlight and rainfall it receives, the gases that surround it, the range of air temperatures, and the nature of the plants that grow nearby, which also serve to block winds, sunlight, and rain.

Just as humans interact with their environment, so do other animals and plants. The study of relationships between organisms, whether animal or plant, and their environments is a science known as **ecology.** Ecological relationships are complex but naturally balanced "webs of life." Disrupting the natural ecology of a community of organisms may have negative results (although this is not always so). For example, filling in or polluting coastal marshlands may disrupt the natural ecology of such areas. As a result, fish spawning grounds may be destroyed, and the food supply of some marine animals and migratory birds could be greatly depleted. The end product is the destruction of valuable plant and animal life.

The word *ecosystem* is a contraction of *ecological system.* An **ecosystem** is a community of organisms and the relationships of those organisms to their environment. An ecosystem is dynamic in that its various parts are always in flux. For instance, plants grow, rain falls, animals eat, and soil matures—all changing the environment of a particular ecosystem. Since each member of the ecosystem belongs to the environment of every other part of that system, a change in one alters the environment for all the others. As those components react to the alteration, they in turn continue to transform the environment for the others. A change in the weather, from sunshine to rain, affects plants, soils, and animals. Heavy rain may carry away soils and plant nutrients so that plants may not be able to grow as well and animals, in turn, may not be able to eat as much. On the other hand, the addition of moisture to the soil may help some plants grow, increasing the amount of shade beneath them and thus keeping other plants from growing.

The ecosystem concept (like other systems models) can be applied on almost any scale from local to global, in a wide variety of geographic locations, and under all environmental conditions in which life is possible. Hence, your backyard, a farm pond, a grass-covered field, a marsh, a forest, or a portion of a desert can be viewed as an ecosystem. Ecosystems are found wherever there is an exchange of materials among living organisms and where there are functional relationships between the organisms and their natural surroundings. Ecosystems are open systems, as both energy and material move across their boundaries. Although some ecosystems, such as a small lake or a desert oasis, have clear-cut boundaries, the limits of many others are not as precisely defined. Often the change from one ecosystem to another is obscure and transitional, occurring slowly over distance.

Ever since human beings first walked on Earth, they have affected each ecosystem they have inhabited, and in modern times, the ability of humans to alter the landscape has been increasing. For example, a century ago the interconnected Kissimmee River–Lake Okeechobee–Everglades ecosystem constituted one of the most productive and stable wetland regions on Earth. But sawgrass marsh and slow-moving water stood in the way of urban and agricultural development. Intricate systems of ditches and canals were built, and since 1900, half of the original 4 million acres of the Everglades has disappeared (Fig. 1.16). The Kissimmee River has been channelized into an arrow-straight ditch, and wetlands along the river have been drained. Levees have prevented water in Lake Okeechobee from contributing sheet flow to the Everglades, and highway construction has divided the region, further disrupting natural drainage patterns. Fires have been more frequent and destructive, and entire biotic communities have been eliminated by lowered water levels. During excessively wet periods, portions of the Everglades are deliberately flooded to prevent drainage canals from overflowing. As a result, animals drown and birds cannot rest and reproduce. South Florida's wading bird population has decreased by 95 percent in the last hundred years. Without the natural purifying effects of wetland systems, water quality in South Florida has deteriorated, and with lower water levels, saltwater encroachment is a serious problem in coastal areas.

Today, backed by regional, state, and federal agencies, scientists are struggling to restore South Florida's ailing ecosystems. There are extensive plans to allow the Kissimmee River to meander again across its former flood plain, to return agricultural land to sawgrass marsh, and to restore historic water-flow patterns through the Everglades. The problems of South Florida should serve as a useful lesson. Alterations of the natural environment should not be undertaken without serious consideration of all the consequences.

The Human-Environment Equation

One lesson is obvious as we examine the varied physical environments that are produced by Earth's many subsystems. Despite the wealth of resources available in the air, water, soil, minerals, vegetation, and animal life on Earth, the capacity of our planet to support the growing numbers of humans may have an ultimate limit, a threshold population. And there are dangerous signs that such a limit may someday be reached. Human numbers on Earth have passed the 6 billion mark, and United Nations estimates indicate the total will reach 9 billion by 2050 if current growth rates continue. Today, more than half the world's people live in countries in which they must tolerate substandard living conditions and insufficient food. A major problem today is the distribution of food supplies, but ultimately, over the long term, the size of the human population cannot exceed the environmental resources needed to sustain them.

Although our current objective is to study physical geography, we should not ignore the information shown in the World Map of Population Density (inside textbook back cover). The map shows the distribution of people over the land areas of Earth and illustrates an important aspect of the human-environment equation. World population distributions are highly irregular; people have chosen to live, and have multiplied rapidly, in some places and not in others. One reason for this uneven distribution is the differing capacities of Earth's varied environments to support humans in large numbers. The World Map of Population Density also shows where most of the problems that threaten Earth's environments originate. The spatial information presented in this map and the world distributional maps in the chapters that follow will also help us identify those locations where large numbers of people live in hazard zones under the constant threat of property damage or death from natural phenomena such as earthquakes, hurricanes, floods, fires, or volcanic eruptions.

The relationships between humans and the environments in which they live will be emphasized throughout this book. Geographers are keenly aware that, as in most relationships, the nature or behavior of each of the parties in the relationship may have direct effects on the other. However, when considering the human-environment equation and the sustaining of humans at acceptable living standards for generations to come, it is most important to note that environments do not change their nature to accommodate humans. Humans should make greater attempts to alter their behavior to accommodate

R. Gabler

EPA, South Florida Water Management Division

Figure 1.16   The mixed tree and grass vegetation characteristics of the Florida Everglades are shown in the top photo. Large areas of this valuable ecosystem have been lost to farmland, industry, and housing developments. Drainage ditches and highways have altered the remainder at the expense of plants, animals, and human water supplies.

As a natural stream channel, the Kissimmee River meandered (flowed in broad, sweeping bends) on its flood plain for a 100-mile stretch from Lake Kissimmee downstream to Lake Okeechobee center photo. In the 1960s and early 1970s, the river was "channelized" (artificially straightened), disrupting the previously existing ecosystem. The artificially straightened channel is shown at bottom. Today, as part of an ongoing project to restore this riparian (river bank) habitat, the Kissimmee is reestablishing its flood plain, associated wetland environments, and its meandering channel. These changes are intended to restore, as much as is possible, the natural environment of the Kissimmee River and its flood-plain habitat.   ■ **What factors should be considered prior to any attempts to return rivers and riparian habitats to their original condition?**

EPA, South Florida Water Management Division

the limitations and potentials of Earth environments. It has been said that humans are not passengers on Spaceship Earth; rather, they are the crew. This means we have the responsibility to maintain our own habitat. Poised at the interface between Earth and human existence, geography has much to offer in helping us understand the factors involved in meeting this responsibility.

Physical Geography and You

Many aspects of the physical environment affect our everyday lives. The principles, processes, and themes of physical geography provide keys that help us to be environmentally aware—that help us assess each environmental situation, analyze the factors involved, and then make an informed choice among possible courses of action. For example, should you plant a new lawn before or after the spring rains? What sort of environmental impacts might be expected from a proposed shopping center? What are the environmental advantages and disadvantages of a particular homesite? What potential impacts of natural hazards—flooding, landslides, earthquakes, hurricanes, and tornadoes—should you be aware of where you live? What can you do to minimize potential damage to your household from a natural hazard? What can you do to assure that both you and your family are as prepared as possible for the kind of natural hazard that might affect your home?

It is apparent, then, that the study of physical geography, and the understanding of our natural environment that it provides, are valuable to all of us. Perhaps you have wondered, however, what do those people who call themselves physical geographers do in the workplace? What kinds of jobs do they hold? Physical geography sounds interesting and exciting, but can I make a living at it?

By applying their knowledge, skills, and techniques to real-world problems, physical geographers make major contributions to human well-being, environmental stewardship, and the economic development of society. Physical geographers emphasize the Earth system, but they do not ignore the effect of people on that system or the impact that our environment may have on people and the way they live. A knowledge of physical geography can help us analyze and solve environmental problems, such as whether we should continue to build nuclear power plants, allow offshore oil development, or drain coastal marshlands. Each of these questions may generate a different answer depending on the environmental conditions and geography of the location in question. Intelligence efforts by the U.S. Department of Defense must predict the effects that weather and terrain may have on military or naval operations. Industries must evaluate how the development of a proposed plant site may alter the surrounding environment.

Applied physical geography takes many forms, and the Career Vision Series in this book will introduce you to physical geographers in the workplace. The geographers presented in these brief biographies each have a geography background, and they share a common experience with you—each began their geographic education as an undergraduate student in an introductory geography course like the one that you are now taking.

Finally, knowledge of physical geography can not only provide opportunities for personal enrichment and possible employment, it can also be a source of perpetual enjoyment. Geography is a visual science, and it is really more than just a subject. Geography is a way of looking at the world and of observing its features. It involves asking questions about the nature of those features and appreciating their beauty and complexity. It encourages you to seek explanations, gather information, and use geographic skills, tools, and knowledge to solve problems. Even if you forget many of the facts discussed in this book, you will have been shown new ways to consider, to see, and to evaluate the world around you. Just as you see a painting differently after an art course, so too will you see sunsets, waves, storms, deserts, rivers, forests, prairies, and mountains with an "educated eye." You should retain knowledge of geography for life. You will see greater variety in the landscape, not because there is any more variety there but because, like an astronaut in orbit, you will have been trained to observe Earth differently and with deeper understanding.

DEFINE & RECALL

human geography	spatial pattern	hydrosphere	pictorial/graphic model
regional geography	spatial interaction	biosphere	mathematical/statistical
physical geography	system	life-support system	model
spatial science	variable	natural resources	conceptual model
absolute location	Earth system	pollution	mental map
relative location	atmosphere	model	systems theory
spatial distribution	lithosphere	physical model	subsystem

inputs

outputs

open system

closed system

equilibrium

dynamic equilibrium

feedback

negative feedback

positive feedback

threshold

feedback loop

environment

ecology

ecosystem

DISCUSS & REVIEW

1. What steps do scientists take before their discoveries become accepted as scientific laws?

2. What characteristics might early humans have had in common with the physical geographers of today?

3. How is the discipline of geography unique among the sciences?

4. Why can geography be considered both a physical and a social science? What are some of the subfields of physical geography, and what do geographers study in those areas of specialization?

5. Why is geography known as the spatial science? What are some topics that illustrate the role of geography as the spatial science?

6. How does the study of systems relate to the role of geography as a physical science? What are the four major divisions of the Earth system, and how do the divisions interact with one another?

7. How do open and closed systems differ? How does feedback affect the dynamic equilibrium of a system?

8. How does negative feedback operate to maintain a tendency toward balance in a system? What is a threshold in a system?

9. How does the examination of human-environment relationships in ecosystems serve to illustrate the role of geography as an environmental science?

10. What is meant by the human-environment equation? Why is the equation falling further out of balance?

CONSIDER & RESPOND

1. One of the best ways to understand the nature of geography is found in its definition as the spatial science. How does this description of geography make it unique among the sciences?

2. There are advantages and disadvantages to the use of models and the study of systems by scientists. List and compare the advantages and disadvantages from the point of view of a physical geographer.

3. How have various kinds of pollution affected your life? List some sources of pollution in your city or town.

4. Give one example of an ecosystem in your local area that has been affected by human activity. In your opinion, was the change good or bad? What values are you using in making such a judgement?

5. How can a knowledge of physical geography be of value to you now and in the future? What steps should you take if you wish to seek employment as a physical geographer? What advantages might you have when applying for a job?

REPRESENTATIONS OF EARTH

Maps and other graphic representations of Earth are essential tools to geographers, and they are also important to everyday life.

In what ways are maps important in daily life or in the workplace?

What career fields make particularly intense use of maps?

Why are geographers major users of maps?

Why are maps an important means of communication?

The Earth's shape is generally referred to as spherical.

How does the shape of Earth deviate from a sphere?

Why, and by how much?

The world globe is the only representation of the entire Earth with little or no distortion.

Why? What does this statement imply about maps?

The geographic grid is an arbitrary coordinate system for accurately describing a location on Earth.

In what ways is it an arbitrary system of location?

How was the geographic grid developed?

How is it used?

How is the Earth grid associated with navigation and time zones?

The Public Lands Survey System has had a significant impact on land-use patterns in much of the United States.

In what ways? Why was the system developed?

How is this system different from and similar to the geographic grid?

Every map, projection, and image of Earth has both beneficial and detrimental properties that affect the accuracy of the spatial or environmental information they portray.

What does this mean?

What does this suggest in relation to the development of a "perfect" map?

Why should representations of Earth be carefully selected to fit an intended purpose?

Topographic maps, computer maps, geographic information systems, global positioning systems, and remotely sensed images are useful tools for the physical geographer.

Why? What are they?

How can maps, aerial photographs, and remotely sensed images provide complementary information about a place?

Geographic information systems allow the direct comparison and combination of many map information layers.

How can GIS help us to understand spatial relationships in complex environmental systems?

Why is it useful to be able to simultaneously compare the locations and distributions of two or more environmental variables?

A variety of satellite imagery and computer techniques were used to produce this beautiful and accurate digital model of the Grand Canyon.

The principle of location is essential to geographers as they describe and analyze aspects of the Earth system. Although many of the basic principles used in dealing with locational problems have been known for centuries, the technologies and methods applied to these tasks are ever-changing. Today, the tools used by geographers and other scientists to find, describe, record, and display locations are becoming more precise and accurate through the application of computers and space-age technologies.

Location on Earth

Perhaps as soon as people began to communicate with language, they also began to develop a language of location. They used landscape features as directional cues, saying, for example, "Go along the river until it forks, then follow the widest branch until you come to the campsite." Today, in built-up areas, we use streets, traffic lights, and city blocks to find our way. In ancient times, however, humans used rivers, trees, hills, and other natural landmarks to find their way and to locate points of interest.

When ancient peoples began to sail the open ocean, they recognized the need for new ways of finding directions and describing locations. Long before the first compass was developed, they discovered that positions of the sun and stars—rising, setting, or circling in the heavens—could provide accurate directions. Observing relationships between the sun and the stars to a position on Earth is a basic skill in **navigation,** the science of location and wayfinding. Navigation has been called the process of getting from where you are to where you want to go.

Maps and Mapmaking

No one knows when or where the first maps were made, as their origin is lost in antiquity. Early humans certainly drew locational diagrams on rock surfaces or in the soil. Permanent maps were created by ancient cultures in many parts of the world, such as China, Egypt, the Pacific Islands, Greece, Mexico, and Peru. Early maps were constructed of sticks or were drawn on clay tablets, stone slabs, metal plates, papyrus, linen, or silk. These ancestral maps were fundamental to the beginnings of geography because they helped humans to communicate spatial thinking and direction finding (Fig. 2.1).

Most maps and globes are visual representations of all or part of Earth, but they can also be produced to show extraterrestrial features such as the Moon and other planets. These representations convey spatial information through graphic symbols, a "language of location" that must be understood in order to comprehend and use the rich store of information that maps and globes can display.

Cartography is the science of mapmaking. Geographers who specialize in cartography supervise the development of maps and globes to ensure that mapped information and data are accurate and presented in an effective manner. Most cartographers would agree that a map is primarily a tool for communicating spatial information. In recent years, cartography has been revolutionized through the application of computer technology.

The changes in map data collection and display that have occurred in the 20th century are comparable to the change from pedestrian to astronaut. Information that used to be collected little by little from ground observations can now be collected instantly by satellites hurtling through space, and recorded data can be flashed back to Earth at the speed of light.

*Cartographers can now gather spatial data and make maps faster than ever before—within hours—and the accuracy of these maps is excellent. Moreover, **digital mapping** enables mapmakers to experiment with a map's basic characteristics (for example, scale or projections), to combine and manipulate map data, to transmit entire maps electronically, and to produce unique maps on demand.*

United States Geological Survey (USGS)
***Exploring Maps,* page 1**

Digital measurements and images from aircraft and satellites, automated methods of data collection, and computer-assisted cartography have contributed to a map "explosion." Maps are found everywhere. We can all think of many fields, such as navigation, political science, community planning, surveying, history, meteorology, and geology, in which representations of Earth, including globes or

Figure 2.1   When was the first map made by humans? Cave paintings by Cro-Magnon people in France depict the animals they were hunting, sometime between 17,000 and 35,000 years ago. Although this view shows detail of stags crossing a river, experts suggest that some of the artwork may represent a rudimentary map. The paintings include lines that apparently represent migration routes, and other marks appear to represent locational information. If so, this is the earliest known example of humans recording their spatial knowledge.

■ **Why would these prehistoric humans want to record locational information?**

maps, are vital. In our everyday lives, through education, travel, television, recreation, and reading, we have all had experience with maps. Examine the maps in one of your daily newspapers. How do they contribute to your understanding of the news? How many are there? How many would that equal in a year (365 daily papers)?

Size and Shape of Earth

For most of their history, humans have pondered the size and shape of their world. Although it was not until the 1960s that we were able to image the shape of Earth from space, ancient Greeks as early as Pythagoras in 540 B.C. theorized that Earth was a sphere. In 200 B.C., a philosopher-geographer named Eratosthenes estimated the circumference of Earth within a few hundred miles of its actual size.

For most purposes, Earth can be considered a perfect sphere with an equatorial circumference of 39,840 kilometers (24,900 mi). However, because of forces associated with Earth rotation, the area near the equator actually bulges, and the polar regions are flattened slightly. So instead of a perfect sphere, Earth is more properly an **oblate spheroid.** The apparent boundary line between the sky and Earth is called the *horizon*. If you hold a basketball to represent Earth in front of you with the sky as a background, you can see a similar boundary line, which is curved no matter which way you turn the ball.

Earth's deviations from a perfect sphere are relatively minor. Nevertheless, these shape irregularities do affect navigation, mapping, and distance accuracies. People working in navigation, surveying, aeronautics, and cartography must include in their calculations the deviations of Earth's shape from true sphericity. Scientists have been able to confirm the extent of such deviations from measurements of variations in gravitational pull acting on satellites. The diameter of Earth at the equator is 12,758 kilometers (7927 mi), while from pole to pole it is 12,714 kilometers (7900 mi). On a globe with a diameter of 12 inches, this difference of 44 kilometers would be 4/100 of an inch, a deviation of about one-third of 1 percent and not noticeable to the naked eye.

Landforms also cause Earth to deviate from true sphericity. Mount Everest in the Himalayas is the highest point on Earth's surface at 8847 meters (29,028 ft) above sea level. The lowest known point on Earth's surface is in the Challenger Deep, in the Mariana Trench of the Pacific Ocean southwest of Guam. This spot is 11,033 meters (36,198 ft) below sea level. The difference between these two points, 19,880 meters, or just over 12 miles, would also be insignificant and invisible when reduced in scale to a globe with a 12-inch diameter.

Globes and Great Circles

The world globe is a nearly perfect model of our planet. It shows the spherical shape of Earth and accurately displays spatial relationships between landforms and water bodies, comparative distances between locations, relative sizes and shapes of Earth's features, and accurate compass directions. A world globe is an extremely useful tool because it enables us to *see,* rather than merely imagine, many geographic aspects of our planet. For example, learning about the seasons or understanding how the length of daylight hours in one place changes annually is made a great deal easier. Because a globe has essentially the same shape as our planet, it can represent without distortion the various features and relationships associated with Earth (Fig. 2.2).

Figure 2.2 Earth as photographed from space by Apollo astronauts, showing most of Africa, the surrounding oceans, storm systems in the Southern Hemisphere, and the relative thinness of the atmosphere (left). A manufactured globe represents the same general viewpoint (right).

Figure 2.3   Planes slicing through a globe's center (a) at the equator and (b) obliquely. In either case, the globe is divided into equal halves, and the line where the plane intersects the surface of the globe is a great circle with the same circumference as the globe itself. In (c), the plane slices the globe into unequal parts. The line of intersection of such a plane with the globe is a small circle.

Globes also have specific limitations. A world globe would not help us find our way on a hiking or backpacking trail. It would be bulky and awkward to carry, and our location would appear as a tiny pinpoint, with little, if any, local information. We would need a map that clearly showed elevations, trails, and rivers—a map that could be folded and put in a pocket or pack. However, despite the limitations of globes, if we want to view a representation of the entire Earth, a world globe provides the most accurate representation. Also, because maps attempt to depict a spherical Earth, or a part of Earth, on a flat surface, being familiar with the characteristics of a globe helps us to understand maps and how they are constructed.

An imaginary circle drawn in any direction around Earth's surface and whose plane passes through the center of Earth is called a **great circle** (Fig. 2.3a, b). It is "great" because it is the largest circle that can be drawn around Earth through two particular points. Great circles have several useful characteristics: (1) Every great circle divides Earth into equal halves called **hemispheres;** (2) every great circle is a circumference of Earth; and, perhaps most important, (3) great circles mark the shortest travel routes between locations on Earth's surface. Circles whose planes do not pass through the center of Earth are called **small circles** (Fig. 2.3c); any circle on Earth's surface that does not divide the planet into equal halves is a small circle.

The shortest route between two places can be located by finding the great circle that connects them. Put a rubber band (or string) around a globe to help visualize this spatial relationship. Connect any two cities, such as Moscow and New York, San Francisco and Tokyo, New Orleans and Paris, or Kansas City and Singapore, by stretching the rubber band around the globe so that it touches both cities and divides the globe in half. The rubber band then marks the shortest route between the two cities. Navigators chart these *great circle routes* for aircraft and ships because traveling the shortest distance saves time and fuel. The farther away two points are on Earth, the larger the distance savings will be by following the great circle route that connects them. The *arc of a great circle* is the shortest distance between two points on any spherical surface. Another important example of a great circle is the *circle of illumination,* which divides Earth into light and dark halves—a day hemisphere and a night hemisphere.

Latitude and Longitude

Imagine you are traveling by car and you want to visit the Football Hall of Fame in Canton, Ohio. Using the Ohio road map, you look up Canton in the map index and find that it is located at "G-6." On the map, you look down the side until you find 6, and then along the top until you reach G. The letter G and the number 6 meet in a box marked on the map. Scanning the area within box G-6, you locate Canton (Fig. 2.4).

What you have used is a **coordinate system** of intersecting lines, which on the map make up a system of *grid cells.* Without a coordinate system that provides locational reference points, it would be difficult to describe a location on a sphere like Earth. The problem is deciding, determining, or finding where the starting points would be for a grid system on a sphere. Without points of reference, either natural or arbitrary, a sphere is a geometric form that appears the same from any direction.

Measurement of Latitude   Fortunately, Earth has two natural reference points. As Earth spins on its axis, the ends of that axis are points called the **North Pole** and the **South Pole.** The location of the poles can be used to define a great circle, the **equator,** that lies exactly between them. Now we can measure the angle between the equator and any position north or south of the equator. This angular distance is called **latitude** and is measured in degrees north or degrees south of the equator.

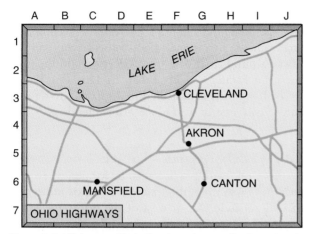

Figure 2.4 Using one kind of rectangular coordinate system to locate a position accurately. This map employs an alphanumeric location system, similar to that used on many road maps.
■ **What are the rectangular coordinates of Mansfield? What is at location F-3?**

In order to locate the latitude of Los Angeles, imagine a line that goes from the center of Earth to Los Angeles and another line that goes from Earth's center to a point on the equator directly south of the city. These two lines form an angle, and the arc of this angle is the latitudinal distance (in degrees) that Los Angeles lies north of the equator (see Fig. 2.5a). The angle made by these two imaginary lines is just over 34°—so the latitude of Los Angeles is about 34°N (north of the equator).

The equator is designated as 0° latitude. As we go farther north or south of the equator, the angles and their arcs increase until we reach the North or South Pole and the maximum latitudes of 90° north or 90° south.

Because Earth's circumference is approximately 40,000 kilometers (25,000 mi) and there are 360 degrees in a circle, we can use division (40,000 km/360°) to find that a degree of latitude is equal to about 111 kilometers (69 mi). One degree of latitude covers a relatively large distance, so degrees are further divided into minutes (') and seconds (") of arc. Actually, Los Angeles is located at 34°03'N (34 degrees, 3 minutes north latitude).

There are 60 minutes of arc in a degree. We can get even more precise: 1 minute is equal to 60 seconds of arc. So we could locate a position at latitude 23° 34'12"S, which we would read as 23 degrees, 34 minutes, 12 seconds south latitude. A minute of latitude equals 1.85 kilometers (1.15 mi), and a second is about 31 meters (102 ft).

A **sextant** can be used to visually determine latitude for celestial navigation (Fig. 2.6). This instrument measures the angle between the horizon and a celestial body, such as the noonday sun or the North Star (Polaris). However, the latitude of a location is only half of its global address. Los Angeles is located approximately 34° north of the equator, but an infinite number of points exist on the same line of latitude.

Measurement of Longitude To fully describe the location of Los Angeles, we must also determine where along the line of 34°N latitude it is situated. However, to describe an east or west position, we must have a starting line, just as we used the equator as our latitudinal reference. Actually, any half of a great circle, running from pole to pole, could serve this function. The global position of the 0° east-west reference point for longitude is arbitrary and in the past was a matter of national pride.

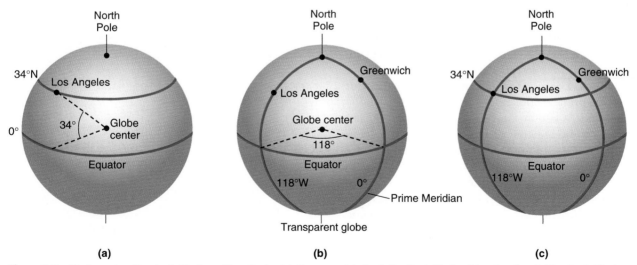

(a)   **(b)**   **(c)**

Figure 2.5 Finding a location by latitude and longitude. (a) The geometric basis for the latitude of Los Angeles, California. Latitude is the angular distance in degrees either north or south of the equator. (b) The geometric basis for the longitude of Los Angeles. Longitude is the angular distance in degrees either east or west of the prime meridian, which passes through Greenwich, England. (c) The location of Los Angeles is 34°N, 118°W. ■ **What is the latitude of the North Pole?**

R. Sager

Figure 2.6 Finding latitude for celestial navigation. A traditional way to determine latitude is by measuring the angle between the horizon and a celestial body with a sextant. Today, most air and sea navigation (as well as certain land travel) is supported by a satellite-assisted technology called the Global Positioning System (GPS).
■ **With high-tech location systems like GPS available, why might understanding how to use a sextant still be important?**

As we go farther east or west from 0° longitude at the prime meridian, our longitude increases. Traveling eastward from the prime meridian, we will eventually be halfway around the world from Greenwich in the middle of the Pacific Ocean at 180°E. This line is also 180°W. Longitude is measured in degrees up to a maximum of 180° east or west of the prime meridian.

The Geographic Grid

Any point on Earth's surface can be located by its latitude north or south of the equator, measured in degrees, and its longitude east or west of the prime meridian, measured in degrees. Similar geographic grids are used for determining locations on the Moon and the planets.

Parallels and Meridians

The geographic grid consists of lines that run east and west around the globe and lines that run north and south from pole to pole (Fig. 2.7). East-west lines mark latitudes north or south of the equator. Latitude lines circle the globe, are evenly spaced, and are parallel to the equator and each other. Hence, they are known as **parallels.** The only parallel that is a great circle is the equator; all other lines of latitude are small circles. One degree of latitude equals about 111 kilometers (69 mi) anywhere on Earth.

Lines of longitude run north and south, converge at the poles, and measure longitudinal distances east or west of the prime meridian. Each **meridian** of longitude, when joined with its mate on the opposite side of Earth,

Some countries designated their own zero line, passing through the country's capital, which caused confusion because the longitudes indicated on maps could vary with the map's country of origin. This difficulty was resolved by international agreement in 1884, when the longitude line that passes through Greenwich, England (near London), was accepted as the **prime meridian,** or 0° longitude. **Longitude** is the angular distance east or west of the prime meridian.

Longitude is also measured in degrees, minutes, and seconds. Returning to our example of Los Angeles, let us imagine a line drawn from the Earth's center to the point where the north-south running half-circle that passes through Los Angeles crosses the equator. This time, a second imaginary line will go from the center of Earth to a point where the prime meridian crosses the equator. Figure 2.5b shows that these two lines drawn from Earth's center define an angle, the arc of which is the angular distance that Los Angeles lies west of the prime meridian. Los Angeles has a longitude of 118° west of Greenwich. Figure 2.5c provides the global address of Los Angeles by combining latitude and longitude.

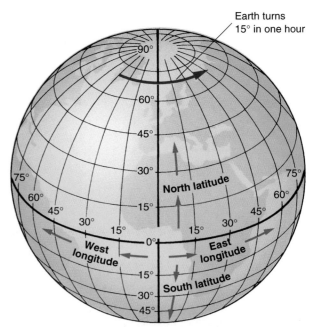

Figure 2.7 A globelike representation of Earth, showing the geographic grid with parallels and meridians at 15° intervals.
■ **How do parallels and meridians differ?**

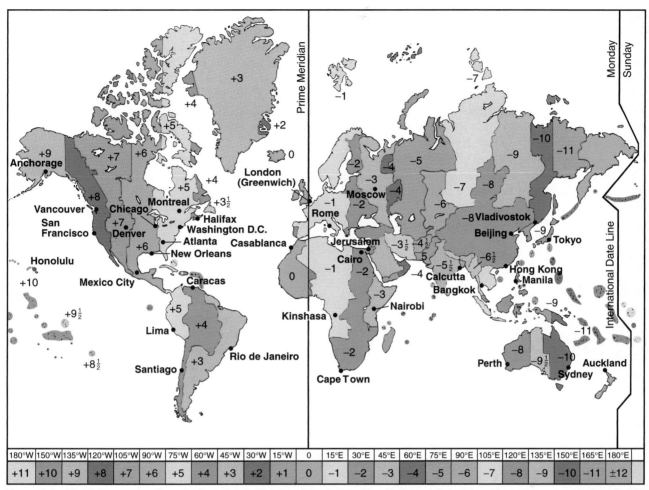

Figure 2.8 World time zones reflect the fact that Earth turns through 15° of longitude each hour. Thus, time zones are approximately 15° wide. Political boundaries usually prevent the time zones from following a meridian perfectly. ■ **What is the time difference between the time zone where you live and Greenwich, England?**

forms a great circle. Meridians at any given latitude are evenly spaced around the globe, although meridians get closer together as they move poleward from the equator. At the equator, meridians separated by 1° of longitude are about 111 kilometers (69 mi) apart, but at 60°N or 60°S latitude, they are only half that distance apart, about 56 kilometers (35 mi).

Longitude and Time

The relationship between longitude and time was used to establish the **time zones** that we know today. Until about 120 years ago, each town or area used what was known as *local time*. **Solar noon** was determined by finding the precise moment in a day when a vertical stake cast its shortest shadow. This meant that the sun had reached its highest angle in the sky for that day at that location—noon—and all local clocks were set to that time. Because of Earth's *rotation* on its axis, noon in a town to the east took place a little earlier, and towns to the west experienced noon a little later.

Using local time did not cause much difficulty until the late 1800s, with the development of railroads and advances in communication such as the telegraph. Increasing long-distance travel and technological changes made the use of local time by each community impractical. In 1884, the International Meridian Conference was held in Washington, D.C., to set up standardized time zones and to establish Greenwich as the location for the prime meridian (0° longitude). Earth was divided into 24 time zones, one for each hour of the day. Ideally, each time zone spans 15° of longitude (360° divided by 24 hours). The prime meridian was made the *central meridian* of its time zone, and the time when solar noon occurred at the prime meridian was established as noon for all places between 7 ½° east and 7 ½° west of that meridian. The same pattern was followed around Earth. Every line of longitude evenly divisible by 15° is the central meridian for a time zone of 7 ½° of longitude on either side. However, as shown in Figure 2.8, the time zones boundaries do not follow meridians exactly. Having a time zone boundary divide cities, towns, or states would be inconvenient, so jogs in the lines have been established to avoid most of these problems. In the United States, time zones often follow state boundaries. Imagine, for instance, the confusion that would result if Chicago were in one time zone and its suburbs were in another.

The time of day at the prime meridian, known as *Greenwich Mean Time* (sometimes called GMT, Universal Time, UTC, or Zulu Time), is used as a worldwide reference, and times to the east or west can be easily determined by comparing them to GMT. Time zones west of the prime meridian are said to be on *slow time* and places to the east are on *fast time*. Thus, a place 90° east of the prime meridian would be 6 hours fast (meaning that the current time is 6 hours later than the time in Greenwich), while time in the Pacific Time Zone of the United States and Canada, whose central meridian is 120°W, is said to be 8 hours slow (local time is 8 hours earlier than GMT).

Figure 2.9   The International Date Line. The new day officially begins at the International Date Line (IDL) and then sweeps westward around the Earth to disappear when it again reaches the IDL. Thus, west of the line is always a day later than east of the line. Maps and globes often have either "Monday|Sunday" or "M|S" shown on opposite sides of the line to indicate the direction of the day change.

■   **Why does the International Date Line deviate from the 180° meridian in some places?**

For navigation, longitude can be determined with a *chronometer,* an extremely accurate clock. Two chronometers are used, one set on Greenwich time, and the other on local time. The number of hours between them, fast or slow, determines longitude (1 hour equals 15° of longitude). Until the advent of electronic navigation by ground- and satellite-based systems, the sextant and chronometer were a navigator's basic tools for determining location.

The International Date Line

On the opposite side of Earth from the prime meridian is the **International Date Line.** It is an imaginary line that follows the 180th meridian, except for jogs to separate Alaska and Siberia and to skirt some Pacific island groups (Fig. 2.9).

At the International Date Line, we turn our calendar back a full day if we are traveling east and a full day forward if we are traveling west. Thus, if we are traveling east from Tokyo to San Francisco and it is 4:30 P.M. Monday just before we cross the International Date Line, it will be 4:30 P.M. Sunday on the other side. If we are traveling west from Alaska to Siberia and it is 10:00 A.M. Wednesday just before we reach the International Date Line, it will be 10:00 A.M. Thursday once we cross it. As a way of remembering this relationship, many globes have Monday and Sunday (or M|S) labeled in that order on the opposite sides of the International Date Line. To find the correct day, you just substitute the current day for Monday or Sunday, and use the same relationship.

Although the International Date Line was not established officially until the 1880s, the need for such a line on Earth to adjust the day was inadvertently discovered by Magellan's crew, who circumnavigated Earth in the years 1519–1521. Sailing westward from Spain around the world, the crew noticed when they returned that one day had apparently been missed in the ship's log. What had actually happened is that, in going around the world in a westward direction, the crew had experienced one less sunset and one less sunrise than had occurred in Spain during their absence.

The U.S. Public Lands Survey System

The geographic grid system of longitude and latitude was designed to locate the *points* where these lines intersect. A different system is used in much of the United States to define and locate land *areas*. This is the **U.S. Public Lands Survey System,** or the *Township and Range System.* Thought to have been suggested by Thomas Jefferson, it was proposed as a method for parceling public lands west of Pennsylvania for sale.

Using the Township and Range System, land was divided into parcels based on selected north-south lines called **principal meridians** and east-west lines called

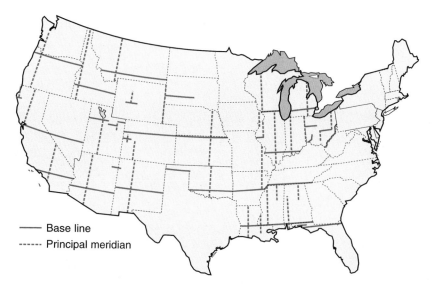

— Base line
----- Principal meridian

Figure 2.10   Principal meridians and base lines of the United States Public Lands Survey System (Township and Range System). ■ **Why wasn't the Township and Range System applied throughout the eastern United States?**

base lines (Fig. 2.10). The base lines were surveyed along parallels of latitude. The north-south meridians, though perpendicular to the base lines, had to be adjusted (jogged) along their length to counteract Earth's curvature. If adjustments were not made, the north-south lines would tend to converge and land parcels defined by this system would be smaller in northern regions of the United States.

The Township and Range System forms a pattern of nearly square parcels called townships laid out in horizontal *tiers* north and south of the base lines and in vertical *columns* ranging east and west of the principal meridians. A **township** is a square plot 6 miles on a side (36 sq mi, or 93 sq km). As illustrated in Figure 2.11, townships are first labeled by their position north or south of a base line; thus, a township in the third tier south of a base line will be labeled Township 3 South, which is abbreviated T3S.

However, labeling a township only by its tier position, north or south, is as insufficient as labeling a global location solely by its latitude. We must also name a township according to its *range*—its location east or west of the principal meridian for the survey area. Thus, if Township 3 South is in the second range east of the principal meridian, its full location can be given as T3S, R2E (Range 2 East).

The Public Lands Survey System further divides townships into 36 **sections** of 1 square mile or 640 acres (2.6 sq km, or 259 ha). Sections are designated by numbers from 1 to 36 that begin in the northeasternmost section with number 1 and, going back and forth across the township, end in the southeast corner with number 36. Each section is divided into four quarters (quarter-sections), named by their location in the section—northeast, northwest, southeast, and southwest, each with 160 acres (65 ha). Quarter-sections are further subdivided into

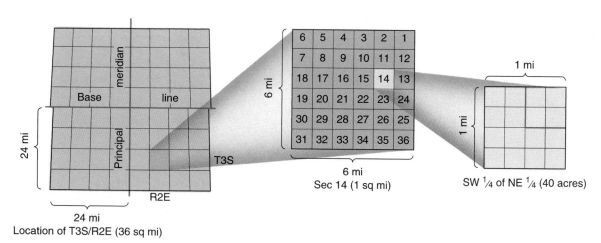

Location of T3S/R2E (36 sq mi)

T3S

R2E

6 mi
Sec 14 (1 sq mi)

1 mi
SW ¼ of NE ¼ (40 acres)

Figure 2.11   Method of location according to the Public Lands Survey System. ■ **How would you describe the extreme southeastern 40 acres of section 20 in the middle diagram?**

Grant Heilman/Grant Heilman Photography

Figure 2.12 Rectangular field patterns resulting from the Public Lands Survey System in the western United States. Note the slight jog in the field pattern to the right of the farm buildings in the center of the photo. ■ **How do you know this photo was not taken in the midwestern United States?**

four quarters (quarter-quarter sections) of 40 acres each (16.25 ha), sometimes known as *forties*. These quarter-quarter sections, or 40-acre plots, are named after their position in the quarter: the northeast, northwest, southeast, and southwest forties. Thus, we can describe the location of the 40 acres that are shaded in Figure 2.11 as being in the SW ¼ of the NE ¼ of Sec. 14, T3S, R2E, which we can find if we locate the principal meridian and the base line for the survey. The order is consistent from smaller division to larger, and township location is always listed before range (T3S, R2E).

The Township and Range system has exerted an enormous influence on landscapes in many areas of the United States; it is what gives most of the Midwest and West its regular checkerboard appearance from the air or from space (Fig. 2.12). The land was surveyed, staked, fenced, and farmed or ranched in parcels formed by the divisions of the Public Lands Survey System. Road maps

in states that use this land survey system reflect the use of this grid, with many roads following angular boundaries surveyed using this government system.

The Global Positioning System

The newest way to determine a geographic location employs the **Global Positioning System (GPS).** The GPS uses a network of satellites orbiting 11,000 miles above Earth. This high-tech locational system was originally created for military applications but today has a host of civilian uses, from surveying to navigation.

The GPS relies on the ability to measure precise distances from a location on Earth to several orbiting satellites. The satellites transmit radio signals that are detected by a ground receiver. Global positioning receivers vary in size, and handheld units are common (Fig. 2.13). The distance from receiver to satellite is calculated by measuring the time it takes for the radio signals, traveling at the speed of light, to arrive. A GPS receiver performs these calculations and displays a locational readout in longitude and latitude. Many receivers also display a digital map that marks the locational position of the receiver.

Global Positioning is based on the principle of triangulation. Assume that we determine that our distance from a GPS satellite is 20,000 kilometers. Thus, we are now certain that our location is somewhere in space on an imaginary sphere with a radius of 20,000 kilometers centered on that satellite (Fig. 2.14). This is an enormous area. However, if we simultaneously determine that we are 19,000 kilometers from a second satellite, our locational possibilities will be greatly reduced. Our location now involves the intersection of two spheres, one with a radius of 20,000 kilometers, the other with a radius of 19,000 kilometers. The only common surface between two intersecting spheres is a circle formed where the two spheres intersect (Fig. 2.14). A simultaneous measurement from a third satellite, 21,000 kilometers away, results in the intersection of a third sphere with the first two. The original sphere of possible locations, which was reduced to a circle of possible locations by the addition of a second satellite, is brought down to just two possible points by the addition of a third satellite (Fig. 2.14c). A fourth satellite allows us to determine which of the two points is our location.

The accuracy of GPS depends on several factors. One is the number of satellites used, as three or more satellite signals are needed to determine your position. Using signals from additional satellites can yield a more precise location. Satellite angles high above the horizon and low levels of interference between you and the satellites (open prairie versus dense forest) are other factors that yield more accurate measurements. With sophisticated GPS equipment and techniques, it is possible to find locational coordinates within small fractions of a meter.

Maps and Map Projections

Maps—representations of Earth or parts of Earth on flat paper—have many advantages. They can be reproduced easily and inexpensively, can depict the entire Earth or show a small area in great detail, and are easy to handle and transport. Maps also have limitations that vary with the type of map, but these limitations can be overcome or minimized by choosing the right kind of map based on an understanding of maps and cartography. There are many kinds of maps, and the key is knowing what kind of map is best for a particular task. No map fits all possible applications.

Advantages of Maps

It has been said that if a picture is worth a thousand words, then a map is worth a million. As visual representations of Earth, maps supply an enormous amount of information that would take pages and pages to describe—probably less successfully—in words. Because they are graphic representations and use symbolic language, maps show spatial relationships with great efficiency. Imagine trying to verbally explain all the information that a map of your city, county, state, or campus provides: sizes, areas, distances, directions, street patterns, railroads, bus routes, hospitals, schools, libraries, museums, highway routes, business districts, residential areas, population centers, and so forth. Maps can show true courses for navigation and true shapes of

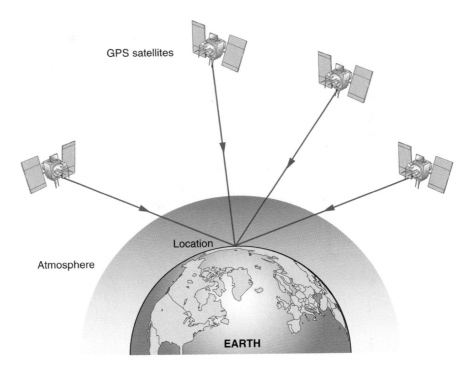

Figure 2.14 The Global Positioning System uses signals from a network of satellites to determine a position on Earth. The distances from several satellites (a minimum of three) are calculated by the GPS receiver to find the location in terms of longitude and latitude and elevation. With the distance from three satellites, a position can be located within meters, and with more satellite signals and sophisticated GPS equipment the position can be located very precisely.

Earth features. They can be used to measure areas or distances, and they can show the best route from one place to another.

Geographers can produce maps to show almost any relationship in the environment. As the interest of physical geographers is frequently spatial—the relationships among environmental subsystems and their changes through time and space—geographers use maps to develop visual or mathematical descriptions of these relationships. For many reasons, then—whether it is presented on paper, on a computer screen, or in the mind—the map is the geographer's most important tool.

In addition to being vital tools for professional geographers, maps are useful to people in every walk of life. They are valuable to tourists, political scientists, historians, geologists, pilots, soldiers, sailors, hikers, and even burglars (police departments also use maps to plot the spatial distribution of certain crimes). The Centers for Disease Control use maps to monitor the geographic distributions of illness outbreaks worldwide. The potential applications of maps are practically infinite, and we could even say "out of this world," because our space programs have produced maps of the Moon, Mars, and other extraterrestrial features.

Limitations of Maps

Despite the benefits and positive aspects of maps, they can never depict Earth with complete accuracy. Unlike globes, every map distorts its portrayal of Earth in some way. It is impossible to present a spherical planet on a flat (two-dimensional) surface and accurately maintain all of its geometric properties. This process has been likened to trying to flatten out an eggshell. Distortion is an unavoidable problem of representing a sphere on a flat map.

Using a globe, we can directly compare the size, shape, and area of Earth features, and we can measure distance, direction, shortest routes, and true directions, all with great accuracy. But because of the inherent distortion of maps, we can never compare or measure all of these properties on one map. There is no such thing as a perfect map, at least not the perfect map for all uses.

We must understand that all map depictions of Earth are distorted in some way. We should also realize, however, that when a map depicts only a small area, the distortion will be so slight as to be negligible. If we use a map of a state park for a day hike, the distortion will be too small to affect us, and we can have confidence in the map information. It is when we attempt to show large regions, or entire Earth, that our planet's curvature causes the map distortion to be apparent and pronounced.

In order to be skilled map users, we must know which properties a map depicts accurately, which features it distorts, and for what purpose a map is best suited. If we are aware of this information, we can make accurate comparisons and measurements on maps, and better understand the information that the map is conveying.

The global grid has four important geometric properties: (1) parallels of latitude are always parallel; (2) parallels are evenly spaced; (3) meridians of longitude converge at the poles; and (4) meridians and parallels always cross at right angles. There are thousands of ways to transfer a spherical grid onto a flat surface to make a **map projection**, but no map projection will maintain all four of these

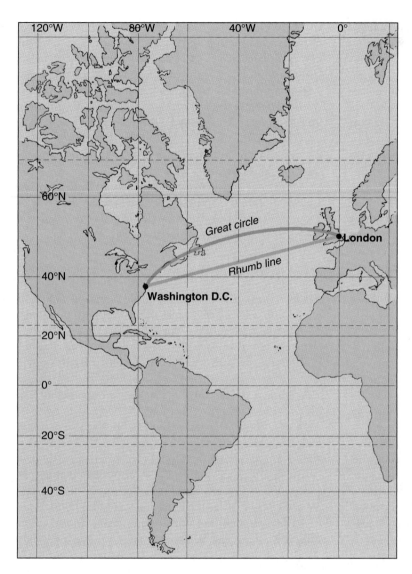

Figure 2.15 The Mercator projection, often misused as a general-purpose world map, was designed for the purpose of navigation. Its most useful property is that lines of constant compass heading, called rhumb lines, are straight lines. The Mercator is developed from a cylindrical projection. ■ **Compare the sizes of Greenland and South America on this map to their proportional sizes on a globe. Is the distortion great or small?**

properties. Checking the grid system of a map for these four properties will help us discover areas of greatest and least distortion on a particular map projection.

Properties of Map Projections

Shape   A flat map cannot depict large areas of the globe without distorting either shape or area. But small areas—regions, lakes, islands, bays—can be depicted in their true shape, with little apparent distortion. Maps that maintain the true shape of areas are known as **conformal maps.**

In order to preserve the shape of Earth features, parallels and meridians transferred to a map must maintain their global relationships. On conformal maps, meridians and parallels always cross at right angles, just as they do on the globe.

Area   Cartographers are able to create a world map that maintains true area relationships for the entire Earth. That is, areas on the map have the same proportions to each other as they have in reality. Thus, if we cover any two parts of the map with, say, a nickel, no matter where the nickel is placed, it will cover equivalent areas on Earth. Maps drawn with this property are called **equal-area maps.** Whenever size comparisons are being made between two or more areas shown on a map, it is important to use one that is equal area. The property of equal area is also useful when examining spatial distributions. As long as the map displays equal area and a symbol represents the same quantity throughout the map, we can get a good idea of the distribution of any feature—for example, people, churches, cornfields, hog farms, or volcanoes.

The only way to show equal area on a flat map of the world or of a large region, however, is to pull areas out of shape. It is impossible to show both equal area and true shape on a flat map. Most of us are familiar with the **Mercator projection** (Fig. 2.15), commonly used in schools and textbooks, although less so in recent years. In the Mercator projection, areas have true shape, so it is conformal and cannot have the property of equal area. Areas are greatly out of proportion. For many years, the Mercator projection was widely used in schools, leading generations of students to believe incorrectly that Greenland is as large as South America. Mercator's projection portrays Greenland as about the size of South America (see Fig. 2.15), but South America is actually, about eight times larger! A nickel placed on Greenland in this conformal map would cover an area of land much smaller than that covered by the same coin placed on South America.

Distortion is a major challenge in mapmaking. Because it is not possible to show both true shape and equal area, cartographers must decide which property they want to preserve at the expense of the other. Another cartographic strategy is to compromise and create a map that does a fairly good job of showing both properties but is not really correct for either (look ahead to Fig. 2.20b).

Distance   Just as no flat map can show true shape for all of Earth or for large areas, neither can it maintain a constant scale of distance over all Earth's surface. The scale on a map depicting a large area cannot be applied equally everywhere on that map. On a map of a small area, distance distortions will be insignificant, and the accuracy will usually be sufficient for most purposes.

It is possible for a map to have the property of *equidistance* in specific instances. That is, on a world map, the equator may have equidistance (a constant scale) along its length. Or all meridians, but not the parallels, may have equidistance. On another map all straight lines drawn from the center may have equidistance, but the scale will not be constant unless lines are drawn from the center.

Direction   Because the compass directions on Earth curve around the sphere, not all flat maps can show true compass directions. Thus, a given map may be able to show true north, south, east, and west, but the directions between those points may not be accurate in terms of the angle between them. So, if we are sailing toward an island, its location may be shown correctly according to its longitude and latitude, but the direction in which we must sail to get there may not be accurate, and we may pass right by it.

Maps that show true directions as straight lines are called *azimuthal* map projections. These are drawn with a center or focus, and all straight lines drawn from that center are true compass directions (Fig. 2.16).

Map Projections

We have been discussing map projections without fully explaining what is meant by this term. A map projection is a representation of the curved surface of a globe on a flat surface. As we have seen, some of the properties of a globe must be compromised in order to flatten it out into a map.

All map projections maintain one aspect of the globe—the property of location. Every place shown on a map must be in the same location with respect to latitude and longitude as it is on the globe. No matter how the arrangement of the global grid is changed by projecting parallels and meridians onto a flat surface, all places must still have their proper location—that is, they must display their accurate latitude and longitude.

Although maps are not actually made this way, certain projections can be demonstrated by putting a light

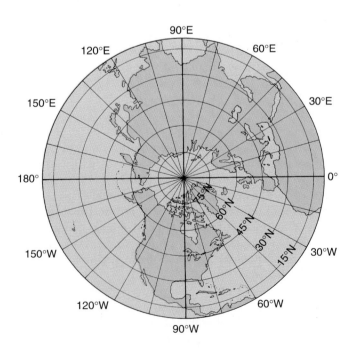

Figure 2.16 Azimuthal map centered on the North Pole. Although this is the conventional orientation of such a map, it could be centered anywhere on Earth. Azimuthal maps show true direction between all points, and can only show half of Earth on a single map.

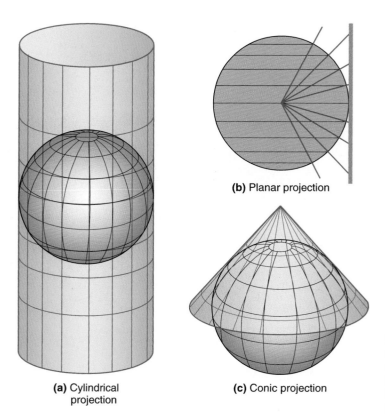

(a) Cylindrical projection

(b) Planar projection

(c) Conic projection

Figure 2.17 The theory behind the development of the (a) cylindrical, (b) planar, and (c) conic projections. Although projections are not actually produced in this manner, they can be demonstrated by projecting light from a transparent globe. ■ **Why do we use different map projections?**

inside a transparent globe so that the grid lines are projected onto a plane or surface (*planar projection*), a cylinder (*cylindrical projection*), or a cone (*conic projection*), which can be flattened (Fig. 2.17). Map projections today are developed by computers using mathematical computations involving the geographic grid.

The Mercator Projection As previously mentioned, one of the best-known conformal projections is the Mercator, named for Gerhardus Mercator, who devised this map in 1569. The Mercator projection, shown in Figure 2.15, is actually a mathematically adjusted cylindrical projection. Imagine a transparent globe with a light inside along its equator that casts shadows of the grid onto a sheet of paper wrapped around the globe to form a cylinder. If the paper touches the globe everywhere at the equator, then the equator on the flat map will be in scale. The rest of the grid, however, will not be to scale (Fig. 2.17a). Meridians will appear as parallel lines instead of converging at the poles. Obviously, there is enormous east-west distortion for the polar regions because the distances between meridians are stretched to the same width that they are at the equator. The spacing of parallels on a Mercator projection is not even (equal) as it is on the globe, but it is proportional to the increased spacing of meridians as they approach the poles. The resulting grid consists of rectangles that increase in size toward the poles. Obviously, this projection does not display equal area, and the size distortion increases toward the poles.

The Mercator projection was developed to provide an important property that no other world projection has. A straight line drawn anywhere on a Mercator projection is a true compass heading. A line of constant direction, called a **rhumb line,** has great value to navigators (see Fig. 2.15). By connecting where they were to where they wanted to go with a straight line on Mercator's map, navigators could follow a constant compass direction to get to their destination.

Gnomonic Projections **Gnomonic projections** are planar projections, made by projecting the grid lines onto a plane, or flat surface (Fig. 2.17b). If we put a flat sheet of paper tangent to (touching) the globe at the equator, the grid will be projected with great distortion. Parallels and meridians will be unevenly spaced, and the shapes and areas of land and water bodies will also be badly distorted.

Despite their distortion, gnomonic projections have a valuable characteristic. In Figure 2.18, we can see that all the parallels appear curved except the equator, which appears as a straight line. In addition, all meridians appear as straight lines. We have learned that the equator and all meridians follow great circles. Thus, if the equator and meridians appear as straight lines on a gnomonic projection, then all great circles must appear as straight lines on maps using this projection. Gnomonic projections are the only maps that display arcs of great circles as straight lines, a characteristic that is important to

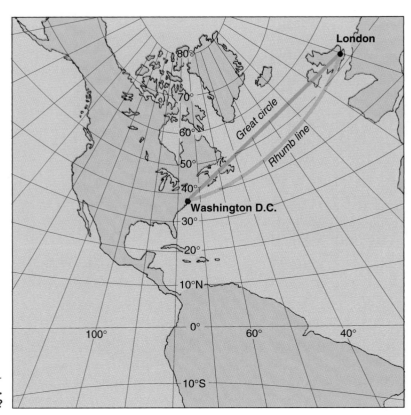

Figure 2.18 The gnomonic projection produces extreme distortion of distances, shapes, and areas. Yet it is extremely valuable for navigation, for it is the only projection that shows all great circles as straight lines. It is developed from a planar projection. ■ **Compare this figure with Figure 2.15. How do the projections in the two figures differ?**

navigation. Navigators can draw a straight line between their location and where they want to go, and this line will be a great circle route—the shortest route between the two places.

An interesting relationship exists between gnomonic and Mercator projections. Great circles on the Mercator projection appear as curved lines, while rhumb lines appear straight (Fig. 2.15). On the gnomonic projection the situation is reversed—great circles appear as straight lines, while rhumb lines are curves (Fig. 2.18).

Conic Projections   Conic projections are used for maps of middle-latitude regions, such as the United States (other than Alaska and Hawaii), because they show regions in these latitudes with minimal distortion. In a simple conic projection, a cone is fitted over the globe with its pointed top centered over a pole. The sides of the cone are tangent to Earth at a predetermined parallel, known as the standard parallel (Fig. 2.17c). The parallels of a simple conic projection, like the lines of latitude they represent, are concentric circles that become smaller the closer they are to the pole. The meridians on a conic projection appear as straight lines radiating from the pole. Distortions are slight, although they increase with distance from the standard parallel. Two standard parallels are used to make a *Lambert conformal conic projection* of the United States (Fig. 2.19),

and the result is a highly accurate map. The distortion on this map is only 1 percent from the east coast to the west coast and from the Canadian to the Mexican border.

Distortion of land masses can also be reduced by using an *interrupted projection* to produce several smaller segments (Fig. 2.20a). Each major continental area is based on its own central meridian, so no part of the projection suffers from extreme shape distortion. If our interest were centered on the world oceans, the projection could be interrupted in the continental areas to minimize distortion of the ocean basins. Some world maps are *compromise projections* that are neither conformal nor equal area, but an effort is made to balance and minimize the distortion to produce an "accurate looking" map, particularly in the areas where most of the world's population is located (Fig. 2.20b).

Map Basics

Grid coordinates are only one important element of a map. Every map is designed to meet a specific need or set of needs and should include the important information necessary to accomplish its intended goal. Among the essential items are title, date, legend, scale, and direction. Cartographers work to select the best way to depict spatial phenomena that the map is meant to show.

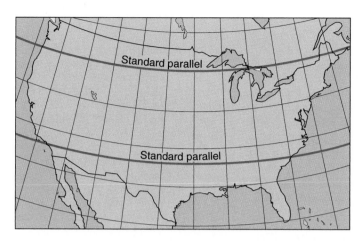

Figure 2.19 The Lambert conformal conic projection is a mathematical projection having two standard parallels. This projection is used when angles and shapes of middle-latitude areas are to be kept as accurate as possible. ■ **What is meant by a mathematical projection?**

Title, Date, and Legend   A map should have a title that tells what area is depicted, what the map is about, and what relationships it shows. For example, a map drawn to provide information for hiking and camping in Yellowstone National Park should have a title like "Yellowstone National Park: Trails and Camping Sites."

Each map should indicate its publication date or the date to which its information applies. For instance, a map of the population distribution in the United States should tell when the census was taken, to let us know how current or outdated the information is, or whether the map is intended to show historical data.

A map should also have a **legend**—a key to symbols used on the map. For example, if one dot represents 1000 people, or the symbol of a pine tree represents a roadside park, the legend should explain this information. If color

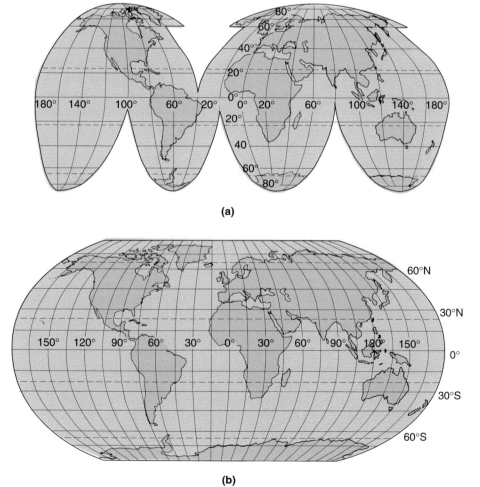

(a)

(b)

Figure 2.20 Distortion in projections can be reduced by interruption (a)—that is, by having a central meridian for each segment of the map. The Robinson projection (b) is considered a compromise projection because it departs from equal area to better depict the shape of the continents. ■ **Compare the distortion of these maps with the Mercator projection (Fig. 2.15). What is a disadvantage of (a) in terms of usage?**

Robert Iantria B.Sc, Physical Geographer, Terra Remote Sensing
Sidney, British Columbia

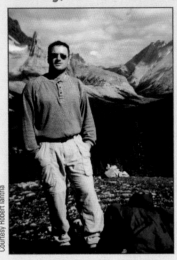

Courtesy Robert Iantria

After two years off to travel, I completed a two-year community college degree, then earned a BSc in Physical Geography from the University of Toronto. In addition, I have taken CAD, MS Project, and management courses.

I liked math, I liked that geography could be both science and humanities, and I liked the fieldwork. I love the outdoors, and the ability to travel was very attractive. I went in knowing that I wanted to do geography. I chose geography because it was all-encompassing in geosciences, and physical geography offered opportunities for fieldwork.

Two months before I graduated, I got a position with a company that was building an 80,000-mile global fiber-optic network.

I was brought on board to oversee the GPS data collection of where we placed the conduit, the splice locations, and the cable routing on a small section of this network in southern Ontario. I developed and revised the specifications for the project and used the final data set to develop "final as-builts."

After a year, I was transferred to Denver as Manager of Engineering Records, GPS, to come up with mapping solutions for our construction, engineering, GIS, and external clients. I did travel extensively for this job, I was able to deliver the best (accurate, geo-referenced, and detailed) telecom as-builts on our fiber-optic lines from Des Moines to Chicago, Chicago to New Orleans, Detroit to Chicago, and Minneapolis to Portal, North Dakota.

I now work for Terra Remote Sensing. We specialize in marine geophysics, hydrography, photogrammetry, LIDAR, and GIS services. I market technology to economic sectors that still use old ways of gathering geo-referenced data. I also research new ways and sectors in which our technology can be applied. I travel extensively, doing presentations at conferences and to industries. People skills, presentation skills, and written and oral communication skills have helped me succeed in this position.

We use LIDAR technology and high-resolution imagery, such as from a digital camera or false infrared, and market them to utility companies. Typical linear corridors are railroad lines, electrical lines, pipelines, fiber-optic lines, and so on. The mapping data we provide enable these companies to run their businesses more effectively.

I'm in charge of railroads, electrical transmission corridors, and flood plains. My knowledge of geomorphology, GIS, surveying, and statistics, along with my knowledge of what we do, helps me to provide them with a solution— why they should install a fiber-optic cable on one side of the track as opposed to the other—and I know how to document that in a GIS world.

Things that I learned in school that helped me succeed include:

- Research and synthesis skills
- Writing and communication skills, especially presentations.
- Statistics
- GIS and spatial theories related to projections and co-ordinate systems.
- Surveying theory

Statistics, for sure. I hated it, but it turns out it pops up in everyday applications in my field. Take any GIS course and as many field courses as possible.

I had the surveying background, I knew the software applications, and I was able to present the solution in having us manage 30,000 miles of fiber-optic network. I was able to use communication, research, and the GIS skills to sell a solution internally. It was cost effective, it was accurate, it was detailed, and it met everyone's needs.

Physical geography helps people think on a broader scale and forces them to examine the forces that they interact with, whether it be spatially or in business. It helped me look at the process, as opposed to cause and effect. I can identify various processes and their interactions. After seeing it in a physical geography course, you can take those skills and apply them to real situations. These processes have social applications as well, such as processes in socio-economic situations and how they relate to one another.

Physical geography is the study of all the physical processes on earth and their interactions. Most of the study of physical geography entails the study of Earth features and the processes that formed them. In all cases, it is a study of a combination of "spheres" that influenced the land features we are looking at. We study features at various scales, from a sand grain to a mountain to an entire mountain range, to figure out why and how they have evolved to what we see today, what we might see in the future, and how these changes affect the different "spheres." Physical geography helps us look at the big and small picture and recognize how different processes throughout time influence what we see today. In life, these skills of looking for and identifying processes and recognizing their interactions can be applied to many everyday social and business problems. The processes and the interactions may be different, but researching and studying them are the same.

My experience has been that an MS or a PhD puts you outside of the trainable scope that most employers are looking for. With a bachelor's, you're more trainable in the employer's eyes. You're more adaptable. If you do some follow-up—specialized education and training with a bachelor's— you're very, very marketable in a business environment.

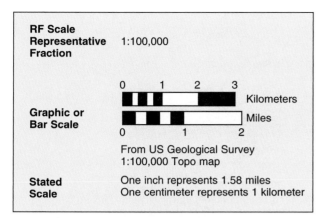

RF Scale Representative Fraction	1:100,000

Graphic or Bar Scale

From US Geological Survey
1:100,000 Topo map

Stated Scale	One inch represents 1.58 miles One centimeter represents 1 kilometer

Figure 2.21 Kinds of map scales. A *representative fraction* scale is a ratio between a distance as represented on a map (1 unit) and its actual length on the ground (here, 100,000 units). Each measurement must be in the same units of measure. A *graphic scale* is a device used for making distance measurements on the map in terms of distances on the ground. A *verbal scale* provides a statement of the relationship between a measurement on a map and the corresponding distance that it represents on the map. Verbal scales generally mix units (kilometer/cm or inches/mile).

shading is used on the map to represent elevations, different climatic regions, or other factors, then a key to the color coding should be provided. Map symbols can be designed to represent virtually any feature (see Appendix A).

Scale   Obviously, maps depict features in a smaller size than they actually are. In order to measure size or distances, we need to know the map **scale** that we are using (Fig. 2.21). That is, we need to know the relationship or ratio between an actual distance on Earth and the same distance as it appears on the map. Most maps indicate the scale at which the map has been drawn and to what parts of the map that scale is applicable. This information is essential in order to use the map for making measurements. Map scales enable us to accurately and precisely measure distances, determine areas, and compare sizes.

On a world globe, one scale is accurate everywhere on its surface, but because making a flat map involves stretching some places more than others, the scale given can never be totally accurate all over the map. However, if the map is of a small area, such as a city or a state, the distortion will be so slight that the scale can be accepted as applicable everywhere on the map. The scale of a map can be conveyed in three basic ways.

A **verbal scale** is a statement on the map that says something like "1 inch to 1 mile" (1 inch on the map represents 1 mile on the ground), or "1 inch to 1000 miles" (1 inch represents 1000 miles). A verbal scale will no longer be correct if the original map is reduced or enlarged.

A **Representative fraction (RF)** scale is presented as a ratio that is free of units of measurement. Thus, it can be used with any unit of linear measurement—feet, inches, meters, or centimeters—as long as the same unit

is used on both sides of the ratio. As an example, a map may have an RF scale of 1:63,360, which can also be expressed 1/63,360. This means that 1 inch on the map represents 63,360 inches on the ground. It also means that 1 cm on the map represents 63,360 cm on the ground. Knowing that 1 inch on the map represents 63,360 inches on the ground may be difficult to conceptualize unless we realize that 63,360 inches is the same as 1 mile. Thus, the representative fraction 1:63,360 means the map has the same scale as a map with a verbal scale of 1 inch to 1 mile.

A **graphic scale** (or **bar scale**) is useful for directly measuring distances from a map. A graphic scale is a graduated line marked with specific map distances proportional to distances on Earth. To use a graphic scale, take a straight edge, such as the edge of a piece of paper, and on it mark the distance between any two points on the map. Then use the graphic scale to measure the equivalent distance on Earth's surface. Graphic scales have two major advantages: (1) They are applicable even if the map is reduced or enlarged, because the graphic scale will also change proportionally in size. (2) No mathematical calculations or unit conversions are necessary to determine true distances on the map, because the graphic scale can be used like a ruler to make direct measurements.

Maps are often described as being of small, medium, or large scale (Fig. 2.22). *Small-scale* maps show large areas in a relatively small size, include little detail, and have large denominators in their representative fractions. *Large-scale* maps show small areas of Earth's surface in greater detail and have smaller denominators in their representative fractions. To help avoid confusion, remember that 1/2 is a *larger* fraction than 1/100. Maps with representative fractions larger than 1:25,000 are large-scale. Medium-scale maps have representative fractions between 1:25,000 and 1:250,000. Small-scale maps have representative fractions less than 1:250,000. This classification follows the guidelines of the United States Geological Survey, publisher of many maps for the federal government and the public.

Direction   Direction can be shown on a map by the orientation of the global grid, because parallels of latitude are east-west lines and meridians of longitude run directly north-south. In addition to, or perhaps instead of, these lines, many maps have an arrow pointing to north (north on the map) to provide an orientation to any compass direction. A north arrow may indicate either *true north* or *magnetic north*—or two north arrows may be given, one for true north and one for magnetic north.

Earth has a magnetic field that emanates from its core, making the planet act like a giant bar magnet. Our planet has a magnetic north pole and a magnetic south pole, with opposite charges, and a magnetic force field that radiates in a distinctive pattern into space. Although the magnetic poles shift position slightly from time to

(a)

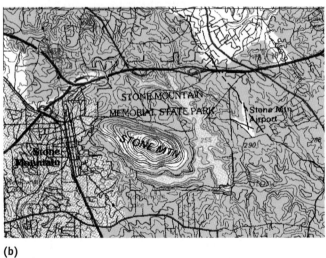

(b)

Figure 2.22 Map scales: Larger versus smaller. The designations *small scale* and *large scale* are related to a map's *representative fraction* scale (RF). For example, consider the question "Which of these maps of Stone Mountain has the smaller scale, (a) at 1:24,000 or (b) at 1:100,000?" It is important to remember that an RF scale is a *fraction* that represents the proportion between a length on the map and the true distance it represents on the ground. So one centimeter on the map would equal the number of centimeters in the denominator of the RF on the ground.
■ **Which number is smaller 1/24,000 or 1/100,000? Which map shows more land area—the larger scale map or the smaller-scale map?**

time, they are located in the Arctic and Antarctic regions. The north magnetic pole is located in northern Canada at 79°N, 103°W, near Ellef Ringnes Island, and the south magnetic pole is at 65°S, 139°E, in Commonwealth Bay, Antarctica.

A compass needle aligns itself with Earth's magnetic field, and the north-seeking end of the needle points toward the magnetic north pole. If we know the **magnetic declination**—the angular difference between magnetic north and true geographic north—we can then adjust our direction accordingly (Fig. 2.23). Thus, if our compass points north and we know that the magnetic declination for our location is 20°, we can adjust our course knowing that our compass is pointing 20° east of true north. To do this, we should turn 20° west from the direction indicated by our compass in order to face true north.

Figure 2.23 Map symbol showing true north, symbolized by a star representing Polaris (the North Star), and magnetic north, symbolized by an arrow. The example indicates 20°E magnetic declination. ■ **In what circumstances would we need to know the magnetic declination of our location?**

Magnetic declination varies from place to place over Earth's surface and also changes through time. For this reason, maps of magnetic declination are revised periodically, and using a recent map is very important. A map of magnetic declination is an *isogonic map* (Fig. 2.24). On such a map, *isogonic lines* connect locations that have the same magnetic declination. It is possible to be in a position where magnetic north and true north are aligned. On an isogonic map, the line that connects

these points is called the *agonic line* (line of zero magnetic declination).

Having examined globes and maps and learned some of their characteristics and symbols, we now have the ability to comprehend and communicate the language of location. We can report a location on a map or globe by latitude and longitude. We can also show location by our compass direction from a known position to our position and by the distance from that known position to where we are. Compass directions can be given by using either the azimuth system or the bearing system. In the **azimuth** system, direction is given in degrees of a full circle (360°) with respect to north. That is, if we imagine a circle of 360° with north at 0° (and at 360°) and read the degrees in a clockwise manner, we can describe a direction by its number of degrees away from north. For instance, straight east would have an azimuth of 90°, and due south would be 180°. The **bearing** system divides compass directions into four quadrants of 90° (N, E, S, W), each numbered by directions in degrees away from either north or south. Using this system, an azimuth of 20° would be North, 20°East (20° east of due north), and an azimuth of 210° would be South, 30°West (30° west of due south). Both azimuths and bearings are used for giving precise directional instructions for mapping, surveying, military, and navigational purposes (see Appendix A).

Figure 2.24 Isogonic map of North America, showing the magnetic declination that must be added (west declination) or subtracted (east declination) from a compass reading to determine true directions. ■ **What is the magnetic declination of your hometown to the nearest degree?**

Figure 2.25 Continuous and Discrete Variables. Continuous means that every location possesses a characteristic that can be measured; for example, everywhere on Earth has an elevation, even if it is zero (at sea level), or below (a negative value). Changes in a continuous variable over an area can be represented in many ways, but isolines, shading, and colors can be used in combination or separately. The map **(a)** shows the continuous distribution of temperature variation for a given day in part of eastern North America.

Discrete variables are features that are present at certain locations, but not found everywhere. Discrete features can be analyzed spatially by their locations, size, shape, extent, number present, arrangement, distribution density, and other measurable characteristics. The locations, distributions, and pattern of discrete features are of great interest in understanding spatial relationships. Discrete features can be **(b)** *areas* like the lands burned by wildfire (or clear-cuts in a forest, or the area where an earthquake was felt), **(c)** *lines* as in the path taken by hurricane Andrew (or river channels, or fault lines), or **(d)** *points* as shown by locations of large earthquakes in Hawaii (or places where lightning has struck, or locations of water pollution sources). **Can you name other environmental examples of continuous and discrete variables?**

Displaying Spatial Data and Information on Maps

Any information or data that vary spatially can be shown on a map. **Thematic maps** are designed to focus attention on the distribution of one feature (or a few related ones). Examples include maps of climate, vegetation, soils, earthquake epicenters, or tornadoes. To fulfill a particular purpose, mapmakers choose methods of cartographic representation that best display and communicate the desired information.

Discrete and Continuous Data

There are two important types of spatial data or information, discrete and continuous (see Fig. 2.25). **Continuous data** means that some numerical value of that characteristic exists everywhere on Earth (or at least within the area of interest displayed on the map); for example, every location on Earth has an elevation (or air pressure value or population density). **Discrete data** means that the

phenomenon is either located at a particular place or it is not—for example, hot springs, tropical rainforests, or earthquake faults. Discrete data are represented on maps by point, area, or line symbols to show their locations and distributions (Fig. 2.25).

The *region* is an important geographic concept that refers to discrete areas that exhibit a common characteristic or set of characteristics within their boundaries. Different colors, line patterns, or shading differentiate adjacent regions from one another. This kind of representation is used to show the areal distribution of soil, climate, and vegetation regions (see the world maps throughout this book).

The distribution of continuous data is often shown using **isolines**—lines on a map that connect points with the same numerical value. We have already encountered one example of an isoline—the isogonic line, which connects points with the same magnetic declination (see Fig. 2.24). Isolines that we will be using later on include *isotherms,* which connect points of equal temperature; *isobars,* which connect points of equal barometric pressure; *isobaths*

Figure 2.26 (Top) A bird's-eye view of a river valley and surrounding hills, shown on a shaded-relief diagram. Note that the river flows into a bay partly enclosed by a hooked sand spit. The hill on the right has a smoothly eroded form, whereas the one on the left rises to a cliff at the edge of an inclined tableland. (Bottom) The same features represented on a contour map.
■ **If you had only a contour map, could you visualize the terrain shown in the shaded-relief diagram?**

(also called bathymetric contours), which connect points with equal water depth; and *isohyets,* which connect points receiving equal amounts of precipitation.

Topographic Maps

Topographic contour lines—isolines connecting points on a map that are at the same elevation above mean sea level (or below, as in Death Valley, California)—are of special interest to physical geographers. For example, if we walk around a hill along the 1200-foot contour line shown on the map, we would be 1200 feet above sea level at all times and be walking on a level line, maintaining a constant elevation.

Contour lines provide an excellent means for depicting the land surface configuration on a map (see Map Interpretation: Topographic Maps). The arrangement, spacing, and changing shapes of the contours give a map reader an accurate idea of what the topography (the "lay of the land") of the map area looks like (Fig. 2.26).

Figure 2.27 illustrates how contour lines show the geometric character of the land surface. The bottom portion of the diagram is a simple contour map of an asymmetrical island. Note that the difference in elevation between adjacent contour lines is a constant interval of 20 feet. The constant

Figure 2.27 A topographic profile and contour map. Topographic contour lines connect points of equal elevation in relation to mean sea level. The upper part of the figure shows the topographic profile (side view) of an island. Horizontal lines mark 20-foot intervals of elevation above sea level. The lower portion of the figure shows how these contour lines look in map view.
■ **Study this figure and the maps in Figure 2.22. What is the relationship between the spacing of contour lines and steepness of slope?**

TOPOGRAPHIC MAPS

The Map

A topographic map is the most widely used tool for graphically depicting elevational variation within an area. A contour line connects points of equal elevation above some reference datum, usually mean sea level. A vast storehouse of information about the relief and the terrain can be interpreted from these maps by understanding the spacing and configuration of contours. For example, elevations of mountains and valleys, steepness of slopes, and the direction of stream flow can be determined by studying a topographic map. In addition to contour lines, many standard symbols are used on topographic maps to represent mapped features, data, and information (a guide to these symbols is in Appendix B).

The elevation difference represented by adjacent contour lines depends on the map scale and the relief in the mapped area, and is called the contour interval. Contour intervals on topographic maps are typically in elevation measurements divisible by ten. In mountainous areas wider intervals are needed to keep the contours from crowding and visually merging together. A flatter locality may require a smaller contour interval to display subtle relief features. It is good practice to note both the map scale and the contour interval when first examining a topographic map.

There are several important rules to keep in mind when interpreting contours:

⊕ Closely spaced contours indicate a steep, and widely spaced contours indicate a gentle slope.

⊕ Evenly spaced contours indicate a uniform slope.

⊕ Closed contour lines represent a hill or a depression.

⊕ Contour lines never cross but may converge along a vertical cliff.

⊕ A contour line will bend upstream when it crosses a valley.

Interpreting the Map

1. What is the contour interval on this map?

2. The map scale is 1:24,000. One inch on the map represents how many feet on the Earth's surface?

3. What is the highest elevation on the map? Where is it located?

4. What is the lowest elevation on the map? Where is it located?

5. Note the mountain ridge between Boat and Emerald Canyons (C-4). Is it steeper on its east side or its west side? What led you to your conclusion?

6. In what direction does the stream in Boat Canyon flow? What led you to your conclusion?

7. The aerial photograph at left depicts a portion of the topographic map on the opposite page. What area of the air photo does the map depict? How well do the contours represent the physical features seen on the air photo?

8. Identify some cultural features on the map. Describe the symbols used to depict these features. The map shown is older than the aerial photograph. Can you identify some cultural features on the aerial photograph not depicted on the contour map?

Aerial Eye, Inc., Irvine, California

Aerial photograph of the coast at Laguna Beach, California

Opposite:
Laguna Beach, California
Scale 1:24,000
Contour interval = 20 feet.
U.S. Geological Survey

vertical difference in elevation between adjacent contour lines is called the **contour interval.**

What kind of terrain would we cover if we were to walk across this island from point A to point B? We start from point A at sea level but immediately begin to climb. We soon cross the 20-foot contour line, then the 40-foot, the 60-foot, and, near the top of our island, the 80-foot level. After walking over a relatively broad summit that is above 80 feet but not as high as 100 feet (or we would cross another contour line), we once again cross the 80-foot contour line, which means we must be starting down. During our descent, we cross each lower level in turn until we arrive back at sea level (point B).

In the top portion of Figure 2.27, a **profile** (side view) helps us to visualize the ground we covered in our walk. We can see why the trip up the mountain was so much more difficult than the trip down. The closely spaced contour lines near point A represent a steeper slope than the more widely spaced contour lines near point B. Actually, we have discovered something that is true of all isoline maps: The closer together the lines are on the map, the steeper the **gradient** (the greater the rate of change per unit of horizontal distance).

When studying a contour map, we should understand that the slope of Earth's surface almost always changes gradually. Thus, the land between two contour lines nearly always slopes gradually toward the line with the lower value; it is unlikely that the land drops off in steps downslope as the contour lines might suggest.

Topographic maps show other features in addition to elevations (see Appendix A). For instance, a contour map may show water bodies such as streams, lakes, rivers, and oceans, or cultural features such as towns, cities, bridges, and railroads. The U.S. Geological Survey (USGS) produces topographic maps of the United States at several different scales. Some of these maps—1:24,000, 1:62,500, 1:63,360 (1 inch to 1 mile), and 1:250,000—use English units for their contour intervals. Many recent maps are produced at scales of 1:25,000 and 1:100,000 and use metric units. Contour maps that show undersea topography are called *bathymetric charts.* In the United States they are produced by the National Ocean Service.

Modern Mapping Technology

Technology has revolutionized cartography from a manual process, slow and sometimes tedious, to an automated process, using computers to store, analyze, and retrieve data and computer-assisted equipment to draw the final map. For most mapping projects, computer systems are faster, more precise, more efficient, and less expensive than the manual cartographic techniques they have replaced. However, an understanding of basic cartographic principles is still required to make a good map, because the computer mapping system will only draw what an operator instructs it to draw.

Digital Mapmaking

Map data can be stored digitally until a map is ready to be produced or revised. Hundreds of millions of bits of mapping data, such as elevations, depths, temperatures, or populations, can be stored in a digital database. A typical USGS topographic map has more than 100 million bits of information to be stored and thousands of bits of data to be plotted on the finished map. A database for a map may include information on coastlines, political boundaries, city locations, river systems, map projections, and coordinate systems. Today, information and measurements can be entered immediately into a data bank while a field survey is being conducted. The data may even be recorded by voice recognition systems, called in by the field surveyors, and entered directly into the map data system.

Figure 2.28 A digital terrain model is a computer-generated, three-dimensional view of a land surface. Many different kinds of terrain displays can be generated from digital elevation data, and the models can be rotated on a computer screen to be viewed from any angle or direction.

Figure 2.29 A continuous variable displayed in three-dimensional perspective: earthquake hazard in the coterminous United States. Computer visualization techniques can produce a display of the spatial distribution of a spatially continuous variable. Here it is easy to get a mental image, or mental map, of how potential earthquake danger varies across this part of the United States.
■ **Other than where you might expect to find a high level of earthquake hazard, are there locations whose level of this characteristic you find surprising?**

A great advantage that computers offer to the mapping process is that maps can be revised without manually redrawing them. Existing map data and new survey data can be displayed on computer screens and then corrected, changed, or improved until the cartographer decides that the final map is ready to be produced in hard copy. The cartographer may tile together adjacent maps to cover a large area or zoom in on a particular area of the map to give a more detailed, larger-scale view. In addition to scale changes, computer map revision also allows easy metric conversions and rapid changes in projections, contour intervals, symbols, colors, and direction of view (orientation).

Computer map revision is essential for rapidly changing phenomena, such as weather systems, air pollution, ocean currents, volcanic eruptions, and forest fires. Weather maps in particular must be constantly revised, and computer mapping allows rapidly changing atmospheric conditions to be monitored and plotted continually.

Of particular interest to physical geographers, geologists, and engineers are **digital terrain models,** three-dimensional (3-D) views of a particular area (Fig. 2.28). A 3-D model is particularly useful for making calculations of the volume of rock material, oil, or water resources. Digital terrain models may be designed to show **vertical exaggeration** by stretching the vertical part of the display to enhance the relief of an area or to show a cross-sectional view of a particular location. Digital elevation data can be translated into many types of terrain displays and maps, including color-scaled contour maps

(where areas between assigned contours are a certain color), conventional contour maps, shaded relief maps, and three-dimensional relief maps.

Computers can facilitate the display of any continuous data, either as a two-dimensional contour map to illustrate the spatial distribution of some feature or as a three-dimensional surface to enhance visualization of the geographic variations in an environmental characteristic. **Visualization** refers to the products of new computer techniques used to generate three-dimensional, sometimes animated, image-models designed to illustrate and explain complex processes and features. For example, the Earth image shown at the beginning of this chapter is a visualization model, based on environmental data sets and remotely sensed imagery that presents several components of the Earth system in a stunning 3-D portrayal. Visualization models help us to understand and conceptualize many environmental processes and features (Figure 2.29).

Although the products and techniques of cartography are now very different from their beginnings and continue to be improved, the goal of making a representation of Earth remains the same: to effectively communicate geographic and spatial knowledge to the user in visual form.

Geographic Information Systems

A tremendously versatile and relatively recent innovation in the computerized handling of maps and spatial data is called a **geographic information system (GIS).** A GIS is a computer-based technology that assists the user in the entry, analysis, manipulation, and display of geographic information derived from map layers. In essence, a GIS blends computer cartography and database management for modeling and analyzing spatial data (Fig. 2.30). It is especially useful to geographers because they often address problems that require large amounts of spatial data from a variety of sources. According to a corporation that develops GIS systems, this technology—software, hardware, service, and applications—represents a multibillion-dollar industry employing hundreds of thousands of people around the world. Geographic information systems (GIS) are widely used by governmental agencies, and many geographers are employed in career fields that apply GIS technology. GIS can not only analyze, manipulate, and display various layers, as in computer cartography, but it can also make the scale and map projection of the layers compatible, thus allowing several or all of the layers to be integrated into new, more meaningful composite maps.

What a GIS Does Imagine you are in a giant map library with thousands of maps, all of the same area but each showing a different aspect of the same place: one map shows roads, another highways, another trails, another rivers (or soils, or vegetation, or slopes, or rainfall, and on almost to infinity). Further, each map is at a different scale,

Relief Presentations and Vertical Exaggeration

Most maps present a landscape as if viewed from directly overhead, looking straight down. In fact, this perspective is sometimes referred to as a *map view* or *plan view* (in the sense of architectural house plans). Measurements of length and distance are accurate, as long as the area depicted is not so large that Earth's curvature becomes a major factor. Topographic maps, for example, show spatial relationships in two dimensions (length and width on the map, called *x* and *y* coordinates in mathematical Cartesian terms). Illustrating terrain, as represented by differences in elevation, requires some sort of symbol to record elevational data on the map. Topographic maps use contour lines, which can also be enhanced by relief shading (see the Map Interpretation, Volcanic Landforms, in Chapter 14 for an example).

For many purposes, though, a side view or an oblique view of what the terrain looks like (also called *perspective*) is important to understanding the landscape. Block diagrams, three-dimensional models of Earth's surface, are very useful for showing the general layout of topography from a perspective view. They provide a perspective with which most of us are familiar, similar to looking out an airplane window or from a high vantage point. Block diagrams are excellent for illustrating three-dimensional relationships in a landscape scene, and information about the subsurface can be included. But such diagrams are not intended for making accurate measurements, and many block diagrams represent hypothetical or stylized, rather than actual, landscapes (see fig. 14.16 for an example).

A topographic *profile* illustrates the shape of a land surface as if viewed directly from the side. It is basically a graph of elevation changes over distance along a transect line. Elevation and distance information collected from a topographic map or from other elevation data in spatial form can be used to draw a topographic profile. Topographic profiles show the terrain. If the geology of the subsurface is represented as well, such profiles are called *geologic cross sections*.

Block diagrams, profiles, and cross sections are typically drawn in a manner that stretches the vertical presentation of the features being depicted. This makes mountains appear taller than they are in comparison to the landscape, valleys deeper, the terrain more rugged, and the slopes steeper. The main reason why vertical exaggeration is used is that it helps to make subtle changes in the terrain more noticeable. In addition, land surfaces are really much flatter than most people think they are. In fact, cartographers have worked with psychologists to determine what degree of vertical exaggeration makes a profile or block diagram appear most "natural" to people viewing a presentation of elevation differences in a landscape. For technical applications, most profiles and block diagrams will indicate how much the vertical presentation has been stretched, so that there is no misunderstanding. Two times vertical exaggeration means that the feature is presented two times higher than it really is, but the horizontal scale is correct.

To illustrate why vertical exaggeration is used, look at the three profiles of a volcano in the Hawaiian Islands. Which do you think shows the true, natural-scale profile of this volcanic mountain? Which one "looks" the most natural to you? What is the true shape of this volcano? After making a guess, check below for the answer and the degree of vertical exaggeration in each of the three profiles. Note that this is a huge volcano—the profile extends horizontally for 100 kilometers.

and there are many different projections (some that do not even preserve shape or direction). These factors will make comparing the different maps a difficult process. You also have digital terrain models and satellite images that you would like to compare directly to the maps. Being interested in geography, you want to be able to put some of this information together, since few aspects of the environment involve only one factor or exist in spatial isolation. You have a problem, and to solve that problem, you need some way of making all of these representations of a part of Earth directly comparable. What you need is a GIS, and the knowledge of how to use this system.

Data and Attribute Entry The first step in solving this problem is to enter the map and image data into a computer system. There are three general ways to do this: (1) Use a device called a *digitizer* to enter the map information into the computer. (2) Use a computer scanner to scan in each map. (3) Directly enter data from a digital source (for example, disks or direct download). Each map is entered and stored as a separate data file that represents an individual thematic map—a separate layer of information (Fig. 2.31). Another step, **geocoding** of the spatial information, is done either simultaneously or afterward. Geocoding means locating all spatial information in relation to some grid coordinates, such as latitude and longitude. Further, a list of *attributes* (information) about each mapped feature or grid cell can be stored and easily accessed.

Registration and Display Each map layer is geometrically registered to the same set of geographic coordinates, as is the geocoded information. The computer will now display any layer (or set of layers) that you wish, stretched to fit any map projection that you specify, at a scale that you specify. The maps, images, and data sets can now be directly compared at the same size and on the same configuration of grid coordinates.

It is also possible to combine the geocoded, registered data and maps in any set of thematic layers that you

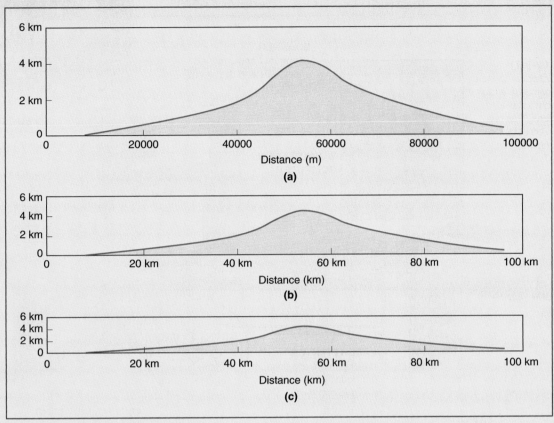

Profiles of Mauna Kea, Hawaii (data from NASA)

(a) 4× vertical exaggeration
(b) 2× vertical exaggeration
(c) natural-scale profile—no vertical exaggeration

wish. If you want to see the location of homes on a river flood plain, the computer will overlay these two thematic layers for you, giving you an instant map of what you want just from entering the commands to retrieve separate homes and flood-plain layers. If you want to see earthquake faults, artificially filled areas, and the locations of fire and police stations, that will require four layers, but this is no problem for a GIS to display.

GIS in the Workplace   As another example of a GIS application, suppose you are a geographer working for the Natural Resources Conservation Service. Your current problem is to control erosion along the banks of a newly constructed reservoir. You know that erosion is a function of many environmental variables, including soil types, slopes, vegetation characteristics, and others. Using a GIS, you could enter data for each of these variables into the computer as a separate thematic layer. These variables could be analyzed individually; however, by integrating information from individual layers (soils,

slope, vegetation, and so on), you could identify the locations most susceptible to erosion. Your resources and personnel could then be directed toward controlling erosion in those target areas.

The capacity of a GIS to integrate and analyze a wide variety of geographic information, from census data to landform characteristics, makes it useful to both human and physical geographers. With many applications in geography and other disciplines, GIS is now, and will continue to be, an important tool for spatial analysis.

Remote Sensing of the Environment

Beginning in the early 1900s, cartographers made extensive use of aerial photographs taken from airplanes. In the 1960s, new technologies for obtaining images from aircraft and satellites were declassified from military applications and made generally available. A geographer,

GPS SATELLITE
A series of satellites, 11,000 miles above Earth, provide the signals that GPS receivers translate.

SCANNER
Scanners transform hard-copy maps and documents into digital format.

GPS RECEIVER
Global Positioning System receivers calculate exact position from satellite transmissions. This has revolutionized data collection for GIS use.

MASS STORAGE DEVICE
Various devices are used to store high-volume data and programs.

WORKSTATION/ COMPUTER

LASER PRINTER
Laser printers are used for high-quality high-volume text and graphics.

PLOTTER
Electrostatic Plotters are able to print large copies of maps, images, and diagrams.

MANUAL DIGITIZING
Manual digitizing is done with a digitizing table and cursor (inset). Lines are traced and cursor buttons pushed to indicate various commands.

Figure 2.30 Components of a geographic information system. A GIS consists of four basic subsystems: (1) Input systems provide spatial data to the GIS through the use of scanners to convert paper images into a digital format, through manual digitizing, and by means of the Global Positioning System, which can feed locational coordinates in digital form directly from the field. (2) Data storage systems include CD-ROMs, computer floppy disks, and hard drives. (3) Computer hardware and software systems are used for data management, access, and analysis, and for updating the database. (4) Output systems include plotters, which have the ability to generate hard-copy maps and images. In addition to these four basic subsystems of a GIS, access to many kinds of spatial data (maps, photographs, satellite imagery, digital elevation data) is important. However, it is often said that people are the most important part of a Geographic Information System.
■ **How can computer technology help us to store, retrieve, and update maps and other spatial information?**

Evelyn Pruitt, coined the term **remote sensing** to refer to the gathering and interpretation of aerial and space imagery. Remote sensing is the collection of information about objects or environments from a distance; a remote sensor is the device that collects, stores, and retrieves this information without being in physical contact with the object of interest. Remote sensing systems can generate images of objects and scenes that would otherwise be invisible to the narrow range of human vision. Remote sensors can "see" types of energy that are nearly or totally invisible and display these scenes as images that humans can visually interpret.

Remote sensing is commonly divided into photographic techniques (using film and cameras to record images of light) and digital image techniques (which can record images from light or other forms of energy). Photographs are made by using cameras to record a picture on film. Digital techniques use image scanners or digital cameras to produce a **digital image**—a visible image derived from numerical data (Fig. 2.32). Digital images consist of **pixels,** a term that is short for "picture element." Pixels are the smallest area resolved in a digital image. We can think of digital images as similar to a mosaic, a grid of cells with varying colors or tones that form a picture. Each cell (pixel) of the grid has a locational address and a value inside the box that represents the brightness of the picture area that the pixel represents. The digital values for the array of grid cells are translated into an image by computer.

Environmental   **Map Layer**

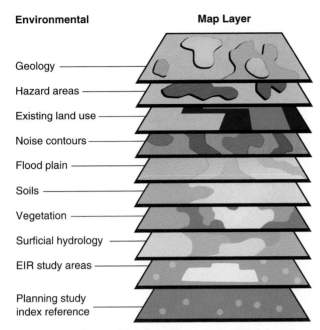

Geology

Hazard areas

Existing land use

Noise contours

Flood plain

Soils

Vegetation

Surficial hydrology

EIR study areas

Planning study
index reference

Figure 2.31 Geographic information systems (GIS) help geographers find solutions to complex problems by storing different information and data files as individual "layers." GIS technology is widely used in environmental studies in which several different variables need to be assessed and compared spatially to solve a problem. This information can then be analyzed and mapped separately or overlaid to provide more detailed information. ■ **Can you think of other applications, for geographic information systems?**

Most images returned from space are digital, because recovering film from space is difficult and digital data can be easily broadcast back to Earth. Digital imagery taken by a camera or scanner also offers the advantage of computer-assisted data processing, image enhancement, and interpretation.

Aerial Photography

Aerial photographs have long provided us with new perspectives on our environment (Fig. 2.33). Most are vertical photographs (looking straight down), although some are oblique (taken at an acute angle to Earth's surface). Air photo interpreters use aerial photographs to examine and describe relationships among objects on Earth's surface. A device called a *stereoscope* allows the overlapping of pairs of aerial photographs (or digital images) taken from different positions to show surface features in three dimensions. Advanced stereographic equipment, supported by computers, has greatly facilitated the production of topographic contour maps.

When astronauts first used cameras on their space missions, many of their first Earth photos were disappointing. One reason was that atmospheric scattering produces haze, so normal photographs taken from high altitudes often appear murky and fuzzy. **Near-infrared (NIR) film** cuts though haze better than normal film

NASA/Earth Observatory/Photo by Mike Abrams/
NASA/GSFC/METI/ERSDAC/JAROS, and U.S/
JAPAN ASTER Science Team

Figure 2.32 This scene, part of Chesapeake Bay, is not a photograph but a digital image. Digital data, beamed back from an orbiting satellite were processed by a computer to produce the image. ■ **What features can you recognize on this image?**

Near-Infrared Sensors: More Than the Eye Can See

Wavelengths of the near-infrared (NIR) band (between 0.7 and 1.5 micrometers) are a bit longer than the wavelength for the color red (from 0.65 to 0.7 micrometers). As such, it lies just outside the range of wavelengths the human eye can discern. This is sad, for if the human eye could see in NIR, people would be able to detect more about our environment. However, we have developed certain NIR sensor systems that can be mounted on airplanes and satellites, allowing us to view Earth's surface with greater accuracy. These NIR sensors have many uses; we discuss two examples here.

The first example concerns Earth's vegetation. When vegetation is exposed to sunlight (or any white light), all the various wavelengths are absorbed, except for a few that reflect off. The wavelength that is reflected with the highest percentage is NIR. Since the eye cannot detect the NIR wavelengths, we must settle for the next highest percentage of wavelengths that are reflected, which are 0.50 to 0.55 micrometers (the color green). Therefore, to the human eye, most vegetation appears in shades of green. Because

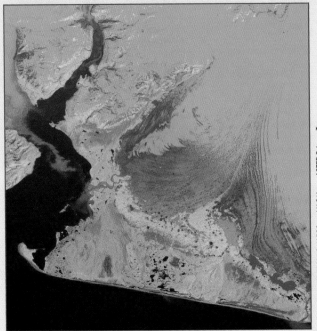

NASA/GSFC/METI/ERSDAC/JAROS, and U.S./Japan ASTER Science Team

The Glacial meltwater helps to keep the vegetation on these slopes quite healthy. The fjord contains some clear water (black) and water with glacial sediment (light blue/gray).

does, giving a clearer, sharper picture. It is sensitive to a portion of visible light as well as to longer-wavelength, near-infrared energy—basically light just beyond red, wavelengths that are too long for our eyes to see.

Color NIR film is sometimes called "false color" film, because strong NIR reflectance shows up as red on the photo. This color was chosen so that a person can instantly tell a color NIR photo or image from a normal color photo. On NIR, healthy grasses, trees, and most plants will show up as bright red, rather than green (Fig. 2.34). A widely held, but incorrect, notion of infrared film is that it images heat. No photographic processes clearly record heat emissions on film.

Remote Sensing through Image Scanning

Image scanners used in remote sensing may use ultraviolet, visible, near-infrared, thermal infrared, and microwave parts of the electromagnetic spectrum. Two of these, **thermal infrared (TIR) scanning** and **imaging radar,** are employed in rather specific applications of environmental remote sensing. **Multispectral scanning,** which senses many kinds of energy at once, can yield an almost unlimited variety of composite images that have very broad uses in studying Earth systems.

Thermal Infrared (TIR) Scanning Images of energy that we sense as heat can be taken either day or night by thermal infrared scanners. How heat is sensed to make an image is complex, but basically the image is scanned line by line with a sensor that records heat variations. The heat data are beamed back, digitally stored, and converted into a visual image. TIR sensors are used on aircraft or spacecraft.

Thermal infrared images record contrasts in temperature, whether they are cold or hot. How well an object will show up on a thermal image depends on how different its temperature is from that of its surroundings. Hot objects show up in light tones and cool objects will be dark, but frequently a computer is used to colorize the gray tones and emphasize heat differences.

Thermal scanning is expensive and tends to be used for specific purposes. Some civilian applications include finding leaks in steam pipes or other pipelines, detecting thermal pollution, finding volcanic hot spots and geothermal sites, finding leaks in building insulation, and locating forest fires through dense smoke.

Weather satellites also use thermal scanners to image atmospheric conditions for forecasting. You have all seen these images on television when the meteorologist says, "Let's see what the weather satellite shows." Clouds are depicted in black on the original thermal image because the eye cannot see NIR, and therefore cannot assign a color to it, the sensor imager automatically assigns shades of red to the intensities of NIR. So, when you look at a NIR satellite or aerial image of the surface and see shades of red and pink, you are looking at vegetation as seen by the NIR sensor. In any such image, brighter and darker shades of red indicate healthy vegetation. As the shades turn lighter toward pink, this indicates vegetation that is under stress. There may be many reasons for this vegetative stress: Perhaps the fields are soaked with too much water, perhaps a drought is in effect, or perhaps a plant disease is causing the vegetation to die off. Whatever the cause, light pink areas show unhealthy plant life, while bright reds show vibrant vegetation.

The second example involves Earth's water bodies. NIR wavelengths tend to be almost totally absorbed by water. Therefore on NIR imagery, water appears to be almost black (or very dark blue) in color, showing almost total absorption of these wavelengths. This fact becomes useful when searching for areas of water pollution, or problems with excessive sediments in the water. Particles floating in the water, such as sediments and certain pollutants, do give off some NIR wavelengths. The sensor system will detect these, and the materials will usually appear as a light blue color mixing with the otherwise black water. When using NIR images to look at lakes, streams, and even the ocean, one can see sediments and foreign materials flowing within the otherwise undisturbed water.

U.S.D.A.

(a)

NASA/Ames Research Center

(b)

Figure 2.33 (a) Oblique photos provide a "natural view," like looking out of an airplane window. This oblique aerial photograph in natural color shows farmland, countryside, and forest. (b) Vertical photos provide a maplike view that is more useful for mapping and making measurements.

Figure 2.34 A comparison of a "false color," near-infrared photograph (left) to the same view in normal color (right). Note that, in general, environmental details can be seen better and are more easily recognized on the near-infrared photograph. ■ **Remembering that red represents healthy growing vegetation and clear or deep water is dark blue, what kinds of details can you recognize more easily on the CIR photo?**

they are colder than their background, the surface of Earth below. Since we don't like to see black clouds, the image is reversed, like a photo negative, so that the clouds will appear white (Fig. 2.35).

Radar *RA*dio *D*etection *A*nd *R*anging is an active sensor that transmits microwave energy and reads the reflected signals to produce an image. Radar imaging operates day or night and also sees through clouds. These are great advantages for mapping hostile regions as well as for gathering information about places that are frequently cloudy, such as tropical rainforests.

Today, specific types of weather radar designed to monitor and track thunderstorms, hurricanes, and tornadoes (Fig. 2.36a) are used all over the world. Weather radar systems produce maplike images of precipitation

Figure 2.35 Thermal infrared scanner images such as this one show temperature differences with gray tones, darker meaning warmer, or with colors. In a sense, they are like a "map" of heat (or cold) patterns. This scene shows part of North America as beamed back from one of a series of U.S. weather satellites called GOES (Geostationary Operational Environmental Satellite). The whiter the cloud tops, the colder (and higher) they are, but here the highest cloud tops have been colorized for emphasis. ■ **What other details can you see in this heat image? The day is April 16, 1998.**

patterns because radar penetrates clouds (day or night) but bounces off raindrops and other precipitation, producing a signal on the radar screen. Precipitation patterns are typically the kind of weather radar image that we see on television. The latest systems include **Doppler radar,** a system that can determine the precipitation pattern, direction of movement, and speed of a storm system (much as police radar measures vehicle speed).

Imaging radar includes several specialized radar systems that sense the ground (topography, rock, water, ice, sand dunes, and so forth) by bouncing microwaves off the surface and converting the reflections into a landscape image. **Side-Looking Airborne Radar (SLAR)** was designed to be airplane-mounted to image areas located to the side of the aircraft (without flying over the region). Other imaging radar systems are mounted on satellites and the space shuttle. Imaging radar generally does not "see" trees (depending on the system), so that it makes an image of

the landscape rather than a crown of trees. Thus, SLAR is excellent for terrain and hydrologic mapping (Fig. 2.36b). SLAR is probably most often used to map remote, inhospitable, inaccessible, cloudy, or heavily forested regions.

Multispectral Analysis

Perhaps the most important development for gaining information about our environment is the ability to use remote sensing from orbiting space satellites. The term *multispectral* really has two related meanings. *Multispectral analysis* means using and comparing more than one type of image of the same place, whether taken from space or not (for example, radar and thermal images, or near-infrared and normal photos). *Multispectral scanning* refers to devices widely used on satellites to simultaneously scan the same area at a variety of wavelengths. Combinations of visible and near-infrared bands are

(a)

(b)

Figure 2.36 (a) NEXRAD (next generation weather radar) image of thunderstorms associated with a storm front. NEXRAD reflections provide detail of the storm (circle is 124 nautical miles across). Blue tones show cloud cover with no precipitation (except in the very center, where blue indicates stray reflections called ground clutter). Other colors show rainfall intensity: green—light rain, yellow—moderate, and orange-red—heavy rain. The storm is moving to the east-southeast. (b) Imaging radar "bounces" a signal off a land surface and monitors variations in the reflected signals to produce a picture of a landscape. Radar images are originally in black and white, but can be colorized to enhance detail. Radar reflections are affected by many factors, particularly the surface materials, as well as steepness and orientation of the terrain. This radar image shows part of Death Valley National Park, California, a desert basin flanked by mountains.

■ **How are weather and imaging radar scenes different in terms of what they record about the environment?**

Polar versus Geostationary Satellite Orbits

When studying images taken from outer space, one can notice some distinct differences in their quality and views of Earth. Some of these differences are related to the type of orbit the satellite imager is using while scanning the surface. There are two distinctively different types of orbits: the *polar orbit* and the *geostationary orbit* (sometimes called a *geosynchronous orbit*).

The polar orbit was developed first; as its name implies, the satellite orbits Earth from pole to pole. This orbit has some distinct advantages. It is typically a lower orbit, usually varying in altitude from 700 kilometers (435 mi) to 900 kilometers (560 mi). At this height, the polar orbiting satellite can produce clear, close-up images of Earth. However, at this distance the satellite must move at a fast speed (it must maintain a fast *orbital velocity*) to overcome the gravitational pull of Earth. This velocity can vary, but for polar orbiters it averages around 27,400 kilometers/hour (17,000 mph). While the satellite orbits from pole to pole, Earth rotates on its axis below, so each orbital path is viewing a different stretch of the surface. Thus, polar orbits will at times cover the dark side of the planet. To adjust for this, a slightly modified polar orbit was developed, called a *sun synchronous orbit*. If the polar orbit is tilted a few degrees off the vertical, then it can remain over the sunlit part of the globe at all times. Most modern polar orbiting satellites are sun synchronous.

The geostationary orbit, developed later, offered some innovations in satellite image gathering. A perfect geostationary orbit must have three characteristics: (1) It must move in the same direction as Earth's rotation. (2) It must orbit exactly over the equator. (3) The orbit must be a perfectly circular. The altitude of the orbit must be exact as well, at 35,900 kilometers (22,300 mi). At this greater height, the orbital velocity is reduced below that for a polar orbit, to 11,120 kilometers/hour (6900 mph), because Earth's gravity is less at that distance. When all these conditions are met, the satellite is perfectly synchronized with Earth's rotation, and the satellite appears to be located in the same spot above Earth all the time. This orbit offers some new advantages. First, at its great distance, a geostationary satellite can view half of the entire Earth (that is, the half it is always facing) in one image. The other great advantage is that it can show movement in our atmosphere and oceans. Keep in mind that the satellite is always above the same point on Earth. Therefore, Earth becomes the background, and any series of images will show movements in the atmosphere and oceans. A film loop of successive geostationary images is what we see on TV weather broadcasts when we see motion in the atmosphere. The disadvantage of the higher geostationary orbit lies in its inability to produce close-up images of the surface. Here the lower polar orbiter has the advantage.

most common, but some satellites also have thermal-infrared capability. Today, many satellites orbiting Earth relay multispectral data back to receiving stations in the form of digital imagery.

Landsat　A series of satellites initially launched in 1972, **Landsat** was the first space program to provide continuous monitoring and imaging of Earth's surface. There have been several Landsat satellites, and configurations vary. Landsat scans a 100-nautical-mile (185 km) swath in several wavelength bands (visible light, near-IR, and thermal IR), and transmits digital image data for later conversion into images.

From 705 kilometers (438 mi) high, the satellite maintains the same north-south orbital path and relies on Earth rotation below to get nearly full planetary coverage. Landsat always overflies and images a place at the same time of day. Every 16 days, a Landsat satellite images the same location. A great benefit of this system is its excellent sequential coverage that is important for monitoring environmental changes.

Many types of images can be generated from multispectral data, but the most familiar is the **color composite** (Fig. 2.37). Blending three images of the same location but using pixel data from three different wavelengths digitally creates a color composite. The usual Landsat color composite was designed to resemble a false-color near-infrared photograph, with the same color assignment as color near-infrared photos. The current Landsat imaging system is called the **Thematic Mapper (TM),** because it was designed to collect image data that can be used to generate thematic maps of a wide variety of environmental variables (for example, water features or vegetation patterns).

On a standard Landsat color composite, red is healthy vegetation; barren areas show up as white or brown; clear, deep water bodies are dark blue; and muddy water appears light blue. Clouds and snow are bright white. Urbanized areas are blue-gray. Although the colors are visually important, the greatest benefit of multispectral imagery is that computers can identify, classify, and map (in a first approximation) these kinds of areas automatically.

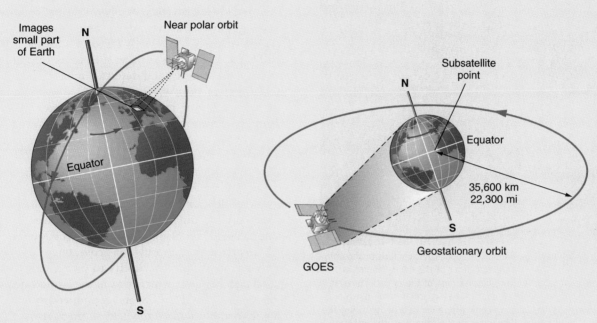

Polar orbits circle Earth approximately from pole to pole and use the movement of Earth as it turns on its axis to image small areas (perhap 100 × 100 kilometers) in order to gain good detail of the surface. This orbital technique yields nearly full Earth coverage in a mosaic of images, and the satellite travels over the same region every few days, always at the same local time. (Not to scale.)

Geostationary orbits are used with satellites orbiting above the equator at a speed that is synchronized with Earth rotation so that the satellite can image the same location continuously. Many weather satellites use this orbit at height that will permit imaging an entire hemisphere of Earth. (Not to scale.)

Multispectral Remote Sensing Applications The future of remote sensing will involve an increased emphasis on computer-automated methods of analyzing imagery, greater spatial resolution of available imagery, and continued use of digital imagery to capture thematic layers for integration into a GIS. Computer analysis permits rapid image processing, although our ability to collect data outstrips our ability to use, apply, and understand the information contained in the images that have already accumulated. This information explosion is complicated by the fact that many geographic and environmental problems are best understood by studying images of several types simultaneously, using multispectral remote sensing. Each part of the spectrum may yield different information about an aspect of the environment, and again, geographic information systems will play a strong role in this kind of analysis.

The first weather satellite was launched by the United States in 1960. Today, National Oceanic and Atmospheric Administration (NOAA) weather satellites observe atmospheric conditions of entire Earth from orbit and return visible and thermal IR coverage of weather and cloud patterns. Geostationary Operational Environmental Satellites (GOES) monitor major storm systems from a **geostationary orbit.** They are positioned 36,000 kilometers (22,300 mi) over the equator and orbit in synch with Earth rotation so that they are always over the same Earth position. GOES images are seen on television weather shows and are reproduced in daily newspapers (see again Fig. 2.34). In addition to NOAA and GOES satellites, the U.S. Air Force, the European Space Agency, Russia, and Japan also have weather satellites in operation.

Weather satellites have saved many lives and millions of dollars by increasing our storm and weather forecasting capabilities. An excellent example of this occurred when Hurricane Andrew struck the south coast of Florida in August 1992. Although this was the most devastating hurricane in U.S. history in terms of property damage ($25–30 billion), the advance warning provided by weather satellites allowed people to evacuate, saving hundreds of lives.

Sonar (*SO*und *NA*vigation and *R*anging) allows us to probe the ocean depths and to map undersea topography.

U.S Geological Survey

Figure 2.37   Landsat Thematic Mapper color composite showing Salt Lake City, Utah; the western front of the Rocky Mountains; the Great Salt Lake; and the deserts west of the city. North is at the top of this image. A railroad causeway isolates the north end from the rest of the lake, and the difference in tone following the causeway results from differences in the salinity of the water on each side. The depiction is intended to be similar to a color infrared photograph, in terms of color representation, ■ **What kinds of features can you identify on this false-color image?**

The use of seismic waves (from natural and artificially generated earthquakes) has given us knowledge of the interior of the planet and helps us find resources such as gas and oil hidden in solid Earth.

A great advantage of remote sensing is that it provides rapid, worldwide coverage of environmental conditions for geographers, cartographers, and other scientists. Satellite coverage is especially important for the less accessible parts of Earth's surface, such as ice caps, mountains, tropical forests, and deserts. Important resources such as mineral ores, water, and potential energy sources can be found more rapidly and with greater accuracy. Through continuous monitoring of Earth's surface, global, regional, and even local changes can be detected and mapped. Geographic information systems offer the ability to match and combine thematic layers of any sort, including digitized maps, scanned photographs, digital terrain models, and multispectral images, instantly accessing any combination of these that we need to solve complex spatial problems related to the environment.

Maps are essential tools for geographers and other Earth scientists, and the ability to gather and present spatial data has greatly increased in the last few decades. Remote sensing, GPS, GIS, and computer-assisted mapping have revolutionized the field of geography.

DEFINE & RECALL

navigation
cartography
digital mapping
oblate spheroid
great circle
hemisphere
small circle
coordinate system
North Pole
South Pole
equator
latitude
sextant
prime meridian
longitude
parallel
meridian
time zone
solar noon
International
   Date Line

U.S Public Lands
   Survey System
principal meridian
base line
township
section
Global Positioning
   System (GPS)
map projection
conformal map
equal-area map
Mercator projection
equidistance
azimuthal map
rhumb line
gnomonic projection
legend
scale
verbal scale
representative fraction (RF)
   scale

graphic (bar) scale
magnetic declination
azimuth
bearing
thematic map
continuous data
discrete data
isoline
topographic contour line
contour interval
profile
gradient
digital terrain model
vertical exaggeration
visualization
geographic information
   system (GIS)
geocoding
remote sensing
digital image
pixels

near-infrared (NIR) film
thermal infrared
   (TIR) scanning
imaging radar
multispectral scanning
Doppler radar
side-looking airborne
   radar (SLAR)
Landsat
color composite
Thematic Mapper (TM)
weather satellite
geostationary orbit
sonar

DISCUSS & REVIEW

1. Why is the concept of a great circle so useful to navigators?

2. What great circle has been chosen as the zero point for latitude?

3. What are the latitude and longitude of your city?

4. Approximately how accurate in meters could you be if you tried to locate a building in your city to the nearest second of latitude and longitude? Using a GPS?

5. What time zone are you in? What is the time difference between Greenwich time and your time zone?

6. If it is 2:00 A.M. Tuesday in New York (EST), what time and day is it in California (PST)? What time is it in London (GMT)?

7. If you fly across the Pacific Ocean from the United States to Japan, how will the International Date Line affect you?

8. How has the use of the Public Lands Survey System affected the landscape of the United States? Has your local area been affected by its use? How?

9. Why can't maps give an accurate representation of Earth's surface? What is the difference between a conformal map and an equal-area map?

10. How does the Mercator projection differ from the gnomonic projection? What advantages does the Mercator offer the navigator?

11. What is the difference between RF and a verbal map scale?

12. Why is the date of publication of a map important? List several reasons, including one related to Earth's magnetic field.

13. Why is the scale of a map so important? What does a small-scale map show in comparison with a large-scale map?

14. Why are contour maps so important to physical geographers? What specific information can be obtained from a contour map that cannot be obtained from other maps?

15. What does the concept of "thematic map layers" mean in a geographic information system? What are attributes?

16. How have computers revolutionized cartography and the handling of spatial data? What specific advantages do computers offer to the mapmaking process?

17. What is the difference between a photograph and a digital image?

18. Describe the difference between imagery from a weather satellite and from Landsat Thematic Mapper. How do the monitoring purposes of these satellites affect the different imagery produced by each?

19. What does a weather radar image show in order to help us understand weather patterns?

CONSIDER & RESPOND

1. Use an atlas and globe to determine what cities are located at the following grid coordinates: 40°N, 75°W; 34°S, 151°E; and 41°N, 112°W.

2. What are the grid coordinates of the following cities: Portland, Oregon; Rio de Janeiro, Brazil; your hometown?

3. You are located at 10°S latitude, 10°E longitude; you travel 30° north and 30° east. What are your new geographic coordinates?

4. You are located at 40°N latitude, 90°W longitude. You travel due north 40°, then due east 60°. What are your new geographic coordinates?

5. Using an atlas and Figure 2.24, identify a city in North America with each of the following magnetic declinations: 20°East; 20°West; 25°East; 0°.

6. Select one place within the United States that you would most like to visit for a vacation. If you could take along only one map and you had your choice of a state highway map, a USGS topographic map, or an official map of a city or town, which would you choose and why would you choose it?

7. If you were an applied geographer and wanted to use a geographic information system to build a database of spatial information about the environment of a park (pick a state or national park near you, for example), what are the seven most important "layers" of mapped information that you would want to have? What combinations of two or more layers would be particularly important to your purpose?

8. GIS has been called a "power tool" for spatial (geographic) analysis. Why are people so important to its effective application?

EARTH IN SPACE AND SOLAR ENERGY

The universe is made up of billions of galaxies, each containing so many stars that the mathematical possibility of other planets containing life as we know it seems unlimited.

What is a galaxy, and what is its relationship to the universe?

How is distance involved in the determination of whether life beyond our solar system does exist in the universe?

Earth is one of nine planets that, together with the sun, comprise the major components of our solar system.

In what ways is the sun of most importance to life on Earth?

What are the chief characteristics of the other planets?

What are the other components of our solar system?

The regular movements of Earth, termed rotation and revolution, are the fundamental elements of Earth–sun relationships, which initially control the dynamics of our atmosphere and the phenomena related to it.

Why is this one of the most important understandings in physical geography?

What other understandings follow from this concept?

The sun is the original and ultimate source of the energy that drives the various components of the Earth system.

How does the sun's energy reach Earth?

How does this energy affect the Earth system?

The types of energy emitted by the sun are represented in the electromagnetic spectrum.

What bands of this spectrum control heating and cooling in the Earth energy system?

What bands of this spectrum affect humans directly?

The relationship of Earth's axis to the plane of Earth's orbit is the key to an explanation of seasons on Earth.

What is the relationship?

How does it operate in conjunction with Earth's revolution to produce seasons?

How does it influence variations in the amounts of insolation reaching different portions of Earth's surface?

The sun affects Earth systems in myriad ways.

With a radius 110 times that of Earth, and a mass 330,000 times greater, the sun reigns as the center of our solar system. The gravitational pull of this fierce, stormy ball of gas holds Earth in orbit and powers the Earth/atmosphere systems on which our lives depend. As the source of almost all the energy in our world, it holds the key to many of our questions about Earth and sky.

Everyone has wondered about environmental changes that take place throughout the year and from place to place over Earth's surface. Perhaps when you were young, you wondered why it got so much warmer in summer than in winter and why some days were long while those in other seasons were much shorter. These questions, and many like them, are probably as old as humanity, and the answers to them help provide us with an understanding of the physical geography of our world.

Physical geographers' concerns take them beyond planet Earth to a consideration of the sun and Earth's position in the solar system. Geographers examine the relationship between the sun and Earth to explain such earthly phenomena as the alternating periods of light and dark that we know as day and night. Other relationships between Earth and sun also help to explain seasonal variations in climate. Although the universe and solar system are not strictly within the province of physical geography, an acquaintance with each can be of help in an examination of Earth as an environment for life as we know it.

The Solar System and Beyond

If you look at the sky on a clear night, all of the stars that you see are part of a single collection of stars called the Milky Way Galaxy. A **galaxy** (Fig. 3.1) is an enormous island in the universe—an almost incomprehensible cluster of stars, dust, and gases. Our sun is one of hundreds of billions of stars that compose the Milky Way Galaxy. In turn, the observable universe appears to contain billions of other galaxies.

Distances within the universe are so vast that it is necessary to use a large unit of measure termed a **light-year**— the distance that light travels in one year. A light-year is equal to 6 trillion miles. Light travels at the amazing speed of 298,000 kilometers per second (186,000 mi/sec). Thus, in 1 second, light could travel seven times around the circumference of Earth. Although that may seem like a great distance, the *closest* star to Earth, other than the sun, is 4.3 light-years away, and the *closest* galaxy to our galaxy is 75,000 light-years away.

The Solar System

The sun is the center of our solar system. A **solar system** can be defined as all the heavenly bodies surrounding a particular star because of the star's dominant mass and gravitational attraction (Fig. 3.2). The principal celestial bodies in our sun's system are the nine major **planets** (celestial bodies that revolve around a star and reflect the star's light rather than producing their own). Our solar system also includes 61 **satellites** (like Earth's moon, these bodies orbit the planets) and numerous **asteroids** (very small planets, usually with a diameter of less than 500 miles), as well as **comets** and **meteors.** A comet is made up of a head—a collection of solid fragments held together by ice—and a tail, sometimes millions of miles long, comprised of gases (Fig. 3.3). Meteors are small, stonelike bodies that, when they enter Earth's atmosphere, burn and often appear as a streak of light, or "shooting star." A meteor that survives the fall through the atmosphere and strikes Earth's surface is called a **meteorite** (Fig 3.4).

The Sun and Its Energy

The sun, like all the other stars in the universe, is a self-luminous sphere of gases that emits radiant energy. A slightly less than average sized star, our sun is the only self-luminous body in our solar system and is the source of almost all the light and heat for the surfaces of the various celestial bodies in our planetary system. It has an estimated surface temperature of between 5500°C and 6100°C (10,000°F and 11,000°F). The energy emitted by the sun comes from **fusion (thermonuclear) reactions** that take place in its interior. There, under high pressure, two hydrogen atoms are fused to form helium

Figure 3.1 Looking toward the center of our galaxy, the Milky Way

NASA

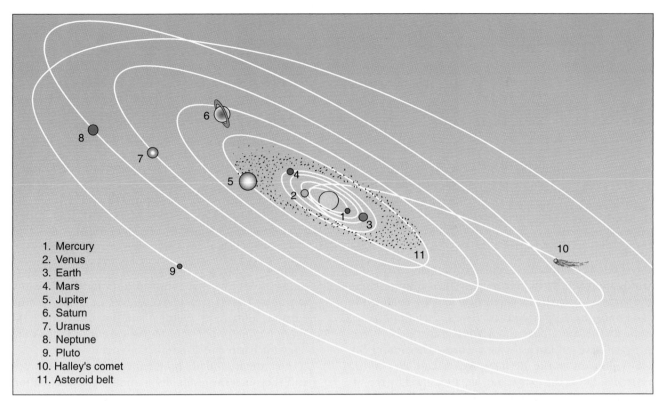

1. Mercury
2. Venus
3. Earth
4. Mars
5. Jupiter
6. Saturn
7. Uranus
8. Neptune
9. Pluto
10. Halley's comet
11. Asteroid belt

Figure 3.2   The solar system, showing the sun and planets in their proper order. The approximate size relationships between the individual planets is shown. However, the planetary orbits are much condensed, and the scale of the sun and planets is greatly exaggerated. The planets would be much too small to be visible at the scale of the orbits shown.

in a process with impact similar to that of a hydrogen bomb explosion (Fig. 3.5). This nuclear reaction releases tremendous amounts of energy that radiate out from the sun in all directions at the speed of light.

Earth receives about $\frac{1}{2,000,000,000}$ of the radiation given off by the sun, but even this tiny amount affects the biological and physical characteristics of Earth's surface.

Other bodies in the solar system receive some of the remainder of the sun's radiant energy, but the vast proportion of it travels out through space unimpeded.

From the outermost layer of the sun, ionized gases acquire enough velocity to escape the gravitational pull of the sun and become **solar wind.** Solar wind is produced by streams of extremely hot protons and electrons that

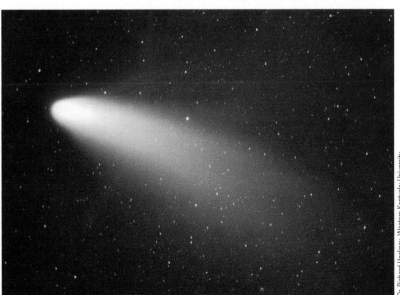

Dr. Richard Hackney, Western Kentucky University

Figure 3.3   The comet Hale–Bopp shows a split tail, because two different types of icy material are emitting different jets of gases.

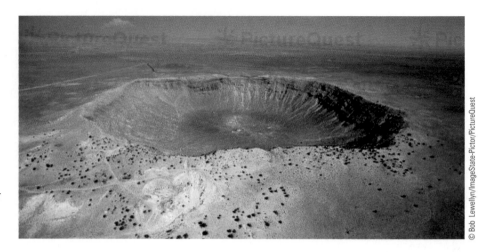

Figure 3.4 Barringer (meteor) Crater in Arizona displays some of the same characteristics as those found on the moon and other planets.

travel out at the speed of light. Occasionally, these solar winds approach Earth, but are prevented from reaching the surface by Earth's magnetic field and are confined to the atmosphere. During these times, they can disrupt radio and television communication, may disable orbiting satellites, and intensify the **auroras.** The Aurora Borealis, known as the *northern lights,* and the Aurora Australis, called the *southern lights,* are caused by the interaction between incoming solar wind and the ions in our upper atmosphere (Fig 3.6).

The intensity of solar winds is influenced by the best-known solar feature, **sunspots.** These dark regions are about 1500°C cooler than the surrounding surface temperature. Galileo began recording sunspots back in the 1600s, and for many years they have been used to indicate solar activity. Sunspots seem to observe an 11-year cycle from one maximum (where 100 or so may be visible) to the next. An individual sunspot may last from a few days to a few months. Just how sunspots might affect Earth's atmosphere is still a matter of controversy. It is difficult to prove the exact connections between sunspot numbers and weather and climate (Fig. 3.7).

The sun's energy is the most important factor determining environmental conditions on Earth. With the exception of geothermal heat sources (such as volcanic eruptions and geyser springs) and heat emitted by radioactive minerals, the sun remains the source of all the energy for Earth and atmospheric systems.

The intimate and life-producing relationship between Earth and sun is the result of the amount and distribution of radiant energy received from the sun. Such factors as our planet's size, its distance from the sun, its atmosphere, the movement of Earth around the sun, and the planet's rotation on an axis all affect the amount of radiant energy that Earth receives. Though some processes of our physical environment result from Earth forces not related to the sun, these processes would have little relevance were it not for the life-giving, life-sustaining energy of the sun.

Earth revolves around the sun at an average distance of 150 million kilometers (93 million mi). The sun's size and its distance from us challenge our comprehension. About 130 million Earths could fit inside the sun, and a plane flying at 500 miles per hour would take 21 years to reach the sun.

Figure 3.5 The fireball explosion of a hydrogen bomb is created by thermonuclear fusion. This same reaction powers the sun.

Figure 3.6 Solar "wind" and Earth's atmosphere interact to produce the Aurora Borealis.

As far as we know with certainty, within our solar system, only on Earth has the energy from the sun been used to create life—to create something that grows, changes, and can reproduce itself. Yet there remains a possibility of life, or at least the basic organic building blocks, on Mars and perhaps even on one or two of the moons of Saturn or Jupiter. What fascinates scientists and philosophers alike, however, is the likelihood that there are millions of planets in the universe like Earth on which forms of life have developed that might be even more sophisticated than humans.

The Planets

The four planets closest to the sun (Mercury, Venus, Earth, and Mars) are called the **terrestrial planets.** They are relatively small, warmed by their proximity to the sun, and composed of rock and metal. They all have solid surfaces that exhibit records of geological forces in the form of craters, mountains, and volcanoes. The next four planets (Jupiter, Saturn, Uranus, and Neptune) are much larger and composed primarily of lighter ices, liquids, and gases. These planets are termed the **giant planets** or **gas planets.** Although they have solid cores at their centers, they are more like huge balls of gas and liquid with no solid surface on which to walk. Finally, at the outer edge of the solar system is Pluto, which is neither a terrestrial nor a giant planet (Fig. 3.8).

Through the years, controversy has arisen as to whether Pluto is officially one of our planets or a large body captured from the Kuiper Belt (a disk-shaped region that lies past the orbit of Neptune containing small icy bodies) by the gravitational pull of the sun. The question has been, Is Pluto a small outermost planet or a large captured member of the Kuiper Belt? For now it has been decided to keep Pluto as one of our family of planets.

The nine major planets that are known to revolve around the sun have several phenomena in common. From a point far out in space above the sun's "north pole," they would all appear to move around the sun in the same counterclockwise direction. Their orbits follow an elliptical, almost circular, path. All the planets also rotate, or spin, on their own axes. With the exception of Venus and Uranus, all rotate in the same direction. All but Pluto lie close to the same plane (the Plane of the Ecliptic) passing through the sun's equator. All the planets have an atmospheric layer of gases with the exception of Pluto and Mercury, which are not dense or heavy enough to hold any appreciable amounts of gases.

Life Beyond Earth?

Although the planets are similar in many ways, they differ in a number of vital characteristics, and the existence of life as we know it beyond Earth in our solar system seems highly unlikely (Table 3.1). Mercury, the smallest terrestrial planet, has a diameter less than half that of Earth, and its **mass** (the total amount of material in the planet) is only 5 percent of Earth's. This means that Mercury does not have sufficient mass and gravitational pull to hold onto a meaningful atmosphere. (**Gravitation** is the attractive force one body has for another. The greater the mass or amount of matter a body has, the greater the gravitational pull it will exert on other bodies.)

Also, partly because of the very thin atmosphere around Mercury, surface temperatures on that planet reach over 425°C (800°F) on the side facing the sun. As we will see later, Earth's atmosphere protects our planet's

Figure 3.7 Sunspots as they appear on the solar surface. Inset shows relative size of Earth.

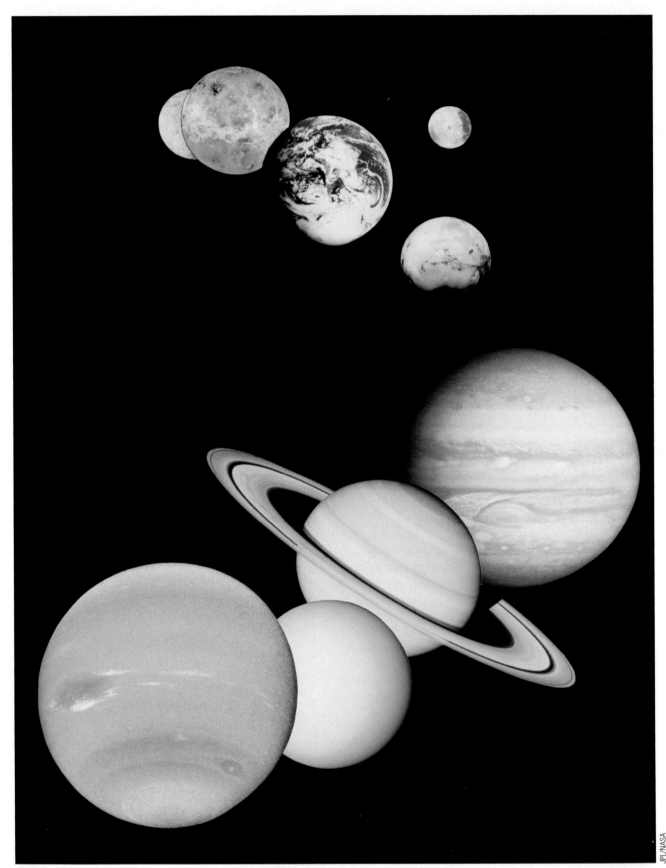

Figure 3.8   The planets in our solar system: Mercury, Venus, Earth, and Mars (the terrestrial planets); Jupiter, Saturn, Uranus, and Neptune (the giant gas planets); and Pluto (which belongs to neither group).

Table 3.1 Comparison of the Planets

Name	Distance from Sun	Revolution Period	Diameter	Mass	Density
	(AU) *	(yrs)	(km)	(10^{23} kg)	(g/cm^3)
Mercury	0.39	0.24	4,878	3.3	5.4
Venus	0.72	0.62	12,102	48.7	5.3
Earth	1.00	1.00	12,756	59.8	5.5
Mars	1.52	1.88	6,787	6.4	3.9
Jupiter	5.20	11.86	142,984	18,991	1.3
Saturn	9.54	29.46	120,536	5,686	0.7
Uranus	19.18	84.07	51,118	866	1.2
Neptune	30.06	164.82	49,660	1,030	1.6
Pluto	39.44	248.60	2,200	0.01	2.1

*An AU (or astronomical unit) is the distance from the Earth to the sun.

surface from much of the sun's radiation, which can be dangerous to life, and keeps temperatures on Earth within a livable range.

Further, because Mercury rotates on its axis only once every 59 days, the same side of the planet is exposed to the sun for long periods, while the opposite side receives none of the sun's energy. This situation compounds the problem of Mercury's surface temperatures. The side facing the sun is much too hot, and the opposite side is much too cold ($-175°$C, or $-280°$F) to support life.

Uranus, Neptune, and Pluto, the three planets farthest from the sun, are too distant to receive enough solar energy to have surface temperatures conducive to life, as we know it. In addition, though Uranus and Neptune are large enough to have atmospheres, theirs are made up primarily of hydrogen and helium; they apparently have no free oxygen, a factor necessary for the development of life like that on Earth.

Jupiter and Saturn are the two largest planets; yet, because significant parts of their masses are gaseous, both have very low densities. For instance, Saturn's density is less than that of water, whereas Earth's density is more than five times as great. The atmospheres of Jupiter and Saturn also have a high proportion of hydrogen and helium, and no free oxygen. Even though Jupiter and Saturn are closer to the sun than Uranus, Neptune, and Pluto, they still do not receive enough solar energy to produce livable surface temperatures; their temperatures are down around $-95°$C to $-150°$C ($-200°$F to $-300°$F). It is unlikely that life exists on the two largest planets, and scientists have begun to look instead at the moons of Saturn and Jupiter for that possibility. In 1997, the spacecraft *Cassini* was launched by the United States and is scheduled to reach Saturn in July 2004. About six months later it will release its piggybacked Huygens probe for descent through the thick atmosphere of Titan, one of Saturn's moons.

Venus and Mars, in the orbits closest to that of Earth, are the planets most similar to our own. Venus, in fact, has been called Earth's twin because it is most like Earth in size, density, and mass. However, we cannot see the surface of Venus, because it is hidden by a permanent layer of thick clouds. Through information gathered by the spacecraft *Magellan* in 1991–1993, we have a much better understanding of surface conditions on Venus. The surface consists primarily of lowland lava plains, much like the basaltic ocean basins of Earth, with two continents rising above the lowlands. The planet has no liquid or frozen water. The atmosphere of Venus is 96 percent carbon dioxide. This thick layer of carbon dioxide allows very little energy to escape from the planet. As a result, the surface temperature of Venus is greater than $450°$C ($850°$F).

Mars is better known to us because thousands of pictures have been taken by numerous spacecraft. In addition, the *Viking* spacecraft, which landed on the surface of Mars, and the *Mars Global Surveyor,* in close orbit, have provided exceptional images of surface features (Fig. 3.9). Since being placed in orbit around Mars in October 2001, the *Mars Odyssey* orbiting spacecraft has relayed new and more revealing images of our neighboring planet and will continue to do so in the future. Like that of Venus, the atmosphere of Mars is 96 percent carbon dioxide. However, it is so thin that energy is not trapped as it is on Venus. Thus, surface temperatures range from $-125°$C ($-190°$F) at the poles to $25°$C ($77°$F) at the equator. Mar has seasonal polar ice caps of frozen carbon dioxide (dry ice), and the 2001 *Mars Odyssey* orbiter detected some water at the Martian south pole. Evidence has been found that suggests that rain once fell and rivers once flowed on Mars. If life has existed on Mars, it is most likely that it existed at a time when water was more abundantly available.

NASA/JPL/MSSS

Figure 3.9   This image taken by Mars Global Surveyor shows a surface cut by ancient streamflow.

The Earth–Sun System

The color views of Earth transmitted by the *Apollo* spacecraft on their lunar missions gave us some unforgettable views of our planet. The cameras showed Earth as a sphere of blue oceans, green and brown landmasses, and swirls of white clouds. One astronaut has described Earth as it appears to someone who has traveled close to the moon:

> *Earth looked so tiny in the heavens that there were times during the Apollo 8 mission when I had trouble finding it. If you can imagine yourself in a darkened room with only one clearly visible object, a small blue-green sphere about the size of a Christmas tree ornament, then you can begin to grasp what Earth looks like from space. I think that all of us subconsciously think that Earth is flat or at least almost infinite. Let me assure you that, rather than a massive giant, it should be thought of as the fragile Christmas tree ball which we should handle with considerable care.*

> **This Island Earth**
> **edited by Oran W. Nicks,**
> **NASA, SP-250, 1970**

As we begin the study of our planet, we should not lose sight of the image of Earth as an exceedingly isolated island in a seemingly endless sea. Scientists have always speculated that there could be a planet in another galaxy that has intelligent life. However, it was not until 1995 that scientists finally proved other planets exist outside our solar system. The fact remains that we have not actually seen other planets but only proved their existence through mathematical modeling. Thus, we should learn as much as we can about our planet and treat it with exceptional care because, in all likelihood, Earth is the only home the human race will ever know.

Solar Energy and Atmospheric Dynamics

As we have previously noted, our sun is the major source of energy, either directly or indirectly, for the entire Earth system. Earth does receive very small proportions of energy from other stars and from the interior of Earth itself (volcanoes and geysers provide certain amounts of heat

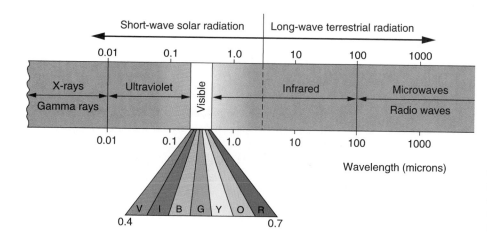

Figure 3.10 Radiation from the sun travels toward Earth in a wide spectrum of wavelengths, which are measured in micrometers (1 micrometer equals one ten-thousandth of a centimeter). Visible light occurs at wavelengths of approximately 0.4 to 0.7 micrometers. Solar radiation is short-wave radiation (less than 4.0 micrometers) whereas terrestrial (Earth) radiation is of long wavelengths (more than 4.0 micrometers).

energy); however, when compared with the amount received from the sun, these other sources are insignificant.

Energy is emitted by the sun in the form of **electromagnetic energy,** which travels at the speed of light in a spectrum of varying wavelengths (Fig. 3.10). It takes about 8⅓ minutes for these waves to reach Earth. About 41 percent of this spectrum comes in the form of visible light rays, but much of the sun's energy cannot be seen by the human eye. About half of the sun's radiant energy is in wavelengths that are longer than visible light rays, including some infrared waves. Although these wavelengths cannot be seen, they can sometimes be sensed by human skin. The longer waves in the infrared part of the spectrum are felt as heat.

The remaining 9 percent of solar energy is made up of *X rays, gamma rays,* and *ultraviolet rays,* all of which are shorter in length than visible light. These also cannot be seen but can affect other tissues of the human body (thus, absorbing too many X rays can be dangerous, and excessive ultraviolet waves gives us sunburned skin). Collectively, visible light, ultraviolet rays, X rays, and gamma rays are known as **short-wave radiation.** We have learned to harness some of these energy waves for communications (radio, microwave transmission, television), health (X rays), and use in the field of remote sensing (photography, radar, infrared imagery).

The sun radiates energy into space at a steady rate. At its outer edge, Earth's atmosphere intercepts an amount of energy equivalent to 1.95 calories per square centimeter per minute. (A **calorie** is the amount of *energy* required to raise the temperature of 1 gram of water 1°C.) This can also be expressed in units of power—in this case, 1380 watts per square meter. The rate of a planet's receipt of solar energy is known as the **solar constant** and has been measured with great precision outside Earth's atmosphere by orbiting satellites. The atmosphere affects the amount of solar radiation received on the surface of Earth, because some energy is absorbed by clouds, some is reflected, and some is refracted. If we

could remove the atmosphere from Earth, we would find that the solar energy striking the surface at a particular location for a particular time would be a constant value determined by the latitude of the location.

Of course, the measured value of the solar constant varies with distance from the sun as the same amount of energy radiates out into larger areas. For example, if we measured the solar constant for the planet Mercury, it would be much higher than that for Earth. When Earth is closest to the sun in its orbit, its solar constant is slightly higher than the yearly average, and when it is farthest away, the solar constant is slightly lower than average. However, this difference does not have a significant effect on Earth's temperatures. When Earth is farthest from the sun in July and the solar constant is lowest because of the distance from the sun, the Northern Hemisphere is in the midst of a summer with temperatures that are not significantly different from those in the Southern Hemisphere six months later. The solar constant also varies slightly with changes in activity on the sun; during intense sunspot or sun storm activity, for example, the solar constant will be slightly higher than usual. However, these variations are not even as great as those caused by Earth's elliptical orbit.

Movements of Earth

Earth has three basic movements: **galactic movement, rotation,** and **revolution.** The first of these is the movement of Earth with the sun and the rest of the solar system in an orbit around the center of the Milky Way Galaxy. This movement has limited effect on the changing environments of Earth and is generally the concern of astronomers rather than of geographers. The other two movements of Earth, rotation on an axis and revolution around the sun, are of vital interest to the physical geographer. The consequences of these movements are the phenomena of day and night, variations in the length of day, and the changing seasons.

The Solar Constant

As you know, the sun is the source of almost all of the energy used to operate Earth systems and sustain life on our planet. It is easy to believe that this ultimate source is everlasting and unchanging. However, this is not the case, and it can be shown with direct measurements.

The **solar constant** is a measure of the amount of solar radiation (expressed as a heat value) that reaches the outer limits of our atmosphere; it amounts to 1.95 calories/square centimeter/minute. This value is often rounded up to 2.0 cal./sq. cm./min. A British radiation scientist by the name of Langley first established the units, "cal./sq. cm." and in his honor we sometimes refer to 1.0 cal./sq. cm. as 1.0 langley (ly). As such, the solar constant is sometimes expressed as 2.0 lys/min. The solar constant can also be expressed in units of power at approximately 1370 watts/square meter.

Regardless of the numerical values and units of measure, we have found incoming solar energy to vary through time. Measurements taken from the Solar Maximum Mission (SMM) and other satellites have shown a variation of as much as (+ or –) 0.2% change through the decades. Some of the reasons for this variation lie within the discussions in this chapter. Activities such as the sunspot cycle, solar flares, and magnetic storms on the solar surface can account for these changes. Though this variation does not seem large, it does exist. How such a change through time might affect Earth systems, is the subject of ongoing research.

Our sun must be viewed as a turbulent entity, an exploding, churning, mass of ignited gases, and to expect such an entity to remain constant through time is unrealistic.

Rotation   Rotation refers to the spin of Earth on its own axis, an imaginary line extending from the North Pole to the South Pole. Earth rotates on its axis at a uniform rate, making one complete turn with respect to the sun in 24 hours.

Earth turns in an eastward direction (Fig. 3.11). The sun "rises" in the east and appears to move westward across the sky, but it is actually Earth, not the sun, that is moving, rotating toward the morning sun (that is, toward the east).

Earth, then, rotates in a direction opposite to the apparent movement of the sun, moon, and stars across the sky. If we look down on a globe from above the North Pole, the direction of rotation is counterclockwise. This eastward direction of rotation not only defines the movement of the zone of daylight on Earth's surface but also helps define the circulatory movements of the atmosphere and oceans.

The velocity of rotation at the Earth's surface varies with the distance of a given place from the equator (the imaginary circle around Earth halfway between the two poles). All points on the globe take 24 hours to make one complete rotation (360°). Thus the *angular velocity* for all locations on Earth's surface is the same—360° per 24 hours, or 15° per hour. However, the *linear velocity* depends on the distance (not the angle) covered in that 24 hours. The linear velocity at the poles is zero. You can see this by spinning a globe with a postage stamp affixed to the North Pole. The stamp rotates 360° but covers no distance and therefore has no linear velocity. If you place the stamp anywhere between the North and South Poles, however, it will cover a measurable distance during one rotation of the globe. The greatest linear velocity is found at the equator, where the distance traveled by a point in 24 hours is largest. At Kampala, Uganda, near the equator, the velocity is about 460 meters (1500 ft) per second, or approximately 1660 kilometers (1038 mi) per hour (Fig. 3.12). In comparison, at St. Petersburg, Russia (60°N latitude), where the distance

North

West

East

South

Figure 3.11   Earth spins around a tilted axis as it follows its orbit around the sun. Earth's rotation is from west to east, making the stationary sun appear to rise in the east and set in the west.

NASA

Surrounding the core of the sun is a radiative zone through which energy passes to the surface. As the energy emerges onto the surface, it creates the grainy pattern we see in this image.

The visible surface of the sun is called the photosphere, a layer about 160 km (100 miles) thick. The lowest part of the transparent atmosphere, lying above the photosphere, is called the chromosphere (note the ragged edges seen around the sun); the upper part, which is seen only during a solar eclipse, is the corona (note the light orange color transition to outer space).

There is a prominent solar flare in the upper right hand corner.

traveled during one complete rotation of Earth is about half that at the equator, Earth rotates about 830 kilometers per hour.

We are unaware of the speed of rotation, because (1) the angular velocity is constant for each place on Earth's surface, (2) the atmosphere rotates with Earth, and (3) there are no nearby objects, either stationary or moving at a different rate with respect to Earth, to which we can relate Earth's movement. Without such references, we are unable to perceive the speed of rotation.

Rotation accounts for our alternating days and nights. This can be demonstrated by shining a light at a globe while rotating the globe slowly toward the east. You can see that half the sphere is always illuminated while the other half is not, and that new points are continually moving into the illuminated section of the globe while

Kampala

Latitude 60°N

St. Petersburg

North Pole

0 kmph

830 kmph

1660 kmph

Equator

Figure 3.12 The speed of rotation of Earth varies with the distance from the equator.

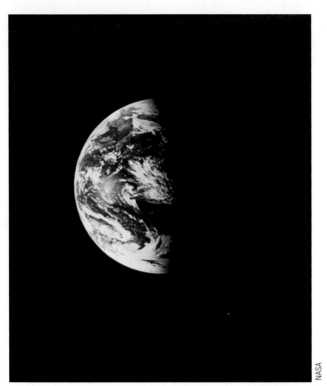

Figure 3.13 The circle of illumination, which separates day from night, is clearly seen on this image of Earth.

others are moving into the darkened sector. This corresponds to Earth's rotation and the sun's energy striking Earth. While one half of Earth receives the light and energy of solar radiation, the other half is in darkness. As noted in Chapter 2, the great circle separating day from night is known as the **circle of illumination** (Fig. 3.13).

Revolution   While Earth rotates on its axis, it also revolves around the sun in a slightly elliptical orbit at an average distance from the sun of about 150,000,000 kilometers (93,000,000 mi) (Fig. 3.14). On about January 3, Earth is closest to the sun and is said to be at **perihelion** (from Greek: *peri,* close to; *helios,* sun); its distance from the sun then is approximately 147,500,000 kilometers. At around July 4, Earth is about 152,500,000 kilometers from the sun. It is then that Earth has reached its farthest

point from the sun and is said to be at **aphelion** (Greek: *ap,* away; *helios,* sun). Five million kilometers is insignificant in space, and these varying distances from Earth to the sun only minimally (3.5 percent) affect the receipt of energy on Earth. Hence, they have no relationship to the seasons.

The period of time Earth takes to make one revolution around the sun determines the length of one year. Earth makes 365¼ rotations on its axis during the time it takes to complete one revolution of the sun; therefore, a year is said to have 365¼ days. Because of the difficulty of dealing with a fraction of a day, it was decided that a year would have 365 days, and every fourth year, called *leap year,* an extra day would be added as February 29th.

Plane of the Ecliptic, Inclination, and Parallelism   In its orbit around the sun, Earth moves in a constant plane, known as the **plane of the ecliptic.** Earth's equator is tilted at an angle of 23½ degrees from the plane of the ecliptic (Fig 3.15). In addition to this constant **angle of inclination,** Earth's axis maintains another characteristic called **parallelism.** As Earth revolves around the sun, Earth's axis remains parallel to its former positions. That is, at every position in Earth's orbit, the axis remains pointed toward the same spot in the sky. For the North Pole, that spot is close to the star that we call the North Star, or Polaris. Thus, Earth's axis is fixed with respect to the stars outside our solar system but not with respect to the sun (see the axis representation in Figure 3.11).

Understanding these relationships with the sun leads us directly into an examination of the seasonal changes on Earth and a discussion of how the intensity of the sun's rays varies from place to place throughout the year.

The Seasons

Many people assume that the seasons must be caused by the changing distance between Earth and the sun during Earth's yearly revolution. As noted earlier, the change in this distance is very small. Further, for people in the Northern Hemisphere, Earth is actually closest to the sun in January and farthest away in July (see Fig. 3.14). This is exactly opposite of that hemisphere's seasonal

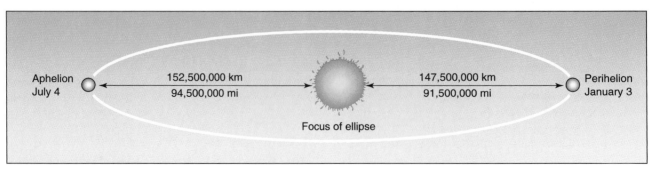

Figure 3.14 Oblique view of the elliptical orbit of Earth around the sun. Earth is closest to the sun at perihelion and farthest away at aphelion. Note that in the Northern Hemisphere summer (July), Earth is farther from the sun than at any other time of year.

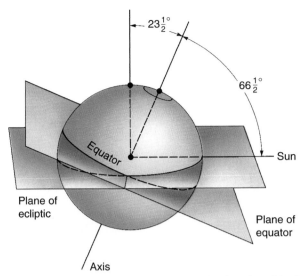

Figure 3.15 The plane of the ecliptic is defined by the orbit of Earth around the sun. The 23½° inclination of Earth's rotational axis causes the plane of the equator to cut across the plane of the ecliptic.

variations. As we will see, seasons are caused by the 23½° tilt of Earth's equator to the plane of the ecliptic (see Fig. 3.15) and the parallelism of the axis that is maintained as Earth orbits the sun. About June 21, Earth is in a position in its orbit so that the northern tip of its axis is inclined toward the sun at an angle of 23½°. In other words, the plane of the ecliptic is directly on 23½°N latitude. This time in Earth's orbit is called the summer **solstice** (from Latin: *sol,* sun; *sistere,* to stand) in the Northern Hemisphere. We can best see what is happening if we refer to Figure 3.16, position A. In that diagram, we can see that the Northern and Southern Hemispheres receive unequal amounts of light from the sun. That is, as we imagine rotating Earth under these conditions, a larger portion of the Northern Hemisphere than the Southern Hemisphere remains in daylight. Conversely, a larger portion of the Southern Hemisphere than the Northern Hemisphere remains in darkness. Thus, a person living at Repulse Bay, Canada, north of the Arctic Circle, experiences a full 24 hours of daylight at the June solstice. On the same day, someone living in New York City will experience a longer period of daylight than of darkness. However, someone living in Buenos Aires, Argentina, will have a longer period of darkness than daylight on that day. This day is called the winter solstice in the Southern Hemisphere. Thus, June 21 is the longest day of the year in the Northern Hemisphere and the shortest day of the year in the Southern Hemisphere.

Now let us imagine the movement of Earth from its position at the June solstice toward a position a quarter of a year later, in September. As Earth moves toward that new position, we can imagine the changes that will be taking place in our three cities. In Repulse Bay, there will be an increasing amount of darkness through July,

August, and September. In New York, sunset will be arriving earlier. In Buenos Aires, the situation will be reversed; as Earth moves toward its position in September, the periods of daylight in the Southern Hemisphere will begin to get longer, the nights shorter.

Finally, on or about September 22, Earth will reach a position known as an **equinox** (Latin: *aequus,* equal; *nox,* night). On this date (the autumnal equinox in the Northern Hemisphere), day and night will be of equal length at all locations on Earth. Thus, on the equinox, conditions are identical for both hemispheres. As you can see in Figure 3.17, position B, Earth's axis points neither toward nor away from the sun (imagine the axis is pointed at the reader); the circle of illumination passes through both poles, and it cuts Earth in half along its axis.

Imagine again the revolution and rotation of Earth while moving from September 22 toward a new position another quarter of a year later in December. We can see that in Repulse Bay the nights will be getting longer until, on the winter solstice, which occurs on or about December 21, this northern town will experience 24 hours of darkness (Fig. 3.16, position C). The only natural light at all in Repulse Bay will be a faint glow at noon refracted from the sun below the horizon. In New York, too, the days will get shorter and the sun will set earlier. Again, we can see that in Buenos Aires the situation is reversed. On December 21, that city will experience its summer solstice; conditions will be much as they were in New York City in June.

Moving from late December through another quarter of a year to late March, Repulse Bay will have longer periods of daylight, as will New York, while in Buenos Aires the nights will be getting longer, though they still will not be as long as the days. Then, on or about March 21, Earth will again be in an equinox position (the vernal equinox in the Northern Hemisphere) similar to the one in September (Fig. 3.17, position D). Again, days and nights will be equal all over Earth (12 hours each). Finally, moving through another quarter of the year toward the June solstice where we began, Repulse Bay and New York City are both experiencing longer periods of daylight than darkness. The sun is setting earlier in Buenos Aires until, on or about June 21, Repulse Bay and New York City will have their longest day of the year and Buenos Aires its shortest. Further, we can see that on June 21, a point on the Antarctic Circle in the Southern Hemisphere will experience a winter solstice similar to that which Repulse Bay had on December 21 (Fig. 3.16). There will be no daylight in 24 hours, except what appears at noon as a glow of twilight in the sky.

Lines on Earth Delimiting Solar Energy

Looking at the diagrams of Earth in its various positions as it revolves around the sun, we can see that the angle of inclination is important. On June 21, the plane of the

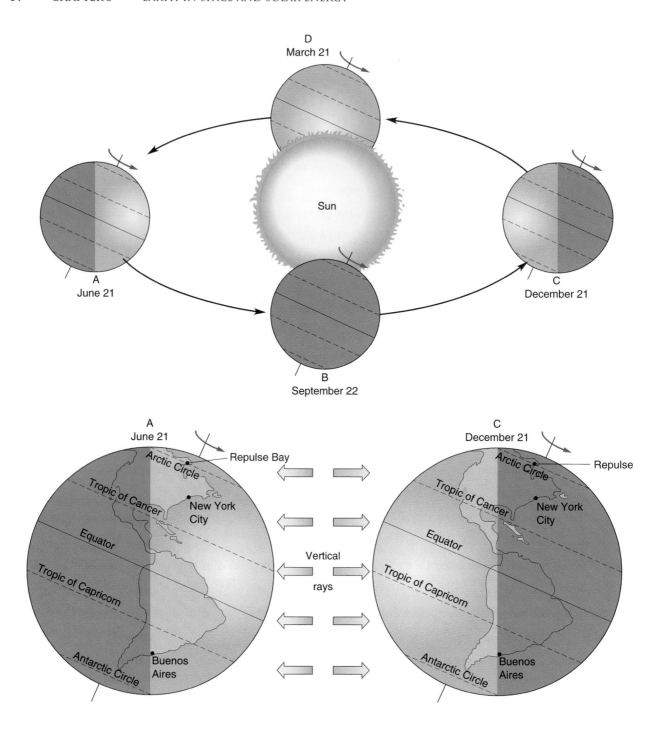

Figure 3.16 Geometric relationships between Earth and the sun at the solstices. Note the differing day lengths at the summer and winter solstices in the Northern and Southern Hemispheres.

ecliptic is directly on 23½°N latitude. The sun's rays can reach 23½° beyond the North Pole, bathing it in sunlight. The **Arctic Circle,** an imaginary line drawn around Earth 23½° from the North Pole (or 66½° north of the equator) marks this limit. We can see from the diagram that all points on or north of the Arctic Circle will experience no darkness on the June sol-

stice, and that all points south of the Arctic Circle will have some darkness on that day. The **Antarctic Circle** in the Southern Hemisphere (23½° north of the South Pole, or 66½° south of the equator) marks a similar limit.

Furthermore, it can be seen from the diagrams that the sun's **vertical,** or **direct rays** (rays that strike Earth's

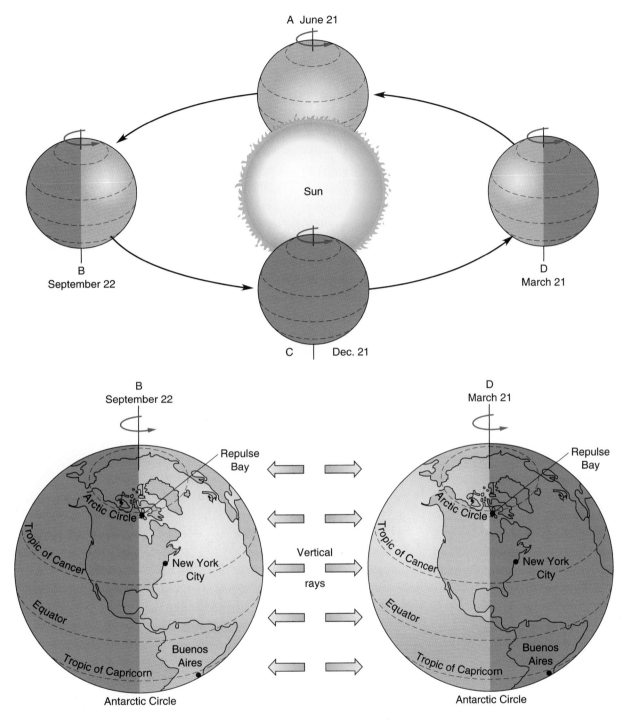

Figure 3.17 Geometric relationships between Earth and the sun at the equinoxes. Day length is 12 hours everywhere, since the circle of illumination crosses the equator at right angles and cuts through both poles. ■ **If earth were not inclined on its axis, would there still be latitudinal temperature variations? Would there be seasons?**

surface at right angles), also shift position in relation to the poles and the equator as Earth revolves around the sun. At the time of the June solstice, the sun's rays are vertical, or directly overhead, at noon at 23½° *north* of the equator. This imaginary line around Earth marks the northernmost position at which the solar rays will ever be directly overhead during a full revolution of our planet

around the sun. The imaginary line marking this limit is called the **Tropic of Cancer.** Six months later, at the time of the December solstice, the solar rays are vertical and the noon sun is directly overhead 23½° *south* of the equator. The imaginary line marking this limit is known as the **Tropic of Capricorn.** At the times of the March and September equinoxes, the vertical solar rays will strike

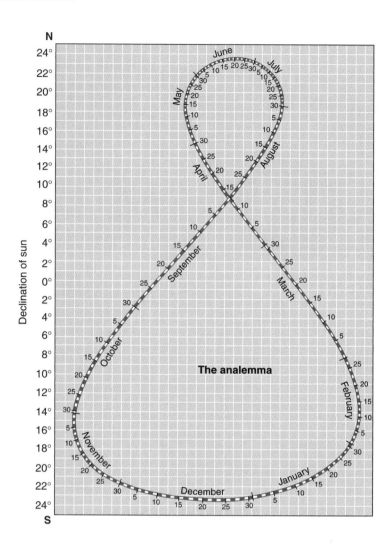

Figure 3.18 An analemma shows the location of the vertical noon sun throughout the year.

directly only at the equator; the noon sun is directly overhead at all points on that line.

Note also that on any day of the year the sun's rays will strike Earth at a 90° angle at only one position, either on or between the two tropics. All other positions that day will receive the sun's rays at an angle of less than 90° (or will receive no sunlight at all).

The Analemma

The latitude at which the noon sun is directly overhead is also known as the sun's **declination.** Thus, if the sun appears directly overhead at 18°S latitude, the sun's declination is 18°S. A figure called an **analemma,** which is often drawn on globes as a big-bottomed "figure 8," shows the declination of the sun throughout the year. A modified analemma is presented in Figure 3.18. Thus, if you would like to know where the sun will be directly overhead on April 25, you can look on the analemma and see that it will be at 13°N. The analemma actually charts the passage of the direct rays of the sun over the 47 degrees of latitude that they cover during a year.

Insolation and the Seasons

Solar radiation received by the Earth system, known as **insolation** (for *in*coming *sol*ar radi*ation*), is the main source of energy on our planet. The seasonal variations in temperature that we experience are due primarily to fluctuations in insolation.

What causes these variations in insolation and brings about seasonal changes? It is true that Earth's atmosphere affects the amount of insolation received. Heavy cloud cover, for instance, will keep more solar radiation from reaching Earth's surface than will a clear blue sky. However, cloud cover is irregular and unpredictable, and it affects total insolation to only a minor degree over long periods of time.

The real answer to the question of what causes variations in insolation can be found by once again studying Figures 3.16 and 3.17, which illustrate the inclination and parallelism of Earth's axis as it revolves around the sun. Two major phenomena vary regularly for a given position on Earth as our planet rotates on its axis and revolves around the sun: the duration of daylight and the angle of

(a)

(b)

(c)

Figure 3.19   (a) The angle at which the sun's rays strike Earth determines the amount of solar energy received per unit of surface area. This amount in turn controls the seasons. The diagram represents the June condition, in which solar radiation strikes the surface perpendicularly only in the Northern Hemisphere, creating summer conditions there. In the Southern Hemisphere, oblique rays are spread over large areas, producing less receipt of energy per unit of area and making this the winter hemisphere. ■ **How would this figure differ in December?** The sun's rays in summer (b) and winter(c). In summer, the sun appears high in the sky, and its rays hit Earth more directly, spreading out less. In winter, the sun is low in the sky, and its rays spread out over a much wider area, becoming less effective at heating the ground.

the solar rays. The amount of daylight controls the duration of solar radiation, and the angle of the sun's rays directly affects the intensity of the solar radiation received. Together, the intensity and the duration of radiation are the major factors that affect the amount of insolation available at any location on Earth's surface (Table 3.2).

Therefore, a location on Earth will receive more insolation if (1) the sun shines more directly, or (2) the sun shines longer, or (3) both. One reason that locations along the Tropics of Cancer and Capricorn are so hot

during their summer solstice is that the sun's rays are intense and the day is long.

The intensity of solar radiation received at any one time varies from place to place because Earth presents a spherical surface to insolation. Therefore, only one line on the Earth's rotating surface can receive radiation at right angles, while the rest receive varying oblique angles (Fig. 3.19a). As we can see from Figure 3.19b, solar energy that strikes Earth at a vertical angle covers less area than an equal amount striking Earth at an oblique angle.

Table 3.2	Radiation Intensity for Certain Solar Angles Expressed as a Percentage of the Maximum Possible (perpendicular beam)						
Solar Angle (degrees above the horizon)	0.0	10.0	20.0	23.5	26.5	30.0	40.0
% of Maximum	00.0	17.4	34.2	39.9	44.6	50.0	64.3
Solar Angle	50.0	60.0	63.5	66.5	70.0	80.0	90.0
% of Maximum	76.6	86.6	89.5	91.7	94.0	98.5	100.0

Solar Energy

In the United States today, a growing number of dedicated scientists, environmentalists, government administrators, politicians, and business executives are convinced that solar energy provides the best alternative to fossil fuels as a power source. Power from solar radiation is nonpolluting and limitless. Earth receives, in just two weeks time, an amount of solar energy that would equal our entire known global supply of fossil fuels.

However, a number of serious obstacles stand in the way of those who see solar energy as the power source of the future. Solar radiation is intermittent (no radiation at night, cloud cover), so a means of storing the power produced, or linking the power to an existing grid system that can accommodate large swings in production, must be indentified. In addition, while the cost of producing energy from solar radi-

tion is reduced by new technologies, additional incentives must be provided for a switch to solar energy.

Two major types of solar technology are being actively developed in various parts of the world, especially in the United States, France, Israel, and Australia. One type involves solar thermal towers, in which racks of tracking mirrors (a heliostat field) follow the sun and focus its heat on a steam boiler perched on a high tower. Temperatures in the boiler may be raised more than 500°C (900°F). No different in principle from our youthful experiments with a magnifying glass to set fire to paper, this device is already operative at several sites. In California, Solar One in the Mojave Desert is the world's largest solar thermal electric power plant of this kind. However, development is slow because additional solar thermal towers must be considered in terms of economic feasibility.

The other major type of solar technology depends on photovoltaic (PV) cells that convert sunlight directly into electrical power (not unlike a camera's light meter).

We can see that the closer to a right angle that the sun's rays strike Earth, the smaller the area covered. Since the amount of energy is the same no matter what kind of angle the rays make with the surface, it follows that the smaller the area that is struck, the greater will be the intensity per unit area. Conversely, the more oblique the angle at which the sun's rays strike Earth, the greater the

area over which those rays will be spread, and the lesser the energy that will strike Earth per unit area.

In addition, the atmosphere acts to diminish, to some extent, the amount of insolation that reaches Earth's surface. Since oblique rays must pass through a greater thickness of atmosphere than vertical rays, more insolation will be lost in the process.

Table 3.3   Duration of Daylight for Certain Latitudes

Latitude (in degrees)	Length of Day (N. Hemisphere) (read down)		
	March 21/Sept. 22	June 21	Dec. 21
0.0	12 hr.	12 hr. 0 min.	12 hr. 0 min.
10.0	12 hr.	12 hr. 35 min.	11 hr. 25 min.
20.0	12 hr.	13 hr. 12 min.	10 hr. 48 min.
23.5	12 hr.	13 hr. 35 min.	10 hr. 41 min.
30.0	12 hr.	13 hr. 56 min.	10 hr. 4 min.
40.0	12 hr.	14 hr. 52 min.	9 hr. 8 min.
50.0	12 hr.	16 hr. 18 min.	7 hr. 42 min.
60.0	12 hr.	18 hr. 27 min.	5 hr. 33 min.
66.5	12 hr.	24 hr.	0 hr.
70.0	12 hr.	24 hr.	0 hr.
80.0	12 hr.	24 hr.	0 hr.
90.0	12 hr.	24 hr.	0 hr.
Latitude	Equinox	Dec. 21	June 21
	Length of Day (S. Hemisphere) (read up)		

One of three solar energy complexes in the Mojave Desert of California that together produce 90 percent of the world's grid-connected solar energy.

Although with PV technology the production of power is limited to daylight hours, this type of solar energy conversion is being given serious consideration as a supplementary source of energy for overtaxed power grids in high-pollution areas. For example, in Las Vegas, Nevada, there are plans to construct a major PV facility on the roof of the Las Vegas Convention Center. The solar PV system will cost between $.17 and $.25 per kwh, which is two to three times as much as the Convention Center currently pays for electricity, but about the same as a customer pays in Marin County in energy-starved California. Although the cost will be high, there is a special reason for considering the project. Nevada law requires that, starting in 2003, a certain percentage of all energy sold in the state must come from solar energy. And Nevada is not alone. Twelve other states have enacted such provisions in their laws. Solar energy may never fully replace fossil fuel as a source of power, but clearly the process has already begun.

Since no insolation is received at night, the duration of solar energy is related to the length of daylight received at a particular point on Earth, (Table 3.3). Obviously, the longer the period of daylight, the greater the amount of solar radiation that will be received at that location. As we have seen, periods of daylight vary in length through the seasons of the year as well as from place to place on Earth's surface.

Variations of Insolation with Latitude

Neglecting for the moment the influence of the atmosphere on variations in insolation during a 24-hour period, a place will receive its greatest insolation at solar noon when the sun has reached its zenith or highest point in the sky for that day. At any location, no insolation will be received during the hours of darkness. The amount of energy received after daybreak increases as Earth rotates toward the time of solar noon. The amount of insolation then decreases until the next period of darkness begins.

We also know that the amount of daily insolation received at any one location on Earth varies with the seasons. There are three distinct patterns in the distribution of the seasonal receipt of solar energy in each hemisphere. These patterns serve as the basis for recognizing six latitudinal zones, or bands, of insolation and temperature that circle Earth (Fig. 3.20).

If we look first at the Northern Hemisphere, we may take the Tropic of Cancer and the Arctic Circle as the dividing lines for three of these distinctive zones. The area between the equator and the Tropic of Cancer can be called the north *tropical zone*. Here, insolation is always high but is greatest at the two times during the year that the sun is directly overhead at noon. These dates vary according to latitude (see again Fig. 3.18). The north *middle-latitude zone* is the wide band between the Tropic of Cancer and the Arctic Circle. In this belt, insolation is greatest on the June solstice, when the sun reaches its highest noon angle and the period of daylight is longest. Insolation is least at the December solstice when the sun is lowest in the sky and the period of daylight the shortest. The north *polar zone,* or *Arctic zone,* extends from the Arctic Circle to the pole. In this region, as in the middle-latitude zone, insolation is greatest at the June solstice, but it ceases during the period that the sun's rays are blocked entirely by the tilt of Earth's axis. This period lasts for six months at the North Pole but is as short as one day directly on the Arctic Circle.

Similarly, there is a south tropical zone, a south middle-latitude zone, and a south polar or *Antarctic zone,* all separated by the Tropic of Capricorn and the Antarctic Circle in the Southern Hemisphere. These areas get their greatest amounts of insolation at opposite times of the year from the northern zones.

Despite various patterns in the amount of insolation received in these zones, we can make some generalizations. For example, total annual insolation at the top of the atmosphere over a particular latitude remains nearly constant from year to year. Furthermore, annual insolation tends to decrease from lower latitudes to higher latitudes. The closer to the poles a place is located, the greater will be its seasonal variations caused by fluctuations in insolation.

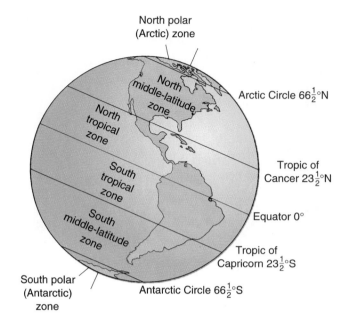

Figure 3.20 The line of the equator, the Tropics of Cancer and Capricorn, and the Arctic and Antarctic Circle define six latitudinal zones that have distinctive insolation characteristics.
■ **Which zone(s) would have the least annual variation in insolation? Why?**

The amount of insolation received by Earth systems is an important concept in understanding atmospheric dynamics and the distribution of climate and vegetation. Such climatic elements as temperature, precipitation, and winds are controlled in part by the amount of insolation received by Earth. People depend on certain levels of insolation for physical comfort, and plant life is especially sensitive to the amount of available insolation. You may have noticed plants that have wilted in too much sunlight or that have grown brown in a dark corner away from a window. Over a longer period of time, deciduous plants have an annual cycle of budding, flowering, leafing, and losing their leaves. This cycle is apparently determined by the fluctuations of increasing and decreasing solar radiation that mark the changing seasons. Even animals respond to seasonal changes. Some animals hibernate; many North American birds fly south toward warmer weather as winter approaches; and many animals breed at such a time that their offspring will be born in the spring, when warm weather is approaching.

DEFINE&RECALL

galaxy
light-year
solar system
planet
satellite
asteroid
comet
meteor
meteorite
fusion (thermonuclear
   reaction)

solar wind
auroras
sunspots
terrestrial planet
giant planet (gas planet)
mass
gravitation
electromagnetic energy
short-wave radiation
calorie
solar constant

galactic movement
rotation
revolution
circle of illumination
perihelion
aphelion
plane of the ecliptic
angle of inclination
parallelism
solstice
equinox

Arctic Circle
Antarctic Circle
vertical rays
Tropic of Cancer
Tropic of Capricorn
declination
analemma
insolation

DISCUSS & REVIEW

1. What is a solar system? What bodies constitute our solar system?

2. How is the energy emitted from the sun produced?

3. Name the terrestrial planets. What do they have in common? Name the giant planets. What do they have in common?

4. Give a unique characteristic for each of the nine planets.

5. The electromagnetic spectrum displays various types of energy by their wavelengths. Where is the division between long-wave and short-wave energy? In what ways do humans use electromagnetic energy?

6. Is the amount of solar energy reaching Earth's outer atmosphere constant? What might make it change?

7. Describe briefly how Earth's rotation and revolution affect life on Earth.

8. If the sun is closest to Earth on January 3, why isn't winter in the Northern Hemisphere warmer than winter in the Southern Hemisphere?

9. Identify the two major factors that cause regular variation in insolation throughout the year. How do they combine to cause the seasons?

CONSIDER & RESPOND

1. Do you think we have discovered all the galaxies in the universe? Why?

2. Given what you know of the sun's relation to life on Earth, explain why the solstices and equinoxes have been so important to cultures all over the world. What are some of the major festivals associated with these times of the year?

3. Use the analemma presented in Figure 3.18 to determine the latitude where the noon sun will be directly overhead on:

   a. February 12

   b. July 30

   c. November 2

   d. December 30

4. Use the discussion of solar angle, including Figure 3.19, to explain why we can look directly at the sun at sunrise and sunset but not at the noon hour.

5. Imagine you are at the equator on March 21. The noon sun would be directly overhead. However, for every degree of latitude that you travel to the north or south, the noon solar angle would decrease by the same amount. For example, if you travel to 40°N latitude, the solar angle would be 50°. Can you explain this relationship? Can you develop a formula or set of instructions to generalize this relationship? What would be the solar angle at 40°N on June 21? on December 21?

6. Describe in your own words the relationship between insolation and latitude.

THE ATMOSPHERE, TEMPERATURE, AND THE HEAT BUDGET

CHAPTER PREVIEW

Our planet's atmosphere is essential to life as we know it here on Earth.

How is this true?

What is the significance of this statement for humans?

How should this fact affect human behavior?

Earth has an energy budget with a multiplicity of inputs and outputs (exchanges) that ultimately remains in balance despite recurring deficits and surpluses from time to time and from place to place.

How is the budget concept useful to an understanding of atmospheric heating and cooling?

How can we tell that the budget remains in balance?

Water plays the single most important role in the exchanges of energy that fuel atmospheric dynamics.

What characteristics of water are responsible for its importance in energy exchange?

In what ways is water involved in the heating of the atmosphere?

As a direct result of differences in insolation and the mechanics of atmospheric heating, air temperature varies over time and both horizontally and vertically through space.

What are the most obvious variations?

Why do they occur?

How do temperatures stay within the ranges suitable for life if there are such great differences in the amounts of insolation received?

Atmospheric elements are affected by atmospheric controls to produce weather and climate.

How do the elements differ from the controls?

How does weather differ from climate?

Blue Skies CD

Chapter 1: Atmospheric Basics, *Layers of Atmosphere*
Chapter 1: Atmospheric Basics, *Energy Balance*
Chapter 2: Sky Identification, *Atmospheric Optics*

Earth Systems Today CD

Chapter 5: Waves, Tides, & Currents, *World Currents*

In actuality, the atmosphere is a very thin layer of gases held to the surface by Earth's gravity.

Water and oxygen are vital for animals and humans to survive. Plant life requires carbon dioxide as well as a sufficient water supply. Most living things we know cannot survive extreme temperatures, nor can they live long if exposed to large doses of harmful radiation. It is the atmosphere, the layer of air that surrounds Earth, that supplies most of the oxygen and carbon dioxide and that helps maintain a constant level of water and radiation in the Earth system.

Though actually a thin film of air, the atmosphere serves as an insulator, maintaining the viable temperatures we find on Earth. Without the atmosphere, Earth would experience temperature extremes of as much as 260°C (500°F) between day and night. The atmosphere also serves as a shield, blocking out much of the sun's ultraviolet radiation and protecting us from showers of meteors. The atmosphere is also described as an ocean of air surrounding Earth. This description reminds us of the currents and circulation of the atmosphere—its dynamic movements—which create the changing conditions on Earth that we know as weather.

For comparison, we can look at our moon—a celestial body with virtually no atmosphere—in order to see the importance of our own atmosphere. Most obviously, a person standing on the moon without a space suit would immediately die for lack of oxygen. Our lunar astronauts recorded temperatures of up to 204°C (400°F) on the hot, sunlit side of the moon and, on the dark side, temperatures approaching –121°C (–250°F). These temperature extremes would certainly kill an unprotected human.

The next thing any astronaut on the moon would notice is the "unearthly" silence. On Earth, we hear sounds because sound waves move by vibrating the molecules in the air. Since the moon has no atmosphere and no molecules to carry the sound waves, the lunar visitor is not able to hear any sounds; only radio communications are possible. Also, because there is no atmosphere, an astronaut cannot fly aircraft or helicopters, and it would be fatal to try to use a parachute. In addition, lack of atmosphere means no protection from the bombardment of meteors that fly through space and collide with the moon. Nearing Earth, most meteors burn up before reaching the surface because of friction within the atmosphere. Without an atmosphere for protection, a visitor to the moon would also be burned by the ultraviolet rays of the sun. On Earth, we are protected to a large degree from ultraviolet radiation because the ozone layer of the upper atmosphere absorbs the major portion of this harmful radiation.

We can see that, in contrast to our stark, lifeless moon, Earth presents a hospitable environment for life almost solely because of its atmosphere. All living things are adapted to its presence. For example, many plants reproduce by pollen and spores that are carried by winds. Birds can fly only because of the air. The water cycle of Earth is maintained through the atmosphere, as are the heat and radiation "budgets." The atmosphere diffuses sunlight as well, giving us our blue skies and the fantastic reds, pinks, oranges, and purples of sunrise and sunset. Without this diffusion, the sky would appear black, as it does from the moon.

Further, the atmosphere provides a means by which the systems of Earth attempt to reach equilibrium. Changes in weather are ultimately the result of the atmospheric effects that equalize temperature and pressure differences on Earth's surface by transferring heat and moisture through atmospheric and oceanic circulation systems.

Characteristics of the Atmosphere

The atmosphere extends to approximately 480 kilometers (300 mi) above Earth's surface. Its density decreases rapidly with altitude; in fact, 97 percent of the air is concentrated in the first 25 kilometers (16 mi) or so. Since air has mass, the atmosphere exerts pressure on Earth's surface. At sea level, this pressure is about 1034 grams per square centimeter (14.7 lb/sq in.), but the higher the elevation, the lower the atmospheric pressure. In Chapter 5, we will examine the relationship between atmospheric pressure and elevation in more detail.

Composition of the Atmosphere

The atmosphere is composed of numerous gases (Table 4.1). Most of these gases remain in the same proportions regardless of the density of the atmosphere. A bit more

Table 4.1 Composition of the Atmosphere Near the Earth's Surface						
Permanent Gases				**Variable Gases**		
Gas	Symbol	Percent (by Volume) Dry Air	Gas (and Particles)	Symbol	Percent (by Volume)	Parts per Million (ppm)*
Nitrogen	N_2	78.08	Water vapor	H_2O	0 to 4	
Oxygen	O_2	20.95	Carbon dioxide	CO_2	0.037	368*
Argon	Ar	0.93	Methane	CH_4	0.00017	1.7
Neon	Ne	0.0018	Nitrous oxide	N_2O	0.00003	0.3
Helium	He	0.0005	Ozone	O_3	0.000004	0.04†
Hydrogen	H_2	0.00006	Particles (dust, soot, etc.)		0.000001	0.01–0.15
Xenon	Xe	0.000009	Chlorofluorocarbons (CFCs)		0.00000002	0.0002

*For CO_2, 368 parts per million means that out of every million air molecules, 368 are CO_2 molecules.
†Stratospheric values at altitudes between 11 km and 50 km are about 5 to 12 ppm.

Figure 4.1 Since 1958, these measurements of carbon dioxide taken at Mauna Loa, Hawaii, have shown an upward trend.

than 78 percent of the atmosphere's volume is made up of nitrogen, and nearly 21 percent consists of oxygen. Argon comprises most of the remaining 1 percent. The percentage of carbon dioxide in the atmosphere has risen through time, but is a little over 0.03 percent by volume. There are traces of other gases as well: ozone, hydrogen, neon, xenon, helium, methane, nitrous oxide, krypton, and others.

Nitrogen, Oxygen, and Carbon Dioxide Nitrogen, which makes up the largest proportion of air, is an important element supporting plant growth. In addition, some of the other gases in the atmosphere are vital to the development and maintenance of life on Earth. One of the most important of the atmospheric gases is, of course, oxygen, which humans and all other animals use to breathe and oxidize (burn) the food they eat. *Oxidation,* which is technically the chemical combination of oxygen with other materials to create new products, occurs in situations outside animal life as well. Rapid oxidation takes place, for instance, when we burn fossil fuels or wood and thus release large amounts of heat energy. The decay of certain rocks or organic debris and the development of rust are examples of slow oxidation. All these processes depend on the presence of oxygen in the atmosphere.

Carbon dioxide is also an important atmospheric gas, because it absorbs heat from Earth. The atmosphere then emits about half of that absorbed heat energy back to Earth. This process helps maintain the warmth of the planet and is a factor in Earth's heat energy budget. However, human activity—largely the burning of fossil fuels and deforestation of the planet—has greatly increased the amount of carbon dioxide in the atmosphere (Fig. 4.1). There is concern that this will upset our current temperature structure.

Carbon dioxide is also involved in the system known as the *carbon cycle.* Plants, through a process known as **photosynthesis,** use carbon dioxide and water to produce carbohydrates (sugars and starches), in which amounts of energy, derived originally from the sun, are stored (Fig. 4.2). Oxygen is given off as a by-product. Animals then use the oxygen to oxidize the carbohydrates, releasing the stored energy. A by-product of this process in animals is the release of carbon dioxide, which completes the cycle when it is in turn used by plants in photosynthesis.

Ozone Another vital gas in Earth's atmosphere is ozone. The ozone molecule (O_3) is related to the oxygen molecule (O_2); it is made up of three oxygen atoms, whereas regular oxygen gas is made up of only two. Ozone is formed in the upper atmosphere when an oxygen molecule is split into two atoms by short-wave solar radiation. Then the free unstable atoms join two other oxygen molecules to form two molecules of ozone gas consisting of three oxygen atoms each. In the lower atmosphere, ozone is a toxic pollutant and is a major component of urban smog.

In the upper atmosphere, however, ozone is vital to terrestrial and marine life. Ozone is important to life forms because it is capable of absorbing large amounts of the sun's ultraviolet radiation. Without the ozone layer of the upper atmosphere, the ultraviolet radiation reaching Earth would severely burn human skin, increase the incidence of skin cancer and optical cataracts, destroy certain microscopic forms of life, and damage plants. At the very least, ultraviolet radiation is responsible for painful sunburns or sensible suntans, depending on individual skin tolerance and exposure time. There is, therefore, increasing concern that human activity, especially the addition of chlorofluorocarbons (CFCs) to the atmosphere, may permanently damage this fragile ozone layer. CFCs, known commercially as freon, are used extensively in refrigeration and air conditioning.

Figure 4.2 The equation of photosynthesis shows how solar energy (light) is used by plants to manufacture sugars and starches from atmospheric carbon dioxide and water, liberating oxygen in the process. The stored food energy is then eaten by animals, which also breathe the oxygen released by photosynthesis.

Through the worldwide use of refrigeration, CFCs have entered the upper atmosphere and continue to pose this threat to our natural ultraviolet filter.

The small proportion of ultraviolet radiation that the ozone layer allows to reach Earth does serve useful purposes. For instance, it is important in the production of certain vitamins (especially Vitamin D), and it helps the growth of some beneficial viruses and bacteria. It also has a function in the process of photosynthesis.

Water Vapor, Liquids, and Solids   Water vapor is always mixed in some proportion with the dry air of the lower part of the atmosphere; it varies from 0.02 percent by volume in a cold, dry climate to a high of more than 4 percent in the humid tropics. The percentage of water vapor in the air will be discussed later under the broad topic of humidity, but it is important to note here that the variations in this percentage over time and place are an important consideration in the examination and comparison of climates.

Water vapor also absorbs heat in the lower atmosphere and so prevents its rapid escape from Earth. Thus, like carbon dioxide, water vapor plays a large role in the insulating action of the atmosphere. In addition to gaseous water vapor, liquid water also exists in the atmosphere as rain and as fine droplets in clouds, mist, and fog. Solid water is found in the atmosphere in the form of ice crystals, snow, and hail.

Suspended in the atmosphere are many other solids. These are not necessarily pollutants from transportation and industry, but rather natural particles that have always existed in our atmosphere. Particles such as dust, smoke, pollen and spores, volcanic ash, bacteria, and salts from ocean spray can all play an important role in absorption of energy and in the formation of raindrops.

 ## Vertical Layers of the Atmosphere

Though people function primarily in the lowest level of the atmosphere, there are times, such as when we fly in aircraft or climb a mountain, when we leave our normal altitude. The thinner atmosphere at these higher altitudes may affect us if we are not accustomed to it. Visitors to Inca ruins in the Andes or high-altitude Himalayan climbers may experience "altitude sickness," and even skiers in the Rockies near "mile-high" Denver may need time to adjust. The air at these levels is much *thinner* than most of us are used to; by this we mean there is more empty space between air molecules, and thus less oxygen and other gases in each breath of air.

The atmosphere can be divided into four layers according to differences in temperature and rates of temperature change (Fig. 4.3). The first of these layers, lying closest to Earth's surface, is the **troposphere** (from Greek: *tropo,* turn—the turning or mixing zone), which extends about 8 to 16 kilometers (5 to 10 mi) above

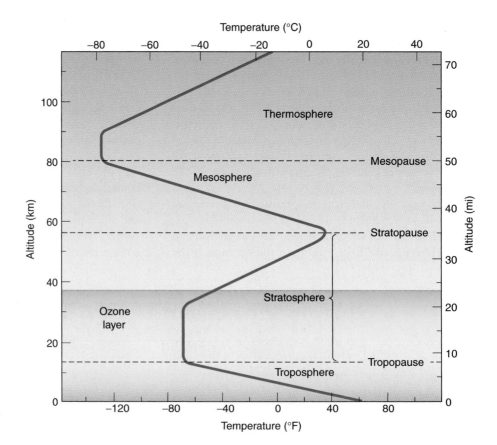

Figure 4.3 Vertical temperature changes in Earth's atmosphere are the basis for its subdivision into the troposphere, stratosphere, mesosphere, and thermosphere.

■ **At what altitude is our atmosphere the coldest?**

Earth. Its thickness, which tends to vary seasonally, is least at the poles and greatest at the equator. It is within the troposphere that people live and work, plants grow, and virtually all Earth's weather takes place.

The troposphere has two distinct characteristics that differentiate it from other layers of the atmosphere. One is that the water vapor and dust particles of the atmosphere are concentrated in this one layer; they are rarely found in the atmospheric layers above the troposphere. The other characteristic of this layer is that temperature normally decreases with increased altitude. The average rate at which temperatures within the troposphere decrease with altitude is called the **normal lapse rate** (sometimes referred to as the environmental lapse rate); it amounts to 6.5°C per 1000 meters (3.6°F/1000 ft).

The altitude at which the temperature ceases to drop with increased altitude is called the *tropopause.* It separates the troposphere from the **stratosphere**—the second layer of the atmosphere. The temperature of the lower part of the stratosphere remains fairly constant (about –57°C, or –70°F) to an altitude of about 32 kilometers (20 mi). It is in the stratosphere that we find the ozone layer that does so much to protect life on Earth from the sun's ultraviolet radiation. As the ozone layer absorbs ultraviolet radiation, this absorbed energy results in the release of heat, and thus temperatures increase in the upper parts of the stratosphere. Temperatures at the *stratopause,* which is about 50 kilometers (30 mi) above Earth, are about the same as temperatures found on Earth's surface, although little of that heat can be transferred because the air is so thin.

Above the stratopause are the **mesosphere,** in which temperatures tend to drop with increased altitude; the *mesopause,* which separates the mesosphere from the thermosphere; and the **thermosphere,** where temperatures

increase until they approach 1100°C (2000°F) at noon. However, the air is so thin at this altitude that there is practically a vacuum and little heat can be transferred.

Astronomers and communications experts use a different method of layering the atmosphere, in which the same region that constitutes the thermosphere is called the *ionosphere.* This name denotes the ionization of molecules and atoms that occurs in this layer, mostly as a result of ultraviolet rays, X rays, and gamma radiation. (Ionization refers to the process whereby atoms are changed to ions through the removal or addition of electrons, giving them an electrical charge.) It is in the ionosphere that the auroras (defined in Chapter 3) occur. The thermosphere gradually gives way to interplanetary space.

Controls of Temperature

Variations in temperatures over Earth's surface are caused by several **controls.** The major controls are (1) latitude, (2) land and water distribution, (3) ocean currents, (4) altitude, (5) landform barriers, and (6) human activity.

Latitude   Latitude is the most important control of temperature variation involved in weather and climate. Recall that, because of the inclination and parallelism of Earth's axis as it revolves around the sun (Chapter 3), there are distinct patterns in the latitudinal distribution of the seasonal and annual receipt of solar energy over Earth's surface. This has a direct effect on temperatures. In general, annual insolation tends to decrease from lower latitudes to higher latitudes (see again Fig. 3.16 and 3.17). Table 4.2 shows the average annual temperatures for several locations in the Northern Hemisphere. We can see that, responding to insolation (with one

Table 4.2 **Temperatures in the Northern Hemisphere**			
		Average Annual Temperature	
Location	*Latitude*	*(°C)*	*(°F)*
Libreville, Gabon	0°23′N	26.5	80
Ciudad Bolivar, Venezuela	8°19′N	27.5	82
Bombay, India	18°58′N	26.5	80
Amoy, China	24°26′N	22.0	72
Raleigh, North Carolina	35°50′N	18.0	66
Bordeaux, France	44°50′N	12.5	55
Goose Bay, Labrador, Canada	53°19′N	–1.0	31
Markova, Russia	64°45′N	–9.0	15
Point Barrow, Alaska	71°18′N	–12.0	10
Mould Bay, NWT, Canada	76°17′N	–17.5	0

exception), a poleward decrease in temperature is true for these locations. The exception is near the equator itself. Because of the heavy cloud cover in the equatorial regions, average annual temperatures there tend to be lower than at places slightly to the north or south, where skies are clearer.

Another very simple way to see this general trend of decreasing temperatures as we move toward the poles is to think about the kinds of clothes we would take along for one month—say, January—if we were to visit Ciudad Bolivar, Venezuela; Raleigh, North Carolina; or Point Barrow, Alaska.

Land and Water Distribution   Not only do the oceans and seas of Earth serve as storehouses of water for the whole system, but they also store tremendous amounts of heat energy. Their widespread distribution makes them an important atmospheric control that does much to modify the atmospheric elements.

All things heat and cool at different rates. On Earth's surface, bodies of water heat and cool more slowly than land surfaces. First, the *specific heat* of water is greater than that of land. Specific heat refers to the amount of heat necessary to raise the temperature of any substance 1°C. Therefore, water must absorb more heat energy than land in order to be raised the same number of degrees in temperature. In fact, the specific heat of water is more than four times that of soil or rock. In addition, water is *transparent,* and solar energy passes through the surface into the layers below, whereas in opaque materials like soil and rock, the energy is concentrated on the surface. Thus, a given unit of heat energy will spread through a

greater volume of water than land. Also, because liquid water flows and *mixes,* it is able to transfer heat to deeper layers within its mass. The result is that as summer changes to winter, the land cools more rapidly than bodies of water, and as winter becomes summer, the land heats more rapidly. Since the air gets much of its heat from the surface with which it is in contact, the differential heating of land and water surfaces produces inequalities in the temperature of the atmosphere above these two surfaces.

The mean temperature in Seattle, Washington, in July is 18°C (64°F), while the mean temperature during the same month in Minneapolis, Minnesota, is 21°C (70°F). Since the two cities are at similar latitudes, their annual pattern and receipt of solar energy are also similar. Therefore, their different temperatures in July must be related to a control other than latitude. Much of this difference in temperature can be attributed to the fact that Seattle is near the Pacific coast, while Minneapolis is in the heart of a large continent, far from the moderating influence of an ocean. Seattle stays cooler than Minneapolis in the summer because the surrounding water warms up slowly, keeping the air relatively cool. Minneapolis, on the other hand, is in the center of a large landmass that warms very quickly. In the winter, the opposite is true. Seattle is warmed by the water while Minneapolis is not. The mean temperature in January is 4.5°C (40°F) in Seattle and –15.5°C (4°F) in Minneapolis.

Not only do water and land heat and cool at different rates, but so do various land surface materials. Soil, forest, grass, and rock surfaces all heat and cool differentially and thus vary the temperatures of the overlying air.

Figure 4.4   A highly simplified map of currents in the Pacific Ocean to show their basic rotary pattern. The major currents move clockwise in the Northern Hemisphere and counterclockwise in the Southern Hemisphere. A similar pattern exists in the Atlantic.

■   **What path would a hurricane forming off western Africa take as it approached the United States?**

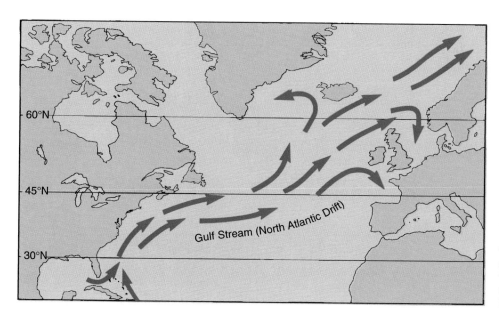

Figure 4.5 The Gulf Stream (the North Atlantic Drift farther eastward) is a warm current that greatly moderates the climate of northern Europe. ■ **Use this figure and the information gained in Figure 4.4 to discuss the route sailing ships would follow from the United States to England and back.**

Ocean Currents Surface ocean currents are large movements of water pushed by the winds. They may flow from a place of warm temperatures to one of cooler temperatures and vice versa. These movements result, as we saw in Chapter 1, from the attempt of Earth systems to reach a balance—in this instance, a balance of temperature and density.

The rotation of Earth affects the movements of the winds, which in turn affect the movement of the ocean currents. In general, the currents move in a clockwise direction in the Northern Hemisphere and in a counterclockwise direction in the Southern Hemisphere (Fig. 4.4).

Since the temperature of the ocean greatly affects the temperature of the air above it, an ocean current that moves warm equatorial water toward the poles (a warm current) or cold polar water toward the equator (a cold current) can significantly modify the air temperatures of those locations. If the currents pass close to land and are accompanied by onshore winds, they can have a significant impact on the coastal climate.

The Gulf Stream, with its extension, the North Atlantic Drift, is an example of an ocean current that moves warm water northward. This warm water keeps the coasts of Great Britain and Norway ice-free in wintertime and moderates the climates of nearby land areas (Fig. 4.5). We can see the effects of the Gulf Stream if we compare the winter conditions of the British Isles with those of Labrador in northeastern Canada. Though both are at the same latitude, the average temperature in Glasgow, Scotland, in January is 4°C (39°F), while during the same month it is −21.5°C (−7°F) in Nain, Labrador.

The California Current off the west coast of the United States helps moderate the climate of that coast as it brings cold water south. As the current swings southwest away from the coast of central California, cold bottom water is drawn to the surface, causing further chilling of the air above. San Francisco's cool summers (July average: 14°C, or 58°F) show the effect of this current.

Altitude As we have seen, temperatures within the troposphere decrease with increasing altitude. In Southern California, you can find snow for skiing if you go to an altitude of 2400 to 3000 meters (8000 to 10,000 ft). Mount Kenya, 5199 meters (17,058 ft) high and located at the equator, is cold enough to have glaciers. Anyone who has hiked upward 500, 1000, or 1500 meters in midsummer has experienced a decline in temperature with increasing altitude. Even if it is hot on the valley floor, you may need a sweater once you climb a few thousand meters. The city of Quito, Ecuador, only 1° south of the equator, has an average temperature of only 13°C (55°F) because it is located at an altitude of about 2900 meters (9500 ft).

Landform Barriers Landform barriers, especially large mountain ranges, can block movements of air from one place to another and thus affect the temperatures of an area. For example, the Himalayas keep cold, wintertime Asiatic air out of India, giving the Indian subcontinent a year-round tropical climate. Mountain orientation can create some significant differences as well. In North America, for example, southern slopes face the sun and tend to be warmer than the shady north-facing slopes. Snowcaps on the south-facing slopes may have less snow and may exist at a higher elevation. North-facing slopes usually have more snow, and it extends to lower elevations.

Human Activities Human beings, too, may be considered "controls" of temperature. Such activities as building cities and road systems, developing industries that

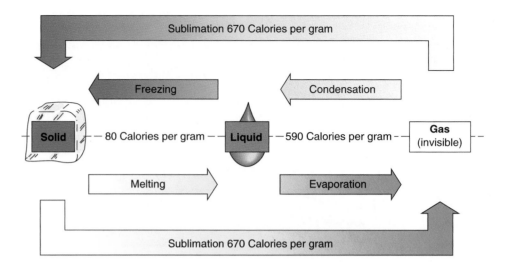

Figure 4.6 The three physical states of water and energy exchanges between them.

burn fossil fuels, and creating places with large population densities can result in an *urban heat island.* This is a measurable pocket of warm air produced by a large urban area. As a result, cities are generally warmer than the surrounding countryside. In addition, human activities like destroying forests, draining swamps, or creating large reservoirs can significantly affect local climatic patterns and, possibly, world temperature patterns as well.

The Role of Water

As it penetrates our atmosphere, some of the incoming solar radiation is involved in energy exchanges. One such exchange involves how water in the Earth system is altered from one state to another. Water is the only substance that can exist in all three states of matter—as a solid, a liquid, and a gas—within the normal temperature range of Earth's surface. In the atmosphere, water exists as a clear, odorless gas called *water vapor.* It is also a liquid in the atmosphere, in the oceans, and in other water bodies on and beneath the surface of Earth. Water is also found within vegetation and animals. Water is a solid in snow and ice in the atmosphere, as well as on and under the surface of the colder parts of Earth.

Not only is water stored in all three states of matter, but it can change from one state to another, as illustrated in Figure 4.6. In doing so, it is involved in the heat energy system of Earth. The molecules of a gas move faster than do those of a liquid. Thus, during the process of *condensation,* when water vapor changes to liquid water, its molecules slow down and some of their energy is released (about 590 calories per gram). The molecules of a solid move even more slowly than those of a liquid, so during the process of *freezing,* when water changes to ice, additional energy is released (80 calories per gram). When the process is reversed, heat must be added. Thus, *melting* requires the addition of 80 calories per gram, and *evaporation* requires the addition of 590 calories per gram. This added energy is stored as *latent* (or hidden) *heat.*

Some of these energy exchanges can be easily demonstrated. For example, if you hold an ice cube in your hand, your hand feels cold because it is giving off the heat needed to melt the ice. We are cooled by perspiration evaporating from our skin, because heat must be absorbed both from our skin and from the remaining perspiration, thereby lowering the temperature of both.

Scientists have become increasingly aware of the importance of the energy exchange between the atmosphere and the hydrosphere, especially at the ocean's surface. Approximately three-fourths of Earth's surface is covered by water. Since water heats and cools more slowly than land, the oceans act as huge reservoirs of heat energy. This energy helps to power the atmosphere, which directly influences our weather and climate on land.

Effects of the Atmosphere on Solar Radiation

As the sun's energy passes through Earth's atmosphere, more than half its intensity is lost through various processes. In fact, the amount of insolation actually received at a particular location depends not only on latitude, time of day, and time of year (all of which are related to the angle at which the sun's rays strike Earth). The transparency of the atmosphere (or the amount of cloud cover, moisture, carbon dioxide, and solid particles in the air) also plays a vital role.

When the sun's energy passes through the atmosphere, several things happen to it (the following figures represent approximate averages for entire Earth; at any one location or time they may differ): (1) 26 percent of the energy is reflected directly back to space by clouds and the ground; (2) 8 percent is *scattered* by minute atmospheric particles and returned to space as diffuse radiation; (3) 20 percent reaches Earth's surface as diffuse radiation after being scattered; (4) 27 percent reaches Earth's surface as direct radiation; and (5) 19 percent is *absorbed* by the ozone layer and by water vapor in the

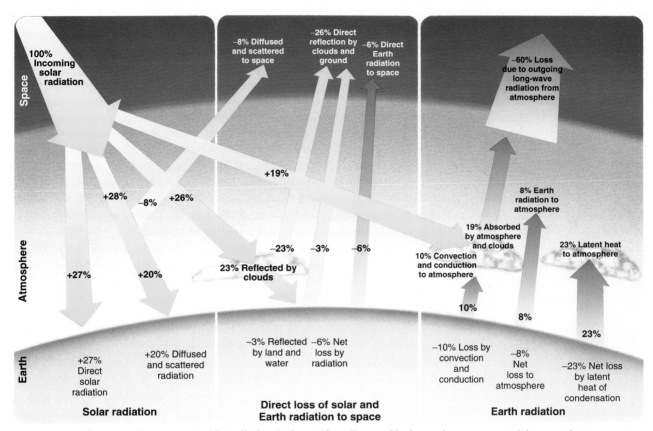

Figure 4.7 **Environmental Systems: Earth's Radiation Budget** If you live outside the tropics, you are certainly aware that temperatures can vary greatly from day to day, month to month, or year to year. However, from one year to the next, Earth's overall average temperature varies very little. This fact indicates that a long-term global balance, or equilibrium, must exist between the energy received and emitted by the Earth system.

   The source of all energy for the Earth system is the sun. As the energy from the sun enters our atmosphere and flows toward Earth's surface, several things happen to it. Some of the solar energy is reflected back to space by clouds and Earth's surface. Additional solar energy is scattered by dust, water, and other small particles in the atmosphere or absorbed by the atmosphere and clouds. Note in the diagram that only 47 percent of the incoming solar energy reaches and is absorbed by Earth's surface.

   The energy absorbed by Earth's surface heats the atmosphere through various processes such as conduction, convection, or latent heat of condensation. The atmosphere readily absorbs Earth-emitted energy. Eventually, all the energy gained by the atmosphere is lost to space. However, the radiation budget is a dynamic one. In other words, alterations to one element affect the other elements. As a result, there is growing concern that one of the elements, human activity, will cause the atmosphere to absorb more Earth-emitted energy, thus raising global temperatures.

clouds of the atmosphere (Fig. 4.7). In other words, on a worldwide average, 47 percent of the incoming solar radiation eventually reaches the surface, 19 percent is retained in the atmosphere, and 34 percent is returned to space. Since Earth's energy budget is in equilibrium, the 47 percent received at the surface is ultimately returned to the atmosphere by processes that we will now examine.

Heating the Atmosphere

The 19 percent of direct solar radiation that is retained by the atmosphere is "locked up" in the clouds and the ozone layer and thus is not available to heat the troposphere. Other sources must be found to explain the creation of atmospheric warmth. The explanation lies in the 47 percent of incoming solar energy reaching Earth's surface (on both land and water) and in the transfer of heat energy from Earth back to the atmosphere. This is accomplished

through such physical processes as (1) radiation, (2) conduction, (3) convection (along with the related phenomenon, advection), and (4) the latent heat of condensation.

Processes of Heat Energy Transfer

Radiation The process by which electromagnetic energy is transferred from the sun to Earth is called **radiation.** We should be aware that all objects emit electromagnetic radiation. The characteristics of that radiation depend on the temperature of the radiating body. In general, the warmer the object, the more energy it will emit and the shorter the wavelengths of peak emission. Since the sun's absolute temperature is 20 times that of Earth, we can predict that the sun will emit more energy, and at shorter wavelengths, than Earth. This is borne out by the facts: The energy output per square meter by the sun is approximately 160,000 times that of Earth! Further, the

Colors of the Atmosphere

When we look to the skies, we are used to seeing certain colors: the brilliant blue of a clear day, the vibrant reds and oranges of a sunset, clouds ranging from pure white to ominous gray/black, and sometimes even a colorful rainbow. These are wondrous colors, and we tend to take them for granted. Why they exist can be explained by the concept of *atmospheric scattering*. This process is explained in the text, but let us examine how it affects the colors of our atmosphere.

It is important to realize first that as sunlight is scattered in our atmosphere, certain wavelengths are scattered more than others. Each wavelength of visible light corresponds to one of the colors of the light spectrum. For example, a wavelength of 0.45 to 0.50 micrometers (1 micrometer = 1/1000 of a millimeter) corresponds to the color blue, and 0.66 to 0.70 micrometers to the color red. The wavelengths that are scattered and then enter the human eye are transmitted to the human brain and translated into the colors. In other words, the wavelengths that are scattered are the colors that we see.

In 1871, Lord Rayleigh, a radiation scientist, explained that our atmospheric gases tend to scatter the shorter wave-lengths of visible light. In other words, *Rayleigh scattering* makes a clear sky appear blue. In 1908, another radiation specialist, Gustav Mie, explained that oblique sun angles coming through the atmosphere tend to cause the scattering of the longer wavelengths of visible light. In other words, *Mie scattering* explains why sunsets bring about a red sky. Another form of scattering, called *indiscriminant scattering*, is caused by tiny water droplets and ice crystals in our atmosphere that tend to scatter all the wavelengths of visible light with equal intensity. When all the wavelengths are scattered equally, white light emerges, and this explains the white clouds we see. (Keep in mind, all clouds are white. When we see dark gray to black storm clouds, we are actually seeing white clouds plus shadows cast by the height and thickness of these massive clouds.)

Rainbows are much more complicated, and their full explanation involves a number of trigonometric functions. Rainbows are caused by water droplets suspended in the atmosphere. The droplets act like tiny glass prisms that take the incoming white sunlight and project the colors of the spectrum. Key elements necessary to observe a rainbow involve the sun angle, water droplet density, and orientation,

majority of solar energy is emitted at wavelengths shorter than 4.0 micrometers, whereas most of Earth's energy is radiated at wavelengths much longer than 4.0 micrometers (see again Fig. 3.10). Thus, *short-wave radiation* from the sun reaches Earth and heats its surface, which, being cooler than the sun, gives off energy in the form of long waves. It is this *long-wave radiation* from Earth's surface that heats the lower layers of the atmosphere and accounts for the heat of the day.

Conduction   The means by which heat is transferred from one part of a body to another or between two touching objects is called **conduction.** Heat flows from the warmer to the cooler part of a body in order to equalize temperature. Conduction actually occurs as heat is passed from one molecule to another in chainlike fashion. It is conduction that makes the bottom of your soup bowl too hot to touch. Conduction also occurs when a spoon left in your coffee gets hot.

Atmospheric conduction occurs at the interface of (zone of contact between) the atmosphere and Earth's surface. However, it is actually a minor method of heat transfer in terms of warming the atmosphere because it affects only the layers of air closest to Earth's surface. This is because air (unlike certain metals) is a poor conductor of heat. In fact, air is just the opposite of a good conductor; it is a good insulator. This property of air is why a layer of air is sometimes put between two panes of glass to help insulate a home. Air is also used as a layer of insulation in sleeping bags and ski parkas. In fact, if air were a good conductor of heat, our kitchens would become unbearable every time we turned on the stove or oven.

Convection   In the atmosphere, as pockets of air near the surface are heated, they expand in volume, become less dense than the surrounding air, and therefore rise. This vertical transfer of heat through the atmosphere is called **convection;** it is the same type of process by which boiling water circulates in a pot on the stove. The water near the bottom is heated first, becoming lighter and less dense as it is heated. As this water tends to rise, colder, denser surface water flows down to replace it. As this new water is warmed, it too flows up, while additional colder water moves downward. This movement within the fluid is called a *convective current.* These currents that are set into motion by the heating of a fluid (liquid or gas) make up a *convectional system.* Such systems account for much of the vertical transfer of heat within the atmosphere and the oceans and are a major cause of clouds and precipitation.

Advection   **Advection** is the term applied to horizontal heat transfer. There are two major advection agents within the Earth-atmosphere system: winds and ocean currents. Both agents help to transfer energy horias well as the observer angle. You will notice when you see a rainbow from a moving vehicle, it is only a matter of time until the angles change and the rainbow disappears.

Incidentally, the colors described above are those distinguishable by the human eye in regard to Earth's atmosphere. Humans visiting another planet with an atmosphere may never see the color blue at all. Furthermore, animals (those that are not color-blind) may see colors other than those described here.

Sunset over the Antarctic Ocean

M. Trapasso

Rainbow

D. Falconer/Photolink/Getty Images

zontally between the equatorial and polar regions, thus maintaining the energy balance in the Earth-atmosphere system.

Latent Heat of Condensation   As we have seen, when water evaporates, a significant amount of energy is stored in the water vapor as latent heat (see again Fig. 4.6). This water vapor is then transported by advection or convection to new locations, where condensation takes place and the stored energy is released. This process plays a major role in the transfer of energy within the Earth system: The **latent heat of evaporation** helps cool the atmosphere while the **latent heat of condensation** helps warm the atmosphere and, in addition, is a source of energy for storms. The power of all severe weather is supplied by the latent heat of condensation.

The Heat Energy Budget

The Budget at Earth's Surface   Now that we know the various means of heat transfer, we are in a position to examine what happens to the 47 percent of solar energy that reaches Earth's surface (look again at Fig. 4.7). Approximately 14 percent of this energy is emitted by the Earth in the form of long-wave radiation. This 14 percent includes a net loss of 6 percent (of the total originally received by the atmosphere) directly to outer space; the other 8 percent is captured by the atmosphere. In addition, there is a net transfer back to the atmosphere (by conduction and convection) of 10 of the 47 percent that reached Earth. The remaining 23 percent returns to the atmosphere through the release of latent heat of condensation. Thus, the 47 percent of the sun's original insolation that reached Earth's surface is all returned to other segments of the system. There has been no long-term gain or loss. Therefore, at Earth's surface, the heat energy budget is in balance.

Examination of the heat energy budget of Earth's surface helps us to understand the open *energy system* that is involved in the heating of the atmosphere. The *input* in the system is the incoming short-wave solar radiation that reaches Earth's surface; this is balanced by the *output* of long-wave terrestrial (Earth's) radiation back to the atmosphere and to space. Since these are in balance, we can say that the overall temperature of Earth's surface is in a state of *dynamic equilibrium*.

Of course, it should be noted that the percentages mentioned earlier represent an oversimplification in that they refer to *net* losses that occur over a long period of time. In the shorter term, heat may be passed from Earth to the atmosphere and then back to Earth in a chain of cycles before it is finally released into space.

We are all familiar with what happens to the inside of a car on a sunny day if all the windows are left closed. Short-wave radiation from the sun is able to penetrate the glass windows (Fig. 4.8). When insolation strikes the

The Energy Budget and Global Climatic Change

Earth's energy budget (or energy balance), as discussed in this chapter, is important to understand for a number of reasons. The budget explains how energy moves through our environment. It shows us that solar radiation (our ultimate energy source) can be gained and lost. Our energy sources can remain untouched or can be converted into some other form. The budget is a means by which we can understand where our energy comes from and where it goes. There are energy budget variations that occur at different latitudes (more so at some latitudes and less at others), but when the planet is taken as a whole the energy flow tends to balance out. When energy flow remains in balance, all Earth systems seem to maintain a balance as well. However, when the energy flow is out of balance, we can expect worldwide changes to occur.

Such is the case with climatic changes. Whenever a theory for worldwide climatic change is proposed, invariably Earth's energy balance is brought into play. Let us consider two examples. The first is probably the most well-known theory for global climatic change, the Greenhouse Effect. Gases in our atmosphere, such as carbon dioxide, methane, ozone, nitrogen oxides, and several others, act like the glass in a greenhouse. They allow solar radiation to reach Earth's surface relatively unhampered but retard the escape of Earth's

interior of the car it is absorbed and heats up the exposed surfaces. Energy, emitted from the surfaces as long-wave radiation, cannot escape through the glass as freely. The result is that the interior of the vehicle gets hotter throughout the day. In extreme cases, windows in some cars have cracked as a result of differential expansion, or, what is more serious, temperatures have become so great in automobiles with closed windows as to pose a serious threat to small children and pets left behind.

A similar phenomenon also occurs in the atmosphere. Like glass, carbon dioxide and water vapor (and other so-called greenhouse gases) can impede the escape of long-wave radiation by absorbing it and then radiating it back to Earth. Termed the **greenhouse effect** this is the primary reason for the moderate temperatures observed on Earth.

The Heat Budget in the Atmosphere   At one time or another, about 60 percent of the solar energy intercepted by the Earth system is temporarily retained by the atmosphere. This includes 19 percent of direct solar radiation absorbed by the clouds and the ozone layer, 8 percent emitted by *long-wave radiation* from the Earth's surface, 10 percent transferred from the surface by *conduction* and *convection,* and 23 percent released by the *latent heat of condensation.* Some of this energy is recycled back to the surface for short periods of time, but eventually all of it is lost into outer space as more solar energy is received. Hence, just as was the case at Earth's surface, the heat energy budget in the atmosphere is in balance over long periods of time—a dynamically stable system.

Many scientists believe that an imbalance in the heat energy budget, with possible negative effects, is developing. Since the Industrial Revolution, human beings have been adding more and more carbon dioxide to the atmosphere through their burning of fossil (carbon) fuels (see again Fig. 4.1). Since carbon dioxide absorbs the long-wave radiation from Earth's surface, restricting its escape to space, this increase in carbon dioxide could increase the greenhouse effect. The rising temperatures would then have significant effects on other Earth features, such as the extent of the polar ice sheets and world sea level.

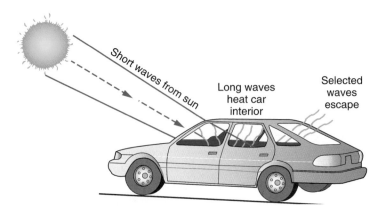

Figure 4.8   Heat buildup in a closed car results from penetration of short-wave radiation through the car window and inability of long-wave radiation to return outward through the glass.

radiated energy to outer space. In essence, this keeps some of Earth's heat energy (thermal infrared energy) from leaving the Earth system. As a result, the energy budget is tilted out of balance. Solar radiation comes in as before, but not all Earth's heat energy escapes. The result is a global warming.

The second example involves an asteroid or comet (the size of Manhattan Island), which struck Earth about 65 million years ago (during a geologic time period known as the Cretaceous-Tertiary boundary). The crater suspected to be the result of this gigantic impact (called Chicxulub) lies off the coast of the Yucatan Peninsula of Mexico. The impact caused an explosion of catastrophic proportions, bringing about earthquakes, tidal waves, and a "nuclear-winter" effect. The atmosphere was dark with pulverized earthen material and vaporized ocean. The resulting darkness cooled Earth so much that it killed much of the vegetation the dinosaurs fed on. In addition, the cold-blooded dinosaurs were vulnerable to the cold, dark conditions. According to the theory, as a result of the cold temperatures and the lack of vegetation, the reign of dinosaurs on Earth came to an end. The immense pall of particulates in the atmosphere had caused incoming solar radiation to decrease dramatically. The amount of solar radiation reflected from and absorbed by the atmosphere had greatly increased, and less energy reached Earth's surface. In other words, Earth's energy flow was out of balance. Once again, when proposing a theory of climatic change on a global scale, the scientist is drawn back to a consideration of Earth's energy budget.

Variations in the Heat Energy Budget Remember that the figures we have seen for the heat energy budget are averages for the *whole* Earth over many years. For any *particular* location, the heat energy budget is most likely not balanced. Some places have a surplus of incoming solar energy over outgoing energy loss, while others have a deficit. The main causes of these variations are differences in latitude and seasonal fluctuations.

As we have noted previously, the amount of insolation received is directly related to latitude (Fig. 4.9). In the tropical zones, where insolation is high throughout the year, more solar energy is received at Earth's surface and in the atmosphere than can be emitted back into space. In the Arctic and Antarctic zones, on the other hand, there is so little insolation during the winter, when Earth is still emitting long-wave radiation, that there is a large deficit for the year. Locations in the middle-latitude zones have lower deficits or surpluses, but only at about latitude 38° is the budget balanced. If it were not for the heat transfers within the atmosphere and the oceans, the tropical zones would get hotter and the polar zones would get colder through time.

At any location, the heat energy budget varies throughout the year according to the seasons, with a tendency toward a surplus in the summer or high-sun season, and a tendency toward a deficit six months later. Seasonal differences may be small near the equator, but they are great in the middle-latitude and polar zones.

Air Temperature

Temperature and Heat

Although heat and temperature are highly related, they are not the same. **Heat** is a form of energy—the total kinetic energy of all the atoms that make up a substance. All substances are made up of atoms that are constantly in motion (vibrating and colliding) and therefore possess kinetic energy—the energy of motion. This energy is manifested as heat. **Temperature,** on the other hand, is the average kinetic energy of the individual atoms of a substance. When something is heated, its atoms vibrate faster and its temperature increases. It is important to remember that the amount of heat energy depends on the mass of the substance under discussion, whereas the temperature refers to the energy of individual molecules. Thus, a burning match has a high temperature but minimal heat energy; the oceans have moderate temperatures but high heat energy content.

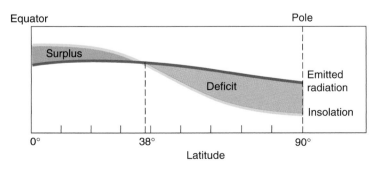

Figure 4.9 Latitudinal variation in the energy budget. Low latitudes receive more insolation than they lose by reradiation and have an energy surplus. High latitudes receive less energy than they lose outward and therefore have an energy deficit. ■ **How do you think the surplus energy in the low latitudes is transferred to higher latitudes?**

Scales

Three different scales are generally used in measuring temperature. The one with which Americans are most familiar is the **Fahrenheit scale,** devised in 1714 by Daniel Fahrenheit, a German scientist. On this scale, the temperature at which water boils at sea level is 212°F, and the temperature at which water freezes is 32°F. This scale is used in the English system of measurements.

The **Celsius scale** (also called the **centigrade scale**) was devised in 1742 by Anders Celsius, a Swedish astronomer. It is part of the metric system. The temperature at which water freezes at sea level on this scale was arbitrarily set at 0°C, and the temperature at which water boils was designated as 100°C.

The Celsius scale is used nearly everywhere but in the United States, and even in the United States the Celsius scale is used by the majority of the scientific community. By this time, you have undoubtedly noticed that throughout this book comparable figures in both the Celsius (centigrade) and Fahrenheit scales are given side by side for temperatures. Similarly, whenever important figures for distance, area, weight, or speed are given, we use the metric system followed by the English system.

Figure 4.10 can help you compare the Fahrenheit and Celsius systems as you encounter temperature figures outside this book. In addition, the following formulas can be used for conversion from Fahrenheit to Celsius or vice versa:

$$C = (F - 32) \div 1.8$$
$$F = C \times 1.8 \ (+32)$$

The third temperature scale, used primarily by scientists, is the **Kelvin scale.** Lord Kelvin, a British radiation scientist, felt that negative temperatures were not proper, and should not be used. In his mind, no temperature should ever go below zero. This scale is based on the fact that the temperature of a gas is related to the molecular movement within the gas. As the temperature of a gas is reduced, the molecular motion within the gas slows. There is a temperature at which all molecular motion stops and no further cooling is possible. This temperature, approximately −273°C, is termed *absolute zero.* The Kelvin scale uses absolute zero as its starting point. Thus 0°K equals −273°C. Conversion of Celsius to Kelvin is expressed by the following formula:

$$K = C + 273$$

Short-Term Variations in Temperature

Local changes in atmospheric temperature can have a number of causes. These are related to the mechanics of the receipt and dissipation of energy from the sun and to various properties of Earth's surface and the atmosphere.

The Daily Effects of Insolation   As we noted earlier, the amount of insolation at any particular location varies both throughout the year (annually) and throughout the day (diurnally). Annual fluctuations are associated with the sun's changing declination and hence with the seasons. Diurnal changes are related to the rotation of Earth about its axis. Each day, insolation receipt begins at sunrise, reaches its maximum at noon (local solar time), and returns to zero at sunset.

Although insolation is greatest at noon, you may have noticed that temperatures usually do not reach their maximum until 3 or 4 o'clock in the afternoon (Fig. 4.11). This is because the insolation received by Earth from shortly after sunrise until the afternoon hours exceeds the energy being lost through Earth radiation. Hence,

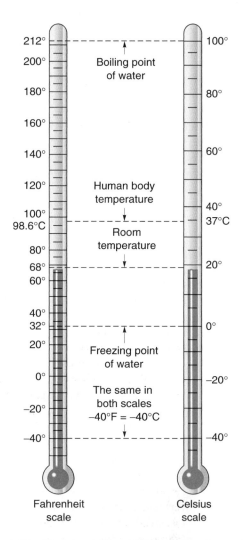

Figure 4.10   The Fahrenheit and Celsius temperature scales. The scales are aligned to permit direct conversion of readings from one to the other.   ■ **When it is 70°F what is the temperature in Celsius degrees?**

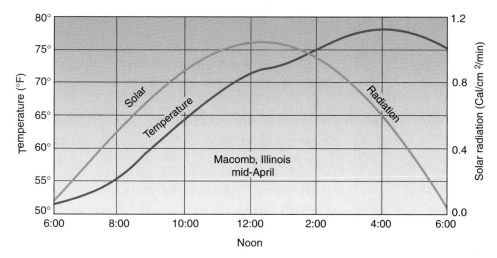

Figure 4.11 Diurnal changes in insolation and temperature in mid-April for Macomb, Illinois.
■ **Why does temperature rise even after solar energy declines?**

during that period, as Earth and atmosphere continue to gain energy, temperatures normally show a gradual increase. Sometime around 3 P.M., when outgoing Earth radiation begins to exceed insolation, temperatures start to fall. The daily lag of Earth radiation and temperature behind insolation is accounted for by the time it takes for Earth's surface to be heated to its maximum and for this energy to be radiated to the atmosphere.

Insolation receipt ends at sunset, but energy that has been stored in Earth's surface layer during the day continues to be lost throughout the night and the ability to heat the atmosphere decreases. The lowest temperatures occur just before dawn, when the maximum amount of energy has been emitted and before replenishment from the sun can occur. Thus, if we disregard other factors for the moment, we can see that there is a predictable hourly change in temperature called the **daily march of temperature.** There is a gentle decline from midafternoon until dawn and a rapid increase in the eight hours or so from dawn until the next maximum is reached.

Cloud Cover The extent of cloud cover is another factor that affects the temperature of Earth's surface and the atmosphere (Fig. 4.12). Weather satellites have shown that, at any time, about 50 percent of Earth is covered by clouds (Fig. 4.13). This is important because a heavy cloud cover can reduce the amount of insolation a place receives, thereby causing daytime temperatures to be lower on a cloudy day. On the other hand, we also have the greenhouse effect, in which clouds, composed in large part of water droplets, are capable of absorbing heat energy radiating from Earth. Clouds therefore keep temperatures near Earth's surface warmer than they would otherwise be, especially at night. The general effect of cloud cover, then, is to moderate temperature by lowering the potential maximum and raising the potential minimum temperatures. In other words, cloud cover makes for cooler days and warmer nights.

Differential Heating of Land and Water Earlier we saw that bodies of water heat and cool more slowly than the land. The air above Earth's surface is heated or cooled in part by what is beneath it. Therefore, temperatures over bodies of water or on land subjected to ocean winds (**maritime** locations) tend to be more moderate than those of land-locked places at the same latitude. Thus, the greater the **continentality** of a location (the distance removed from a large body of water), the less its temperature pattern will be modified.

Reflection The capacity of a surface to reflect the sun's energy is called its **albedo;** a surface with a high albedo has a high percentage of reflection. The more solar energy that is reflected back into space by an Earth surface, the less that is absorbed for heating the atmosphere. Temperatures will be higher at a given location if its surface has a low albedo rather than a high albedo.

As you may know from experience, snow and ice are good reflectors; they have an albedo of 90 to 95 percent. This is one reason why glaciers on high mountains do not melt away in the summer or why there may still be snow on the ground on a sunny day in the spring: solar energy is reflected away. A forest, on the other hand, has an albedo of only 10 to 12 percent, which is good for the trees because they need solar energy for photosynthesis. The albedo of cloud cover varies, from 40 percent to 80 percent, according to the thickness of the clouds. The high albedo of many clouds is why much solar radiation is reflected directly back into space by the atmosphere. Cities have an albedo of only about 10 percent. This is one reason why summer days can be hotter in the city relative to the surrounding countryside, which may be several degrees cooler. Albedo of surface cover is one cause for the *urban heat island* effect.

The albedo of water varies greatly, depending on the depth of the water body and the angle of the sun's rays. If the angle of the sun's rays is high, smooth water will

(a) Day

(b) Night

Figure 4.12 The effect of cloud cover on temperatures. (a) By intercepting insolation, clouds produce lower air temperatures during the day. (b) By trapping long-wave radiation from Earth, clouds increase air temperatures at night. The overall effect is a great reduction in the diurnal temperature range. ■ **Desert regions have large diurnal variations in temperature. Why is this so?**

reflect little. In fact, if the sun is vertical over a calm ocean, the albedo will be only about 2 percent. However, a low sun angle, such as just before sunset, causes an albedo of more than 90 percent from the same ocean surface. Likewise, a snow surface in winter, when solar angles are lower, can reflect up to 95 percent of the energy striking it, and skiers must constantly be aware of the danger of severe burns and possible snow blindness from reflected solar radiation. In a similar fashion, the high albedo of sand causes the sides of sunbathers' legs to burn faster when they lie on the beach.

Horizontal Air Movement We have already seen that advection is the major mode of horizontal transfer of heat and energy over Earth's surface. Any movement of air due to the wind, whether on a large or small scale, can have a significant short-term effect on the temperatures of a given location. Thus, wind blowing from an ocean to

land will generally bring cooler temperatures in summer and warmer temperatures in winter. Large quantities of air moving from polar regions into the middle latitudes can cause sharp drops in temperature, whereas air moving poleward will usually bring warmer temperatures.

Vertical Distribution of Temperature

Normal Lapse Rates We have learned that Earth's atmosphere is primarily heated from the ground up as a result of long-wave terrestrial radiation, conduction, and convection. Thus, temperatures in the troposphere are usually highest at ground level and decrease with increasing altitude. As noted earlier in the chapter, this decrease in the free air of approximately 6.5°C per 1000 meters (3.6°F/1000 ft) is known as the *normal lapse rate*.

The lapse rate at a particular place can vary for a variety of reasons (Fig. 4.14). Low lapse rates can exist if denser and colder air is drained into a valley from a higher elevation or if advectional winds bring air in from a cooler region at the same altitude. In each case, the surface is cooled so that its temperature is closer to that at higher elevations directly above it. On the other hand, if the surface is heated strongly by the sun's rays on a hot summer afternoon, the air near Earth will be disproportionately warm, and the lapse rate will be steep. Fluctuations in lapse rates due to abnormal temperature conditions at various altitudes can play an important role in the weather a place may have on a given day.

Figure 4.13 This GOES satellite image shows a section of North America and a variety of types of cloud cover.

Figure 4.14 Steep, normal, and low atmospheric lapse rates.

Temperature Inversions   Under certain circumstances, the normal observed decrease of temperature with increased altitude may be reversed; temperature may actually *increase* for several hundred meters. This is called a **temperature inversion.**

Some inversions take place 1000 or 2000 meters above the surface of Earth where a layer of warmer air interrupts the normal decrease in temperature with altitude (Fig. 4.15). Such inversions tend to stabilize the air, causing less turbulence and discouraging both precipitation and the development of storms. Upper-air inversions may occur when air settles slowly from the upper atmosphere. Such air is compressed as it sinks and rises in temperature, becoming more stable and less buoyant. Inversions caused by descending air are common at about 30° to 35° north and south latitudes.

An upper-air inversion common to the coastal area of California results when cool marine air blowing in from the Pacific Ocean moves under stable, warmer, and lighter air aloft created by subsidence and compression. Such an inversion layer tends to maintain itself; that is, the cold underlying air is heavier and cannot rise through the warmer air above. Not only does the cold air resist rising or moving, but pollutants, such as smoke, dust particles, and automobile exhaust, that are created at Earth's surface also fail to disperse. They therefore accumulate, filling the lower atmosphere. This situation is particularly acute in the Los Angeles area, which is a basin surrounded by higher mountainous areas (Fig. 4.16). Cooler air blows into the basin from the ocean and then cannot escape either horizontally, because of the landform barriers, or vertically, because of the inversion.

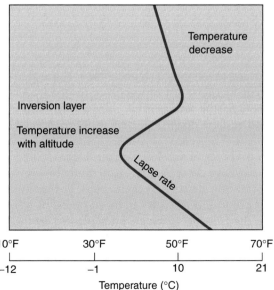

Figure 4.15   (Left) Temperature inversion caused by subsidence of air. (Right) Lapse rate associated with the column of air (A) in left-hand drawing.

Figure 4.16 Conditions producing smog-trapping inversion in the Los Angeles area.

Some of the most noticeable temperature inversions are those that occur near the surface when Earth cools the lowest layer of air through conduction and radiation (Fig. 4.17). In this situation, the coldest air is nearest the surface and the temperature rises with altitude. Inversions near the surface most often occur on clear nights in the middle latitudes. They may be enhanced by snow cover or the recent advection of cool, dry air into the area. Such conditions produce extremely rapid cooling of Earth's surface at night as it loses the day's insolation through radiation. Then the layers of the atmosphere that are closest to Earth are cooled by radiation and conduction more than those at higher altitudes. Calm air conditions near the surface help produce, and partially result from, these temperature inversions.

Surface Inversions: Fog and Frost Fog and frost often occur as the result of a surface inversion. Especially where Earth's surface is hilly, dense cold surface air will tend to flow downslope and accumulate in the lower valleys. The colder air on the valley floors sometimes produces fog or, in more extreme cases, a killing frost.

Farmers use a variety of methods to prevent such frosts from destroying their crops. For example, fruit trees in California that can be destroyed by a frost during the growing season are often planted on the warmer hillsides instead of in the valleys. Farmers may also put blankets of straw, cloth, or some other poor conductor over their plants. These take the place of the missing water vapor in the clear atmosphere, preventing the escape of Earth's radiation to outer space and thereby keeping the plants warmer.

Large fans and helicopters are sometimes used to stir the air in an effort to mix the layers and disturb the inversion (Fig. 4.18a). Another device used to prevent frost is huge orchard heaters that heat the air, disturbing the temperature layers. Smudge pots, an older method of preventing frost, pour smoke into the air, which provides an insulation blanket much like the straw or blankets just mentioned, preventing the escape of terrestrial radiation (Fig. 4.18b). However, these devices have declined in favor because of their air pollution potential.

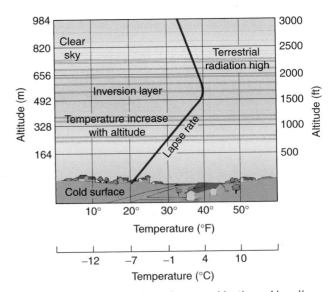

Figure 4.17 Temperature inversion caused by the rapid cooling of the air above the cold surface of Earth at night. ■ **What is the significance of an inversion?**

Figure 4.18 (left) Fans (propellers) are used to protect Washington apple orchards from frost. (right) Smudge pots were an older method of trying to keep surface temperatures above freezing.

Temperature Distribution at Earth's Surface

Displaying the distribution of temperatures over the surface of the Earth requires a mapping device called **isotherms.** Isotherms (from Greek: *isos,* equal; *therm, heat*) are defined as lines that connect places of equal temperature. When constructing isothermal maps showing temperature distribution, we need to account for elevation by adjusting temperature readings to what they would be at sea level. This adjustment means adding 6.5°C for every 1000 meters of elevation (the normal lapse rate). The rate of temperature change on an isothermal map is called the **temperature gradient.** Closely spaced isotherms indicate a steep temperature gradient (a rapid temperature change over a shorter distance), and widely spaced lines indicate a weak one (a slight temperature change over a longer distance).

Figures 4.19 and 4.20 show the horizontal distribution of temperatures for Earth at two critical times, during January and July, when the seasonal extremes of high and low temperatures are most obvious in the Northern and Southern Hemispheres. The easiest feature to recognize on both maps is the general orientation of the isotherms; they run nearly east-west around Earth, as do the parallels of latitude.

A more detailed study of Figures 4.19 and 4.20 and a comparison of the two maps reveal some additional important features. The highest temperatures in January are in the Southern Hemisphere; in July, they are in the Northern Hemisphere. Comparing the latitudes of Portugal and southern Australia can demonstrate this point. Note on the July map that Portugal in the Northern Hemisphere is nearly on the 70°F isotherm, while at southern Australia in the Southern Hemisphere the average July temperature is less than 50°F, even though the two cities are approximately the same distance from the equator. The temperature differences between the two hemispheres are again a product of insolation, this time changing as the sun shifts north and south across the equator between its positions at the two solstices.

Note that the greatest deviation from the east-west trend of isotherms occurs where the isotherms leave large landmasses to cross the oceans. As the isotherms leave the land, they usually bend rather sharply toward the pole in the hemisphere experiencing winter and toward the equator in the summer hemisphere. This behavior of the isotherms is a direct reaction to the differential heating and cooling of land and water. The continents are hotter than the oceans in the summer and colder in the winter.

Other interesting features on the January and July maps can be mentioned briefly. Note that the isotherms

Average sea-level temperature °F (January)

Figure 4.19   Average sea-level temperatures in January (°F).

poleward of 40° latitude are much more regular in their east-west orientation in the Southern than in the Northern Hemisphere. This is because in the Southern Hemisphere (often called the "water hemisphere"), there is little land south of 40°S latitude to produce land and water contrasts. Note also that the temperature gradients are much steeper in winter than in summer in both hemi-spheres. The reason for this can be understood when you recall that the tropical zones have high temperatures throughout the year, whereas the polar zones have large seasonal differences. Hence, the difference in temperature between tropical and polar zones is much greater in winter than in summer. As a final point, observe the especially sharp swing of the isotherms off the coasts of

Average sea-level temperature °F (July)

Figure 4.20   Average sea-level temperatures in July (°F). ■ **Observe the temperature gradients between the equator and northern Canada in July and January. Which is greater? Why?**

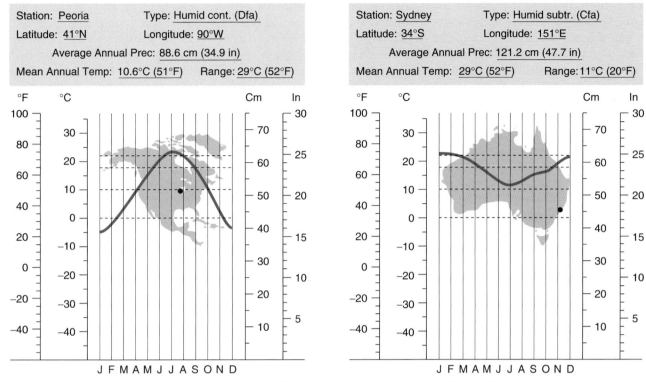

Station: Peoria	Type: Humid cont. (Dfa)
Latitude: 41°N	Longitude: 90°W
Average Annual Prec: 88.6 cm (34.9 in)	
Mean Annual Temp: 10.6°C (51°F)	Range: 29°C (52°F)

Station: Sydney	Type: Humid subtr. (Cfa)
Latitude: 34°S	Longitude: 151°E
Average Annual Prec: 121.2 cm (47.7 in)	
Mean Annual Temp: 29°C (52°F)	Range: 11°C (20°F)

Figure 4.21   The annual march of temperature at Peoria, Illinois, and Sydney, Australia.   ▪ **Why do these two locations have opposite temperature curves?**

eastern North America, southwestern South America, and southwestern Africa in January, and off southern California in July. In these locations, the normal bending of the isotherms due to land-water differences is augmented by the presence of warm or cool ocean currents.

Annual March of Temperature

Isothermal maps are commonly plotted for January and July because there is a lag of about 30 to 40 days from the solstices, when the amount of insolation is at a minimum or maximum (depending on the hemisphere), to the time minimum or maximum temperatures are reached. This **annual lag of temperature** behind insolation is similar to the daily lag of temperature explained previously. It is a result of the changing relationship between incoming insolation and outgoing Earth radiation. Temperatures continue to rise for a month or more after the summer solstice because insolation continues to exceed radiation loss. Temperatures continue to fall after the winter solstice until the increase in insolation finally matches Earth's radiation. In short, the lag exists because it takes time for Earth to heat or cool and for those temperature changes to be transferred to the atmosphere.

The annual changes of temperature for a location can be plotted in a graph (Fig. 4.21). The mean temperature for each month in a place such as Peoria, Illinois, is recorded and a line drawn connecting the 12 temperatures. The mean monthly temperature is the average of the daily mean temperatures recorded at a weather station during a month. The daily mean temperature is the average of the maximum and minimum temperatures for a 24-hour period. Such a temperature graph, depicting the **annual march of temperature,** shows both the decrease in solar radiation, as reflected by a decrease in temperature, from midsummer to midwinter and the increase in temperature from midwinter to midsummer caused by the increase in solar radiation.

Weather and Climate

We frequently use the words *weather* and *climate* in general conversation, but it is important in a science course to carefully distinguish between the terms. **Weather** refers to the condition of atmospheric elements at a given time and for a specific area. That area could be as large as the New York metropolitan area or as small and specific as a weather observation station. Higher atmospheric layers are important in such fields as space research, remote sensing, and telecommunications. However, it is the lowest layer, the troposphere, that exhibits Earth's weather and is of the greatest interest to physical geographers and weather forecasters who survey the changing conditions of the atmosphere in the field of study known as **meteorology.**

Many observations of the weather of a place over a period of at least 30 years provide us with a description of its climate. **Climate** describes an area's average weather, but it also includes those common deviations from the normal or average that are likely to occur. Climates also include extreme situations, which can be very significant. Thus, we could describe the climate of the southeastern United States in terms of average temperatures and precipitation through a year, but we would also have to include mention of the likelihood of events such as hurricanes and snowstorms during certain periods of the year. **Climatology** is the study of the varieties of climates, both past and present, found on our planet and their distribution over its surface.

Weather and climate are of prime interest to the physical geographer because they affect and are interrelated with other parts of the Earth system. The changing conditions of atmospheric elements such as temperature, rainfall, and wind affect soils and vegetation, modify landforms, cause flooding of towns and farms, and, in a multitude of other ways, influence the function of systems in each of Earth's varied physical environments.

Five basic **elements** of the atmosphere serve as the "ingredients" of weather and climate: (1) solar energy (or insolation), (2) temperature, (3) pressure, (4) wind, and (5) precipitation. We must examine these elements in order to understand and categorize weather and climate. Thus, a weather forecast will generally include the present temperature, the probable temperature range, a description of the cloud cover, the chance of precipitation, the speed and direction of the winds, and air pressure.

In Chapter 3, we noted that the amount of solar energy received at one place on Earth's surface varies during a day and throughout the year. The amount of insolation a place receives is the most important weather element; the other four elements depend, in part, on the intensity and duration of solar energy.

The temperature of the atmosphere at a given place on or near the surface of Earth is largely a function of the insolation received at that location. It is also influenced by many other factors, such as land and water distribution and altitude. Unless there is some form of precipitation occurring, the temperature of the air may be the first element of weather we describe when someone asks us what it is like outside.

However, if it is raining or foggy or snowing, we will probably notice and mention that condition first. We are less aware of the amount of water vapor or moisture in the air (except in very arid or humid areas). However, moisture in the air is a vital weather element in the atmosphere, and its variations play an important role in the likelihood of precipitation.

We are probably least aware of variations in air pressure, although the fluctuations in air pressure are basic to the development of winds and storms. However, some people say they can feel a change in the weather. Some weather-sensitive people feel it "in their bones" because they have arthritis and can probably sense the movement of fluids under pressure in their joints; others may feel changing pressure in their sinus passages.

We all know that weather varies. Since it is the momentary state of the atmosphere at a given location, it varies in both time and place. There are even variations in the amount that weather varies. In some places and at some times of the year, the weather changes almost daily. In other places, there may be weeks of uninterrupted sunshine, blue skies, and moderate temperatures followed by weeks of persistent rain. A few places experience only minor differences in the weather throughout the year. The language of the original people of Hawaii is said to have no word for weather because conditions there varied so little.

Complexity of Earth's Energy Systems

This chapter was designed to show the variations of Earth's energy systems and dynamic balances. These variations are the result of complex interrelationships between the characteristics of Earth and its atmosphere, and the energy gained and lost by Earth's environments. The variations are both horizontal across the surface and vertical through our atmosphere. Further, they vary on both daily and seasonal time frames.

Variations of Earth's energy systems impose both diurnal and annual rhythms on our agricultural activities, our recreational pursuits, our clothing styles, our architecture, and our energy bills. Human activities are constantly influenced by temperature changes, which reflect the input-output patterns of Earth's energy systems.

DEFINE & RECALL

photosynthesis	controls (weather and	convection	greenhouse effect
troposphere	climate)	advection	heat
normal lapse rate	solar constant	latent heat of	temperature
stratosphere	radiation	evaporation	Fahrenheit scale
mesosphere	long-wave radiation	latent heat of	Celsius (centigrade)
thermosphere	conduction	condensation	scale

Kelvin scale

daily march of
   temperature

maritime

continentality

albedo

temperature inversion

isotherm

temperature gradient

annual lag of temperature

annual march of
   temperature

weather

meteorology

climate

climatology

elements (weather and
   climate)

DISCUSS & REVIEW

1. Why is it useful to think of the atmosphere as a thin film of air? As an ocean of air?

2. Name the gases of which the atmosphere is composed, and give the percentage of the total that each supplies.

3. What function does ozone play in the support of life on Earth? Where and how is ozone formed?

4. How is the atmosphere subdivided? In what level do you live? Have you been in any of the other levels?

5. Would you expect an area like Seattle to have a milder or a harsher winter than Grand Forks, North Dakota? Why?

6. Discuss the role of water in energy exchange. What characteristics of water make it so important?

7. How does Earth's atmosphere affect incoming solar radiation (insolation)? By what processes is insolation prevented from reaching Earth's surface? What percentages are involved in a generalized situation? What percentages reach the surface, and by what processes?

8. How is the atmosphere heated from Earth's surface? What processes and percentages are involved in a generalized situation?

9. What is meant by Earth's energy budget? List and define the important energy exchanges that keep it in balance.

10. What is the temperature in Fahrenheit degrees today in your area? In Celsius degrees?

11. At what time of day does insolation reach its maximum? Its minimum? Compare this to the daily temperature maximum and minimum.

12. How is albedo a factor in your selection of outdoor clothes on a hot, sunny day? On a cold, clear winter day?

13. What is a temperature inversion? Give several reasons why temperature inversions occur.

14. Why do citrus growers use wind machines and heaters? Describe any techniques you are familiar with to prevent frost damage to plants in your area.

15. Describe the behavior of the isotherms in Figures 4.19 and 4.20. What factors cause the greatest deviation from an east-west trend? What factors cause the greatest differences between the January and July maps?

16. What is the difference between meteorology and climatology?

17. What are the basic characteristics that we call *elements* of weather and climate?

18. What factors cause variation in the elements of weather and climate?

CONSIDER & RESPOND

1. Convert the following temperatures to Fahrenheit: 20°C, 30°C, 15°C.

2. Convert the following temperatures to Celsius: 60°F, 15°F, 90°F.

3. Convert the temperatures in (1) and (2) to Kelvin.

4. Refer to Figure 4.7. List the major means by which the atmosphere gains heat; loses heat.

5. What are the major weather and climate controls that operate in your area?

6. The normal lapse rate is 6.5°C/1000 meters. If the surface temperature is 25°C, what is the air temperature at 10,000 meters above the Earth's surface? Convert your answer to degrees Fahrenheit.

7. Refer to Figures 4.19 and 4.20. What location on the Earth's surface exhibits the greatest annual range of temperature? Why?

ATMOSPHERIC PRESSURE, WINDS, AND CIRCULATION PATTERNS

CHAPTER PREVIEW

Latitudinal differences in temperature (as a result of differential receipt of insolation) provide a partial explanation for latitudinal differences in pressure.

What is the relation between temperature and pressure? Why is this only a partial explanation?

The fact that land heats and cools more rapidly than water is of significance not only to world patterns of temperature but also to world patterns of pressure, winds, and precipitation.

How can you explain this fact? What effect does this fact have on world patterns? In what ways can it affect human beings?

The horizontal transfer of air (wind) is a direct result of atmospheric movement to adjust for pressure inequalities.

How do these inequalities develop? How does wind adjust for high and low pressure?

Convergent and divergent wind systems (cyclones and anticyclones) in the Northern Hemisphere are characterized by air systems spiraling in opposite directions from one another and in opposite directions from their counterparts in the Southern Hemisphere.

In what directions (clockwise or counterclockwise) do these systems spiral in the Northern and Southern Hemispheres?

Planetary wind systems in association with global pressure patterns play a major role in global circulation.

What are the primary sources of the planetary winds? What are the six major planetary wind belts or zones, and what are their chief characteristics? Why do the wind belts migrate with the seasons?

Ocean currents constitute a system of global circulation that can exert significant influence on the atmospheric conditions of nearby land areas.

What is the driving force behind the ocean current circulation patterns? How do ocean currents affect atmospheric conditions of land areas?

El Niños can have a devastating impact on our global weather.

What is an El Niño? How does it influence global weather? Why have scientists recently become better able to predict the onset of an El Niño?

Upper air winds and atmospheric circulation play a major role in controlling surface weather and climatic conditions.

What is upper air circulation like? How does it affect surface conditions? How are human beings affected?

Blue Skies CD

Chapter 4: Atmospheric Forces, *Coriolis Force*
Chapter 4: Atmospheric Forces, *Winds in 2 Hemispheres*
Chapter 5: Atmospheric Circulation, *Global Atmosphere*
Chapter 5: Atmospheric Circulation, *Global Ocean*
Chapter 5: Atmospheric Circulation, *Southern Oscillation*

Earth Systems Today CD

Chapter 1: Weather & Climate, *El Niño*
Chapter 5: Waves, Tides, & Currents, *World Currents*

Laboratory for Atmospheres at NASA Goddard Space Flight Center

Hurricanes, the largest storms on Earth, show distinctive circulation patterns.

An individual gas molecule weighs almost nothing; however, the atmosphere as a whole has considerable weight and exerts an average pressure of 1034 grams per square centimeter (14.7 lb/sq in.) on Earth's surface. The reason why people are not crushed by this atmospheric pressure is that we have air and water inside us—in our blood, tissues, and cells—exerting an equal outward pressure that balances the inward pressure of the atmosphere. Atmospheric pressure is important because variation in pressure within the Earth-atmosphere system causes our atmospheric circulation and thus plays a major role in determining our weather and climate.

In 1643, Evangelista Torricelli, a student of Galileo, performed an experiment that was the basis for the invention of the *mercury barometer,* an instrument that measures atmospheric (also called barometric) pressure. Torricelli took a tube filled with mercury and inverted it in an open pan of mercury. The mercury inside the tube fell until it was at a height of about 76 centimeters (29.92 in.) above the mercury in the pan, leaving a vacuum bubble at the closed end of the tube (Fig. 5.1). At this point, the pressure exerted by the atmosphere on the open pan of mercury was equal to the pressure from the mercury in the tube. Torricelli observed that as the air pressure increased, it pushed the mercury up into the tube, increasing the height of the mercury until the pressure exerted by the mercury (under the pull of gravity) would equal the pressure of the air. On the other hand, as the air pressure decreased, the mercury column fell.

In the strictest sense, a mercury barometer does not actually measure the pressure exerted by the atmosphere on Earth's surface, but instead measures the *response* to that pressure. That is, when the atmosphere exerts a specific pressure, the mercury will respond by rising to a specific height. Meteorologists usually prefer to work with actual pressure units. The unit most often used is the millibar (mb). Standard sea-level pressure of 1013.2 millibars will cause the mercury to rise 76 centimeters (29.92 in.).

Our study of the atmospheric elements that combine to produce weather and climate has to this point focused on the fundamental influence of solar energy on the global distributional patterns of temperature. The unequal receipt of insolation by latitude over Earth's surface produces temperature patterns that vary from the equator to the poles. In this chapter, we will learn that these temperature differences are one of the major causes of the development of patterns of higher and lower pressure

that also vary with latitude. In addition, we will examine patterns of another kind—patterns of movement or, more properly, circulation, in which both energy and matter travel cyclically through Earth subsystems.

Geographers are particularly interested in circulation patterns because they illustrate spatial interaction, one of geography's major themes introduced in Chapter 1. Patterns of movement between one place and another reveal that the two places have a relationship and prompt geographers to seek both the nature and effect of that relationship. It is also important to understand the causes of the spatial interaction taking place. As we examine the circulation patterns featured in the latter portion of the chapter, you should make a special effort once again to trace each pattern back to the fundamental influence of solar energy.

Variations in Atmospheric Pressure

Vertical Variations in Pressure

Imagine a pileup of football players during a game. The player on the bottom gets squeezed more than a player near the top because he has the weight of all the others on top of him. Similarly, air pressure decreases with elevation, for the higher we go, the more diffused the air molecules become. The increased intermolecular space results in less pressure. In fact, at the top of Mount Everest (elevation 8848 m or 29,028 ft), the air pressure is only about one-third the pressure at sea level.

Humans are usually not sensitive to small, everyday variations in air pressure. However, when we climb or fly to altitudes significantly above sea level, we become aware of the effects of air pressure on our system. When jet aircraft fly at 10,000 meters (33,000 ft), they have to be pressurized and nearly airtight so that a near-sea-level pressure can be maintained. Even then, the pressurization does not work perfectly, so our ears may pop as they adjust to a rapid change in pressure when ascending or descending. Hiking or skiing at heights that are a few thousand meters in elevation will affect us if we are used to the air pressure at sea level. The reduced air pressure means less oxygen is contained in each breath of air. Thus, we sometimes find that we get out of breath far more easily at high elevations until our bodies adjust to the reduced air pressure.

Changes in air pressure are not solely related to altitude. At Earth's surface, small but important variations in pressure are related to the intensity of radiation, the general movement of global circulation, and local humidity and precipitation. Consequently, a change in air pressure at a given locality often indicates a change in the weather. Weather systems themselves can be classified by the structure and tendency toward change of their pressure.

Figure 5.1 A simple mercury barometer. Standard sea-level pressure of 1013.2 millibars will cause the mercury to rise 76 centimeters (29.92 in.) in the tube. ■ **When air pressure increases, what happens to the mercury in the tube?**

Horizontal Variations in Pressure

The causes of horizontal variation in air pressure are grouped into two types: thermal (determined by temperature) and dynamic (related to motion).

We will look at the simpler thermal type first. In Chapter 4, we saw that Earth is heated unevenly because of unequal distribution of insolation, differential heating of land and water surfaces, and different albedos of surfaces. One of the basic laws of gases is that the pressure and density of a given gas vary inversely with temperature. Thus, during the day, as Earth heats the air above it, the air expands in volume and decreases in density. Such air has a tendency to rise as its density decreases. When the warmed air rises, there is less air near the surface, with a consequent decrease in surface pressure. The equator is an area where such low pressure occurs.

In an area with cold air, there is an increase in density and a decrease in volume. This causes the air to sink and pressure to increase. The poles are an area where such high pressure occurs. Thus, the constant low pressure in the equatorial zone and the high pressure at the poles are thermally induced.

From this we might expect a gradual increase in pressure from the equator to the poles to accompany the gradual decrease in average annual temperature. However, actual readings taken at Earth's surface indicate that pressure does not increase in a regular fashion latitudinally poleward from the equator. Instead, there are regions of high pressure in the subtropics and regions of low pressure in the subpolar regions. The dynamic causes of these zones, or *belts,* of high and low pressure are more complex than the thermal causes.

These dynamic causes are related to the rotation of Earth and the broad patterns of circulation. For example, as air rises steadily at the equator, it moves toward the poles. Earth's rotation, however, causes the poleward-flowing air to drift to the east. In fact, by the time it is over the subtropical regions, the air is flowing from west

to east. This bending of the flow as it moves poleward impedes movement and causes the air to pile up over the subtropics, which results in increased pressure at Earth's surface there.

With high pressure over the polar and subtropical regions, dynamically induced areas of low pressure are created between them, in the subpolar region. As a result, air flows from the highs to the lows, where it rises. Thus, both the subtropical and subpolar pressure regions are dynamically induced. This example describes horizontal pressure variations on a global scale. We will concentrate on this scale later in this chapter.

Basic Pressure Systems

Before we begin our discussion of circulation patterns leading up to the global scale, we must start by describing the two basic types of pressure systems: the **low,** or **cyclone,** and the **high,** or **anticyclone.** These are represented by the letters **L** and **H** that we commonly see on TV, newspaper, and official weather maps.

A low, or cyclone, is an area where air is ascending. As air moves upward away from the surface, it relieves pressure from that surface. In this case, barometer readings will begin to fall. A high, or anticyclone, is just the opposite. In a high, air is descending toward the surface and thus barometer readings will begin to rise, indicating an increased pressure on the surface. Lows and highs are illustrated in Figure 5.2.

Convergent and Divergent Circulation

As we have just seen, winds blow toward the center of a cyclone and can be said to *converge* toward it. Hence, a cyclone is a closed pressure system whose center serves as the focus for **convergent wind circulation.** The winds of an anticyclone blow away from the center of high pressure and are said to be *diverging.* In the case of an anticyclone, the center of the system serves as the source

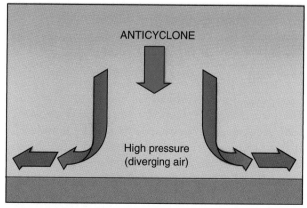

Figure 5.2 Winds converge and ascend in cyclones (low-pressure centers) and descend and diverge from anticyclones (high-pressure centers). ■ **How is temperature related to the density of air?**

Effective Temperatures

Effective temperatures (formerly known as *sensible temperatures*) are temperatures as they might be experienced by a person at rest, in ordinary clothing, in a motionless atmosphere. In other words, at any given time, how comfortable does a temperature feel to the individual experiencing it? This temperature value cannot be obtained by just reading a thermometer.

Several factors come into play when considering effective temperatures. These factors can be divided into two cate-gories: atmospheric factors and human factors. Of the atmospheric factors, four are most important. First is the actual temperature; obviously, thermometers will help distinguish between cold and warm days. Second is humidity; since the evaporation of sweat is a cooling process for the human body, humid days feel warmer than dry days. Third is wind speed; winds not only carry heat from the body but can also accelerate the evaporation of sweat. Fourth is the percentage of clear sky; shady areas are cooler than sunny areas.

Air temperature (°F)	Relative humidity (%)																				
	0	**5**	**10**	**15**	**20**	**25**	**30**	**35**	**40**	**45**	**50**	**55**	**60**	**65**	**70**	**75**	**80**	**85**	**90**	**95**	**100**
140	125																				
135	120	128																			
130	117	122	131																		
125	111	116	123	131	141																
120	107	111	116	123	130	139	148														
115	103	107	111	115	120	127	135	143	151												
110	99	102	105	108	112	117	123	130	137	143	150										
105	95	97	100	102	105	109	113	118	123	129	135	142	149								
100	91	93	95	97	99	101	104	107	110	115	120	126	132	138	144						
95	87	88	90	91	93	94	96	98	101	104	107	110	114	119	124	130	136				
90	83	84	85	86	87	88	90	91	93	95	96	98	100	102	106	109	113	117	122		
85	78	79	80	81	82	83	84	85	86	87	88	89	90	91	93	95	97	99	102	105	108
80	73	74	75	76	77	77	78	79	79	80	81	81	82	83	85	86	86	87	88	89	91
75	69	69	70	71	72	72	73	73	74	74	75	75	76	76	77	77	78	78	79	79	80
70	64	64	65	65	66	66	67	67	68	68	69	69	70	70	70	70	71	71	71	71	72

Heat index (or apparent temperature)

(a)

Two examples of effective temperatures are (a) Heat Index and (b) Wind Chill factors.

for **divergent wind circulation.** Figure 5.2 shows converging and diverging winds moving in straight paths. This is not a true picture of reality. In fact, winds moving out of a high and into a low do so in a spiraling motion created by another force we will cover in the chapter section on wind.

Mapping Pressure Distribution

Geographers and meteorologists can best study pressure systems when they are mapped. In mapping air pressure, we reduce all pressures to what they would be at sea level, just as we changed temperature to sea level in order to eliminate altitude as a factor. The adjustment to sea level is especially important for atmospheric pressure because the variations due to altitude are far greater than those due to atmospheric dynamics and would tend to mask the more meteorologically important regional differences.

Isobars (from Greek: *isos,* equal; *baros,* weight) are lines drawn on maps to connect places of equal pressure. When the isobars appear close together, they portray a significant difference in pressure between places, hence a strong **pressure gradient.** When the isobars are far apart, a weak pressure gradient is indicated. When depicted on a map, high and low pressure cells are outlined by concentric isobars that form a closed system around centers of high or low pressure.

Wind

Wind is the horizontal movement of air in response to differences in pressure. Winds are the means by which the atmosphere attempts to balance the uneven distribution of pressure over Earth's surface. The movements of the wind also play a major role in correcting the imbalances in radiational heating and cooling that occur over

Of the many human factors, the following stand out as important. First is respiration; breathing in a lung full of cold air will make one feel colder. Second is perspiration; the evaporative cooling process is quite efficient for the human body, but it differs from one individual to another. Third is the amount of activity involved; physical work or playing a physical sport can heat the body rapidly. Fourth is the amount of exposed skin; tank tops versus sweatshirts can make a world of difference.

Effective temperatures are established by considering the interplay between these two sets of factors. For example, the well-known Heat Index (also called Apparent Temperature) takes into account the temperature and humidity of a summer day and calculates how it might feel. The equally well known Wind Chill Index considers both temperature and wind speed to establish how cold it might feel on a winter day. Keep in mind that these are only theoretical values. No one can predict exactly how comfortable a particular person will feel on a given day. However, these temperature indexes can help guide us when dealing with seasonally extreme days.

Wind Chill Chart

Temperature (°F)

Calm	40	35	30	25	20	15	10	5	0	−5	−10	−15	−20	−25	−30	−35	−40	−45
5	36	31	25	19	13	7	1	-5	-11	-16	-22	-28	-34	-40	-46	-52	-57	-63
10	34	27	21	15	9	3	-4	-10	-16	-22	-28	-35	-41	-47	-53	-59	-66	-72
15	32	25	19	13	6	0	-7	-13	-19	-26	-32	-39	-45	-51	-58	-64	-71	-77
20	30	24	17	11	4	-2	-9	-15	-22	-29	-35	-42	-48	-55	-61	-68	-74	-81
25	29	23	16	9	3	-4	-11	-17	-24	-31	-37	-44	-51	-58	-64	-71	-78	-84
30	28	22	15	8	1	-5	-12	-19	-26	-33	-39	-46	-53	-60	-67	-73	-80	-87
35	28	21	14	7	0	-7	-14	-21	-27	-34	-41	-48	-55	-62	-69	-76	-82	-89
40	27	20	13	6	-1	-8	-15	-22	-29	-36	-43	-50	-57	-64	-71	-78	-84	-91
45	26	19	12	5	-2	-9	-16	-23	-30	-37	-44	-51	-58	-65	-72	-79	-86	-93
50	26	19	12	4	-3	-10	-17	-24	-31	-38	-45	-52	-60	-67	-74	-81	-88	-95
55	25	18	11	4	-3	-11	-18	-25	-32	-39	-46	-54	-61	-68	-75	-82	-89	-97
60	25	17	10	3	-4	-11	-19	-26	-33	-40	-48	-55	-62	-69	-76	-84	-91	-98

Wind (mph)

Frostbite occurs in 15 minutes or less

(b)

Earth's surface. On average, locations below 38° latitude receive more radiant energy than they lose, while locations poleward of 38° lose more than they gain (see again Fig. 4.9). Our global wind system transports energy poleward to help maintain an energy balance. The global wind system also gives rise to the ocean currents, which are another significant factor in equalizing the energy imbalance. Thus, without winds and their associated ocean currents, the equatorial regions would get hotter and hotter and the polar regions colder and colder.

Besides serving a vital function in the advectional (horizontal) transport of heat energy over Earth's surface, winds also transport water vapor from the air above bodies of water, where it has evaporated, to land surfaces, where it condenses and releases latent energy. This allows greater precipitation over land surfaces than could otherwise occur. In addition, winds exert influence on the rate of evaporation itself. Furthermore, as we become more aware and concerned about the effect that the burning of fossil fuels has on our atmosphere, we look for alternate energy sources. Natural sources, such as water, solar energy, and wind, become increasingly attractive alternatives to fossil fuels. They are clean, abundant, and renewable.

Pressure Gradients and Winds

Winds vary widely in speed, intensity, duration, and direction. Much of their strength depends on the size or strength of the pressure gradient to which they are responding. As we noted previously, *pressure gradient* is the term applied to the rate of change of atmospheric pressure between two points (at the same elevation). The greater this change—that is, the steeper the pressure gradient—the greater will be the wind response (Fig. 5.3). Winds tend to flow down a pressure gradient from high

Figure 5.3 The relationship of wind to the pressure gradient: The steeper the pressure gradient, the stronger the resulting wind.

■ **Where else on this figure (other than the area indicated) would winds be strong?**

pressure to low pressure, just as water flows down a slope from a high point to a low one. A useful little rhyme, "Winds always blow, from high to low," will always remind you of the direction of surface winds. The steeper the pressure gradients involved, the faster and stronger the winds. Yet wind does not flow directly from high to low, as we might expect, because other factors also affect the direction of wind.

The Coriolis Effect and Wind

Two factors, both related to our Earth's rotation, greatly influence wind direction. First, our fixed grid system of latitude and longitude is constantly rotating. Thus, our frame of reference for tracking the path of any free-moving object—whether it is an air mass or a missile—is constantly changing its position. Second, the speed of rotation of Earth's surface increases as we move equatorward and decreases as we move toward the poles (see again Fig. 3.12). Thus, to use our previous example, someone in St. Petersburg (60° north latitude), where the distance around a parallel of latitude is about half that at the equator, moves at about 840 kilometers per hour (525 mph) as Earth rotates, while someone in Kampala, Uganda, near the equator, moves at about 1680 kilometers per hour (1050 mph).

Because of these Earth rotation factors, anything moving horizontally appears to be deflected to the right of the direction in which it is traveling in the Northern Hemisphere and to the left in the Southern Hemisphere. This apparent deflection is termed the **Coriolis effect.** The degree of deflection, or curvature, is a function of the speed of the object in motion and the latitudinal location of the object. The higher the latitude, the greater will be the Coriolis effect (Fig. 5.4). In fact, not only

does the Coriolis effect decrease at lower latitudes, but it does not exist at the equator. Also, the faster the object is moving, the greater will be the apparent deflection, and the greater the distance something must travel, the greater will be the Coriolis effect.

As we have said, anything that moves horizontally over Earth's surface exhibits the Coriolis effect. Thus, both the atmosphere and the oceans are deflected in their movements. Winds in the Northern Hemisphere moving across a gradient from high to low pressure are apparently deflected to the right of their expected path (and to the left in the Southern Hemisphere). In addition, when considering winds at Earth's surface, we must take into account another force. This force, **friction,** interacts with the pressure gradient and the Coriolis effect.

Friction and Wind

Above Earth's surface, frictional drag is of little consequence to wind development. At this level, the wind starts down the pressure gradient and turns 90 degrees in response to the Coriolis effect. At this point, the pressure gradient is balanced by the Coriolis effect, and the wind, termed a **geostrophic wind,** flows parallel to the isobars (Fig. 5.5a).

However, at or near Earth's surface (up to about 1000 meters above the surface), frictional drag is important because it reduces the wind speed. A reduced wind speed, in turn, reduces the Coriolis effect, while the pressure gradient is not affected. With the pressure gradient and Coriolis effect no longer in balance, the resultant surface wind does not flow between the isobars like its upper-level counterpart. Instead, a surface wind flows obliquely across the isobars toward the low-pressure area (Fig. 5.5b).

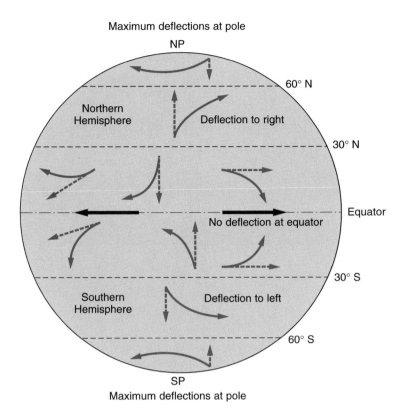

Maximum deflections at pole
NP

60° N

Northern
Hemisphere   Deflection to right

30° N

Equator
No deflection at equator

30° S

Southern
Hemisphere   Deflection to left

60° S

SP
Maximum deflections at pole

Figure 5.4 Schematic illustration of the apparent deflection (Coriolis effect) of an object caused by Earth's rotation when an object (or the wind) moves north, south, east, or west in both hemispheres.

Wind Terminology

Winds are named after their source. Thus, a wind that comes out of the northeast is called a northeast wind. One coming from the south, even though going toward the north, is called a south wind.

Windward refers to the direction from which the wind blows. The side of something that faces the direction from which the wind is coming is called the *wind-*ward side. Thus, a windward slope is the side of a mountain against which the wind blows (Fig. 5.6). **Leeward,** on the other hand, means the direction toward which the wind is blowing. Thus, when the winds are coming out of the west, the *leeward* slope of a mountain would be the east slope. We know that winds can blow from any direction, yet in some places winds may tend to blow more from one direction than any other. We speak of these as the **prevailing winds.**

GEOSTROPHIC WIND

High pressure — Coriolis effect — Pressure gradient — Low pressure

Geostrophic wind

(a)

SURFACE WIND

Friction

High pressure — Pressure gradient — Low pressure

Coriolis effect — Surface wind

(b)

Figure 5.5 These Northern Hemisphere examples illustrate that (a) in a geostrophic wind, as a parcel of air starts to flow down the pressure gradient, the Coriolis effect causes it to veer to the right until the pressure gradient and Coriolis effect reach an equilibrium and the wind flows between isobars; (b) in a surface wind, this equilibrium is upset by friction, which reduces the wind speed. Since the Coriolis effect is a function of wind speed, it also is reduced. With the Coriolis effect reduced, the pressure gradient dominates and the wind now flows across isobars in the direction of low pressure. ■ **If the amount of friction increased, would the surface wind be closer to the pressure gradient, closer to the Coriolis effect, or unchanged?**

Figure 5.6 Illustration of the meaning of windward (facing into the wind) and leeward (facing away from the wind). ■ **How might vegetation differ on the windward and leeward sides of an island?**

Cyclones, Anticyclones, and Winds

Imagine a high pressure cell (anticyclone) in the Northern Hemisphere in which the air is moving from the center in all directions down pressure gradients. As it moves, the air will be deflected to the right, no matter which direction it was originally going. Therefore, the wind moving out of an anticyclone in the Northern Hemisphere will move from the center of high pressure in a clockwise spiral (Fig. 5.7).

Air tends to move down pressure gradients from all directions toward the center of a low-pressure area (cyclone). However, since the air is apparently deflected to the right in the Northern Hemisphere, the winds move into the cyclone in a counterclockwise spiral. Because all objects including air and water are apparently deflected to the left in the Southern Hemisphere, spirals there are reversed. Thus, in the Southern Hemisphere, winds moving away from an anticyclone do so in a counterclockwise spiral and winds moving into a cyclone move in a clockwise spiral.

Subglobal Surface Wind Systems

As we have seen, winds develop whenever differential heating causes differences in pressure. The global wind system is a response to the constant temperature imbalance between tropical and polar regions. On a smaller, or subglobal, scale, additional wind systems develop. We will begin with a discussion of small pressure and wind systems, then move to global systems. Monsoon winds are continental in size and develop in response to the seasonal variations in temperature between large landmasses and adjacent oceans. On the smallest scale are local winds, which develop in response to diurnal (daily) variation in heating.

Local Winds

Later in this chapter, we will discuss the major circulation patterns of Earth's atmosphere, a knowledge of which is vital to understanding the climatic regions of Earth and the fundamental climatic differences between those regions. Yet we are all aware that there are winds that affect weather on a far smaller scale. These local winds are often a response to local landform configurations and add further complexity to the problem of understanding the dynamics of weather.

Land Breeze–Sea Breeze The **land breeze–sea breeze cycle** is a diurnal (daily) one in which the differential heating of land and water again plays a role (Fig. 5.8). During the day, when the land—and consequently the air above it—is heated more quickly and to a higher temperature than the nearby ocean (or large lake or sea), the air above the land expands and rises. This process creates a local area of low pressure, and the rising air tends to be replaced by the denser, cooler air from over the ocean. Thus, a sea breeze of cool, moist air blows in over the land during the day. This sea breeze helps explain why seashores are so popular in summer, because cooling winds help alleviate the heat. At times, however, sea breezes are responsible for afternoon cloud cover and light rain, spoiling an otherwise sunny day at the shore. These winds can mean a 5° to 9°C (9°–16°F) reduction in temperature along the coast, as well as a lesser influence on land perhaps as far from the sea as 15 to 50 kilometers (9–30 mi). During hot summer days, such winds cool cities like Chicago, Milwaukee, and Los Angeles. At night, the land and the air above it cool more quickly and to a cooler temperature than the nearby water body and the air above it. Consequently, the pressure builds up over the land and air flows out toward the lower pressure over the water, creating a land breeze. For thousands of years, sailboats have left their coasts at dawn, when

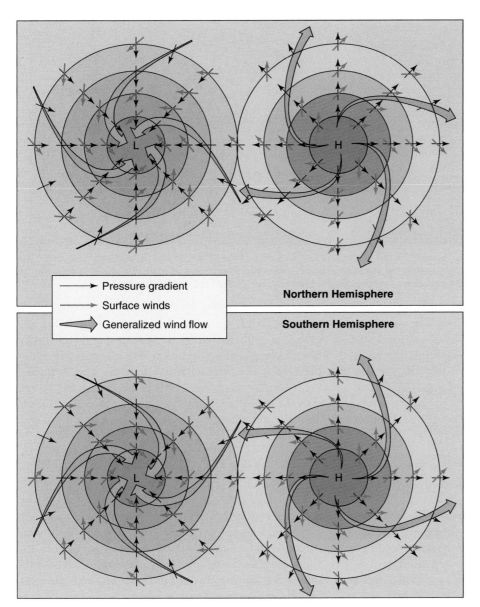

Pressure gradient
Surface winds
Generalized wind flow

Northern Hemisphere

Southern Hemisphere

Figure 5.7 Movement of surface winds associated with low-pressure centers (cyclones) and high-pressure centers (anticyclones) in the Northern and Southern Hemispheres. Note that the surface winds are to the right of the pressure gradient in the Northern Hemisphere and to the left of the pressure gradient in the Southern Hemisphere. ■ **What do you think might happen to the diverging air of an anticyclone if there is a cyclone nearby?**

there is still a land breeze, and have returned with the sea breeze of the late afternoon.

Mountain Breeze–Valley Breeze Under the calming influence of a high pressure system, there is a daily **mountain breeze–valley breeze cycle** (Fig. 5.9) that is somewhat similar in mechanism to the land breeze–sea breeze cycle just discussed. During the day, when the valleys and slopes of mountains are heated by the sun, the air expands and rises up the sides of the mountains. This warm daytime breeze is the valley breeze, named for its place of origin. Clouds, which can often be seen hiding mountain peaks, are actually the visible evidence of condensation in the warm air rising from the valleys. At night, when the valley and slopes are cooled because Earth is giving off more radiation than it is receiving, the air cools and sinks once again into the valley as a cool mountain breeze.

Drainage Winds Also known as **katabatic winds, drainage winds** are local to mountainous regions and can occur only under calm, clear conditions. Cold, dense air will accumulate in a high valley, plateau, or snowfield within a mountainous area. Because the cold air is very dense, it tends to flow downward, escaping through passes and pouring out onto the land below. Drainage winds can be extremely cold and strong, especially when they result from cold air accumulating over ice sheets such as Greenland and Antarctica. These winds are known by many local names; for example, on the Adriatic coast they are called the *bora*, in France the *mistral*, and in Alaska the *Taku*.

Chinook-type Winds A fourth type of local wind is also known by several names in different parts of the world—for example, **Chinook** in the Rocky Mountain area and **foehn** (pronounced "fern") in the Alps. Chinook-type

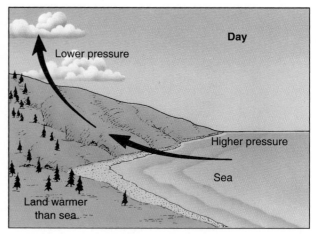

Figure 5.8 Land and sea breezes. This day-to-night reversal of winds is a consequence of the different rates of heating and cooling of land and water areas. The land becomes warmer than the sea during the day and colder than the sea at night; the air flows from the cooler to the warmer area. ■ **What is the impact on daytime coastal temperatures of the land and sea breeze?**

winds occur when air originating elsewhere must pass over a mountain range. As these winds flow down the leeward slope after crossing the mountains, the air is compressed and heated at a greater rate than it was cooled when it ascended the windward slope (Fig. 5.10). Thus, the air enters the valley below as warm, dry winds. The rapid temperature rise brought about by such winds has been known to damage crops, increase forest-fire hazard, and set off avalanches.

An especially dry foehn-type wind is the **Santa Ana** of southern California. It forms when high pressure develops over the interior desert regions of southern California. The clockwise circulation of the high drives the air of the desert southwest over the mountains of eastern California, accentuating the dry conditions as the air moves down the western slopes. The hot, dry Santa Ana winds are notorious for fanning forest and brush fires which plague the southwestern United States, especially in California.

There is no question that winds, both local and global, are effective elements of atmospheric dynamics. We all know that a hot, breezy day is not nearly as unpleasant as a hot day without any wind. This difference exists because winds increase the rate of evaporation and thus the rate of removal of heat from our bodies, the air, animals, and plants. For the same reason, the wind on a cold day increases our discomfort.

Monsoon Winds

The term *monsoon* comes from the Arabic word *mausim*, meaning season. This word has been used by Arab sailors for many centuries to describe seasonal changes in wind direction across the Arabian Sea between Arabia and India. As a meteorological term, **monsoon** refers to the directional shifting of winds from one season to the next. Usually, the monsoon occurs when a humid wind blow-

Figure 5.9 Mountain and valley breezes. This daily reversal of winds results from heating of mountain slopes during the day and their cooling at night. Warm air is drawn up slopes during the day, and cold air drains down the slopes at night. ■ **Do you think fog formation in valleys is more likely during the day or at night? Why?**

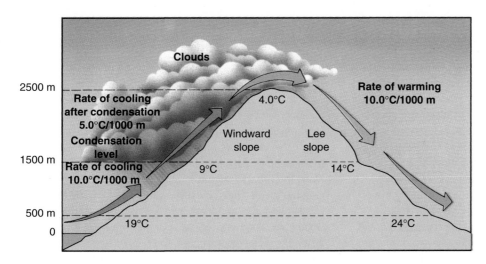

Figure 5.10 Foehn winds result when air ascends a mountain range and undergoes condensation that dries it out and adds latent heat that slows its rate of cooling. As the dry air descends the leeward side of the range, it is compressed and heated at a greater rate than it was initially cooled. This produces the relatively warm, dry conditions with which foehn winds are associated.
■ **The term Chinook, a type of foehn wind, means "snow eater." Can you offer an explanation for how this name came about?**

ing from the ocean toward the land in the summer shifts to a dry, cooler wind blowing seaward off the land in the winter, and it involves a full 180-degree direction change in the wind.

The monsoon is most characteristic of southern Asia, although it occurs on other continents as well. As the large landmass of Asia cools more quickly than the surrounding oceans, the continent develops a strong center of high pressure from which there must be an outflow of air in winter (Fig. 5.11). This outflow blows across much land toward the tropical low before reaching the oceans. It brings cold, dry air south.

In summer the Asian continent heats quickly and develops a large low-pressure center. This development is reinforced by a poleward shift of the warm, moist tropi-

cal air to a position over southern Asia. Warm, moist air from the oceans is attracted into this low. Though full of water vapor, this air does not in itself cause the wet summers with which the monsoon is associated. However, any turbulence or landform barrier that makes this moist air rise and, as a result, cool off will bring about precipitation. This precipitation is particularly noticeable in the foothills of the Himalayas, the western Ghats of India, and the Annamese Highlands of Vietnam. This is the time of year when the rice crop is planted in many parts of Asia.

In the lower latitudes, a monsoonal shift in winds can come about as a reaction to the migration of the direct rays of the sun. For example, the winds of the equatorial zone migrate during the summer months northward

July

January

Figure 5.11 Seasonal changes in surface wind direction that create the Asiatic monsoon system. The "burst" of the "wet monsoon," or the sudden onshore flow of tropical humid air in July, is apparently triggered by changes in the upper air circulation, resulting in heavy precipitation. The offshore flow of dry continental air in winter creates the "dry monsoon" and drought conditions in southern Asia.
■ **How do the seasonal changes of wind direction in Asia differ from those of the southern United States?**

The Santa Ana Winds and Fire

Wildfires require three factors to occur: *oxygen, fuel,* and an *ignition source.* The conditions for all three of these factors vary geographically, so their spatial distributions are not equal everywhere. In locations where all of these three factors have the potential to exist, the danger from wildfires is high. *Oxygen* in the atmosphere is a constant, but winds, which supply more oxygen as a fire consumes it, vary with location, weather, climate, and terrain. High winds cause fires to spread faster and make them difficult to control or extinguish. *Fuel* in wildland fires is usually supplied by dry vegetative litter (leaves, branches, and dry annual grasses), and certain environments have more of this fuel than others. Dense vegetation tends to support the spread of fires. Growing vegetation can also become dessicated—dried out by transpiration losses during a drought or an annual season of dryness. In addition, once a fire becomes large, extreme heat in the areas where it is spreading causes vegetation along the edges of the burning area to lose their moisture through evaporation. *Ignition sources* are the means by which a fire is started.

Lightning and human causes such as campfires provide the main ignition sources for wildfires.

Southern California offers a regional example of how conditions related to these three factors combine with the local physical geography to create an environment that is conducive to wildfire hazard. This is also a region where many people live in forested or chaparral-covered locales or along the urban-wildland fringe—areas that are very susceptible to fire. High pressure, warm weather, and low relative humidity dominate the Mediterranean climate of Southern California's coastal region for much of the year. When these conditions occur, the region experiences high fire potential, because of the warm dry air and the vegetation that has dried out during the arid summer season with long rainless periods.

The most dangerous circumstances for wildfires in Southern California occur when high winds are sweeping the region. When a strong cell of high pressure forms east of Southern California, the clockwise anticyclonic circulation directs winds from the north and east toward the coast. These warm, dry winds (called Santa Ana winds) blow

---

toward the southern coast of Asia, bringing with them warm, moist, turbulent air. The winds of the Southern Hemisphere also migrate north with the sun, some crossing the equator. They also bring warm, moist air (from their travels over the ocean) to the southern and especially the southeastern coasts of India. In the winter months, the equatorial and tropical winds migrate south, leaving southern Asia under the influence of the dry, calm winds of the tropical Northern Hemisphere. Asia and northern Australia are true monsoon areas, with a full 180-degree wind shift with changes from summer to winter. Other regions, like the southern United States and West Africa, have "monsoonal tendencies," but are not monsoons in the true meaning of the term.

The phenomenon of monsoon winds and their characteristic seasonal shifting cannot be fully explained by the differential heating of land and water, however, or by the seasonal shifting of tropical and subtropical wind belts. Some aspects of the monsoon system—for example, its "burst" or sudden transition between dry and wet in southern Asia—must have other causes. Meteorologists looking for a more complete explanation of the monsoon are examining the role played by the jet stream (described later in this chapter) and other wind movements of the upper atmosphere.

Global Pressure Belts

Idealized Global Pressure Belts

Using what we have learned about pressure on Earth's surface, we can construct a theoretical model of the pressure belts of the world (Fig. 5.12). Later, we will see how real conditions depart from our model and examine why these differences occur.

Centered approximately over the equator in our model is a belt of low pressure, or a **trough.** Because this is the region on Earth of greatest annual heating, we can conclude that the low pressure of this area, the **equatorial low (equatorial trough),** is determined primarily by thermal factors which cause the air to rise.

North and south of the equatorial low and centered on the so-called "horse latitudes," about 30°N and 30°S, are cells of relatively high pressure. These are the **subtropical highs,** which are the result of dynamic factors related to the sinking of convectional cells initiated at the equatorial low.

Poleward of the subtropical highs in both the Northern and Southern Hemispheres are large belts of low pressure that extend through the upper middle latitudes. Pressure decreases through these **subpolar lows** until

Geographic setting and wind direction for Santa Ana winds.

down from nearby high desert regions, becoming adiabatically warmer and drier as they descend into the coastal lowlands. The Santa Ana winds are most common in fall and winter, and wind speeds can be 50–90 kilometers per hour with stronger local wind gusts reaching 160 kmh (100 mph). Just like using a bellows, or blowing on a campfire to get it started, the Santa Ana winds produce fire weather that can cause the spread of a wildfire to be extremely rapid after ignition. Most people take great care during these times to avoid or strictly control any activities that could cause a fire to start, but occasionally accidents, acts of arson, or lightning strikes ignite a wildfire. Given the physical geography of the Los Angeles region, when the Santa Ana winds are blowing the fire danger is especially extreme. Ironically, although the Santa Ana winds create dangerous fire conditions, they also provide some benefits to local residents, as they tend to blow air pollutants offshore and out of the urban region. In addition, because they are strong winds flowing opposite to the direction of ocean waves, experienced surfers can enjoy higher than normal waves during those periods when Santa Ana winds are present.

about 65° latitude. Again, dynamic factors play a role in the existence of subpolar lows.

In the polar regions are high-pressure systems called the **polar highs.** These are determined by the cold temperatures and consequent sinking of the dense air in those regions.

This system of pressure belts that we have just developed is a generalized picture. Just as temperatures change from month to month, day to day, and hour to hour, so do pressures vary from time to time at any one place. Our model disguises these changes, but it does give an idea of broad pressure patterns on the surface of Earth.

The Global Pattern of Atmospheric Pressure

As our idealized model suggests, the atmosphere tends to form belts of high and low pressure along east-west axes in areas where there are no large bodies of land. These belts are latitudinally arranged and generally maintain their bandlike pattern. However, where there are continental landmasses, belts of pressure are broken and tend to form cellular pressure systems. The landmasses affect the development of belts of atmospheric pressure in several ways. Most influential is the effect of

the differential heating of land and water surfaces. In addition, landmasses affect the movement of air, and consequently the development of pressure systems, through friction with their surfaces. Landform barriers such as mountain ranges also block the movement of air and thereby affect atmospheric pressure.

Seasonal Variations in the Pattern

In general, the global atmospheric pressure belts shift northward in July and southward in January, following the changing position of the sun's direct rays as they migrate between the Tropics of Cancer and Capricorn. Thus, there are thermally induced seasonal variations in the pressure patterns, as seen in Figures 5.13a and 5.13b. These seasonal variations tend to be small at low latitudes, where there is little temperature variation, and large at high latitudes, where there is an increasing contrast in length of daylight and angle of the sun's rays. Furthermore, landmasses tend to alter the general pattern of seasonal variation. This is an especially important factor in the Northern Hemisphere, where land accounts for 40 percent of the total Earth surface, as opposed to less than 20 percent in the Southern Hemisphere.

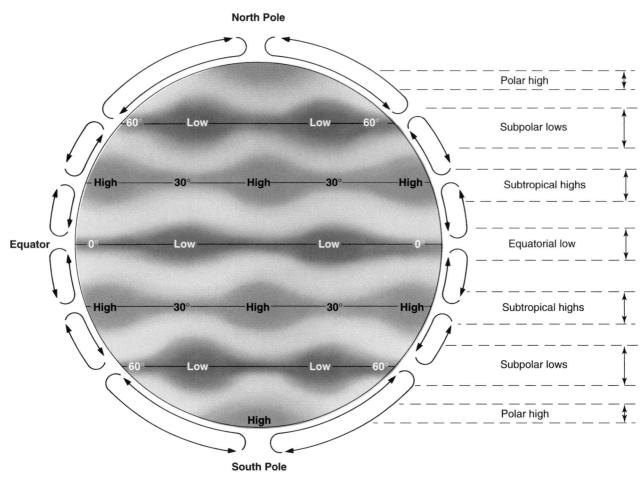

Figure 5.12 Idealized world pressure belts. Note the arrows on the perimeter of the globe that illustrate the cross-sectional flow associated with the surface pressure belts.

January Because land cools more quickly than the oceans, its temperatures will be lower in winter than those of the surrounding seas. Figure 5.13a shows that in the middle latitudes of the Northern Hemisphere, this variation leads to the development of cells of high pressure over the land areas. In contrast, the subpolar lows develop over the oceans because they are comparatively warmer. Over eastern Asia, there is a strongly developed anticyclone during the winter months that is known as the **Siberian High.** Its equivalent in North America, known as the **Canadian High,** is not nearly so well developed, because North America is a smaller land mass than the Eurasian continent.

In addition to the Canadian High and the Siberian High, two low-pressure centers develop: one in the North Atlantic, called the **Icelandic Low,** and the other in the North Pacific, called the **Aleutian Low.** The air in them has relatively lower pressure than either the subtropical or the polar high systems. Consequently, air moves toward these low-pressure areas from both north and south. Such low-pressure regions are associated with cloudy, unstable weather and are a major source of winter storms, whereas

high-pressure areas are associated with clear, blue-sky days, calm, starry nights, and cold, stable weather. Therefore, during the winter months, cloudy, stormy weather tends to be associated with the two oceanic lows and clear weather with the continental highs.

We can also see that the polar high in the Northern Hemisphere is well developed. This development is due primarily to thermal factors, because January is the coldest time of the year. The subpolar lows have developed into the Aleutian and Icelandic cells described earlier. At the same time, the subtropical highs of the Northern Hemisphere appear slightly south of their average annual position because of the migration of the sun toward the Tropic of Capricorn. The equatorial trough also appears centered south of its average annual position over the geographic equator.

In January in the Southern Hemisphere, the subtropical belt of high pressure appears as three cells centered over the oceans because the belt of high pressure has been interrupted by the continental landmasses where temperatures are much higher and pressure tends to be lower than over the oceans. Because there is virtually no

Average sea-level pressure (January)

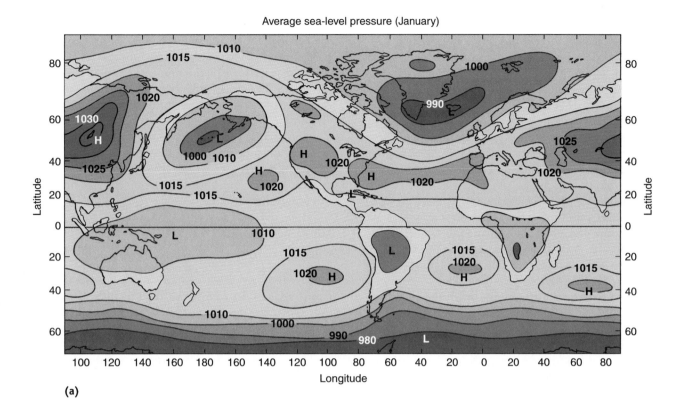

(a)

Average sea-level pressure (July)

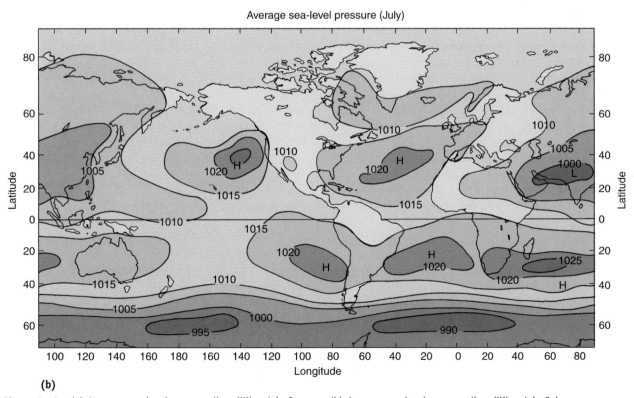

(b)

Figure 5.13   (a) Average sea-level pressure (in millibars) in January. (b) Average sea-level pressure (in millibars) in July.
■ **What is the difference between the January and July average sea-level pressures at your location? Why do they vary?**

land between 45°S and 70°S latitude, the subpolar low circles Earth as a real belt of low pressure and is not divided into cells by interrupting landmasses. There is little seasonal change in this belt of low pressure other than in January (summer in the Southern Hemisphere), when it lies a few degrees north of its July position.

July   The anticyclone over the North Pole is greatly weakened during the summer months in the Northern Hemisphere, primarily because of the lengthy (24 hours) heating of the oceans and landmasses in that region (Fig. 5.13b). The Aleutian and Icelandic Lows nearly disappear from the oceans, while the landmasses, which developed high pressure cells during the cold winter months, have extensive low pressure cells slightly to the south during the summer. In Asia, a low pressure system develops, but it is divided into two separate cells by the Himalayas (see again Fig. 5.11). The low pressure cell over northwest India is so strong that it combines with the equatorial trough, which has moved north of its position six months earlier. The subtropical highs of the Northern Hemisphere are more highly developed over the oceans than over the landmasses. In addition, they migrate northward and are highly influential factors in the climate of landmasses nearby. In the Pacific, this subtropical high is termed the **Pacific High;** this system of pressure plays an important role in moderating the temperatures of the west coast of the United States. In the Atlantic Ocean, the corresponding cell of high pressure is known as the **Bermuda High** to North Americans and as the **Azores High** to Europeans. As we have already mentioned, the equatorial trough of low pressure moves north in July, following the migration of the sun's vertical rays, and the subtropical highs of the Southern Hemisphere lie slightly north of their January locations.

In examining pressure systems at Earth's surface, we have seen that there are essentially seven belts of pressure (two polar highs, two subpolar lows, two subtropical highs, and one tropical low), which are broken into cells of pressure in some places, primarily because of the influence of certain large landmasses. We have also seen that these belts and cells vary in size, intensity, and location with the seasons and with the migration of the sun's vertical rays over Earth's surface. Since these global-scale pressure systems migrate by latitude with the position of the direct sun angle, they are sometimes referred to as *semipermanent pressure systems*, because they are never permanently fixed in the same location.

Global Surface Wind Systems

The planetary or global wind system that is a response to the global pressure patterns also plays a role in the maintenance of those same pressures. This wind system, which is the major means of transport for energy and moisture through Earth's atmosphere, can be examined

in an idealized state. To do so, however, we must ignore the influences of landmasses and seasonal variations in solar energy. By assuming, for the sake of discussion, that Earth has a homogeneous surface and that there are no seasonal variations in the amount of solar energy received at different latitudes, we can examine a theoretical model of the atmosphere's planetary circulation. Such an understanding will help to explain specific features of climate, such as the rain and snow of the Sierra Nevada and Cascade mountains and the existence of arid regions farther to the east. It will also account for the movement of great surface currents in our oceans that are driven by this atmospheric engine.

Idealized Model of Atmospheric Circulation

Since winds are caused by pressure, various types of winds are associated with different kinds of pressure cells. Therefore, a system of global winds can be demonstrated using the model of pressures that we previously developed (see again Fig. 5.12).

The characteristics of convergence and divergence are very important to our understanding of global wind patterns. We know that because of the pressure gradient, surface winds always blow from high pressure cells to low pressure cells. In other words, surface air diverges from zones of high pressure and converges on areas of low pressure.

Knowing that surface winds originate in areas of high pressure, and taking into account the global system of pressure cells, we can develop our model of the wind systems of the world (Fig. 5.14). This model takes into account differential heating, Earth rotation, and atmospheric dynamics. Note that the winds do not blow in a straight north-south line. The variation is due, of course, to the Coriolis effect, which causes an apparent deflection to the right in the Northern Hemisphere and to the left in the Southern Hemisphere.

Our idealized model of global atmospheric circulation includes six wind belts or zones in addition to the seven pressure zones we have previously identified. Two wind belts, one in each hemisphere, are located where winds move out of the polar highs and down the pressure gradients toward the subpolar lows. As these winds are deflected to the right in the Northern Hemisphere and to the left in the Southern, they become the **polar easterlies.**

The remaining four wind belts are closely associated with the divergent winds of the subtropical highs. In each hemisphere, winds flow out of the poleward portions of these highs toward the subpolar lows. Because of their general movement from the west, the winds of the upper middle latitudes are labeled the **westerlies.** The winds blowing from the highs toward the equator have been called the **trade winds.** Because of the Coriolis effect,

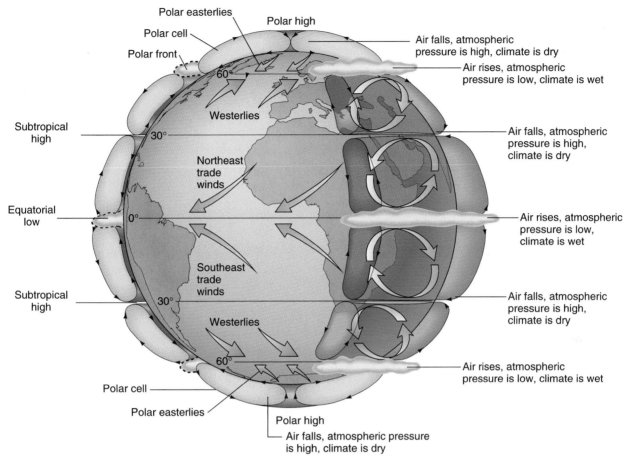

Polar easterlies

Polar high

Polar cell

Polar front

Air falls, atmospheric pressure is high, climate is dry

Air rises, atmospheric pressure is low, climate is wet

60°

Westerlies

Subtropical high

30°

Northeast trade winds

Air falls, atmospheric pressure is high, climate is dry

Equatorial low

0°

Air rises, atmospheric pressure is low, climate is wet

Southeast trade winds

Subtropical high

30°

Air falls, atmospheric pressure is high, climate is dry

Westerlies

60°

Air rises, atmospheric pressure is low, climate is wet

Polar cell

Polar easterlies

Polar high

Air falls, atmospheric pressure is high, climate is dry

Figure 5.14 The general circulation of Earth's atmosphere.

they are the **northeast trades** in the Northern Hemisphere and the **southeast trades** south of the equator.

Our model does not conform exactly to actual conditions. First, as we know, the vertical rays of the sun do not stay precisely over the equator but move north and south during the year as far as the Tropics of Cancer and Capricorn. Therefore, the pressure systems, and consequently the winds, must move in order to adjust to the change in the position of the sun. Then, as we have already discovered, the existence of the continents, especially in the Northern Hemisphere, causes longitudinal pressure differentials that affect the zones of high and low pressure.

Conditions within Latitudinal Zones

Trade Winds A good place to begin our examination of winds and associated weather patterns as they actually occur is in the vicinity of the subtropical highs. On Earth's surface, the trade winds, which blow out of the subtropical highs toward the equatorial trough in both the Northern and Southern Hemispheres, can be identified between latitudes 5° and 25°. Because of the Coriolis effect, the northern trades move away from the subtropical high in a clockwise direction out of the northeast. In the Southern Hemisphere, the trades diverge out of the subtropical high toward the tropical low from the southeast, as their movement is counterclockwise. Because the trades tend to blow out of the east, they are also known as the **tropical easterlies.**

The trade winds tend to be constant, steady winds, consistent in their direction. This is most true when they cross the eastern sides of the oceans (near the eastern portion of the subtropical high). The area of the trades varies somewhat during the solar year, moving north and south a few degrees of latitude with the sun. Near their source in the subtropical highs, the weather of the trades is clear and dry, but after crossing large expanses of ocean, the trades have a high potential for stormy weather.

Early Spanish sea captains depended on the northeast trade winds to drive their galleons to destinations in Central and South America in search of gold, spices, and new lands. Going eastward toward home, navigators usually tried to plot a course using the westerlies to the north. The trade winds are one of the reasons that the Hawaiian Islands are so popular with tourists; the steady winds help to keep temperatures pleasant, even though Hawaii is located south of the Tropic of Cancer.

Sandra Diaz

Courtesy Sandra Diaz

Bachelor of Science, Physical and Applied Geography, Southwest Texas State University, San Marcos, Texas, 1977; Distinguished Alumni Award, 2002

I am a broadcast meteorologist at the Weather Channel. My job involves understanding and explaining weather patterns for the world. A strong foundation in physical geography is essential because we cover atmospheric topics, topography, wind patterns, seasons, and oceanic issues.

Why did you decide to major in geography?

I found every class interesting and fun. I loved the combination of science, nature and math. I have always enjoyed exploring the world around me.

Coursework in physical geography provided a strong foundation for a career in meteorology. I was then able to build on that foundation with specialized meteorological courses. Almost everything I studied comes into play during my daily work now.

Location is very important for forecasting weather patterns. Knowledge of an area's geography can help me better understand how dangerous weather might impact that community—that is how we save lives.

When I was learning the Köppen Climate Classification System, I was not sure if I would ever really need to know this extensive information in the future. Nonetheless, my instructor wanted us to know this forward and backward. At the time, I thought this was a tedious exercise, but now I am very happy he did because it has been very useful in understanding the climate and vegetation of places around the world.

To prepare for a career like mine, take as many geography courses related to meteorology as possible. Be sure to graduate with a BS, not a BA. Take a public speaking course to help you get over any fear of speaking to a group of people. Lastly, I would highly recommend doing an internship with a local television station in order to get a better understanding about what a broadcast meteorologist does.

Why do you think that physical geography is an important course for those who are majoring in other fields?

We are all affected by the natural world around us. The more we understand our environment, the better we can face challenges in nature.

In my spare time, I enjoy rock collecting, star gazing, outdoor activities (such as hiking the Franklin Mountains in West Texas), church, and golf.

Doldrums Where the trade winds converge in the equatorial trough (or tropical low) lies a zone of calm and weak winds of no prevailing direction. Here the air, which is very moist and heated by the sun, tends to expand and rise, maintaining the low pressure of the area. This zone, which is roughly between 5°N and 5°S, is generally known as the **doldrums.** It is also called the **intertropical convergence zone (ITCZ),** or the "equatorial belt of variable winds and calms." Because of the converging moist air and high potential for rainfall in the doldrums, this region coincides with the world's latitudinal belt of heaviest precipitation and most persistent cloud cover.

Old sailing ships often remained becalmed in the doldrums for days at a time. A description of a ship becalmed in the doldrums appears in *The Rime of the Ancient Mariner* by Samuel Taylor Coleridge (lines 103–118). The ship is sailing northward from the tropical southeasterlies (trades) when it gets to the doldrums.

The fair breeze blew, the white foam flew,
The furrow followed free;
We were the first that ever burst
Into that silent sea.

Down dropt the breeze, the sails dropt down,
'Twas sad as sad could be;
And we did speak only to break
The silence of the sea!

All in a hot and copper sky,
The bloody Sun, at noon,
Right up above the mast did stand,
No bigger than the Moon.

Day after day, day after day,
We stuck, nor breath nor motion;
As idle as a painted ship
Upon a painted ocean.

This is obviously not a happy poem! It is interesting to note that the word *doldrums* in the English language means a bored or depressed state of mind.

Subtropical Highs   The areas of subtropical high pressure, generally located between latitudes 25° and 35°N and S, and from which winds blow equatorward as the trades, are often called the subtropical belts of variable winds, or the "horse latitudes." This name comes from the occasional need by the Spanish conquistadors to eat their horses or throw them overboard in order to conserve drinking water and lighten the weight when their ships were becalmed in these latitudes. The subtropical highs are areas, like the doldrums, in which there are no strong prevailing winds. However, unlike the doldrums, which are characterized by convergence, rising air, and heavy rainfall, the subtropical highs are areas of sinking and settling air from higher altitudes, which tend to build up the atmospheric pressure. Weather conditions are typically clear, sunny, and rainless, especially over the eastern portions of the oceans where the high-pressure cells are strongest.

Westerlies   The winds that flow poleward out of the subtropical high pressure cells in the Northern Hemisphere are deflected to the right and thus blow from the southwest. Those in the Southern Hemisphere are deflected to the left and blow out of the northwest. Thus, these winds have been correctly labeled the westerlies. They tend to be less consistent in direction than the trades, but they are usually stronger winds and may be associated with stormy weather. The westerlies occur between about 35° and 65°N and S latitudes. In the Southern Hemisphere, where there is less land than in the Northern Hemisphere to affect the development of winds, the westerlies attain their greatest consistency and strength. Most of the United States, except Florida, Hawaii, and Alaska, is under the influence of the westerlies.

Polar Winds   Accurate observations of pressure and wind are sparse in the two polar regions; therefore, we must rely on remotely sensed information. Our best estimate is that pressures are consistently high throughout the year at the poles and that prevailing easterly winds blow from the polar regions to the subpolar low-pressure systems.

Polar Front   Despite our limited knowledge of the wind systems of the polar regions, we do know that the winds can be highly variable, blowing at times with great speed and intensity. When the cold air flowing out of the polar regions and the warmer air moving in the path of the westerlies meet, they do so like two warring armies: One does not absorb the other. Instead, the denser, heavier cold air pushes the warm air upward, forcing it to rise rapidly. The line along which these two great wind systems battle is appropriately known as the **polar front.** The weather that results from the meeting of the cold polar air and the warmer air from the subtropics can be very stormy. In fact, most of the storms that move slowly through the middle latitudes in the path of the prevailing westerlies are born at the polar front.

The Effects of Seasonal Migration

Just as insolation, temperature, and pressure systems migrate north and south as Earth revolves around the sun, Earth's wind systems also migrate with the seasons. During the summer months in the Northern Hemisphere, maximum insolation is received north of the equator. This condition causes the pressure belts to move north as well, and the wind belts of both hemispheres shift accordingly. Six months later, when maximum heating is taking place south of the equator, the various wind systems have migrated south in response to the migration of the pressure systems. Thus, seasonal variation in wind and pressure conditions is one important way in which actual atmospheric circulation differs from our idealized model.

The seasonal migration will most affect those regions near the boundary zone between two wind or pressure systems. During the winter months, such a region will be subject to the impact of one system. Then, as summer approaches, that system will migrate poleward and the next equatorward system will move in to influence the region. Two such zones in each hemisphere have a major effect on climate. The first lies between latitudes 5° and 15°, where the wet equatorial low of the high-sun season (summer) alternates with the dry subtropical high and trade winds of the low-sun season (winter). The second occurs between 30° and 40°, where the subtropical high dominates in summer but is replaced by the wetter westerlies in winter.

California is an example of a region located within a zone of transition between two wind or pressure systems (Fig. 5.15). During the winter, this region is under the influence of the westerlies blowing out of the Pacific High. These winds, turbulent and full of moisture from the ocean, bring winter rains and storms to "sunny" California. As summer approaches, however, the subtropical high, and its associated westerlies, moves north. As California comes under the influence of the calm and steady high pressure system, it experiences again the climate for which it is famous: day after day of warm, clear, blue, cloudless skies. This alternation of moist winters and dry summers is typical of the western sides of all landmasses between 30° and 40° latitude.

Longitudinal Differences in Winds

We have seen that there are sizable latitudinal differences in pressure and winds. In addition, there are significant longitudinal variations, especially in the zone of the subtropical highs.

Harnessing the Wind

For centuries, windmills provided the power to pump water and grind grain in rural areas throughout the world. But the widespread availability of inexpensive electricity changed the role of most windmills to that of a nostalgic tourist attraction. Should we then conclude that energy from the wind is only a footnote in the history of power? In no way is that a reasonable assumption if mounting needs for electricity and increasing problems from atmospheric pollution associated with fossil fuels are taken into consideration.

Wind power is an inexhaustible source of clean energy. Although the cost of electrical energy produced by the wind depends on favorable sites for the location of wind turbines, wind power is already cost competitive with power produced from fossil fuels. One expert calls wind generation the fastest-growing electricity producing technology in the world. During the last decade, power production from the wind increased more than 25 percent. Much of the growth was in Europe, where most of the world's 17,000 megawatts of wind power is generated. As examples, 13 percent of

Denmark's power and more than 20 percent of the power in the Netherlands, Spain, and Germany is supplied by the wind.

Two criteria are more important than others in the location of wind turbines. The site must have persistent strong winds, and it must be in an already developed region so that the power from the turbines can be linked directly to an existing electrical grid system. Although individual wind turbines (such as those located on farms scattered throughout the Midwest and Great Plains of the United States) can be found producing electricity, most wind power is generated from wind farms. These are long rows, or more concentrated groups, of as many as 50 or more turbines. Each turbine can economically extract up to 60 percent of the wind's energy at minimum wind speeds of 20 kilometers (12 mi) per hour, although higher wind speeds are desirable. Since the power generated is proportional to the cube of the wind speed, a doubling of wind velocity increases energy production eight times.

Although North America currently lags far behind Europe in the production of energy from the wind, the continent has great potential. Excellent sites for the location of wind farms exist throughout the open plains of North America's interior

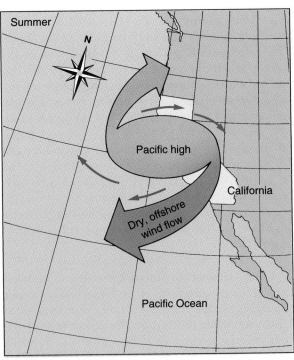

Figure 5.15 Winter and summer positions of the Pacific anticyclone in relation to California. In the winter, the anticyclone lies well to the south and feeds the westerlies that bring the cyclonic storms and rain from the North Pacific to California. The influence of the anticyclone dominates during the summer. The high pressure blocks out cyclonic storms and produces warm, sunny, and dry conditions.

■ **In what ways would the seasonal migration of the Pacific anticyclone affect agriculture in California?**

and along its coasts from the Maritime Provinces of Canada to Texas and from California to the Pacific Northwest. In addition the newest wind power technology places wind farms out of sight and sound in offshore locations that avoid navigation routes and marine-life sanctuaries. And North Amer-

ica has some of the longest coastlines in the world with major adjacent power needs. The sites are available, the technology is developed, the costs are competitive, and the resolve to shift from fossil fuels is growing. Is it not time for power from the wind to come of age in North America?

Centuries-old wind power still in use in the Dutch Polders.

Wind farm close to the North Sea in the Weiringermeer Polder of the Netherlands. Why is a coastal location an advantage when choosing a site for a wind farm?

As was previously noted, the subtropical high pressure cells, which are generally centered over the oceans, are much stronger on their eastern sides than on their western sides. Thus, over the eastern portions of the oceans (west coasts of the continents) in the subtropics, subsidence and divergence are especially noticeable. The above-surface temperature inversions so typical of anticyclonic circulation are close to the surface, and the air is calm and clear. The air moving equatorward from this portion of the high produces the classic picture of the steady trade winds with clear, dry weather.

Over the western portions of the oceans (eastern sides of the continents), conditions are markedly different. In its passage over the ocean, the diverging air is gradually warmed and moistened; turbulent and stormy weather conditions are likely to develop. As indicated in Figure 5.16, wind movement in the western portions of the anticyclones may actually be poleward and directed toward landmasses. Hence, the trade winds in these areas are especially weak or nonexistent much of the year.

As we have pointed out in discussing Figures 5.13 and 5.14, there are great land-sea contrasts in temperature and pressure throughout the year farther toward the poles, especially in the Northern Hemisphere. In the cold continental winters, the land is associated with pressures that are higher than those over the oceans, and thus

there are strong, cold winds from the land to the sea. In the summer, the situation changes, with relatively low pressure existing over the continents because of higher temperatures. Wind directions are thus greatly affected, and the pattern is reversed so that winds flow from the sea toward the land. In North America, this is sometimes referred to as the "North American monsoon."

Upper Air Winds

Thus far we have closely examined the wind patterns near Earth's surface. Of equal, or perhaps even greater, importance is the flow of air above Earth's surface—in particular, the flow of air at altitudes above 5000 meters (16,500 ft), in the upper troposphere. The formation, movement, and decay of surface cyclones and anticyclones in the middle latitudes depend to a great extent on the flow of air high above Earth's surface.

The circulation of the upper air winds is a far less complex phenomenon than surface wind circulation. In the upper troposphere, an average westerly flow, the *upper air westerlies,* is maintained poleward of about 15° to 20° latitude in both hemispheres. Because of the reduced frictional drag, the upper air westerlies move much more rapidly than their surface counterparts. Between 15° and 20°N and S latitudes are the *upper air*

Figure 5.16 Circulation pattern in a Northern Hemisphere subtropical anticyclone. Subsidence of air is strongest in the eastern part of the anticyclone, producing calm air and arid conditions over adjacent land areas. The southern margin of the anticyclone feeds the persistent northeast trade winds. ■ **What wind system is fed by the northern margin?**

easterlies, which can be considered the upper air extension of the trade winds. The flow of the upper air winds became very apparent during World War II, when high-altitude bombers moving eastward were found to cover similar distances faster than those flying westward. Pilots had encountered the upper air westerlies, or perhaps even the **jet streams**—very strong air currents embedded within the upper air westerlies.

The upper air westerlies form as a response to the temperature difference between warm tropical air and cold polar air. The air in the equatorial latitudes is warmed, rises convectively to high altitudes, and then flows toward the polar regions. At first this seems to contradict our previous statement, relative to surface winds, that air flows from cold areas (high pressure) toward warm areas (low pressure). This apparent discrepancy disappears, however, if you recall that the pressure gradient, down which the flow takes place, must be assessed between two points *at the same elevation.* A column of

cold air will exert a higher pressure at Earth's surface than a column of warm air. Consequently, the pressure gradient established at Earth's surface will result in a flow from the cooler air toward the warmer air. However, cold air is denser and more compact than warm air. Thus pressure decreases with height more rapidly in cold air than in warm air. As a result, at a specific height above Earth's surface, a lower pressure will be encountered above cold surface air than above warm surface air. This will result in a flow (pressure gradient) from the warmer surface air toward the colder surface air at that height. Figure 5.17 illustrates this concept.

Returning to our real-world situation, as the upper air winds flow from the equator toward the poles (down the pressure gradient), they are turned eastward because of the Coriolis effect. The net result is a broad circumpolar flow of westerly winds throughout most of the upper atmosphere (Fig. 5.18). Since the upper air westerlies form in response to the thermal gradient between

Figure 5.17 Variation of pressure surfaces with height. Note that the horizontal pressure gradient is from cold to warm air at the surface and in the opposite direction at higher elevations (such as 400 meters). ■ **In what direction would the winds flow at 300 meters?**

Figure 5.18 The upper-air westerlies form a broad circumpolar flow throughout most of the upper atmosphere.

tropical and polar areas, it is not surprising that they are strongest in winter (the low-sun season), when the thermal contrast is greatest. On the other hand, during the summer (the high-sun season), when the contrast in temperature over the hemisphere is much reduced, the upper air westerlies move more slowly.

The temperature gradient between tropical and polar air, especially in winter, is not uniform but rather is concentrated where the warm tropical air meets cold polar air. This boundary, called the polar front, with its stronger pressure gradient, marks the location of the **polar front jet stream.** Ranging from 40 to 160 kilometers (25–100 mi) in width and up to 2 or 3 kilometers (1–2 mi) in depth, the polar front jet stream can be thought of as a faster, internal ribbon of air within the upper air westerlies. While the polar front jet stream flows over the middle latitudes, another westerly **subtropical jet stream** flows above the sinking air of the subtropical highs in the lower

middle latitudes. Like the upper air westerlies, both jets are best developed in winter, when hemispherical temperatures exhibit their steepest gradient (Fig 5.19). During the summer, both jets weaken in intensity. The subtropical jet stream frequently disappears completely, and the polar front jet tends to migrate northward.

We can now go one step further and combine our knowledge of the circulation of the upper air and surface to yield a more realistic portrayal of the vertical circulation pattern of our atmosphere (Fig. 5.20). In general, the upper air westerlies and the associated polar jet stream flow in a fairly smooth pattern (Fig. 5.21a). At times, however, the upper air westerlies develop oscillations, termed *long* waves or **Rossby waves,** after the Swedish meteorologist Carl Rossby who proposed and proved their existence (Fig. 5.21b). Rossby waves result in cold polar air pushing into the lower latitudes and forming *troughs* of low pressure, while warm tropical air moves into higher latitudes, forming *ridges* of high pressure. It is when the upper air circulation is in this configuration that surface weather is most influenced. We will examine this influence in more detail in Chapter 7.

Eventually, the upper air oscillations become so extreme that the "tongues" of displaced air are cut off, forming upper air cells of warm and cold air (Fig. 5.21c). This process helps to maintain a net poleward flow of energy from equatorial and tropical areas. The cells eventually dissipate, and the pattern returns to normal (Fig. 5.21a). The complete cycle takes four to eight weeks. Although it is not completely clear why the upper atmosphere goes

Figure 5.19 Approximate location of the subtropical jet stream and area of activity of the polar front jet stream (shaded) in the Northern Hemisphere winter.

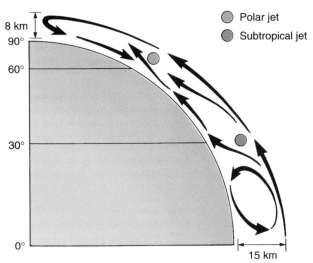

Figure 5.20 A more realistic schematic cross section of the average circulation in the atmosphere.

into these oscillating patterns, we are currently gaining additional insights. One possible cause is variation in ocean surface temperatures. If the oceans in, say, the northern Pacific or near the equator become unusually cold or warm (e.g., El Niño or La Niña), this apparently triggers oscillations, which continue until the ocean surface temperature returns to normal. Other causes are also likely.

In addition to this influence on weather, jet streams are important to study for other reasons. They can carry pollutants, such as radioactive wastes or volcanic dust, over great distances and at relatively rapid rates. It was the polar jet stream that carried ash from the Mount St. Helens eruption eastward into Idaho and Montana. Nuclear fallout from the Chernobyl incident in the former Soviet Union could be monitored in succeeding days as it crossed the Pacific, and later the United States, in the jet streams. Pilots flying eastward—for example, from North America to Europe—take advantage of the

(a)

(b)

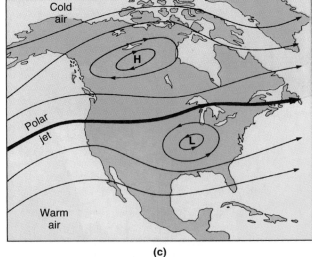

(c)

Figure 5.21 Development and dissipation of Rossby waves in the upper air westerlies. (a) A fairly smooth flow prevails. (b) Rossby waves form, with a ridge of warm air extending into Canada and a trough of cold air extending down to Texas. (c) The trough and ridge are cut off and will soon dissipate. The flow will then return to a pattern similar to (a). ■ **How are Rossby waves closely associated with the changeable weather of the central and eastern United States?**

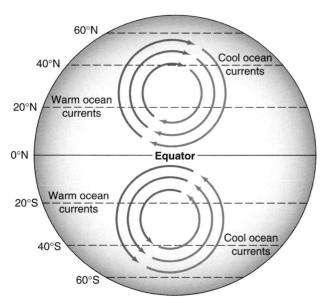

Figure 5.22 The major ocean currents flow in broad gyres in opposite directions in the Northern and Southern Hemispheres.

jet stream, so the flying times in this direction may be significantly shorter than those in the reverse direction.

Ocean Currents

Like the planetary wind system, surface ocean currents play a significant role in helping to equalize the energy imbalance between the tropical and polar regions. In addition, surface ocean currents greatly influence the climate of coastal locations.

Earth's surface wind system is the primary control of the major surface currents and drifts. Other controls are the Coriolis effect and the size, shape, and depth of the sea or ocean basin. Other currents may be caused by differences in density due to variations in temperature and salinity, tides, and wave action.

The major surface currents move in broad circulatory patterns, called **gyres,** around the subtropical highs. Because of the Coriolis effect, the gyres flow clockwise in the Northern Hemisphere and counterclockwise in the Southern Hemisphere (Fig. 5.22). As a general rule, the surface currents do not cross the equator.

Waters near the equator in both hemispheres are driven west by the tropical easterlies or the trade winds. The current thus produced is called the Equatorial Current. At the western margin of the ocean, its warm tropical waters are deflected poleward along the coastline. As these warm waters move into higher latitudes, they move through waters cooler than themselves and are identified as warm currents (Fig. 5.23).

In the Northern Hemisphere, warm currents, such as the Gulf Stream and the Kuroshio Current, are deflected more and more to the right (or east) because of the Coriolis effect. At about 40°N, the westerlies begin to drive these warm waters eastward across the ocean, as in the North Atlantic Drift and the North Pacific Drift. Eventually, these currents run into the land at the eastern margin of the ocean, and most of the waters are deflected toward the equator. By this time, these waters have lost much of their warmth, and as they move equatorward into the subtropical latitudes, they are cooler than the adjacent waters. They have become cool currents. These waters complete the circulation pattern when they rejoin the westward-moving Equatorial Current.

On the eastern side of the North Atlantic, the North Atlantic Drift moves into the seas north of the British Isles and around Scandinavia, keeping those areas warmer than their latitudes would suggest. Some Norwegian ports north of the Arctic Circle remain ice-free because of this warm water. Cold polar water—the Labrador and Oyashio Currents—flows southward into the Atlantic and Pacific oceans along their western margins.

The circulation in the Southern Hemisphere is comparable to that in the Northern except that it is counterclockwise. Also, because there is little land poleward of 40°S, the West Wind Drift (or Antarctic Circumpolar Drift) circles Earth as a cool current across all three major oceans almost without interruption. It is cooled by the influence of the Antarctic ice sheet (Table 5.1).

In general, then, warm currents move poleward as they carry tropical waters into the cooler waters of higher latitudes, as in the case of the Gulf Stream or the Brazil Current. Cool currents deflect water equatorward, as in the California Current and the Humboldt Current. Warm currents tend to have a humidifying and warming effect on the east coasts of continents along which they flow, while cool currents tend to have a drying and cooling effect on the west coasts of the land masses. The contact between the atmosphere and ocean currents is one reason why subtropical highs have a strong side and a weak side. Subtropical highs on the west coast of continents are in contact with cold ocean currents, which cools the air and makes the eastern side of a subtropical high more stable and stronger. On the east coasts of continents, contact with warm ocean currents cause the western side of subtropical highs to be less stable and weaker.

The general circulation just described is consistent throughout the year, although the position of the currents follows seasonal shifts in atmospheric circulation. In addition, in the North Indian Ocean, the direction of circulation reverses seasonally according to the monsoon winds.

The cold currents along west coasts in subtropical latitudes are frequently reinforced by **upwelling.** As the trade winds in these latitudes drive the surface waters offshore, the wind's frictional drag on the ocean surface displaces the water to the west. As surface waters are dragged away, deeper, colder water rises to the surface to replace them. This upwelling of cold waters adds to the strength and effect of the California, Humboldt (Peru), Canary, and Benguela Currents.

Warm water current
Cool water current

Figure 5.23 Map of the major world ocean currents, showing warm and cool currents. ■ **How does this map of ocean currents help explain the mild winters in London, England?**

 El Niño

As you can see in Figure 5.23, the cold Humboldt Current flows equatorward along the coasts of Ecuador and Peru. When the current approaches the equator, the westward-flowing trade winds cause upwelling of nutrient-rich cold water along the coast. Fishing, especially for anchovies, is a major local industry.

Every year during the months of December and January, a weak warm countercurrent replaces the normally cold coastal waters. Without the upwelling of nutrients

Table 5.1 Primary Ocean Currents and Temperature Characteristics					
Pacific Ocean		**Atlantic Ocean**		**Indian Ocean**	
Oyashio	Cool	East Greenland	Cool	North Indian monsoon	
Bering	Cool	Labrador	Cool	currents (reverse seasonally	
North Pacific Drift	Warm	North Atlantic Drift	Warm	with the monsoon winds)	
Kuroshio (Japan)	Warm	Gulf Stream	Warm	North Indian Equatorial	Warm
Alaska	Warm	Canary	Cool	Indian Equatorial	
California	Cool	Guinea	Warm	Counter Current	Warm
North Pacific Equatorial	Warm	North Atlantic		South Indian Equatorial	Warm
Pacific Equatorial		Equatorial	Warm	Agulhas (Mozambique)	Warm
Counter Current	Warm	North Atlantic		West Australian	Cool
South Pacific Equatorial	Warm	Equatorial Counter			
Humboldt (Peru)	Cool	Current	Warm		
East Australian	Warm	South Atlantic			
West Wind Drift	Cool	Equatorial	Warm		
(Antarctic Circumpolar		Brazil	Warm		
Drift; also present in		Benguela	Cool		
South Atlantic and		Falkland	Cool		
South Indian Oceans)					

25 MAR 97

25 APR 97

25 MAY 97

25 JUN 97

Courtesy NASA/JPL/Caltech

Figure 5.24 Enhanced satellite images show the development of an El Niño episode. Top left: Patches of cold water (blue color) are seen in the Tropical Pacific Ocean. This situation more closely resembles the normal situation. Bottom right: A strip of abnormally warm water (reds and whites) stretch across the Tropical Pacific from the West Coast of South America.

from below to feed the fish, fishing comes to a standstill. Fishermen in this region have known of the phenomenon for hundreds of years. In fact, this is the time of year they traditionally set aside to tend to their equipment and await the return of cold water. The residents of the region have given this phenomenon the name **El Niño,** which is Spanish for "The Child," because it occurs about the time of the celebration of the birth of the Christ Child.

The warm-water current usually lasts for two months or less, but there are occasions when the disruption to the normal flow lasts for many months. In these situations, water temperatures are raised not just along the coast, but for thousands of kilometers offshore (Fig. 5.24). Over the past decade, the term *El Niño* has come to describe these exceptionally strong episodes and not the annual event. During the past 50 years, ten El Niños

Figure 5.25 During El Niño, the easterly surface winds weaken and retreat to the eastern Pacific, allowing the central Pacific to warm and the rain area to migrate eastward.

have been observed. Not only do the El Niños affect the temperature of the equatorial Pacific, but the strongest of them impact worldwide weather.

El Niño and the Southern Oscillation

To completely understand the processes that interact to produce an El Niño requires that we study conditions all across the Pacific, not just in the waters off South America. More than 50 years ago, Sir Gilbert Walker, a British scientist, discovered a connection between surface pressure readings at weather stations on the eastern and western sides of the Pacific. He noted that a rise in pressure in the eastern Pacific is usually accompanied by a fall in pressure in the western Pacific, and vice versa. He called this seesaw pattern the **Southern Oscillation.** The link between El Niño and the Southern Oscillation is so great that they are often referred to jointly as ENSO (*El Niño/Southern Oscillation*).

During a typical year, the eastern Pacific has a higher pressure than the western Pacific. This east-to-west pressure gradient enhances the trade winds over the equatorial Pacific waters. This results in a warm surface current that moves from east to west at the equator. The western Pacific develops a thick, warm layer of water while the eastern Pacific has the cold Humboldt Current enhanced by upwelling.

Then, for unknown reasons, the Southern Oscillation swings in the opposite direction, dramatically changing the usual conditions described above, with pressure increasing in the western Pacific and decreasing in the eastern Pacific. This change in the pressure gradient causes the trade winds to weaken or, in some cases, to reverse. This causes the warm water in the western Pacific to flow eastward, increasing sea surface temperatures in the central and eastern Pacific. This eastward shift signals the beginning of El Niño.

In contrast, at times and for reasons we do not fully know, the trade winds will intensify. These more powerful trade winds will cause even stronger upwelling than usual to occur. As a result, sea surface temperatures will be even colder than normal. This condition is known as **La Niña** (in Spanish, "Little Girl," but scientifically, simply the opposite of El Niño). La Niña episodes will at times, but not always, bring about the opposite effects of an El Niño episode.

El Niño and Global Weather

Cold ocean waters impede cloud formation. Thus, under normal conditions, clouds tend to develop over the warm waters of the western Pacific but not over the cold waters of the eastern Pacific. However, during an El Niño, when warm water migrates eastward, clouds develop over the entire equatorial water of the Pacific (Fig. 5.25). These clouds can build to heights of 18,000 meters (59,000 ft). Clouds of this magnitude can disrupt the high-altitude wind flow above the equator. As we have seen, a change in the upper air wind flow in one portion of the atmosphere will trigger wind flow changes in other portions of the atmosphere. Alterations in the upper air winds result in alterations to surface weather.

Scientists have tried to document as many past El Niño events as possible by piecing together bits of historical evidence, such as sea surface temperature records, daily observations of atmospheric pressure and rainfall, fisheries' records from South America, and the writings of Spanish colonists living along the coasts of Peru and Ecuador dating back to the fifteenth century. Additional evidence comes from the growth patterns of coral and trees in the region.

Based on this historical evidence, we know that El Niños have occurred as far back as records go. One disturbing fact is that they are occurring more often. Records indicate that during the sixteenth century, an El Niño occurred, on average, every six years. Evidence gathered over the past few decades indicates El Niños are now occurring, on average, every 2.2 years. Even more alarming is the fact that they appear to be getting stronger. The record-setting El Niño of 1982–83 was recently surpassed by the one in 1997–98.

The 1997–98 El Niño brought copious and damaging rainfall to the southern United States, from California to Florida. Snowstorms in the northeast portion of the United States were more frequent and stronger than in most years. The warm El Niño winters fueled Hurricane Linda, which devastated the western coast of Mexico. Linda was the strongest hurricane ever recorded in the eastern Pacific.

In recent years, scientists have become better able to monitor and forecast El Niño and La Niña events. An elaborate network of ocean-anchored weather buoys plus satellite observations provide an enormous amount of data that can be analyzed by computer to help predict the formation and strength of El Niño and La Niña

events. Our improved observation skills have led to the discovery of the **North Atlantic Oscillation**—a strengthening of the subtropical (Azores) high and subpolar (Icelandic) low over the Atlantic Ocean. The North Atlantic Oscillation (NAO) is not as well understood as ENSO. One phase of the NAO has been associated with increased precipitation and temperatures in the southeastern United States, while cold, dry conditions occur over the northeastern United States. ENSO and the NAO both require more research in the future if scientists are to better understand how these ocean phenomena affect weather and climate.

Will scientists ever be able to predict the occurrence of such phenomena as ENSO or the NAO? No one can answer that question, but as our technology improves, our forecasting ability will also increase. We have made tremendous progress: In the past few decades, we have come to recognize the close association between the atmosphere and hydrosphere as well as to better understand the complex relationship between these Earth systems.

This chapter began with an examination of the behavior of atmospheric gases as they respond to solar radiation and other dynamic forces. This information enabled a definition and thorough discussion of global pressure systems and their accompanying winds. This discussion, in turn, permitted a description of atmospheric circulation patterns that began with the smallest and simplest (diurnal) systems and progressed to seasonal systems and then systems at a global scale. Once again, we can recognize the interactions among Earth's systems. Earth's radiation budget helps create movements in our atmosphere, which in turn help drive ocean circulation, which in turn creates feedback with the atmosphere:

Solar radiation→Atmosphere→Hydrosphere→Back to atmosphere

In following chapters, we will examine the role of the atmosphere in controlling variations in weather and climate and, later, weather and climate systems as they affect surface landforms.

DEFINE & RECALL

cyclone (low)	friction	Chinook	Siberian High
anticyclone (high)	geostrophic wind	foehn	Canadian High
convergent wind circulation	windward	Santa Ana	Icelandic Low
divergent wind circulation	leeward	monsoon	Aleutian Low
isobar	prevailing winds	trough	Pacific High
pressure gradient	land breeze–sea breeze cycle	equatorial low (equatorial trough)	Bermuda High (Azores High)
wind	mountain breeze–valley breeze cycle	subtropical high	polar easterlies
Coriolis effect	drainage wind (katabatic wind)	subpolar low	westerlies
		polar high	trade winds

northeast trades
southeast trades
tropical easterlies
doldrums

intertropical convergence
    zone (ITCZ)
polar front
polar front jet stream

subtropical jet stream
Rossby waves
gyre
upwelling

El Niño
Southern Oscillation
La Niña
North Atlantic Oscillation

DISCUSS & REVIEW

1. What is atmospheric pressure at sea level? How do you suppose Earth's gravity is related to atmospheric pressure?

2. Horizontal variations in air pressure are caused by thermal or dynamic factors. How do these two factors differ?

3. How does incoming insolation affect pressure in the atmosphere? Give an example of an area where incoming insolation would create a pressure system. Would high or low pressure occur?

4. What is the difference between a cell and a belt of pressure?

5. What kind of pressure (high or low) would you expect to find in the center of an anticyclone? Describe and diagram the wind patterns of anticyclones in the Northern and Southern Hemispheres.

6. What is the circulation pattern around a center of low pressure (cyclone) in the Northern Hemisphere? In the Southern Hemisphere? Draw diagrams to illustrate these circulation patterns.

7. Explain how surface friction causes the surface winds to flow *across* isobars rather than *parallel to* the isobars, as in the case of a geostrophic wind.

8. Explain how water and land surfaces affect the pressure overhead during summer and winter. How does this relate to the afternoon sea breeze?

9. How are the land breeze–sea breeze and monsoon circulations similar? How are they different?

10. What effect on valley farms could a strong drainage wind have?

11. What effect would foehn-type winds have on farming, forestry, and ski resorts?

12. What are monsoons? Have you ever experienced one? What causes them? Name some nations that are concerned with the arrival of the "wet monsoon."

13. How do landmasses affect the development of belts of atmospheric pressure over Earth's surface?

14. What are the horse latitudes? The doldrums?

15. Why do Earth's wind systems migrate with the seasons?

16. Map the trade winds of the Atlantic Ocean, and compare your map with one of trade routes in the nineteenth century or earlier.

17. Describe the movements of the upper air. How have pilots applied their experience of the upper air to their flying patterns?

18. What is the relationship between the polar front jet stream and upper air westerlies?

19. What is the relationship between ocean currents and global surface wind systems? How does the gyre in the Northern Hemisphere differ from the one in the Southern Hemisphere?

20. Where are the major warm and cool ocean currents located in respect to Earth's continents? Which currents have the greatest effects on North America?

21. What roles do ocean currents and atmospheric pressure differences play in the development of an El Niño? How might an El Niño affect global weather?

22. How can you apply knowledge of pressure and winds in your everyday life?

CONSIDER & RESPOND

1. If you wished to sail from London to New York and back, what route would you take? Why?

2. Is the polar front jet stream stronger in the summer or the winter? Why?

3. The amount of power that can be generated by wind is determined by the equation

$$P = \frac{1}{2}D \times S^3$$

where P is the power in watts, D is the density, and S is the wind speed in meters per second (m/s). Since $D = 1.293$ kg/m^3, we can rewrite the equation as

$$P = 0.65 \times S^3$$

a. How much power (in watts) is generated by the following wind speeds: 2 m/s, 6 m/s, 10 m/s, 12 m/s?

b. Since wind power increases significantly with increased wind speed, very windy areas are ideal locations for "wind farms." Cities A and B both have average wind speeds of 6 m/s. However, City A tends to have very consistent winds; in City B, half of its winds tend to be at 2 m/s and the other half at 10 m/s. Which site would be the better location for a wind generation plant?

4. Atmospheric pressure decreases at the rate of 0.036 mb/ft as one ascends through the lower portion of the atmosphere.

   a. The Sears Tower in Chicago, Illinois, is one of the world's tallest buildings at 1450 ft. If the street-level pressure is 1020.4 mb, what is the pressure at the top of the Sears Tower?

   b. If the difference in atmospheric pressure between the top and ground floor of an office building is 13.5 mb, how tall is the building?

   c. A single story of a building is 12 ft. You enter an elevator on the top floor of the building and want to descend five floors. The elevator has no floor markings—only a barometer? If the initial reading was 1003.2, at what pressure reading would you want to get off?

5. Look at the January (Fig. 5.13a) and July (Fig. 5.13b) maps of average sea-level pressure. Answer the following questions:

   a. Why is the subtropical high-pressure belt more continuous (linear, not cellular) in the Southern Hemisphere than in the Northern Hemisphere in July?

   b. During July, what area of the United States exhibits the lowest average pressure? Why?

MOISTURE, CONDENSATION, AND PRECIPITATION

The hydrologic cycle involves the circulation of water throughout all the major Earth spheres; it is therefore fundamental to the nature and operation of the entire Earth system.

How does the hydrologic cycle involve all of Earth's spheres? In what ways does the hydrologic cycle affect the Earth system?

There is a direct relationship between the temperature of the atmosphere and the amount of water vapor the atmosphere can hold.

What is this relationship? Why is it important?

Although water may evaporate from all Earth surfaces and transpiration may add considerable moisture to the air, the oceans are the most important source of water vapor in the atmosphere.

Why is this so? What portion (latitudes) of the oceans would have the highest evaporation rates, and why?

There is only one way for significant condensation, cloud formation, and precipitation to occur: Air must be forced to rise so that sufficient adiabatic cooling will take place.

How are condensation, clouds, and precipitation related?

Why does adiabatic cooling take place in rising air? Why might the rate of cooling differ in wet and dry air?

Stability is associated with low (or weak) environmental lapse rates; instability is associated with high (or steep) environmental lapse rates.

What is stability? Instability? Why does the lapse rate affect the tendency of air to rise?

The horizontal distribution of precipitation over Earth can be explained by either one or a combination of the following: frontal, cyclonic, orographic, or convectional precipitation.

Which is most important in your community? Which is most closely related to landform? How is this statement related to the fourth statement above?

As a general rule, the less precipitation a region receives, the greater will be the variability of precipitation in that region from year to year.

What is the effect of this variability on human beings? Where do you think the effect would be greater—in deserts or in semiarid grasslands?

Blue Skies CD

Chapter 3: Atmospheric Moisture & Stability, *Moisture Graph*
Chapter 3: Atmospheric Moisture & Stability, *Adiabatic*
Chapter 2: Sky Identification, *Name that Cloud*

A winter storm hits Washington, D.C.

Water is vital to all life on Earth. Although some living organisms can survive without air, nothing can survive without water. Water is necessary for photosynthesis, cell growth, protein formation, soil formation, and the absorption of nutrients by plants and animals.

Water affects Earth's surface in innumerable ways. The structure of the water molecule is such that water is able to dissolve an enormous number of substances—so many, in fact, that it has been called the universal solvent. Because water acts as a solvent for so many substances, it is almost never found in a pure state. Even rain is filled with impurities picked up in the atmosphere. Indeed, without these impurities to condense around, neither clouds nor precipitation could occur. In addition, rainwater usually contains some dissolved carbon dioxide from the air. Therefore, rain is a very weak form of carbonic acid. We shall see later (Chapter 17) that this fact affects how water shapes certain landforms. The weak acidity of rainwater should not be confused with the environmentally damaging *acid rain,* which is at least ten times more acidic.

Not only can water dissolve and transport many minerals, it also can transport solid particles in suspension. These characteristics make water a unique transportation system for Earth. Water supplies nutrients that would not otherwise be available to plants. Water carries minerals and nutrients down streams, through the soil, through the openings in subsurface rocks, and through plants and animals. It deposits solid matter on stream floodplains, in river deltas, and on the ocean floor.

The surface tension of water and the behavior of water molecules make possible **capillary action**—the ability of water to pull itself upward through small openings against gravity. Capillary action also permits transport of dissolved material in an upward direction—that is, against the pull of gravity. Capillary action moves water into the stems and leaves of plants—even to the topmost needles of the great California redwoods and the top leaves of the rainforest trees. Capillary action is also important in the movement of blood through our bodies. Without it, many of our cells could not receive the necessary nutrients carried by the blood.

Another important and highly unusual property of water is that it expands when it freezes. Most substances contract when cooled and expand when heated. Water follows these rules until it is cooled below 4°C (39°F); then it begins to expand. Ice is therefore less dense than water and consequently will float on water, as do ice floes and icebergs.

Finally, compared with solids, water is slow to heat and slow to cool. Therefore, as we saw in Chapter 4, large bodies of water on Earth act as reservoirs of heat during winter and have a cooling effect in summer. This moderating effect on temperature can be seen in the vicinity of lakes as well as on seacoasts.

Earth's water—the *hydrosphere* (from Latin: *hydros,* water)—is found in all three states: as a liquid in rivers, lakes, oceans, and rain; as a solid in the form of snow and ice; and as a vapor in our atmosphere. Even the water temporarily stored in living things can be considered part of the hydrosphere. About 73 percent of Earth's surface is covered by water, with the largest proportion contained within the world's oceans (Fig. 6.1). In all, the total water content of the Earth system, whether liquid, solid, or vapor, is about 1.33 billion cubic kilometers (326 million cubic mi). Although water cycles in and out of the atmosphere, lithosphere, and biosphere, the total amount of water in the hydrosphere remains constant.

The Hydrologic Cycle

The circulation of water from one part of the general Earth system to another is known as the **hydrologic cycle.** The air contains water vapor that has entered the atmosphere through evaporation from Earth's surface. When water vapor condenses and falls as precipitation, several things may happen to it. First, it may go directly into a body of water, where it is immediately available for evaporation back into the atmosphere. Alternatively, it may fall onto the land, where it may run off the sur-

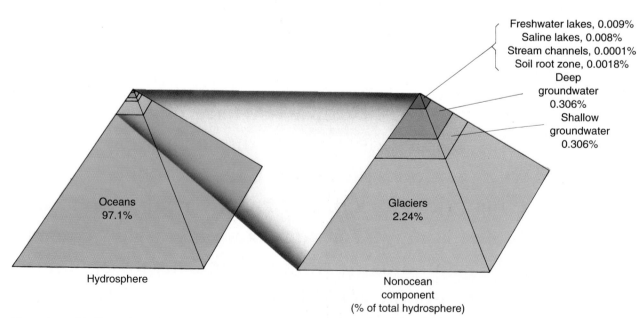

Freshwater lakes, 0.009%
Saline lakes, 0.008%
Stream channels, 0.0001%
Soil root zone, 0.0018%
Deep groundwater 0.306%
Shallow groundwater 0.306%

Oceans 97.1%

Glaciers 2.24%

Hydrosphere

Nonocean component (% of total hydrosphere)

Figure 6.1 This illustration emphasizes the fact that the vast majority of water in the hydrosphere is salt water, stored in the world's oceans. The bulk of the supply of fresh water is relatively unavailable because it is stored in polar ice sheets. ■ **How might global warming or cooling alter this figure?**

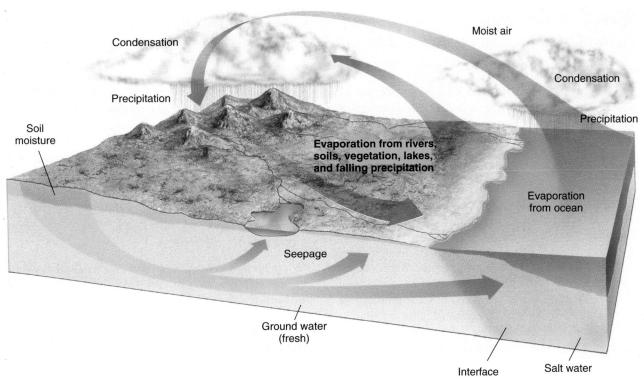

Figure 6.2 Environmental Systems: The Hydrologic System—The Water Cycle The hydrologic system is concerned with the circulation of water from one part of the Earth system to another. The subsystem of the hydrologic system illustrated in this diagram is referred to as the hydrologic cycle. Largely through condensation, precipitation, and evaporation, water is cycled endlessly between the atmosphere, the soil, subsurface storage, lakes and streams, plants and animals, glacial ice, and the principal reservoir—the oceans.

face to form streams, ponds, or lakes. Or it may be absorbed into the ground, where it can either be contained by the soil or flow through open spaces, called *interstices,* that exist in loose sand, gravel, clay, and voids in solid rock. Ultimately, much of the water in or on Earth's surface reaches the oceans. Some water that reaches the surface as snow becomes a part of massive ice over Greenland and Antarctica as well as in high mountain glaciers, while some snow remains frozen until spring thaw, then reenters the hydrologic cycle as a liquid. Other water is used by plants and animals and temporarily becomes a part of living things. Thus, there are six storage areas for water in the hydrologic cycle: the atmosphere, the oceans, bodies of fresh water on the surface, plants and animals, open spaces beneath Earth's surface, and glacial ice.

Liquid water is returned to the atmosphere as a gas through evaporation. Water evaporates from all bodies of water on Earth, from plants and animals, and from soils; it can even evaporate from falling precipitation. Once the water is an atmospheric gas again, the cycle can be repeated.

The hydrologic cycle is basically one of evaporation, condensation, and precipitation, and to some extent

transportation, since these processes are in constant motion. The hydrologic cycle is a *closed system;* that is, water may appear in the system in all three major states—as a liquid, vapor, or solid—and may be transferred from one state to another, but there is no gain or loss of water by the system. This is unlike the radiation budget depicted in Figure 4.7, which is an *open system,* where energy can flow both into and out of the system. Although the hydrologic cycle is a closed system, it is not static but exceedingly dynamic. The percentage of water associated with any one component of the system changes constantly from place to place as well as over time. For example, during the last ice age, evaporation and precipitation were greatly reduced. Also, some changes are human induced; the cutting down of a forest or the damming of a river will cause adjustments among the components.

The hydrologic cycle is one of the most important subsystems of the larger Earth system. It is linked to numerous other subsystems that rely on water as an agent of movement. For example, it plays a major role in the redistribution of energy over Earth's surface. Figure 6.2 provides a schematic illustration of the circulation of water in the hydrologic cycle.

Water in the Atmosphere

The Water Budget and Its Relation to the Heat Budget

We are most familiar with, and most often take notice of, water in its liquid form, as it pours from a tap or as it exists in fine droplets within clouds or fog. Most commonly, however, water exists as a tasteless, odorless, transparent gas known as water vapor, which is mixed with the other gases of the atmosphere in varying proportions. Water vapor is found within approximately the first 5500 meters (18,000 ft) of the troposphere and makes up a small but highly variable percentage of the atmosphere by volume. Atmospheric water is the source of all condensation and precipitation. Through these processes, as well as through evaporation, water plays a significant role as Earth's temperature regulator and modifier. In addition, as we noted in Chapter 4, water vapor in the atmosphere absorbs and reflects a significant portion of both incoming solar energy and outgoing Earth radiation. By preventing great losses of heat from Earth's surface, water vapor helps to maintain the moderate range of temperature found on this planet.

As we noted previously, Earth's hydrosphere is a closed system; that is, water is neither received from outside the Earth system nor given off from it. Thus, an increase in water within one subsystem must be accounted for by a loss in another. Put another way, we say that the Earth system operates with a *water budget,* in which the total quantity of water remains the same and in which the deficits must balance the gains throughout the entire system.

We know that the atmosphere gives up a great deal of water, most obviously by condensation into clouds, fog, and dew, and through several forms of precipitation (rain, snow, hail, sleet). If the quantity of water in the atmosphere remains at the same level through time, the atmosphere must be absorbing from other parts of the system an amount of water equal to that which it is giving up. During 1 minute, as many as 1 billion tons of water are given up by the atmosphere through some form of precipitation or condensation, while another billion tons are evaporated and absorbed as water vapor by the atmosphere.

If we look again at our discussion of the heat energy budget in Chapter 4, we can see that a part of that budget is the latent heat of condensation. Of course, this energy is originally derived from the sun. The sun's energy is used in evaporation and is then stored in the molecules of water vapor, to be released only during condensation. Although the heat transfers involved in evaporation and condensation within the total heat energy budget are proportionately small, the actual energy is significant. Imagine the amount of energy released every minute when a billion tons of water condense out of the atmosphere. It is this vast storehouse of energy,

the latent heat of condensation, that provides the major source of power for Earth's storms: hurricanes, tornadoes, and thunderstorms.

There are limits to the amount of water vapor that can be held by any parcel of air. The most important determinant of the amount of water vapor that can be held by the air is temperature. The warmer air is, the greater the quantity of water vapor it can hold. Therefore, we can make a generalization that air in the polar regions can hold far less water vapor (approximately 0.2 percent by volume) than the hot air of the tropics and equatorial regions of Earth, where the air can contain as much as 5 percent by volume.

Saturation and Dew Point

When air of a given temperature holds all of the water vapor that it possibly can, it is said to be in a state of **saturation** and to have reached its **capacity.** If a constant temperature is maintained in a quantity of air, there will come a point, as more water vapor is added, when the air will be saturated and unable to hold any more water vapor. For example, when you take a shower, the air in the room becomes increasingly humid until a point is reached at which the air cannot contain more water. Then, excess water vapor condenses onto the colder mirrors and walls.

We know that the capacity of air to hold water vapor varies with temperature. In fact, as we can see in Figure 6.3, this capacity of air to contain moisture increases with rising temperatures. Some examples will help illustrate the relationship between temperature and water vapor capacity. If we assume that a parcel of air at 30°C is saturated, then it will contain 30 grams of water vapor in each cubic meter of air (30 g/m^3). Now suppose we increase the temperature of the air to 40°C *without* increasing the water vapor content. The parcel is no longer saturated, since air at 40°C can hold more than 30 g/m^3 of water vapor (actually, 50 g/m^3). Conversely, if we decrease the temperature of saturated air from 30°C (which contains 30 g/m^3 of water vapor) to 20°C (which has a water vapor capacity of only 17 g/m^3), some 13 grams of the water vapor will be forced to condense out of the air because of the reduced capacity.

It is also evident that if an unsaturated parcel of air is cooled, it will eventually reach a temperature where the air will become saturated. This critical temperature is known as the **dew point**—the temperature at which condensation takes place. For example, we know that if a parcel of air at 30°C contains 20 g/m^3 of water vapor, it is not saturated, since it can hold 30 g/m^3. However, if we cool that parcel of air to 21°C, it would become saturated, since the capacity of air at 21°C is 20 g/m^3. Thus, that parcel of air at 30°C has a dew-point temperature of 21°C. It is the cooling of air to below its dew-point temperature that brings about the condensation that must precede precipitation.

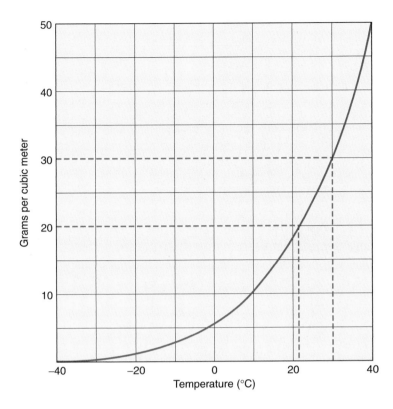

Figure 6.3 The graph shows the maximum amount of water vapor that can be contained in a cubic meter of air over a wide range of temperatures. ■ **Compare the change in capacity when the air temperature is raised from 0° to 10°C with a change from 20° to 30°C. What does this indicate about the relationship between temperature and capacity?**

Because the capacity of air to hold water vapor increases with rising temperatures, air in the equatorial regions has a higher dew point than does air in polar regions. Thus, because the atmosphere can hold more water in these equatorial regions, there is greater potential for large quantities of precipitation than in polar regions. Likewise, in the middle latitudes, summer months, because of their higher temperatures, have more potential for large-scale precipitation than do winter months.

 Humidity

The amount of water vapor in the air at any one time and place is called **humidity.** There are three common ways to express the humidity content of the air. Each method provides information that contributes to our discussion of weather and climate.

Absolute and Specific Humidity **Absolute humidity** is the measure of the mass of water vapor that exists within a given volume of air. It is expressed either in the metric system as the number of grams per cubic meter (g/m^3), or in the English system as grains per cubic foot (gr/ft^3). **Specific humidity** is the mass of water vapor (given in grams) per mass of air (given in kilograms). Obviously, both are measures of the actual amount of water vapor in the air. Since most water vapor gets into the air through the evaporation of water from Earth's surface, it stands to reason that absolute and specific humidity will decrease with height from Earth.

We have also learned that air is compressed as it sinks and expands as it rises. Thus, a given parcel of air changes its volume as it moves vertically, although its weight remains the same and there may be no change in the amount of water vapor in that quantity of air. We can see, then, that absolute humidity, although it measures the amount of water vapor, can vary simply as a result of the vertical movement of a parcel of air. Specific humidity, on the other hand, changes *only* as the quantity of the water vapor changes. For this reason, when assessing the changes of water vapor content in large masses of air, which often have vertical movement, specific humidity is the preferred measurement among geographers and meteorologists.

Relative Humidity Probably the best-known means of describing the content of water vapor in the atmosphere—the one commonly given on television and radio weather reports—is **relative humidity.** It is simply the ratio between the amount of water vapor in air of a given temperature and the maximum amount of vapor that the air could hold at that temperature; it is reported as a percentage that expresses how close the air is to saturation. This method of describing humidity has its strengths and weaknesses. Its strength lies in the ease with which it is communicated. Most people can understand the concept of "percent" and are unlikely to understand "g/m^3." However, relative humidity percentages can vary widely through the day, even when the water vapor content of the air remains the same. The unsteady nature of relative humidity is its weakness.

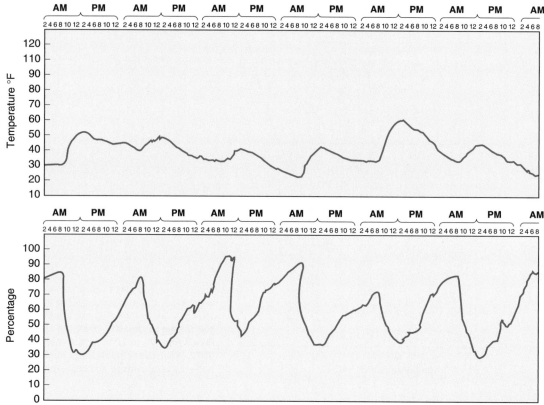

Figure 6.4 This hygrothermograph illustrates the relationship between air temperature (top half) and relative humidity (bottom half) throughout a few days in May in Bowling Green, Kentucky. ■ **What do you notice about these lines?**

If the temperature and absolute humidity of an air parcel are known, its relative humidity can be determined by using Figure 6.3. For instance, if we know that a parcel of air has a temperature of 30°C and an absolute humidity of 20 g/m³, we can look at the graph and determine that if it were saturated, its absolute humidity would be 30 g/m³. To determine relative humidity, all we do is divide 20 grams (actual content) by 30 grams (potential content) and multiply by 100 (to get an answer in percentage):

$$(20 \text{ grams} \div 30 \text{ grams}) \times 100 = 67\%$$

The relative humidity in this case is 67 percent. In other words, the air is holding only two-thirds of the water vapor it could contain at 30°C; it is only at 67 percent of its capacity.

Two important factors are involved in the horizontal distribution and variation of relative humidity. One of these is the availability of moisture. For example, air above bodies of water is apt to contain more moisture than similar air over land surfaces because there is simply more water available for evaporation. Conversely, the air overlying a region like the central Sahara Desert is usually very dry because it is far from the oceans and little water is available to be evaporated. The second factor in the horizontal variation of relative humidity is temperature. In regions of higher temperature, relative humidity for the same amount of water vapor will be lower than it would be in a cooler region.

At any one point in the atmosphere, relative humidity varies if the amount of water vapor increases as a result of evaporation *or* if the temperature increases or decreases. Thus, although the quantity of water vapor may not change through a day, the relative humidity will vary with the daily temperature cycle. As air temperature increases from shortly after sunrise to its maximum in midafternoon, the relative humidity decreases as the air becomes capable of holding greater and greater quantities of water vapor. Then, as the air becomes cooler, decreasing toward its minimum temperature right after sunrise, the relative humidity increases (Fig. 6.4).

Relative humidity affects our comfort through its relationship to the rate of evaporation. Perspiration evaporates into the air, leaving behind a salty residue, which you can taste if you lick your lips after perspiring a great deal. Evaporation is a cooling process, because the heat used to change the perspiration to water vapor (which becomes locked in the water vapor as latent heat) is subtracted from your skin. This is the reason that, on a hot August day when the temperature approaches 35°C (95°F), you will be far more uncomfortable in Atlanta, Georgia, where the relative humidity is 90 percent, than in Tucson, Arizona, where it may be only 15 percent at the same temperature. At 15 percent, your perspiration will be evaporated at a faster rate than at a higher relative humidity, and you will benefit from the resultant cooling effects. When the relative humidity is 90 percent, the air is nearly saturated, and far less evaporation can take place.

Sources of Atmospheric Moisture

In our earlier discussion of the hydrologic cycle, we saw that the atmosphere receives water vapor through the process of evaporation. Water evaporates into the atmosphere from many different places, most important of which are the surfaces of Earth's bodies of water. Water also evaporates from wet ground surfaces and soils, from droplets of moisture on vegetation, from city pavements, building roofs, cars and other surfaces, and even from falling precipitation.

Vegetation provides another source of water vapor. Plants give up water in a complex process known as **transpiration,** which can be a significant source of atmospheric moisture. A mature oak tree, for instance, can give off 400 liters (105 gal) of water per day, and a cornfield may add 11,000 to 15,000 liters (2900–4000 gal) of water to the atmosphere per day for each acre under cultivation. In some parts of the world—notably tropical rainforests of heavy, lush vegetation—transpiration accounts for a significant amount of atmospheric humidity. Together, evaporation and transpiration, or **evapotranspiration,** account for virtually all the water vapor in the air.

Rate of Evaporation

The rate of evaporation is affected by several factors. First, it is affected by the amount of accessible water. Thus, as Table 6.1 shows, the rate of evapotranspiration tends to be greater over the oceans than over the continents. The only place this generalization is not true is in equatorial regions between 0° and 10°N and S, where the vegetation is so lush on the land that transpiration provides a large amount of water for the air.

Second is the degree to which the air is saturated with water vapor. The drier the air and the lower the relative humidity, the greater the rate of evaporation can be. Some of us have had direct experience with this principle. Compare the length of time it takes your bathing suit to dry on a hot, humid day with how long it takes on a day when the air is dry.

Third, temperature also affects the rate of evaporation. As the temperature of the air increases, its capacity to contain moisture also increases, providing room for additional water in the atmosphere. Also, as air temperature increases, so does the temperature of the water at the evaporation source. Such increases in temperature assure that more energy is available to the water molecules for their escape from a liquid state to a gaseous one. Consequently, more molecules can make the transition.

Wind, too, affects the rate of evaporation. If there is no wind, the air that overlies a water surface will approach saturation as more and more molecules of liquid water change to water vapor. Once saturation is reached, evaporation will cease. However, if there is a wind, it will blow the saturated or nearly saturated air away from the evaporating surface, replacing it with air of a lower humidity. This allows evaporation to continue as long as the wind keeps blowing saturated air away and bringing in drier air. Anyone who has gone swimming on a windy day has experienced the chilling effects of rapid evaporation.

Potential Evapotranspiration

So far, we have discussed actual evaporation and transpiration (evapotranspiration). However, geographers and meteorologists are also concerned with **potential evapotranspiration** (Fig. 6.5). This term refers to the idealized conditions in an area under which there would be sufficient moisture for all possible evapotranspiration to occur. Various formulas have been derived for estimating the potential evapotranspiration at a location because it is difficult to measure directly. These formulas commonly use temperature, latitude, vegetation, and soil character (permeability, water retention ability) as factors that could affect the potential evapotranspiration.

Table 6.1	Distribution of Actual Mean Evapotranspiration					
Zone	**Latitude**					
	60°–50°	50°–40°	40°–30°	30°–20°	20°–10°	10°–0°
	Northern Hemisphere					
Continents	36.6 cm(14.2 in.)	33.0(13.0)	38.0(15.0)	50.0(19.7)	79.0(31.1)	115.0(45.3)
Oceans	40.0(15.7)	70.0(27.6)	96.0(37.8)	115.0(45.3)	120.0(47.2)	100.0(39.4)
Mean	38.0(15.0)	51.0(20.1)	71.0(28.0)	91.0(35.8)	109.0(42.9)	103.0(40.6)
	Southern Hemisphere					
Continents	20.0 cm(7.9 in.)	NA	51.0(20.1)	41.0(16.1)	90.0(35.4)	122.0(48.0)
Oceans	23.0(9.1)	58.0(22.8)	89.0(35.0)	112.0(44.1)	119.0(47.2)	114.0(44.9)
Mean	22.5(8.8)	NA	NA	99.0(39.0)	113.0(44.5)	116.0(45.7)

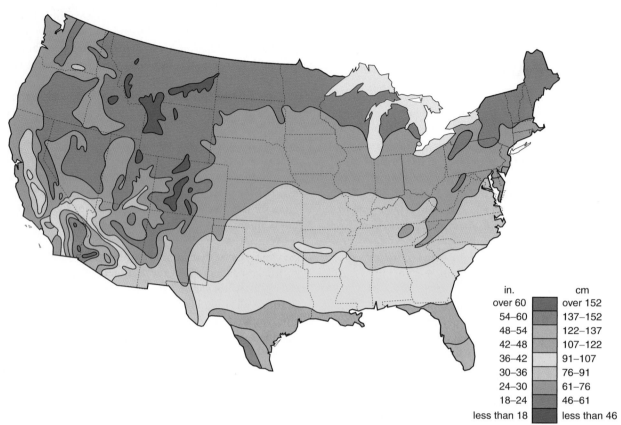

in.		cm
over 60		over 152
54–60		137–152
48–54		122–137
42–48		107–122
36–42		91–107
30–36		76–91
24–30		61–76
18–24		46–61
less than 18		less than 46

Figure 6.5 Potential evapotranspiration for the contiguous 48 states
desert southwest? ■ **Why is potential evapotranspiration so high over the**

In places where precipitation exceeds potential evapotranspiration, there is a surplus of water for storage in the ground and in bodies of water, and water can even be exported to other places if canal construction is feasible. When potential evapotranspiration exceeds precipitation, as it does during the dry summer months in California, then there is no water available for storage; in fact, the water stored during previous rainy months evaporates quickly into the warm, dry air (Fig. 6.6). Soil becomes dry and vegetation turns brown as any available water is soaked up by the atmosphere. For this reason, fires become a potential hazard during the late summer months in California.

Knowledge of potential evapotranspiration is used by irrigation engineers to learn how much water will be lost through evaporation so that they can determine whether the water that is left is enough to justify a drainage canal. Farmers, by assessing the daily or weekly relationship between potential evapotranspiration and precipitation, can determine when and how much to irrigate their crops.

Condensation

Condensation is the process by which a gas is changed to a liquid. In our present discussion of atmospheric moisture and precipitation, condensation refers to the change of water vapor to liquid water.

Condensation occurs when air saturated with water vapor is cooled. Viewed in another way, we can say that if we lower the temperature of air until it has a relative humidity of 100 percent (the air has reached the dew point), condensation will occur with additional cooling.

It follows, then, that condensation depends on (1) the relative humidity of the air and (2) the degree of cooling. In the arid air of Death Valley, California, a huge amount of cooling must take place before the dew point is reached. In contrast, on a humid summer afternoon in Mississippi, a minimal amount of cooling will bring on condensation.

This is the principle behind the formation of droplets of water on the side of a glass containing a cold drink on a warm afternoon: The temperature of the air is lowered when it comes in contact with the cold glass. Consequently, the air's capacity to hold water vapor is diminished. If air touching the glass is cooled sufficiently, its relative humidity will reach 100 percent. Any cooling beyond that point will result in condensation in the form of water droplets on the glass.

Condensation Nuclei

For condensation to occur, one other factor is necessary: the presence of **condensation nuclei.** These are minute particles in the atmosphere that provide a surface upon

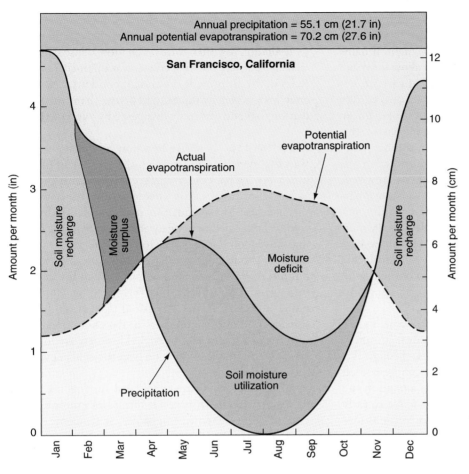

Figure 6.6 This is an example of the Thornthwaite water budget system, which "keeps score" of the balance between water input by precipitation and water loss to evaporation and transpiration, permitting month-by-month estimates of both runoff and soil moisture. ■ **When would irrigation be necessary at this site?**

which condensation can take place. Condensation nuclei are most often sea-salt particles in the air from the evaporation of seawater. They also can be particles of dust, smoke, dirt, pollen, or volcanic material. Some particles are said to be more *hygroscopic* (the property to attract moisture) than others. More and more commonly, they are chemical particles that are the by-products of industrialization. The condensation that takes place on such chemical nuclei is often corrosive and dangerous to human health; when it is, we know it as *smog* (an invented term derived by combining the words *smoke* and *fog*).

Theoretically, if all such particles were removed from a volume of air, we could cool that air below its dew point without condensation occurring. Conversely, if there is a superabundance of such particles, condensation may take place at relative humidities just below 100 percent. For example, ocean fogs, which are an accumulation of condensation droplets formed on sea-salt particles, can form when the relative humidity is as low as 92 percent.

In nature, condensation appears in a number of forms. Fog, clouds, frost, and dew are all the results of condensation of water vapor in the atmosphere. The type of condensation produced depends on a number of factors, including the cooling process itself. The cooling that produces condensation in one form or another can occur as a result of radiation cooling, through advection, through convection, or through a combination of these processes.

Fog

Fogs and clouds appear when water vapor condenses on nuclei and a large number of these droplets form a mass. Not being transparent to light in the way that water vapor is, these masses of condensed water droplets appear to us as fog or clouds, in any of a number of shapes and forms, usually in shades of white or gray.

In the water budget and the hydrologic cycle, fog is a minor form of condensation. Yet, in certain areas of the world, it has important climatic effects. The "drip factor" helps to sustain vegetation along desert coastlines where fog occurs. Fog also plays havoc with our modern transportation systems. Navigation on the seas is made more difficult by fog, and air travel can be greatly impeded. In fact, fog sometimes causes major airports to shut down until visibility improves. Highway travel is also greatly hampered by heavy fogs, which can lead to huge, chain-reaction pileups of cars.

Radiation Fog Radiation cooling can lead to **radiation fog,** also called **surface-inversion fog.** This kind of fog is likely to occur on a cold, clear, calm night, usually in the middle latitudes. These conditions allow for maximum outgoing radiation from the ground with no incoming radiation. The ground gets colder and colder during the night as it gives up more and more of the heat that it has received during the day. As time passes, the air directly above is cooled by conduction through contact with the ground. Since the cold ground can cool only the lower few meters of the atmosphere, an inversion is created in which cold air at the surface is overlain by warmer air above. If this cold layer of air at the surface is cooled to a temperature below its dew point, then condensation will occur, usually in the form of a low-lying fog. However, wind strong enough to disturb this inversion layer can prevent the formation of the fog by not allowing the air to stay at the surface long enough to become cooled below its dew point.

The chances of a surface-inversion fog occurring are increased by certain types of land formations. In valleys and depressions, cold air accumulates through air drainage. During a cold night, this air can be cooled below its dew point, and a fog forms like a pond in the valley. It is common in mountainous areas to see an early morning radiation fog on the valley floor while snow-capped mountaintops shine against a clear blue sky. Radiation fogs have a diurnal cycle. They form during the night and are usually the densest right after sunrise, when temperatures are lowest. They then "burn off" during the day, when the heat from the sun slowly penetrates the fog and warms Earth. Earth in turn warms the air directly above it, increasing its temperature and consequently its capacity to hold water vapor. This greater capacity allows the fog to evaporate into the air. As Earth's heat penetrates to higher and higher layers of air, the fog continues to burn off—from the ground up!

Radiation fog often forms even more densely in industrial areas, where the high concentration of chemical particles in the air provides abundant condensation nuclei. Such a fog is usually thicker and denser than "natural" radiation fogs and less easily dissipated by wind or sun.

Advection Fog Another common type of fog is **advection fog,** which occurs through the movement of warm, moist air over a colder surface, either land or water. When the warm air is cooled below its dew point through heat loss by radiation and conduction from the colder surface below, condensation occurs in the form of fog. Advection fog is usually less localized than radiation fog. It is also less likely to have a diurnal cycle, though if not too thick, it can be burned off early in the day to return again in the afternoon or early evening. More common, however, is the persistent advection fog that spreads itself over a large area for days at a time. Advection fog is a major reason why ski resorts are forced to close. Warm, moist air moving over the cold snow causes the dense fog.

Advection fog forms over land during the winter months in middle latitudes. It forms, for example, in the United States when warm, moist air from the Gulf of Mexico flows northward over the cold, frozen, and sometimes snow-covered upper Mississippi Valley.

During the summer months, advection fog may form over large lakes or over the oceans. Formation over lakes occurs when warm continental air flows over a colder water surface, such as when a warm air mass passes over the cool surface of Lake Michigan. An advection fog also can be formed when a warm air mass moves over a cold ocean current and the air is cooled sufficiently to bring

(a)

© Richard Hamilton Smith / CORBIS

Figure 6.7 (a) Radiation fog forms when cold air drains to a valley floor. (b) Advection (sea) fog is caused by warm moist air passing over colder water. (c) Upslope fog is caused by moist air, adiabatically cooling as it rises up the mountain face.

■ **What unique problems might coastal residents face as a result of these fogs?**

about condensation. This variety of advection fog is known as a sea fog. Such a situation accounts for fogs along the west coast of the United States. During the summer months, the Pacific subtropical high moves north with the sun, and winds flow out of the high toward the coast, where they pass over the cold California Current. When condensation occurs, fogs form that flow in over the shore, pushed from behind by the eastward movement of air and pulled by the low pressure of the warmer land (Fig. 6.7). Advection fogs also occur in New England, especially along the coasts of Maine and the Canadian Maritime Provinces, when warm, moist air from above the Gulf Stream flows north over the colder waters of the Labrador Current. Advection fog over the Grand Banks off Newfoundland has long been a hazard for cod fishermen there.

Upslope Fog   Another type of fog clings to windward sides of mountain slopes and is known as **upslope fog.** Its appearance is sometimes the source of geographic place names—for example, the Great Smoky Mountains, where this type of fog is quite common. During early morning hours in middle-latitude locations, a light, moist breeze may ascend a slope and cool to the dew point, leaving a blanket of fog behind. In tropical rain-forest regions, mountain slopes may be covered in a misty fog any time of day. Because the atmosphere is much more humid, reaching the dew point is easier.

(b)

© Craig Aurness / CORBIS

(c)

NASA / Space Shuttle Image/ SRTM

Other Minor Forms of Condensation

Dew, which is made up of tiny droplets of water, is formed by the condensation of water vapor at or near the surface of Earth. Dew collects on surfaces that are good radiators of heat (such as your car, or blades of grass). These good radiators give up large amounts of heat during the night hours when there is no incoming solar radiation. When the air comes in contact with these cold surfaces, it cools; if it is cooled to the dew point, droplets of water will form as beads on the surface. When the temperature of the air is below 0°C (32°F), **white frost** forms. It is important to note that frost is not frozen dew but instead represents a sublimation process—water vapor changing directly to the frozen state.

Sometimes, under very still conditions with low air pressure, though air temperatures may be below 0°C (32°F), the liquid droplets that make up clouds or fogs are not frozen into solid particles. When such *supercooled* water droplets come in contact with a surface, such as the edge of an airplane wing or a tree branch or a window, ice crystals are created on that surface in a formation known as **rime.** Icing on the wings, nose, and tail of an airplane is extremely hazardous and has been the cause of numerous aviation disasters through time.

Clouds

Clouds are the most common form of condensation and are important for several reasons. First, they are the source of all precipitation. **Precipitation** is made up of condensed water particles, either liquid or solid, that fall to Earth. Obviously, not all clouds result in precipitation, but we cannot have precipitation without the formation of a cloud first. Clouds also serve an important function in the heat energy budget. We have already noted that clouds absorb some of the incoming solar energy. They also reflect some of that energy back to space and scatter and diffuse other parts of the incoming energy before it strikes Earth as diffuse radiation. In addition, clouds absorb some of Earth's radiation so that it is not lost to space, and then reradiate it back to the surface. Finally, clouds are a beautiful and ever-changing aspect of our environment. The colors of the sky and the variations in the shapes and hues of clouds have provided us all with a beautiful backdrop to the natural scenery here on Earth.

Adiabatic Heating and Cooling   The cooling process that leads to cloud formation is quite different from that associated with the other condensation forms we have already examined. The cooling process that produces fog, frost, and dew is either radiation or advection. On the other hand, clouds usually develop from a cooling process that results when a parcel of air on Earth's surface is lifted into the atmosphere.

The rising parcel of air will expand as it encounters decreasing atmospheric pressure with height. This expansion allows the air molecules to spread out, which causes the parcel's temperature to decrease. This is known as **adiabatic cooling** and occurs at the constant lapse rate of approximately 10°C/1000 meters (5.6°F/1000 ft). By the same token, air descending through the atmosphere is compressed by the increasing pressure and undergoes **adiabatic heating** of the same magnitude.

However, the rising and cooling parcel of air will eventually reach its dew point—the temperature at which water vapor begins to condense out, forming cloud droplets. From this point on, the adiabatic cooling of the rising parcel will decrease, as latent energy released by the condensation process is added to the air. To differentiate between these two adiabatic cooling rates, we refer to the precondensation rate (10°C/1000 m) as the **dry adiabatic lapse rate** and the lower, postcondensation rate as the **wet adiabatic lapse rate.** The latter rate averages 5°C/1000 meters (3.2°F/1000 ft), but varies according to the amount of water vapor that condenses out of the air.

A rising air parcel will cool at one of these two adiabatic rates. Which rate is in operation depends on whether condensation is (wet adiabatic rate) or is not (dry adiabatic rate) occurring. On the other hand, the warming temperatures of descending air allow it to hold greater quantities of water vapor. Consequently, condensation will not occur, so the heat of condensation will not affect the rate of rise in temperature. Thus, the temperature of air that is descending and being compressed always increases at the dry adiabatic rate.

It is important to note that adiabatic temperature changes are the result of changes in volume and do not involve the addition or subtraction of heat from external sources.

It is also extremely important to differentiate between the *environmental lapse rate* and *adiabatic lapse rates*. In Chapter 4, we found that, in general, the temperature of our atmosphere decreases with increasing height above Earth's surface; this is known as the environmental lapse rate, or the normal lapse rate. Although it averages 6.5°C/1000 meters (3.6°F/1000 ft), this rate is quite variable and must be measured through the use of meteorological instruments sent aloft. Whereas the environmental lapse rate reflects nothing more than the vertical temperature structure of the atmosphere, the adiabatic lapse rates are concerned with temperature changes as a parcel of air moves through the atmospheric layers (Fig. 6.8).

Stability and Instability   Although adiabatic cooling results in the development of clouds, the various forms of clouds are related to differing degrees of vertical air movement. Some clouds are associated with rapidly rising, buoyant air, while other forms result when air resists vertical movement.

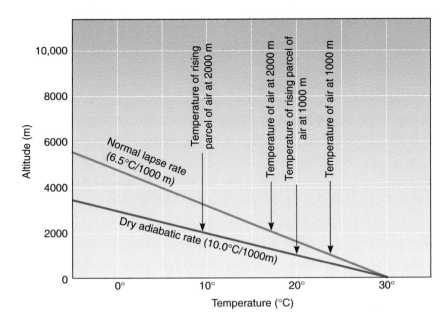

Figure 6.8 Comparison of the dry adiabatic lapse rate and the environmental lapse rate. The environmental lapse rate is the average vertical change in temperature. Air displaced upward will cool (at the dry adiabatic rate) because of expansion. ■ **In this example, what is the temperature of the layer of air at 4000 meters?**

An air parcel will rise of its own accord as long as it is warmer than the surrounding atmospheric air. When it reaches a layer of the atmosphere that is the same temperature as itself, it will stop rising. Thus, an air parcel warmer than the surrounding atmospheric air will rise and is said to be *unstable*. On the other hand, an air parcel that is colder than the surrounding atmospheric air will resist any upward movement and is said to be *stable*.

Determining the stability or instability of an air parcel involves nothing more than asking the question, If an air parcel were lifted to a specific elevation (cooling at an adiabatic lapse rate), would it be warmer, colder, or the same temperature as the atmospheric air (determined by the environmental lapse rate at that time) at that same elevation?

If the air parcel is warmer than the atmospheric air at the selected elevation, then the parcel would be unstable and would continue to rise, since warmer air is less dense and, therefore, buoyant. Thus, under conditions of **instability,** the environmental lapse rate must be *greater than* the adiabatic lapse rate in operation. For example, if the environmental lapse rate is 12°C/1000 meters and the ground temperature is 30°C, then the atmospheric air temperature at 2000 meters would be 6°C. On the other hand, an air parcel (assuming no condensation occurs) lifted to 2000 meters would have a temperature of 10°C. Since the air parcel is warmer than the atmospheric air around it, it is unstable and will continue to rise (Fig. 6.9).

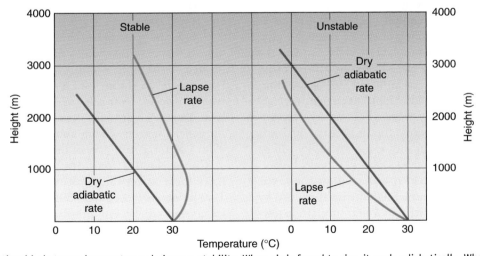

Figure 6.9 Relationship between lapse rates and air mass stability. When air is forced to rise, it cools adiabatically. Whether it continues to rise or resists vertical motion depends on whether adiabatic cooling is less rapid or more rapid than the prevailing vertical temperature lapse rate. If the adiabatic cooling rate exceeds the lapse rate, the lifted air will be colder than its surroundings and will tend to sink when the lifting force is removed. If the adiabatic cooling rate is less than the lapse rate, the lifted air will be warmer than its surroundings and will therefore be buoyant, continuing to rise even after the original lifting force is removed. ■ **In this example, what would be the temperature of the air if it rose to 2000 meters?**

The Lifting Condensation Level (LCL)

When you look at clouds in our atmosphere, it is often quite easy to see their relatively flat bases. Cloud tops may appear quite irregular, but cloud bases are often flat. If the cloud bases do not seem flat, it will be obvious that the clouds you see all seem to be formed at the same level above the surface. This level represents the altitude to which the air must be lifted (and cooled at the dry adiabatic rate) before saturation is

reached. Any additional lifting, and the clouds will form and build. Therefore, the height at which clouds form from lifting is called the lifting condensation level (LCL) and can be estimated by the equation:

LCL (in meters) = 125 m × (Celsius temperature – Celsius dew point)

For example, if the surface temperature is 7.2°C (45°F) and the dew-point temperature is 4.4°C (40°F), then

Now let us assume that it is another day and all the conditions are the same, except that measurements indicate the environmental lapse rate on this day is 2°C/1000 meters. Consequently, although our air parcel if lifted to 2000 meters would still have a temperature of 10°C, the temperature of the atmosphere at 2000 meters would now be 26°C. Thus, the air parcel would be colder and would sink back toward Earth as a result of its greater density (see Fig.6.9). As you can see, under conditions of **stability,** the environmental lapse rate is *less than* the adiabatic lapse rate in operation. If an air parcel, upon being lifted to a specific elevation, has the same temperature as the atmospheric air surrounding it, it is neither stable nor unstable. Instead, it is considered *neutral;* it will neither rise nor sink but will remain at that elevation.

Whether an air parcel will be stable or unstable is related to the amount of cooling and heating of air at Earth's surface. With chilling of the air through radiation and conduction on a cool, clear night, air near the surface will be relatively close in temperature to that aloft, and the environmental lapse rate will be low, thus enhancing stability. With the rapid heating of the surface on a hot summer day, there will be a very steep environmental lapse rate because the air near the surface is so much warmer than that above, and instability will be enhanced.

Pressure zones can also be related to atmospheric stability. In areas of high pressure, stability is maintained by the slow subsiding of relatively cool air from aloft. In low-pressure regions, on the other hand, instability is promoted by the tendency for air to converge and then rise.

Cloud Forms Clouds are composed of billions of tiny water droplets and/or ice crystals so small (some measured in 1000ths of a millimeter) that they can remain suspended in the atmosphere. As you will

learn in this and later chapters, clouds are a manifestation of various movements in our atmosphere. Clouds appear white or in shades of gray, even deep gray approaching black. They differ in color depending on how thick or dense they are and if the sun is shining on the surface that we see. The thicker a cloud is, the more light it is able to absorb and thus block from our view, and the darker it will appear. Clouds also seem dark when we are seeing their shaded side instead of their sunlit side.

In 1803, Luke Howard proposed the first cloud classification scheme, which has been modified through time into the current system. Cloud names may (but not always) consist of two parts. The first part of the name refers to the cloud's height: low-level clouds, below 2000 meters (6500 ft), are called **strato;** middle-level clouds from 2000 to 6000 meters (6500–19,700 ft), are named **alto;** and high-level clouds, above 6000 meters (19,700 ft), are termed **cirro.**

The second part of the name concerns the morphology, or shape, of the clouds. The three basic shapes are termed *cirrus, stratus,* and *cumulus.* Classification systems categorize these cloud formations into many subtypes; however, most subtypes are overlaps of the three basic shapes. Figure 6.10 illustrates the appearance and the general heights of common clouds; Figure 6.11 provides a pictorial summary of the major cloud types.

Cirrus clouds (from Latin: *cirrus,* a lock or wisp of hair) form at very high altitudes, normally 6000 to 10,000 meters (19,800–36,300 ft), and are made up of ice crystals rather than droplets of water. They are thin, stringy, white clouds that trail like feathers across the sky. When associated with fair weather, cirrus clouds are scattered white patches in a clear blue sky.

Stratus clouds (from Latin: *stratus,* layer) can appear anywhere from near the surface of Earth to almost 6000 meters (19,800 ft). The variations of stratus clouds are based in part on their altitude. The basic characteristic of the LCL is estimated at 350 meters (1148 ft) above the surface.

Caution: Keep in mind that different layers of clouds may exist at the same time. Low, middle, and high clouds as defined in this chapter may all appear on the same afternoon. These clouds may have formed in other regions and be only passing overhead. The formula presented here is best used with the lowest level of cloud cover that appears overhead.

This cloud layer shows its development at the LCL.

stratus clouds is their horizontal sheetlike appearance, lying in layers with fairly uniform thickness. The horizontal configuration indicates that they form in stable atmospheric conditions, which inhibit vertical development.

Often stratus clouds cover the entire sky with a gray cloud layer. It is stratus clouds that make up the dull, gray, overcast sky common to winter days in much of the midwestern and eastern United States. The stratus

Figure 6.10 Clouds are named based on their form and height. ■ **Observe this figure and Figure 6.11; what cloud type is present in your area today?**

Cirrocumulus

C. Donald Ahrens

Cirrostratus

Steve McCutcheon/ Visuals Unlimited

Altocumulus

Mark A. Schneider/ Visuals Unlimited

Altostratus

Mark A. Schneider/ Visuals Unlimited

Stratocumulus

Michael Trapasso

Stratus

Ralph F. Kresge/ NOAA

Figure 6.11   Cloud types.

cloud formation may overlie an area for days, and any precipitation will be steady and persistent.

Cumulus clouds (from Latin: *cumulus,* heap or pile) develop vertically rather than forming the more horizontal structures of the cirrus and stratus forms. Cumulus are massive piles of clouds, rounded or cauliflower in appearance, usually with a flat base, which can be anywhere from 500 to 12,000 meters (1650–39,600 ft)

above sea level. From this base, they pile up into great rounded structures, often with tops like cauliflowers. The cumulus cloud is the visible evidence of an unstable atmosphere; its base is the point where condensation has begun in a column of air as it moves upward.

Examine Figures 6.10 and 6.11 to familiarize yourself with the basic cloud types and their names. Keep in mind that some cloud shapes exist in all three levels—

Cirrus

William Weber/ Visuals Unlimited

Cumulonimbus

John Cunningham/ Visuals Unlimited

Nimbostratus

Martin Miller/ Visuals Unlimited

Cumulus

Yva Momatiuk and John Eastcott/ Photo Researchers, Inc.

for example, *stratocumulus* (*strato* = low-level + *cumulus* = a rounded shape), *altocumulus,* and *cirrocumulus.* These three share the similar rounded or cauliflower appearance of cumulus clouds, which can exist at all three levels. You may notice that *altostratus* (*alto* = middle level + *stratus* = layered shape) and *cirrostratus* have two part names, but low-level layered clouds are called *stratus* only. Lastly, thin, stringy *cirrus* clouds are found only as high-level clouds, so the term *cirro* (meaning high-level cloud) is not necessary here.

Other terms used in describing clouds are **nimbo** or **nimbus,** meaning precipitation (rain is falling). Thus, the *nimbostratus* cloud may bring a long-lasting drizzle, and the *cumulonimbus* is the thunderstorm cloud. This latter cloud has a flat top, called an anvil head, as well as a flat base, and it becomes darker as condensation within it increases and blocks the sun. The cumulonimbus is the source of the gusty winds, torrential rain, and lightning common on hot afternoons in humid regions.

Precipitation

Condensed droplets within cloud formations stay in the air and do not fall to Earth because of their general buoyancy and the upward movement of the air. These droplets of condensation are so minute that they are kept floating in the cloud formation; their mass and the consequent pull of gravity are insufficient to overcome the buoyant effects of air and the vertical currents, or updrafts, within the clouds.

Precipitation occurs when the droplets of water, ice, or frozen water vapor coalesce and develop masses too great to be held above Earth. They then fall to Earth as rain, snow, hail, or sleet. The form that precipitation takes depends largely on the method of formation and the temperature during formation.

Forms of Precipitation

Rain, consisting of droplets of liquid water, is by far the most common form of precipitation. In most cases, raindrops develop through the grouping of water droplets around ice crystals formed within cumulus clouds that reach into altitudes with subfreezing temperatures. However, rain is also produced by clouds that do not reach such high altitudes. To explain the development of raindrops within these clouds, meterologists suggest the fusion of small droplets into bigger drops, which in beginning their fall attract even more droplets. It has been estimated that one raindrop is due to the coalescence of

GEOGRAPHY AS **ENVIRONMENTAL** SCIENCE

Desertification

Although desertification is sometimes confused with drought, the two terms describe distinctly different processes. Drought— a longer than normal period of little or no rainfall—is a naturally recurring climatic event. It is especially common in arid and semiarid regions, but can occur in subhumid and sometimes even humid climates. As a general rule, the drier a climate is, the greater the variability of rainfall and the greater the risk of drought.

Desertification, by contrast, is the natural process of desert expansion caused by climatic change, but accelerated by human activities. It is a more serious situation than drought because it involves long-term environmental and human consequences. Desertification expands the margins of the desert when rare rains cause gully erosion, sheet erosion, and loss of soil. It also increases wind erosion, causing dust storms and sand dune movement into grassland and

Map of the Sahel region in northern Africa

about 1 million cloud droplets. Raindrops vary in size but are generally about 2.5 to 6 millimeters (0.1–0.25 in.) in diameter. As we all know, rain can come in many ways: as a brief afternoon shower, a steady three-day rainfall, or the deluge of a tropical rainstorm. When the temperature of an air mass is only slightly below the dew point, the raindrops may be very small (about 0.5 millimeters or less in diameter) and close together. The result is a fine mist or haze called **drizzle.** Drizzle is so light that it is greatly affected by the direction of air currents and the variability of winds. Consequently, drizzle seldom falls vertically.

Snow is the second most common form of precipitation. When water vapor is frozen directly into a solid without first passing through a stage as liquid water (sublimation), it forms minute ice crystals around certain types of nuclei, and these crystals are usually in six-sided, symmetric shapes. Combinations of these ice-crystal shapes make up the intricate and delicate patterns of snowflakes.

Sleet is frozen rain, formed when rain, in falling to Earth, passes through a cold layer of air and freezes. The re-

sult is the creation of solid particles of clear ice. In English-speaking countries outside the United States, sleet refers not to this phenomenon of frozen rain but rather to a mixture of rain and snow.

Hail is a less common form of precipitation than the three just described. It occurs most often during summer months and is the result of certain phenomena that are peculiar to the cumulonimbus cloud form. Hail appears as rounded lumps of ice, called hailstones, which can vary in size from 5 millimeters (0.2 in.) in diameter up to the size of a baseball. The world record is a hailstone 30 centimeters (12 in.) in diameter that fell in Australia. Dropping from the sky, hailstones can be highly destructive to crops and other vegetation, as well as to cars and buildings. Hailstones have even been known to kill animals and humans. Children think it is strange that they must leave a pool or lake where they are swimming simply because hailstones are falling, but this is a sensible precaution because the atmospheric conditions that produce hailstones also produce thunder and lightning. In fact, these phenomena often occur in conjunction with one another.

166 **CHAPTER 6** MOISTURE, CONDENSATION, AND PRECIPITATION farming areas. Desertification is pronounced in regions of the world where humans have accelerated the expansion of desert climate and landform features into former grassland and woodland regions. Although climate change may be the trigger, the process is accelerated by deforestation, overcultivation, soil salinization due to irrigation, and overgrazing by cattle, sheep, and goats.

Desertification is not new. Archeological evidence from Israel and Jordan indicates that as far back as 4000 B.C., early farming communities may have destroyed the soil and deforested the hills, causing desertification. Recent research into ancient environmental catastrophes has shown a similar pattern of denudation of the hilly landscape of Greece as early as 3000 B.C. Today, evidence of desertification is visible in areas of Spain that exhibit deep gully erosion, in northwestern India as the Thar desert expands into Rajasthan's farming areas, and throughout much of the Middle East, northern China, and Africa. Along with the threat to the human population, desertification endangers habitats for wildlife.

It was not until the 1970s, however, that desertification became well known, as television revealed starving and suffering citizens of the nations of the African Sahel. It showed bone-thin cattle trying to find a blade of grass in a barren landscape. The TV also revealed villages being invaded by sand dunes. The Sahel is the semiarid zone bordering the southern margin of the Sahara. It extends across northern African from Mauritania on the Atlantic coast to Somalia on the Indian Ocean. The term *desertification* was popularized at a United Nations conference dealing with problems like those of the Sahel, and most people associate the term with the continuing plight of the people in the region.

The United Nations Environment Program (UNEP) includes a cost estimate of up to $20 billion annually for 20 years in order to successfully fight worldwide desertification. In 1994, 87 nations signed the Desertification Convention in Paris. When ratified by 50 nations, this treaty will budget funds to help protect the fertility of lands that are at the greatest risk of desertification. It will take the support of all the world's nations for antidesertification programs to be successful. Only a major international effort can deal with a natural hazard that causes such large-scale environmental deterioration and human suffering.

Desertification in the Sahel region in the west African country of Mali

Hail forms when ice crystals are lifted by strong updrafts in a cumulonimbus (thunderstorm) cloud. Then, as these ice crystals fall through the cloud, supercooled water droplets attach themselves and are frozen in a layer. Sometimes these pellets are lifted up into the cold layer of air and then dropped again and again. The resulting hailstone, made up of concentric layers of ice, has a frosty, opaque appearance when it finally breaks out of the strong updrafts of the cloud formation and falls to Earth.

On occasion, a raindrop can form and have a temperature below 0°C (32°F). These supercooled droplets will freeze the instant they fall onto a cold surface. The resulting icy covering on trees, plants, and wires is known as **glaze** (or **freezing rain**). People usually call the rain and its blanket of ice an "ice storm." Because of the weight of ice, glazing can break off large branches of trees, bringing down telephone and power lines. It can also make roads practically impassable. A small counterbalance against the negative effects of glazing is the beauty of the natural landscape after an ice storm. Against the background of a clear blue sky, sunlight catches on the ice, reflecting and making diamonds out of the most ordinary weeds and tree branches.

Conditions Causing Precipitation

Three factors are necessary for the formation of any type of precipitation on Earth. The first is the presence of *moist air* on the surface. This air obviously represents the source of moisture (for the precipitation) and energy (in the form of latent heat of condensation). Second are the *condensation nuclei,* around which the water vapor can condense, discussed previously in this chapter. Third is a *mechanism of uplift.* These **uplift mechanisms** are responsible for forcing the air higher into the atmosphere so it can cool down (by the dry adiabatic rate) to the dew point. These uplift mechanisms are vital to the process of precipitation.

A parcel of air can be forced to rise in four major ways, each of which produces its own characteristic types of precipitation (Fig. 6.12). **Convectional precipitation** results from the displacement of warm air upward in a convectional system. **Frontal precipitation** and **cyclonic precipitation** can both take place when a warm air mass

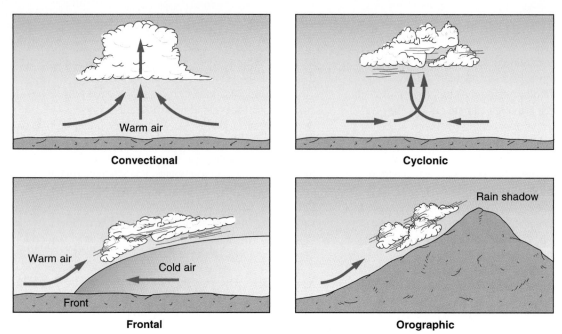

Figure 6.12   The principal cause of precipitation is upward movement of moist air resulting from convectional, frontal, cyclonic, or orographic lifting.   ■   **What do all four diagrams have in common?**

rises after encountering a colder and denser air mass. In **orographic precipitation,** an air mass encounters a land barrier, usually a mountain, and must rise above it in order to pass.

Convectional Precipitation   The simple explanation of convection is that when air is heated near the surface, it expands, becomes lighter, and rises. It is displaced by the cooler, denser air around it to complete the convection cycle. The important factor in convection for our discussion of precipitation is that the heated air rises and thus fulfills the one essential criterion for significant condensation and, ultimately, precipitation.

To enlarge our understanding of convectional precipitation, let us apply what we have learned about instability and stability. Figure 6.13 illustrates two different cases in which air rises due to convection. In both, the lapse rate in the free atmosphere is the same; it is especially high during the first few thousand meters, but slows after that (as on a hot summer day).

In the first case (Fig. 6.13a), the air parcel is not very humid, and thus the dry adiabatic rate applies throughout its ascent. By the time the air reaches 3000 meters (9900 ft), its temperature and density are the same as those of the surrounding atmospheric air. At this point, convectional lifting ceases.

In the second case (Fig. 6.13b), we have introduced the latent heat of condensation. Here again, the as yet unsaturated rising column of air cools at the dry adiabatic rate of 10°C/1000 meters (5.6°F/1000 ft) for the first 1000 meters (3300 ft). However, since the air parcel is humid, the rising air column soon reaches the dew

point, condensation takes place, and cumulus clouds begin to form. As condensation occurs, the heat locked up in the water vapor is released and heats the moving parcel of air, retarding the adiabatic rate of cooling so that the rising air is now cooling at the wet adiabatic rate (5°C/1000 meters). Hence, the temperature of the rising air parcel remains warmer than that of the atmospheric air it is passing through, and the air parcel will continue to rise and rise and rise. In this case, which incorporates the latent heat of condensation, we have massive condensation, towering cumulus clouds, and a thunderstorm potential.

Convectional precipitation is most common in the humid equatorial and tropical areas that receive much of the sun's energy and in summer in middle latitudes. Though differential heating of land surfaces plays an important role in convectional precipitation, it is not the sole factor. Other factors, such as surface topography and atmospheric dynamics associated with the upper air winds, may provide the initial upward push for air that is potentially unstable. Once condensation begins in a convectional column, additional energy is available from the latent heat of condensation for further lifting.

This convectional lifting can result in the heavy precipitation, thunder, lightning, and tornadoes of summer afternoon thunderstorms. When the convectional currents are strong in the characteristic cumulus clouds, hail can result.

Frontal Precipitation   The zones of contact between relatively warm and relatively cold bodies of air are known as **fronts.** When two large bodies of air that

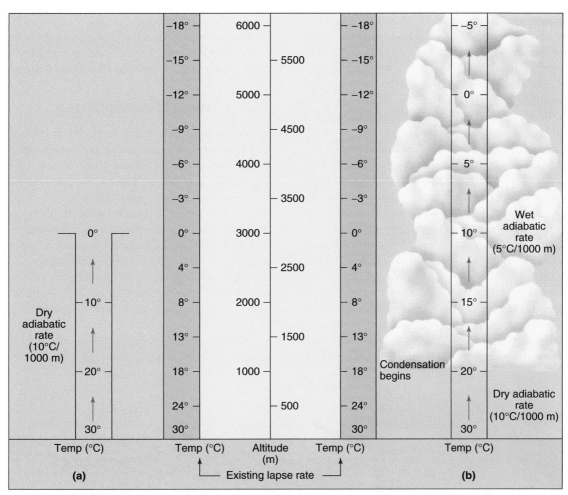

Figure 6.13 Effect of humidity on air mass stability. (a) Warm, dry air rises and cools at the dry adiabatic rate, soon becoming the same temperature as the surrounding air, at which point convectional uplift terminates. Since the rising dry air did not cool to its dew-point temperature by the time convectional lifting ended, no cloud formed. (b) Rising warm, moist air soon cools to its dew-point temperature. The upward-moving air subsequently cools at the wet adiabatic rate, which keeps the air warmer than the surrounding atmosphere so that the uplift continues. Only when all moisture is removed by condensation will the air cool rapidly enough at the dry adiabatic rate to become stable. ■ **What would be necessary for the cloud in (b) to stop its upward growth at 4500 meters?**

differ in density, humidity, and temperature meet, the warmer one is lifted above the colder. When this happens, the major criterion for large-scale condensation and precipitation is once again met. Frontal precipitation thus occurs as the moisture-laden warm air rises above the front caused by contact with the cold air. Continuous frontal precipitation caused the devastating floods of the Mississippi River in 1993.

To fully understand fronts, we must examine what causes unlike bodies of air to come together, and what happens when they do. This will be discussed in Chapter 7, where we will take a more detailed look at frontal disturbances and precipitation.

Cyclonic Precipitation   The third mechanism, the cyclonic, was first introduced in Chapter 5 (see Fig. 5.2). When air enters a low pressure system, or *cyclone,* it does so (in a counterclockwise fashion in the Northern Hemisphere) from all directions. When air converges on a low

pressure system, it has little option but to rise. Therefore, clouds and possible precipitation are common around the center of a cyclone.

Orographic Precipitation   As was the case with convectional rainfall, orographic rainfall has a simple definition and a somewhat more complex explanation. When land barriers—such as mountain ranges, hilly regions, or even the escarpments (steep edges) of plateaus or tablelands—lie in the path of prevailing winds, large portions of the atmosphere are forced to rise above these barriers. This fills the one main criterion for significant precipitation—that large masses of air are cooled by ascent and expansion until large-scale condensation takes place. The resultant precipitation is termed *orographic* (from Greek: *oros,* mountains). As long as the air parcel rising up the mountainside remains stable (cooling at a greater rate than the environmental lapse rate), any resulting cloud cover will be a type of stratus cloud. However, the situation can be complicated by the same circumstances illustrated in Figure 6.13b. A potentially unstable air parcel may need only the initial lift provided by the orographic barrier to set it in motion. In this case, it will continue to rise of its own accord (no longer forced) as it seeks air of its own temperature and density. Once the land barrier provides the initial thrust, it has performed its function as a lifting mechanism.

Because the air deposits most of its moisture on the windward side of a mountain, there will normally be a great deal less precipitation on the leeward side; on this side, the air will be much drier and the dew point consequently much lower. Also, as air descends the leeward slope, its temperature warms (at the dry adiabatic rate) and condensation ceases. The leeward side of the mountain is thus said to be in the **rain shadow** (Fig. 6.14a). Just as being in the shade, or in shadow, means that you are not receiving any direct sun, so being in the rain shadow means that you do not receive much rain. If you live near a mountain range, you can see the effects of orographic precipitation and the rain shadow in the pattern of vegetation (Figs. 6.14b and 6.14c). The wind-

ward side of the mountains (say, the Sierra Nevada in California) will be heavily forested and thick with vegetation. The opposite slopes in the rain shadow will usually be drier and the cover of vegetation sparser.

Distribution of Precipitation

The precipitation a region receives can be described in different ways. We can look at average annual precipitation to get an overall picture of the amount of moisture the region gets during a year. We can also look at its number of raindays—days on which 1.0 millimeter (0.01 in.) or more of rain is received during a 24-hour period. Less than this amount is known as a **trace** of rain. If we divide the number of raindays in a month or year by the total number of days in that period, the resulting figure represents the probability of rain. Such a measure is important to farmers and to ski or summer resort owners whose incomes may depend on precipitation or the lack of it.

We can also look at the average monthly precipitation. This provides a picture of the seasonal variations in

Figure 6.14 (a) Orographic uplift over the windward (western) slope of the Sierras produces condensation, cloud formation, precipitation, resulting in (b) dense stands of forest. (c) Semiarid or rain-shadow conditions occur on the leeward (eastern) slope of the Sierras.
■ **Can you identify a mountain range in Eurasia in which the leeward side of that range is in the rain shadow?**

Station:	San Francisco		
Latitude:	38°N	Longitude:	122°W
	Average annual prec.:	55 cm (21.7 in.)	
Mean annual temp.:	12.8°C (55°F)	Range:	7.2°C (13°F)

Figure 6.15 Average monthly precipitation in San Francisco, California, is represented by colored bars along the bottom of the graph. Such a graph of monthly precipitation figures gives a much more accurate picture than the annual precipitation total, which does not tell us that nearly all the precipitation occurs in only half of the year. ■ **How would this rainfall pattern affect agriculture?**

precipitation (Fig. 6.15). For instance, in describing the climate of the west coast of California, average annual precipitation would not give the full story because this figure would not show the distinct wet and dry seasons that characterize this region.

Horizontal Distribution of Precipitation   Figure 6.16 shows average annual precipitation for the world's continents. We can see that there is great variability in the distribution of precipitation over Earth's surface. Although there is a zonal distribution of precipitation related to latitude, this distribution is obviously not the only factor involved in the amount of precipitation an area receives.

The likelihood and amount of precipitation are based on two factors. First, precipitation depends on the degree of lifting that occurs in air of a particular region. This lifting, as we have already seen, may be due to the convergence of different air masses, to differential heating of Earth's surface, to the lifting that results when an air mass encounters a rise in Earth's surface, or to a combination of these factors. Also, precipitation depends on the internal characteristics of the air itself, including its degree of instability, its temperature, and its humidity.

Since higher temperatures, as we have seen, allow air masses to hold greater amounts of water vapor, and since, conversely, cold air masses can hold less water vapor, we can expect a general decrease of precipitation from the equator to the poles that is related to the un-

equal zonal distribution of incoming solar energy discussed in Chapter 3.

However, if we look again at Figure 6.16, we see a great deal of variability in average annual precipitation beyond the general pattern of a decrease with increased latitude. In the following discussion, we will examine some of these variations and give the reasons for them. We will be applying what we have already learned about temperature, pressure systems, wind belts, and precipitation.

Distribution within Latitudinal Zones   The equatorial zone is generally an area of high precipitation—more than 200 centimeters (79 in.) annually—largely due to the zone's high temperatures, high humidity, and the instability of its air. High temperatures and instability lead to a general pattern of rising air, which in turn allows for precipitation. This tendency is strongly reinforced by the convergence of the trades as they move toward the equator from opposite hemispheres. In fact, the Intertropical Convergence Zone is one of the two great zones where air masses converge. (The other is the polar front within the westerlies zone.)

In general, the air of the trade wind zones is stable compared with the instability of the equatorial zone. Under the control of these steady winds, there is little in the way of atmospheric disturbances to lead to convergent or convectional lifting. However, since the trade winds are basically easterlies, when they move onshore

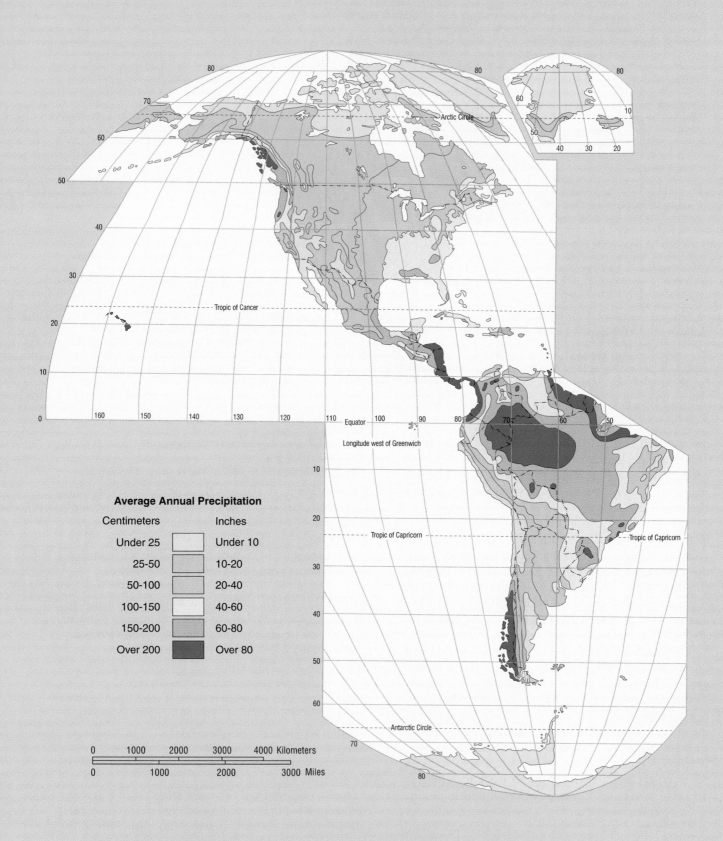

Average Annual Precipitation

Centimeters		Inches
Under 25		Under 10
25-50		10-20
50-100		20-40
100-150		40-60
150-200		60-80
Over 200		Over 80

Figure 6.16 World map of average annual precipitation. ■ **In general, where on Earth's surface does the heaviest rainfall occur? Why?**

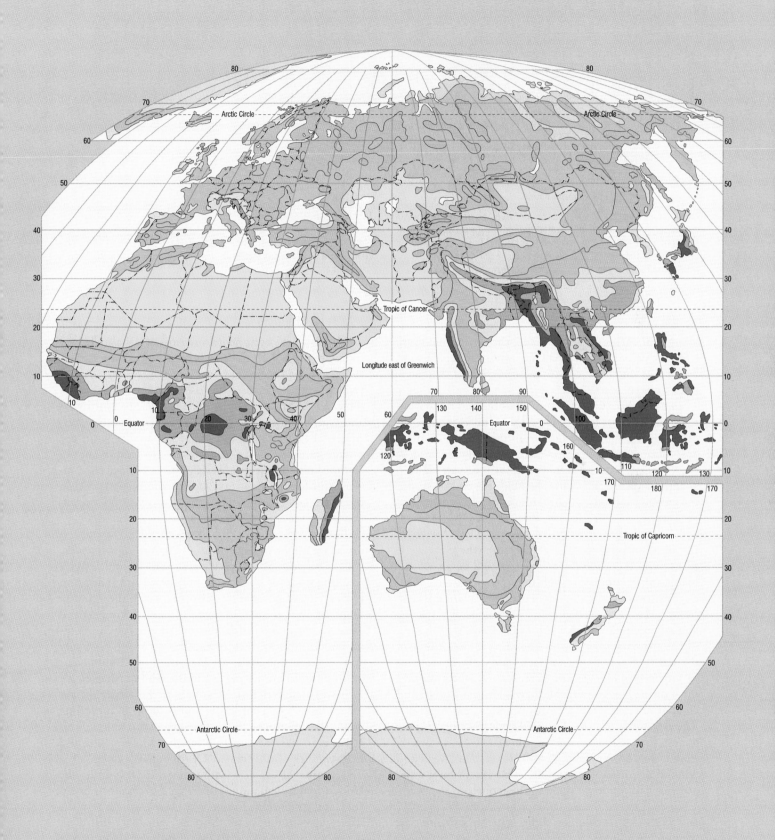

80
70
Arctic Circle
60
50
40
30
Tropic of Cancer
20
Longitude east of Greenwich
10
10
10
10
0 Equator
20
30 40 50
10
20
30
40
50
60
Antarctic Circle
70
80

60
70
130
140
150
90
Equator 0
120
160
10
170
110
120
130
170
180
Tropic of Capricorn
20
30
40
50
60
Antarctic Circle
70
80

100

A Western Paragraphic Projection
developed at Western Illinois University

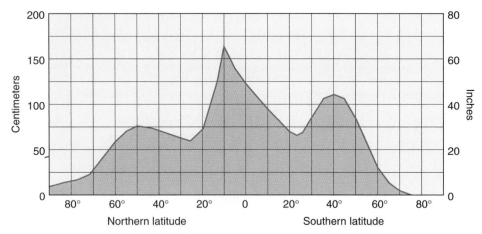

Figure 6.17 Latitudinal distribution of average annual precipitation. This figure illustrates the four distinctive precipitation zones: high precipitation caused by convergence of air in the tropics and in the middle latitudes along the polar front, and low precipitation caused by subsidence and divergence of air in the subtropical and polar regions. ■ **Compare this figure with Figure 5.12. What is the relationship between world rainfall patterns and world pressure distribution?**

along east coasts or high islands, they bring moisture from the oceans with them. Thus, within the trade wind belt, continental east coasts tend to be wetter than continental west coasts.

In fact, where the air of the equatorial and trade wind regions—with its high temperatures and vast amounts of moisture—moves onshore from the ocean and meets a landform barrier, record rainfalls can be measured. The windward slope of Mount Waialeale on Kauai, Hawaii, at approximately 22°N latitude, holds the world's record for greatest average annual rainfall—1146 centimeters (451 in.).

Moving poleward from the trade wind belts, we enter the zones of subtropical high pressure where the air is subsiding. As it sinks lower, it is warmed adiabatically, increasing its moisture-holding capacity and consequently reducing the amount of precipitation in this area. In fact, if we look at Figure 6.17, which shows average annual precipitation on a latitudinal basis, we can see a dip in precipitation level corresponding to the latitude of the subtropical high pressure belts and cells. These areas of subtropical high pressure are, in fact, where we find most of the great deserts of the world: in northern and southern Africa, Arabia, North America, and Australia. The exceptions to this subtropical aridity occur along the eastern sides of the landmasses, where, as we have already noted, the subtropical high pressure cells are weak and wind direction is often onshore. This exception is especially true of regions affected by the monsoons.

In the zones of the westerlies, from about 35° to 65°N and S latitude, precipitation occurs largely as a result of the meeting of cold, dry polar air masses and warm, humid subtropical air masses along the polar front. Thus, there is much cyclonic and frontal precipitation in this zone.

Naturally, the continental interiors of the middle latitudes are drier than the coasts because they are farther away from the oceans. Furthermore, where air in the prevailing westerlies is forced to rise, as it is when it crosses the Cascades and Sierra Nevada of the Pacific Northwest and California, especially during the winter months, there is heavy orographic precipitation. Thus, in the middle latitudes, continental west coasts tend to be wet, and precipitation decreases with movement eastward toward continental interiors. Along eastern coasts within the westerlies, precipitation usually increases once again because of proximity to humid air from the oceans.

In the United States, the interior lowlands are not as dry as we might expect within the prevailing westerlies. This is because of the great amount of frontal activity resulting from the conflicting northward and southward movements of polar and subtropical air. If there were a high east-west mountain range extending from central Texas to northern Florida, the lowlands of the continental United States north of that range would be much drier than they actually are because they would be cut off from moist air originating in the Gulf of Mexico.

Also characteristic of the belt of the westerlies are desert areas that occur in the rain shadows of prevailing winds that are forced to rise over mountain ranges. This effect is in part responsible for the development and maintenance of California's Death Valley, as well as the desert zone of eastern California and Nevada in the United States, the mountain-ringed deserts of eastern Asia, and Argentina's Patagonian Desert, which is in the rain shadow of the Andes. Note in Figure 6.17 that there is greater precipitation in the middle latitudes of the Southern Hemisphere than there is in the Northern. This occurs largely because there is a lot more ocean and less landmass in the Southern Hemisphere westerlies than in the corresponding zone of the Northern Hemisphere.

Moving poleward, we find that temperatures decrease, along with the moisture-holding capacity of the air. The low temperatures also lead to low evaporation rates. In addition, the air in the polar regions shows a general pattern of subsidence that yields areas of high pressure. This settling of the air in the polar regions is the opposite of the lifting needed for precipitation. All of these factors combine to cause low precipitation values in the polar zones.

Variability of Precipitation

The rainfall depicted in Figure 6.16 is an annual average. It should be remembered, however, that in many parts of the world, there are significant variations in precipitation, both within any one year and between years. For example, areas like the Mediterranean region, California, Chile, South Africa, and Australia, which are on the west sides of the continents and roughly between 30° and 40° latitude, get much more rain in the winter than in the summer. There are also areas between 10° and 20° latitude that get much more of their precipitation in the summer (high-sun season) than in the winter (low-sun season).

Rainfall totals can change markedly from one year to the next, and, tragically for many of the world's people, the drier a place is on the average, the greater will be the statistical variability in its precipitation (compare

Fig. 6.18 with Fig. 6.16). To make matters worse for people in dry areas, a year with a particularly high amount of rainfall may be balanced with several years of below-average precipitation. This situation has occurred recently in West Africa's Sahel, the Russian steppe, and the American Great Plains.

Thus, there are years of drought and years of flood, each bringing its own kind of disaster upon the land. Farmers, resort owners, construction workers, and others whose economic well-being depends in one way or another on weather can determine only a probability of rainfall on an annual, monthly, or even a seasonal basis.

Meteorologists are not able to predict rainfall with 100 percent accuracy. This inability is due to the many factors involved in causing precipitation—temperature, available moisture, atmospheric disturbances, landform barriers, frontal activity, air mass movement, upper air winds, and differential surface heating, among others. In addition, the interaction of these factors in the development of precipitation is very complex and not completely understood.

Study of the hydrologic system helps scientists understand how changes in one subsystem can greatly impact other subsystems. For example, it is theorized that changes in the radiation budget of the Earth system can produce changes in the hydrologic cycle that, in turn, can explain changes in glacial subsystems and bring about continental glaciation. If the energy available to heat

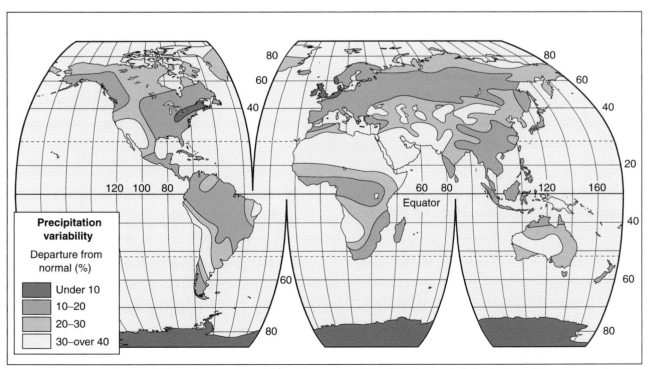

Precipitation variability

Departure from normal (%)

- Under 10
- 10–20
- 20–30
- 30–over 40

Figure 6.18 Precipitation variability. The greatest variability in year-to-year precipitation totals is in the dry regions, accentuating the critical problem of moisture supply in those parts of the world. ■ **Compare this figure with figure 6.16. What are some of the similarities and differences?**

Earth's atmosphere were reduced, the colder temperatures would affect the hydrologic cycle. A colder atmosphere would store less water vapor. Thus, there would be less precipitation, but the colder temperatures would result in more snowfall. With time, glacial ice would increase, and sea level would drop. The high albedo of the ice would further reduce the energy available to heat our at-mosphere, which would allow glacial ice to increase and eventually result in continental ice sheets covering large portions of Earth.

Even without a dramatic example like the one described above, our planet's hydrologic system relates to all other Earth systems. Later chapters will demonstrate other aspects of this theme, which runs throughout this textbook.

DEFINE & RECALL

capillary action
hydrologic cycle
saturation
capacity
dew point
humidity
absolute humidity
specific humidity
relative humidity
transpiration
evapotranspiration
potential
   evapotranspiration

condensation nuclei
radiation fog
   (surface-inversion fog)
advection fog
upslope fog
dew
white frost
rime
precipitation
adiabatic cooling
   (adiabatic heating)
dry adiabatic lapse rate
wet adiabatic lapse

rate
instability
stability
strato
alto
cirro
cirrus
stratus
cumulus
nimbus (nimbo)
rain
drizzle
snow

sleet
hail
glaze (freezing rain)
uplift mechanisms
convectional
   precipitation
frontal precipitation
cyclonic precipitation
orographic precipitation
front
rain shadow
trace

DISCUSS & REVIEW

1. How is the hydrologic cycle related to Earth's water budget?

2. Is the saturation or dew point in your area generally higher or lower than that in Siberia? Honduras?

3. What is the difference between absolute and specific humidity? What is relative humidity?

4. Why is the inside of a greenhouse generally more humid than an ordinary room?

5. Why is wearing wet clothes sometimes bad for your health? Under what conditions might wearing wet clothes be good for your health?

6. Imagine that you are deciding when, in your daily schedule, to water the garden. What time of day would be best for conserving water? Why?

7. What factors must be taken into consideration when calculating potential evapotranspiration? How is evapotranspiration related to the water budget of a region?

8. What factors affect the formation of surface-inversion fogs?

9. What kinds of fogs occur in the area where you live?

10. What causes adiabatic cooling? Differentiate between the environmental lapse rate and adiabatic lapse rates.

11. How and why does the wet adiabatic lapse rate differ from the dry adiabatic lapse rate?

12. How is atmospheric stability related to the adiabatic lapse rates?

13. Describe the most recent cloud formations in your area. With what kinds of weather (temperature, humidity, precipitation) were the cloud types associated?

14. What atmospheric conditions are necessary for precipitation to occur?

15. Find out how many inches of precipitation have fallen in your area this year. Is that average or unusually high or low?

16. Compare and contrast convectional, orographic, cyclonic, and frontal precipitation.

17. How is rainfall variability related to total annual rainfall? How might this relationship be considered a double problem for people?

CONSIDER & RESPOND

1. Refer to Figure 6.3.

    a. What is the water vapor capacity of air at 0°C? 20°C? 30°C?

    b. If a parcel of air at 30°C has an absolute humidity (actual water vapor content) of 20.5 g/m^3, what is the parcel's relative humidity?

    c. If the relative humidity of a parcel of air is 33 percent and the air temperature is 15°C, what is the absolute humidity of the air in g/m^3?

    d. A major concern of northern climate residents is the low relative humidity within their homes during the winter. Low relative humidity is not healthy, and it has an adverse effect on the homes' furnishings. The problem results when cold air, which can hold little water vapor, is brought indoors and heated up. The following example will illustrate the problem: Assume the air outside is 5°C and has a relative humidity of 60 percent. What is the actual water vapor content of this air? If it is brought indoors (through the doors, windows, and cracks in the home) and heated to 20°C, with no increase in water vapor content, what is the new relative humidity?

2. Recall that as a parcel of air rises, it expands and cools. The rate of cooling, termed the dry adiabatic lapse rate, is 10°C/1000 meters. (A descending parcel of air will always warm at this rate.) In addition, the dew-point temperature decreases 2°C/1000 meters within a rising parcel of air. At the height at which the dew-point temperature is reached, condensation begins and, as discussed in the text, the wet adiabatic lapse rate of 5°C/1000 meters becomes operational. When the wet adiabatic lapse rate is in operation, the dew-point temperature will be the same as the air temperature. When a parcel descends through the atmosphere, its dew-point temperature increases 2°C/1000 meters. The height at which condensation begins, termed the lifting condensation level (LCL), can be determined by using the formula found in Geography As a Physical Science: The Lifting Condensation Level, pages 162–163.

    a. A parcel of air has a temperature of 25°C and a dew-point temperature of 14°C. What is the height of the LCL? If that parcel were to rise to 4000 meters, what would be its temperature?

    b. A parcel of air at 6000 meters has a temperature of –5°C and a dew point of –10°C. If it descended to 2000 meters, what would be its temperature and dew-point temperature?

AIR MASSES AND WEATHER SYSTEMS

The movement of relatively large bodies of the air (air masses) is responsible for the transportation of distinguishable characteristics of temperature and humidity to regions far from their original sources.

How is this important to the operation of the Earth system?

What is the significance of air mass movement to human beings?

The meeting of the leading edges of two unlike air masses occurs along a sloping surface of discontinuity called a front.

How do air masses differ?

What kinds of air masses meet along the polar front?

Why are fronts important in explaining middle-latitude weather?

The major explanation for the variable and nearly unpredictable weather of the middle latitudes may be found in the irregular migration of relatively short-lived low pressure systems (cyclones) in the path of the prevailing westerlies.

Why do cyclones play such a significant role?

In which direction would you normally look to obtain some forewarning of future weather?

What are the human consequences of variable and unpredictable weather?

Meteorology is an inexact science, and there is much yet to be learned about the behavior of air masses, fronts, and pressure systems. We should therefore anticipate that weather forecasting will remain a complicated art.

What questions about the weather remain unanswered?

How accurate is weather prediction?

How successful are humans at altering the weather?

The latent heat of condensation is the major source of energy behind the violent atmospheric disturbances of the middle latitudes, such as thunderstorms, tornadoes, and hurricanes.

How might this information be useful?

What have we done in the United States to protect against violent storms?

Blue Skies CD

Earth Systems Today CD

A disastrous tornado near Dimmitt, Texas on 10 April 1999

If we are to understand the types of weather systems that rule the middle latitudes, we must first come to grips with the vital parts of our basic weather systems. In the previous four chapters, we have looked at the elements of the atmosphere and investigated some of the controls that act upon those elements, causing them to vary from place to place and through time. However, even more is involved in the examination of weather. We have not yet looked at storms (atmospheric disturbances)—their types and characteristics, their origin, and their development. Storms are an important part of the weather story. They help illustrate the interactions among the weather elements. Further, they represent a major means of energy exchange within the atmosphere.

Air Masses

Before we begin to study weather systems, we should understand the nature and significance of air masses. In themselves, air masses provide a straightforward way of looking at the weather. An **air mass** is a large body of air, at times subcontinental in size, that moves over Earth's surface with distinguishable characteristics. An air mass is relatively homogeneous in temperature and humidity. That is, at approximately the same altitude within the air mass, the temperature and humidity will be similar. As a result of this temperature and moisture uniformity, the density of air will be much the same throughout any one level within an air mass. Of course, since an air mass may extend over 20 or 30 degrees of latitude, we can expect some slight variations due to changes in sun angle and its corresponding insolation, which are significant over that distance. Changes caused by contact with differing land and ocean surfaces also affect the characteristics of air masses.

The similar characteristics of temperature and humidity within an air mass are determined by the nature of its **source region**—the place where the air mass originates. Only a few areas on Earth make good source regions. In order for the air mass to have similar characteristics throughout, the source region must have a nearly homogeneous surface. For example, it can be a desert, an ice sheet, or an ocean body, but not a combination of surfaces. In addition, the air mass must have sufficient time to acquire the characteristics of the source region. Hence, gently settling, slowly diverging air will mimic a source region, whereas converging, rising air will not.

Air masses are identified by a simple letter code. The first is always a lowercase letter. There are only two choices: The letter *m,* for maritime, means the air mass originates over the sea and is therefore relatively moist. The letter *c,* for continental, means the air mass originates over land and is therefore dry. The second letter is always a capital. These help to locate the latitude of the

Table 7.1   Types of Air Masses

	Source Region	Usual Characteristics at Source	Accompanying Weather
Maritime Equatorial (*mE*)	Equatorial oceans	Ascending air, very high moisture content	High temperature and humidity, heavy rainfall; never reaches the United States
Maritime Tropical (*mT*)	Tropical and subtropical oceans	Subsiding air; fairly stable, but some instability on western side of oceans; warm and humid	High temperatures and humidity, cumulus clouds, convectional rain in summer; mild temperatures, overcast skies, fog, drizzle, and occasional snowfall in winter; heavy precipitation along *mT/cP* fronts in all seasons
Continental Tropical (*cT*)	Deserts and dry plateaus of subtropical latitudes	Subsiding air aloft; generally stable, but some local instability at surface; hot and very dry	High temperatures, low humidity, clear skies, rare precipitation
Maritime Polar (*mP*)	Oceans between 40° and 60° latitude	Ascending air and general instability, especially in winter; mild and moist	Mild temperatures, high humidity; overcast skies and frequent fogs and precipitation, especially during winter; clear skies and fair weather common in summer; heavy orographic precipitation, including snow, in mountainous areas
Continental Polar (*cP*)	Plains and plateaus of subpolar and polar latitudes	Subsiding and stable air, especially in winter; cold and dry	Cool (summer) to very cold (winter) temperatures, low humidity; clear skies except along fronts; heavy precipitation, including winter snow, along *cP/mT* fronts
Continental Arctic (*cA*)	Arctic Ocean, Greenland, and Antarctica	Subsiding very stable air; very cold and very dry	Seldom reaches United States, but when it does, bitter cold, subzero temperatures, clear skies, often calm conditions

source region. *E* stands for Equatorial; this air is very warm. The letter *T* identifies a Tropical origin and is therefore warm air. A *P* represents Polar; this air can be quite cold. Lastly, an *A* identifies Arctic air, which is very cold. These six letters can be combined to give us the classification of air masses first described in 1928 and still used today: **Maritime Equatorial (*mE*), Maritime Tropical (*mT*), Continental Tropical (*cT*), Continental Polar (*cP*), Maritime Polar (*mP*), and Continental Arctic (*cA*).** These six types are described more fully in Table 7.1. From now on, we will use the symbols rather than the full names as we discuss each type of air mass.

Modification and Stability of Air Masses

As a result of the general circulation patterns within the atmosphere, air masses do not remain stationary over their source regions indefinitely. When an air mass begins to move over Earth's surface along a path known as a trajectory, for the most part it retains its distinct and homogeneous characteristics. However, modification does occur as the air mass gains or loses some of its thermal energy and moisture content to the surface below. Although this modification is generally slight, the gain or loss of thermal energy can make an air mass more stable or unstable.

An air mass is further classified by whether it is warmer or colder than the surface over which it travels, because this has a bearing on its stability. If an air mass is colder than the surface over which it passes, then the surface will heat the air mass from below. This will, in turn, increase the environmental lapse rate, enhancing the prospect of instability. To describe such a situation, the letter *k* (from German: *kalt,* cold) is added to the other letters that symbolize the air mass. For example, an *mT* air mass originating over the Gulf of Mexico in summer that moves onshore over the warm land would be denoted *mTk*. Such an air mass is often unstable and can produce copious convective precipitation. On the other hand, this same *mT* air mass moving onshore during the winter would be warmer than the land surface. Consequently, the air mass would be cooled from below, decreasing its environmental lapse rate, which enhances the prospect of stability. We describe this situation with the letter *w* (from German: *warm,* warm), and the air mass would be denoted *mTw*. In this case, stratiform, not convective, precipitation is most likely.

North American Air Masses

Most of us are familiar with the weather in at least one region of the United States or Canada; therefore, this chapter will concentrate on the air masses of North America and their effects on weather. What we learn will be applicable to the rest of the world, and as we examine climatic regions in some of the following chapters, we will be able to understand that weather everywhere is most often affected by the movements of air masses. Especially in middle-latitude regions, the majority of atmospheric disturbances result from the confrontations of different air masses.

Five types of air masses (*cA, cP, mP, mT,* and *cT*) influence the weather of North America, some more than others. Air masses assume characteristics of their source regions (Fig. 7.1). Consequently, as the source regions change with the seasons, primarily because of changing insolation, the air masses also will vary.

Continental Arctic Air Masses The frigid, frozen surface of the Arctic Ocean serves as the source region for this air mass. It is extremely cold, very dry, and very stable. Though it will affect parts of Canada, even during the winter, when this air mass is best developed, it seldom travels far enough south to affect the United States. However, on those few occasions when it does extend down into the midwestern and southeastern United States, its impact is awesome. Record-setting cold temperatures often result. If the *cA* air mass remains in the Midwest for an extended period, vegetation—not accustomed to the extreme cold—can be severely damaged or killed.

Continental Polar Air Masses At its source in north-central North America, a continental polar air mass is cold, dry, and stable because it is warmer than the surface beneath it; the weather of a *cP* air mass is cold, crisp, and clear. Since there are no east-west landform barriers in North America, *cP* air can migrate south across Canada and the United States. A tongue of *cP* air can sometimes reach as far south as the Gulf of Mexico or Florida. When winter *cP* air extends into the United States, its temperature and humidity are raised only slightly. The movement of such an air mass into the Midwest and South brings with it a cold wave characterized by colder-than-average temperatures and clear, dry air, causing freezing temperatures as far south as Florida and Texas.

The general westerly direction of atmospheric circulation in the middle latitudes, rarely allows a *cP* air mass to break through the great western mountain ranges to the west coast of the United States. When such an air mass does reach the Washington, Oregon, and California coasts, it brings with it unusual freezing temperatures that do great damage to agriculture.

Maritime Polar Air Masses During winter months, the oceans tend to be warmer than the land, so an *mP* air mass tends to be warmer than its counterpart on land (the *cP* air mass). Much *mP* air is originally cold, dry, *cP* air that has moved to a position over the ocean. There, it is heated by the warmer water, and collects moisture. Thus, *mP* air is cold (although not as cold as *cP* air) and damp, with a tendency toward instability. The northern Pacific Ocean serves as the source region for maritime

Figure 7.1 Source regions of North American air masses. Air mass movements import the temperature and moisture characteristics of these source regions into far distant areas. ■ **Use Table 7.1 and this figure to determine which air masses affect your location. Are there seasonal variations?**

polar air masses, which, because of the general westerly circulation of the atmosphere in the middle latitudes, affect the weather of the northwestern United States and southwestern Canada. When this *mP* air meets an uplift mechanism (such as a mass of colder, denser air or coastal mountain ranges), the result is usually very cloudy weather with a great deal of precipitation. An *mP* air mass is still the source of many midwestern snowstorms even after crossing the western mountain ranges.

Generally, an *mP* air mass that develops over the northern Atlantic Ocean does not affect the weather of the United States because such an air mass tends to flow eastward toward Europe. However, on some occasions, there may be a reversal of the dominant wind direction accompanying a low pressure system, and New England can be made miserable by the cool, damp winds, rain, and snow of a weather system called a *nor'easter*.

Maritime Tropical Air Masses   The Gulf of Mexico and subtropical Atlantic Ocean serve as a source region for maritime tropical air masses that have a great influence on the weather of the United States, and at times

southeastern Canada. During winter, the waters are warm and the air above is warm and moist. As the warm, moist air moves northward up the Mississippi Lowlands, it travels over increasingly cooler land surfaces. The lower layers of air are chilled, and dense advection fog often results. When it reaches the *cP* air migrating southward from Canada, the warm *mT* air is forced to rise over the colder, drier *cP* air, and significant precipitation can occur.

The longer days and more intense insolation of summer months modify a Gulf and Atlantic air mass at its source region by increasing its temperature and moisture content. However, during summer, the land is warmer than the nearby waters, and as the *mT* air mass moves onto the land, the instability of the air mass is increased. This air mass is the source of great thunderstorms and convective precipitation on hot, humid days, and it is also responsible for much of the hot, humid weather of the southeastern and eastern United States.

Other *mT* air masses form over the Pacific Ocean in the subtropical latitudes. These air masses tend to be slightly cooler than those that form over the Gulf and the Atlantic, partly because of their passage over the cooler California Current. A Pacific *mT* air mass is also more stable because of the strong subsidence associated with the eastern portion of the Pacific subtropical high. This air mass contributes to the dry summers of southern California and occasionally brings moisture in winter as it rises over the mountains of the Pacific Coast.

Continental Tropical Air Masses   There is a fifth type of air mass that may affect North America, but it is the least important to the weather of the United States and Canada. This is the continental tropical air mass that develops over large, homogeneous land surfaces in the subtropics. The Sahara Desert of North Africa is a prime example of a source region for this type of air mass. The weather typical of the *cT* air mass is usually very hot and dry, with clear skies and major heating from the sun during daytime.

In North America, there is little land in the correct latitudes to serve as a source region for a *cT* air mass of any significant proportion. A small *cT* air mass can form over the deserts of the southwestern United States and central Mexico in the summer. In its source region, a *cT* air mass provides hot, dry, clear weather. When it moves eastward, however, it is usually greatly modified as it comes in contact with larger and stronger air masses of different temperatures, humidities, and densities. At times, *cT* air from Mexico and Texas meets with *mT* air from the Gulf of Mexico. This boundary is known as a *dry line*. Here the drier air is denser and will lift the moister air over it. This mechanism of uplift may act as a trigger for precipitation episodes and perhaps thunderstorm activity.

Fronts

We have seen that air masses migrate with the general circulation of the atmosphere. Over the United States, which is influenced primarily by the westerlies, there is a general eastward flow of the air masses. In addition, air masses tend to diverge from areas of high pressure and converge toward areas of low pressure. This tendency means that the tropical and polar air masses, formed within systems of divergence, tend to flow toward areas of convergence within the United States. As previously noted, an important feature of an air mass is that it maintains the primary characteristics first imparted to it by its source region, although some slight modification may occur during its migration.

When air masses differ, they do so primarily in their temperature and in their moisture content, which in turn affect the air masses' density and atmospheric pressure. As we saw in Chapter 6, when different air masses come together, they do not mix easily but instead come in contact along sloping boundaries called *fronts*. Although usually depicted on maps as a one-dimensional boundary line separating two different air masses, a front is actually a three-dimensional surface with length, width, and height. To emphasize this concept, a front is sometimes referred to as a **surface of discontinuity.** This surface of discontinuity is a zone that can cover an area from 2 or 3 kilometers (1–2 mi) wide to as wide as 150 kilometers (90 mi). Hence, it is more accurate to speak of a frontal zone rather than a frontal line.

The sloping surface of a front is created as the warmer and lighter of the two contrasting air masses is lifted or rises above the cooler and denser air mass. Such rising, known as *frontal uplift,* is a major source of precipitation in middle-latitude countries like the United States and Canada, where contrasting air masses are most likely to converge.

The steepness of the frontal surface is governed primarily by the degree of difference between the two converging air masses. When there is a sharp difference between the two air masses, as when an *mT* air mass of high temperature and moisture content meets a *cP* air mass with its cold, dry characteristics, the slope of the frontal surface will be steep. With a steep slope, there will be greater frontal uplift. Provided other conditions (for example, temperature and moisture content) are equal, a steep slope, with its greater frontal uplift, will produce heavier precipitation than will a gentler slope.

Fronts are differentiated by determining whether the colder air mass is moving in on the warmer one, or vice versa. The weather that occurs along a front also depends on which air mass is the "aggressor."

Cold Front

A **cold front** occurs when a cold air mass actively moves in on a warmer air mass and pushes it upward. The colder air, denser and heavier than the warm air it is displacing, stays at the surface while forcing the warmer air to rise. As we can see in Figure 7.2, a cold front usually results in a relatively steep slope in which the warm air may rise 1 meter in the vertical for every 40 to 80 meters of horizontal distance. If the warm air mass is unstable and has a high moisture content, heavy precipitation can result, sometimes in the form of violent thunderstorms. A **squall line** may result when several storms align themselves on (or in advance of) a cold front. In any case, cold fronts are usually associated with strong weather disturbances or sharp changes in temperature, air pressure, and wind.

Warm Front

When a warmer air mass is the aggressor and invades a region occupied by a colder air mass, a **warm front** results. In a warm front, the warmer air, as it slowly pushes against the cold air, also rises over the colder, denser air mass, which again stays in contact with Earth's surface. The slope of the surface of discontinuity that results is

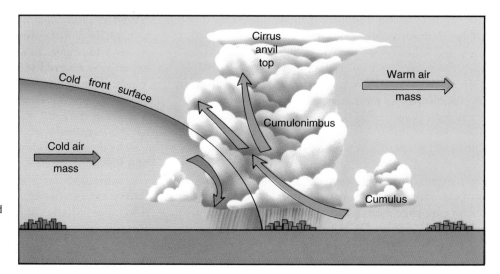

Figure 7.2 Cross section of a cold front. Cold fronts generally move rapidly, with a blunt forward edge that drives adjacent warmer air upward. This can produce violent precipitation from the warmer air.

usually far gentler than that occurring in a cold front. In fact, the warm air may rise only 1 meter in the vertical for every 100 or even 200 meters of horizontal distance. Thus, the frontal uplift that develops will not be as great as that occurring along a cold front. The result is that the warmer weather associated with the passage of a warm front tends to be less violent and the changes less abrupt than those associated with cold fronts.

If we look at Figure 7.3, we can see why the advancing warm front affects the weather of areas ahead of the actual surface location of the frontal zone. Changes in the weather from approaching fronts can sometimes be indicated by the series of cloud types that precede them.

Stationary and Occluded Fronts

When two air masses have converged and formed a frontal boundary, but then neither moves, we have a situation known as a *quasi-stationary* or, as it is more commonly called, a **stationary front.** Locations under the

influence of a stationary front are apt to experience clouds, drizzle, and rain (or possible thunderstorms) for several days. In fact, a stationary front and its accompanying weather will remain until the contrasts between the two air masses are reduced or the circulation of the atmosphere finally causes one of the air masses to move. If a stationary front holds a position for a length of time, then regional flooding is possible.

An **occluded front** occurs when a faster-moving cold front overtakes a warm front, pushing all of the warm air aloft. This frontal situation usually occurs in the latter stages of a middle-latitude cyclonic storm, which will be discussed next. Map symbols for the four frontal types are shown in Figure 7.4.

Atmospheric Disturbances

Imbedded within the wind belts of the general atmospheric circulation (see Chapter 5) are secondary circulations. These are made up of storms and other atmospheric

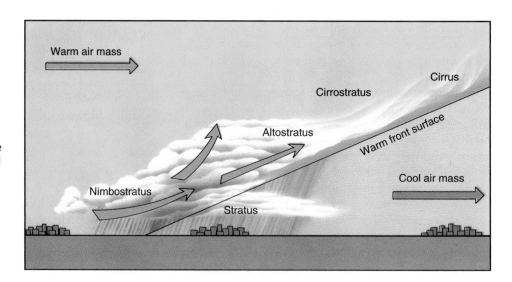

Figure 7.3 Cross section of a warm front. Warm fronts advance more slowly than cold fronts and replace rather than displace) cold air by sliding upward over it. The gentle rise of the warm air produces stratus clouds and gentle drizzles. ■ **Compare Figures 7.2 and 7.3. How are they different? How are they similar?**

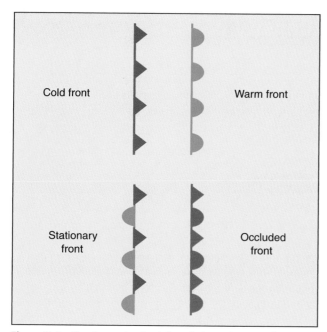

Figure 7.4 The four major frontal symbols used on weather maps

disturbances. We use the term **atmospheric disturbance** because it is more general than a storm and includes variations in the secondary circulation of the atmosphere that cannot be correctly classified as storms.

Partly because our primary interest is in the weather of North America, we will concentrate on an examination of **middle-latitude disturbances,** sometimes known as **extratropical disturbances.**

Shortly after World War I, Norwegian meteorologists Vilhelm and Jacob Bjerknes (a father and son) put forth the *polar front theory,* which provided insight into the development, movement, and dissipation of middle-latitude storms. They recognized the middle latitudes as an area of convergence where unlike air masses, such as cold polar air and warm subtropical air, commonly meet at a boundary called the *polar front.* Though the polar front may be a continuous boundary circling the entire globe, it is most often fragmented into several individual line segments. Furthermore, the polar front tends to move north and south with the seasons and is apt to be stronger in winter than in summer. It is along this wavy polar front that the upper air westerlies, also known as the *Polar Front Jet Stream,* develop and flow.

Middle-latitude storms develop at the front and then travel along it. These migrating storms, with their opposing cold, dry polar air and warm, humid tropical air, can cause significant variation in the day-to-day weather of the locations over which they pass. It is not unusual in some parts of the United States and Canada for people to go to bed at the end of a beautiful warm day in early spring and wake up to falling snow the next morning. Such variability is common for middle-latitude weather, especially during certain times of the year when

the weather changes from a period of cold, clear, dry days to a period of snow, only to be followed by one or two more moderate but humid days.

Cyclones and Anticyclones

Nature, Size, and Appearance on Maps We have previously distinguished cyclones and anticyclones according to differences in pressure and wind direction. Also, when studying maps of world pressure distribution, we identified large areas of semipermanent cyclonic and anticyclonic circulation in Earth's atmosphere (the subtropical high, for example). Now, when examining middle-latitude atmospheric disturbances, we will use the terms *cyclone* and *anticyclone* to describe the moving cells of low and high pressure, respectively, that drift with varying regularity in the path of the prevailing westerly winds. As systems of higher pressure, anticyclones are usually characterized by clear skies, gentle winds, and a general lack of precipitation. As centers for converging, rising air, the cyclones are the true storms of the middle latitudes, with associated fronts of various types.

As we know from experience, no middle-latitude cyclonic storm is ever exactly like any other. The storms vary in their intensity, their longevity, their speed, the strength of their winds, their amount and type of cloud cover, the quantity and kind of their precipitation, and the surface area they affect.

Since there is a seemingly endless variety of cyclones, we will describe "model cyclones" in the following discussions. Not every storm will act in the way we describe, but certain generalizations will be helpful in understanding middle-latitude cyclones.

A cyclone has a low pressure center; thus, winds tend to converge toward that center in an attempt to equalize pressure. If we visualize air moving in toward the center of the low pressure system, we can see that the air that is already at the center must be displaced upward. The lifting that occurs in a cyclone results in clouds and precipitation, and the warm air rises like a corkscrew.

Anticyclones are high pressure systems in which atmospheric pressure decreases toward the outer limits of the system. Visualizing an anticyclone, or high, we can see that air in the center of the system must be subsiding, in turn displacing surface air outward, away from the center of the system. Hence, an anticyclone has diverging winds. In addition, an anticyclone tends to be a fair-weather system; the subsiding air in its center increases in temperature and stability, reducing the opportunity for condensation.

We should note here that the pressures we are referring to in these two systems are relative. What is important is that in a cyclone, pressure decreases toward the center, and in an anticyclone, pressure increases toward the center. Furthermore, the intensities of the winds involved in these systems depend on the steepness of the

The Density of Air

Density is defined as the ratio of the mass of any substance to the volume it occupies. When dealing with the atmosphere, density can be explained as the number of air molecules held within a given volume of atmosphere. This can be expressed as grams/cubic centimeter or kilograms/cubic meter. For the purposes of our meteorological discussions, as the density of the atmosphere increases, so does its weight. In other words, the more molecules that are packed into a given volume of atmosphere, the heavier that volume becomes.

Density is often (but not always) dependent on the temperature of the air. According to Boyle's Law:

$$pV = f(T)$$

where:

$$p = \text{pressure (held constant)}$$

$$V = \text{volume}$$

$$f(T) = \text{a function of temperature}$$

In other words, volume and temperature are directly related: As temperature increases, so does the volume of air. If the volume of the air increases, the density of gas molecules within that air should decrease. Therefore warm air tends to have less density (and corresponding weight) than cold air.

Here is a case in which the way your body feels can deceive you! On a hot, humid day, the atmosphere feels oppressive. The air feels very heavy as you sluggishly move through the day. In actuality, the opposite is true. Warm, humid air is less dense and lighter than cold, dry air. When the air is cold, the gas molecules are closely packed within a given volume, and the overall density (and weight) increases. With warm air, on the other hand, the gas molecules are

pressure gradients (the change in pressure over a horizontal distance) involved. Thus, if there is a steep pressure gradient in a cyclone, with the pressure much lower at the center than at the outer portions of the system, the winds will converge toward the center with considerable velocity.

The situation is easier to visualize if we imagine these pressure systems as landforms. A cyclone is shaped like a basin (Fig. 7.5). If we are filling the basin with water, we know that the water will flow in faster the steeper the

sides and the deeper the depression. If we visualize an anticyclone as a hill or mountain, then we can also see that just as water flowing down the sides of such landforms will flow faster with increased height and steepness, so will the air blowing out of an area of very high pressure move rapidly.

On a surface weather map, cyclones and anticyclones are depicted by concentric isobars of increasing pressure toward the center of a high and of decreasing pressure toward the center of a low. Usually a high will cover a

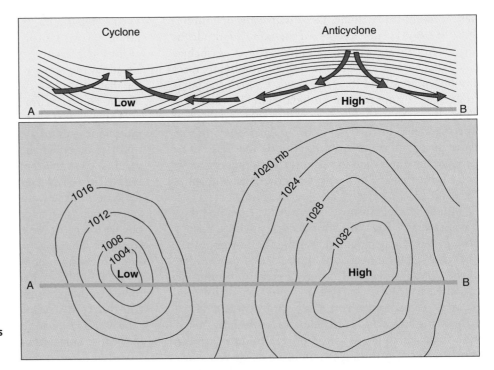

Figure 7.5 Close spacing of isobars around a cyclone or anticyclone indicates a steep pressure gradient that will produce strong winds. Wide spacing of isobars indicates a weaker system. ■
Where would the strongest winds be in this figure? Where would the weakest winds be?

o = Gas molecule

(a) With warmer air, the gas molecules are spread farther apart; the density is less.
(b) Cold air has greater density and corresponding weight.

spaced farther apart, there are fewer molecules per unit volume, and the density (and weight) of the air decreases.

Even when humidity is considered, the addition of moisture to the air adds little to the overall weight. Water (H_2O) is a remarkably small, lightweight molecule. Hydrogen, of course, is the lightest element known to exist, and oxygen (with an atomic mass of 16) is rather lightweight as well. Thus, humid air is not as heavy and as oppressive as it seems to be.

Remember our discussions of air masses and fronts? Boyle's law explains why cold air masses and dry air masses are heavier than warm air masses and moist air masses. As a result, when air masses meet along a frontal (or dry line) boundary, the less dense (warmer and lighter) air masses will override and rise above the denser (colder and heavier) air masses.

larger area than a low, but both pressure systems are capable of covering and affecting extensive areas. There are times when nearly the entire midwestern United States is under the influence of the same system. The average diameter of an anticyclone is about 1500 kilometers (900 mi); that of a cyclone is about 1000 kilometers (600 mi).

General Movement   The cyclones and anticyclones of the middle latitudes are steered, or guided, along a path reflecting the configuration and speed of the upper air westerlies (or the jet stream). The upper air flow can be quite variable, with wild oscillations. However, a general west-to-east pattern does prevail. Consequently, people in most of the eastern United States look at the weather occurring to the west to see what they might expect in

the next few days. Most storms that develop in the Great Plains or Far West move across the United States during a period of a few days at an average speed of about 36 kilometers per hour (23 mph), and then travel on into the North Atlantic before occluding.

Although neither cyclones nor anticyclones develop in exactly the same places at the same times each year, they do tend to develop in certain areas or regions more frequently than in others. They also follow the same general paths, known as **storm tracks** (Fig. 7.6). These storm tracks vary with the seasons. In addition, because the temperature variations between the air masses are stronger during the winter months, the atmospheric disturbances that develop in the middle latitudes during those months are greater in number and intensity.

Figure 7.6   Common storm tracks for the United States. Virtually all cyclonic storms move from west to east in the prevailing westerlies and swing northeastward across the Atlantic coast. Storm tracks originating in the Gulf of Mexico represent tropical hurricanes.   ■ **What storm tracks influence your location?**

187

Cyclones Now let us look more closely at cyclones—their origin, development, and characteristics. Warm and cold air masses meet at the polar front, where most cyclones develop. These two contrasting air masses do not merge but may move in opposite directions along the frontal zone. Although there may be some slight uplift of the warmer air along the edge of the denser, colder air, the uplift will not be significant. There may be some cloudiness and precipitation along such a frontal zone, though not of storm caliber.

For reasons not completely understood, but certainly related to the wind flow in the upper troposphere, a wavelike kink may develop along the polar front. This is the initial step in the formation of a fully grown middle-latitude cyclone (Fig. 7.7). At this bend in the polar front, we now have warm air pushing poleward (a warm front) and cold air pushing equatorward (a cold front), with a center of low pressure at the location where the two fronts are joined.

As the contrasting air masses jockey for position, the clouds and precipitation that exist along the fronts are greatly intensified, and the area affected by the storm is much greater. Along the warm front, precipitation will be more widespread but less intense than along the cold front. One factor that can vary the kind of precipitation occurring at the warm front is the stability of the warm air mass. If it is stable, then its uplifting over the cold air mass may cause only a fine drizzle—or a gentle, powdery snow if the temperatures are cold enough. On the other hand, if the warm air mass is moist and unstable, the uplifting may set off heavier precipitation. As you can see by referring again to Figure 7.3, the precipitation that falls at the warm front may *appear* to be coming from the colder air. Though weather may feel cold and damp, the precipitation is actually coming from the overriding warmer air mass above, then falling through the colder air mass to reach Earth's surface.

Because the cold front moves faster, it will eventually overtake the warm front. This produces the situation we previously identified as an *occluded front*. Since additional warm, moist air will not be lifted after occlusion, condensation and the release of latent heat energy will diminish, and the storm will soon die. Occlusions are usually accompanied by rain and are the major process by which middle-latitude cyclones dissipate.

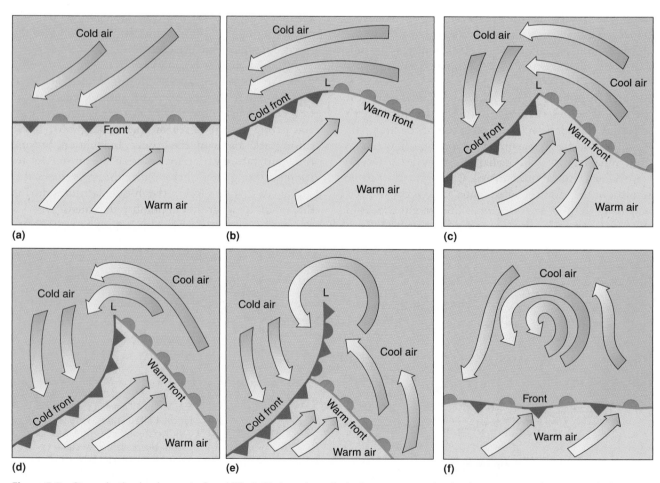

Figure 7.7 Stages in the development of a middle-latitude cyclone. Each view represents the development somewhat eastward of the preceding view as the cyclone travels along its storm track. Note the occlusion in (e). ■ **In (c), where would you expect rain to develop? Why?**

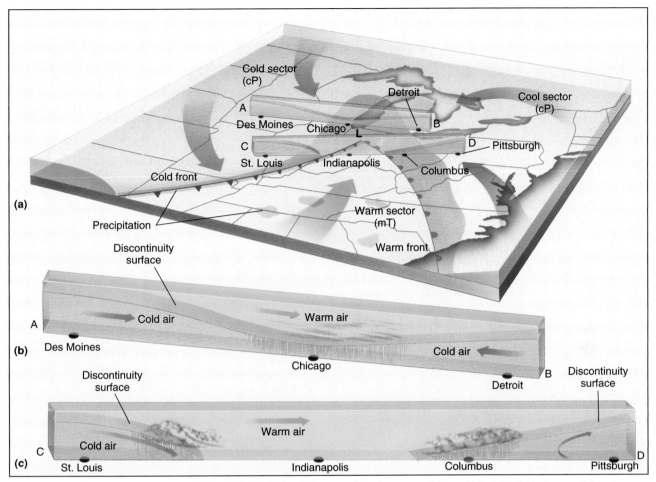

Figure 7.8   Environmental Systems: Middle-latitude Cyclonic Systems This diagram models a middle-latitude cyclone positioned over the Midwest as the system moves eastward: (a) a map view of the weather system; (b) a cross section along line AB north of the center of low pressure; (c) a cross section along line CD south of the center of low pressure.

Middle-latitude cyclones are dynamic systems created when air masses with different temperatures and humidity characteristics converge along the polar front. As the contrasting air masses seek equilibrium, fronts form between the warm and cold air. Along the fronts, warm, moist air is forced to rise and cools adiabatically, leading to condensation and often precipitation. Condensation supplies energy to the storm system by the release of latent heat. Much of the water vapor in this model was initially evaporated from the tropics. Middle-latitude storms play a significant role in transferring energy from lower latitudes to higher latitudes.

Typically, a cold front moves faster than a warm front and eventually overtakes it. At that time the air becomes stratified, with the warm air aloft and the cold air near the surface (part b). Because the air is now more or less in equilibrium, no additional warm air is rising, so both condensation and release of latent heat diminish, and the storm system soon dies.

Cyclones and Local Weather   Different parts of a middle-latitude cyclone exhibit different weather. Therefore, the weather that a location experiences at a particular time depends on which portion of the middle-latitude cyclone is over the location. Also, since the entire cyclonic system tends to travel from west to east, a specific sequence of weather can be expected at a given location as the cyclone passes over that location.

Let us assume that it is late spring. A cyclonic storm has originated in the southeast corner of Nebraska and is following a track (refer again to Fig. 7.6) across northern Illinois, northern Indiana, northern Ohio, through Pennsylvania, and finally out over the Atlantic Ocean. A view of this storm on a weather map, at a specific time in its journey, is presented in Figure 7.8a. Figure 7.8b

shows a cross-sectional view north of the center of the cyclone, and Figure 7.8c a cross-sectional view south of the center of the cyclone. As the storm continues eastward, at 9 to 13 meters per second (20–30 mph), the sequence of weather will be different for Detroit, where the warm and cold fronts will pass just to the south, than for Pittsburgh, where both fronts will pass overhead. To illustrate this point, let us examine, element by element, the variation in weather that will occur in Pittsburgh, with reference, where appropriate, to the differences that occur in Detroit as the cyclonic system moves east.

As we have previously stated, atmospheric temperature and pressure are closely related. As temperature increases, air expands and pressure decreases. Therefore, these two elements will be discussed together. Since a

Thomas G. Novak, Computer Training Consultant, Meteorlogix, Minneapolis, Minnesota

Early Interests: weather, geography, history, and sports. I remember hand-drawing weather maps when I was 8 years old. I could draw the United States in my sleep or with my eyes closed!

Education: 1997 graduate of the University of Minnesota, where I received my Bachelor's of Arts degree in Geography

No schools in the Twin Cities offered a degree in meteorology. Weather is my first love; my second love is maps and making maps. I didn't want to move, so I decided to go with geography. It was a can't lose situation. What many people don't understand is that geography offers you a well-rounded education. Geography gives you the whole gamut—meteorology, geology, climatology, cartography.

Prior Work Experience: Internship with the National Weather Service; President of the Upper Midwest Weather Consultants; Weather Broadcaster for KAAL TV

Meteorlogix is one of the largest commercial weather data providers in the world. We provide our weather data to anyone who needs them, such as broadcasting companies that display our data on beautifully textured maps that we provide. We provide our weather data to a variety of industries, including broadcast companies, TV stations, cable companies, farmers, and aviation.

My job is to go to each customer who needs training on our equipment and provide expert training. Some of the training involves computer data management, and the rest focuses on our software and how to create an aesthetically pleasing weather map. That's exactly where geography has helped me out. My job actually focuses more on geography than meteorology.

Today I've gotten several calls from weather broadcasters wanting to know how to construct a great-looking map that features Hurricane Isidore, currently in the Gulf of Mexico. If I'm not traveling around the country or world, I'm fielding calls in reference to our company's product.

Meanwhile, we've recently started implementing our weather data with GIS software. Being able to incorporate our weather data within a GIS system is going to be key to our future. That's where my geography background has come in extremely handy. One of the eye-opening aspects of my resume is that I have GIS experience and I understand what it is about.

Geography is playing a huge role in what I do now. In geography, I learned about meteorology, climatology, and cartography, which is extremely important in what I do. And just as important was GIS. What was stressed to me, as I was leaving the University of Minnesota, was that GIS will play a significant technological role in the future. In general,

people like to visualize data through maps, instead of just having statistics and numbers in front of them. GIS gives people the opportunity to visually see what is happening in their industry and in the market around them.

The one aspect of my geography degree that has helped me the most at Meteorlogix is knowing how to make an aesthetically pleasing map. It's a talent that most people don't possess. Creating an informative and aesthetically pleasing weather map usually incorporates knowledge of cartography, meteorology, and climatology. A Geography degree puts all these backgrounds together; it's like having a full basket rather than just a partial basket of knowledge. I often hear a lot of professional individuals say "How did you make such a nice-looking map?" Well, it comes down to years upon years of studying this kind of stuff, seeing what actually looks good. I wouldn't have that knowledge if it weren't for geography.

Number one, I would tell people to get some GIS knowledge and then wrap all the other geography knowledge base (cartography, meteorology, climatology, hydrology) in with it. A degree in Geography offers many different courses to choose from, and it is your responsibility to correctly navigate through them. I headed more for a physical geography background, but others may choose the human side of geography by getting into statistics, demographics, and the like.

Tell people to turn on the television and watch their local news. The map that the weather broadcasters are pointing to, I likely had a hand in creating that map and animation.

A degree in Geography offers a well-rounded education. I've always believed in being well rounded rather than having knowledge in only one thing. It can certainly open doors for you in several different career paths.

cyclonic storm is composed of two dissimilar air masses, there are usually important temperature contrasts. The sector of warm, humid *mT* air between the two fronts of the cyclone is usually considerably warmer than the cold *cP* air surrounding it. The temperature contrast is accentuated in the winter, when the source region for *cP* air is the cold cell of high pressure normally found in Canada at that time of year. During the summer, the contrast between these air masses is greatly reduced.

As a consequence of the temperature difference, the atmospheric pressure in the warm sector is considerably lower than the atmospheric pressure in the cold air behind the cold front. Far in advance of the warm front, the pressure is also high, but as the warm front (see Fig. 7.3) approaches, increasingly more cold air is replaced by overriding warm air, thus steadily reducing the surface pressure.

Therefore, as the warm front of this late-spring cyclonic storm approaches Pittsburgh, the pressure will decrease. After the warm front passes through Pittsburgh, where the temperature may have been 8°C (46°F) or more, the pressure will stop falling, and the temperature may easily rise to 18° to 20°C (64°–69°F) as *mT* air invades the area. At this point, Indianapolis has already experienced the passage of the warm front. After the cold front passes, the pressure will rise rapidly and the temperature will drop. In this late-spring storm, the *cP* air temperature behind the cold front might be 2° to 5°C (35°–40°F). Detroit, which is to the north of the center of the cyclone, will miss the warm air sector entirely and therefore will experience a slight increase in pressure and a temperature change from cool to cold as the cyclone moves to the east.

Changes in wind direction are one signal of the approach and passing of a cyclonic storm. Since a cyclone is a center of low pressure, winds flow counterclockwise toward its center. Also, winds are caused by differences in pressure. Therefore, the winds associated with a cyclonic storm are stronger in winter when the pressure (and temperature) differences between air masses are greatest.

In our example, Pittsburgh is located to the south and east of the center of low pressure and ahead of the warm front, and it is experiencing winds from the southeast. As the entire cyclonic system moves east, the winds in Pittsburgh will shift to the south-southwest after the warm front passes. Indianapolis is currently in this position. After the cold front passes, the winds in Pittsburgh will be out of the north-northwest. St. Louis has already experienced the passage of the cold front and currently has winds from the northwest. The changing direction of wind, clockwise around the compass from east to southeast to south to southwest to west and northwest, is called a **veering wind shift** and indicates that the center of a low has passed to the north of your position. On the other hand, Detroit, which is also experiencing winds from the southeast, will undergo a completely different sequence of directional wind changes as the cyclonic storm moves eastward. Detroit's winds will shift to the northeast as the center of the storm passes to the south. Chicago has just undergone this shift. Finally, after the storm has passed, the winds will blow from the northwest. Des Moines, to the west of the storm, currently has northwest winds. Such a change of wind direction, from east to northeast to north to northwest, is called a **backing wind shift,** as the wind "backs" counterclockwise around the compass. A backing wind shift indicates that you were north of the cyclone's center.

The type and intensity of precipitation and cloud cover also vary as a cyclonic disturbance moves through a location. In Pittsburgh, the first sign of the approaching warm front will be high cirrus clouds. As the warm front continues to approach, the clouds will thicken and lower. When the warm front is within 150 to 300 kilometers (90–180 mi) of Pittsburgh, light rain and drizzle may begin, and stratus clouds will blanket the sky.

After the warm front has passed, precipitation will stop and the skies will clear. However, if the warm, moist *mT* air is unstable, convective showers may result, especially with afternoon sun angles.

As the cold front passes, warm air in its path will be forced to move aloft rapidly. This may mean that there will be a cold, hard rain, but the band of precipitation normally will not be very wide because of the steep angle of the surface of discontinuity along a cold front. In our example, the cold front and the band of precipitation have just passed St. Louis.

Thus, Pittsburgh can expect three zones of precipitation as the cyclonic system passes over its location: (1) a broad area of cold showers and drizzle in advance of the warm front; (2) a zone within the moist, subtropical air from the south where there can be scattered convectional showers; (3) and a narrow band of hard rainfall associated with the cold front (Fig. 7.8c). However, locations to the north of the center of the cyclonic storm, such as Detroit, will usually experience a single, broad band of light rains resulting from the lifting of warm air above cold air from the north (Fig. 7.8b). In winter, the precipitation is likely to be snow, especially in locations just to the northwest of the center of the storm, where the humid *mT* air overlies extremely cold northern air.

As you can see, different portions of a middle-latitude cyclone are accompanied by different weather. If we know where the cyclone will pass relative to our location, we can make a fairly accurate forecast of what our weather will be like as the storm moves east along its track (see Map Interpretation: Weather Maps).

Cyclones and the Upper Air Flow   The upper air wind flow greatly influences our surface weather. We have already discussed the role of these upper air winds in the steering of surface storm systems. Another less obvious

WEATHER MAPS

The Map

One of the most widely used weather maps is a *surface weather map*. These maps, which portray meteorological conditions over a large area at a given moment in time, are important for current weather descriptions and forecasting. Simultaneous observations of atmospheric pressure, temperature, humidity, precipitation measurements, cloud cover, visibility, and wind velocity are taken at locations across the United States and worldwide. This information is electronically relayed to the National Meteorological Center near Washington, D.C., where the surface data are analyzed and mapped.

The surface map provides a wealth of information. Atmospheric conditions observed at specified weather stations are displayed by using an internationally accepted symbol code. Meteorologists at the Center then use the individual pieces of information to develop and depict the general weather picture over a larger area. For example, isobars (lines of equal air pressure) are drawn to reveal the locations of cyclones (L) and anticyclones (H). Patterns of temperature and wind directions clearly indicate frontal boundaries. Areas that were receiving precipitation at the time the map depicts are shaded so that these patterns are highlighted. The end result is a map of surface weather conditions that can be used in forecasting future changes in weather patterns.

Interpreting the Map

1. Isobars are lines of equal atmospheric pressure expressed in millibars. What is the interval (in millibars) between adjacent isobars on this weather map?

2. Based on the depicted pattern of isobars, are surface winds stronger over Iowa or Georgia? Explain your choice.

3. Study the map of upper air flow (below, left). What path will the center of low pressure likely follow in the next 24 hours?

4. Circulation around a Northern Hemisphere low-pressure area is inward, upward, and counterclockwise. Identify the wind direction at several stations to confirm this circulation pattern.

5. What air mass is influencing Atlanta, Georgia; Washington, D.C.; Wichita, Kansas?

6. Does the surface map accurately depict the cloud cover indicated on the accompanying satellite image?

7. At a location where precipitation is not occurring, how could you use the station information to forecast the *type* of precipitation that might occur in the near future?

8. Describe the current weather in Indiana. In a *relative* sense, focusing on changes in wind speed and direction, atmospheric pressure, cloud cover, precipitation, and temperature, how do you expect the weather will change in the next 24 to 36 hours?

9. Do the locations of the fronts and areas of precipitation depicted on the map agree with the idealized relationship represented in Figure 7.8 (Middle-latitude Cyclonic Systems).

NOAA/ GOES satellite image

A satellite image of the atmospheric conditions shown on the accompanying weather map.

**Opposite:
(Top) Surface weather maps depicting conditions on a mid-April day at 7 AM EST. (Lower left) The upper air wind flow at 500 millibars (approximately 18,000 ft) for the same period. (Lower right) Station model illustrating the type of information reported at each weather station.**

1012   1008   1008   1012   1016

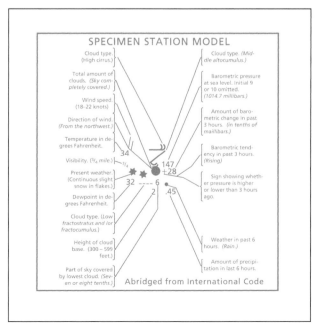

SPECIMEN STATION MODEL

Cloud type. (High cirrus.)

Total amount of clouds. (Sky completely covered.)

Wind speed. (18–22 knots.)

Direction of wind. (From the northwest.)

Temperature in degrees Fahrenheit.

Visibility. (3/4 mile.)

Present weather. (Continuous slight snow in flakes.)

Dewpoint in degrees Fahrenheit.

Cloud type. (Low fractostratus and/or fractocumulus.)

Height of cloud base. (300 – 599 feet.)

Part of sky covered by lowest cloud. (Seven or eight tenths.)

Cloud type. (Middle altocumulus.)

Barometric pressure at sea level. Initial 9 or 10 omitted. (1014.7 millibars.)

Amount of barometric change in past 3 hours. (in tenths of millibars.)

Barometric tendency in past 3 hours. (Rising)

Sign showing whether pressure is higher or lower than 3 hours ago.

Weather in past 6 hours. (Rain.)

Amount of precipitation in last 6 hours.

34

3/4

32   6   147   +28

2   .45

Abridged from International Code

Figure 7.9 The upper air wind pattern, such as that depicted here can have a significant influence on temperatures and precipitation on Earth's surface. ■ **Where would you expect storms to develop?**

influence of the upper air flow is related to the undulating, wavelike flow so often exhibited by the upper air. As the air moves its way through these waves, it undergoes divergence or convergence because of the rather complex dynamics associated with curved flow. This upper air convergence and divergence greatly influence the surface storms below.

The region between a ridge and the next downwind trough (A–B in Fig. 7.9) is an area of upper-level convergence. In our atmosphere, an action taken in one part of the atmosphere is compensated for by an opposite reaction somewhere else. In this case, the upper air convergence is compensated for by divergence at the surface. In this area, anticyclonic circulation is promoted. This pattern will inhibit the formation of a middle-latitude storm or cause an existing storm to weaken or even dissipate. On the other hand, the region between a trough and the next downwind ridge (B–C in Fig. 7.9) is an area of upper-level divergence, which in turn is compensated

for by surface convergence. This is an area where cyclonic circulation is encouraged. Convergence at the surface will certainly enhance the prospects of storm development or strengthen an already existing storm.

In addition to storm development or dissipation, upper air flow will have an impact on temperatures as well. If we assume that our "average" upper air flow is from west to east, then any deviation from that pattern will cause either colder air from the north or warmer air from the south to be advected into an area. Thus, after the atmosphere has been in a wavelike pattern for a few days, the areas in the vicinity of a trough (area B in Fig. 7.9) will be colder than normal as polar air from higher latitudes is brought into that area. Just the opposite occurs at locations near a ridge (area C in Fig. 7.9). In this case, warmer air from more southerly latitudes than would be the case with west-to-east flow is advected into the area near the ridge.

Weather Forecasting Weather forecasting, at least in principle, is fairly straightforward. Meteorological observations are made, collected, and mapped to depict the current state of the atmosphere. From this information, the probable movement, as well as any anticipated growth or decay, of the current weather systems is projected for a specific amount of time into the future.

When a forecast is wrong—which we all know occurs—it is usually either because limited or erroneous information has been collected and processed in the first place or, more likely, because errors have been made in anticipating the path or growth of the storm systems. Little errors will compound themselves over time. For example, a few degrees' shift in a storm's path may result in an error of a few miles in the projected location of that storm in a 2-hour forecast. However, this same few-degree error may result in a projected locational error of hundreds of miles in a 48-hour forecast. Consequently, the further into the future one tries to forecast, the greater the chance of error.

Figure 7.10 GOES satellite image of a low pressure system entering the West Coast of North America.

Although forecasts are not perfect, they are much better today than in the past. Much of this improvement can be attributed to our current sophisticated technology and equipment. Increased knowledge and surveillance of the upper atmosphere have improved the accuracy of weather prediction. Weather satellites have helped tremendously by providing meteorologists with a better understanding of weather and weather systems. They have been of particular value to forecasters on the west coast of the United States (Fig. 7.10). Before the advent of weather satellites, these forecasters had to rely on information relayed from ships, leaving enormous areas of the Pacific unobserved. Thus, forecasters were often caught off-guard by unexpected weather events.

In addition, high-speed computers allow rapid processing and mapping of observed weather conditions. Computers also allow the processing of numerical forecasts, which are based on the solution of physical equations that govern our atmosphere. Numerical forecasts and long-term forecasts based on solving statistical relationships and equations would not be possible without computers. In fact, computers now play such an important role in forecasting that some of the world's largest and fastest computers are used to forecast the weather.

Though forecasters now possess a great deal of knowledge and a variety of highly sophisticated devices that were previously unavailable, these devices are not foolproof. Understanding some of the problems the weather forecaster faces may make us more understanding when a forecast fails. No one can promise a sunny day. Nor can anyone say that it will definitely rain tomorrow, for no one can truly predict the future. The weather forecaster combines science and art, fact and interpretation, data and intuition, to come up with some probabilities about future weather conditions.

Anticyclones Just as cyclones are centers of low pressure that are typified by the convergence of air, so anticyclones are cells of high pressure in which air diverges. The subsidence of air in the center of an anticyclone encourages stability, as the air is warmed adiabatically while sinking toward the surface. Consequently, the air is able to hold additional moisture as its capacity increases with increasing temperatures. The weather resulting from the influence of an anticyclone is often clear, with no rainfall. There are, however, certain conditions under which there can be some precipitation within a high pressure system. When such a system passes near or crosses a large body of water, the resulting evaporation can cause variations in humidity significant enough to result in precipitation.

There are two sources for the relatively high pressures that are associated with anticyclones in the middle latitudes of North America. Some anticyclones move into the middle latitudes from northern Canada and the Arctic Ocean, in what are called outbreaks of cold arctic air. These outbreaks can be quite extensive, covering much

of the midwestern and eastern United States. The temperatures in an anticyclone that has developed in a *cA* air mass can be markedly lower than those expected for any given time of year. They may be far below freezing in the winter. Other anticyclones are generated in zones of high pressure in the subtropics. When they move across the United States toward the north and northeast, they bring waves of hot, clear weather in summer and unseasonably warm days in the winter months.

Thunderstorms

Thunderstorms are common local storms of the middle and lower latitudes. Very simply, a thunderstorm is a storm accompanied by thunder and lightning. Lightning is an intense discharge of electricity. For lightning to occur, positive and negative electrical charges must be generated within a cloud. It is believed that the intense friction of the air on moving ice particles within a cumulonimbus cloud generates these charges. A clustering of positive charges tends to occur in the upper portion of the cloud, with negative charges clustering in the lower portion. When the potential difference between these charges becomes large enough to overcome the natural insulating effect of the air, a lightning flash, or discharge, takes place. These discharges, which often involve over a million volts, can occur within the cloud, between two clouds, or from cloud to ground. The air immediately around the discharge is momentarily heated to temperatures in excess of 25,000°C (45,000°F)! The heated air expands explosively, creating the shock wave we call thunder (Fig. 7.11).

Figure 7.11 A cross-sectional view of a thunderstorm showing the distribution of electrical charges

(a) Convectional uplift   **(b) Orographic uplift**

Figure 7.12   Thermal convection and orographic uplift.

Thunderstorms usually cover a small area of a few miles, although there may be a series of related thunderstorms covering a larger region. The intensity of a thunderstorm depends on the degree of instability of the air and the amount of water vapor it holds. A thunderstorm will die out when most of its water vapor has condensed, for there will no longer be energy available for continued vertical movement. In fact, most thunderstorms last less than an hour.

As an intense form of precipitation, thunderstorms result from the uplift of moist air. As is the case for other types of precipitation, the trigger mechanism causing that uplift can be thermal convection (warm unstable air rising on a warm afternoon, Fig. 7.12a), orographic uplift (warm moist air ramping up a mountain side, Fig. 7.12b), or frontal uplift (see again Figs. 7.2 and 7.3).

Convective thunderstorms are most common in lower latitudes during the warmer months of the year and during the warmer hours of the day. It is apparent, then, that the amount of solar heating affects the development of thunderstorms. This is true because the intense heating of the surface steepens the environmental lapse rate, which in turn leads to increased instability of the air, allowing for greater moisture-holding capacity and adding to the buoyancy of the air.

Orographic thunderstorms occur when air is forced to rise over land barriers, providing the necessary initial trigger action leading to the development of thunderstorm cells. Thunderstorms of orographic origin play a large role in the tremendous precipitation of the monsoons of South and Southeast Asia. In North America, they occur over the mountains in the West (the Rockies and the Sierra Nevada), especially during summer afternoons, when heating of south-facing slopes increases the air's instability. For this reason, pilots of small planes try to avoid flying in the mountains during summer afternoons for fear of getting caught in the turbulence of a thunderstorm.

Frontal thunderstorms are often associated with cold fronts, where a cooler air mass forces a warmer air mass to rise. This action can bring about the strong, vertical updrafts necessary for precipitation. In fact, a cold front may be immediately preceded by a line of thunderstorms (a squall line) resulting from such frontal uplift.

As we mentioned in the discussion of precipitation types in Chapter 6, hail can be a product of thunderstorms when the vertical updrafts of the cells are sufficiently intense to carry water droplets repeatedly into a freezing layer of air. Fortunately, since thunderstorms are primarily associated with warm-weather areas, only a very small percentage of storms around the world produce hail. In fact, hail seldom occurs in thunderstorms in the lower latitudes. In the United States, there is little hail along the Gulf of Mexico, where thunderstorms are most common.

Tornadoes

Tornadoes are the most violent storms on the face of Earth (Fig. 7.13). They can occur almost anywhere, but are far more common in the interior of North America than any other country in the world (Fig. 7.14). In fact, Oklahoma and Kansas lie in the path of so many "twisters" that together they are sometimes referred to as "tornado alley."

Systematic government documentation of tornado activity, such as that depicted in Figure 7.14a, began in 1875. Accounts of tornadoes occurring prior to 1875 must be tracked down through other sources. These accounts, though often unverifiable and vague, do offer interesting and informative insights into our forebears' perceptions of tornadoes. The accounts below describe a tornado that killed several people as it swept across several counties in western Illinois on May 21, 1859.

It was "a violent storm or hurricane [which] did immense damage to houses, barns, fences, and also caused some destruction of life." It was described as having a "frightful, . . . balloon or funnel shape, and appeared . . . peculiarly bright and luminous, not at all black or dark in any of its parts, except its base or bottom." A vivid account of what surely must be related to the output of static electricity associated with a tornado is given in this account

AP/ Wide World Photos

Figure 7.13 Tornado damage at a trailer park near Chesapeake, Missouri in December, 2002.

of the same tornado as it swept across Morgan county: "Mr. Cow-ell was plowing his field. . . . He saw the frightful cloud approaching . . . and at once attempted to drive his horses and plow to the house. . . . The horses suddenly took fright . . . their manes and tails and all their hair 'stood right out straight' as he expressed it, and . . . the iron in the harness . . . and plow, in his language 'seemed all covered with fire.' He felt a violent pulling of his own hair which left 'his head sore for some days' and the hair itself rigid and inflexible." In addition, although uncon-firmed by others, Mr. Cowell was one of the few individuals to have a tornado pass directly over him and live to tell about it. He described the light in the center of the tornado as being "so bril-liant that he could not endure it with his eyes open, and for the most part kept them shut. . . . Yet [inside the tornado] there was no wind, no thunder and no noise whatever. . . ." Another inter-esting feature of this same tornado can be attributed to the low pressure of the vortex: "When the terrific whirl struck . . . [it] stripped all of the feathers off from the hens and turkeys, as per-fectly clean as if picked for the table. Some, though badly plucked, and made entirely blind, still lived." Such a bizarre occurrence probably resulted when the hollow quills of the feathers expanded so suddenly — as the low pressure vortex moved over the area — that the birds' feathers "exploded."

Transactions of the Illinois Natural History Society
Phillips Bros., 1861

A **tornado** is actually a small, intense cyclonic storm of very low pressure, violent updrafts, and converging winds of enormous contrast. Fortunately, they are small and short-lived. Even in tornado alley, a tornado is likely to strike a given locale only once in 250 years.

Although only 1 percent of all thunderstorms pro-duce a tornado, 80 percent of all tornadoes are associ-ated with thunderstorms and middle-latitude cyclones.

The remaining 20 percent of tornadoes are spawned by hurricanes that make landfall. Approximately 700 to 1000 tornadoes occur each year in the United States, most of them from March to July in the late afternoon or early evening in the central part of the country.

Because of their small size and limited life span, tor-nadoes are extremely difficult to detect and forecast. However, relatively new radar technology, **Doppler radar,** improves tornado detection and forecasting sig-nificantly. Doppler radar has more power concentrated in a narrower beam than previous radar units. This al-lows meteorologists to assess storms in much greater de-tail (Fig. 7.15). Even more important is a Doppler radar's ability to measure the speed of the wind flow toward or away from the radar site, as well as any turbulence or sheer within the wind flow.

When the energy emitted by radar strikes precipita-tion, a small portion is scattered back to the radar. De-pending on whether the precipitation is moving toward or away from the radar site, the wavelength of the re-turned energy is either compressed or elongated. The faster the movement, the greater the change. Previous radars could not measure this change; however, Doppler radar does and uses it to estimate the wind circulation within the storm.

In fact, Doppler radar is so sensitive that it can detect the wind pattern in clear air by detecting the back-scattered energy from clouds, pollution, insects, and so forth. This is something previous radars could not achieve. It allows meteorologists to see the formation of a tornado, thus increasing warning time to the public. It also permits the detection, around airports, of clear air turbulence (CAT), a major factor in airline accidents. The United States government, through the NEXRAD

(a)

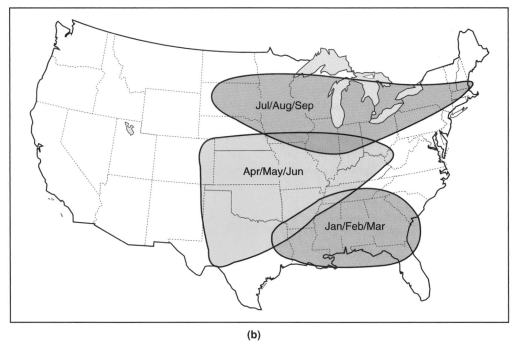

(b)

Figure 7.14 (a) Average number of tornadoes per 26,000 square kilometers (10,000 sq mi). (b) Seasonal march of peak tornado activity.

program (NEXt-generation weather RADar), has installed 151 Doppler radar sites across the country.

Doppler radar observations indicate that most tornadoes (63 percent) are fairly weak, with wind speeds of 45 meters per second (100 mph) or less. About 35 percent of tornadoes can be classified as strong, with wind speeds reaching 90 meters per second (200 mph). Nearly 70 percent of all tornado fatalities result from violent tornadoes. Although very rare (only 2 percent of all tornadoes reach this stage), these may last for hours and have wind speeds in excess of 135 meters per second (300 mph).

Ron Holmes/ National Weather Service/ NOAA

Figure 7.15 Doppler radar image taken at Missoula, Montana shows various levels of precipitation.

Before Doppler radar, wind speeds within a tornado could not be measured directly. Therefore, tornado intensity was estimated from the damage produced by the storm. The most commonly used scale of tornado intensity was developed by the late T. Theodore Fujita, a former professor at the University of Chicago. The scale is termed the Fujita Intensity Scale or, more commonly, the F-scale (Table 7.2).

A tornado first appears as a swirling, twisting funnel cloud that moves across the landscape at 10 to 15 meters per second (22–32 mph). Its narrow end may be only 100 meters (330 ft) across. The funnel cloud becomes a tornado when its narrow end is in contact with the ground, where the greatest damage is done. Above the ground, the end can swirl and twist, but little or nothing is done to the ground below. The color of a tornado

Table 7.2 Fujita Tornado Intensity Scale

(F Scale)	Wind Speed (KPH)	(MPH)	Expected Damage
F–0	<116	<72	Light Damage Damage to chimneys and billboards; broken branches; shallow-rooted trees pushed over.
F–1	116–180	72–112	Moderate Damage The lower limit is near the beginning of hurricane wind speed. Surfaces peeled off roofs; mobile homes pushed off foundations or overturned; moving autos pushed off the road.
F–2	181–253	113–157	Considerable Damage Roofs torn off houses; mobile homes demolished; boxcars pushed over, large trees snapped or uprooted; light-object missiles generated.
F–3	254–332	158–206	Severe Damage Roofs and some walls torn off well-constructed houses; trains overturned; most trees in forest uprooted; heavy cars lifted off ground and thrown.
F–4	333–419	207–260	Devastating Damage Well-constructed houses leveled; structures with weak foundations blown ome distance; cars thrown and large missiles generated.
F–5	>419	>260	Incredible Damage Strong frame houses lifted off foundations and carried considerable distance to disintegrate; automobile-sized missiles fly through the air farther than 100 m; trees debarked; incredible phenomena occur.

NOAA/ NOSC/ HSEI

Figure 7.16   Enhanced GOES image of Hurricane Fran as it moves up the East Coast of the United States during October 1996.

can be milky white to black, depending on the amount and direction of sunlight and the type of debris being picked up by the storm as it travels across the land. Although most tornado damage is caused by the violent winds, most tornado injuries and deaths result from flying debris. The small size and short duration of a tornado greatly limit the number of deaths caused by tornadoes. In fact, more people die from lightning strikes each year than from tornadoes.

Weak Tropical Disturbances

Until World War II, the weather of tropical regions was described as hot and humid, generally fair, but basically pretty monotonous. The only tropical disturbance given any attention was the tropical cyclone (also called a *hurricane* or, in other parts of the world, a *typhoon*), a spectacular but relatively uncommon storm that affects only islands, coastal lands, and ships at sea (Fig. 7.16).

Even a few decades ago, an aura of mystery remained about the weather of the tropics. One reason for this lack of information was that the few weather stations located in tropical areas were widely scattered and often poorly equipped. As a result, it was difficult to understand completely the passing weather disturbances in the tropics.

Largely through satellite technology and computer analysis, it is now known that a variety of weak atmo-spheric disturbances affect the weather and relieve the monotony, although it is likely that the full number of these disturbances has not yet been recognized. The primary impact of these weak tropical disturbances on the weather of tropical regions is not on the temperature but rather on the cloud cover and the amount of precipitation. Temperatures in the tropics are largely unaffected during the passage of a tropical storm, except that as the cloud cover increases, temperature extremes are reduced.

Easterly Wave   The best known of the weak tropical disturbances is the **easterly wave.** It shows up in Figure 7.17 as a trough-shaped, weak low pressure region that is generally aligned on an approximate north-south axis. Traveling slowly in the trade wind belt from east to west, it is preceded by fair, dry weather and followed by cloudy, showery weather. This occurs because air tends to converge into the low from its rear, or the east, causing lifting and convectional showers. The resulting divergence and subsidence to the west account for the fair weather. Meteorologists believe that this type of disturbance can, on occasion, develop into a tropical hurricane.

Polar Outbreak   Occasionally, an outbreak of polar air may follow a low into the subtropics and tropics. Such an outbreak would, of course, be preceded by the squalls

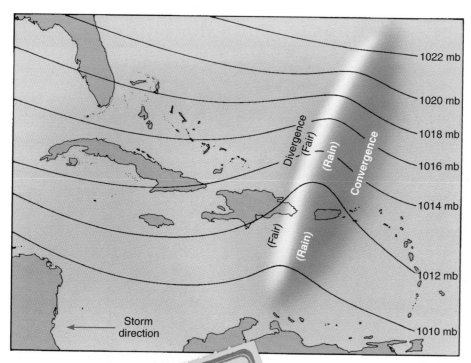

1022 mb
1020 mb
1018 mb
1016 mb
1014 mb
1012 mb
1010 mb

Divergence (Fair)
(Rain)
Convergence
(Fair)
(Rain)

Storm direction

Figure 7.17 A typical easterly wave in the tropics. Note that the isobars (and resulting winds) do not close in a circle but merely make a poleward "kink," indicating a low pressure trough rather than a closed cell. The resulting weather is a consequence of convergence of air coming into the trough, producing rains, and divergence of air coming out of the trough, producing clear skies. ■ **Why do easterly waves move toward the west?**

and clouds and rain. . . . . Following, however, would . . . . eather as the modified polar . . . . occasions near the equator . . . . an Antarctic outburst, kno . . . . ng freezing temperatures ar. . . . . tion. Farther to the sou . . . . e coffee crop can be ruined, causin. . . . . s in North America to rise.

Hurricanes

Hurricanes are severe tropical cyclones that receive a great deal of attention from scientists and laypeople alike, primarily because of their great destructive powers (Fig. 7.18). Abundant, even torrential, rains and winds often exceeding 45 meters per second (100 mph) characterize hurricanes. Though hurricanes develop over the oceans, their paths at times do take them over islands

AP/Wide World Photos

Figure 7.18 Damage to beachfront homes in North Carolina by Hurricane Fran in October 1996.

Hurricane Paths and Landfall Probability Maps

Hurricanes (typhoons) are generated over tropical or subtropical oceans and then tend to build strength as they move over regions of warm ocean water. Ships and aircraft regularly avoid hurricane paths by navigating away from these huge violent storms. Residents of landmasses in the path of an oncoming hurricane try to prepare their homes and other structures, and may have to evacuate if the hazard potential of the impending storm is great enough.

Landfall refers to the location where the storm, particularly the most severe part of the hurricane, encounters (or will encounter) the coastline. Generally in the Northern Hemisphere, the location where the northeast quadrant of these circular storms strikes the coast is the area most at risk for the highest amount of damage. This is because the winds are usually strongest in that quadrant, and they tend to blow directly toward the coastline, driving ocean water toward

the shoreline. Here the potential is greatest for a storm surge, where the ocean violently washes over and floods low-lying coastal areas. Storm surges are the most dangerous condition associated with hurricanes. In 1900, some 6000 residents of Galveston Island in Texas were killed when a hurricane pushed a 7-meter high wall of water over the island. Much of the city was destroyed by this storm,

Figure 7.19 As a hurricane moves ashore, a storm surge combines with the normal high tide to create a storm tide. This mound of water, topped by battering waves, moves ashore along an area of the coastline as much as 100 miles wide. The combination of the storm surge, battering waves, and high winds is deadly.

the worst natural disaster ever to occur in the United States in terms of lives lost.

With today's better technology for tracking and evaluating tropical storms, providing adequate warnings to coastal residents might seem like a simple process. Maps of prior hurricane paths, however, show that predicting exactly which way a hurricane will move based on its previous movement is very difficult. Computer models, developed from maps of past storm behavior, are used to indicate the storms' most likely path and landfall location, as well as the chance that it may strike the coast at different locations. The nearer the storm is to the coast, the more accurate the predicted landfall site should be, but in some cases the storm may begin to move in a completely different direction. Nature still remains unpredictable, so potential landfall sites are shown on probability maps, to help local authorities and residents decide what course of action is best to take in preparing for the storm. A probability of 90 percent means that similar storms, 9 times out of 10, have moved onshore in that direction. A probability of 60 percent means that six of ten moved as indicated, and so forth. In recent years the National Weather Service has worked hard to develop computer models that yield better predictions of hurricane paths and landfalls.

Here is an example of a landfall probability map for hurricane Isidore. If you live in a coastal area affected by tropical storms, understanding landfall probability maps may be very important to your safety and ability to prepare for a potential storm.

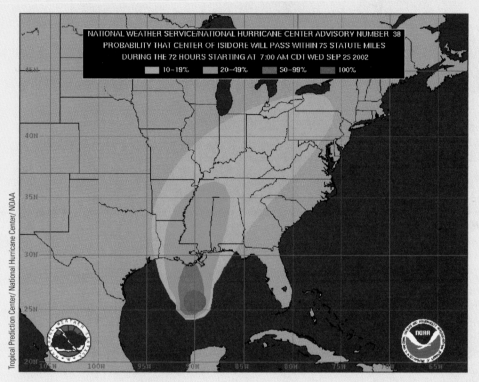

NATIONAL WEATHER SERVICE/NATIONAL HURRICANE CENTER ADVISORY NUMBER 38
PROBABILITY THAT CENTER OF ISIDORE WILL PASS WITHIN 75 STATUTE MILES
DURING THE 72 HOURS STARTING AT 7:00 AM CDT WED SEP 25 2002
10–19%   20–49%   50–99%   100%

Tropical Prediction Center/ National Hurricane Center/ NOAA

and coastal lands. The results can be devastating destruction of property and loss of life. It is not just the rains and winds that cause damage. Accompanying the hurricane are unusually high seas, called **storm surges,** which can flood entire coastal communities (Fig. 7.19).

A **hurricane** is a circular, cyclonic system with wind speeds in excess of 33 meters per second (74 mph). It has a diameter of 160 to 640 kilometers (100–400 mi). Extending upward to heights of 12 to 14 kilometers (40,000–45,000 ft), the hurricane is a towering column of spiraling air (Fig. 7.20). At its base, air is sucked in by the very low pressure at the center and then spirals inward. Once within the hurricane structure, air rises rapidly to the top and spirals outward. This rapid upward movement of moisture-laden air produces enormous amounts of rain. Furthermore, the release of latent heat energy provides the power to drive the storm.

At the center of the hurricane lies the eye of the storm, an area of calm, clear, usually warm and humid but rainless air. Sailors traveling through the eye have been surprised to see birds flying there. Unable to leave the eye because of the strong winds surrounding it, these birds will often alight on the passing ship as a resting spot.

Hurricanes have very strong pressure gradients because of the extreme low pressure at their centers. The strong pressure gradients in turn cause the powerful winds of the hurricane. In contrast to the middle-latitude cyclone, a hurricane is formed from a single air

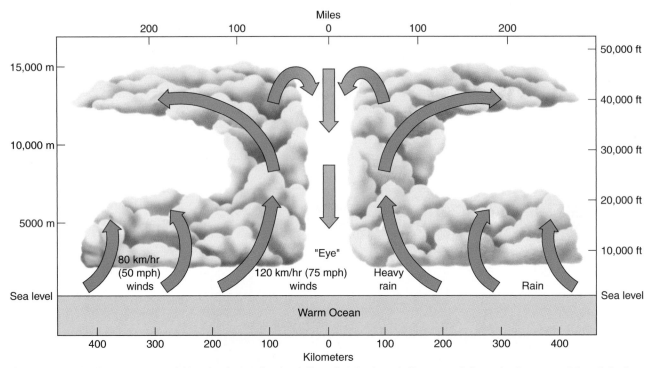

Figure 7.20   Circulation patterns within a hurricane, showing inflow of air in the spiraling arms of the cyclonic system, rising air in the towering circular wall cloud, and outflow in the upper atmosphere. Subsidence of air in the storm's center produces the distinctive calm, cloudless "eye" of the hurricane. ■ **Why is this so?**

mass and does not have the different temperature sectors like a frontal system. Rather, a hurricane has a fairly even, circular temperature distribution, which we might have been led to expect from its circular winds.

The Saffir-Simpson hurricane scale provides a means of classifying hurricane intensity, and potential damage, by assigning a number from 1 to 5 based on a combination of central pressure, wind speed, and the height of the storm surge (Table 7.3).

Although a great deal of time, effort, and money has been spent on studying the development, growth, and paths of hurricanes, much is still not known. For example, it is not yet possible to predict the path of a hurri-

cane, even though it can be tracked with radar and studied by planes and weather satellites. In addition, meteorologists can list factors favorable for the development of a hurricane but cannot say that in a certain situation a hurricane will definitely develop and travel along a particular path. As with tornadoes, there are also "hurricane alleys," or areas where their development is more likely to occur (Fig. 7.21).

Among the factors leading to hurricane development are a warm ocean surface of about 25°C (77°F) and warm, moist overlying air. These factors are probably the reasons why hurricanes occur most often in the late summer and early fall, when air masses have maximum hu-

Table 7.3	Saffir-Simpson Hurricane Scale					
Scale Number	Central Pressure	Wind Speed		Storm Surge		Damage
(Category)	(Millibars)	(KPH)	(MPH)	(Meters)	(Feet)	
1	≥980	119–153	74–95	1.2–1.5	4–5	Minimal
2	965–979	154–177	96–110	1.6–2.4	6–8	Moderate
3	945–964	178–209	111–130	2.5–3.6	9–12	Extensive
4	920–944	210–250	131–155	3.7–5.4	13–18	Extreme
5	<920	>250	>155	>5.4	>18	Catastrophic

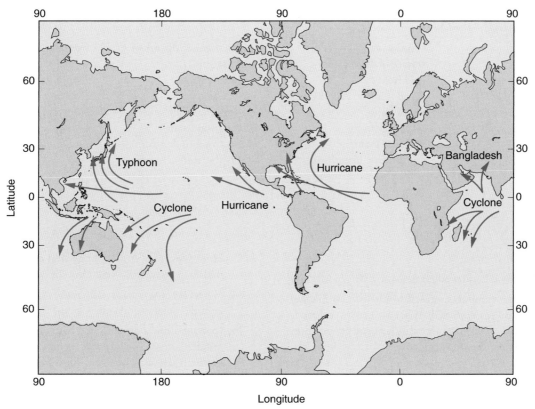

Figure 7.21 A world map showing major "Hurricane Alleys"

midity and ocean surface temperatures are highest. Also, the Coriolis effect must be sufficient to support the rapid spiraling of the hurricane. Therefore, hurricanes neither develop nor survive in the equatorial zone from about 8°S to 8°N, for there the Coriolis effect is far too weak. Hurricanes begin as weak tropical disturbances such as the easterly wave and, in fact, will not develop without the impetus of such a disturbance. It is further speculated that some sort of turbulence in the upper air may play a part in a hurricane's initial development.

Names are assigned to storms once they reach tropical storm status, with winds greater than 17 meters per second (38 mph). Each year the names are selected from a different list of alternating female and male names—one list for the North Atlantic and one for the North Pacific. If a hurricane is especially destructive, its name is retired and never used again.

Hurricanes do not last long over land, because their source of moisture (and consequently their source of energy) is cut off. Also, friction with the land surface produces a drag on the whole system. North Atlantic hurricanes that move first toward the west with the trades, and then north and northeast as they intrude into the westerlies, become polar cyclones if they remain over the colder region of the ocean and eventually die out.

Over land, they will also become simple cyclonic storms, but even when they have lost some of their power, hurricanes can still do great damage.

Hurricanes can occur over most subtropical and tropical oceans and seas (the South Atlantic is an exception, though it is not known why). In Australia and the South Pacific, they are called tropical cyclones (or *willy-willies*). Near the Philippines, they are known as *bagyos*, but in most of East Asia they are called **typhoons.** In the Bay of Bengal, they are referred to as *cyclones*.

Two hurricanes caused extensive damage in the United States during 1992. Hurricane Iniki struck the Hawaiian Islands, doing extensive damage to the island of Kauai. On August 24, 1992, Hurricane Andrew crashed into the populated areas of southern Florida with wind speeds of 150 miles per hour and then proceeded across the Gulf of Mexico to Louisiana. Andrew was the costliest natural disaster in U.S. history, causing more than $20 billion in damage.

Some people have suggested that we seek ways to control these destructive storms. On the other hand, hurricanes are a major source of rainfall and an important means of transferring energy within Earth's system away from the tropics. Eliminating them might cause unwanted and unforeseen climate changes.

DEFINE & RECALL

air mass
source region
Maritime Equatorial (*mE*)
Maritime Tropical (*mT*)
Continental Tropical (*cT*)
Continental Polar (*cP*)
Maritime Polar (*mP*)

Continental Arctic (*cA*)
surface of discontinuity
cold front
squall line
warm front
stationary front
occluded front
atmospheric disturbance

middle-latitude disturbance
  (extratropical
  disturbance)
storm track
veering wind shift
backing wind shift
convective thunderstorm
orographic thunderstorm

frontal thunderstorm
tornado
Doppler radar
easterly wave
storm surge
hurricane
typhoon

DISCUSS & REVIEW

1. What is an air mass?

2. Do all areas on Earth produce air masses? Why or why not?

3. What letter symbols are used to identify air masses? How are these combined? What air masses influence the weather of North America? Where and at what time of the year are they most effective?

4. Use Table 7.1 and Figure 7.1 to find out what kinds of air masses are most likely to affect your local area. How do they affect weather in your area?

5. What forces modify the behavior of air masses? What kinds of weather may be produced when an air mass begins to move?

6. Why do you suppose air masses can be classified by whether they develop over water or over land?

7. Why does *mP* air affect the United States? Are there any deviations from this tendency?

8. What kind of air mass forms over the southwestern United States in summer? Have you ever experienced weather in such an air mass? What was it like? What kind of weather might you expect to experience if such an air mass met an *mP* air mass?

9. What is a front? How does it occur?

10. In a meeting of two contrasting air masses, how can the aggressor be determined?

11. Compare warm and cold fronts. How do they differ in duration and precipitation characteristics?

12. What kind of weather often results from a stationary front? What kinds of forces, do you suppose, tend to break up stationary fronts?

13. How does the westerly circulation of winds affect air masses in your area? What kinds of weather result?

14. What are the major differences between middle-latitude cyclones and anticyclones?

15. Can you draw a diagram of a mature (fully developed) middle-latitude cyclone that includes the center of the low with several isobars, the warm front, the cold front, wind direction arrows, appropriate labeling of warm and cold air masses, and zones of precipitation?

16. If a wind changes to a clockwise direction, what is the shift called? Where does it locate you in relation to the center of a low pressure system? Explain why this happens.

17. How does the configuration of the upper air wind patterns play a role in the surface weather conditions?

18. How can a knowledge of cyclones be used to help forecast weather changes? For how long in advance? What changes might occur to spoil your forecast?

19. Describe the sequence of weather events over a 48-hour period in St. Louis, Missouri, if a typical low pressure system (cyclone) passes 300 kilometers (180 mi) north of that location in the spring.

20. List three major causes of thunderstorms. How might the storms that develop from each of these causes differ?

21. Have you ever experienced a tornado or hurricane? Describe your feelings during it and the events surrounding it. How do the news media prepare us for such natural disasters?

CONSIDER & RESPOND

1. Collect a three-day series of weather maps from your local newspapers. Based upon the migration of high and low pressure systems during that period, discuss the likely pattern of the upper air winds.

2. List the ideal conditions for the development of a hurricane.

3. Look at Figure 7.8a. Assume you are driving from Point A to Point B. Describe the changes in weather (temperature, wind speed and direction, barometric pressure, precipitation, and cloud cover) you would encounter on your trip. Do the same analysis for a trip from Point C to Point D.

4. The location of the polar front changes with seasons. Why? In what way is Figure 7.14b related to the seasonal migration of the polar front?

5. Redraw Figure 7.7 so that it depicts a Southern Hemisphere example.

GLOBAL CLIMATES AND CLIMATE CHANGE

Atmospheric elements vary so greatly from place to place over Earth's landmasses, that scientists have classified climates by combining those elements with similar statistics to identify a manageable number of groups or types.

What two atmospheric elements are most often used when classifying climates, and why were these elements selected?

Why is there a need for more than one system of climate classification?

On a global or macro scale, the Köppen system of climate classification is one of the most widely used.

What are some of the advantages and disadvantages of the Köppen system?

What are some advantages of the Thornthwaite system, and how does it differ from the Köppen system?

At what other scales can climate be studied?

A knowledge of past climates is critical to an understanding of the nature of present-day climates and to the attempted prediction of future climates.

What important climatic changes have occurred over the past 2.4 million years?

How has the Holocene, or most recent climatic epoch, differed from much of the rest of the Pleistocene?

What recent evidence has been most important to the deciphering of climatic changes during the Pleistocene?

Scientists today believe that dramatic changes in climate can occur over surprisingly short periods of time.

On what evidence are climatologists basing this belief?

What type of feedback system is probably responsible for such rapid climatic change?

What major climatic changes have occurred over the past 18,000 years?

There are a number of plausible causes of climatic change, and it is therefore difficult for scientists to determine which cause is responsible for a specific climatic event.

What are some of the major causes?

Which causes seem to best explain some of the most important climatic events of the past?

In which of the causes is human activity most likely involved?

Despite the best scientific methods and the most modern technology, accurately predicting future climate remains exceedingly difficult.

Why is this so?

What methods and technology are most commonly used?

What climate changes are most likely in the immediate future? In the more distant future?

The glaciated valley leading to the terminus of the Franz-Josef Glacier, New Zealand. About 150 years ago, the entire valley was filled with glacial ice.

If someone asked you "What's the weather like where you live?" how might you respond? Would you talk about the storm that occurred last week, or say that summers are very warm where you live? You may find that a question dealing with local atmospheric conditions is difficult to answer. Is the question referring to *weather* or *climate?* It is essential that you can distinguish between the two. Weather and climate were defined briefly in Chapter 4. In Chapter 7, we thoroughly discussed weather. In this chapter, we will begin the study of climate in much greater detail.

Unlike weather, which describes the state of the atmosphere over short periods of time, climatic analysis relies heavily on averages and statistical probabilities involving data accumulated for the atmospheric elements over periods of many years. Climatic descriptions include such things as average annual precipitation and temperature; patterns of change in wind direction, temperature, and precipitation throughout the year; anticipated number of days of sunshine or cloud cover; and ranges between the statistical highs and lows that may be expected for each atmospheric element.

In the first part of this chapter, we introduce the characteristics and classification of modern climates. Since climate can be defined at different scales, from a single hillside to a region as large as the Sahara stretching across much of northern Africa, two systems of describing and classifying climates will be discussed. The first of those systems has been widely adopted by physical geographers and other scientists; in a modified version, it will be the basis for the worldwide regional study of present-day climates in Chapters 9 and 10. The second system is introduced because it is one of the most effective available to scientists for classification and study of climates on a local scale.

The remainder of the chapter will focus on climatic change. Climates in the past were not the same as they are today, and there is every reason to believe that future climates will be different as well.

It is now also recognized that humans can alter Earth's climate. The most direct human impact on climate results from a change in the composition and relative abundance of atmospheric gases—for example, the amount of carbon dioxide (CO_2). The concentrations of gases can change either directly, through the release of enormous quantities of CO_2 and other "greenhouse" gases, or indirectly, from such actions as paving over fields or forests, thus reducing the number of plants that absorb CO_2 from the atmosphere.

For decades scientists have realized that Earth has experienced major climate shifts during the past 30,000 years. It was believed that these shifts were gradual and could not be detected by humans during a lifetime or two. However, recent research into the record of the deep sea sediments and polar ice sheets reveals that climate has shifted repeatedly between extremes over some exceedingly short intervals. Moreover, the research has revealed that climate during the most recent 10,000 years has been extraordinarily stable compared to similar intervals in the past.

In order to predict future climates, it is critical that we examine the details of past climate changes, including both the magnitude and rates of prehistoric climate change. During the past 18,000 years, Earth has experienced both ice ages and lengthy periods that were warmer than today. These fluctuations serve as indicators of the natural variability of climate in the absence of significant human impact. Using knowledge of present and past climates, as well as models of how and why climate changes, we will conclude this chapter with some predictions of future climate trends.

Classifying Climates

Knowledge that climate varies from region to region dates to ancient times. The Ancient Greeks (such as Aristotle, ca. 350 B.C.) classified the known world into Torrid, Temperate, and Frigid zones based on average temperatures. It was also recognized that these zones varied systematically with latitude, and that the flora and fauna reflected these changes as well. With the further exploration of the world by Europeans, naturalists noticed that the distribution of climates could be explained

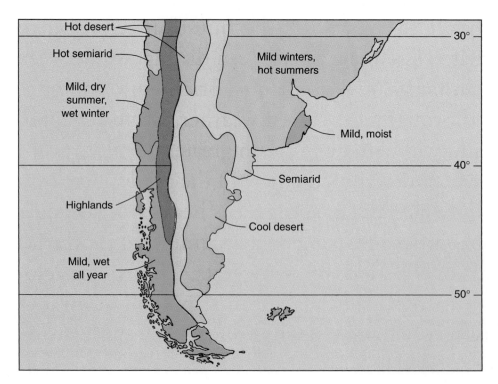

(a)

Figure 8.1   (a) This map shows the diversity of climates possible in a relatively small area, including portions of Chile, Argentina, Uruguay, and Brazil. The climates range from dry to wet and from hot to cold, with many possible combinations of temperature and moisture characteristics. (b) The Argentine Patagonian Steppe. (c) A meadow in the Argentine, Tierra del Fuego.   ■   **What can you suggest as the causes for the major changes in climate as you follow the 40° south latitude line from west to east across South America?**

Michael Trapasso

(b)

Michael Trapasso

(c)

using factors such as sun angles, prevailing winds, elevation, and proximity to large water bodies.

The two weather variables used most often as indicators of climate are temperature and precipitation. To classify climates accurately, climatologists need many years of records of these two atmospheric elements. In most cases, a minimum of 30 years is required to describe the climate of an area. The invention of an instrument to reliably measure temperature—the thermometer—dates only to Galileo in the early seventeenth century. European settlement of and sporadic collection of temperature and precipitation data from distant colonies began in the 1700s but was not routine until the mid-nineteenth century.

This was soon followed in the early twentieth century by some of the first attempts to classify global climates using actual temperature and precipitation data.

As we have seen in earlier chapters, temperature and precipitation vary greatly over Earth. Climatologists have worked to reduce the infinite number of worldwide variations in atmospheric elements to a comprehensible number of groups or varieties by combining elements with similar statistics (Fig. 8.1). That is, they can classify climates strictly on the basis of atmospheric elements, ignoring the causes of those variations (such as the frequency of air mass movements). This type of classification, based on statistical parameters or physical characteristics, is called an **empirical classification.** A classification based on the causes, or *genesis,* of climatic variation, is known as a **genetic classification.**

Ordering the vast wealth of available climatic data into descriptions of major climatic groups, on either an empirical or a genetic basis, enables geographers to concentrate on the larger-scale causes of climatic differentiation. In addition, they can examine exceptions to the general relationships, the causes of which are often one or more of the other atmospheric controls. Finally, differentiating climates helps to explain the distribution of other climate-related phenomena of importance to humans.

Despite its value, climate classification is not without its problems. Climate is a generalization about observed facts based on the averages and probabilities of weather. It does not describe a real weather situation; instead, it presents a composite weather picture. Within such a generalization, it is impossible to include the many variations that actually exist. On a global scale, generalizations, simplifications, and compromises are made to distinguish among climatic types and regions.

The Köppen System

The most widely used climate classification is based on temperature and precipitation patterns. It is referred to as the **Köppen system,** after the German botanist and climatologist who developed it. Dr. Wladimir Köppen recognized that major vegetation associations reflect the local climate. Hence, his climatic regions were formulated to coincide with well-defined vegetation regions, and each climatic region can be described by the natural vegetation most often found there. Evidence of the strong influence of Köppen's system is seen in the wide usage of his climatic terminology, even in nonscientific literature (for example, steppe climate, tundra climate, rainforest climate).

Advantages and Limitations of Köppen's System Not only are temperature and precipitation two of the easiest weather elements to measure, they are also measured more often and in more parts of the world than any other variables. By using temperature and precipitation statistics to define his boundaries, Köppen was able to develop precise definitions for each climatic region, eliminating the imprecision that can develop in verbal and sometimes in genetic classifications.

Moreover, temperature and precipitation are the most important and effective weather elements. Variations caused by the atmospheric controls will show up most obviously in temperature and precipitation statistics. At the same time, temperature and precipitation are the weather elements that most directly affect humans, other animals, vegetation, soils, and the form of the landscape. Tying classification to these two elements, Köppen's system is closely related to the visible aspects of our environment.

Köppen's climatic boundaries were designed to define the vegetation regions included in a classification of world vegetation produced earlier by the Swiss botanist Alphonse de Candolle. Thus, Köppen's climatic boundaries reflect "vegetation lines." For example, the Köppen classification uses the 10°C (50°F) monthly isotherm because of its relevance to the timberline—the line beyond which it is too cold for trees to thrive. Trees need a certain amount of warmth to grow, and unless at least one month of the year has an average temperature exceeding 10°C, trees will ordinarily be unable to survive. For this reason, Köppen defined the treeless polar climates as those areas where the mean temperature of the warmest month is below 10°C. Clearly, if climates are divided according to associated vegetation types, and if the division is based on the atmospheric elements of temperature and precipitation, then the result will be a visible association of vegetation with climatic types. The relationship with the visible world in Köppen's climate classification system is one of its most appealing features to geographers.

There are, of course, limitations to Köppen's system. For example, Köppen considered only average monthly temperature and precipitation in making his climatic classifications. These two elements permit estimates of precipitation effectiveness but do not measure it with enough precision to permit comparison from one locality to another. In addition, for the purposes of generalization and simplification, Köppen ignored winds, cloud cover, intensity of precipitation, humidity, and daily temperature extremes—much, in fact, of what makes local weather and climate distinctive.

Simplified Köppen Classification The Köppen system, as modified by later climatologists, divides the world into six major climate categories. The first four are based on the annual range of temperatures: humid tropical climates (*A*), humid mesothermal (mild winter) climates (*C*), humid microthermal (severe winter) climates (*D*), and polar climates (*E*). Another category, the arid and semiarid climates (*BW* and *BS*), identifies regions that are characteristically dry based on both temperature and precipitation values. Since plants need more moisture to survive as the temperature increases, the arid and semiarid climates include regions where the temperatures range from cold to very hot. The final category, highland climates (*H*), identifies mountainous regions where vegetation and climate vary rapidly as a result of changes in elevation and exposure.

Within each of the first five major categories, individual climatic types and subtypes are differentiated from one another by specific parameters of temperature and precipitation. Some are characterized by seasonal rainfall in either summer or winter, whereas others exhibit evenly distributed monthly precipitation. Other climatic types are differentiated by aspects of their annual temperature ranges that mostly reflect the degree of conti-

Station: Nashville, Tenn.
Latitude: 36°N

Type: Humid Subtropical (Cfa)
Longitude: 88°W

Av. annual prec.: 119.6 cm (47.1 in.)

Av. Annual temp.: 15.2°C (59.5°F)

Range: 22.5°C (40.5°F)

Figure 8.2 A standard climograph showing average monthly temperature (curve) and rainfall (bars). The horizontal index lines at 0°C (32°F), 10°C (50°C), 18°C (64.4°F), and 22°C (71.6°F) are the Köppen temperature parameters by which the station is classified. ■ **What specific information can you read from the graph that identifies Nashville as a specific climate type (humid subtropical) in the Köppen classification? (See Appendix C.)**

perature changes clearer, we can connect the monthly values with a continuous line, producing an annual temperature curve. To avoid confusion, monthly precipitation amounts are usually shown as bars reaching to various heights above the line of zero precipitation. Such a display of a location's climate is called a **climograph.** To read the graph, one must relate the temperature curve to the values given along the left side, and the precipitation amounts to the scale on the right. Other information may also be displayed, depending on the type of climograph used. Figure 8.2 represents the type we will use in Chapters 9 and 10. This climograph can be used to determine the Köppen classification of the station as well as to show its specific temperature and rainfall regimes.

The Thornthwaite System

For a farmer interested in growing a specific crop in a particular area, the Köppen system is inadequate. Although the Köppen system correctly identifies the major vegetation type of the region and the annual range of both temperature and precipitation, it does not provide a farmer with information concerning the amounts and timing of annual soil moisture surpluses or deficits. From an agricultural perspective, it is much more important to know that moisture will be available in the growing season, whether it comes directly in the form of precipitation or from the soil.

The **Thornthwaite system** was developed by an American climatologist, Dr. C. Warren Thornthwaite, to establish moisture availability at the subregional scale (Fig. 8.3). It is the system preferred by many soil scientists and others examining climates on a local scale. Development of detailed climate classification systems such as the Thornthwaite system became possible only after temperature and precipitation data were widely collected at numerous locations beginning in the latter half of the nineteenth century.

The Thornthwaite system is based on the concept of **potential evapotranspiration (PE),** which approximates the water use of plants with an unlimited water supply. Potential evapotranspiration is a theoretical value that increases with increasing temperature, winds, and length of daylight and decreases with increasing humidity. **Actual evapotranspiration (AE),** which reflects actual water use by plants, can be supplied during the dry season by soil moisture if the soil is saturated, the climate is relatively cool, and/or the daylengths are short. Thus, measurements

nentality associated with the climate (for example, hotter summers or colder winters inland, milder summers or winters in coastal locations).

Köppen gave names to each of his individual climatic types and, as indicated when we introduced his six major categories, he used combinations of letters to symbolize them. The letter symbols provide an international shorthand describing climatic types that are often difficult to characterize in words. In the interest of generalization, we will use only the names when identifying climates in the discussion of climatic regions in Chapters 9 and 10. However, a detailed presentation of Köppen's letter symbols, along with the definitions used to classify climates in the system, can be found in Appendix C.

Climographs

It is possible to summarize the nature of the climate at any point on Earth in graph form, as shown in Figure 8.2. Given information on mean monthly temperature and rainfall, we can express the nature of the changes in these two elements throughout the year simply by plotting their values as points above or below (in the case of temperature) a zero line. To make the pattern of the monthly tem-

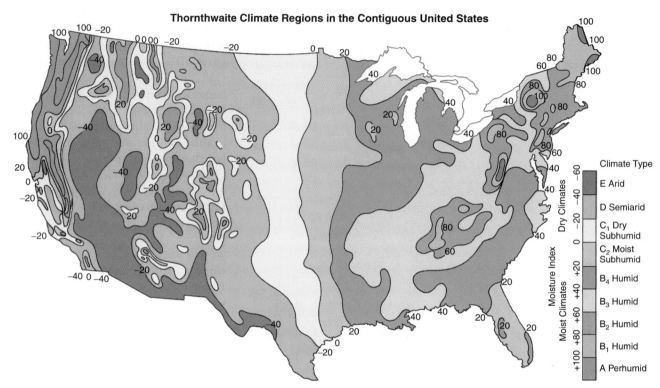

Figure 8.3 Thornthwaite climate regions in the contiguous United States are based on the relationship between precipitation (P) and potential evaporation (PE). The moisture index (MI) for a region is determined by this simple equation: MI = 100 × [(P − PE) ÷ PE]. Where precipitation exceeds potential evaporation, the index is positive; where potential evaporation exceeds precipitation, the index is negative. ■ **What are the moisture index and Thornthwaite climatic type for coastal southern California?**

of AE relative to PE and available soil moisture are the determining factors for most vegetation and crop growth.

The Thornthwaite system recognizes three climate zones based on PE values: low-latitude climates, with PE greater than 130 centimeters (51 in.); middle-latitude climates, with PE less than 130 but greater than 52.5 centimeters (20.5 in.); and high-latitude climates, with PE less than 52.5 centimeters. Climate zones may be subdivided based on how long, and by how much, AE is below PE. Moist climates have either a surplus or a minor deficit of less than 15 centimeters (6 in.). Dry climates have an annual deficit greater than 15 centimeters. In the case of extreme moisture seasons, two classes do not fit into the normal classification system: the Tropical Wet and Dry regions and the Mediterranean climate regions.

Thornthwaite's original equations for PE were based on analyses of data collected in the midwestern and eastern United States. The method was subsequently used with less success in other parts of the world. Over the past few decades, many attempts have been made to improve the accuracy of the Thornthwaite system for regions outside the United States.

Scale and Climate

Climate can be measured at different scales (macro, meso, or micro). The climate of a large (macro) region, such as the Sahara, may be described correctly as hot and dry. Climate can also be described at mesoscale levels; for example, the climate of coastal southern California is sunny and warm, with dry summers and wet winters. Finally, climate can be described at local scales, such as on the slopes of a single hill. This is termed a **microclimate.**

At the microclimate level, many factors will cause the climate to differ from nearby areas. For example, in the United States and other regions north of the Tropic of Cancer, south-facing slopes tend to be warmer and drier than north-facing slopes because they receive more sunlight (Fig. 8.4). This variable is referred to as *slope aspect*—the direction a mountain slope faces in respect to the sun's rays. Microclimatic differences such as slope aspect can cause significant differences in vegetation and soil moisture that over the long term affect soil thickness and composition, and at times may even affect the appearance of the slope itself.

Human activities can influence microclimates as well. Recent research indicates that the construction of a large reservoir leads to greater annual precipitation immediately downwind of the lake. This occurs because the lake supplies additional water vapor to passing storms, which intensifies the rainfall or snows immediately downwind of the lake. These microclimatic effects are similar to the "lake effect" snows that occur downwind of the Great Lakes in the early winter when the lakes are not frozen. Another example of human impact on microclimates is the **urban heat island** effect, which leads to changes in temperature (urban centers tend to be warmer than their outlying rural areas), rainfall, wind speeds, and many other phenomena (Table 8.1).

Past Climates

To try to predict future climates, it is critical to understand the magnitude and frequency of previous climate changes. Knowledge that Earth experienced major climate changes in the past is not new. In 1837, Louis Agassiz, a European naturalist, proposed that Earth had experienced major periods of *glaciation*, periods known as ice ages, when large areas of the continents were covered by sheets of ice. He presented evidence, controversial at the time, that glaciers (flowing ice) had once covered most of England, northern Europe, and Asia, as well as the foothill regions of the Alps. Agassiz' ideas were not readily accepted in Europe, in part because the continental glaciers covering Greenland and Antarctica had yet to be discovered. Agassiz arrived in the United States in 1846, and similar evidence of widespread glaciation was soon recognized throughout North America.

Norman Meek

Figure 8.4 This photograph facing west near Desert View, Arizona, illustrates the significance of slope aspect. The south-facing slopes on the right of the photograph are receiving the full effect of the desert sun's energy. These slopes are clearly hotter and drier than the more shaded north-facing slopes. ■ **Why do the differing angles that the sun's rays strike the two opposite slopes affect temperatures?**

Table 8.1   Effects of Urbanization on Climatic Elements

Elements	Comparison with Rural Environment
Pollutants	
Solid particles	10 times more
Gases	5–25 times more
Cloud cover	5–10 percent greater
Fog, winter	100 percent more
Fog, summer	30 percent more
Precipitation	5–10 percent more
Snowfall	5 percent less
Rain days with less than 5 mm	10 percent more
Relative humidity, winter	2 percent less
Relative humidity, summer	8 percent less
Radiation	15–20 percent less
Ultraviolet radiation, winter	30 percent less
Ultraviolet radiation, summer	5 percent less
Duration of sunshine	5–15 percent less
Annual mean temperature	0.5°–1.0°C higher
Heating degree days	10 percent fewer
Annual mean wind speed	20–30 percent less
Calms	5–20 percent more

Source: World Meteorological Organization, 1970

The Pleistocene Epoch

The nomenclature that defines Earth's geologic history was developed and defined before thorough knowledge of ice ages existed. The **Pleistocene** epoch has become synonymous with the time of the most recent ice age. Unfortunately, the Pleistocene was originally defined on the basis of the first appearance of cold-water marine species in a sequence of Italian sedimentary rocks. Subsequent radiometric dating of the sediments (mineral and organic material can be dated by measuring the extent to which radioactive elements in the material have decayed through time) reveals that this occurred only about 1.6 million years ago in the Mediterranean Sea. Today, it is recognized that major glacial advances occurred as early as 2.4 million years ago, several hundred thousand years before the Pleistocene "officially" began.

Until the 1960s, it was widely believed that Earth had experienced four major glacial advances followed by warmer interglacial periods (glacial cycles) during the Pleistocene. In the foothills of the Alps near the Danube River, evidence of four glacial advances was discovered. These glacial epochs were termed the Günz (oldest), Mindel, Riss, and Würm. Likewise, in the United States, evidence of four glacial periods was recognized. Based on the southward limit of the glacial advances, these were termed the Nebraskan (oldest), Kansan, Illinoian, and Wisconsinan glaciations.

A major problem with studying the advance and retreat of glaciers on land is that each subsequent advance of the glaciers tends to destroy, bury, or greatly disrupt the sedimentary evidence of the previous glacial period.

The evidence of the fourfold record of glacial advances was largely recognized on the basis of minor glacial deposits lying beyond the limit of the two most recent glaciations. Evidence of "average" glacial advances that were subsequently overridden by more recent glaciers was rarely recognized.

Before the advent of radiometric dating techniques following World War II, the timing of the glacial advances in both Europe and the United States was only crudely known. For example, estimates of the age of the last deglaciation were based on the rates at which Niagara Falls and St. Anthony's Falls had eroded headward after the areas were first exposed when the glaciers retreated. Calculations ranging from 8000 to 30,000 years ago—surprisingly recent—were produced. In Sweden, the age of deglaciation was based on the number of sedimentary couplets, or **varves,** that are present in some newly exposed lake basins. A varve is a pairing of organic-rich summer sediments and organic-poor winter sediments, and each couplet represents one year of time. This technique suggested that the region had been deglaciated only about 13,000 years ago. Many scientists were skeptical about such studies because it was difficult to believe that several thousand feet of ice had covered large areas of Europe and North America so recently.

Modern Research

Two major advances in scientific knowledge about climate change occurred in the 1950s. First, radiometric techniques, such as radiocarbon dating, that measured the absolute ages of landforms produced by the glaciers

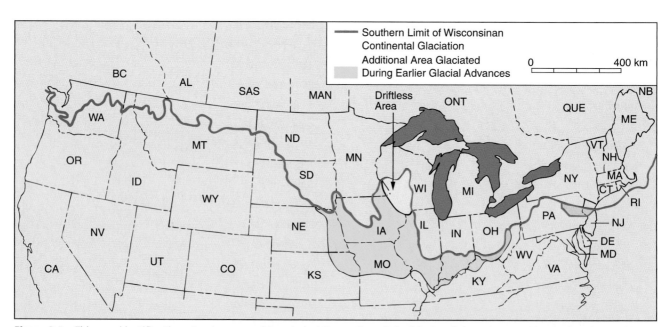

Figure 8.5 This map identifies the extensive areas of Canada and the northern United States that were covered by moving sheets of ice as recently as 18,000 years ago. It also shows that glaciers extended even farther south during earlier glacial advances. ■ **Can you suggest a relationship between glaciation and the location of the Missouri and Ohio Rivers?**

began to be widely used. Radiocarbon dating of Pleistocene lake deposits in the western United States and glacial debris in the midwest conclusively showed that the last ice advance peaked a mere 18,000 years ago (Fig. 8.5). Given that the Pyramids of Egypt were built about 4000 years ago, it was mind-boggling for scientists to realize that major glaciers dominated essentially all of Canada and the United States as far south as the Ohio and Missouri Rivers only 18,000 years ago. This ice sheet covered modern-day city sites such as Boston, New York, Indianapolis, and Des Moines.

The second major discovery was that evidence of detailed climate changes have been recorded in the sediments on the ocean floors. Unlike the continental record, the deep-sea sedimentary record has not been disrupted by subsequent glacial advances. Rather, the slow, continuous sediment record provides a complete history of climate changes during the past several million years. The most important discovery of the deep-sea record is that Earth has experienced numerous major glacial advances during the Pleistocene, not just the four that had been identified previously. Today, the names of only two of the North American glacial periods, the Illinoian and Wisconsinan, have been retained.

Because the deep-sea sedimentary record is so important to climate change studies, it is important to understand how the record is deciphered. The deep-sea mud contains the microscopic record of innumerable surface-dwelling marine animals, collectively known as **foraminifera.** These miniscule creatures built tiny shells for protection. When they died, these tiny shells sank to the seafloor, forming the layers of mud. Different species thrive in different surface water temperatures; therefore, the stratigraphic record of foraminifera fossils (analysis) produces a detailed history of water temperature fluctuations.

Foraminifera shells are composed of calcium carbonate ($CaCO_3$); therefore, the analyses also record the oxygen composition of the seawater in which they lived. One common measurement technique for determining oxygen composition is known as **oxygen-isotope analysis.** Two of the most common isotopes of oxygen are O^{16} and the "heavier" O^{18}. Modern seawater has a fixed ratio of the two oxygen isotopes. When water evaporates from the ocean, slightly more O^{16} than O^{18} evaporates, because water containing the lighter-weight oxygen evaporates more readily. During an ice age, the evaporated water is stored on the continents in the form of glaciers rather than returned to the oceans. Not only does the storage of water on land result in a sea level drop of more than 100 meters (330 ft), but the ratio of O^{16} to O^{18} in the ocean changes slightly to reflect the O^{16}-enriched water being stored in the glaciers. Analysis of the oxygen-isotope ratios recorded in the foraminifera tests reveals a record of sea level changes that reflect the growth and decline of ice sheets. The combined record of cold and warm foraminifera species and the associated oxygen-isotope ratios has produced a detailed reconstruction of climate changes.

Analyses of the deep-sea sedimentary record produced a very complex record of climate changes, and a new system was required to describe the major shifts in global climates. The record of oxygen-isotope changes was divided into intervals, termed "stages," of cold and warm climates. These shifts were numbered back in time, beginning with the number 1 for the current warm interval. Thus, all warm intervals have odd-numbered oxygen-isotope stages (OIS), and all glacial intervals have even-numbered oxygen-isotope stages. A review of the oxygen-isotope record indicates that the last glacial advance about 18,000 years ago was only one of many major glacial advances during the past 2.4 million years (Fig. 8.6).

Today, climatologists are aware that the present climate is but a short interval of relative stability in a time of major climate shifts. Moreover, the modern climatic epoch, known as the **Holocene** (most recent), is a time

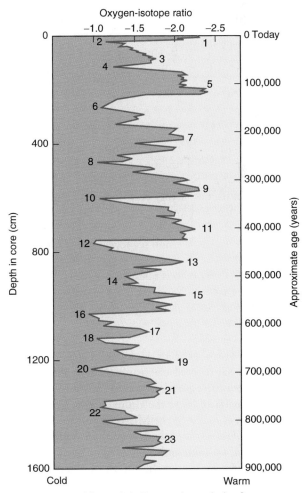

Figure 8.6 As this graph indicates, the analysis of oxygen-isotope ratios in fossils that have accumulated in deep-sea sediments over the past 900,000 years reveals frequent changes between periods of colder and warmer climates. ■ **How does this recent reconstruction of past climates differ from the widely accepted theories of 50 years ago?**

Determining Past Climates

As an aid to understanding oxygen-isotope analysis as a means of reconstructing paleoclimates (ancient climates), it is helpful to review some basic definitions. *Isotopes* are defined as atoms with the same atomic number but different atomic mass. The atomic number is equal to the positive charge of the nucleus—essentially, the number of *protons* present in the nucleus. The atomic mass (or atomic weight) is equal to the number of *protons and neutrons* that comprise the nucleus of the atom. *Electrons,* which orbit a particular atom, are negatively charged particles that possess no appreciable mass (or weight). In a neutral atom, the number of electrons should equal the number of protons. When an electron is lost or gained to the atom, then a net (+) or (–) charge, respectively, will result, and the particle is then classified as an *ion.*

When dealing with isotopes, the atomic number (proton number) must remain the same (giving the atom its identity in the Periodic Table of Elements), but the number of neutrons can vary. In the example of oxygen isotopes, the atomic number is always 8. This is necessary to identify the atom as oxygen. However, the atom may contain 8 neutrons (O^{16}, a lighter isotope), or 10 neutrons (O^{18}, a heavier isotope). In oxygen-isotope analysis, the ratio of the heavier to the lighter isotope is measured and compared to the normal ratio. The offset of this ratio can help determine the temperature at which these ratios were formed. (A larger than normal proportion of the O^{18} isotope in relation to the O^{16} isotope is an indication of colder temperatures.) Thus O^{18}/O^{16} ratios can be an effective way of reconstructing temperature records of the past.

Through time, paleoclimatologists have discovered new and different ways to determine climates of the past. The oxygen-isotope analysis is one of the most widely accepted methods, but there are other well-established methods as well. Two are worth a brief discussion; they are *dendrochronology* and *palynology.*

Dendrochronology (or tree-ring dating) has been used for decades. This analysis calls for the examination of tree rings exposed by cores taken through the middle of certain species of trees. The core (small enough so as not to harm the tree) will reveal each yearly tree ring. The rings are counted back through time to supply a time scale for the analysis. Each ring, by its thickness, color, and texture, can reveal the climate conditions during that particular year of the tree's growth. Thus, a short-term climate record can be determined.

---

of extraordinarily stable, warm temperatures compared to most of the last 2.4 million years. Based on the deep-sea record, it appears that global climates tend to rest at one of two extremes: a very cold interval characterized by major glaciers and lower sea levels, and shorter intervals between the glacial advances marked by unusually warm temperatures and high sea levels. With the realization that global climates have changed dramatically numerous times, two obvious questions arise: How quickly does global climate change from one extreme to the other, and what causes global climate to change so often?

Rates of Climatic Change

Radiocarbon dating of glacial features has confirmed that full-glacial conditions existed about 18,000 years ago. Through Canada and into the United States, glaciers covered most areas north of the Missouri and Ohio Rivers. In the west, freshwater lakes more than 500 feet deep covered much of Utah and Nevada. The same scientific methods that have shown the extent of glacial features also reveal that the United States was mostly glacier-free, and the western lake basins were dry, by about 9000 years ago. Abundant evidence has even been found that the climate about 7000 years ago (a time known as the **Altithermal**) was hotter than today.

For glaciers several thousand feet thick to melt completely, and for deep lakes to evaporate, a substantial increase in insolation is required over a few thousand years. Where did so much extra energy come from?

To answer questions about such rapid rates of climate change requires a more detailed record of climate than the deep-sea sediments can provide. This is because the deep-sea sedimentary record is extraordinarily slow—a few centimeters of sea mud accumulates in a thousand years. Rapid shifts in climate during periods of a few hundred years are not recorded clearly in the seafloor sediments. This problem has been solved by coring the thick glaciers covering Antarctica and Greenland. Glacial ice records yearly amounts of snowfall and is much more likely to provide short-term evidence of climate changes. Analyses of oxygen isotopes in the glacial ice of Antarctica and, most recently, Greenland, have revealed a detailed record of climate changes during the past 250,000 years (Fig. 8.7).

A surprising discovery of the ice-sheet analyses is the speed at which climate changes. Rather than changing gradually from glacial to interglacial conditions over thousands of years, the ice record indicates that the shifts can occur in a few years or decades. Thus, whatever is most responsible for major climate changes can develop rapidly. This probably requires a *positive feedback system,* which means, as explained in Chapter 1, that a change in one variable will cause changes in other variables that magnify the amount of original change. For example, most glaciers have high albedos, reflecting significant

National Ice Corps Laboratory (NICL)

(a)

Akira Kaede/ Getty Images

(b)

Getty Images

(c)

Palynology (or pollen-analysis dating) is also a well-established way of reconstructing past climates. Though pollen samples can be recovered from a variety of environments, organic bogs are the best place to conduct this analysis. A core is removed from a bog showing the layers of organic material all the way to the bottom layer. Then each organic bog layer is radiocarbon dated to identify its age. Afterwards, all the pollen is removed from each layer of organic material, and the pollen is used to identify the numbers, types, and relative distribution of the trees from which the pollen came. It is then left to the paleoclimatologist to determine what type of climate would be required to sustain a forest at the time of the one described by the analysis of each bog layer.

Paleoclimates can be reconstructed using oxygen $^{18}/_{16}$ ratios from layers found in ice cores (a). Other methods include the identification and characterization of tree rings (b) and pollen samples (c).

amounts of sunlight back to space. However, if the ice sheets begin to decline for whatever reason, low-albedo land begins to replace the high-albedo ice, increasing the amount of energy available to melt the ice. Thus, the more ice that melts, the more energy is available to melt the ice further, magnifying the initial glacial retreat.

In contrast, a *negative feedback system,* where changes in one of the variables induce the system to remain stable,

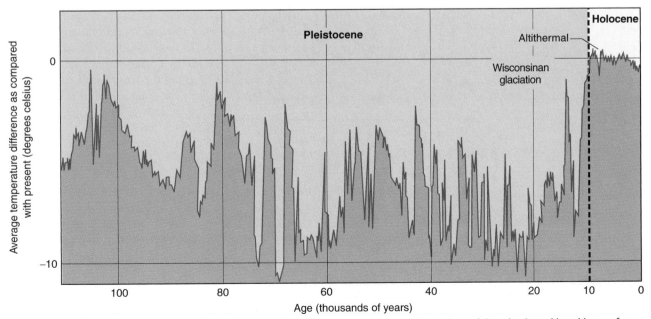

Figure 8.7   Analyses of oxygen-isotope ratios in ice cores taken from the glacial ice of Antarctica and Greenland provide evidence of surprising shifts of climate over short periods of time.   ■   **Has the general trend of temperatures on Earth been warmer or colder during the Holocene?**

will also affect the likelihood or rate of climate change. For example, increasing global temperatures cause evaporation rates to increase. The more water that evaporates from the ocean surface, the more clouds will form. The more clouds that exist, the more insolation is reflected back to space, cooling Earth's surface. (A counterargument to this effect is that clouds also operate as a greenhouse blanket, trapping heat in the lower atmosphere.) Thus, for climate changes to occur rapidly, negative feedback cycles such as this one must be overwhelmed by positive feedback cycles.

Causes of Climatic Change

Although theories about the causes of climatic change are numerous, they can be organized into five broad categories: (1) astronomical variations in Earth's orbit;

(2) changes in the output of solar energy and/or the transparency of outer space; (3) changes in atmospheric gases or dust, causing changes in incoming or outgoing radiation; (4) changes in oceanic circulation; and (5) changes in landmasses that affect albedo and oceanic circulation.

Orbital Variations

Astronomers have detected slow changes in Earth's orbit that affect the distance between the sun and Earth as well as the deviation of Earth's orbit from the plane of the ecliptic. These orbital cycles produce regular changes in the amount of solar energy that reaches Earth. The longest is known as the **eccentricity cycle,** which is a 100,000-year variation in the shape of Earth's orbit around the sun.

In simple terms, Earth's orbit changes from an ellipse, to a more circular orbit, and then back. In addition, the planet slightly changes its position relative to the plane of the ecliptic. A second cycle, termed the **obliquity cycle,** represents a 41,000-year variation in the tilt of Earth's axis from a maximum 24.5° to a minimum of 22.0°, and back. The greater Earth's tilt, the greater the seasonality at middle and high latitudes. Finally, a **precession cycle** has been recognized that has a major periodicity of 21,000 years. The precession cycle determines the time of year that perihelion occurs. Today, Earth is closest to the sun on January 3 and, as a result, receives about 3.5 percent greater insolation than the average in January. According to the precession cycle, this date advances one calendar day each 71 years, thereby altering the timing of maximum insolation receipt. Because the Northern Hemisphere is closest to the sun on January 3, the winter season in the Northern Hemisphere is somewhat warmer than average today. When aphelion occurs on January 3 in about 10,500 years, the Northern Hemisphere winters should be somewhat colder.

These cycles operate collectively, and the combined effect of the three cycles can be calculated (Fig. 8.8). The first person to examine all three of these cycles in detail was a mathematician, Milutin Milankovitch. In 1924, prior to the invention of calculators, Milankovitch completed

Figure 8.8   The bottom line graph in this figure is the most significant because it demonstrates the combined effect that eccentricity, obliquity, and precession cycles have had on the amounts of solar radiation received by Earth during the past 800,000 years. Because of the close correlation between this graph and climate change as indicated by the deep-sea record, many scientists believe that variations in Earth's orbit exert a major influence on glacial cycles.   ■   **Which two of the top three graphs exhibit the greatest similarity?**

the complex mathematical calculations and published graphs showing how these subtle changes in Earth's orbit would affect insolation at three different latitudes (Fig. 8.9). Computers permit the values to be calculated and mapped for all latitudes (Fig. 8.10). Milankovitch's calculations indicated that there should be numerous glacial cycles during a million-year interval. However, many scientists believed that there had been only four major glaciations, and because some early radiocarbon evidence appeared contradictory, Milankovitch's ideas were largely dismissed until the recovery and interpretation of the deep-sea record in the 1950s.

Analysis of the deep-sea sedimentary record indicates that the 41,000-year obliquity cycle is the most important cycle in the climate record from 2,400,000 to 900,000 years ago. But for the past 900,000 years, the 100,000-year eccentricity cycle has dominated the record. By the 1980s, most paleoclimate scientists were convinced that there is an unusually good correlation between the deep-sea record and Milankovitch's predictions. This suggests that the primary driving force behind glacial cycles is regular orbital variations, and it indicates that long-term climate cycles are entirely predictable! Unfortunately for humans, the Milankovitch theory indicates that the warm Holocene interglacial will soon end and that Earth is destined to experience full glacial conditions (glacial ice possibly as far south as the Ohio and Missouri Rivers) in about 23,000 years.

Changes in Solar Output

Astronomers have also detected regular cycles in the energy output of the sun. The most famous of these cycles is the approximately 11-year sunspot cycle, with the greatest solar output correlating with the maximum number of sunspots (discussed in Chapter 3). These changes have been confirmed recently with satellite measurements of solar fluxes. Solar maxima occurred in 1991 and 2000, and a solar minimum occurred in 1997. Therefore, another minimum should occur in the very near future.

Climatic effects associated with the 11-year sunspot cycle are less certain. Deviations from average weather are so difficult to recognize, and detailed meteorological records are so short, that nothing definite can be concluded based on direct measurements of climate variables. However, some indirect records of climate (proxy records), such as tree-ring records, show rather good evidence of an 11-year cycle.

Other cycles related to variations in solar output are more speculative. Cycles of 2 to 2.2, 5.5, 20 to 23, 50, 80, and 100 years have been reported in the scientific literature.

Changes in Earth's Atmosphere

Many theories attribute climate changes to variations in atmospheric dust levels. The primary villain is volcanic activity, which pumps enormous quantities of ash into the stratosphere, where strong winds spread it around the world. The dust reduces the amount of insolation reaching Earth's surface for periods of one to three years (Figs. 8.11 and 8.12).

Volcanic Activity   The climatic cooling effect of volcanic activity is unquestioned: All of the coldest years on record over the past two centuries have occurred in the year following a major eruption. Following the massive eruption of Tambora (in Indonesia) in 1815, 1816 was known as "the year without a summer." Killing frosts in July ruined crops in New England and Europe, resulting in

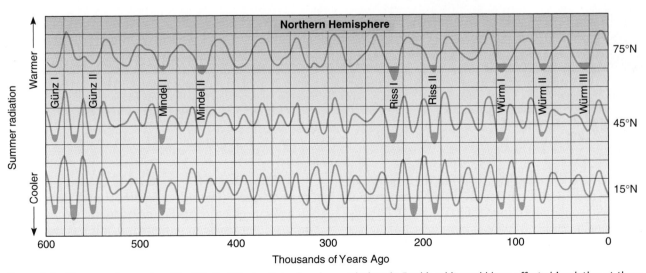

Figure 8.9   These graphs, produced by Milutin Milankovitch, show how variations in Earth's orbit would have affected insolation at three different latitudes over the past 600,000 years.

Courtesy Arizona State University Departments of Geography and Computer Science

Figure 8.10   The nine computer images in this figure are based on the Milankovitch theory that links changes in receipt of insolation on Earth to changes in Earth's orbital characteristics. The blue end of the spectrum indicates values below those of the present time; the red end of the spectrum indicates values above those of the present time.   ■   **What effect should these changes in receipt of insolation have had on global climates?**

famines. Several decades later, following the massive eruption of Krakatoa (also in Indonesia) in 1883, temperatures decreased significantly during 1884 (Fig. 8.13). Although no twentieth-century eruptions have approached the magnitude of these two, the 1991 eruption of Mt. Pinatubo (in the Philippines) produced a substantial respite of cool conditions in an otherwise continuous series of record warm years (see again Fig. 8.13).

Volcanic activity is an important variable in global temperatures. In early climate models attempting to replicate global temperature variations, volcanic activity was found to be so important that it was given twice the weight of any other variable. Interestingly, volcanic ashes that accumulated about 2.6 million years ago are ten times thicker than average on the floor of the North Pacific Ocean. The modern series of major global

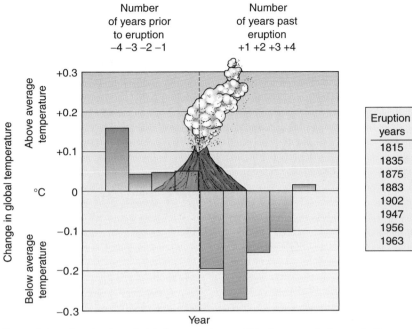

Eruption years
1815
1835
1875
1883
1902
1947
1956
1963

Figure 8.11 Examination of global temperatures within four years before and after major volcanic eruptions provides compelling evidence that volcanic activity can have a direct effect upon the amounts of insolation reaching Earth's surface. ■ **At what period after an eruption year does the effect seem the greatest?**

climate changes began soon thereafter, suggesting that the massive eruptions could have set in motion a series of feedback cycles that continue today.

Atmospheric Gases   Another phenomenon closely correlated with average global temperatures is the composition of atmospheric gases. Scientists have known for many years that CO_2 acts as a "greenhouse gas." Although the analogy to a greenhouse is problematic (because gases are not solids and do not physically restrict air movement), there is no question that CO_2 is transparent to incoming short-wave radiation and blocks outgoing long-wave radiation, similar to the effect of the glass panes in a greenhouse or in your automobile on a sunny day (refer again to Chapter 4). Thus, as the atmospheric content of greenhouse gases rises, so will the amount of heat trapped in the lower atmosphere. Although the amount of

Figure 8.12   A photograph of a volcanic explosion producing clouds of volcanic ash.

Figure 8.13   This graph shows the gradual warming trend in global temperatures since 1880. It also documents the sharp reversal of the trend and the cooling of temperatures after the eruptions of Krakatoa in 1883 and Mt. Pinatubo in 1991.

CO_2 in the atmosphere is about 0.037 percent, its effect on climate is considerable. It has been estimated that the average temperature of Earth would be below freezing, rather than a balmy 16°C (61°F), if CO_2 were not present in the atmosphere.

A consequence of the combustion of fossil fuels is that CO_2 is released in huge quantities. Detailed measurements of atmospheric CO_2 content since 1958 reveal a cyclical and ever-increasing atmospheric CO_2 level (Fig. 8.14). The cyclical nature of the curve is related to the dominance of plant growth in the Northern Hemisphere summer, which temporarily lowers the global CO_2 level.

Trapped in the glacial ice of Antarctica and Greenland are air bubbles containing minor samples of the atmosphere that existed at the time the ice formed. One of the important discoveries of the ice-core projects is that prehistoric atmospheric CO_2 levels increased during interglacials and decreased during major glacial advances. It is still not known why the two are so closely correlated. One explanation involves the oceans, since they serve as a temporary storage site for CO_2. Unlike many gases, the solubility of CO_2 increases as water temperature decreases. This means that cold polar waters absorb CO_2 from the atmosphere. In turn, CO_2 is released into the atmosphere as the water warms in tropical regions. Global cooling leads to colder ocean water, which increases the amount of CO_2 it can hold. The decreasing atmospheric CO_2 levels then cause temperatures to

drop and glaciers to advance, resulting in a positive feedback cycle of global cooling. Operating in the reverse, warming ocean water releases CO_2, which should produce a positive feedback cycle of global warming by enhancing the greenhouse effect.

The fact that average global temperatures and CO_2 levels are so closely correlated suggests that Earth will experience record warmth as the atmospheric level of CO_2 increases. The present level of approximately 370 parts per million of CO_2 is already higher than at any time in the past million years.

Carbon dioxide is not the only greenhouse gas. Molecule for molecule, methane (CH_4) is more than 20 times more effective than CO_2 as a greenhouse gas but is considered less important because the atmospheric concentrations and the lengths of time the molecules of gas remain in the atmosphere (residence times) are much smaller. Widespread publicity about the need to reduce the increase in methane levels has led to proposals to monitor (and perhaps control) dump emissions and termite mounds, both of which produce substantial quantities of CH_4. A much more important source of atmospheric methane may be from the tundra regions or the deep sea, where vast quantities of methyl hydrates could be released if oceanic temperatures rise. If warming tundra or ocean water indeed releases large amounts of methane, the resulting positive feedback cycle of warming could be enormous.

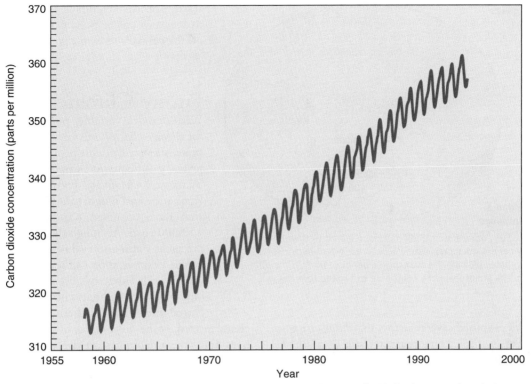

Figure 8.14 The steady increase since 1958 in the concentration of carbon dioxide in the atmosphere is demonstrated by this graph. Although motor vehicle exhaust contributes major quantities of CO_2 to the atmosphere worldwide, other sources of CO_2 such as heavy industry, power production, and burning of wood for heat, cooking, and clearing of land are just as important. ■ **What causes the annual fluctuations of CO_2 levels that are indicated on the graph?**

Other greenhouse gases are CFCs (chlorofluorocarbons) and N_2O (nitrous oxide). The relative greenhouse contribution of common greenhouse gases and their average residence times in the atmosphere are presented in Figure 8.15.

Changes in the Ocean

Oceans cover 70 percent of Earth's surface. Their enormous volume and high heat capacity make the oceans the single largest buffer against changes in Earth's climate. Whenever changes occur in oceanic temperatures, chemistry, or circulation, significant changes in global climate are certain to follow.

Surface oceanic currents are driven mostly by winds. However, a much slower circulation deep below the surface moves large volumes of water between the oceans. A major driving force of the deep circulation appears to be differences in water buoyancy caused by differences in salinity. Where surface evaporation is rapid, the rising salinity content causes the seawater density to increase, inducing subsidence. On the other hand, when major influxes of freshwater flow from adjacent continents or concentrations of melting icebergs, called *Heinrich events,* flood into the North Atlantic, the salinity is reduced, thereby increasing the buoyancy of the water.

When the surface water is buoyant, deep-water circulation slows. In many cases, the freshwater influx is immediately followed by a major flow of warm surface waters into the North Atlantic, causing an abrupt warming of the Northern Hemisphere. Subsurface ocean currents are also affected by water temperature. Extremely cold Arctic and Antarctic waters are quite dense and tend to subside, whereas tropical water is warmer and may tend to rise. Therefore, salinity and temperature taken together bring about rather complex subsurface flows deep within our ocean basins.

In modern times, short-term changes in Pacific circulation are primarily responsible for El Niño and La Niña events. The onset of El Niño/Southern Oscillation (ENSO) climatic events is both rapid and global in extent, and it is widely believed that changes in oceanic circulation may be responsible for similar rapid climate changes during the Pleistocene.

Changes in Landmasses

The final category of climate change theories invokes changes in Earth's surface to explain lengthy periods of cold climates. Ice ages, some with more than 125 glacial advances, occurred during past periods of Earth's history (Fig. 8.16). To explain some of the previous glacial periods,

Greenhouse gases

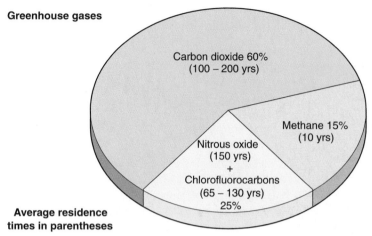

**Average residence
times in parentheses**

Figure 8.15 Gases, other than carbon dioxide, released to the atmosphere by human activity contribute approximately 40% to the greenhouse effect. The figures in parentheses indicate the average number of years that the different gases remain in the atmosphere and contribute to temperature change.

scientists have proposed several factors that might be responsible. For example, one characteristic that all of these glacial periods have in common with the Pleistocene is the presence of a continent in polar latitudes. Polar continents permit glaciers to accumulate on land, which results in lowered sea levels and consequent global effects. For example, when sea level was lowered between 40 and 50 meters (130 and 160 ft) some 10 million years ago (see Fig. 8.16), the Mediterranean Sea was cut off from the Atlantic Ocean. As a result, evaporation of seawater in the Mediterranean basin caused the accumulation of large quantities of salt on the seafloor. When the link between the Mediterranean Sea and the global ocean was reestablished, the average salinity of the oceans was reduced by 6 percent. As oceanic salinity decreases, seawater is more likely to freeze, enhancing global cooling. In short, the presence of a polar continent may set feedback mechanisms in motion that could eventually lead to a series of glacial cycles.

Another geologic factor sometimes invoked as a cause of climate change is the formation of a landmass that restricts oceanic or atmospheric circulation, or the disappearance of such a barrier because of landmass movements. For example, eruptions of volcanoes on the isthmus of Panama would have restricted ocean currents flowing between the Atlantic and Pacific, thereby closing a pathway of significant energy transfers. Another example is the uplift of the Himalayas, altering atmospheric flows, monsoonal effects, and weathering rates in Asia. Both of these events, and several other significant changes, immediately predate the onset of the modern series of glaciations. Which ones caused climate changes and which changes are simply coincidences is yet to be determined.

A final group of theories involve changes in albedo, caused either by major snow accumulations on high-latitude landmasses or by large oceanic ice sheets drifting into tropical regions. The increased reflection of sunlight

starts a positive feedback cycle of cooling that may end when the polar oceans freeze, shutting off the primary moisture source for the polar ice sheets.

Future Climates

With so many variables potentially responsible for climate change, reliably predicting future climate is a very difficult proposition at best. The primary difficulty in climate prediction is posed by natural variability. Figure 8.17 displays the frequency and magnitudes of climate changes that have occurred naturally over the past 150,000 years. Although the Holocene has been the most stable interval of the whole period, a detailed examination of the Holocene record reveals a wide range of climates. For example, a long interval of climates hotter than today's occurred during the Altithermal. This interval was characterized by the dominance of grasslands in the Sahara and severe droughts on the Great Plains. Other warm intervals occurred during the Bronze Age, during the second half of the Roman Empire, and in medieval times. An unusually cold interval began with the eruption of Santorini (the site of a civilization that may have been the basis for the Atlantis myth) in the Aegean Sea. Other cold periods occurred during the Dark Ages and again beginning about 1197 A.D. in the North Atlantic and 1309 A.D. in continental Europe. This last episode has been termed the **Little Ice Age.** The Little Ice Age had major impacts on civilizations—from the isolation of the Greenland settlements established during the medieval warm period to the abandonment of the Colorado Plateau region by the Anasazi cultures. An important point to remember is that, with the exception of the cold interval that began with the eruption of Santorini, climate scientists do not know what variables changed to cause each of these major climate fluctuations.

Attempts to predict future climates are complicated further by the operation of many feedback cycles. Simply increasing the amount of heat that is trapped by atmospheric gases in the lower atmosphere may or may not result in long-term warming. Negative feedback processes such as increased cloud formation and increased plant uptake of CO_2 may operate to counteract the warming. On the other hand, warming of the oceans and tundra may release additional greenhouse gases, setting into motion some significant positive feedback cycles. Which feedback mechanisms will dominate is not certain; therefore, all predictions must be tentative.

There have been numerous attempts to simulate the variables affecting climate. **General circulation models (GCMs)** are complex computer simulations based on the relationships among variables discussed throughout this book: sun angles, temperature, evaporation rates, land versus water effects, energy transfers, and so on (Fig. 8.18).

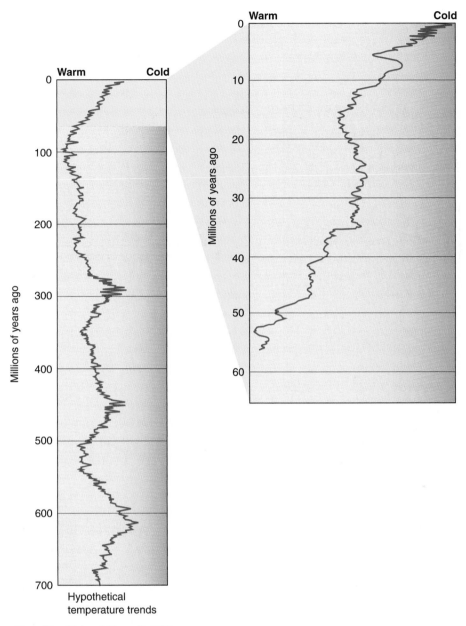

From Sheckleton & Kennett, 1975.

Figure 8.16 These graphs show hypothetical temperature trends and periods of glaciation as suggested by evidence in the rocks and fossils that have accumulated over 700 million years of Earth history. ■ **Why would the simulation of climates of the last few million years be considered more reliable than that of previous periods of Earth history?**

The models often simulate Earth's climate by dividing the surface into grid cells and estimating how all of the variables will respond when a change is made in one of them. Although the complexity and usefulness of GCMs is increasing rapidly, so far none of the models has accurately predicted most previous climate changes, and their ability to predict future climate changes is therefore still suspect. Nonetheless, GCMs appear to do a good job predicting how conditions will change in specific regions as the globe warms or cools, and they have added new insights into how some climatic variables interact.

Based on the record of climate changes during the past, only one thing can be concluded about future climates:

They will change. Looking into the distant future, the Milankovitch cycles indicate that another glacial cycle is probably on the way. The most rapid cooling should occur between 3000 and 7000 years from now. The end result will be glacial ice covering most of Canada and northern Europe, and the United States can expect glacial ice as far south as Long Island, Indianapolis, and Des Moines about 25,000 years from now.

In the near term, however, global warming is most likely. The rise of atmospheric CO_2, the widespread destruction of vegetation, and the feedback cycles that will most likely result are bound to increase the average global temperature for the foreseeable future. An average

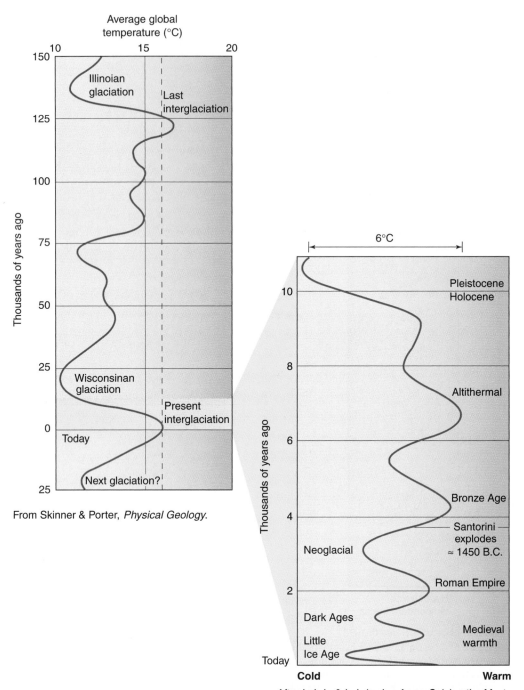

From Skinner & Porter, *Physical Geology*.

After Imbrie & Imbrie, *Ice Ages: Solving the Mystery*
Enslow Publishers, Short Hills, NJ, p. 179.

Figure 8.17 This figure shows broad climatic trends of the past 150,000 years, with significant details for the Holocene. Climatologists have been remarkably successful in dating recent climatic change, but predicting future climates remains extremely difficult. ■ **Why is this so?**

increase of 1°C (nearly 2°F) would be equivalent to the change that has occurred since the end of the Little Ice Age in about 1850. A 2°C warming would be greater than anything that has happened in the Holocene, including the Altithermal. A 3°C warming would exceed anything that has happened in the past million years. Current estimates and the most reliable GCMs predict a 1°C to 3.5°C (2° to 6°F) warming in the twenty-first century.

Detailed studies of conditions during the Altithermal, as well as GCMs, permit climatologists to determine how global warming will likely affect Earth. It is clear that not all areas will be affected equally. One of the most important effects is expected to be a more vigorous hydrologic cycle, fueled largely by increases in evaporation from the ocean. Intense rainfalls will be more likely everywhere, as will droughts in regions such as the Great

(a)

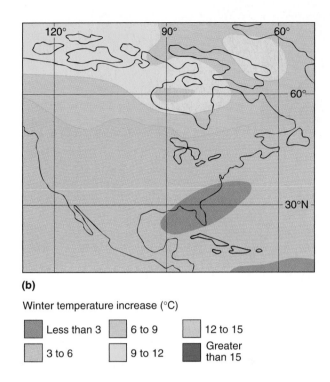

(b)

Winter temperature increase (°C)

- Less than 3
- 3 to 6
- 6 to 9
- 9 to 12
- 12 to 15
- Greater than 15

(c)

Figure 8.18   These three estimates of increasing global temperature in response to a doubling of atmospheric carbon dioxide were produced as a result of different general circulation models (GCMs): (a) National Center for Atmospheric Research, (b) NASA/Goddard Institute for Space Studies, (c) United Kingdom Meteorological Office. All three estimates agree that global warming will occur.   ■ **Why do you think the estimates vary so widely?**

Plains. Temperatures will rise most in the polar regions, during nighttime rather than daytime, and during winter months. As a result of the warming, sea levels will rise, mostly because of the thermal expansion of the oceans. By 2100, sea levels should be between 15 and 95 centimeters (0.5–3.1 ft) higher than today. In addition, the ranges of tropical diseases will expand toward higher latitudes, tree lines will rise, and many alpine glaciers will disappear.

However, some greenhouse effects will be beneficial to humans. Growing seasons in the high latitudes will increase in length. The increase in atmospheric CO_2 will help some crops such as wheat, rice, and soybeans grow larger faster. In the United States, a 1°C increase in average temperature should decrease heating bills by about 11 percent.

Many scientists believe that the warming is already occurring. The five hottest years of the twentieth century occurred during the 1990s, and each subsequent year usually set a new record. Average annual global temperatures have already risen between 0.3° and 0.6°C (0.5–1.1°F), and sea level has risen between 10 and 25 centimeters (4–10 in.), during the past 100 years. Given the long residence times of many greenhouse gases (see again Fig. 8.15) and the thermal inertia of the oceans, some warming is inevitable and will likely continue for hundreds of years no matter what actions are taken to reduce greenhouse gas emissions. On the other hand, the lesson of this chapter is that short-term climate trends, and some longer-term climate trends, will never be reliably predictable. Major volcanic eruptions, changing oceanic circulation, or human impacts (such as massive deforestation and urbanization) can significantly disrupt climate trends at any time. Moreover, because sudden, major shifts in the climate system that were not caused by human activity are abundant in the record of the last 2.4 million years, predicting climate will always be a risky endeavor.

Climatic Change and Its Impact on Coastlines

When we look at a map or a globe, one of the pieces of geographic information that we see is so obvious and basic that we often take it for granted, perhaps failing to recognize that it is spatial information. This is the location of coastlines — the boundaries between land and ocean regions. One of the most basic aspects of our planet that a world map shows us is where landmasses exist and where the oceans are located, as well as the generally familiar shape of these major Earth features. But maps of our planet today only show where the coastline is currently located. We know that sea level has changed over time, and that it rose 20 to 30 centimeters (8–12 in.) during the twentieth century. The hydrologic system on Earth is a closed system because the total amount of water (as a gas, liquid, and solid) on our planet is fixed. When the climate supports more glacial ice, sea level falls. When world climates experience a warming tendency, sea level rises. More ice in glaciers means less water in the oceans, and vice versa. If global warming trends continue at their present rate, the U.S. Environmental Protection Agency predicts that sea level will rise 31 centimeters (1ft) in the next 25 to 50 years. That amount of sea-level rise will cause problems for low-lying coastal areas; the populations of some coral islands in the Pacific are already concerned, as their homelands are barely above the high-tide level.

Looking at a map of world population distribution and comparing it to a world physical map shows a strong link between settlement density and coastal areas. For low-lying coastal regions, sea-level rise is a major concern, and the more gentle the slope of the coast is, the farther inland the inundation would be with every increment of sea-level rise. Scientists at the United States Geological Survey have determined that if all of the glaciers on Earth were to melt, sea level would rise 80 meters (263 ft), and that a 10-meter (33 ft) rise would displace 25 percent of the United States population.

Before the ice ages of the Pleistocene, worldwide climates were generally warmer; through much of Earth's history, no glaciers have existed on the planet. During times like those, when Earth was ice-free, sea level would have been at a maximum, and that might occur again in the distant future under similar climatic conditions. At the time when glaciers were most extensive, during the maximum advance of Pleistocene glaciers, sea level fell to about 100 meters (330 ft) below today's level. Maps that create the positions of coastlines and the shape of continents

DEFINE & RECALL

empirical classification
genetic classification
Köppen system
climograph
Thornthwaite system

potential evapotranspiration (PE)
actual evapotranspiration (AE)
microclimate
urban heat island

Pleistocene
varve
foraminifera
oxygen-isotope analysis
Holocene
Altithermal

eccentricity cycle
obliquity cycle
precession cycle
Little Ice Age
general circulation model (GCM)

DISCUSS & REVIEW

1. Why is it important to study the nature and possible causes of past climates when attempting to predict future climate change?

2. What is the difference between genetic and empirical classification? Is a classification based on statistics empirical or genetic?

3. Why are temperature and precipitation the two atmospheric elements most widely used as the sources of statistics for climatic classification? How are these two elements used in the Köppen system to identify six major climate categories?

4. What are the advantages and disadvantages of the Köppen system for geographers? Why are the Köppen climatic boundaries often referred to as "vegetation lines"?

5. What is a climograph? What is its function?

6. How does the Thornthwaite system of climate classification differ from the Köppen system? What are the advantages of the Thornthwaite system?

7. What examples of microclimates can be found in your local area? How did you decide that these are examples of microclimates?

8. Why is the occurrence, frequency, and dating of glacial advances and retreats so important to the study of past climates? How has modern research changed earlier theories of glacial coverage and associated climate change during the Pleistocene?

during times of major environmental change show how temporary and vulnerable coastal areas can be.

The maps below show the pre-Pleistocene maximum rise of sea level, the Pleistocene drop in sea level, and the impact on North American coastlines. It seems odd to think of world maps as temporary, but they only show us where the coast is located today. Like other natural boundaries, coasts can shift over time.

Maps of coastline change

9. In what way has the Holocene been unusual compared to climatic conditions during most of the rest of the Pleistocene?

10. How have scientists been able to document the rapid shifts of climates that have occurred during the latter part of the Pleistocene?

11. What are the major possible causes of global climatic change? What contribution did the mathematician Milankovitch make to theories regarding glaciation?

12. What is the evidence that volcanic activity can affect global temperatures? How does this occur?

13. What effects can changes in the amounts of CO_2 and other "greenhouse" gases in the atmosphere have on global temperatures? How can past changes in amounts of CO_2 be determined?

14. How might changes in Earth's oceans and landmasses affect global climates?

15. What is the primary difficulty for any climatologist who attempts to predict future climates? How effective has the use of computer simulations been in the prediction of future climate change?

16. What changes are likely to occur in Earth's major subsystems if global warming continues for the near term, as many scientists believe?

CONSIDER & RESPOND

1. Study the definitions for the individual Köppen climatic types described in Appendix C. Why do you think Köppen and later climatologists who modified the system selected the particular temperature and precipitation parameters that separate the individual types from one another?

2. Examine the climograph in Figure 8.2. During what month does Nashville experience the greatest precipitation? What major change would immediately identify this graph as representing a Southern Hemisphere location? What do you think the four horizontal dashed lines represent?

3. Review Figures 8.6, 8.7, 8.16, and 8.17. State in your own words the general conclusions you would draw from a study of all four figures.

4. After studying Chapters 7 and 8 in your textbook, which do you believe is more important to you now and in the future—the subject of weather or of climate? Defend your answer.

LOW-LATITUDE AND ARID REGIONS

Regardless of the chosen system of classification, the world's climates are distributed in patterns that occur on each landmass in similar positions and in similar latitudes.

What system of classification has been modified for the presentation of climatic types in this textbook?

Why are the climatic types of this system so closely related to major vegetation zones?

Knowledge of the controls of weather and climate is sufficient preparation to permit a reasonably accurate prediction of the broad distribution of climatic types over Earth's land surfaces.

What control has the greatest influence over climatic distribution?

What other controls produce significant deviation from the pattern established by the major control?

Geographers use regions for much the same reasons that scholars in other disciplines use arbitrary systems for the organization of information—to create an orderly presentation of diverse phenomena.

How does a geographer identify and define a region?

What type of phenomena can be organized into regions?

Although the humid tropical climates are all characterized by high temperatures throughout the year, they exhibit significant differences based on either the amounts or the distribution of the precipitation they receive.

What temperature parameter do these climates have in common?

How do they differ from one another on the basis of precipitation?

What controlling factors explain these differences?

Except for a few unusual circumstances, the tropical rainforest climatic regions are among the least populated areas of the world, despite being coincident with the belt of heaviest rainfall, insolation, and vegetative growth.

Why are there so few people in these regions?

What are the exceptions?

In what ways are these regions valuable to humankind?

Although rainfall is seasonal in both tropical savanna and tropical monsoon climatic regions, the differences in total rainfall between the two climates cause major dissimilarities in their appearance, resource characteristics, and human use.

What are the chief dissimilarities?

How is rainfall responsible?

How are humans affected?

Knowledge of the location of the world's deserts is similar to an understanding of the distribution of the world's steppe regions.

What is the association between deserts and steppes?

How do the regions differ?

In what ways are they similar?

Typical vegetation of a tropical rainforest climate can be seen along the Madre de Dios River in Peru.

The odds are overwhelming that within your lifetime, if you have not already done so, you will travel. You may travel extensively either within or beyond the borders of North America to destinations far from where you are living today. You may be moving to a new home or place of employment. You may be traveling on business or simply for pleasure. Whatever the reason, it is likely that you will ask the question almost every other traveler asks: "I wonder what the weather will be like?"

Realistically, as you have learned from reading previous chapters, the question should probably be "I wonder what the climate is like?"

The constant variability associated with weather in many areas of the world makes it difficult to predict. However, the long-term ranges and averages upon which climate is based allow geographers to provide the traveler with a general idea of the atmospheric conditions likely to be experienced at specific locations throughout the world during different times of the year.

Of course, there are many reasons other than travel why knowledge of climate and its variation over Earth's surface is a valuable asset. An understanding of climate in other areas of the world helps us to understand the adaptations to atmospheric conditions that have been made by the people who live there. We can better appreciate some of their economic activities and certain aspects of their cultures. In addition, the climate of any place on Earth has a dominant effect on native vegetation and animal life. It influences the rate and manner by which rock is destroyed and soil is formed. It is a contributing factor in the way landforms are reduced and physical landscapes are sculptured. In short, knowledge of climates provides endless clues not only to atmospheric conditions but to numerous other aspects of the physical environment.

In this chapter and the next, you will be provided with a broad descriptive survey of world climates: their locations, distributions, general characteristics, associated features, and related human activities. The information contained in these two chapters can serve as a valuable knowledge base as you prepare for the future. Throughout these chapters, we will use a modified and simplified version of the Köppen climate classification that was introduced in Chapter 8. (A detailed presentation of the Köppen system is included in Appendix C.) It is interesting to note that each of the climates discussed can be found within North America so even if you never travel beyond this continent's borders, you will still find the discussions of all climates valuable preparation as you move about your own country.

The Distribution of Climatic Types

Five of the six major climate categories of the Köppen classification include enough differences in the ranges, total amounts, and seasonality of temperature and precipitation to produce the 13 distinctive climatic types listed in Table 9.1. The tropical and arid climatic types will be discussed in some detail in this chapter; the mesothermal, microthermal, and polar climates will be presented in Chapter 10, along with a brief coverage of undifferentiated highlands.

Before we begin a study of individual climates, it is worthwhile to review what we have already learned about the atmospheric controls that produce these climates (Fig. 9.1). Applying our knowledge of these controls, we should be able to predict, with a reasonable amount of accuracy, the climatic patterns that exist throughout Earth's land areas. For example, suppose a large, unexplored continent divided the vast Pacific Ocean, stretching almost from pole to pole. Based on our knowledge of the atmospheric controls, and assum-

Table 9.1   Simplified Köppen Climate Classes

Climates	Climograph Abbreviation
Humid Tropical Climates (A)	
Tropical Rainforest Climate	Tropical Rf.
Tropical Monsoon Climate	Tropical Mon.
Tropical Savanna Climate	Tropical Sav.
Arid Climates (B)	
Steppe Climate	Low-lat./Mid-lat. Steppe
Desert Climate	Low-lat./Mid-lat. Desert
Humid Mesothermal (Mild Winter) Climates (C)	
Mediterranean Climate	Medit.
Humid Subtropical Climate	Humid Subt.
Marine West Coast Climate	Marine W.C.
Humid Microthermal (Severe Winter) Climates (D)	
Humid Continental, Hot Summer Climate	Humid Cont. H.S.
Humid Continental, Mild Summer Climate	Humid Cont. M.S.
Subarctic Climate	Subarctic
Polar Climates (E)	
Tundra Climate	Tundra
Ice-sheet Climate	Ice-sheet
Highland Climates (H)	

Latitude

Tropical (*A*) climate: Island of Jamaica

Mesothermal, mild winter (*C*) climate: Summer in southern Spain

Microthermal, severe winter (*D*) climate: Winter in Illinois

Polar (*E*) climate: Glacier in the Alaska Mountain Range

Atmospheric Pressure

Altitude

Arid (*B*) climate: Mojave Desert, southern California

Highland (*H*) climate: Teton Range, near Jackson Lake, Wyoming

Figure 9.1 Photographs representing Köppen's six major climate categories. ■ **What advantages or disadvantages for human use does each photograph suggest?**

ing that this continent did not have a mountainous landscape, it would not be difficult to construct a map of its climatic patterns (see Fig. 9.2). Throughout the following section, frequent reference should be made to this figure and to Figure 9.3, which includes photographs related to each climatic type.

Tropical Climates

Near the equator of our hypothetical continent, we would find high temperatures year-round, because the noon sun would never be far from the zenith (at 90°, or directly overhead). Humid climates of this type with no

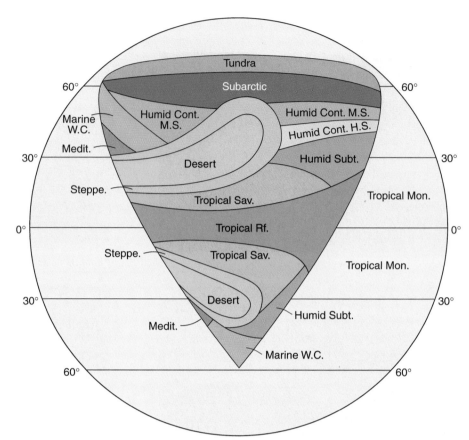

Figure 9.2 A generalized continent showing the characteristic locations of the Köppen climatic regions.
■ **How might a mountain range extending from north to south along the west coast of the northern middle latitudes affect the distribution of climates in this model?**

winter season are Köppen's **tropical** climates. As his boundary for tropical climates, Köppen chose 18°C (64.4°F) for the average temperature of the coldest month because it closely coincides with the geographic limit of certain tropical palms.

Table 9.1 and Figure 9.3 indicate that there are three humid tropical climates, reflecting major differences in the amount and distribution of rainfall within the tropical regions. Tropical climates extend poleward to 30° latitude or higher in the continent's interior but to lower latitudes near the coasts, because of the moderating influence of the oceans on coastal temperatures. Note also that the humid tropical climates extend further poleward along the east coast than along the west coast because cool currents flow along tropical west coasts while warm currents flow along the tropical east coasts.

Regions near the equator are influenced by the Intertropical Convergence Zone (ITCZ). However, the convergent and rising air of the ITCZ, which brings rain to the tropics, is not anchored in one place; instead, it follows the 90° sun angle (see again The Analemma, Chapter 3), migrating with the seasons. Within 5° to 10° latitude of the equator, rainfall occurs year-round because the ITCZ moves through twice a year and is never far away. Likewise, tropical coasts facing the easterly trade winds rarely experience drought because of the steady onshore flow of humid maritime air. Poleward of

this zone, precipitation becomes seasonal. When the ITCZ is over the region during the high-sun period (summer), there is adequate rainfall. However, during the low-sun period (winter), the subtropical highs and trade winds invade the area, bringing clear, dry weather.

Thus, on our hypothetical continent, we find the tropical rainforest climate in the equatorial region flanked both north and south by the dry-winter tropical savanna climate. Finally, along coasts facing the strong, moisture-laden inflow of air associated with the summer monsoon, we find the tropical monsoon climate. This climate owes its existence to the seasonal reversal of atmospheric pressure and winds over both land and water, described in Chapter 5.

Polar Climates

Just as the tropical climates lack winters (cold periods), the polar climates—at least statistically—lack summers. **Polar** climates, as defined by Köppen, are areas in which no month has an average temperature exceeding 10°C (50°F). Poleward of this temperature boundary, trees cannot survive. On our hypothetical continent, the 10°C isotherm for the warmest month more or less coincides with the Arctic Circle, poleward of which the sun does not rise above the horizon in midwinter and in summer strikes at a low angle. Again, the effects of ocean currents prevent this isotherm from being a straight line. It is deflected northward on the west side of our hypothetical continent, where tropical waters move poleward, and dips to the south on the east side, where cold arctic water flows into the general oceanic circulation.

The polar climates are subdivided into tundra and ice-sheet climates. Tundra climate occurs where at least one month averages above 0°C (32°F); in the ice-sheet climate, no month has an average temperature above 0°C.

Mesothermal and Microthermal Climates

Except where arid climates intervene, the lands between the tropical and polar climates are occupied by the transitional middle-latitude mesothermal and microthermal climates. As they are neither tropical nor polar, the mild and severe winter climates must have at least one month averaging below 18°C (64.4°F) and one month averaging above 10°C (50°F). Although both middle-latitude climatic categories have distinct temperature seasons, the **microthermal** climates have severe winters with at least one month averaging below freezing. Once again, vegetation reflects the climatic differences. In the severe-winter climates, all broadleaf and even a species of needle-leaf trees defoliate naturally during the winter (generally, needle-leaf trees do not defoliate in winter) as soil water is temporarily frozen and unavailable. Much of the natural vegetation of the mild-winter **mesothermal** climates retains its foliage throughout the year because liquid water is always present in the soil. The line separating mild from severe winters usually lies in the vicinity of the 40th parallel.

A number of important internal differences within the mesothermal and microthermal climatic groups produce individual climatic types based on precipitation patterns or seasonal temperature contrasts. On our hypothetical continent, the Mediterranean, or dry summer, mesothermal climate appears along west coasts between 30° and 40° latitude, where the dry subtropical highs of summer alternate with the humid conditions of the westerly wind belt in winter. On the east coasts, in generally the same latitudes, the humid subtropical climate is found. The contrast between coasts should be expected, because the subsidence and temperature inversions of the subtropical highs, which promote dry weather during the summer months, are only in evidence along the west coasts of continents. In contrast, continental east coasts receive both cyclonic and convective precipitation year-round.

The distinction between the humid subtropical and marine west coast climates illustrates a second important criterion for the internal subdivisions of middle-latitude climates: seasonal contrasts. Both mesothermal climates have year-round precipitation, but humid subtropical summer temperatures are much higher than those in the marine west coast climate. The humid subtropical climate is located along continental east coasts; it lies directly in the path of maritime tropical air masses moving onshore from source regions over high-temperature tropical oceans. Therefore, summers are hot. In contrast, the mild summers of the marine west coast climate, located poleward of the Mediterranean climate along continental west coasts, often extending beyond 60° latitude, are a direct result of mild, moist conditions carried onshore by the westerly winds in these latitudes. This is but one example of how seasonal temperatures are used to distinguish between climatic types in the Köppen system.

Another example is found among microthermal climates, which usually receive year-round precipitation associated with storms traveling along the polar front. Internal subdivision into climatic types is based on summers that become shorter and cooler and winters that become longer and more severe with increasing latitude and continentality. On our generalized continent, the microthermal climates are found exclusively in the Northern Hemisphere. As in the real world, there is no land in the Southern Hemisphere latitudes that would normally be occupied by the humid microthermal climates. In the Northern Hemisphere, these climates progress poleward through the humid continental, hot-summer climate, to the humid continental, mild-summer climate, and finally, to the subarctic climate.

Arid Climates

Arid climates are dominated by year-round moisture deficiency. These climates will penetrate deep into the continent, interrupting the neat zonation of climates that would otherwise exist. The definition of climatic aridity is that precipitation received is less than potential evapotranspiration (ET). (Evapotranspiration, discussed in Chapter 6, is a combination of evaporation and transpiration, or water loss through vegetation.) Aridity does not depend solely on the amount of precipitation received; ET rates and temperature must also be taken into account. In a low-latitude climate with relatively high temperatures, the ET rate is greater than in a colder, higher-latitude climate. As a result, more rain must fall in the lower latitudes to produce the same effects (especially noticeable on vegetation) that smaller amounts of precipitation produce in areas with lower temperatures and, consequently, lower ET rates.

On our hypothetical continent, the arid climates are concentrated in a zone about 15°N and S latitude to about 30°N and S along the western coasts, expanding much farther poleward over the heart of the landmass. This eastward expansion is a consequence of remoteness from the oceanic moisture supply. The main factor that causes the deficiency of precipitation is air subsidence on the eastern sides of the subtropical high pressure systems. The correspondence between the arid climates and the belt of subtropical high pressure is quite unmistakable.

Tropical rainforest: Windward coast of Jamaica

Desert: Sonoran Desert of Arizona

Tropical monsoon: Himalayan foothills, West Bengal, India

Steppe: Sand Hills of Nebraska

Tropical savanna: East African highlands

Mediterranean: Village in southern Spain

Figure 9.3 Examples of climates in the Köppen classification system.

Ocean currents reinforce the effect of the subtropical anticyclones in suppressing precipitation along west coasts in the subtropical zone. On the eastern sides of the oceanic anticyclones, or the west coasts of continents, cold water from the polar seas moves equatorward. Air moving onshore from the west is cooled from below by these currents, increasing its stability. This allows the zone of deficient precipitation to be accentu-

ated on the west coast itself. In fact, coastal deserts receive the lowest precipitation amounts of any region on Earth.

In desert climates, the annual amount of precipitation is less than half the annual potential ET. Bordering the deserts are steppe climates—semiarid climates that are transitional between the extreme aridity of the deserts and the moisture surplus of the humid climates. The definition of the steppe climate is an area where annual

Humid subtropical: Grapefruit grove in central Florida

Humid continental, mild summer: Fall season in New Hampshire

Marine west coast: North Sea coast of Scotland

Subarctic: Southern edge of the Canadian shield

Humid continental, hot summer: Corn Belt farm in Missouri

Tundra: Ice meets ocean in southern Greenland

precipitation is less than potential ET but more than half the potential ET.

Highland Climates

We have been speaking of our hypothetical continent as if it were a uniform surface with no significant relief features. Of course, the pattern of climates and extent of aridity would be affected by irregularities in Earth's surface, such as the presence of deep gulfs, interior seas, or significant highlands. The climatic patterns of Europe and North America are quite different because of such variations.

Highlands can channel air-mass movements and create abrupt climatic divides. Their own microclimates form an intricate pattern related to elevation, cloud

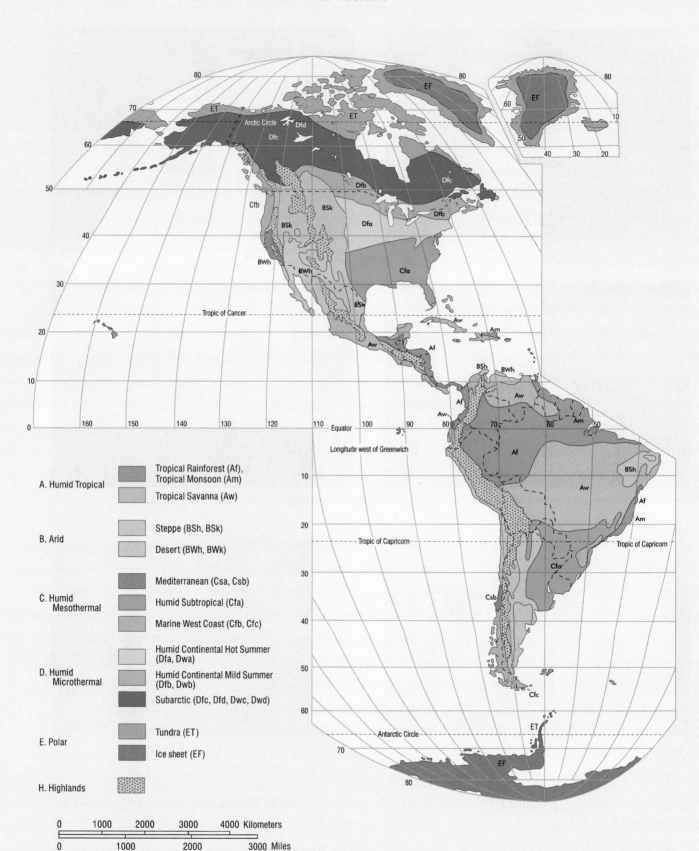

Figure 9.4 World map of climates in the modified Köppen classification system

A Western Paragraphic Projection
developed at Western Illinois University

cover, and exposure. One significant effect of highlands aligned at right angles to the prevailing wind direction is the creation of arid regions extending tens to hundreds of kilometers leeward.

Climatic Regions

Each of our modified Köppen climatic types is defined by specific parameters for monthly averages of temperature and precipitation; thus, it is possible to draw boundaries between these types on a world map. The areas within these boundaries are examples of one type of world region. A *region,* as the term is used by geographers, refers to an area that has recognizably similar internal characteristics that are distinct from those of other areas. A region may be described on any basis that unifies it and differentiates it from others.

As we examine the climatic regions of the world in the pages that follow, you should make frequent reference to the map of world climatic regions (Fig. 9.4). It shows the patterns of Earth's climates as they are distributed over each of the continents. It is not surprising that these patterns closely resemble the distribution of climates on our hypothetical continent, as we applied real-world climatic controls to our model. However, a word of caution is in order. On a map of climatic regions, distinct lines separate one region from another. Obviously, the lines do not mark points where there are abrupt changes in temperature or precipitation conditions. Rather, the lines signify **zones of transition** between different climatic regions. Furthermore, these zones or boundaries between regions are based on monthly and annual averages and may shift as temperature and moisture statistics change over the years.

The actual transition from one climatic region to another is gradual, except in cases where the change is brought about by an unusual climatic control such as a mountain barrier. It would be more accurate to depict climatic regions and their zones of transition on a map by showing one color fading into another. Always keep in mind, as we describe Earth's climates, that it is the core areas of the regions that best exhibit the characteristics that distinguish one climate from another.

Now, let us look more closely at Figure 9.4. One thing that is immediately noticeable is the change in climate with latitude. This is especially apparent in North America. Looking at the east coast of the United States and Canada, we see that the southern tip of Florida has a tropical savanna climate. This gives way almost immediately to the humid subtropical climate of the southern states. As we move poleward from Virginia toward Maryland and southern New Jersey, the humid subtropical climate changes to a humid continental, hot-summer climate. Further north, Vermont, New Hampshire, and Maine have a humid continental climate with mild summers. Moving still further north, we see that much of

Canada has a subarctic climate until we reach latitudes of 55° to 60°N, or 65° to 70°N in the western part of the country. Poleward of these latitudes, the climate is described as polar tundra.

It is also interesting to discover locations that have similar climates. For instance, how many areas in the world have climates similar to the one with which you are most familiar? The Mediterranean climate is found in many parts of the world: California, southern Europe, the Middle East, central Chile, and small sections of the southwest coasts of Africa and Australia. These places are all in similar latitudes (between about 30° and 40°) and receive winds from oceans to the west. The marine west coast climate is often found poleward of the Mediterranean type: in coastal Oregon to British Columbia, southern Chile, and much of northwest Europe, at latitudes between 40° and 60°. This climate is found as well in New Zealand, on the southeast coast of Australia, and in southeast Africa. The subarctic category of the humid microthermal climatic category is found over the vast continental areas of both northern Canada and northern Eurasia. Also, if you follow the equator around the world, you will spend most of your time in a tropical rainforest climate, except for excursions into the mountains and savanna of eastern Africa.

A first look at the world climate map may have given you the impression that climatic regions are scattered about haphazardly. A closer inspection, however, reveals that many areas show a latitudinal progression of climate. Furthermore, we have seen that similar climates usually appear in similar latitudes and/or in similar locations with respect to landmasses, ocean currents, or topography. These climatic patterns emphasize the close relationship among climate, the weather elements, and the climatic controls. As we predicted when we examined the hypothetical continent, there is an order to Earth's atmospheric conditions, and so also to its climatic regions.

Carrying this idea even further, let us compare the world maps of temperature (page 112), precipitation (pages 172–173), and pressure and winds (page 133) with Figure 9.4, which shows world climates. To a large degree, the superposition of the maps showing the individual atmospheric elements or the combination of them has produced the map of climatic regions. We know, of course, that ultimately the order and pattern of the climatic regions are based on the patterns produced by Earth-sun relationships. The exceptions to those patterns are the result of the uneven distribution and irregular sizes and shapes of land and water bodies, of the Coriolis effect on wind and ocean currents, and of interruption of the continents by mountain barriers.

A striking variation in these global climatic patterns becomes apparent when we compare the Northern and Southern Hemispheres. The Southern Hemisphere lacks the large landmasses of the Northern Hemisphere; thus, no climates in the higher latitudes (in land regions)

can be classified as humid microthermal, and only one small peninsula of Antarctica can be said to have a tundra climate.

Humid Tropical Climatic Regions

We have already learned a good deal about the climatic regions of the humid tropics through our study of a hypothetical continent and our examination of the maps throughout this book. It now remains for us to identify the major characteristics of each humid tropical climatic type in turn, along with its associated world regions.

A glance at Figure 9.5 and a careful reading of Table 9.2 will provide the locations of the humid tropical climates and a preview of the significant facts associated with them. The table also reminds us that although each of the three humid tropical climates has high average temperatures throughout the year, they differ greatly in the amount and distribution of precipitation.

Tropical Rainforest Climate

The **tropical rainforest climate** probably comes most readily to mind when someone says the word *tropical*. Hot and wet throughout the year, the tropical rainforest climate has been the stage for many stories of both fact and fiction. One cannot easily forget the classic struggle with the elements portrayed by Bogart and Hepburn in the film *The African Queen*. Upon visiting this type of climate, one would easily feel the high temperatures, oppressive humidity, and the frequent heavy rains for which it is known.

Constant Heat and Humidity Most weather stations in the tropical rainforest climatic regions record average monthly temperatures of 25°C (77°F) or more (Fig. 9.6). Since these regions are usually located within 5° or 10° of the equator, the sun's noon rays are always close to being directly overhead. Days and nights are of almost equal length, and the amount of insolation received remains nearly constant throughout the year. Consequently, there are no appreciable temperature variations that can be linked to the sun angle and therefore considered seasonal.

The **annual temperature range**—the difference between the average temperatures of the warmest and coolest months of the year—reflects the consistently high angle of the sun's rays. As indicated in Figure 9.6, the annual range is seldom more than 2° or 3°C (4°–5°F). In fact, at Ocean Island in the central Pacific, the annual range is 0°C because of the additional moderating influence of a large water body on the already nearly uniform pattern of insolation.

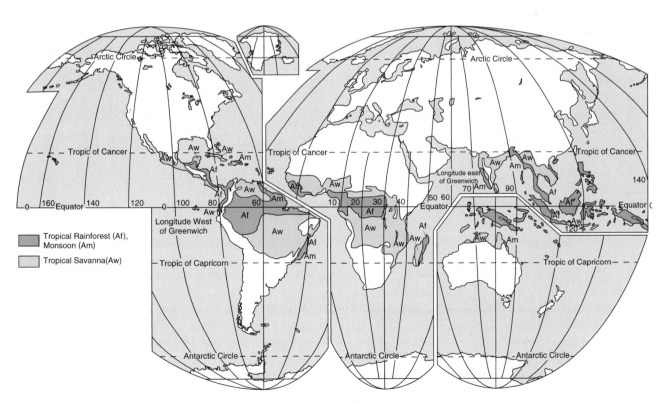

Humid Tropical Climates

Figure 9.5 Index map of humid tropical climates

Table 9.2 The Humid Tropical Climates

Name and Description	Controlling Factors	Geographic Distribution	Distinguishing Characteristics	Related Features
Tropical Rainforest				
Coolest month above 18°C (64.4°F); driest month with at least 6 cm (2.4 in.) of precipitation	High year-round insolation and precipitation of doldrums (ITC); rising air along trade wind coasts	Amazon R. Basin, Congo R. Basin, east coast of Central America, east coast of Brazil, east coast of Madagascar, Malaysia, Indonesia, Philippines	Constant high temperatures; equal length of day and night; lowest (2°–3°C/3°–5°F) annual temperature ranges; evenly distributed heavy precipitation; high amount of cloud cover and humidity	Tropical rainforest vegetation (selva); jungle where light penetrates; tropical iron-rich soils; climbing and flying animals, reptiles, and insects; slash-and-burn agriculture
Tropical Monsoon				
Coolest month above 18°C (64.4°F); one or more months with less than 6 cm (2.4 in.) of precipitation; excessively wet during rainy season	Summer onshore and winter offshore air movement related to shifting ITC and changing pressure conditions over large landmasses; also transitional between rainforest and savanna	Coastal areas of southwest India, Sri Lanka, Bangladesh, Myanmar, southwestern Africa, Guyana, Surinam, French Guiana, northeast and southeast Brazil	Heavy high-sun rainfall (especially with orographic lifting), short low-sun drought; 2°–6°C (3°–10°F) annual temperature range, highest temperature just prior to rainy season	Forest vegetation with fewer species than tropical rainforest; grading to jungle and thorn forest in drier margins; iron-rich soils; rainforest animals with larger leaf-eaters and carnivores near savannas; paddy rice agriculture
Tropical Savanna				
Coolest month above 18°C (64.4°F); wet during high-sun season, dry during lower-sun season	Alternation between high-sun doldrums (ITCZ) and low-sun subtropical highs trades caused by shifting winds and pressure belts	Northern and eastern India, interior Myanmar and Indo-Chinese Peninsula; northern Australia; borderlands of Congo R., south central Africa; llanos of Venezuela, campos of Brazil; western Central America, south Florida, and Caribbean Islands	Distinct high-sun wet and low-sun dry seasons; rainfall averaging 75–150 cm (30–60 in.); highest temperature ranges for humid tropical climates	Grasslands with scattered, drought-resistant trees, scrub, and thorn bushes; poor soils for farming, grazing more common; large herbivores, carnivores, and scavengers

One of the most interesting features of the tropical rainforest climate is that the **daily (diurnal) temperature ranges**—the differences between the highest and lowest temperatures during the day—is usually far greater than the annual range. Highs of 30° to 35°C (86°–95°F) and lows of 20° to 24°C (68°–75°F) produce daily ranges of 10° to 15°C (18°–27°F). However, the drop in temperature at night is small comfort. The high humidity causes even the cooler evenings to seem oppressive.

The climographs of Figure 9.6 illustrate that significant variations in precipitation can occur even within rainforest regions. Although most rainforest locations receive more than 200 centimeters (80 in.) a year of precipitation, and the average is in the neighborhood of 250 centimeters (100 in.), there are some that record an annual precipitation of more than 500 centimeters (200 in.). Ocean locations, near the greatest source of moisture, tend to receive the most rain. Mt. Waialeale,

Iquitos, Peru — Tropical Rf (Af)
4°S — Precip.: 262 cm (103.1 in.) — 73°W
Av. temp.: 25°C (77°F) — Range: 2.2°C (4°F)

Akassa, Nigeria — Tropical Rf (Af)
5°N — Precip.: 365 cm (143.8 in.) — 6°E
Av. temp.: 25.5°C (78°F) — Range: 2.2°C (4°F)

Ocean Is., Gilbert Is. — Tropical Rf (Af)
1°S — Precip.: 213 cm (83.9 in.) — 170°E
Av. temp.: 27°C (80.6°F) — Range: 0°

in the Hawaiian Islands, receives a yearly average of 1146 centimeters (451 in.), making it the wettest spot on Earth. As a group, climate stations in the humid tropics experience much higher annual totals than typical humid middle-latitude stations. Compare, for example, the 365 centimeters in Akassa with the average 112 centimeters received annually in Portland, Oregon, or the 61 centimeters received in London.

We should recall that the heavy precipitation of the tropical rainforest climate is associated with the warm, humid air of the doldrums and the unstable conditions along the ITCZ. Both convection and convergence serve as uplift mechanisms, causing the moist air to rise and resulting in the heavy rains that are characteristic of this climate. There is heavy cloud cover during the warmer, daylight hours when convection is at its peak, although the nights and early mornings can be quite clear. Variations in rainfall can usually be traced to the ITCZ and its weak low pressure systems. Many tropical rainforest locations (Akassa, for example) exhibit two maximum precipitation periods during the year, one during each appearance of the ITCZ as it follows the migration of the sun's direct rays. In addition, although no season can be called dry, during some months it may rain on only 15 or 20 days.

A Delicate Balance   The most common vegetation of tropical rainforest climatic regions is multistoried, broadleaf evergreen forest made up of many species whose tops form a thick, almost continuous cover that blocks out much of the sun's light (Fig. 9.7). This type of rainforest is called **selva**. Within the selva, there is usually little undergrowth on the forest floor because sunlight cannot penetrate enough to create much low-growing vegetation.

The relationship between the soils beneath the selva and the vegetation the soils support is so close that there exists a nearly perfect ecological balance between the two, threatened only by people's efforts to earn a living from the soil. The trees of the selva supply the tropical soils with the nutrients the trees need for growth. As leaves, flowers, and branches fall to the ground, or as roots die, the numerous soil animals and bacteria act on them, transforming the organic matter into usable nutrients. However, if the trees are removed, there is no replenishment of these nutrients and no natural counterbalance to the processes by which large amounts of percolating soil moisture remove the soluble minerals from the soil. The intense activities of microorganisms, worms, termites, ants, and other insects cause rapid deterioration of

Figure 9.6 Climographs for tropical rainforest climate stations.
■ **Why is it difficult, without looking at the climograph key, to determine whether each station is located in the Northern or Southern Hemisphere?**

The Amazon Rainforest

Currently, an area of Earth's virgin tropical forest somewhat larger than a football field is being destroyed every second. During a recent decade, the rate of deforestation doubled, and in the Amazon Basin, where more than half of the rainforest resources are located, the rate nearly tripled. Simple mathematics indicates that even at the present rate of deforestation, the Amazon rainforest will virtually disappear in 150 years.

From the point of view of a developing country, there are basic economic reasons for clearing the Amazon rainforest, and Peru, Ecuador, Colombia, and Brazil (where most of the forest is located) are developing countries. Tropical timber sales provide short-term income to finance national growth and repay staggering debts owed to foreign banks. In addition, Brazil, in particular, has viewed the Amazon rainforest as a frontier land available for agricultural development and for resettlement of the poor from overcrowded urban areas.

No one questions that decisions concerning the future of the Amazon rainforest rest with the governments of nations that control these resources. So why has the rainforest become a serious international issue? The answer is found in the concerns of physical geographers and other environmental scientists throughout the world.

When the tropical rainforests are removed, the hydrologic cycle and energy budgets in the previously forested areas are dramatically altered, often irreversibly. In tropical rainforests, the canopy shades the forest floor, thus helping to keep it cooler. In addition, the huge mass of vegetation provides a tremendous amount of water vapor to the atmosphere through transpiration. The water vapor condenses to

Cattle grazing along the Rio Salimoes in Brazil in an area of former rainforest

Dan Guarvarch/ Photo Researchers, Inc.

the remaining organic debris, and soon all that remains is an infertile mixture of insoluble manganese, aluminum, and iron compounds.

In recent years, there has been large-scale harvesting of the tropical rainforests as land has been cleared for agriculture, especially in the Amazon River basin. Such deforestation can have a significant and, unfortunately, permanent impact on the delicate balance that exists among Earth's systems.

Environmental conditions vary from place to place within climatic regions; therefore, the typical rainforest situation we have just described does not apply everywhere in the tropical rainforest climate. Some regions are covered by true jungle, a term often misused when describing the rainforest. **Jungle** is a dense tangle of vines and smaller trees that develops where direct sunlight does reach the ground, as in clearings and along streams. Other regions have soils that remain fertile or have bedrock that is chemically basic and provides the soils above with a constant supply of soluble nutrients

through the natural weathering processes. Examples of the former region are found along major river floodplains; examples of the latter are the volcanic regions of Indonesia and the limestone areas of Malaysia and Vietnam. Only in such regions of continuous soil fertility can agriculture be intensive and continuous enough to support population centers in the tropical rainforest climate.

Human Activities Throughout much of the tropical rainforest climate, humans are far outnumbered by other forms of animal life. Though there are few large animals of any kind, a great variety of smaller tree-dwelling and aquatic species live in the rainforest. Birds, monkeys, bats, alligators, crocodiles, snakes, and amphibians such as frogs of many varieties abound. Animals that can fly or climb into the food-rich leaf canopy have become the dominant animals in this world of trees.

Most common of all, though, are the insects. Mosquitoes, ants, termites, flies, beetles, grasshoppers, butterflies, and bees live everywhere in the rainforest. Insects form clouds, which in turn provide rainfall to nourish the forest. With the forests removed, transpiration is diminished, which leads to less cloud cover and less rainfall. With fewer clouds and no forest canopy, more solar energy reaches Earth's surface.

The unfortunate outcome is that areas that are deforested soon become hotter and drier, and any ecosystem in place is seriously damaged or destroyed. Once the rainforests are removed, the soils lose their source of plant nutrients, and this precludes the growth of any significant crops or plants. In addition, the rainforest, which was in harmony with the soil, cannot reestablish itself. Thus, the multitude of flora and fauna species indigenous to the rainforest is lost forever. It is impossible to calculate the true cost of this reduction in biodiversity (the total number of different plant and animal species in the Earth system). The lost species may have held secrets to increased food production; a cure for AIDS, cancer, or other health problems; or a base for better insecticides that do not harm the environment. Similar services to humanity already have been provided by tropical forest species.

Tropical deforestation is also threatening the natural chemistry of the atmosphere. The tropical forests are a major source of the atmospheric oxygen so essential to all animal life. And deforestation encourages global warming by enhancing the greenhouse effect, since forests act as a major reservoir of carbon dioxide. It has been estimated that forest clearing since the mid-1800s has contributed more than 130 billion tons of carbon to the atmosphere; more than two-thirds as much as has been added by the burning of coal, oil, and natural gases combined.

What can be done? The reasons for tropical deforestation and the solutions to the problem may be economic, but the issues are extremely complex. It is not sufficient for the rest of the world to point out to governments of tropical nations that their forests are a major key to human survival. It is unacceptable for scientists and politicians from nations where barely one-fourth of the original forests remain to insist that the citizens of the tropics cease cutting trees and establish forest plantations on deforested land. These are desired outcomes, but it is first the responsibility of all the world's people to help resolve the serious economic and social problems that have prevented most tropical nations from considering their forests as a sustainable resource.

Gregory Dimijian/ Photo Researchers, Inc.

A section of Amazon rainforest cleared by slash-and-burn techniques for potential farming or grazing

can breed continuously in this climate without danger from cold or drought.

Besides the insects, there are genuine health hazards for human inhabitants of the tropical rainforest. Not only does the oppressive, sultry weather impose uncomfortable living conditions, it also allows a variety of parasites and disease-carrying insects to threaten human survival. Malaria, yellow fever, dengue fever, and sleeping sickness are all insect-borne diseases of the tropics, frequently fatal, and uncommon in middle-latitude climates.

Whenever native populations have existed in the rainforest, subsistence hunting and gathering of fruits, berries, small animals, and fish have been important. Since the introduction of agriculture, land has been cleared and crops such as manioc, yams, beans, maize (corn), bananas, and sugarcane have been grown. It has been the practice to cut down the smaller trees, burn the resulting debris, and plant the desired crops (Fig. 9.8). With the forest gone, this kind of farming is possible for only two years, or perhaps three, before the soil is completely exhausted of its small supply of nutrients and the surrounding area is depleted of game. At this point, the native population moves to another area of forest to begin the practice over again. This kind of subsistence agriculture is known as **slash-and-burn** or simply **shifting cultivation.** Its impact on the close ecological balance between soil and forest is obvious in many rainforest regions. Sometimes the damage done to the system is irreparable, and only jungle, thornbushes, or scrub vegetation will return to the cleared areas.

In terms of numbers of people supported, the most important agricultural use of the tropical rainforest climate is the wet-field (paddy) rice agriculture on the river floodplains of southeastern Asia. However, this type of agriculture is best developed in the monsoon variant of this climate. Commercial plantation agriculture is also significant. The principal plantation crops are rubber and cacao, both of which originally grew wild in the forests of the Amazon Basin but are now of greatest importance elsewhere—rubber in Malaysia and Indonesia, and cacao in West Africa and the Caribbean area.

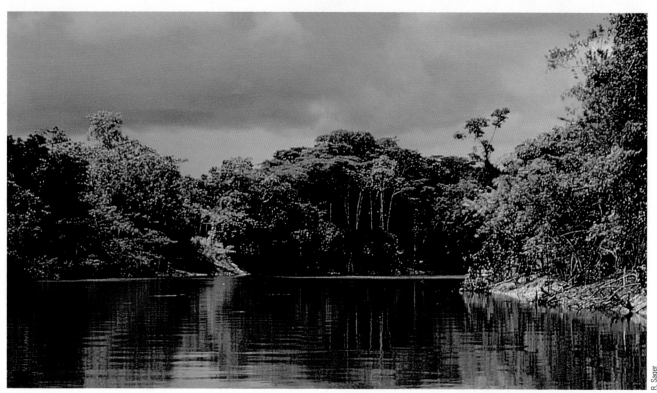

Figure 9.7 The multistoried broad-leaf evergreen forest of the Amazon Basin near Iquitos, Peru. ■ **Why might this photograph, taken along the riverbank, give an incorrect impression of vegetation at ground level in the forest?**

Tropical Monsoon Climate

We associate the **tropical monsoon climate** most closely with the peninsula lands of Southeast Asia. Here the alternating circulation of air (from land to sea in winter, from sea to land in summer) is strongly related to the shifting of the ITCZ. During the summer, the ITCZ moves suddenly north into the Indian subcontinent and adjoining lands to a latitude of 20° or 25°. This is due in part to the attracting force of the deep low pressure area of the Asian continent. However, as we have previously noted, the mechanism is complex and involves changes in the upper air flow as well as in surface currents. Several months later, the moisture-laden summer monsoon is replaced by an outflow of dry air from the massive Siberian high pressure area that develops in the winter season over central Asia. By this time, the ITCZ has shifted far to the south (see again Figure 5.11).

Figures 9.5 and 9.9 confirm that there are climatic regions outside of Asia that fit the simplified Köppen classification of tropical monsoon

as well. A modified version of the monsoonal wind shift occurs at Freetown, Sierra Leone, in Africa, but the climate there might also be described as transitional between the constantly wet rainforest climate and the sharply seasonal wet and dry conditions of the tropical savanna. Cayenne, French Guiana, on the other hand,

Figure 9.8 An example of subsistence slash-and-burn agriculture: preparing an area for planting in Jamaica. ■ **Would you expect shifting cultivation to be on the increase or decrease in tropical rainforests?**

receives much of its rainfall during the low-sun (winter) season, when the trade winds move onshore along the northeast coast of South America, bringing abundant moisture from the nearby ocean.

Distinctions between Rainforest and Monsoon   Whatever the factors are that produce tropical monsoon climatic regions, these regions have strong similarities to those classified as tropical rainforest. In fact, although their core regions are distinctly different, the two climates are often intermixed over transition areas. A major reason for the similarity between monsoon and rainforest climates is that a monsoon area has enough precipitation to allow continuous vegetative growth with no dormant period during the year. Rains are so abundant and intense, and the dry season so short, that the soils usually do not dry out completely. As a result, this climate and its soils support a plant cover much like that of the tropical rainforests.

However, there are clear distinctions between rainforest and monsoon climatic regions. The most important distinction, of course, concerns precipitation, including both distribution and amount. The monsoon climate has a short dry season, whereas the rainforest does not. Perhaps even more interesting, the average rainfall in monsoon regions varies more widely from place to place. It usually totals between 150 and 400 centimeters (60–150 in.) and may be massive where the onshore monsoon winds are forced to rise over mountain barriers. Mahabaleshwar, altitude 1362 meters (4467 ft), on the windward side of India's Western Ghats, averages more than 630 centimeters (250 in.) of rain during the five months of the summer monsoon.

The annual march of temperature of the monsoon climate differs appreciably from the monotony of the rainforest climate. The heavy cloud cover of the rainy monsoon reduces insolation and temperatures during that time of year. During the period of clear skies just prior to the onslaught of the rains, higher temperatures are recorded. As a result, the annual temperature range in a monsoon climate is from 2° to 6°C (compared with 2°–3°C in the tropical rainforest). This is illustrated by the temperature curves on the climographs of Figure 9.9.

Some additional distinctions between monsoon and rainforest regions can be found in vegetation and animal life. Toward the wetter margins, the tropical monsoon forest resembles the tropical rainforest, but fewer species are present and certain ones become dominant. The seasonality of rainfall in the monsoon narrows the range of species that will prosper. Toward the drier margins of the climate, the trees grow farther apart, and the monsoon

Figure 9.9   Climographs for tropical monsoon climate stations.
■ **Why does the distribution of precipitation at Cayenne differ so much from that of the other two stations, although all three are in the Northern Hemisphere?**

Slash-and-Burn Farming Fertilizes the Soils

As stated in the text, the slash-and-burn technique is used for farming in tropical rainforest regions. This seemingly destructive method is necessary for small-scale subsistence farming for two reasons. First, it is necessary to clear a space for cultivation. Second, since the heavy rainfall of these regions washes the nutrients out of the soils, slash-and-burn farming is a means of treating the topsoil. In regions where other fertilizers may not be available, this technique is designed to add nutrients to the soils. The technique is quite simple. First, a small area is slashed and cleared of standing native vegetation. Then, the cut vegetation is laid on the ground and allowed to dry (the drying process is sometimes delayed by afternoon rainfall). The next step is completed when the vegetation is ignited and burned totally to ashes. Finally, the ashes are mixed in with the topsoil.

A standard fertilizer usually (but not always) contains NO_3 (nitrates), K_2O (potassium oxide), and P_2O_5 (diphosphorus pentoxide). When natural rainforest vegetation is burned, its ashes contain, among other chemicals, N (nitrogen), K (potassium), and P (phosphorus). When these ashes are worked into the upper soil layer, they act as a low-grade fertilizer. As also stated in the text, this slightly more fertile soil will grow crops for only two to three years, after which the process must be repeated on another patch of the forest.

If the slash-and-burn fields are small, and if they are kept separated, they will recover from the burning. The surrounding native vegetation will encroach and once again take over. However, when massive burning is used to clear vast acreage of rainforest for mining or for pastureland, then recovery of the grounds is impossible, and the rainforest is lost forever.

---

forest often gives way to jungle or a dwarfed thorn forest. The composition of the animal kingdom here also changes. The climbing and flying species that dominate the forest are joined by larger, hoofed leaf-eaters and by carnivores such as the famous tigers of Bengal.

Effects of Seasonal Change The seasonal precipitation of the tropical monsoon climate is of major importance for economic reasons, especially to the people of

Southeast Asia and India. Most of the people living in those areas are farmers, and their major crop is rice, which is the staple food for millions of Asians. Rice is most often an irrigated crop, so the monsoon rains are very important to its growth. Harvesting, on the other hand, must be done during the dry season (Fig. 9.10).

Each year, an adequate food supply for much of South and Southeast Asia depends on the arrival and departure of the monsoon rains. The difference between

David Watanabe/ Getty Images.

Figure 9.10 Crop selection and agricultural production adjust to the wet (a) and dry (b) seasons throughout Southeast Asia. ■ **Would it be beneficial to the people of Southeast Asia if the traditional rice farming methods were replaced by mechanized rice agriculture as practiced in the United States?**

(a)

It is essential that the ashes from the burning be dug into the topsoil.

Michael Trapasso

famine and survival for many people in these regions is very much associated with the climate.

Tropical Savanna Climate

Located well within the tropics (usually between latitudes 5° and 20° on either side of the equator), the **tropical savanna climate** has much in common with the tropical rainforest and monsoon. The sun's vertical rays at noon are never far from overhead, the receipt of solar energy is nearly at a maximum, and temperatures remain constantly high. Days and nights are of nearly equal length throughout the year, as they are in other tropical regions.

However, as previously noted, its distinct seasonal precipitation pattern identifies the tropical savanna. As the latitudinal wind and pressure belts shift with the sun, savanna regions are under the influence of the rain-producing ITCZ (doldrums) for part of the year and the rain-suppressing subtropical highs for the other part. In fact, the poleward limits of the savanna climate are approximately the poleward limits of migration of the ITCZ, and the equatorward limits of this climate are the equatorward limits of movement by the subtropical high pressure systems.

As you can see in Table 9.2 and Figure 9.5, the greatest areas of savanna climate are found peripheral to the rainforest climates of Central and South America and Africa. Lesser but still important savanna regions occur in India, peninsular Southeast Asia, and Australia. In some instances, the

Armstrong Roberts/ Getty Images

(b)

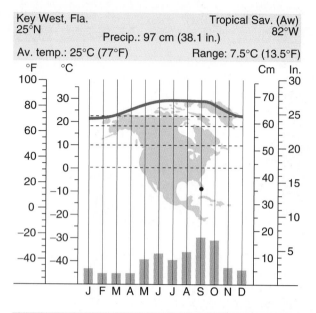

Key West, Fla.   Tropical Sav. (Aw)
25°N   82°W
            Precip.: 97 cm (38.1 in.)
Av. temp.: 25°C (77°F)   Range: 7.5°C (13.5°F)

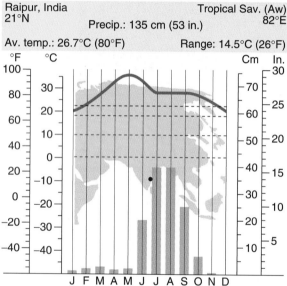

Raipur, India   Tropical Sav. (Aw)
21°N   82°E
            Precip.: 135 cm (53 in.)
Av. temp.: 26.7°C (80°F)   Range: 14.5°C (26°F)

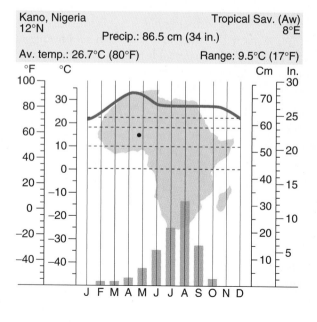

Kano, Nigeria   Tropical Sav. (Aw)
12°N   8°E
            Precip.: 86.5 cm (34 in.)
Av. temp.: 26.7°C (80°F)   Range: 9.5°C (17°F)

climate extends poleward of the tropics, as it does in the southernmost portion of Florida.

Transitional Features of the Savanna   Of particular interest to the geographer is the transitional nature of the tropical savanna. Often situated between the humid rainforest climate on one side and the rain-deficient steppe climate on the other, the savanna experiences some of the characteristics of both. During the rainy, high-sun season, atmospheric conditions resemble those of the rainforest, whereas the low-sun season can be as dry as nearby arid lands are all year. The gradational nature of the climate causes precipitation patterns to vary considerably (Fig. 9.11). Savanna locations close to the rainforest may have rain during every month, and total annual precipitation may exceed 180 centimeters (70 in.). In contrast, the drier margins of the savanna have longer and more intensive periods of drought and lower annual rainfalls, less than 100 centimeters (40 in.).

Other characteristics of the savanna help to demonstrate its transitional nature. The higher temperatures just prior to the arrival of the ITCZ produce annual temperature ranges 3° to 6°C (5°–11°F) wider than those of the rainforest, but still not as wide as those of the steppe and desert. Although the typical savanna vegetation (known as **llanos** in Venezuela and **campos** in Brazil) is a mixture of grassland and trees, scrub, and thorn bushes, there is considerable variation. Near the equatorward margins of these climates, grasses are taller, and trees, where they exist, grow fairly close together. Toward the drier, poleward margins, trees are more widely scattered and smaller, and the grasses are shorter. Soils, too, are affected by the climatic gradation, as the iron-rich reddish soils of the wetter sections are replaced by darker-colored wet and dry tropical soils in the drier regions.

Both vegetation and soils have made special adaptations to the alternating wet-dry seasons of the savanna. During the wet (high-sun) period, the grasslands are green and the trees are covered with foliage. During the dry (low-sun) period, the grass turns dry, brown, and lifeless, and most of the trees lose their leaves as an aid in reducing moisture loss through transpiration. The trees have deep roots that can reach down to water in the soil during the dry season. They are also fire resistant, an advantage for survival in the savanna where the grasses may burn during the winter drought.

Savanna Potential   Conditions within tropical savanna regions are not well suited to agriculture, although many of our domesticated grasses (grains) are presumed to

Figure 9.11 Climographs for tropical savanna climate stations.
■ **Consider the differences in climate, and human use of the environment, between Key West and Raipur. Which are more important in the geography of the two places, the physical or the human factors?**

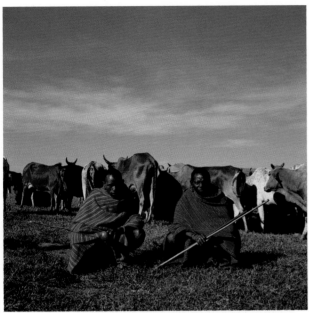

Figure 9.12 The grasslands of the East African savanna are well suited to the support of large numbers of grazing animals. The Masai cattle herders shown here in Kenya count their wealth by the numbers of animals they own. ■ **What problems are created for the Kenyan government as cattle herds grow?**

have grown wild there. Rainfall is far less predictable than in the rainforest or even the monsoon climate. For example, Nairobi, Kenya, has an average rainfall of 86 centimeters (34 in.). Yet from year to year, the amount of rain received may vary from 50 to 150 centimeters (20–60 in.). The drier the savanna station, the more unreliable the rainfall. However, the rains are essential for human and animal survival in savanna regions. When they are late or deficient, as they have been in West Africa in recent years, severe drought and famine result. On the other hand, when the rains last longer than usual or are excessive, they can cause major floods, often followed by outbreaks of disease.

Savanna soils (except in areas of recent stream deposits) also limit productivity. During the rains of the wet season, they become gummy; during the dry season, they are hard and almost impenetrable. Consequently, people in the savannas have often found the soils better suited to grazing than to farming. The Masai cattle herders of East Africa are world-famous examples (Fig. 9.12). However, even animal husbandry has its problems. Many savanna regions make poor pasturelands, at least during the dry part of the year.

The savannas of Africa have exhibited the greatest potential of the world's savanna regions. They have been veritable zoological gardens for the larger tropical animals, to such an extent that the popularity of classic photo safaris has made the African savannas a major center for tourism. The grasslands support many different herbivores (plant eaters), such as the elephant, rhinoceros, giraffe, zebra, and wildebeest (Fig. 9.13). The herbivores,

in turn, are eaten by the carnivores (flesh eaters), such as the lion, leopard, and the cheetah. Lastly, scavengers, such as hyenas, jackals, and vultures, devour what remains of the carnivore's kill. During the dry season, the herbivores find grasses and water along stream banks and forest margins and at isolated water holes. The carnivores follow the herbivores to the water, and a few human hunters and scavengers still follow them both.

Arid Climatic Regions

Arid climatic regions in the simplified Köppen system are widely distributed over Earth's surface. A brief study of Figure 9.14 confirms that they are found from the vicinity of the equator to more than 50°N and S latitude. There are two major concentrations of desert lands, and each illustrates one of the important causes of climatic aridity. The first is centered on the Tropics of Cancer and Capricorn (N and S latitudes) and extends 10° to 15° poleward and equatorward from there. The second is located poleward of the first and occupies continental interiors, particularly in the Northern Hemisphere.

The concentration of deserts in the vicinity of the two Tropic lines is directly related to the subtropical high pressure cells. Although the boundaries of the subtropical highs may migrate north and south with the sun, their influence remains constant in these latitudes. We have already learned that the subsidence and divergence of air associated with these cells is strongest along the eastern portions of the oceans (western portions of the continents). Hence, the clear weather and dry conditions of the subtropical high pressure extend inland from the western coasts of each landmass in the subtropics. The Atacama, Namib, and Kalahari Deserts and the desert of Baja California are restricted in their development by

Figure 9.13 Elephants have always been a majestic sight in the savanna climate of East Africa. ■ **Why are opportunities for photographs such as this in serious danger of disappearing completely?**

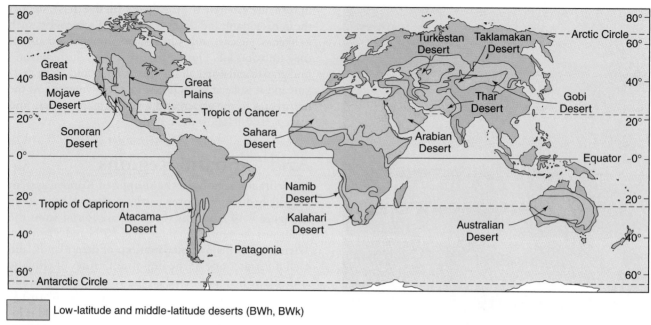

Low-latitude and middle-latitude deserts (BWh, BWk)

Low-latitude and middle-latitude steppes (BSh, BSk)

Figure 9.14   A map of the world's arid lands.   ■   **What does a comparison of this map with the Map of World Population Density (inside back cover) suggest?**

the small size of the landmass or by landform barriers to the interior. However, the western portion of the Afro-Eurasian landmass comprises the greatest stretch of desert in the world and includes the Sahara, Arabian, and Thar Deserts. Similarly, the Australian Desert occupies most of the interior of the Australian continent.

The second concentration of deserts is located within continental interiors remote from moisture-carrying winds. Such arid lands include the vast cold-winter deserts of inner Asia and the Great Basin of the western United States. The dry conditions of the latter region extend northward into the Columbia Plateau and southward into the Colorado Plateau and are increased by the mountain barriers that restrict the movement of rain-bearing air masses from the west. Similar rain shadow conditions help to explain the Patagonia Desert of Argentina and the arid lands in western China (Xinjiang/Sinkiang).

Both wind direction and ocean currents can accentuate aridity in coastal regions. When prevailing winds blow parallel to a coastline instead of onshore, desert conditions are likely to occur because little moisture is brought inland. This seems to be the case in eastern Africa and in northeastern Brazil. Where a cold current flows next to a coastal desert, foggy conditions may develop. Warm, moist air from the ocean may be cooled to its dew point as it passes over the cooler current. A temperature inversion is created, increasing stability and preventing the upward movement of air required for precipitation. The unique, fog-shrouded coastal deserts in Chile (the Atacama), southwest Africa (the Namib), and Baja California have the lowest precipitation of any regions on Earth.

Figure 9.14 shows deserts of the world to be core areas of aridity, usually surrounded by the slightly moister steppe regions. Hence our explanations for the location of deserts hold true for the steppes as well. The steppe climates either are subhumid borderlands of the humid tropical, mesothermal, and microthermal climates, or are transitional between these climates and the deserts. As previously noted, we classify both steppe and desert on the basis of the relation between precipitation and potential ET. In the **desert climate,** the amount of precipitation received is less than half the potential ET. In the **steppe climate,** the precipitation is more than half but less than the total potential ET.

The criterion for determining whether a climate is desert, steppe, or humid is *precipitation effectiveness.* The amount of precipitation actually available for use by plants and animals is the effective precipitation. Precipitation effectiveness is related to temperature. At higher temperatures, it takes more precipitation to have the same effect on vegetation and soils than at lower temperatures. The result is that areas with higher temperatures, and therefore greater potential for ET, can receive more precipitation than cooler regions and yet have a more arid climate.

Because of the temperature influence, precipitation effectiveness depends on the season in which an arid region's meager precipitation is concentrated. Obviously, precipitation received during the low-sun period will be more effective than that received when temperatures are higher, because less will be lost through ET. Graphs based on the concept of precipitation effectiveness are

Table 9.3   The Arid Climates

Name and Description	Controlling Factors	Geographic Distribution	Distinguishing Characteristics	Related Features
Desert				
Precipitation less than half of potential evaporation; mean annual temperature above 18°C (64.4°F) (low-lat.), below (mid-lat.)	Descending, diverging circulation of subtropical highs; continentality often linked with rain-shadow location	Coastal Chile and Peru, southern Argentina, southwest Africa, central Australia, Baja California and interior Mexico, North Africa, Arabia, Iran, Pakistan and western India (low-lat.); inner Asia and western United States (mid-lat.)	Aridity; low relative humidity; irregular and unreliable rainfall; highest percentage of sunshine; highest diurnal temperature range; highest daytime temperatures; windy conditions	Xerophytic vegetation; often barren, rocky, or sandy surface; desert soils; excessive salinity; usually small, nocturnal, or burrowing animals; nomadic herding
Steppe				
Precipitation more than half but less than potential evaporation; mean annual temperature above 18°C (64.4°F) (low-lat.), below (mid-lat.)	Same as deserts; usually transitional between deserts and humid climates	Peripheral to deserts, especially in Argentina, northern and southern Africa, Australia, central and southwest Asia, and western United States	Semiarid conditions, annual rainfall distribution similar to nearest humid climate; temperatures vary with latitude, elevation, and continentality -	Dry savanna (tropics) or short grass vegetation; highly fertile black and brown soils; grazing animals in vast herds, predators and smaller animals; ranching, dry farming

included in Appendix C and may be used to determine whether a particular location has a desert, steppe, or humid climate.

Desert Climates

The deserts of the world extend through such a wide range of latitudes that the simplified Köppen system recognizes two major subdivisions. The first are low-latitude deserts, where temperatures are relatively high year-round and frost is absent or infrequent even along poleward margins; the second are middle-latitude deserts, which have distinct seasons, including below-freezing temperatures during winter (Table 9.3). However, the significant characteristic of all deserts is their aridity. The relative unimportance of temperature is emphasized by the small number of occasions on which we distinguish between low-latitude and middle-latitude deserts in the discussion that follows.

Land of Extremes   By definition, deserts are associated with a minimum of precipitation, but they represent the extremes in other atmospheric conditions as well. With few clouds and little water vapor in the air, as much as 90 percent of insolation reaches Earth in desert regions. This is why the highest insolation and highest temperatures are recorded in low-latitude desert areas and not in the more humid tropical climates that are closer to the equator. Again because of light cloud cover, there is little atmospheric effect, and much of the energy received by Earth during the day is radiated back to the atmosphere at night. Consequently, night temperatures in the desert drop far below their daytime highs. This excessive heating and cooling give low-latitude deserts the greatest diurnal temperature ranges in the world, and middle-latitude deserts are not far behind. In the spring and fall, these ranges may be as great as 40°C (72°F) in a day. More common diurnal temperature ranges in deserts are 22° to 28°C (40°–50°F).

The sun's rays are so intense in the clear, dry desert air that temperatures in shade are much lower than those a few steps away in direct sunlight. (Keep in mind that all temperatures recorded for statistical purposes, including those in the desert, are shade temperatures.) Khartoum (in the Sahara of the Sudan) has an *average* annual temperature of 29.5°C (85°F), which is a *shade* temperature. Temperatures in the bright desert sun under cloudless skies at Khartoum are often 43°C (110°F) or more. Soil temperatures rise to close to 95°C (200°F) in midsummer in the Mojave Desert of southern California.

Aswan, Egypt   Low-lat. Desert (BWh)
24°N   Precip.: <.25 cm (0.1 in.)   33°E
Av. temp.: 27°C (80.5°F)   Range: 17°C (31°F)

Lima, Peru   Low-lat. Desert (BWh)
12°S   Precip.: 4 cm (1.6 in.)   77°W
Av. temp.: 20°C (68°F)   Range: 9°C (15.5°F)

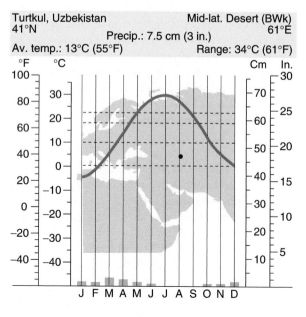

Turtkul, Uzbekistan   Mid-lat. Desert (BWk)
41°N   Precip.: 7.5 cm (3 in.)   61°E
Av. temp.: 13°C (55°F)   Range: 34°C (61°F)

During low-sun or winter months, deserts experience colder temperatures than more humid areas at the same latitude, and in summer they experience hotter temperatures. Just as with the high diurnal ranges in deserts, these high annual temperature ranges can be attributed to the lack of moisture in the air.

Annual temperature ranges are usually greater in middle-latitude deserts, such as the Gobi in Asia, than in low-latitude deserts because of the colder winters experienced at higher latitudes. Compare, for example, the climograph for Aswan in south central Egypt—at 24°N, a low-latitude desert location—with the climograph for Turtkul, Uzbekistan—at 41°N, a middle-latitude desert location (Fig. 9.15). The annual range for Aswan is 17°C (31°F); in Turtkul, it is 34°C (61°F).

Precipitation in the desert climate is irregular and unreliable, but when it comes, it may arrive in an enormous cloudburst, bringing more precipitation in a single rainfall than has been recorded in years. This happened in the extreme at the port of Walvis Bay, on the coast of the Namib Desert, a cold-current coastal desert of southwest Africa. The equivalent of 10 years' rain was received in one night when a freak storm dumped 3.2 centimeters (1.3 in.) of rain.

The actual amount of water vapor in the air may be high in desert areas. However, the hot daytime temperatures increase the capacity of the air, so that the relative humidity during daylight hours is quite low (10 to 30 percent). Desert nights are a different story.

Radiation of energy is rapid in the clear air. As temperature drops, relative humidity increases, and the formation of dew in the cool hours of early morning is common. Where measurements have been made, the amount of dew formed has sometimes considerably exceeded the annual rainfall for that location. It has been suggested, in fact, that dew may be of great importance to plant and animal life in the desert. Studies are now being carried out on the use of dew as a moisture source for certain crops, thereby minimizing the need for large-scale irrigation.

The convection currents set up by the intense heating of the land during the day help to make the desert a windy place. In addition, the sparseness of vegetation and the absence of topographic interruptions in some deserts allow winds to sweep across these arid lands unimpeded. Sand and dust are carried by the desert winds, lowering visibility and irritating eyes and throats.

Adaptations by Plants and Animals   Deserts tend to have sparse vegetation, and large tracts may be barren bedrock, sand, or gravel. The plants that do exist are

Figure 9.15 Climographs for desert climate stations. ■ **If you consider the serious limitations of desert climates, how do you explain why some people choose to live in desert regions?**

R. Gabler

Figure 9.16   This diverse association of vegetation is well adapted to survive the year-round drought conditions prevailing in desert climates.   ■   **Under what circumstances might significant numbers of people be attracted to regions with desert climates?**

xerophytic, or adapted to extreme drought. They may have thick bark, thorns, little foliage, and waxy leaves, all of which reduce loss of water by transpiration (Fig. 9.16). Another characteristic adaptation is the storage of moisture in stem or leaf cells, as in the cactus (barrel cactus, prickly pear, saguaro). Some plants, such as creosote

bush, mesquite, and acacias, have deep root systems to reach water; others, such as the Joshua tree, spread their roots widely near the surface for their moisture supply.

Even humans, the most adaptable of animals, find the desert environment a lasting challenge. For the most part, people have been hunters and gatherers, nomadic herders, and subsistence farmers wherever there was a water supply from wells or exotic streams (streams that bring water from outside the region) such as the Nile, Tigris, Euphrates, Indus, and Colorado Rivers. Desert people have learned to adjust their habits to the environment. For example, they wear loose clothing to protect themselves from the burning rays of the sun and to prevent moisture loss by evaporation from the skin. At night, when the temperatures drop, the clothing keeps them warm by insulating and minimizing the loss of body heat.

Permanent agriculture has been established in desert regions all around the world wherever river or well water is available (Fig. 9.17). Every desert has its irrigated oases, although they vary considerably in size. Some produce mainly subsistence crops, but others have become significant producers of commercial crops for export.

R. Gabler

Figure 9.17   Irrigated cotton, shown here in the Imperial Valley of California, owes its existence to water from the Colorado River.

Salt Lake City, Utah — Mid-lat. Steppe (BSk)
40°N — 112°W
Precip.: 41 cm (16 in.)
Av. temp.: 12°C (54°F) — Range: 27°C (49°F)

Semipalatinsk, Kazakhstan — Mid-lat. Steppe (BSk)
50°N — 80°E
Precip.: 18 cm (7 in.)
Av. temp.: 3°C (37°F) — Range: 37.5°C (68°F)

Daly Waters, Australia — Low-lat. Steppe (BSh)
16°S — 133°E
Precip.: 63 cm (24.7 in.)
Av. temp.: 27°C (80.5°F) — Range: 11°C (19.5°F)

Steppe Climates

Further study of Figure 9.14 and Table 9.3 provides a reminder that the distribution of the world's steppe lands is closely related to the location of deserts. Both moisture-deficient climate types share the controlling factors of continentality, rain-shadow location, the subtropical high pressure cells, or some combination of the three. The transitional nature of the steppes may make them seem like better-watered deserts at one time and like slightly subhumid versions of their humid climate neighbors at another. Herein lies the major problem of steppe regions: How and to what extent should these variable and unpredictable climatic regimes be used by humans?

Similarities to Deserts We are already aware that steppe regions are differentiated from deserts by their greater precipitation. Whereas most low-latitude desert locations receive fewer than 25 centimeters (10 in.) of rain annually, low-latitude steppe regions usually receive between 25 and 50 centimeters (10–20 in.). However, the similarities between deserts and steppes are often greater than the differences. In both climates, the potential ET exceeds the precipitation. As in the deserts, precipitation in steppe regions is unpredictable and varies widely in total amount from year to year. Annual rainfall differs significantly from place to place within both desert and steppe regions, and vegetation varies accordingly.

To be more specific, both the general precipitation pattern and the nature of the vegetation of a steppe region are usually closely related to the more humid climate immediately adjacent to it. Thus, when the steppe is located between desert and tropical savanna, the steppe's rains come with the high-sun season. Next to a Mediterranean climate, the steppe receives primarily winter precipitation. Similarly, the short, shallow-rooted grasses most commonly associated with the steppe climate occur in the areas of transition from mesothermal and microthermal climates to desert. However, in the areas of transition from tropical savanna to desert, the vegetation is the dry savanna type, including scrub tree and bush growth, which becomes more stunted and sparse toward the drier margins until the typical desert shrub vegetation type is dominant.

With similar conditions of atmospheric moisture and cloud cover, temperatures in the steppe regions also have much in common with those of the desert. Both low-latitude and middle-latitude varieties are identified by

Figure 9.18 Climographs for steppe climate stations. ■ **What are the chief differences between the climographs for Salt Lake City and Daly Waters? What are the causes of these differences?**

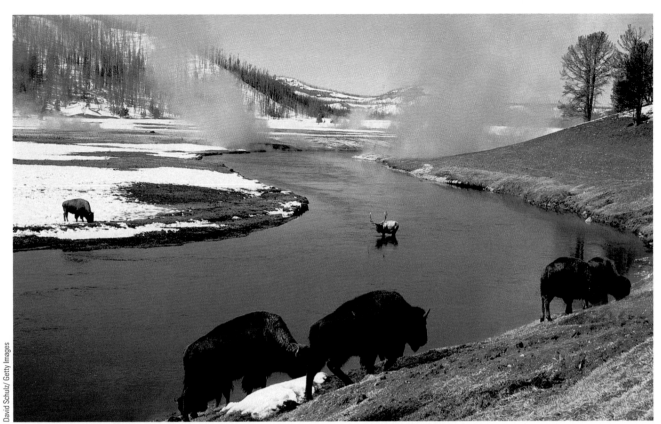

Figure 9.19 It is hard to imagine the magnitude of the animal population on the North American steppes little more than a century ago. One is now fortunate to see a single remnant such as this one of the former herds of millions of bison and other grazing animals.
■ **Where would a person go today to take a similar photograph?**

mean annual temperature (see Table 9.3 and Fig. 9.18). As in the desert, steppe temperatures vary throughout the climate type with latitude, distance from the sea, and elevation. The climographs of Figure 9.18 demonstrate that although summer temperatures are high in all steppe regions, the differences in winter temperatures can produce annual ranges in middle-latitude steppes that are two or three times as great as those in low-latitude steppes.

A Dangerous Appeal Although the surface cover is often incomplete, in more humid regions of the middle-latitude steppes the grasses have been excellent for pasture. In North America, this was the realm of the bison (Fig. 9.19) and antelope; in Asia, it was the domain of wild horses. Steppe soils are usually high in organic matter and soluble minerals. Attributes such as these have attracted farmers and herders alike to the rich grasslands—but not without penalty to both humans and land.

The climate is dangerous for agriculture, and people take a sizable risk when they attempt to farm. Although dry-farmed wheat and drought-resistant barley and sorghum can be successfully raised because both farming methods and crops are adapted to the environment, the use of techniques employed in more humid regions can lead to serious problems. During dry cycles, crops fail year after year, and with the land stripped of its natural sod, the soil is exposed to wind erosion. Even using the grasses for grazing domesticated animals is not always the answer, for overgrazing can just as quickly create "Dust Bowl" conditions.

The difficulties in making steppe regions more productive point out again the sensitive ecological balance of Earth's systems. The natural rains in the steppe are usually sufficient to support a vegetation cover of short grasses that can feed the roaming herbivores that graze on them. The herbivore population, in turn, is kept in check by the carnivores who prey on them. When people enter the scene, however, sending out more animals to graze, plowing the land, or merely killing off the predators, the ecological balance is tipped, and sometimes the results are disastrous.

DEFINE & RECALL

tropical
polar
microthermal
mesothermal
arid
region
zones of transition

tropical rainforest
  climate
annual temperature
  range
daily (diurnal)
  temperature range
selva

jungle
slash-and-burn (shifting)
  cultivation
tropical monsoon
  climate
tropical savanna
  climate

llanos
campos
desert climate
steppe climate
xerophytic

DISCUSS & REVIEW

1. Where would you expect each of the 13 distinctive climatic types of the modified Köppen system used in your text to be located on a hypothetical continent? What reasons explain the location of each?

2. Describe the three tropical (*A*) climates. Why does each exist?

3. How do ocean currents affect tropical and polar climates?

4. How are the mesothermal (*C*) and the microthermal (*D*) climates similar? In what important respect do they differ?

5. In what two important ways do the *C* and *D* climates differ internally?

6. What is the major factor producing the dryness of arid climates? What factors produce the variations in highland climates?

7. What is a region, as defined by geographers? Why is regionalization such a useful concept?

8. How many areas in the world have climates similar to the one with which you are most familiar? (Refer to Figure 9.4 for help in identifying them.)

9. What exceptions are there to the general rule that the pattern of climatic regions is produced by Earth-sun relationships? How do the Northern and Southern Hemispheres differ in climatic patterns, and why?

10. What are the controlling factors that explain the tropical rainforest climate? Give a brief verbal description of this climatic type.

11. What aspects of tropical rainforests are favorable to human use? Unfavorable? Describe the delicate balance between forest and soil. How might humans affect that balance?

12. Explain the seasonal precipitation pattern of the tropical savanna climate. State some of the transitional features of this climate. How have vegetation and soils adapted to the wet-dry seasons?

13. What conditions give rise to desert climates?

14. How do steppes differ from deserts? Why might human use of steppe regions in some ways be more hazardous than use of deserts?

15. Do each of the following for the listed climates: tropical rainforest, tropical monsoon, tropical savanna, desert, steppe.

    a. Identify the climate from a set of data or a climograph indicating average monthly temperature and precipitation for a representative station within a region typical of that climate.

    b. Match the climatic type with a written statement that includes one or more of the following: the statistical parameters of the climate in the modified Köppen classification; the particular climatic controls (controlling factors) that produce the climate; and the geographic distributions of the climate, stated in terms of physical or political location.

    c. Distinguish between the important subtypes (if any) of each climate by identifying the characteristics that separate them from one another.

CONSIDER&RESPOND

1. Based on the classification scheme presented in Appendix C, classify the following climatic stations from the data provided.

		J	F	M	A	M	J	J	A	S	O	N	D	Yr.
a.	Temp. (°C)	20	23	27	31	34	34	34	33	33	31	26	21	29
	Precip. (cm)	0.0	0.0	0.0	0.0	0.2	0.1	1.1	0.2	0.0	0.0	0.0	0.0	1.6
b.	Temp. (°C)	21	24	28	31	30	28	26	25	26	27	25	22	26
	Precip. (cm)	0.0	0.0	0.2	0.8	7.1	11.9	20.9	31.1	13.7	1.4	0.0	0.0	87.2
c.	Temp. (°C)	19	20	21	23	26	27	28	28	27	26	22	20	24
	Precip. (cm)	5.1	4.8	5.8	9.9	16.3	18.0	17.0	17.0	24.0	20.0	7.1	3.0	149.0
d.	Temp. (°C)	2	4	8	13	18	24	26	24	21	14	7	3	14
	Precip. (cm)	1.0	1.0	1.0	1.3	2.0	1.5	3.0	3.3	2.3	2.0	1.0	1.3	20.6
e.	Temp. (°C)	27	26	27	27	27	27	27	27	27	27	27	27	27
	Precip. (cm)	31.8	35.8	35.8	32.0	25.9	17.0	15.0	11.2	8.9	8.4	6.6	15.5	243.8
f.	Temp. (°C)	13	14	17	19	22	24	26	26	26	24	19	15	20
	Precip. (cm)	6.6	4.1	2.0	0.5	0.3	0	0	0	0.3	1.8	4.6	6.6	26.7

2. The following six locations are represented by the data in the previous table, although not in the order listed: Albuquerque, New Mexico; Belém, Brazil; Benghazi, Libya; Faya, Chad; Kano, Nigeria; Miami, Florida. Use an atlas and your knowledge of climates to match the climatic data with the locations.

3. Benghazi, Libya, and Albuquerque, New Mexico, both exhibit steppe climates, but the cause (or control) of their dry conditions is quite different. Describe the primary factor responsible for the dry conditions at each location.

4. Tropical *B* climates are located more poleward than *A* climates, yet their daytime high temperatures are often higher. Why?

5. Kano, Nigeria, has copious amounts of rainfall during the summer season. There are two sources, or causes, of this rainfall. What are they?

MIDDLE-LATITUDE, POLAR, AND HIGHLAND CLIMATIC REGIONS

Perhaps the most distinctive feature of Mediterranean climatic regions is that they receive most of their precipitation during the season opposite that during which greatest plant growth normally occurs.

Why is this so?

What plant adaptations have developed as a result?

What human use patterns have developed in response to these climatic conditions?

The locations of the Mediterranean and the humid subtropical climates on opposite sides of continents in the lower middle latitudes provide convincing evidence of the dominance of the subtropical highs as the climatic control of these latitudes.

In what ways are the subtropical highs responsible for these contrasting climates?

How do the highs influence climate both poleward and equatorward of these latitudes?

Some geographers refer to the marine west coast climate as the "temperate oceanic," or simply the "marine," climate.

Why do we call this climate the marine west coast climate?

Why do all these names emphasize maritime terms?

In what ways does the climate exemplify its names?

What factors produce the climate?

Although humid continental hot-summer and mild-summer regions have more in common than not, there are distinct differences in natural and cultivated vegetation between the two climates, particularly in North America.

What are these differences, and why do they exist?

What does this statement suggest about scientific classification?

The climatic and related physical characteristics that distinguish the subarctic climate and the two polar climates are accompanied by human utilization patterns significantly different from those found in the humid continental climates.

In what ways does the physical environment in each of these high-latitude climates affect human use patterns?

In what ways are cultural factors more influential?

Highlands are occupied by a complex pattern of widely varying microclimates far too intricate to be shown on anything but large-scale maps.

What factors are most important in explaining the existence of a particular highland microclimate?

What other phenomena are most directly affected by the different microclimates?

What are the major peculiarities of mountain climates?

Antarctic island, ice-free during the summer months, displays a tundra climate.

In this chapter, we will further explain the world climates as they dominate higher latitudes and higher altitudes. As we have noted, the tropical climates exhibit the constant characteristic of heat, whereas polar climates exhibit the constant characteristic of cold. The arid climates have inadequate precipitation in common. But what is the constant characteristic of the mesothermal and microthermal middle-latitude climates that we are about to examine? If there is one, then it is the oxymoron "constant inconsistency." Each of the middle-latitude climates is dominated by the changing of the seasons and the variability of atmospheric conditions associated with migrating air masses or cyclonic activity along the polar front.

Middle-Latitude Climates

If change is constant in the mesothermal and microthermal climatic groups, then degree of change is what distinguishes one climatic type from another. In one or another of the middle-latitude climates, summers vary from hot to cool and winters from mild to extremely cold. Some climates receive adequate monthly precipitation year-round; others experience winter drought or, even more challenging to humans who live there, dry months during the normal summer growing season. Despite the changing atmospheric conditions of middle-latitude climates, their regions are home to the majority of Earth's people, and they are major factors in some of Earth's most attractive, interesting, and productive physical environments.

Humid Mesothermal Climatic Regions

When we use the term *mesothermal* (from Greek: *mesos, middle*) in describing climates, we are usually referring to the moderate temperatures that characterize such regions. However, we could also be referring to their middle position between those climates that have high temperature throughout the year and those that experience severe cold. By definition, the mesothermal cli-

mates experience seasonality, with distinct summers and winters that distinguish them from the humid tropics. Their summers are long and their winters mild, and this separates them climatically from the microthermal climates, which lie poleward.

Table 10.1 presents the three distinct mesothermal climates introduced in Chapter 9. In all three, the annual precipitation exceeds the annual potential evapotranspiration, but the Mediterranean climate has a lengthy period of precipitation deficit in the summer season that distinguishes it from the humid subtropical and the marine west coast climates. The latter two are further differentiated by the fact that the humid subtropical regions have hot summers, whereas the marine west coast regions experience mild summers.

Mediterranean Climate

The **Mediterranean climate** is one of the best arguments for organizing a study of the environment, or developing an understanding of world regions, based on climatic classification. Such a climate appears with remarkable regularity in the vicinity of 30° to 40° latitude along the west coasts of each landmass (Fig. 10.1). The alternating controls of subtropical high pressure in summer and westerly wind movement in winter are so predictable that all Mediterranean lands have notably similar and easily

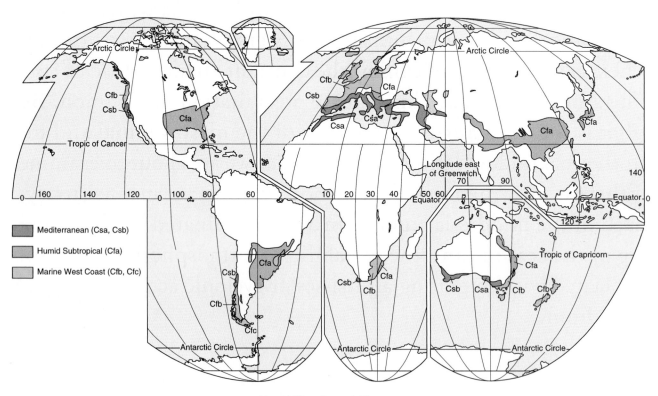

Humid Mesothermal Climates

Figure 10.1   Index map of humid mesothermal climates

Table 10.1   The Mesothermal Climates

Name and Description	Controlling Factors	Geographic Distribution	Distinguishing Characteristics	Related Features
Mediterranean				
Warmest month above 10°C (50°F); coldest month between 18°C (64.4°F) and 0°C (32°F); summer drought; hot summers (inland), mild summers (coastal)	West coast location between 30° and 40°N and S latitudes; alternation between subtropical highs in summer and westerlies in winter	Central California; central Chile; Mediterranean Sea borderlands, Iranian highlands; Capetown area of South Africa; southern and southwestern Australia	Mild, moist winters and hot, dry summers inland with cooler, often foggy coasts; high percentage of sunshine; high summer diurnal temperature range; frost danger	Sclerophyllous vegetation; low, tough brush (chaparral); scrub woodlands; varied soils, erosion in Old World regions; winter-sown grains, olives, grapes, vegetables, citrus, irrigation
Humid Subtropical				
Warmest month above 10°C (50°F); coldest month between 18°C (64.4°F) and 0°C (32°F); hot summers; generally year-round precipitation, winter drought (Asia)	East coast location between 20° and 40°N and S latitudes; humid onshore (monsoonal) air movement in summer, cyclonic storms in winter	Southeastern United States; southeastern South America; coastal southeast South Africa and eastern Australia; eastern Asia from northern India through south China to southern Japan	High humidity; summers like humid tropics; frost with polar air masses in winter; precipitation 62–250 cm (25–100 in.), decreasing inland; monsoon influence in Asia	Mixed forests, some grasslands, pines in sandy areas; soils productive with regular fertilization; rice, wheat, corn, cotton, tobacco, sugarcane, citrus
Marine West Coast				
Warmest month above 10°C (50°F); coldest month between 18°C (64.4°F) and 0°C (32°F); year-round precipitation; mild to cool summers	West coast location under the year-round influence of the westerlies; warm ocean currents along some coasts	Coastal Oregon, Washington, British Columbia, and southern Alaska; southern Chile; interior South Africa; southeast Australia and New Zealand; northwest Europe	Mild winters, mild summers, low annual temperature range; heavy cloud cover, high humidity; frequent cyclonic storms, with prolonged rain, drizzle, or fog; 3- to 4-month frost period	Naturally forested, green year-round; soils require fertilization; root crops, deciduous fruits, winter wheat, rye, pasture and grazing animals; coastal fisheries

recognized temperature and precipitation characteristics (Fig. 10.2). The special appearance, combinations, and climatic adaptations of Mediterranean vegetation not only are unusual but also are clearly distinguishable from those of other climates. Agricultural practices, crops, recreational activities, and architectural styles all exhibit strong similarities within Mediterranean lands.

Warm, Dry Summers; Mild, Moist Winters   The major characteristics of the Mediterranean climate are a dry summer, a mild, moist winter, and abundant sunshine (90 percent of possible sunshine in summer and as much as 50 to 60 percent even during the rainy winter season). Summers are warm throughout the climate, but there are enough differences between the monthly temperatures in coastal and interior locations to recognize two distinct subtypes. The moderate-summer subtype has the lower summer temperatures associated with a strong maritime influence. The hot-summer subtype is located further inland and reflects an increased continental influence. The inland version has higher summer and daytime temperatures and slightly cooler winter and

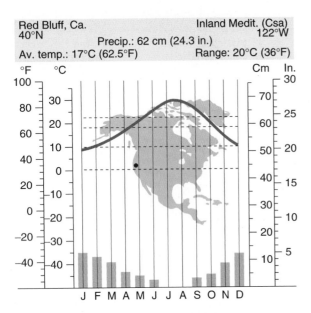

Red Bluff, Ca.   Inland Medit. (Csa)
40°N   122°W
                Precip.: 62 cm (24.3 in.)
Av. temp.: 17°C (62.5°F)   Range: 20°C (36°F)

San Francisco, Ca.   Coastal Medit. (Csb)
38°N   122°W
                Precip.: 56 cm (22.1 in.)
Av. temp.: 13.6°C (56.5°F)   Range: 6.5°C (12°F)

Capetown, S. Africa   Coastal Medit. (Csb)
34°S   19°E
                Precip.: 50.8 cm (20 in.)
Av. temp.: 16.5°C (62°F)   Range: 8.5°C (15.5°F)

nighttime temperatures than its coastal counterpart, and it has greater annual and diurnal temperature ranges as well. Compare, for example, the annual range for Red Bluff, California, an inland station, with that of San Francisco, a coastal station about 240 kilometers (150 mi) further south (see Fig. 10.2).

Whichever the subtype, Mediterranean summers clearly show the influence of the subtropical highs. Weeks go by without a sign of rain, and evapotranspiration rates are high. Effective precipitation is lower than actual precipitation, and the summer drought is as intense as that of the desert. Days are warm to hot, skies are blue and clear, and sunshine is abundant. The high percentage of insolation coupled with nearly vertical rays of the noon sun may drive daytime temperatures as high as 30° to 38°C (86°–100°F), except where moderated by a strong ocean breeze or coastal fog.

Fog is common throughout the year in coastal locations and is especially noticeable during the summer. As moist maritime air moves onshore, it passes over the cold ocean currents that typically parallel west coasts in Mediterranean latitudes. The air is cooled, condensation takes place, and fog regularly creeps in during the late afternoon, remains through the night, and "burns off" during the morning hours. As in the desert, radiation loss is rapid at night, and even in summer, nighttime temperatures are commonly only 10° to 15°C (50°–60°F). People accustomed to the humid summer nights of the eastern United States do not recognize, without experience, the need to bring a sweater for use during the evenings.

Winter is the rainy season in the Mediterranean climate. The average annual rainfall in these regions is usually between 35 and 75 centimeters (15–30 in.), with 75 percent or more of the total rain falling during the winter months. The precipitation results primarily from the cyclonic storms and frontal systems common in the westerlies. Annual amounts increase with elevation and decrease with increased distance from the ocean. Only because the rain comes during the cooler months, when evapotranspiration rates are lower, is there sufficient precipitation to make this a humid climate.

Despite the rain during the winter season, there are often many days of fine, mild weather. Insolation is still usually above 50 percent, and the average temperature of the coldest month rarely falls below 4° to 10°C (40°–50°F). Frost is uncommon, and because of its rarity, many less hardy tropical varieties of fruits and vegetables are grown in these regions. However, when frost does occur, it can do great damage.

Figure 10.2 Climographs for Mediterranean climate stations.
■ **In what ways is the climograph of Capetown clearly distinguishable from the two California climographs? What causes the differences?**

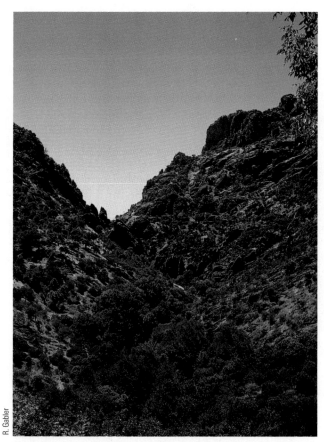

R. Gabler

Figure 10.3 Typical remnant chaparral vegetation in southern Spain.   ■ **Why would you expect to find little "native" vegetation left in Old World Mediterranean lands?**

Special Adaptations   The summer drought, not frost, is the great challenge to vegetation in Mediterranean regions. The natural vegetation reflects the wet-dry seasonal pattern of the climate. During the rainy season, the land is covered with lush, green grasses that turn golden and then brown under the summer drought. Only with winter and the return of the rains does the landscape become green again. Much of the natural vegetation is **sclerophyllous** (hard-leafed) and drought resistant. Like xerophytes, these plants have tough surfaces, shiny, thick leaves that resist moisture loss, and deep roots to help combat aridity.

One of the most familiar plant communities is made up of many low, scrubby bushes that grow together in a thick tangle. In the western United States, this is called **chaparral** (Fig. 10.3). A common low bush in California is the manzanita, a tough, red-stemmed shrub with crooked limbs that interlock to form an almost impenetrable barrier. Most chaparral is less than 25 years old because of the frequent fires that occur in this dry brush. The fires help to perpetuate the chaparral because the associated heat is required to open seedpods and allow many chaparral species to reproduce. People often remove chaparral as a preventive measure against fires, but the removal can have disastrous results because the chaparral acts as a check against erosion of soils during the

rainy season. With the chaparral removed, soils wash or slide down hillsides during the heavy rains of winter, frequently taking homes with them.

Brush similar to the chaparral of California is called *mallee* in Australia, *mattoral* in Chile, and *maquis* in France. In fact, the "maquis" French Underground that fought against Nazi occupation in World War II literally meant "underbrush."

Trees that appear in the Mediterranean climate also respond to moisture conditions. Because of their drought-resistant qualities, the needle-leaf pines are among the more common species. Groves of deciduous and evergreen oaks appear in depressions where moisture collects and on the shady north sides of hills where evapotranspiration rates are lower. Where the summer drought is more distinct, the scrub and woodlands open up to parklands of grasses and scattered oak trees. Even the great redwood forests of northern California probably could not survive without the heavy fogs that regularly invade the coastal lands in summer (Fig. 10.4).

© Lake County Museum/CORBIS

Figure 10.4 California redwoods (*Sequoia sempervirens*) may reach heights of 100 meters (330 ft) and live for thousands of years.   ■ **Why is it considered unusual to find redwoods growing in a Mediterranean climate?**

Figure 10.5 Grapes adapt well to the Mediterranean climate. This vineyard is located in California.

of nuts, and countless vegetables. California, blessed both with fertile valleys for growing fruits, vegetables, and flowers and with snow meltwater for irrigation, is probably the most agriculturally productive of the Mediterranean regions.

Even the houses of these regions show people's adaptation to the climate. Usually white or pastel in color, they gleam in the brilliant sunshine against clear blue skies. Many have shuttered windows to cut the glare of the sun and to keep the houses cool in summer. On the other hand, much less attention is paid to keeping homes warm during the cooler winter months. This is true even in the United States, where many Midwesterners and Easterners are surprised at the lack of insulation in California homes and the small number and size of heating devices.

In response to the dry summers, most soils in Mediterranean regions are high in soluble nutrients, but the potential for agriculture differs widely from one region to another. The lime-rich soils around the Mediterranean Sea originally were highly productive, but destructive agricultural practices and overgrazing over thousands of years of human use have caused serious erosion problems. Today, bare white limestone is widely visible on hillsides in Spain, Italy, Greece, Crete, Syria, Lebanon, Israel, and Jordan. In the Mediterranean regions of California, the soils are dense clays, gluelike when wet and hard as concrete when dry. During the Spanish period in California, the clay was formed into adobe bricks and used for building material. These clay soils are hazardous on slopes, for they absorb so much water during the wet season that they can become mudflows destructive to roads and homes in areas where vegetation has been removed by fire or human interference.

In all Mediterranean regions, the most productive areas are the lowlands covered by stream deposits. Here farmers have made special adaptations to climatic conditions. There is sufficient rainfall in the cool season to permit fall planting and spring harvesting of winter wheat and barley. These grasses originally grew wild in the eastern Mediterranean region. Grapevines, fig and olive trees, and the cork oak, which undoubtedly were also native to Mediterranean lands, are especially well adapted to the dry summers because of their deep roots and thick, well-insulated stems or bark (Fig. 10.5). Where water for irrigation is available, an incredible diversity of crops may be seen. These include, in addition to those already mentioned, oranges, lemons, limes, melons, dates, rice, cotton, deciduous fruits, various types

Humid Subtropical Climate

The **humid subtropical climate** extends inland from continental east coasts between 15° to 20° and 40°N and S latitude (refer again to Fig. 10.1 and Table 10.1). Thus, it is located within approximately the same latitudes and in a similar transitional position as the Mediterranean climate, but on the eastern instead of the western continental margins. There is ample evidence of this climatic transition. Summers in the humid subtropics are similar to the humid tropical climates further equatorward. When the noon sun is nearly overhead, these regions are subject to the importation of moist tropical air masses. High temperatures, high relative humidity, and frequent convectional showers are all characteristics that they share with the tropical climates. In contrast, during the winter months, when the pressure and wind belts have shifted equatorward, the humid subtropical regions are more commonly under the influence of the cyclonic systems of the continental middle latitudes. Polar air masses can bring colder temperatures and occasional frost.

Comparison with the Mediterranean Climate Like the inland version of the Mediterranean climate, the humid subtropical climate has mild winters and hot summers. But it has no dry season. Whereas the Mediterranean lands are under the drought-producing eastern flank of the subtropical highs, the humid subtropical regions are located on the weak western sides of the subtropical high pressure cells. Subsidence and stability are greatly reduced or absent, even during the summer months. Here again, ocean temperatures play a significant role. The warm ocean currents that are commonly found along continental east coasts in these latitudes also moderate the winter temperatures and warm the lower atmosphere, thus increasing lapse rates, which enhances instability. Furthermore, a modified monsoon effect (especially in Asia, and to some extent in the southern United States) increases summer precipitation as the moist, unstable tropical air is drawn in over the land.

As might be expected from the year-round rainfall, average annual precipitation for humid subtropical locations usually exceeds that for Mediterranean stations and may vary more widely as well. Humid subtropical regions receive anywhere from 60 to 250 centimeters (25–100 in.) a year. Precipitation generally decreases inland toward continental interiors and away from the oceanic sources of moisture. It is not surprising that these regions are noticeably drier the closer they are to steppe regions inland toward their western margins.

Both the Mediterranean and humid subtropical climates receive winter moisture from cyclonic storms, which travel along the polar front. As we have noted, the great contrast occurs in the summer when the humid subtropics receive substantial precipitation from convectional showers, supplemented in certain regions by a modified monsoon effect. In addition, because of the shift in the sun and wind belts during the summer months, the humid subtropical climates are subject to tropical storms, some of which develop into hurricanes (or typhoons), especially in late summer. These three factors—the modified monsoon effect, convectional activity, and tropical storms—combine in most of these regions to produce a precipitation maximum in late summer. The climographs for New Orleans, Louisiana, and Brisbane, Australia, illustrate these effects (Fig. 10.6).

A subtype of the humid subtropical climate is found most often on the Asian continent, where the monsoon effect is most pronounced because of the magnitude of the seasonal pressure changes over this immense landmass. There, the low-sun period or winter season is noticeably drier than the high-sun period. High pressure over the continent blocks the importation of moist air, so that some months receive less than 3 centimeters (1.2 in.) of precipitation. The climograph for Lashio, Myanmar, illustrates the subtropical wet-dry climate (refer back to Fig. 10.6).

Temperatures in the humid subtropics are much like those of the Mediterranean regions. Annual ranges are similar, despite a greater variation among climate stations in the humid subtropical climate, primarily because the climate covers a far larger land area. Mediterranean stations record higher summer daytime temperatures, but summer months in both climates average around 25°C (77°F), increasing to as much as 32°C (90°F) as

Figure 10.6 Climographs for humid subtropical climate stations.
■ **What causes the atypical humid subtropical precipitation characteristic of Lashio?**

New Orleans, La.   Humid Subt. (Cfa)
30°N   90°W
Precip.: 146 cm (57.4 in.)
Av. temp.: 21°C (69.5°F)   Range: 16°C (28.5°F)

Brisbane, Australia   Humid Subt. (Cfa)
27°S   153°E
Precip.: 113.5 cm (44.7 in.)
Av. temp.: 20.5°C (69°F)   Range: 10°C (18.5°F)

Lashio, Myanmar   Humid Subt. (Cfa)
23°N   98°E
Precip.: 157 cm (62 in.)
Av. temp.: 22°C (71.5°F)   Range: 9.5°C (17°F)

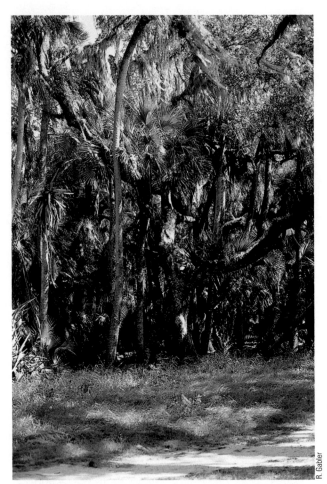

R. Gabler

Figure 10.7 Forest vegetation similar to this in Myakka State Park originally covered much of humid subtropical central Florida.
■ **How has the physical landscape of central Florida been changed by human occupancy?**

maritime influence decreases inland. Winter months in both climates average around 7° to 14°C (45°–57°F). Frost is a similar problem. The long growing season in the warmest humid subtropical regions enables farmers to grow such delicate crops as oranges, grapefruit, and lemons, but, as in the Mediterranean climates, farmers must be prepared with various means to protect their more sensitive crops from the danger of freezing. The growers of citrus crops in Florida are concentrated in the central lake district to take advantage of the moderating influence of nearby bodies of water.

Variation in humidity greatly affects the effective temperatures—the temperatures we feel—in the humid subtropical and Mediterranean climates. The summer temperatures in humid subtropical regions feel far warmer than they are because of the high humidity there. In fact, summers in this climate are oppressively hot, sultry, and uncomfortable. Nor is there the relief of lower night temperatures, as in the Mediterranean regions. The high humidity of the humid subtropical climate prevents much radiative loss of heat at night. Consequently, the air remains hot and sticky. Diurnal temperature ranges,

in winter as well as in summer, are far smaller in the humid subtropical than in the drier Mediterranean regions. Despite the relatively mild temperatures, humid subtropical winters seem cold and damp, again because of the high humidity.

A Productive Climate Vegetation generally thrives in humid subtropical regions, with their abundant rainfall, high temperatures, and long growing season. The wetter portions support forests of broad-leaf deciduous trees, pine forests on sandy soils, and mixed forests (Fig. 10.7). In the drier interiors near the steppe regions, forests give way to grasslands, which require less moisture. There is an abundant and varied fauna. A few of the common species are deer, bears, foxes, rabbits, squirrels, opossums, raccoons, skunks, and birds of many sizes and species. Bird life in lake and marsh areas is incredibly rich. Alligators inhabit the American swamps, such as the Dismal Swamp of North Carolina and the Okefenokee in Georgia.

As in the tropics, soils tend to have limited fertility because of rapid removal of soluble nutrients. However, there are exceptions in drier grassland areas, such as the pampas of Argentina and Uruguay that constitute South America's "bread basket." Whatever the soil resource, the humid subtropical regions are of enormous agricultural value because of their favorable temperature and moisture characteristics. They have been used intensively for both subsistence crops, such as rice and wheat in Asia, and commercial crops, such as cotton and tobacco in the United States. When we consider that this climate (with its monsoon phase) is characteristic of south China as well as of the most densely populated portions of both India and Japan, we realize that this climatic regime contains and feeds far more human beings than any other type (Fig. 10.8). The care with which agriculture has been practiced and the soil resource conserved over thousands of years of intensive use in eastern Asia is in sharp contrast to the agricultural exploitation of the past 200 years in the corresponding area of the United States. The traditional system of cotton and tobacco farming, in particular, devastated the land by exhausting the soil and triggering massive sheet and gully erosion (Fig. 10.9). Over much of the old cotton and tobacco belt, extending from the Carolinas through Georgia and Alabama to the Mississippi delta, all of the topsoil or a significant part of it has been lost, and we see only the red clay subsoil and occasionally bare rock where crops formerly flourished. In the remaining areas, practices have had to change to conserve the soil that is left. Heavy applications of fertilizer, scientific crop rotation, and careful tilling of the land are now the rule.

Where forests still form the major natural vegetation, forest products, such as lumber, pulpwood, and turpentine, are important commercially (Fig. 10.10). The long-leaf and slash pines of the southeastern United States are a source of lumber, as well as the resinous products of the pine tree (pitch, tar, resin, and turpentine). The absence

Robert Essel, NYC/CORBIS

Figure 10.8 The terraced fields in this humid subtropical climatic region near Wakayama, Japan, are ideally suited for rice production.
■ **Why is rice a preferred crop in Japan?**

of temperature and moisture limitations strongly favors forest growth. In Georgia, for example, trees may grow two to four times faster than in colder regions such as New England. This means that trees can be planted and harvested in much less time than in cooler forested regions, offering distinct commercial advantages.

The fact that living things thrive in the humid subtropical climate presents certain problems, however, for parasites and disease-carrying insects thrive along with other forms of life. African "killer bees," for example, have migrated through Mexico and now pose a serious new threat to humans and livestock in the southern United States.

Agriculture and lumbering are not the only important industries in the humid subtropical climates. In the southeastern United States, livestock raising has been increasing greatly in importance. Despite the commercial advantages of this climate, people often find it an uncomfortable one in which to live. The development and spread of air conditioning helps to mitigate this problem. Where the ocean offers relief from the summer heat, as in Florida, the humid subtropical climate is an attractive recreation and retirement region. The beauty of its more unusual features, such as its cypress swamps and forests draped with Spanish moss, has to be experienced to be fully appreciated.

Marine West Coast Climate

Proximity to the sea and prevailing onshore winds make the **marine west coast climate** one of the most temperate in the world. Thus, it is sometimes known as the *temperate oceanic climate*. Found in those middle-latitude regions (between 40° and 65°) that are continuously influenced by the westerlies, the marine west coast climate receives ample precipitation throughout the year. However, unlike the humid subtropical climate just discussed, it has mild to cool summers. The climographs for Bordeaux, France, and Stuttgart Germany, in Figure 10.11 are representative of the mild summer marine west coast climate, and the climograph for Reykjavik, Iceland, represents the cool summer variety.

Oceanic Influences As they travel onshore, the westerlies carry with them the moderating marine influence on temperature, as well as much moisture. In addition, warm ocean currents, such as the North Atlantic Drift, bathe some of the coastal lands in the latitudes of the marine west coast climate, further moderating climatic conditions and accentuating humidity. This latter influence is particularly noticeable in Europe, where the marine west coast climate extends along the coast of Norway to beyond the Arctic Circle (see Fig. 10.1).

In this climatic zone, the marine influence is so strong that temperatures decrease little with poleward movement. Thus, the influence of the oceans is even stronger than latitude in determining these temperatures. This is obvious when we examine isotherms in the areas of marine west coast climates on a map of world temperatures (see again Figs. 4.19 and 4.20). Wherever the marine west coast climate prevails, the isotherms swing poleward, parallel to the coast, clearly demonstrating the dominant marine influence.

Figure 10.9 Providence Canyon, Georgia, pictured here, did not exist at the time of the Civil War. It began as a system of gulleys when the original vegetative cover was removed during the most active period of farming and forestry in the postwar South. The original fertile soils are exposed near the top of the canyon walls. ■ **What is being done throughout Georgia today to prevent such serious soil erosion from occurring again?**

Another result of the ocean's moderating effect is that the annual temperature ranges in the marine west coast climates are relatively small, considering the latitude. For an illustration of this, compare the monthly temperature graphs for Portland, Oregon, and Eau Claire, Wisconsin (Fig. 10.12). Though these two cities are at the same latitude, the annual range at Portland is 15.5°C (28°F), while at Eau Claire it is 31.5°C (57°F). The moderating effect of the ocean on Portland's temperatures is clearly contrasted to the effect of continentality on the temperatures in Eau Claire.

Diurnal temperature ranges are also smaller than they are in other climatic regions at similar latitudes and in more arid climates. Heavy cloud cover and high humidity in both summer and winter diminish daytime heating and prevent much radiational cooling at night. Consequently, the difference between the daily maximum and minimum temperatures is small.

Of course, these climographs and climate statistics represent averages, which can be misleading. Marine west coast climatic regions

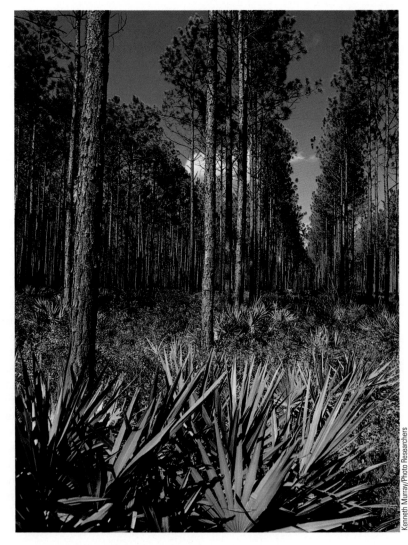

Figure 10.10 A commercial tree farm near Waycross, Georgia. Note that the pine trees are planted in orderly rows to expedite cultivation and harvest. ■ **Why are tree farms common in the U.S. Southeast and Pacific Northwest but not in New England or the Upper Midwest?**

experience the unpredictable weather conditions associated with the polar front. Occasional invasions of a tropical air mass in summer or a polar air mass in winter can move against the general westerly flow of air in these latitudes and produce surprising results. For example, under just such weather conditions, temperatures in Seattle, Washington, have reached a high of 38°C (100°F) and a low of −19°C (−3°F).

Despite the insulating effect of cloudy skies and high moisture content of the air, slowing heat loss at night, frost is a significant factor in the marine west coast climate. It occurs more often, may last longer, and is more intense than in other mesothermal regions. The growing season is limited to eight months or less, but even during the months when freezing temperatures may occur, only half the nights or fewer may experience them. The possibility of frost and the frequency of its occurrence increase inland far more rapidly than they do poleward, once more illustrating the importance of the marine influence.

As final evidence of oceanic influences, study the distribution of the marine west coast climate in Figure 10.1. Where mountain barriers prevent the movement of maritime air inland, this climate is restricted to a narrow coastal strip, as in the Pacific Northwest of North America and in Chile. Where the land is surrounded by water, as in New Zealand, or where the air masses move across broad plains, as in much of northwestern Europe, the climate extends well into the interior of the landmass.

Clouds and Precipitation   The marine west coast has a justly deserved reputation as one of the cloudiest, foggiest, rainiest, and stormiest climates in the world. This is particularly true during the winter season. Rain or drizzle may last off and on for days, though the amount of rain received is small for the number of rainy days recorded. Even when rain is not falling, the weather is apt to be cloudy or foggy. Advection fog may be especially common and long-lasting in the winter months, when air masses pass over warm ocean currents and pick up considerable moisture, which is then condensed as a fog when the air masses move over colder land. The cyclonic storms and frontal systems are also strongest in the winter, when the subtropical highs have shifted equatorward. Conspicuous winter maximums in rainfall occur near the coasts and near boundaries with the Mediterranean climate. However, further inland a summer maximum may occur, as in Stuttgart, Germany (see Fig. 10.11).

Though all parts of this climatic type receive ample precipitation, there is much greater station-to-station variation in precipitation averages than in temperature statistics. Precipitation tends to decrease very gradually

Figure 10.11   Climographs for marine west coast climate stations. ■ **How do you explain the fact that Reykjavik has the lowest temperature range of the three stations?**

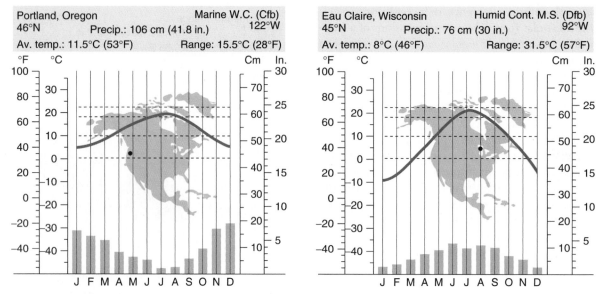

Figure 10.12 Effect of maritime influence on climates of two stations at the same latitude. Portland, Oregon, exemplifies the maritime influences dominating marine west coast climates. Eau Claire, Wisconsin, shows the effect of location in the continental interior. The difference in temperature ranges for the two stations are significant, but note also the interesting differences in precipitation distribution. ■ **How do you explain these differences?**

as one moves inland, away from the oceanic source of moisture. It also decreases equatorward, especially during summer months, as the influence of the subtropical highs increases and the influence of the westerlies decreases. This can bring about periods of beautiful, clear weather, something rarely associated with this climate but not uncommon in our Pacific Northwest.

The most important factor in the amount of precipitation received is local topography. When a mountain barrier such as the Cascades in the Pacific Northwest or the Andes in Chile parallels the coast, abundant precipitation, both cyclonic and orographic, falls on the windward side of the mountains. Valdivia, Chile, located windward of the Andes, receives an average of 267 centimeters (105 in.) of precipitation a year. A similar location in Canada, Henderson Lake, British Columbia, averages 666 centimeters (262 in.) of rain a year, the highest figure for the entire North American continent. During the colder Pleistocene Epoch, these high precipitation amounts, falling largely as winter snow, produced large mountain glaciers. In many cases, these came down to the sea, excavating deep troughs that now appear as steep-walled inlets or fjords. Fjord coasts are present in Norway, British Columbia, Chile, and New Zealand—all areas of marine west coast climates today (Fig. 10.13). In contrast, where there are lowlands and no major landforms of high elevation, precipitation is spread more evenly over a wide area, and the amount received at individual stations is more moderate, around 50 to 75 centimeters (20–30 in.) annually. This is the situation in much of the Northern European Plain, extending from western France to eastern Poland.

Two aspects of precipitation are directly related to the moderate temperatures of this climatic regime. Snow falls infrequently, and when it does, it melts or turns to slush as soon as it hits the ground. Snow is especially rare in lowland regions of this climatic zone. Paris averages only 14 snow days a year; London, 13; and Seattle, 10. In addition, thunderstorms and convectional showers are uncommon, although they occur occasionally. Even in summer, surface heating is rarely sufficient to produce the towering cumulonimbus clouds.

Resource Potential   There is little doubt that this climate offers certain advantages for agriculture. The small annual temperature ranges, mild winters, long growing seasons, and abundant precipitation all favor plant growth. Many crops, such as wheat, barley, and rye, can be grown further poleward than in more continental regions. Although the soils common to these regions are not naturally rich in soluble nutrients, highly successful agriculture is possible with the application of natural or commercial fertilizers (Fig. 10.14). Root crops, such as potatoes, beets, and turnips; deciduous fruits, such as apples and pears; berries; and grapes join the grains previously mentioned as important farm products. Grass in particular requires little sunshine, and pastures are always lush. The greenness of Ireland—the Emerald Isle—is evidence of these favorable conditions, as is the abundance of herds of beef and dairy cattle.

The magnificent forests that form the natural vegetation of the marine west coast regions have been a readily available resource. Some of the finest stands of commercial timber in the world are found along the Pacific coast

Figure 10.13   The scenic Kenai Fjords of Alaska, shown here, were produced by glacial erosion during the Pleistocene ice advance.
■ **In what other areas of the world are fjords common?**

Figure 10.14   Reliable precipitation makes a diversified type of agriculture possible in the marine west coast climatic areas, with emphasis on grain, orchard crops, vineyards, vegetables, and dairying. The village in the photograph is Iphofen, Germany.   ■ **Although the climate is similar, why would a photo taken in a marine west coast agricultural region of the United States depict a scene that is significantly different from this one?**

Figure 10.15 Commercial timber is the outstanding natural resource of the marine west coast climatic region in North America. ■ **Why is this not the case in other marine west coast regions?**

George Herbert/Alaska Stock Images

of North America, where pines, firs, and spruces abound, commonly exceeding 30 meters (100 ft) in height (Fig. 10.15). Europe and the British Isles were once heavily forested, but most of those forests (even the famous Sherwood Forest) have been cut down for building material and have been replaced by agricultural lands and urbanization.

Humid Microthermal Climatic Regions

Our definition of humid microthermal includes temperatures high enough during part of the year to have a recognizable summer and cold enough six months later to have a distinct winter. In between are two periods, called spring and fall, when all life, and especially vegetation, makes preparations for the temperature extremes. Thus, in this section we will be talking about climatic regions that clearly display four readily identifiable seasons.

Seasonality is not the only reason we often use the word *variable* when describing the humid microthermal climates. As Figure 10.16 indicates, these climates are generally located between 35°N and 75°N on the North American and Eurasian landmasses. Thus, they share the westerlies and the storms of the polar front with the marine west coast climate. However, their position in the continental interiors and in high latitudes prevents them from experiencing the moderating influence of the

oceans. In fact, the dominance of continentality in these climates is best demonstrated by the fact that they do not exist in the Southern Hemisphere, where there are no large landmasses in the appropriate latitudes.

The recognition of three separate microthermal climates is based mainly on latitude and the resulting differences in the length and severity of the seasons (Table 10.2). Winters tend to be longer and colder toward the poleward margins because of latitude and toward interiors throughout the microthermal climates because of the continental influence. Summers inland are also inclined to be hotter, but they become progressively shorter as the winter season lengthens poleward. Thus, the three microthermal climates can be defined as humid continental, hot summer; humid continental, mild summer; and subarctic, with a cool summer and, in extreme cases, a long, bitterly cold winter.

All microthermal climates have several features in common. By definition, they all experience a surplus of precipitation over potential evapotranspiration, and they have year-round precipitation. An exception to this rule lies in an area of Asia where the Siberian High causes winter droughts to occur. The greater frequency of maritime tropical air masses in summer and continental polar air masses in winter, combined with the monsoonal effect and strong summer convection, produce a precipitation maximum in the summer. Although the length of time snow remains on the ground increases poleward and toward the continental interior (Fig. 10.17),

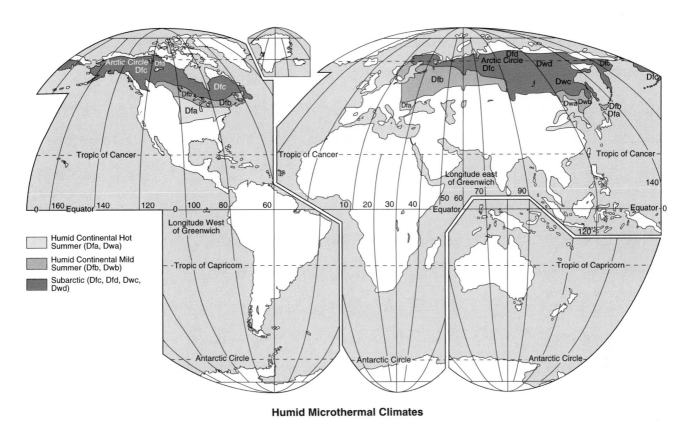

Humid Microthermal Climates

Figure 10.16   Index map of humid microthermal climates

all three microthermal climates experience significant snow cover. This decreases the effectiveness of insolation and helps to explain their cold winter temperatures. Finally, the unpredictable and variable nature of the weather is especially apparent in the humid microthermal climates and is present throughout all microthermal regions.

With these generalizations in mind, let's compare the microthermal climates with the mesothermal climates we have just examined. Regions with microthermal climates have more severe winters, a lasting snow cover, shorter summers, shorter growing seasons, shorter frost-free seasons, a truer four-season development, lower nighttime temperatures, greater average annual temperature ranges, lower relative humidity, and much more variable weather than do the mesothermal climatic regions.

Humid Continental, Hot Summer Climate

Unlike the other two microthermal climates, the **humid continental, hot summer climate** is relatively limited in its distribution on the Eurasian landmass (see Table 10.2). This is unfortunate for the people of Europe and Asia because it has by far the greatest agricultural potential and is the most productive of the microthermal climates. In the United States, this climate is distributed over a wide area that begins with the eastern seaboard of New York, New

Jersey, and southern New England and stretches continuously across the heartland of the eastern United States to encompass much of the American Midwest. It is one of the most densely populated, highly developed, and agriculturally productive regions in the world.

In terms of environmental conditions, the hot-summer variety of microthermal climate has some obvious advantages over its poleward counterparts. Its higher summer temperatures and longer growing season permit farmers to produce a wide variety of crops. Those lands within the hot-summer region that were covered by ice sheets are far enough equatorward that there has been sufficient time for most negative effects of continental glaciation to be removed, and primarily positive effects remain. Soils are inclined to be more fertile, especially under forest cover, where the typical soil-forming processes are not as extreme and where deciduous trees are more common than the acid-associated pine. Of course, some advantages are matched by liabilities. The lower fuel bills of winter in the humid continental, hot summer climate are often more than offset by the cost of air conditioning during the long, hot summers not found in other microthermal climates.

Internal Variations   From place to place within the humid continental, hot summer climate, there are significant differences in temperature characteristics. The length of the growing season is directly related to latitude,

Table 10.2 The Microthermal Climates

Name and Description	Controlling Factors	Geographic Distribution	Distinguishing Characteristics	Related Features
Humid Continental, Hot Summer				
Warmest month above 10°C (50°F); coldest month below 0°C (32°F); hot summers; usually year-round precipitation, winter drought (Asia)	Location in the lower middle latitudes (35° to 45°); cyclonic storms along the polar front; prevailing westerlies; continentality; polar anticyclone in winter (Asia)	Eastern and mid-western U.S. from Atlantic coast to 100°W longitude; east central Europe; northern China, Manchuria, northern Korea, and Honshu (Japan)	Hot, often humid summers; occasional winter cold waves; rather large annual temperature ranges; weather variability; precipitation 50–115 cm (20–45 in.), decreasing inland and poleward; 140- to 200-day growing season	Broad-leaf deciduous and mixed forest; moderately fertile soils with fertilization in wetter areas; highly fertile grassland and prairie soils in drier areas; "corn belt," soybeans, hay, oats, winter wheat
Humid Continental, Mild Summer				
Warmest month above 10°C (50°F); coldest month below 0°C (32°F); mild summers; usually year-round precipitation, winter drought (Asia)	Location in the middle latitudes (45° to 55°); cyclonic storms along the polar front; prevailing westerlies; continentality; polar anticyclone in winter (Asia)	New England, the Great Lakes region, and south central Canada; southeastern Scandinavia; eastern Europe, west central Asia; eastern Manchuria and Hokkaido (Japan)	Moderate summers; long winters with frequent spells of clear, cold weather; large annual temperature ranges; variable weather with less total precipitation than further south; 90- to 130-day growing season	Mixed or coniferous forest; moderately fertile soils with fertilization in wetter areas; highly fertile grassland and prairie soils in drier areas; spring wheat, corn for fodder, root crops, hay, dairying
Subarctic				
Warmest month above 10°C (50°F); coldest month below 0°C (32°F); cool summers, cold winters poleward; usually year-round precipitation, winter drought (Asia)	Location in the higher middle latitudes (50° to 70°); westerlies in summer, strong polar anticyclone in winter (Asia); occasional cyclonic storms; extreme continentality	Northern North America from Newfoundland to Alaska; northern Eurasia from Scandinavia through most of Siberia to the Bering Sea and the Sea of Okhotsk	Brief, cool summers; long, bitterly cold winters; largest annual temperature ranges; lowest temperatures outside Antarctica; low precipitation, 20–50 cm (10–20 in.); unreliable 50- to 80-day growing season; permafrost common	Northern coniferous forest (taiga); strongly acidic soils; poor drainage and swampy conditions in warm season; experimental vegetables and root crops

varying from 200 days equatorward to as little as 140 days along poleward margins of the climate. In addition, the degree of continentality can have an effect on both summer and winter temperatures and, as a result, on temperature range. Ranges are consistently large, but they become progressively larger toward continental interiors. Especially near the coasts in this climatic region, temperatures may be modified by a slight marine influence, so that temperatures are milder, in both summer and winter, than those at inland locations at comparable latitudes. Large lakes may cause a similar effect. Even the size of the continent exerts an influence. Galesburg, Illinois, a typi-

cal station in the United States, has a significantly lower temperature range than Mukden, Manchuria, which is located at almost the same latitude but which experiences the greater seasonal contrasts of the Eurasian landmass (Fig. 10.18).

The amount and distribution of precipitation are also variable from station to station. As in all microthermal climates, the total precipitation received decreases both poleward and inland (Fig. 10.19). A move in either direction is a move away from the source regions of warm maritime air masses that provide much of the moisture for cyclonic storms and convectional showers. This decrease

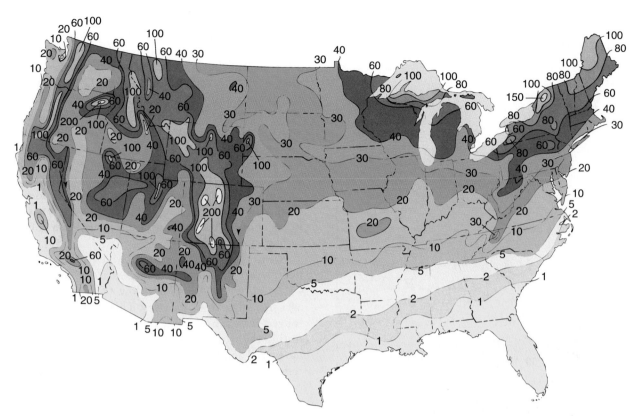

Figure 10.17 Map of the contiguous United States showing average annual number of days of snow cover. ■ **What areas of the United States average the greatest number of days of snow cover?**

can be seen in the average annual precipitation figures for the following cities, all at a latitude of about 40°N: New York (longitude 74°W), 115 centimeters (45 in.); Indianapolis, Indiana (86°W), 100 centimeters (40 in.); Hannibal, Missouri (92°W), 90 centimeters (35 in.); and Grand Island, Nebraska (98°W), 60 centimeters (24 in.).

Most stations have a precipitation maximum in summer, when the warm, moist air masses dominate. In certain regions of Asia, not only does a summer maximum of precipitation exist, but the monsoon circulation inhibits winter precipitation to such a degree that they experience winter drought (see Mukden, Fig. 10.18).

Figure 10.18 Climographs for humid continental, hot summer climate stations. ■ **What are the reasons for the differences in temperature and precipitation between the two stations?**

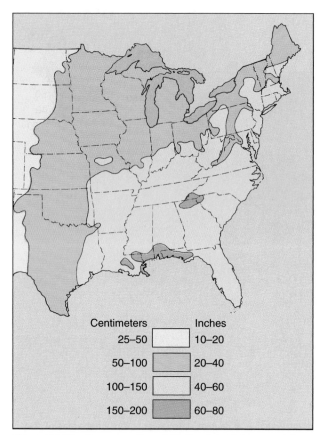

Centimeters		Inches
25–50		10–20
50–100		20–40
100–150		40–60
150–200		60–80

Figure 10.19 Decreases of precipitation inland from coastal regions are clearly evident in this map of the eastern United States. ■ **Why does precipitation decrease inland and poleward?**

As might be expected, vegetation and soils vary with the climatic elements, especially precipitation. In the wetter regions, forests and forest soils predominate. At one time, in certain sections of the American Midwest, tall prairie grasses grew where precipitation was sufficient to support forests; in all the drier portions of the climate, grasslands are the natural vegetation. The soils that developed under these grasslands are among the richest in the world.

Seasonal Changes The four seasons are highly developed in the humid microthermal, hot summer climate. Each is distinct from the other three and has a character all its own. The winter is cold and often snowy; the spring is warmer, with frequent showers that produce flowers, budding leaves, and green grasses; the hot, humid summer brings occasional violent thunderstorms; and the fall has spells of both clear and rainy weather, with mild days and frosty nights in which the green leaves of summer turn to colorful reds, oranges, yellows, and browns before falling to the ground.

Of course, the most significant differences are between summer and winter. In all regions, the summers are long, humid, and hot. The centers of the migrating low pressure systems usually pass poleward of these regions, and they are dominated by tropical maritime air. So-called "hot spells" can go on day and night for a week or more, with only temporary relief available from convectional thunderstorms or an occasional cold front. Asia, in particular, experiences the heavy summer precipitation associated with the monsoonal effect. Conditions are usually ideal for vigorous vegetative growth. In fact, it is said the corn grows so fast in Iowa that you can hear it! Even if this seems an exaggeration, one thing is certain: The summer heat and humidity are the ideal formula for insects; mosquitoes, flies, gnats, and bugs of all kinds abound. "Putting up the screens" is a common early summer project.

Winters are not as severe as those further poleward, but average January temperatures are usually between –5° and 0°C (23°–32°F) or below. Once again, the averages tell only part of the story. There is invariably a prolonged invasion of cold, dry, arctic air once or twice during the winter. This often occurs just after a storm has passed and the ground is covered with snow. The sky remains clear and blue for days at a time; the temperatures will stay near –18°C (0°F) and may occasionally dip to 30° or 35°C (20°or 30°F) *below zero* at night. The ground remains frozen for long periods, and there may be snow cover for several days, even weeks, at a time. However, these characteristics do not last continuously, because the greatest frequency of cyclones occurs during winter and sudden weather contrasts are common. Cold air precedes warmer air, and thaw follows freeze. Vegetation remains dormant throughout the winter season but bursts into life again with the return of consistently warmer temperatures. Throughout its early growth, vegetation is in constant danger of late spring frost.

As should be apparent from this description, the atmospheric changes within seasons are just as significant as those between seasons. The humid continental, hot summer climate is the classic example of variable middle-latitude weather. This is the domain of the polar front. Cyclonic storms are born as tropical air masses move northward and confront polar air masses migrating to the south. The daily weather in these regions is dominated by days of stormy frontal activity followed by the clear conditions of a following anticyclone. Above the land is a battlefield in which storms mark the struggles of air masses for dominance, and as in battle, the conflict occurs at the *front*. The general circulation of the atmosphere in these latitudes continuously carries the cyclones and anticyclones toward the east along the polar front. When the polar front is most directly over these regions, as it is in winter and spring, one storm and its associated fronts seem to follow directly behind another with such speed and regularity that the only safe weather prediction is that the weather will change.

Figure 10.20 Climographs for humid continental, mild summer climate stations. ■ **What characteristics of these climographs distinguish them from the climographs of Figure 10.18?**

Humid Continental, Mild Summer Climate

If you review the relative distributions of the humid continental, hot summer and mild summer climates in Figure 10.16, the close relationship between the two is unmistakable. Where one is found, the other is found as well; in each situation, the mild summer climate invariably lies adjacent to and poleward from the hot summer climate.

In most instances, the **humid continental, mild summer climate** is a more continental or severe-winter version of its equatorward counterpart. It is characterized by distinct seasonality. There is significant climatic variation, particularly with respect to precipitation, from place to place within the climate. Variable weather is the rule, and storms along the polar front provide most of the precipitation within this climatic type. Of course, there are differences between the neighboring climates. These are especially apparent when we examine certain aspects of temperature, growing season, vegetation, and human activity. A brief comparison of the humid continental, mild summer climate with the humid continental, hot summer climate should help to point out important similarities and differences.

Mild Summer–Hot Summer Comparison In the microthermal climates, precipitation tends to decrease poleward; therefore, the humid continental, mild summer climate tends to have less precipitation than the hot-summer regions closer to the equator. Precipitation continues to decrease throughout this climatic type toward the poleward margins and from the coasts toward the arid continental interiors. As in its hot-summer

counterpart, the monsoon effect in the mild-summer climate is strong enough in Asia to produce a dry winter season (see Vladivostok, Fig. 10.20).

Winters in the mild-summer climate are more severe and longer than in its neighbor to the south. Summers, on the other hand, are not as long or hot. The combination of more severe winter and shorter summers makes for a growing season of between 90 and 130 days, which is one to three months shorter than in the hot-summer climate. In addition, although overall precipitation totals—50 to 100 centimeters (20–40 in.)—are generally lower, snowfall is greater, and snow cover is both thicker and longer lasting (Fig. 10.21).

The humid continental, mild-summer regions exhibit seasonal changes just as clearly as the hot-summer regions. Annual temperature ranges are generally larger. Vigorous polar and tropical air mass interaction makes weather change a common occurrence. However, the more poleward position of the mild-summer climate brings about a greater dominance of the colder air masses and explains why temperature variability is not as abrupt or as great as it is further south. Under normal conditions, tropical air is strongly modified by the time it reaches mild-summer regions: Even in the high-sun season, intrusions of warm, humid air rarely last more than a few days at a time. By contrast, winter invasions of cold arctic air periodically bring several successive days or weeks of clear skies and frigid temperatures.

As in the hot-summer climate, the wetter regions of the mild-summer climate are associated with a natural forest vegetation. However, many trees common in the hot-summer climate, such as oaks, hickories, and maples, find it difficult to compete with firs, pines, and spruces toward the colder, polar margins of these regions. Fortunately for

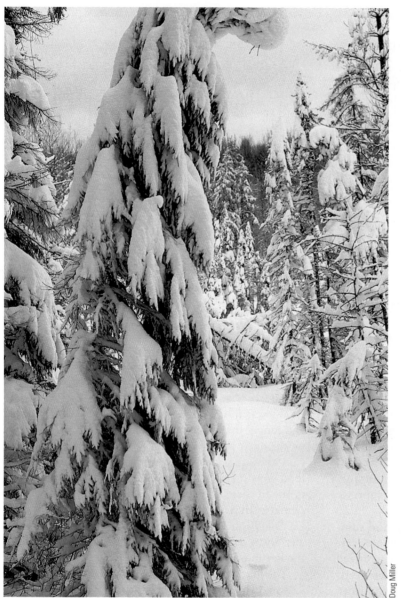

Figure 10.21 People, animals, and plants living in the humid continental, mild summer regions have learned to cope with, as well as enjoy, the beauty of abundant snow, which is present continuously for many months at a time, as in this example in northern Michigan. ■ **In what ways might snow cover be considered an asset for some people living in humid continental, mild summer regions?**

agriculture, northward extensions of the grasslands and the rich soils that accompany them in the hot-summer regions are found in the drier portions of the mild-summer climate.

Human Activity in the Humid Continental Climates

Perhaps the greatest contrast between the hot-summer and mild-summer humid continental regions is exhibited in agriculture. Despite the unpredictability of the weather, the humid continental hot-summer agricultural regions are among the finest in the world. The favorable combination of long, hot summers, ample rainfall, and highly fertile soils has made the American Midwest a

leading producer of corn, beef cattle, and hogs (Fig. 10.22). Soybeans, which are native to similar climatic regions in northern China, are now second to corn throughout the Midwest as feed for animals and as a raw material for the food-processing, plastics, and vegetable oil industries. Wheat, barley, and other grains are especially important in European and Asian regions, and winters are sufficiently mild that fall-sown varieties can be raised in the United States. In the mild-summer climate, on the other hand, a shorter growing season imposes certain limitations on agriculture and restricts the crops that can be grown. Farmers rely more on quick-ripening varieties, grazing animals, orchard products, and root crops. Dairy products—milk, cheese, butter, cream—are mainstays in the economies of Wisconsin, New York, and northern New England. The moderating effect of the Great Lakes or other water bodies permits the growth of deciduous fruits, such as apples, plums, and cherries.

The length of the growing season is the most obvious reason for the differences in agriculture between the two humid continental climates, but there is another climate-related reason as well. The great ice sheets of the Pleistocene Epoch have had significant but different effects on mild-summer and hot-summer regions, especially in North America. In the hot-summer regions, the ice sheets thinned and receded, releasing the enormous load of soil and solid rock debris they had stripped off the lands nearer to their centers of origin. The material was laid down in a blanket hundreds of feet thick in the areas of maximum glacial advance. As the ice retreated northward, less and less debris was deposited, much of it flushed away by meltwater streams. The more southerly, hot-summer region consequently has an undulating topography underlain by thick masses of glacial debris. The soils formed on this debris are well developed and fertile, and plant nutrients are more likely to be evenly distributed because steep slopes are lacking. The more northerly, mild-summer region, on the other hand, mainly shows the effects of glacial erosion (Fig. 10.23). Rockbound lakes and marshy lowlands alternate with ice-scoured rock hills. Soils are either thin and stony or waterlogged. Because of its lower agricultural potential, large sections of this area remain in forest.

However, because of its wilderness character and the abundance of lakes in basins produced by glacial erosion, recreational possibilities in a mild-summer region

Figure 10.22 (Top) Corn harvested in the United States; (bottom) hog production in the United States. Both maps are overlaid by boundaries of the humid continental, hot summer climate. ■ **From an examination of the maps, which two states do you believe would lead the United States in production of corn and hogs each year?**

far exceed those of a more subdued hot-summer region. Minnesota calls itself the "Land of 10,000 Lakes," and in New England, lakes, rough mountains, and forest combine to produce some of the most spectacular scenery east of the Rocky Mountains.

Subarctic Climate

The **subarctic climate** is the furthest poleward and most extreme of the microthermal climates. By definition, it has at least one month with an average temperature above 10°C (50°F), and its poleward limit roughly coincides with the 10°C isotherm for the warmest month of the year. As you may recall from our earlier discussion of the simplified Köppen system, forests cannot survive where at least one month does not have an average temperature over 10°C. Thus, the poleward boundary of the subarctic climate is the poleward limit of forest growth as well.

As Figure 10.16 indicates, the subarctic climate, like the other microthermal climates, is found exclusively in the Northern Hemisphere. It covers vast areas of subpolar Eurasia and North America. Conditions vary widely over such great distances. Extremely severe winter regions are located along the polar margins or deep in the interior of the Asian landmass, and climatic subtypes with winter drought are found in association with the Siberian high and its clear skies, bitter cold, and strong subsidence of air over interior Asia during winter. Other subarctic regions experience less severe winters or year-round precipitation.

Further study of Figure 10.16 suggests two additional observations. First, ocean currents tend to influence the distribution of the subarctic climate. Along the west coasts of the continents, especially in North America, the warm ocean currents modify temperatures sufficiently to permit the marine west coast climate to extend into latitudes normally occupied by the subarctic and to cause the subarctic to be found well beyond the Arctic Circle. Along east coasts, where cold ocean currents help to reduce winter temperatures, the subarctic is situated further south. Second, the development of the subarctic climate is not as extensive in North America as it is in Eurasia. This is because (1) the Eurasian continent is a larger landmass, which increases the effect of continentality,

Figure 10.23 Although the effect of glacial action further south was to deposit material, here in New Hampshire we see an area of glacial erosion—Newfound Lake, a former glacial trough. ■ **How might this area be considered an economic resource?**

R.Gabler

and (2) the large water surface of Hudson Bay in Canada provides a modifying marine influence inland, which tends to counter the effect of continentality there.

The Effects of High Latitude and Continentality Subarctic regions experience short, cool summers and long, bitterly cold winters (Fig. 10.24). The rapid heating and cooling associated with continental interiors in the higher latitudes allow little time for the in-between seasons of spring and fall. At Eagle, Alaska, a station in the Klondike region of the Yukon River Valley, the temperature climbs 8° to 10°C (15° to 20°F) per month as summer approaches and drops just as rapidly prior to the next winter season. At Verkhoyansk in Siberia, the change between the seasonal extremes is even more rapid, averaging 15° to 20°C (30° to 40°F) per month.

Because of the high latitudes of these regions, summer days are quite long and nights are short. The noon sun is as high in the sky during a subarctic summer as during a subtropical winter. The combination of a moderately high angle of the sun's rays and many hours of daylight means that some subarctic locations receive as much insolation at the time of the summer solstice as the equator does. As a result, temperatures during the one to three months of the subarctic summer usually average 10° to 15°C (50°–60°F), and on some days they

Eagle, Alaska
65°N
Precip.: 27.5 cm (10.9 in.)
Subarctic (Dfc)
141°W
Av. temp.: −4.5°C (24°F)
Range: 41°C (74°F)

Verkhoyansk, Russia
68°N
Precip.: 13.5 cm (5.3 in.)
Subarctic (Dwd)
134°E
Av. temp.: −17°C (1°F)
Range: 64°C (115°F)

Figure 10.24 Climographs of the subarctic climate stations. ■ **Why would people settle in such severe-winter climatic regions?**

may even approach 30°C (86°F). Thus, the brief summer in the subarctic climate is often pleasantly warm, even hot, on some days.

The winter season in the subarctic is bitter, intense, and lasts for as long as eight months. Eagle has eight months with average temperatures below freezing. In the Siberian subarctic, the January temperatures regularly *average* −40° to −50°C (−40° to −60°F). The coldest temperatures in the Northern Hemisphere—officially, −68°C (−90°F) at both Verkhoyansk and Oymyakon; unofficially, −78°C (−108°F) at Oymyakon—have been recorded there. In addition, the winter nights, with an average 18 to 20 hours of darkness that extend well into one's working hours, can be mentally depressing and can increase the impression of climatic severity.

As a direct result of the intense heating and cooling of the land, the subarctic has the largest annual temperature ranges of any climate. Average annual ranges in equatorward margins of the climate vary from near 40°C (72°F) to more than 45°C (80°F). The exceptions are near western coasts, where warm ocean currents and the marine influence may significantly modify winter temperatures. Average annual ranges for poleward stations are even greater. The climograph for Verkhoyansk, which indicates a range of 64°C (115°F), is an extreme example.

As with subarctic temperature, latitude and continentality influence subarctic precipitation. These climatic controls combine to limit annual precipitation amounts to less than 50 centimeters (20 in.) for most regions and to 25 centimeters (10 in.) or less in northern and interior locations. The low temperatures in the subarctic reduce the moisture-holding capacity of the air, thus minimizing precipitation during the occasional passage of cyclonic storms. Location toward the center of large landmasses or near lee coasts increases distance from oceanic sources of moisture. Finally, the higher latitudes

occupied by the subarctic climate are dominated by the polar anticyclone, especially in the winter season. Much as the subtropical high pressure systems restrict precipitation in the subtropical arid regions, subsidence and divergence of air in the polar anticyclone limit the opportunity for lifting and hence for precipitation in the subarctic regions. This high pressure system also blocks the entry of moist air from warmer areas to the south.

Subarctic precipitation is cyclonic or frontal, and because the polar anticyclone is weaker and further north during the warmer summer months, more precipitation comes during that season. The meager winter precipitation falls as fine, dry snow. Though there is not as much snowfall as in less severe climates, the temperatures remain so cold for so long that the snow cover lasts for as long as seven or eight months. During this time, there is almost no melting of snow, especially in the dark shadows of the forest.

A Limiting Environment The climatic restrictions of subarctic regions place distinct limitations on plant and animal life and on human activity. The characteristic vegetation is coniferous forest, adapted to the severe temperatures, the physiologic drought associated with frozen soil water, and the infertile soils. Seemingly endless tracts of spruce, fir, and pine thrive over enormous areas, untouched by humans (Fig. 10.25). In Russia, the forest is called the *taiga* (or *Boreal forests* in other regions), and this name is sometimes given to the subarctic climate type itself.

The brief summers and long, cold winters severely limit the growth of vegetation in subarctic regions. Even the trees are shorter and more slender than comparable species in less severe climatic regimes. There is little hope for agriculture. The growing season averages 50 to 75 days, and frost may occur even during June, July, or August. Thus,

Georg Gerster/Comstock

Figure 10.25 Seemingly endless tracts of taiga (boreal forest) are typical of the vegetation throughout much of the Siberian Subarctic. This photo was taken between the Russian towns of Vanavara and Tura. ■ **Why are these virgin forests currently of little economic value?**

M. Trapasso

Figure 10.26 Frost polygons, or patterned ground, as seen in the subarctic region of Alaska. Repeated freezing and thawing causes the soil to produce polygonal shapes.

in some years, a subarctic location may have no truly frost-free season. Although scientists are working to develop plant species that can take advantage of the long hours of daylight in summer, only minimal success has been achieved, in southern parts of the climate, with certain vegetables such as cabbage and root crops such as potatoes.

A particularly vexing problem to people in subarctic (as well as tundra) regions is **permafrost,** a permanently frozen layer of subsoil and underlying rock that may extend to a depth of 300 meters (1000 ft) or more in the northernmost sections of the climate. Permafrost is present over much of the subarctic climate, but it varies greatly in thickness and is often discontinuous. Where it

occurs, the land is frozen completely from the surface down in winter. The warm temperatures of spring and summer cause the top few feet to thaw out, but because the land beneath this thawed top layer remains frozen, water cannot percolate downward, and the thawed soil becomes sodden with moisture, especially in spring, when there is an abundant supply of water from the melting snow. Permafrost poses a problem to agriculture by preventing proper soil drainage. The seasonal freeze and thaw of the surface layer above the permafrost also poses special problems for construction engineers. The cycle causes repeated expansion and contraction, heaving the surface up and then letting it sag down. The effects of this cycle break up roads, force buried pipelines out of the ground, cause walls and bridge piers to collapse, and even swivel buildings off their foundations. Landscapes called **patterned ground** or **frost polygons** are commonly found in regions of subarctic yearly freezing and thawing (Figure 10.26).

With agriculture a questionable occupation at best, there is little economic incentive to draw humans to subarctic regions. Logging is unimportant because of small tree size. Even use of the vast forests for paper, pulp, and wood products is restricted by their interior location far from world markets. Miners occasionally exploit rich ore deposits; others, many of them native to the subarctic regions, pursue hunting, trapping, and fishing of the relatively limited wildlife. Fur-bearing animals such as mink, fox, wolf, ermine, otter, and muskrat are of greatest value.

Table 10.3	The Polar Climates			
Name and Description	**Controlling Factors**	**Geographic Distribution**	**Distinguishing Characteristics**	**Related Features**
Tundra				
Warmest month between 0°C (32°F) and 10°C (50°F); precipitation exceeds potential evaporation	Location in the high latitudes; subsidence and divergence of the polar anticyclone; proximity to coasts	Arctic Ocean borderlands of North America, Greenland, and Eurasia; Antarctic Peninsula; some polar islands	At least 9 months average below freezing; low evaporation; precipitation usually below 25.5 cm (10 in.); coastal fog; strong winds	Tundra vegetation; tundra soils; permafrost; swamps and bogs during melting period; life most common in nearby seas; Inuit; mineral and oil resources; defense industry
Ice-sheet				
Warmest month below 0°C (32°F); precipitation exceeds potential evaporation	Location in the high latitudes and interior of landmasses; year-round influence of the polar anticyclone; ice cover; elevation	Antarctica; interior Greenland; permanently frozen portions of the Arctic Ocean and associated islands	Summerless; all months average below freezing; world's coldest temperature; extremely meager precipitation in the form of snow, evaporation even less; gale-force winds	Ice- and snow-covered surface; no vegetation; no exposed soils; only sea life or aquatic birds; scientific exploration

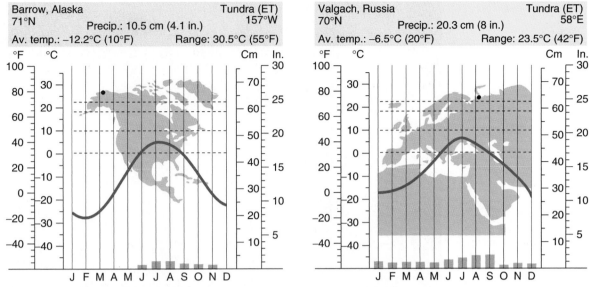

Figure 10.27 Climographs for tundra climate stations. ■ **Why is it not surprising that both stations are located in the Northern Hemisphere?**

Polar Climatic Regions

The polar climates are the last of Köppen's humid climatic subdivisions to be differentiated on the basis of temperature. These climatic regions are situated at the greatest distance from the equator, and they owe their existence primarily to the low annual amounts of insolation they receive. No polar station experiences a month with average temperatures as high as 10°C (50°F), and hence all are without a warm summer (Table 10.3). Trees cannot survive in such a regimen. In the regions where at least one month averages above 0°C (32°F), they are replaced by tundra vegetation. Elsewhere, the surface is covered by great expanses of frozen ice. Thus, there are two polar climatic types, tundra and ice-sheet.

There are two important points to keep in mind in the discussion of polar climatic regions. First, these regions have a large net annual radiation loss; that is, they give up much more radiation or energy than they receive from the sun during a year, resulting in a major radiation deficiency. The transfer of heat from lower to higher latitudes to make up this deficiency is the driving force of the general atmospheric circulation. Without this compensating poleward transfer of heat from the lower latitudes, the polar regions would become too cold to permit any form of life, and the equatorial regions would heat to temperatures no organism could survive.

A second and equally important characteristic of polar climates is the unique pattern of day and night. At the poles, six months of relative darkness, caused when the sun is positioned below the horizon, alternate with six months of daylight during which the sun is above the horizon. Even when the sun is above the horizon, however, the sun's rays are at a sharply oblique angle, and little insolation is received for the number of hours of daylight. Moving outward from the poles, the lengths of

periods of continuous winter night and continuous summer day decrease rapidly from six months at the poles to 24 hours at the Arctic and Antarctic Circles (66° N and S). Here the 24-hour night or day occurs only at the winter and summer solstices, respectively.

Tundra Climate

Compare the location of the **tundra climate** with that of the subarctic climate in Figure 9.4. You can see that although the tundra climate is situated closer to the poles, it is also along the periphery of landmasses, and, with the exception of the Antarctic Peninsula, it is everywhere adjacent to the Arctic Ocean. Even though temperature ranges in the tundra are large, they are not as large as in the subarctic because of the maritime influence. Winter temperatures, in particular, are not as severe in the tundra as they are inland (Fig. 10.27).

It almost seems inappropriate to call the unpleasantly chilly and damp conditions of the tundra's warmer season "summer." Temperatures average around 4°C (40°F) to 10°C (50°F) for the warmest month, and frosts occur regularly. The air does warm sufficiently to melt the thin snow cover and the ice on small bodies of water, but this only causes marshes, swamps, and bogs to form across the land because drainage is blocked by permafrost (Fig. 10.28). Out of this soggy landscape, known as **muskeg** in Canada and Alaska, swarm clouds of black flies, mosquitoes, and gnats. The one bright note in the landscape is provided by the enormous number of migratory birds that nest in the arctic regions at this time of year and feed on the insects. However, as soon as the shrinking days of autumn approach, these birds depart for warmer climates.

Winters are cold and seem to last forever, especially in tundra locations where the sun is below the horizon

Figure 10.28 Permafrost regions, such as this area at the base of the Alaska Range, become almost impenetrable swampland during the brief Alaskan summer. Travel over land is feasible only in the winter season. ■ **What is the preferred means of travel in the summer?**

R.Gabler

for days at a time. The climograph for Barrow, Alaska, illustrates the low temperatures of this climate. Note that average monthly temperatures are *below freezing* nine months of the year. The average annual temperature is −12°C (10°F). The low-growing tundra vegetation survives despite the forbidding environment. It consists of lichens, mosses, sedges, flowering herbaceous plants, small shrubs, and grasses. In particular, the plants have adjusted to the conditions associated with nearly universal permafrost.

The tundra regions exhibit several other significant climatic characteristics. Diurnal temperature ranges are small because insolation is uniformly high during the long summer days and uniformly low during the long winter nights. Precipitation is generally low, except in eastern Canada and Greenland, because of exceedingly low absolute humidity and the influence of the polar anticyclone. Icy winds sweep across the open land surface and are an added factor in eliminating the trees that might impede their progress. Coastal fog is characteristic in marine locations, where cool polar maritime air drifts onshore and is chilled below the dew point by contact with the even colder land.

Ice-Sheet Climate

The **ice-sheet climate** is the most severe and restrictive climate on Earth. As Table 10.3 indicates, it covers large areas in both the Northern and Southern Hemispheres, a total of about 16 million square kilometers (6 million sq mi)—nearly the same area as the United States and Canada combined. All average monthly temperatures are below freezing, and because most surfaces are covered with glacial ice, no vegetation can survive in this climate. It is a virtually lifeless region of perpetual frost.

Antarctica is the coldest place on Earth (although Siberia sometimes has longer and more severe periods of cold in winter). The world's coldest temperature, −88°C (−127°F), was recorded at Vostok, Antarctica. Consider the climographs for Little America, Antarctica, and Eismitte, Greenland, for a fuller picture of the cold ice-sheet temperatures (Fig. 10.29).

The primary reason for the low temperatures of ice-sheet climates is the minimal insolation received in these regions. Not only is little or no insolation received during half the year, but the sun's radiant energy that is received arrives at sharply oblique angles. In addition, the perpetual snow and ice cover of this climate reflects nearly all of the incoming radiation. A further factor, in both Greenland and Antarctica, is elevation. The ice sheets covering both regions rise more than 3000 meters (10,000 ft) above sea level (Fig. 10.30). Naturally, this elevation contributes to the cold temperatures.

The polar anticyclone severely limits precipitation in the ice-sheet climate to the fine, dry snow associated with occasional cyclonic storms. Precipitation is so meager in this climate that regions within this regime are sometimes incorrectly referred to as "polar deserts." However, because of the exceedingly low evaporation rates (evapotranspiration rates are not considered in this climate because there is no plant life to account for any transpiration) associated with the severely cold temperatures, precipitation still exceeds potential evaporation and the climate can be classified as humid. The annual precipitation surplus produces glaciers, which export snowfall similar to the way rivers export rainfall.

The strong and persistent polar winds are another staple of the harsh ice-sheet climate. Mawson Base, Antarctica, for example, has approximately 340 days a year with gale-force winds of 15 meters per second (33 mph) or

Figure 10.29 Climographs for ice-sheet climate stations. ■ **If you were to accept an offer for an all-expense-paid trip to visit either Greenland or Antarctica, which would you choose, and why would you go?**

Human Activity in Polar Regions

The climatic severity that limits animal life in polar regions to a few scattered species in the tundra is just as restrictive on human settlement. The celebrated Lapps of northern Europe migrate with their reindeer to the tundra from the adjacent forest during warmer months. They join the musk ox, arctic hare, fox, wolf, and polar bear that manage to make a home there despite the prohibitive environment. Only the Inuit (Eskimos) of Alaska, northern Canada, and Greenland have in the past succeeded in developing a year-round lifestyle adapted to the tundra regime. Yet even this group relies less on the resources of the tundra than on the large variety of fish and sea mammals, such as cod, salmon, halibut, seal, walrus, and whale, that occupy the adjacent seas.

As their communication with the rest of the world has increased and they have become acquainted with alternative lifestyles, the permanent population of Inuit living in the tundra has greatly diminished, and life for those re-

more. *Katabatic winds,* which are caused by the downslope drainage of heavy cold air accumulated over ice sheets, are common along the edges of the polar ice. The winds of these regions can result in "whiteouts"—periods of zero visibility due to blowing fine snow and ice crystals.

maining has changed drastically. Some have gained new economic security through employment at defense installations or at sites where they join other skilled workers from outside the region to exploit mineral or energy resources. However, the new population centers based on the construction and maintenance of radar and missile defense stations or, as in the case of Alaska's North Slope, on the production and transportation of oil, cannot be considered permanent (Fig. 10.31). Workers depend on other regions for support and often inhabit this region only temporarily.

The ice-sheet climate cannot serve as a home for humans or other animals. Even the penguins, gulls, leopard seals, and polar bears are coastal inhabitants. It is

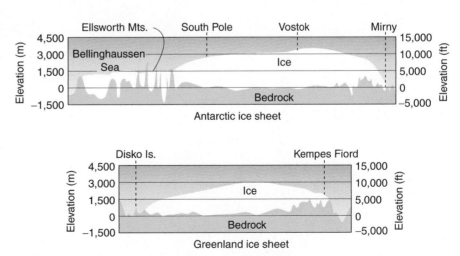

Figure 10.30 The Antarctic ice sheet resulting from the ice-sheet climate of the south polar regions is as much as 4000 meters (13,200 ft) thick locally. Where it is thinnest, it is floated by seawater to produce an ice shelf. The smaller Greenland ice sheet is about 3000 meters (10,000 ft) thick. ■ **What reasons might be given for the fact that more land in Greenland than in Antarctica is free of glacial ice?**

Why Are "Ozone Holes" Located at the Poles?

First of all, it is important to ask, is there indeed a "hole" in our protective ozone layer? The correct answer is no. From Earth's surface to the outer atmosphere there is always some ozone present (keeping in mind, of course, that ozone is only a trace gas). There have been years when all the ozone was missing from certain levels above the surface, but there has always been some ozone above us. If all our ozone were forced down to sea level, the atmospheric pressure present there would squeeze the ozone into a worldwide layer of only about 3mm in thickness.

When we see a satellite image of the infamous "ozone hole," we see an elliptical area centered on the polar regions (especially the South Pole). The color pattern of the images corresponds to the amount of ozone in the air using Dobson units (du). Dobson units in the 250du to 350du range indicates a sufficient amount of ozone to prevent serious damage to Earth life forms. G.M.B. Dobson developed these units, along with the instrument (spectrophotometer) used to measure ozone, in the late 1920s. One hundred du in the strastosphere translates to 1-mm of thickness that ozone would have at sea level. The ozone hole then really indicates an area where ozone is considerably less than it should be. Ozone measured inside the "hole" has dropped as low as 95 du in recent years. The areal extent of this ozone-deficient region (a more accurate description) has exceeded that of the North American continent.

People have wondered why "ozone holes" form at the South Pole (and to a lesser extent, the North Pole). Ozone in the upper atmosphere is destroyed by chemical reaction to chlorofluorocarbons (CFCs) and other gases that are products of pollution by industrial nations. With the exception of scientific research bases, no one lives near the South Pole. There are certainly no big cities or any widespread industrial activity present. Why should the hole appear there instead of over major urbanized countries?

Research through the decades has shown three factors necessary for ozone destruction and the creation of the "holes." First, there are circulation patterns in the stratosphere, which move from the tropics to the poles and back again. This slow, conveyor belt-like movement transports pollution from the tropics, through populated middle latitudes, and on to the poles. Therefore it is quite possible to bring pollutants to polar regions.

Second, during the *austral winter* (June, July, and August) the South Pole experiences total darkness and is colder than anyplace on Earth. A closed circulation pattern of extremely cold air called the *circumpolar vortex,* forms above the dark pole. Ozone is trapped, along with other pollutants (like CFCs and NO_x), within this vortex, until the return of sunlight energizes the destructive reaction. During the *austral spring* (September, October, and November) the incoming solar radiation dissolves the vortex and begins the destructive process. Usually the "ozone hole" reaches its greatest extent each year within the first two weeks of October.

Figure 10.31 One result of North Slope oil production that had a decidedly negative effect outside the polar regions: Cleanup crews work to remove oil that washed onshore after the oil tanker *Exxon Valdez* spilled 11 million gallons of crude oil into Prince William Sound, Alaska.
■ **Considering the vulnerability of Alaska's physical environment, should development of the North Slope oil fields have been permitted?**

Ron Levy/Gamma Liaison

The third factor involves Stratospheric Ice Clouds (SIC) found many miles above the poles. These clouds, located in the stratosphere, do not affect the weather, but they act as the perfect laboratory for the ozone destructive reaction to take place.

$$Cl + O_3 \rightarrow ClO + O_2$$

This ozone-destructive reaction takes place most efficiently on an ice crystal surface.

In light of these three factors, it becomes clear that if an "ozone hole" were going to appear in our atmosphere, it would necessarily appear at the poles first.

Image produced by Fritz Hasler and Hal Pierce; Laboratory for Atmospheres, NASA Goddard Space Flight Center for the *National Geographic Satellite Atlas of the World*.

NASA

Areas of ozone depletion are found in both the Antarctic (a) and the Arctic (b).

without question the harshest, most restrictive, most nearly lifeless climatic zone on Earth. Yet, especially in Antarctica, it is of strategic importance and of great scientific interest. Scientists study the oxygen-isotope ratios of the Antarctic ice cores and the gas bubbles trapped within them to help them reconstruct past climates. They have also noted that ozone concentrations have decreased over Antarctica every fall for several decades. The result is a hole in the ozone layer above Antarctica that is as large as the North American continent. This decrease in the ozone layer is a major environmental concern. Antarctica's strategic value is so widely recognized that the world's nations have voluntarily given up claims to territorial rights on the continent in exchange for cooperative scientific exploration on behalf of all humankind.

Highland Climatic Regions

As we saw in Chapter 4, temperature decreases with increasing altitude at the rate of about 6.5°C per 1000 meters (3.6°F per 1000 ft). Thus, you might suspect that highland regions exhibit broad zones of climate based on changes in temperature with elevation that roughly correspond to Köppen's climatic zones based on change of temperature with latitude (Fig. 10.32). This is indeed the case, with one important exception: Seasons only exist in highlands if they also exist in the nearby lowland regions. For example, although zones of increasingly cooler temperature occur at progressively higher elevations in the tropical climatic regions, the seasonal changes of Köppen's middle-latitude climates are not present.

Elevation is only one of several controls of highland climates; **exposure** is another. Just as some continental coasts face the prevailing wind, so do some mountain slopes; others are lee slopes or are sheltered behind higher topography. The nature of the wind, its temperature, and its moisture content depend on whether the mountain is (1) in a coastal location or deep in a continental interior and (2) at a high or low latitude within or beyond the reaches of cyclonic storms and monsoon circulation. In the middle and high latitudes, mountain

The Effects of Altitude on the Human Body

The changes in climate, soils, and vegetation as an individual climbs into higher altitudes are discussed elsewhere in the text. However, it is interesting to note also some of the ways in which high altitudes affect the human body. Besides food and water, the metabolism of a human being also depends on the consumption of oxygen. A lessening of the normal amount of oxygen intake can have some profound physiological effects. One way to look at the amount of oxygen available for human use is to consider something called the *partial pressure of oxygen gas,* or p(O_2). Partial pressure, in this case, refers to the portion of total atmospheric pressure attributed to oxygen alone. At sea level, atmospheric pressure is 1013.2 mb, and the pO_2 is about 212 mb. In other words, 212 of the 1013.2 mb is attributed to oxygen gas. At an altitude of 10 km (6.2 miles), the pO_2 drops to only 55 mb! Even at moderate altitudes, the effects of *hypoxia* (oxygen starvation) can cause headaches and nausea. Above 6 km (3.7 miles), this lack of oxygen can seriously affect the brain. Since the body's need for oxygen does not alter with major changes in altitude, any significant drop in pO_2 with height can cause severe body stress. Pressure-controlled cabins of high-flying aircraft and the use of oxygen by mountain climbers provide dramatic examples of ways to meet the vital need for oxygen.

Fortunately, the human body can acclimatize to moderate changes in altitude, but it takes time (days to sometimes weeks, depending on the altitude). During this time, however, one may experience some of the following symptoms: sleeplessness, headaches, loss of weight, thyroid deficiency, increased excitability, muscle pain, gastrointestinal disturbances, swelling of the lungs, severe infections, psychological and mental disturbances, and others. At the very least, when visiting higher altitude locations, most people can expect headaches (perhaps leading to nausea), a drop in physical endurance (climbing several flights of stairs will be more difficult), and hampered mental function (solutions to simple problems may be difficult to grasp). There is little need to worry, though; humans do acclimatize, and these symptoms will pass.

It is also important to remember that, in addition to hypoxia, higher altitudes may cause increased susceptibility to severe sunburn. If you climb to a higher altitude, this means that more of Earth's atmosphere is below you. Correspondingly, there is less of the protective atmosphere above to filter harmful UV radiation. Keeping latitude constant, sunburn of exposed skin is more likely on a high-altitude mountainside than on a sea-level beach.

---

slopes and valley walls that face the equator receive the direct rays of the sun and are warm; poleward-facing slopes are shadowed and cool. West-facing slopes feel the hot afternoon sun, whereas east-facing slopes are sunlit only in the cool of the morning. This factor, known as **slope aspect,** affects where people live in the mountains and where particular crops will do best. The higher one rises in the mountains, the more important direct sunlight is as a source of warmth and energy for plant and animal life processes.

(a)

Figure 10.32 Note the similarities between (a) Slovar, Norway, 68°N (an Arctic location), at sea level, and (b) Machu Picchu, 13°S latitude (a tropical location), at 2590 meters (8500 feet) above sea level.

M. Trapasso

AP/Wide World Photos

Both air pressure and air density decrease rapidly with increasing altitude.

(b)

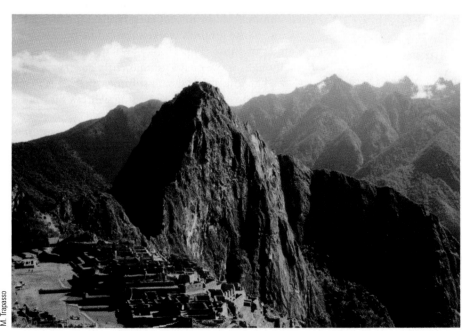

M. Trapasso

Complexity is the hallmark of highland climates. Every mountain range of significance is composed of a mosaic of climates far too intricate to differentiate on a world map, or even on a map of a single continent. Highland climates are therefore undifferentiated, signi-fying climate complexity. Highland climates are indi-cated on Figure 9.4 wherever there is marked local varia-tion in climate as a consequence of elevation, exposure, and slope aspect. We can see that these regions are distributed widely over Earth but are particularly concentrated in Asia, central Europe, and western North and South America.

The areas of highland climate on the world map are cool, moist islands in the midst of the zonal climates that dominate the areas around them. Consequently, highland areas are also biotic is-lands, supporting a flora and fauna adapted to cooler and wet-ter conditions than those of the surrounding lowlands. This coolness is part of the highland charm, particularly where moun-tains rise cloaked with forests above arid plains, as do the Cana-dian Rocky Mountains and Cali-fornia's Sierra Nevada.

Highlands stimulate moisture condensation and precipitation by forcing moving air masses

Figure 10.33 Variation in precipitation caused by uplift of air crossing the Sierra Nevada range of California from west to east. The maximum precipitation occurs on the windward slope, because air in the summit region is too cool to retain a large supply of moisture. Note the strong rain shadow to the lee that gives Reno a desert climate. ■ **Taking into consideration the locations of the recording stations, during what season of the year does the maximum precipitation on the windward slope occur?**

to rise over them (Fig. 10.33). Where mountain slopes are rocky and forest-free, their surfaces grow warm during the day, causing upward convection, which often produces afternoon thundershowers. Mountains receive abundant precipitation and are the source area for multitudes of streams that join to form the great rivers of all of the continents.

There are few streams of significance whose headwaters do not lie in rugged highlands. Much of the stream flow on all continents is produced by the summer melting of mountain snowfields. Thus, the mountains not only wring water from the atmosphere but also store much of it in a form that gradually releases it throughout summer droughts, when water is most needed for irrigation and for municipal and domestic uses.

Peculiarities of Mountain Climates

A general characteristic of mountain weather is its variability from hour to hour as well as from place to place. Strong orographic flow over mountains often causes clouds to form very quickly, leading to thunderstorms and longer rains that do not affect surrounding cloud-free lowlands. Where the cloud cover is diminished, diurnal temperature ranges over mountains are far greater than those over lowlands. Since mountains penetrate upward beyond the densest part of the atmosphere, the greenhouse effect is less developed there than anywhere else on Earth. The thinner layer of low-density air above a mountain site does not greatly impede insolation, thus allowing surfaces to warm dramatically during the daytime. By the same token, the atmosphere in these areas does little to impede long-wave radiation loss at night.

Consequently, air temperatures overnight are cooler than the elevation alone would indicate. Since the atmospheric shield is thinnest at high elevations, plants, animals, and humans receive proportionately more of the sun's short-wave radiation at high altitudes. Violet and ultraviolet radiation are particularly noticeable; severe sunburn is one of the real hazards of a day in the high country.

In the middle and high latitudes, mountains rise from mesothermal and microthermal climates into tundra and snow-covered zones. The lower slopes of mountains are commonly forested with conifers, which become more stunted as one moves upward, until the last dwarfed tree is passed at the **tree line.** This is the line beyond which low winter temperatures and severe wind stress eliminate all forms of vegetation except those that grow low to the ground, where they can be protected by a blanket of snow (Fig. 10.34). Where mountains are high enough, snow or ice permanently covers the land surface. The line above which summer melting is insufficient to remove all of the preceding winter's snowfall is called the **snow line.**

In tropical mountain regions, the vertical zonation of climate is even more pronounced. Both tree line and snow line occur at higher elevations than in middle latitudes. Any seasonal change is mainly restricted to rainfall; temperatures are stable year-round, regardless of elevation. Each climatic zone has its own particular association of natural vegetation and has given rise to a distinctive crop combination where agriculture is practiced (Fig. 10.35). In South America, four vertical climatic zones are recognized: *tierra caliente* (hot lands); *tierra templada* (temperate lands); *tierra fría* (cool lands), and *tierra helada* (frozen lands).

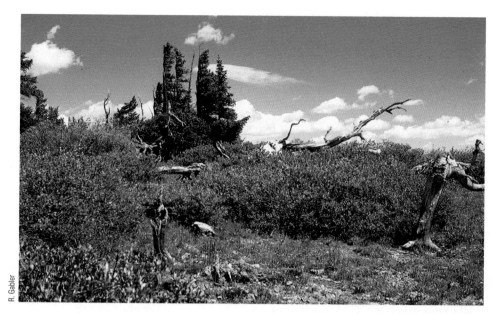

Figure 10.34 As in this example in the Colorado Rocky Mountains, the last tree species found at the tree line are stunted, prostrate forms, which often produce an elfin forest. Where the trees are especially gnarled and misshapen by wind stress, the vegetation is called *krummholz* (crooked wood).
■ **What do you see in the photograph that indicates prevailing wind direction?**

Highland Climates and Human Activity

In middle-latitude highlands, soils are poor, the growing season is short, and the winter snow cover is heavy in the conifer zone, which dominates the lower and middle mountain slopes. Therefore, little agriculture is practiced, and permanent settlements in the mountains are few.

However, as the winter snow melts off the high ground just below the bare rocky peaks, grass springs into life, and humans drive herds of cattle and flocks of sheep and goats up from the warmer valleys. The high pastures are lush throughout the summer, but in early fall they are once again vacated by the animals and their keepers, who return to the valleys. This seasonal movement of herds and herders between alpine pastures and villages in the

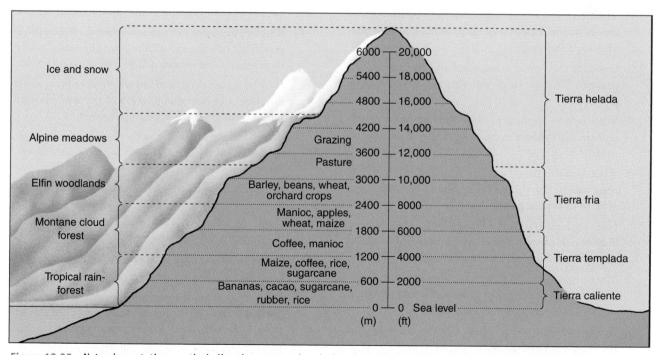

Figure 10.35 Natural vegetation, vertical climatic zones, and agricultural products in tropical mountains. Note that this example extends from tropical life zones to the zone of permanent snow and ice. There is little seasonal temperature change in tropical mountains, which allows life forms sensitive to low temperatures to survive at relatively high elevations. ■ **When Europeans first settled in the highlands of tropical South America, in which vertical climatic zone did they prefer to live?**

valleys, termed *transhumance,* was once common in the European highlands (the Alps, Pyrenees, Carpathian Mountains, and mountains throughout Scandinavia) and is still practiced there on a reduced scale.

Otherwise, the middle-latitude highlands serve mainly as sources of timber and of minerals formed by the same geologic forces that elevated the mountains, and as arenas for recreation—both summer and winter. Recreational use of the highlands is a relatively recent phenomenon, resulting both from new interest in mountain areas and from new access routes by road, rail, and air that did not exist a century ago.

In contrast to poleward mountain regions, tropical highlands may actually experience more favorable cli-matic conditions and are often a greater attraction to human settlement than adjacent lowlands. In fact, large permanent populations are supported throughout the tropics where topography and soil favor agriculture in the vertical climatic zones. Highland climates are at such a premium in many areas that steep mountain slopes have been extensively terraced to produce level land for agricultural use. Spectacular agricultural terraces can be seen in Peru, Yemen, the Philippines, and many other tropical highlands. Where the climate is appropriate and population pressure is high, people have created a topography to suit their needs, although they have had to hack it out of mountainsides.

DEFINE & RECALL

Mediterranean climate
sclerophyllous
chaparral
humid subtropical climate
marine west coast climate

humid continental, hot
   summer climate
humid continental, mild
   summer climate
subarctic climate

permafrost
patterned ground
   (frost polygons)
tundra climate
muskeg

ice-sheet climate
exposure
slope aspect
tree line
snow line

DISCUSS & REVIEW

1. What characteristics make the Mediterranean climate readily distinguishable from all others? What controls are responsible for producing these characteristics?

2. Summarize the special adaptations of vegetation and soils in Mediterranean regions. How have humans also adapted to these regions?

3. Compare the humid subtropical and Mediterranean climates. What are their most obvious similarities and differences?

4. What factors combine to cause a precipitation maximum in late summer in most of the humid subtropical regions?

5. What factors serve to make the humid subtropical climate one of the world's most productive? What are some of its handicaps?

6. How are temperature, precipitation, and geographic distribution of marine west coast regions linked to the controlling factors for this climate?

7. The marine west coast climate has long had its supporters and critics. Give reasons why you would or would not wish to live in such a climatic region.

8. Explain why the microthermal climates are limited to the Northern Hemisphere.

9. List several features that all humid microthermal climates have in common. How do these features differ from those displayed by the humid mesothermal climates?

10. Why is weather in the humid continental, hot summer climate so variable?

11. Describe the relationship between vegetation and climate in the humid continental, mild summer regions.

12. What are the major differences between the humid continental, hot summer and mild summer climates? Contrast the human use of regions occupied by the two climates.

13. How have past climatic changes helped to bring about differences in the configuration of the land between some hot-summer and mild-summer regions?

14. Refer to Figure 10.24. Using the climographs for Eagle, Alaska, and Verkhoyansk, Russia, describe the temperature patterns of the subarctic regions.

15. What factors limit precipitation in the subarctic regions?

16. How does permafrost affect human activity in both the subarctic and tundra regions?

17. Identify and compare the controlling factors of the tundra and ice-sheet regions. How do these controlling factors affect the distribution of these climates?

18. What kind of plant and animal life can survive in the polar climates? What special adaptations must this life make to the harsh conditions of these regions?

19. How do elevation, exposure, and slope aspect affect the microclimates of highland regions? What are the major climatic differences between highland regions and nearby lowlands?

20. Describe and compare the vertical zonation of highland climates in the tropics and in the middle latitudes. How are the climatic zones in tropical highlands related to agriculture?

21. How does human use of highland regions in the middle latitudes differ from that in the tropics? What special human adaptations have aided utilization of these regions?

22. Do each of the following for these climates: Mediterranean; humid subtropical; marine west coast; humid continental, hot summer; humid continental, mild summer; subarctic; tundra; ice-sheet.

   a. Identify the climate from a set of data or a climograph indicating average monthly temperature and precipitation for a representative station within a region of that climate.

   b. Match the climatic type with a written statement that includes one or more of the following: the statistical parameters of the climate in the modified Köppen classification; the particular climatic controls (controlling factors) that produce the climate; the geographic distribution of the climate as stated in terms of physical or political location; the unique climatic characteristics or combination of characteristics that distinguishes the climate from others; types of plants, animals, and soils associated with the climate; and the human utilization typical of the climate.

   c. Distinguish between the important subtypes (if any) of each climate by identifying the characteristics that separate them from one another.

CONSIDER & RESPOND

1. Based on the classification scheme presented in Appendix C, classify the following climatic stations from the data provided.

		J	F	M	A	M	J	J	A	S	O	N	D	Yr.
a.	Temp. (°C)	−42	−47	−40	−31	−20	15	−11	−18	−22	−36	−43	−39	−30
	Precip. (cm)	0.3	0.3	0.5	0.3	0.5	0.8	2.0	1.8	0.8	0.3	0.8	0.5	8.6
b.	Temp. (°C)	3	3	5	8	10	13	15	14	13	10	7	5	9
	Precip. (cm)	4.8	3.6	3.3	3.3	4.8	4.6	8.9	9.1	4.8	5.1	6.1	7.4	65.8
c.	Temp. (°C)	23	23	22	19	16	14	13	13	14	17	19	22	18
	Precip. (cm)	0.8	1.0	2.0	4.3	13.0	18.0	17.0	14.5	8.6	5.6	2.0	1.3	88.1
d.	Temp. (°C)	−27	−28	−26	−18	−8	1	4	3	−1	−8	−18	−24	12
	Precip. (cm)	0.5	0.5	0.3	0.3	0.3	1.0	2.0	2.3	1.5	1.3	0.5	0.5	10.9
e.	Temp. (°C)	−4	−2	5	14	20	24	26	25	20	13	3	−2	12
	Precip. (cm)	0.5	0.5	0.8	1.8	3.6	7.9	24.4	14.2	5.8	1.5	1.0	0.3	62.2
f.	Temp. (°C)	9	9	9	10	12	13	14	14	14	12	11	9	11
	Precip. (cm)	17.0	14.8	13.3	6.8	5.5	1.9	0.3	0.3	1.6	8.1	11.7	17	97.6
g.	Temp. (°C)	−3	−2	2	9	16	21	24	23	19	13	4	−2	11
	Precip. (cm)	4.8	4.1	6.9	7.6	9.4	10.4	8.6	8.1	6.9	7.1	5.6	4.8	84.8
h.	Temp. (°C)	0	0	4	9	16	21	24	23	20	14	8	2	12
	Precip. (cm)	8.1	7.4	10.7	8.9	9.4	8.6	10.2	12.7	10.7	8.1	8.9	8.1	111.5

2. The data in the previous table represent the following eight locations, although not in this order: Beijing, China; Point Barrow, Alaska; Chicago, Illinois; Eismitte, Greenland; Eureka, California; Edinburgh, Scotland; New York, New York; Perth, Australia. Use an atlas and your knowledge of climates to match the climatic data with the locations.

3. Eureka, Chicago, and New York are located within a few degrees latitude of one another, yet they represent three distinctly different climatic types. Discuss these differences, and identify the primary cause, or source, of the differences.

4. The precipitation recorded at Albuquerque, New Mexico (see Consider and Respond, Chapter 9), is almost twice that recorded at Point Barrow, Alaska, yet Albuquerque is considered a dry climate and Point Barrow a humid climate. Why?

5. What differentiates the *Dw* climate from the other *D* climates? Why is the *Dw* climate type only found in Asia?

6. *Csa* climatic regions and *Cfa* climatic regions are both under the influence of subtropical high pressure cells during the summer, yet *Csa* climates are dry during the summer and *Cfa* climates are wet. Why?

BIOGEOGRAPHY

Plants, animals, and the environments in which they live are interdependent, each affecting the others.

In what ways are plants as a group and animals as a group mutually dependent on one another? In what ways do humans have a much greater impact on ecosystems than all other life forms?

As the basic producers, autotrophs are generally considered to be the most important component of an ecosystem.

What are the other components? In what ways are the autotrophs affected by the other components?

The large majority of Earth's net primary productivity is from terrestrial ecosystems, and humans use less than 1 percent for plant food.

What does this suggest concerning the oceans as a source of food?

Could humans alter these figures if they changed their eating habits?

The most common species in ecosystems are generalists.

Why is this so?

What effect does a species' eating habits have on its ecological niche?

How can the ecological niche of a generalist change from one ecosystem to another?

Plant and animal distributions present a mosaic on the landscape, with the most broadly distributed plant associations typically comprising the matrix of the landscape.

Why isn't a matrix based on animal distributions? What are patches and corridors within the matrix? How do successional processes relate to the concept of patches within a mosaic?

Although other environmental controls may be more important on a local scale, climate has the greatest influence over ecosystems on a worldwide basis.

What is the evidence that this is the case?

Which climatic factors have the greatest effect on plants and animals?

What environmental controls other than climate are important?

How does climate influence the shape and size of animals and their appendages?

Earth's major terrestrial ecosystems (biomes) are classified on the basis of the dominant vegetation types that occupy the ecosystems.

What are these vegetation types? Why not base the classification on animals? How does this statement relate to the second statement above?

Freely floating plants called phytoplankton are the most important link in the ocean food chain.

What role do they play? Phytoplankton perform what function in relation to the atmosphere? Where in the ocean are phytoplankton concentrated?

Fall season in the woodlands of New Hampshire

As was pointed out in Chapter 1, biogeographers are those physical geographers who specialize in the study of natural and human-modified environments and the ecological processes that influence each environment's nature and distribution. In company with ecologists from a host of other science disciplines, biogeographers often focus their research on *ecosystems*—communities of organisms that function together in an interdependent relationship with the environments they occupy.

In almost all respects, the study of ecosystems provides the ideal opportunity to demonstrate the multiple roles of physical geography among the sciences: geography as the spatial science, geography as a physical science, and geography as environmental science. Biogeographers examine the locations, distribution, spatial patterns, and spatial interactions of all plant and animal life.

They delineate the boundaries and study the characteristics of the ecosystems they identify and also monitor the flow of energy and material through each system. In addition, they pay particular attention to the impact of humans on each ecosystem's living and nonliving physical environment.

Animal life would not exist without plants as basic food, and most plants could not survive without some animals. Together, plants and animals must adapt to their physical environment. Humans alone have the intelligence and the capacity to alter, either carelessly or deliberately, the plant–animal–physical environment relationship.

What lasting effects will increased industrialization have on the water that animals in the twenty-first century will drink or on the atmosphere in which plants will grow? What will be the consequences for life forms if concrete and steel replace additional square kilometers of forest? What could happen to marine life if toxic wastes accumulate in the world's oceans? The understandings that come from the study of ecosystems by biogeographers can provide answers to these questions and can help humans learn to work with, and not against, nature to sustain and improve life as we know it on planet Earth today.

Organization within Ecosystems

It can be said that ecology is an old science. The great voyages of exploration that began in the fifteenth century carried colonists and adventurers to uncharted lands with exotic environments. The more scholarly observers within each group made careful note of the flora and fauna found in each new part of the world. It soon became apparent that certain plants and animals were found together and that they bore a direct relationship to the climate in which they lived. As information about various world environments became more reliable and readily available, early biologists began to study plant communities and classify vegetation types. As the relationships of animals to these plant communities were recognized, naturalists in the early twentieth century began dividing Earth's life forms into *biotic associations*. Recently, the functional relationships of plants, animals, and their physical environment have been the primary focus of attention, and the concept of the ecosystem has become widely used.

Our definition of an ecosystem is both broad and flexible. The term can be used in reference to the Earth system in its entirety (the ecosphere) or to any group of organisms occupying a given area and functioning together with their nonliving environment (Fig. 11.1). An ecosystem may be large or small, marine or terrestrial (on land), short-lived or long-lasting. It may even be an artificial ecosystem, such as a farmer's field. When a farmer plants crops, spreads fertilizer, practices weed control, and sprays insecticides, a new ecosystem is created, but this does not alter the fact that plants and animals are living together in an interdependent relationship with the soil, rainfall, temperatures, sunshine, and other factors that constitute the physical environment (Fig. 11.2).

As noted in Chapter 1, ecosystems are *open systems*. There is free movement of both energy and materials across their boundaries. They are usually so closely related to nearby ecosystems or so integrated with the larger ecosystems of which they are a part that they are not isolated in nature or readily delimited. Nevertheless, the concept of the ecosystem is a valuable model for examining the structure and function of life on Earth.

Figure 11.1 Relatively small ecosystems, such as the one associated with this pond in Richmond Park, London, England, were among the first to be studied because they have fairly distinct boundaries and less complicated relationships between the nonliving environment and the organisms that occupy it. ■ **Why are the boundaries of ecosystems almost always arbitrary and determined by the ecologist who studies them?**

R. Gabler

Figure 11.2 Hybrid seed, fertilizers, insecticides, and, on occasion, irrigation systems may be used by the farmer to insure the success of this artificial ecosystem in the Corn Belt. ■ **How does the role of humans in an artificial ecosystem differ from that in a natural one?**

Major Components

Ecosystems may be many and varied, but the typical ecosystem has four basic components. The first of these is the nonliving, or **abiotic,** part of the system. This is the physical environment in which the plants and animals of the system live. In an aquatic ecosystem (a pond, for example), the abiotic component would include such inorganic substances as calcium, mineral salts, oxygen, carbon dioxide, and water. Some of these would be dissolved in the water, but the majority would lie at the bottom as sediments—a natural reservoir of nutrients for both plants and animals. In a terrestrial ecosystem, the abiotic component provides life-supporting elements and compounds in the soil, groundwater, and atmosphere.

The second and perhaps most important component of an ecosystem consists of the basic producers, or **autotrophs** (meaning *self-nourished*). Plants, the most important autotrophs, are essential to virtually all life on Earth because they are capable of using energy from sunlight to convert water and carbon dioxide into organic molecules through the process known as photosynthesis. The sugars, fats, and proteins produced by plants through photosynthesis are the foundation for the food supply that supports other forms of life. It should be noted that some bacteria are also capable of photosynthesis and hence are classed as autotrophs along with plants. Sulfur-dependent organisms that dwell at ocean-bottom thermal vents are also classified as autotrophs.

A third component of most ecosystems consists of consumers, or **heterotrophs** (meaning *other-nourished*). These are animals that survive by eating plants or other

animals. Heterotrophs are classified on the basis of their feeding habits. **Herbivores** eat only living plant material; **carnivores** eat other animals; **omnivores** feed on both plants and animals. Animals make an essential contribution to the Earth ecosystem of which they are a part. They use oxygen in their respiration and return as an end product to the atmosphere the carbon dioxide that is required for photosynthesis by plants. They can influence soil development through their digging and trampling activities, and those activities in turn may affect local plant distributions.

We might assume that plants, animals, and a supporting environment are all that are required for a functioning ecosystem, but such is not the case. Without the fourth component of ecosystems, the decomposers, plant growth would soon come to a halt. The **decomposers,** or **detritivores,** feed on dead plant and animal material and waste products. They promote decay and return mineral nutrients to the soil and sea in a form that plants can use.

Trophic Structure

From the discussion of the autotrophs and heterotrophs, it becomes apparent that there is a definite arrangement of the major components of an ecosystem. The components form a sequence in their levels of eating: Herbivores eat plants, carnivores may eat herbivores or other carnivores, and decomposers feed on dead plants and animals and their waste products. The pattern of feeding in an ecosystem is called the **trophic structure,** and the sequence of levels in the feeding pattern is referred to as a **food chain.** The simplest food chain would include only plants and decomposers. However, the chain usually includes at least four steps—for example, grass–field mouse–owl–fungi (plants–herbivore–carnivore–decomposer). More complex food chains may include six or more levels as carnivores feed on other carnivores—for example, zooplankton eats plants, small fish eats zooplankton, larger fish eats small fish, bear eats larger fish.

Organisms within a food chain are often identified by their **trophic level,** or the number of steps they are removed from the autotrophs or plants in a food chain (Table 11.1). Plants occupy the first trophic level, herbivores the second, carnivores feeding on herbivores the third, and so forth until the last level, the decomposers, is reached. Omnivores may belong to several trophic levels because they eat both plants and animals.

In reality, linear food chains do not operate in isolation; they overlap and interact to form a feeding mosaic

Table 11.1	Trophic Structure of Ecosystems	
Ecosystem Component	**Trophic Level**	**Examples**
Autotroph	First	Trees, shrubs, grass
Heterotroph	Second	Locust, rabbit, field mouse, deer, cow, bear
	Third	Praying mantis, owl, hawk, coyote, wolf, bear
	Fourth, etc.	Bobcat, wolf, hawk, bear
Decomposer	Last	Fungi, bacteria

within an ecosystem called a **food web** (Fig. 11.3). Both food chains and food webs merit careful study because they can be used to trace the movement of food and energy from one level to another in the ecosystem.

Energy Flow

When physical geographers study ecosystems, they trace the flow of energy through the system just as they do when they study energy flow in other systems, such as streams or glaciers. And just as in other systems, the laws of thermodynamics apply to ecosystems. For example, as the first law of thermodynamics states, energy cannot be created or destroyed; it can only be changed from one form to another. Energy comes to the ecosystem in the form of sunlight, which is used by plants in photosynthesis. This energy is stored in the system in the form of organic material in plants and animals. It flows through the system along food chains and webs from one trophic level to the next. It is finally released from the system when oxygen is combined with the chemical compounds of the organic material through the process of oxidation. Fire is one form of oxidation. Respiration, which involves the combination of oxygen with chemical com-

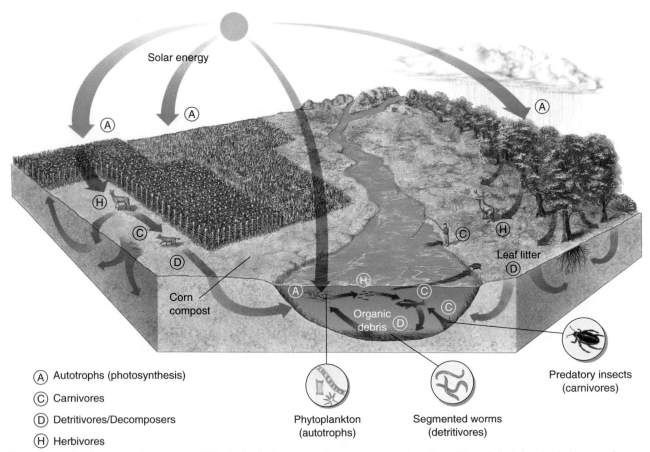

(A) Autotrophs (photosynthesis)
(C) Carnivores
(D) Detritivores/Decomposers
(H) Herbivores

Phytoplankton (autotrophs)

Segmented worms (detritivores)

Predatory insects (carnivores)

Figure 11.3   Environmental Systems: Middle-latitude Ecosystem. Ecosystems are a favorite subject of study by physical geographers for several reasons. They clearly illustrate the interdependence of the variables in systems, especially the close relationships between the living components of systems (the plants and animals, particularly humans, of the biosphere) and the nonliving or abiotic components in systems (the atmosphere, hydrosphere, and lithosphere). Plants and animals depend on the energy from the sun and the nutrients in the atmosphere, dissolved in the water, and in the soil for the creation of new organic material. Annual or seasonal changes in temperature and precipitation or alterations of the soil or water quality bring about changes in the energy and food supplies of plants and animals throughout the food webs of an ecosystem.

A major key to understanding ecosystems is revealed by an examination of energy as it flows through ecosystem food webs from one trophic level to the next. Note the movement of solar energy as it is converted to organic material through photosynthesis by the autotrophs and then moves through successively higher levels in the ecosystem food webs—from autotrophs to herbivores to carnivores. Energy is consumed and biomass decreases at each level until the organic material is recycled by the detritivores and decomposers after death.

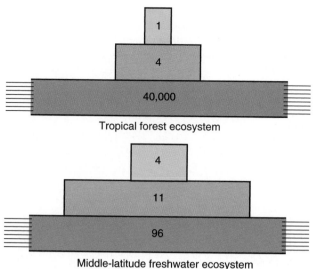

Figure 11.4 Trophic pyramids showing biomass of organisms at various trophic levels in two contrasting ecosystems. Dry weight is used to measure biomass because the proportion of water to total mass differs from one organism to another. ■ **How can you explain the exceptionally large loss of biomass between the first and second trophic levels of the tropical forest ecosystem?**

pounds in living cells, and which can occur at any trophic level, is another.

The total amount of living material in an ecosystem is referred to as the **biomass.** Because the energy of a system is stored in the biomass, scientists measure the biomass at each trophic level to trace the energy flow through the system. They usually find that the biomass decreases with each successive trophic level (Fig. 11.4). There are a number of explanations for this, each involving loss of energy. The first instance occurs between trophic levels. The second law of thermodynamics states that whenever energy is transformed from one state to another there will be a loss of energy through heat. Hence, when an organism at one trophic level feeds on an organism at another, not all of the food energy is used. Some is lost to the system. Additional energy is lost through respiration and movement. At each successive trophic level, the amount of energy required is greater. A deer may graze in a limited area, but the wolf that preys on the deer must hunt over a much larger territory. Whatever the reason for energy loss, it follows that as the flow of energy decreases with each successive trophic level, the biomass also decreases. This principle also applies to agriculture. There is a great deal more biomass (and food energy) available in a field of corn than there is in the cattle that eat the corn.

Productivity

Productivity in an ecosystem is defined as the rate at which new organic material is created at a particular trophic level. **Primary productivity** refers to the formation of new organic matter through photosynthesis by autotrophs; **secondary productivity** refers to the rate of formation of new organic material at the heterotroph level.

Primary Productivity   Just how efficient are plants at producing new organic matter through photosynthesis? The answer to this question depends on a number of variables. Photosynthesis requires sunlight, the amount of which depends on the length of day and the angle of the sun's rays, which in turn differ widely with latitude. Photosynthesis is also affected by factors such as soil moisture, temperature, the availability of mineral nutrients, the carbon dioxide content of the atmosphere, and the age and species of the individual plants.

Most studies of productivity in ecosystems have been concerned with measuring the net biomass at the autotroph level (Table 11.2). Wherever figures have been compiled on the efficiency of photosynthesis, the efficiency has been surprisingly low. Most studies indicate that less than 5 percent of the available sunlight is used to produce new biomass in ecosystems. For Earth as a whole, the figure is probably less than 1 percent. Nonetheless, the net primary productivity of the ecosphere is enormous. It is estimated to be in the range of 170 billion metric tonnes (a metric tonne is about 10 percent greater than a U.S. ton) of organic matter annually. Even though oceans cover approximately 70 percent of Earth's surface, slightly more than two-thirds of net annual productivity is from terrestrial ecosystems and less than one-third is from marine ecosystems. Perhaps even more surprising is the fact that humans consume less than 1 percent of Earth's primary productivity as plant food. However, humans also use biomass, in a variety of other ways—for example, using lumber for construction and paper production and using biomass energy for feedstocks and as fodder for range animals.

Table 11.2 illustrates the wide range of net primary productivity displayed by various ecosystems. The latitudinal control of insolation and the subsequent effect on photosynthesis can be easily recognized when comparing figures for terrestrial ecosystems. There is a noticeable decrease in terrestrial productivity from tropical ecosystems to those in middle and higher latitudes. Even the tropical savannas, which are dominated by grasses, produce more biomass in a year than the boreal forests, which are found in the colder climates. Today satellites monitor Earth's biological productivity and give us a global perspective on our biosphere (Fig. 11.5).

The reasons for differences among aquatic or water-controlled ecosystems are not quite as apparent. Swamps and marshes are especially well supplied with plant nutrients and therefore have a relatively large biomass at the first trophic level. On the other hand, depth of water has the greatest impact on ocean ecosystems. Most nutrients in the open ocean sink to the bottom, beyond the depth where sunlight can penetrate and make

Table 11.2 Net Primary Productivity of Selected Ecosystems

Type of Ecosystem	Net Primary Productivity, gm² per year	
	Normal Range	Mean
Tropical rainforest	1000–3500	2200
Tropical seasonal forest	1000–2500	1600
Middle-latitude evergreen forest	600–2500	1300
Middle-latitude deciduous forest	600–2500	1200
Boreal forest (taiga)	400–2000	800
Woodland and shrubland	250–1200	700
Savanna	200–2000	900
Middle-latitude grassland	200–1500	600
Tundra and alpine	10–400	140
Desert and semidesert scrub	10–250	90
Extreme desert, rock, sand, and ice	0–10	3
Cultivated land	100–3500	650
Swamp and marsh	800–3500	2000
Lake and stream	100–1500	250
Open ocean	2–400	125
Upwelling zones	400–1000	500
Continental shelf	200–600	360
Algal beds and reefs	500–4000	2500
Estuaries	200–3500	1500

Source: R. H. Whittaker, *Communities and Ecosystems* (2nd ed.). New York: Macmillan, 1975.

photosynthesis possible. Hence the most productive marine ecosystems are found in the sunlit, shallow waters of estuaries, continental shelves, or coral reefs, or in areas where ocean upwelling carries nutrients nearer to the surface (Fig. 11.6).

Some artificial ecosystems associated with agriculture can be fairly productive when compared with the natural ecosystems they have replaced. This is especially true in the warmer latitudes, where farmers may raise two or more crops in a year, or in arid lands where irrigation

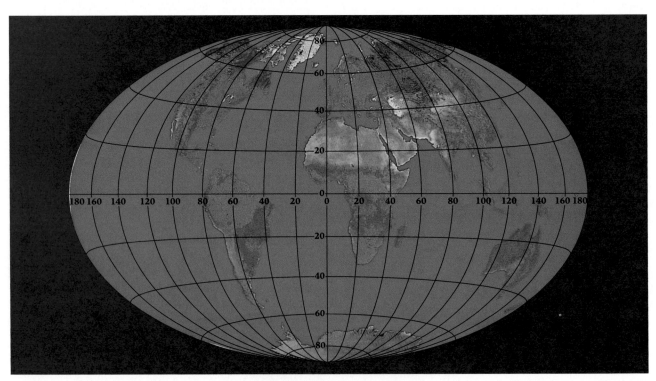

Figure 11.5 Worldwide vegetation patterns revealed through a color index derived from environmental satellite observations. Compare this image with the world map of natural vegetation in Figure 11.21.

© Bill Ross/CORBIS

Figure 11.6 The nutrient-rich waters surrounding Green Island, Great Barrier Reef Marine Park, Australia. ■ **Why are the most productive marine ecosystems found in the shallow waters bordering the world's continents?**

supplies the water essential to growth. However, Table 11.2 indicates that mean productivity for cultivated land does not approach that of forested land and is just about the same as that of middle-latitude grasslands. Most quantitative studies have shown that agricultural ecosystems are significantly less productive than natural systems in the same environment.

Secondary Productivity As we have seen, secondary productivity results from the conversion of plant materials to animal substances. We have also noted that the ecological efficiency, or the rate of energy transfer from one trophic level to the next, is low (Fig. 11.7). The efficiency of transfer from autotrophs to heterotrophs varies widely from one ecosystem to another. The amount of net primary productivity actually eaten by herbivores may range from as high as 15 percent in some grassland areas to as low as 1 or 2 percent in certain forested regions. In ocean ecosystems, the figure may be much higher, but there is a greater loss during the digestion process. Once the food is eaten, energy loss through respiration or body movement reduces secondary productivity to a small fraction of the biomass available as net primary productivity.

Most authorities consider 10 percent to be a reasonable estimate of ecological efficiency for both herbivores and carnivores. If both herbivores and carnivores have ecological efficiencies of only 10 percent, the ratio of biomass at the first trophic level to biomass of carnivores at the third trophic level is several thousand to one. It obviously requires a huge biomass at the autotroph level to support one animal that eats only meat. As human

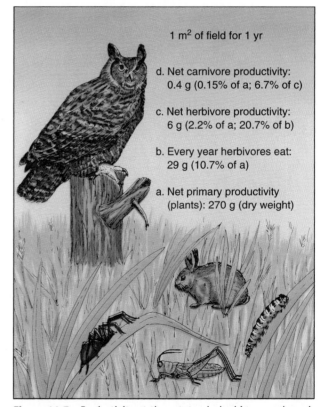

1 m² of field for 1 yr

d. Net carnivore productivity: 0.4 g (0.15% of a; 6.7% of c)

c. Net herbivore productivity: 6 g (2.2% of a; 20.7% of b)

b. Every year herbivores eat: 29 g (10.7% of a)

a. Net primary productivity (plants): 270 g (dry weight)

Figure 11.7 Productivity at the autotroph, herbivore, and carnivore trophic levels as measured in a Tennessee field. The figures represent productivity for one square meter of field in one year. Note the extremely small proportion of primary productivity that reaches the carnivore level of the food chain. ■ **In this example, which group—carnivores or herbivores—is more efficient (produces the greater percentage of energy available at the trophic level immediately below it in the food chain)?**

populations grow at increasing rates and agricultural production lags behind, it is indeed fortunate that human beings are omnivores and can adopt a more vegetarian diet (Fig. 11.8).

Ecological Niche

There are a surprising number of species in each ecosystem, except for those ecosystems severely restricted by adverse environmental conditions. Yet each organism performs a specific role in the system and lives in a certain location, described as its **habitat.** The combination of role and habitat for a particular species is referred to

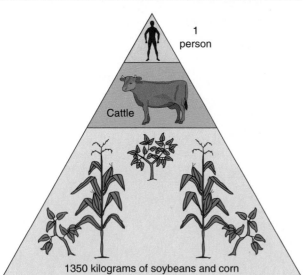

Figure 11.8 The triangles illustrate the advantages of a vegetarian diet as we contemplate another century of rapid population growth. It is fortunate that humans are omnivores and can choose to eat grain products. The same 1350 kilograms of grain that will support, if converted to meat, only one person will support 22 people if cattle or other animals are omitted from the food chain.

■ **In what areas of the world today do grain products constitute nearly all of the total food supply?**

as its **ecological niche.** A number of factors influence the ecological niche of an organism. Some species are **generalists** and can survive on a wide variety of food. The North American brown, or grizzly, bear, as an omnivore, will eat berries, honey, and fish. On the other hand, the koala of Australia is a specialist and eats only the leaves of certain eucalyptus trees. Specialists do well when their particular food supply is abundant, but they cannot adapt to changing environmental conditions. The generalists are in the majority in most ecosystems, as their broader ecological niche allows survival on alternative food supplies.

Some generalists among species occupy an ecological niche in one ecosystem that is quite different from the niche they occupy in another. As food supply varies with habitat, so varies the ecological niche. Humans are the extreme example of the generalist; in some parts of Earth they are carnivores, in some parts herbivores, and in some parts omnivores. It is also true that different species may occupy the same ecological niche in habitats that are similar but located in separate ecosystems. As an example, native horses grazed the steppes of Eurasia when bison roamed the grasslands of North America. Yet when horses were introduced to North America, they did well in the local grasslands.

Succession and Climax Communities

Up to this point, we have been discussing ecosystems in general terms. In the remainder of this chapter, we will note that it is the species that occupy the ecosystem that give the ecosystem its character. At least for terrestrial ecosystems, it is the autotrophs—the plant species at the first trophic level—that most easily distinguish one ecosystem from another. All other species in an ecosystem depend on the autotrophs for food, and the association of all living organisms determines the energy flow and the trophic structure of the ecosystem. It should also be noted that the species that occupy the first trophic level are greatly influenced by climate; again we see the interconnections between the major Earth subsystems.

If the plants that comprise the biomass of the first trophic level are allowed to develop naturally without obvious interference from or modification by humans, the resulting association of plants is called **natural vegetation.** These plant associations, or **plant communities,** are compatible because each species within the community has different requirements in relation to major environmental factors such as light, moisture, and mineral nutrients. If two species within a community were to compete, one would eventually eliminate the other. The species forming a community at any specific place and time will be an aggregation of those that together can adapt to the prevailing environmental conditions.

Succession

Natural vegetation of a particular location develops in a sequence of stages involving different plant communities. This developmental process, known as **succession,** usually begins with a relatively simple plant community. Two types of succession, *primary* and *secondary,* are recognized. In primary succession, a bare substrate is the beginning point. No soil or seedbed exists at this point. A *pioneer* community invades the bare substrate (whether it be volcanic lava, glacially deposited sediment, or a bare beach, among others) and begins to alter the environment. As a result, the species structure of the ecosystem does not remain constant. In time, the alterations of the environment become sufficient to allow a new plant community (a community that could not have survived under the original conditions) to appear and eventually to dominate the original community. The process continues, with each succeeding community rendering further changes to the environment. Because of the initial absence of soil, primary succession can take hundreds or even a few thousand years. Secondary succession begins when some natural process, such as a forest fire, tornado, or landslide, has destroyed or damaged a great deal of the existing vegetation. Ecologists refer to this process as **gap** creation. Even with such damage, however, soil still typically exists, and seeds may be lying dormant in that soil ready to invade the newly opened gap. Secondary succession, therefore, can occur much more quickly than primary succession.

A common form of secondary succession associated with agriculture in the southeastern United States is depicted in Figure 11.9. After agriculture has ceased, weeds and grasses are the first vegetative types to adapt to the somewhat adverse conditions associated with bare fields.

These low-growing plants will stabilize the topsoil, add organic matter, and in general pave the way for the development of hardwood brush such as sassafras, persimmon, and sweet gum. During the brush stage, the soil will become richer in nutrients and organic matter, and its ability to retain water will increase. These conditions encourage the development of pine forests, the next stage in this vegetative evolutionary process. Pine forests thrive in the newly created environment and will eventually dwarf and dominate the weeds, grasses, and brush that preceded them.

Ironically, the dominance of the pine forest leads to its demise. Pine trees require much sunlight if their seeds are to germinate. When competing with low-lying brush, grasses, and weedy annuals, there is no problem in getting enough sunlight, but once the pines dominate the landscape, their seeds will not germinate in the shade and litter that their dense foliage creates. Thus, pines eventually will give way to hardwoods, such as oak and hickory, whose seeds can germinate under those conditions. These seeds may have been present and dormant in the soil, or blown into the forest, or carried into it by animals. In this example, then, the end result is an oak-hickory forest. If the changes continue uninterrupted, it is estimated that the complete succession will take 100 to 200 years. In other ecosystems, such as a tropical rainforest, this complete process may take many hundreds of years.

The Climax Community

The theory of plant succession was introduced early in the twentieth century. However, some of the original ideas have undergone considerable modification. Succession was considered to be an orderly process that included

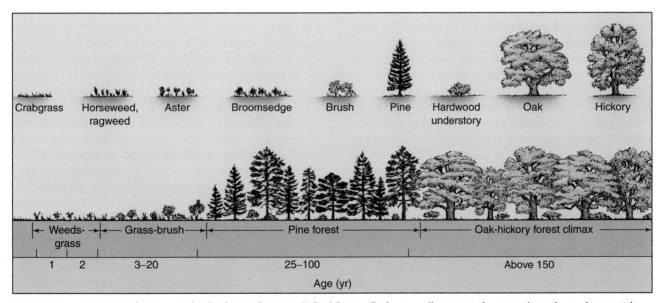

Figure 11.9 A common plant succession in the southeastern United States. Each succeeding vegetation type alters the environment in such a way that species having more stringent environmental requirements can develop. ■ **Why would plant succession be quite different in another region of the United States?**

The Theory of Island Biogeography

Biogeographers are intrigued by the life forms and the diversity of species that exist on islands—areas that are isolated from larger landmasses by ocean environments that are not inhabited by these life forms. How can there be land plants and animals living on an island surrounded by a wide expanse of sea? How did this assortment of flora and fauna become established and flourish on these often distant and geologically recent terrains (volcanic islands, for example, which were barren after they formed)? The farther an island is from the nearest landmasses, the more difficult it is for species to migrate to the island and to establish a viable population there. The seeds of some plants are carried by the wind, by birds, or by currents to islands, and germinate to develop the vegetative environments on these isolated landmasses. Many other plant and animal species living on islands were introduced to these remote locations by humans as they migrated to these islands. But why did the species adapt and survive?

The theory of island biogeography offers an explanation for how natural factors interact to affect successful colonization or extinction of species that initially come to live on an island. The theory considers the degree of isolation of the island (the distance from a mainland source of migrating species), the size of the island, and the number of species living on an island. Generally, the diversity of life forms on islands is low compared to mainland locations with a similar climate and other environmental characteristics. Low diversity of species typically also means that the floral and faunal populations of that place exist in an environmentally sensitive location. Many extinctions have occurred on islands because of the introduction of some factor that made the habitat nonviable for that species to survive.

Several major factors affect the species diversity on islands (as long as other environmental conditions such as climate are comparable):

1. The farther an island is from the area from which species must migrate, the lower the species diversity. Islands nearer to large landmasses tend to have higher diversity than those that are more distant.
2. The larger the island, the greater the species diversity. This is partly because larger islands tend to offer a wider variety of environments to colonizing organisms than smaller islands do. Larger islands also offer more space for species to occupy.
3. The species diversity of an island results from an equilibrium between the rates of extinction of species on the island and the colonization of species. If the islands' extinction rate is higher, only a few hardy species will live there; if the extinction rate is lower compared to the colonization rate, then more species will thrive, and the diversity will be higher.

The theory of island biogeography has also been useful in understanding the ecology and biota of many other kinds of isolated environments, such as high mountain areas that stand, much like islands, above surrounding deserts. In those regions, plants and animals adapted to cool wet environments live in isolation from similar populations on nearby mountains, separated by inhospitable arid environments.

---

various *predictable* steps or phases and ended with a dominant vegetative cover that would remain in balance with the environment until disturbed by human activity—or until there were major changes in the environment. The final step in the process of succession has been referred to as the **climax community.** It was generally agreed that such a community was self-perpetuating and had reached a state of equilibrium or stability with the environment. In our illustration of plant succession in the southeastern United States, the oak-hickory forest would be considered the climax community.

Succession is still a useful model in the study of ecosystems. Why, then, have some of the original ideas been challenged, and what is the most recent thought on the subject? For one thing, early proponents of succession emphasized the predictable nature of the theory. One plant community would follow another in regular order as the species structure of the ecosystem evolved. But it has been demonstrated on many occasions that changes in ecosystems do not occur in such a rigid fash-ion. More often than not, the movement of species into an area to form a new community occurs in a random fashion and may be largely a function of chance events.

Many scientists today no longer believe that only one type of climax vegetation is possible for each major climatic region of the world. Some suggest that one of several different climax communities might develop within a given area, influenced not only by climate but also by drainage conditions, nutrients, soil, or topography. The dynamic nature of climate is now also much more fully understood than it was when the theories of succession and climax were developed, so it is now seen that by the time the species structure of a plant community has adjusted to new climatic conditions, the climate may change again. Because of the dynamic nature of each habitat, no one climax community can exist in equilibrium with the environment for an indefinite period of time. Many biogeographers and ecologists today view plant communities, and the ecosystems upon which they are based, as a *landscape* that is an expression of all the

Diverse species despite serious isolation: Moorea Island in French Polynesia.

various environmental factors functioning together. They view the landscape of an area as a **mosaic** of interlocking parts, much like the tiles in a mosaic artwork. In a pine forest, for example, other plants also exist, and some areas do not support pine trees. The dominant area of the mosaic—that is, the pine forest—is referred to as the **matrix.** Gaps within the matrix, resulting from areas of different soil conditions or gaps created by natural processes, are referred to as **patches** within the matrix (Fig. 11.10). Relatively linear features cutting across the mosaic, including natural features such as rivers and human-created structures such as roads, fence lines, power lines, and hedgerows, are termed **corridors** (Fig. 11.11). Each particular habitat is unique and constantly changing, and resultant plant and animal communities must constantly adjust to these changes. The dominant environmental influence is climate.

As alluded to above, however, climates are also changing. Over both relatively short time periods, on the scales of decades and centuries, and much longer times mea-

sured in millennia, climate changes occur. They may be subtle, or they may be sufficiently drastic to create ice ages or periods of warmth between ice ages. Plant and animal communities must be able to respond to these ongoing changes, or they will not survive. Many modern biogeographers are deeply involved in the reconstruction of the vegetation communities of past climatic periods, through the examination of evidence such as tree rings, pollen, insect fossils, and plant fossils. By determining how past climate changes affected and induced change in Earth's ecosystems, biogeographers hope to be able to suggest how future changes may develop as climate continues to change.

Environmental Controls

It is apparent from the preceding paragraphs that the plants and animals occupying a particular ecosystem at a given time are those species that are most successful in adjusting to the unique environmental conditions that

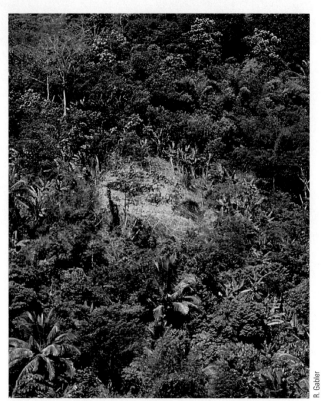

Figure 11.10 This patch in the tropical rain forest matrix of Jamaica is the result of land cleared for shifting (slash-and-burn) cultivation. ■ **What types of human activity might be responsible for patches in the matrix of a middle-latitude forest?**

constitute their habitat. Each species of living organism has a range within which it can adapt to environmental factors. For example, some plants can survive under a wide range of temperature conditions, whereas others have narrow temperature requirements. Biogeographers and ecologists refer to this characteristic as an organism's range of **tolerance** for a particular environmental condition. The ranges of tolerance for a species will determine where on Earth that species may be found, and species with wide ranges of tolerance will be the most widely distributed. The *ecological optimum* refers to the environmental conditions under which a species will flourish (Fig. 11.12). As a species moves away from its ecological optimum, or as one moves away from the geographic core of a plant or animal community, the environmental conditions become increasingly difficult for that species or community to survive. At the same time, those conditions may be more amenable for another species or community. The **ecotone** is the overlap, or zone of transition, between two plant or animal communities (see Fig. 11.12).

Climate has the greatest influence over natural vegetation when we observe plant communities on a worldwide basis. The major types of terrestrial ecosystems, or **biomes,** are each associated with specific ranges of temperature and critical precipitation characteristics, such as

annual amounts and seasonal distribution. Climate influences leaf shape and size in trees and determines if trees will exist in a region. At the local scale, however, other environmental factors can be as important as climate. A plant's range of tolerance for the acidity of the soil, the drainage of the land, or the salinity of the water may be the critical environmental factor in determining whether that plant is a part of the ecosystem. The discussion that follows serves to illustrate how the major environmental factors influence the organization and structure of ecosystems.

Climatic Factors

Of all the various climatic factors that influence the ecosystem, sunlight conditions are often the most critical. Sunlight is the vital source of energy for photosynthesis in plants and a control of life patterns for animals. The competition for light makes forest trees grow taller and limits growth on the forest floor to plants, such as ferns, that can tolerate shade conditions. Leaf sizes and shapes may reflect this variation in light reception, with

Figure 11.11 This aerial view of Taskinas Creek, Virginia shows riparian corridors passing through a forest matrix on the Atlantic coastal plain. ■ **How does a corridor differ from a patch?**

Figure 11.12 The concepts of *ecotone, ecological optimum, range of tolerance, mosaic, matrix, patch,* and *corridor* are illustrated in the diagram. Two ecosystems exist adjacent to each other, a pine forest on the left and a grassland on the right. Between the two is an area of overlap, where grasses and pine trees intermingle. This overlap is the *ecotone*. Toward the left of the overlap, the area would be more pine-forest-like, but with elements of grassland as well; on the right of the overlap, the reverse would occur. The forest landscape achieves its *ecological optimum* some distance away from the ecotone. Within the ecotone, the pine forest (or the grassland) is reaching the outer geographic and ecological limits of its *range of tolerance*. The pine forest landscape is a *mosaic* of a number of landscape elements: The pine forest itself, which is the dominant landscape element in the area, comprises the landscape *matrix*. Smaller *patches* of non–pine forest, created by both natural and human forces, differ from the surrounding matrix. Examples include a beaver pond, a forest clear-cut, and a burned area. A *corridor*, a relatively straight-line disruption to the matrix, is created by a power-line right-of-way running through the forest. In the adjacent grassland matrix to the right of the ecotone, a *riparian corridor*, a specialized form of corridor associated with river channels, cuts through the grassland. ■ **What effect would a change to drier climatic conditions throughout the area have on the relative sizes of the two ecosystems, as well as the position of the ecotone?**

large leaves in areas of limited light reception. The *quality* of light is important, especially in mountain areas, where plant growth may be severely retarded by excess ultraviolet radiation. This radiation does significant damage in the thin air at higher elevations but is effectively screened out by the denser atmosphere at lower elevations. Light *intensity* affects the rate of photosynthesis and hence the rate of primary productivity in an ecosystem. The more intense light of the low latitudes produces a higher energy input and greater biomass in the tropical forest than does the less intense light of the higher latitudes in Arctic regions. The *duration* of daylight, in association with the changing seasons, has a profound effect on the flowering of plants, the activity patterns of insects, and the migration and mating habits of animals.

Introduction and Diffusion of an Exotic Species: Fire Ants

Exotic species are plants or animals that have been introduced to a new environment from where they originated, usually by human activities. Some exotics are regarded as beneficial; many landscaping plants used in North America were brought in from other continents or regions and planted as decorative foliage. Problems have occurred, however, with many other exotics, introduced either purposefully or inadvertently, that have adapted to their new environment to the detriment of the native plants or animals. If the exotic is unable to adapt to its new home, then it dies off and will not be a problem. Many exotics, however, thrive at the expense of native populations. One example of a destructive and harmful exotic is the imported red fire ant, which causes a painful bite and kills off many native insect and other populations (often beneficial insects) in the areas they invade. Fire ants came to the United States in the 1930s from Brazil, as "hitchhikers" on a ship that docked at Mobile, Alabama. Since that time, the geographic distribution of fire ant colonies has been closely documented as they have spread throughout the southern states from their point of introduction.

A map of the expansion of fire ants outward from this location illustrates the rapid impact that the introduction of a species to a new environment can have. This spread of fire ants is an example of spatial diffusion—the expansion of a distribution over an area. Climatic factors will influence or limit the eventual distribution of fire ant populations in the United States, but as of now they are still expanding their geographic range. Cold to the north and aridity to the west are the limiting environmental factors, but these ants also have been discovered in Southern California and in settled areas of the arid West, where lawn or agricultural irrigation provides adequate moisture for their survival. The fire ants may have been spread to these locations by potted nursery plants or turf grasses shipped in from areas already infested with this pest.

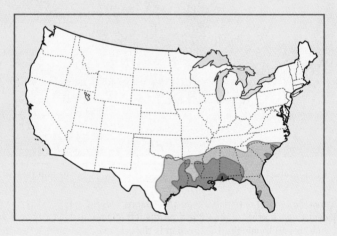

Invasion of Fire Ants 1931–1998

A second important climatic control of ecosystems is the availability of water. Virtually all organisms require water to survive. Plants require water for germination, growth, and reproduction, and most plant nutrients are dissolved in soil water before they are absorbed by plants. Marine plants are adapted to living completely in water (seaweed); some plants, such as mangrove (Fig. 11.13) and bald cypress (Fig. 11.14), rise from coastal marshes and inland swamps; others thrive in the constantly wet rainforest. Certain tropical plants drop their leaves and become dormant during dry seasons; others store water received during periods of rain in order to survive seasons of drought. Desert plants, such as cacti, are especially adapted to obtaining and storing water when it is available while minimizing their water loss from transpiration.

Animals, too, are severely restricted when water is in short supply. In arid regions, animals must make special adaptations to environmental conditions. Many become inactive during the hottest and driest seasons, and most leave their burrows or the shade of plants and rocks only at night. Others, like the camel, can travel for great distances and live for extended periods without a water supply.

Organisms are affected less by temperature variations than by sunlight and water availability. Many plants can tolerate a wide range of temperatures, although each species has optimum conditions for germination, growth, and reproduction. These functions can be impeded by excessively high and excessively low temperatures. Temperatures may also have indirect effects on vegetation. For example, high temperatures will lower the relative

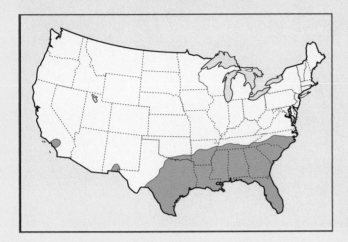

Invasion of Fire Ants 2001. Note the small incursions in the arid southern regions of California and New Mexico.

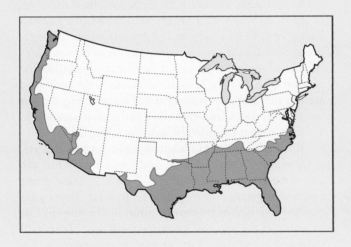

Possible Future Fire Ant Infestation Area. Suburban and agricultural irrigation provides adequate moisture for the potential spread of this invasive exotic species. However, winter temperature colder than 10°F is a limiting factor.

humidity, thus increasing transpiration. If a plant's root system cannot extract enough moisture from the soil to meet this increase in transpiration, the plant will wilt and eventually die.

Because of their mobility, animals are not as dependent on the vagaries of climate as are plants. In spite of the great advantage afforded by mobility, however, animals are nevertheless subject to climatic stress. The geographic distribution of some groups of animals reflects this sensitivity to climate. Cold-blooded animals are, for example, more widespread in warmer climates and more restricted in colder climates. Some warm-blooded animals develop a layer of fat or fur and are able to shiver to protect themselves against the cold. In hot periods, they may sweat or lick their fur in an attempt to stay cool. In

extremely cold or arid regions, animals may hibernate. During hibernation, the body temperature of the animal changes roughly in response to outside and ground temperatures. Cold-blooded animals such as the desert rattlesnake move in and out of shade in response to temperature change. Warm-blooded animals may migrate great distances out of environmentally harsh areas.

Some warm-blooded animals exhibit an interesting linkage between body shape and size and variations in average environmental temperature. These adaptations have been described by biologists as *Bergmann's Rule* and *Allen's Rule*. Bergmann's Rule states that, within a warm-blooded species, the body size of the subspecies usually increases with the decreasing mean temperature of its habitat; Allen's Rule notes that, in warm-blooded

Figure 11.13 Mangrove thicket along the Gulf of Mexico coast of southern Florida. ■ **How did this vegetation influence the routes that were followed by the Spanish adventurers who first explored Florida?**

species, the relative size of exposed portions of the body decreases with the decrease of mean temperature. These rules essentially boil down to the fact that members of the same species living in colder climates eventually evolve shorter or smaller appendages (ears, noses, arms, legs, and so on) than their relatives in warmer climates (Allen's Rule), and that in cold climates, body size will

be larger, with more mass to provide the body heat needed for survival and for protection of the main trunk of the body where vital organs are located (Bergmann's Rule). In cold climates, small appendages are advantageous because they reduce the amount of exposed area subject to temperature loss, frostbite, and cellular disruption (Fig. 11.15). In warm climates, large body sizes are not necessary for protection of internal organs, but long limbs, noses, and ears allow for heat dissipation in addition to that provided by panting or fur-licking.

Although most significant in areas such as deserts, polar regions, coastal zones, and highlands, wind can also serve as a climatic control. Wind may cause direct injury to vegetation or may have an indirect effect by increasing the rate of evapotranspiration. To prevent loss of water in the areas of severe wind stress, plants will twist and grow close to the ground to minimize the degree of their exposure (Fig. 11.16). During severe winters, they are better off buried by snow than exposed to icy gales. In some coastal regions, the shoreline may be devoid of trees or other tall plants; where trees do grow, they are often misshapen or swept bare of leaves and branches on their windward sides (Fig. 11.17).

Soil and Topography

In terrestrial ecosystems, the soils in which plants grow supply much of the moisture and minerals that are transformed into plant tissues. Soil variations are among the most conspicuous influences on plant distribution and often produce sharp boundaries in vegetation type. This is partly a consequence of the varying chemical requirements of different plant species and partly a reflection of other factors, such as soil texture. In a particular area,

Figure 11.14 This stand of bald cypress trees is located in extreme southern Illinois, at the poleward limit of growth for this type of vegetation. ■ **What Köppen climatic type does this site represent?**

Figure 11.15 The Arctic polar bear provides an excellent example of both Allen's and Bergmann's Rules. ■ **What warm climate animal might you suggest as a prime example of Allen's Rule?**

clay soil may retain too much moisture for certain plants, whereas sandy soil retains too little. It is well known that pines thrive in sandy soils, grasses in clays, cranberries in acid soils, and wheat and chili peppers in alkaline soils. The subject of soils and their influence on vegetation will be explored more thoroughly in Chapter 12.

In the discussion of highland climatic regions in Chapter 10, we learned that topography influences ecosystems indirectly by providing many microclimates within a relatively small area. Plant communities vary significantly from place to place in highland areas in response to the differing nature of the climatic conditions. Some plants thrive on the sunny south-facing slopes of highland areas in the Northern Hemisphere; others survive on the colder, shaded, north-facing slopes. The steepness and shape of a slope also affect the amount of time water is present there before draining downslope.

Luxuriant forests tower above the well-watered windward sides of mountain ranges such as the Sierra Nevada and Cascades, while semiarid grasslands and sparse forests cover the leeward sides. Spatial variations in precipitation drainage and resulting vegetation differences would not exist were it not for the presence of the topographic barrier inducing orographic uplift. Each major increase in elevation also produces a different mixture of plant species that can tolerate the lower ranges of temperature found at the higher elevation.

Natural Catastrophes

The distributions of plants and animals are also affected by a diversity of natural processes frequently termed *catastrophes*. It should be noted, however, that this term is applied from a strictly human perspective. What may be catastrophic to a human, such as a hurricane, tornado, forest fire, landslide, or avalanche (Fig. 11.18), is

Figure 11.16 Krummholz vegetation at the upper reaches of the subalpine zone on Pennsylvania Mountain in the Mosquito Range of the Colorado Rockies. The healthy green vegetation has been covered by snow much of the year and has been protected from the bitterly cold temperatures associated with gale-force winter winds. Note the flag trees, which give a clear indication of wind direction. ■ **What type of vegetation would be found at elevations higher than the one depicted in this photograph?**

R.Gabler

R.Gabler

Figure 11.17 Windswept trees along the California coast near Carmel. Strong winds from the Pacific Ocean have bowed trees inland and polished limbs on their exposed windward sides. ■ **The photographer was facing south and a little east; what is the prevailing wind direction in the area?**

simply a natural process operating to produce openings (gaps) in the prevailing vegetative mosaic of a region. The resulting successional processes, whether primary or secondary in nature, produce a variety of patch habitats within the broader regional matrix of vegetation. The study of natural catastrophes and the resulting patch dynamics among the plant and animal residents of an area is a topic of strong interest and ongoing research in modern landscape biogeography.

Biotic Factors

Although they might tend to be overlooked as environmental controls, other plants and animals may be the critical factors in determining whether a given organism is a part of an ecosystem. Some interactions between organisms may be beneficial to both species involved, whereas others may have an adverse effect on one or both. Because most ecosystems are suitable to a wide variety of plants and animals, there is always competition between species and among members of a given species to determine which organisms will survive. The greatest competition occurs between species that occupy the same ecological niche, especially during the earliest stages of life cycles when organisms are most vulnerable. Among plants, the greatest competition is for light. Those trees that become dominant in the forest are those that grow the tallest and partially shade the plants growing beneath them. Other competition occurs underground, where the roots compete for soil water and plant nutrients.

Interactions between plants and animals and competition both within and among animal species also have significant effects on the nature of an ecosystem. Animals are often helpful to plants during pollination or the

dispersal of seeds, and plants are the basic food supply for many animals. The simple act of grazing may help to determine the species that make up a plant community. During dry periods, herbivores may be forced to graze an area more closely than usual, with the result that the taller plants are quickly grazed out. Plants that grow close to the ground, that are nonpalatable, or that have the strongest root development are the ones that survive. Hence, grazing is a part of the natural selection process, but serious overgrazing rarely results under natural conditions because wild animal populations increase or decrease with the available food supply. To be more precise, the number of animals of a given species will fluctuate between the maximum number that can be supported when its food supply is greatest and the minimum number required for reproduction of the species. For most animals, predators are also an important control of numbers. Fortunately, when the predator's favorite species is scarce, it will seek an alternative species for its food supply, and it would be rare for an animal to be excluded from an ecosystem through predation.

Human Impact on Ecosystems

Throughout human history, we have modified the natural development of ecosystems. Except in regions too remote to be altered significantly by civilization, humans have eliminated much of Earth's natural vegetation. Farm-

David Butler

Figure 11.18 The vegetation mosaic of this area is coniferous forest, but frequent snow avalanches passing down the gully keep rigid-stemmed conifers from invading the patch of open, low shrubs and grasses. The effects of attempted colonization are seen at right, where a swath of conifers has been destroyed by a recent snow avalanche. ■ **What other environmental factors might account for the limited amount of coniferous vegetation on the left, south-facing slope?**

ing, fire, grazing of domesticated animals, afforestation and deforestation, road building, urban development, dam building and irrigation, raising and lowering of water tables, mining, and the filling in or draining of wetlands are just a few ways in which humans have modified the plant communities around them. Overgrazing by domesticated animals can seriously harm marginal environments in semiarid climates. Trampling and compaction of the soil by grazing herbivores may reduce the soil's ability to absorb moisture, leading to increased surface runoff of precipitation. In turn, the decreased absorption and increased runoff may respectively lead to *land degradation* and gully erosion.

It should be noted that ecosystems are not the only victims as humans alter natural environments; the changes can often produce long-term negative effects on humans themselves. The desertification of large tracts of semiarid portions of east Africa has resulted periodically in widespread famine in countries such as Ethiopia and Somalia (Fig. 11.19). Elsewhere, the continuing destruction of wetlands not only eliminates valuable plant and animal communities, it often seriously threatens the quality and reliability of the water supplies for the people who drained the land.

We have, in fact, so changed the vegetation in some parts of the world that we can characterize classes of cultivated vegetation cared for by humans—for example, flowers, shrubs, and grasses to decorate our living areas, and grains, vegetables, and fruits that we raise for our own food and to feed the animals we eat. Our focus in the remainder of this chapter, however, will be on the major ecosystems of Earth as we assume they would appear without human modification.

Classification of Terrestrial Ecosystems

The geographic classification of natural vegetation is as difficult as the classification of any other complex phenomena influenced by a variety of factors. However, plant communities are among the most highly visible of natural phenomena, so they can be categorized on the basis of form and structure or gross physical characteristics. Of course, the composition of the natural vegetation changes from place to place in a gradational manner, just as temperature and rainfall do, and although distinctly different types are apparent, there may be broad transition zones (ecotones) between them. Nevertheless, over the world there are distinctive recurring plant communities, indicating a consistent botanical response to systematic con-

trols that are essentially climatic. It is the dominant vegetation of these plant communities that we recognize when we classify Earth's major terrestrial ecosystems (biomes).

All of Earth's biomes can be categorized into one of four easily recognized types: forest, grassland, desert, and tundra (Fig. 11.20). Because vegetation adapted to cold climates may occur not only in high-latitude regions but also at high elevations at any latitude, biogeographers often refer to the last of the major types as arctic and alpine tundra. However, the forests of the equatorial lowlands are an entirely different world from those of Siberia or of New England, and the original grasslands of Kansas bore little resemblance to those of the Sudan or Kenya. Hence, the four major types of ecosystems can be subdivided into distinctive biomes, each of which is an association of plants and animals of many different species. Pure stands of particular trees, shrubs, or even grasses are extremely rare and are limited to small areas having peculiar soil or drainage conditions.

Earth's major biomes are mapped in Figure 11.21 on the basis of the dominant associations of natural vegetation that give each its distinctive character and appearance. The direct influence of climate on the distribution of these biomes is immediately apparent. Temperature (or latitudinal effect on temperature and insolation) and the availability of moisture are the key factors in determining the location of major biomes on the world scale (Fig. 11.22).

Forest Biomes

Forests are easily recognized as associations of large, woody, perennial tree species, generally several times the height of a human, and with a more or less closed

Figure 11.19 Overgrazing by cattle has helped to promote this example of desertification in the Lake Chad Basin near the Sahara. ■ **What are some of the other causes of desertification?**

© Bernard and Catherine Desjeux/CORBIS

Roxie Anderson

Courtesy Roxie Anderson

My early interests growing up in Wisconsin were camping, cross-country skiing, biking, anything that had me outdoors. I loved the change of seasons (autumn is my favorite) and the landscape of western Wisconsin. In school my interests were math and physics; I only recall having one geography class throughout my high school career. But I've always loved maps and cartography (I used topographic maps for hiking, biking, and camping).

I've been with Ducks Unlimited since graduating. I also work on campus as an adventure trip leader with the outdoor recreation department.

I have a variety of interests, and that's one of the reasons I chose to become a geographer. There are so many different routes you can take, and so many different experiences you can gain. I was one of those people who didn't know what I wanted to do for a long time because I wanted to do everything! When I finally found geography, I knew that this field would allow me to expand my horizons and become a jack of all trades, so to speak.

Where did you get your education?
University of Wisconsin, Eau Claire

Why did you decide to major in geography?
Because geography, whether it be land use planning, physical, social, or urban, allows you to choose from a diverse scope of opportunities. Geography can be used to tie together and integrate the environmental sciences and can offer a larger-scale view or "big picture" of the effects of human/environment interaction.

Describe what you now do in your career field. Describe what your company does. How do you, as a physical geographer, fit in with what your company's mission?
The mission of Ducks Unlimited is to conserve, restore, and manage wetlands and associated habitats for North America's waterfowl. These habitats also benefit other wildlife and people.

As a physical geographer, I use tools such as remote sensing and GIS to do earth cover mapping using Landsat satellite imagery. These maps help identify important habitats that need to be conserved, restored, or managed. To date, Ducks Unlimited has mapped about 162 million acres (almost 50% of Alaska); these areas can now be managed more effectively by determining areas of disturbances or important wildlife habitat that needs to be monitored. Ducks Unlimited, as a cooperator in this effort, makes sure that the breeding and nesting habitat for the Pacific flyway waterfowl can now be monitored and managed properly. Recently we have carried these efforts over to the Western Boreal Forests of Canada.

On a smaller scale, Ducks Unlimited works with government agencies and local agencies to restore/manage wetlands, improve water quality, and monitor waterfowl populations. For these efforts, as a physical geographer, I provide GIS support for the biologists and engineers. GIS support often includes creating project area maps, managing and collecting GIS data, and using GIS suitability analysis models.

What did you learn in your physical geography education that is now important to you in your career?

Learning the geomorphology of different landscapes has been the most important lesson for me. My earth cover mapping depends heavily on what I know about the landscape and how I interpret it. Alaska's and Canada's landscapes are heavily influenced by glaciation. So I guess my most important lesson was glacial landforms.

How do you apply your knowledge of geography and geographic concepts to your work?
I think geographers have a broader sense of the environment. I, as a geographer, am able to work within different disciplines and with different professionals to get a project done.

What academic preparation would your recommend to students who are interested in a career similar to yours?
Take as much physical geography as possible! At UWEC I took a remote sensing course, GIS courses, soils, surveying, and GPS/cartography courses. But if I could go back, I would have taken an additional soils class and switched my math minor to a biology or ecology minor.

Also get outside, observe the landscape around you, and try to interpret what the geologic/geomorphic history of the area was to make it look the way it does today. Field trips are an excellent way to get outside and see real-life examples of geomorphology and also have an experienced physical geographer there to explain the landscape to you.

Do you have some interesting examples about how you've used your training in physical geography in your career?
In Alaska, I need to understand the physical geography and landscape to be able to interpret the satellite imagery, in order to do an earth cover classification. So I've used my training in glaciation to help me interpret the different ecological regions of an area. Also important is understanding vegetation and plant distribution within these ecological regions.

Why do you think that physical geography is an important course for those who are majoring in other fields?
Physical geography, especially GIS, is becoming essential throughout all agencies and disciplines. In my four-year experience, I've worked jointly with biologists, archaeologists, agriculture specialists, water specialists, geologists, and engineers.

Figure 11.20 The four major types of Earth biomes: (a) forest biome near Dornbirn, Australia; (b) grassland biome of the Florida Everglades; (c) desert biome at the base of Red Rock Canyon escarpment, Nevada; (d) tundra biome in northern Alaska.

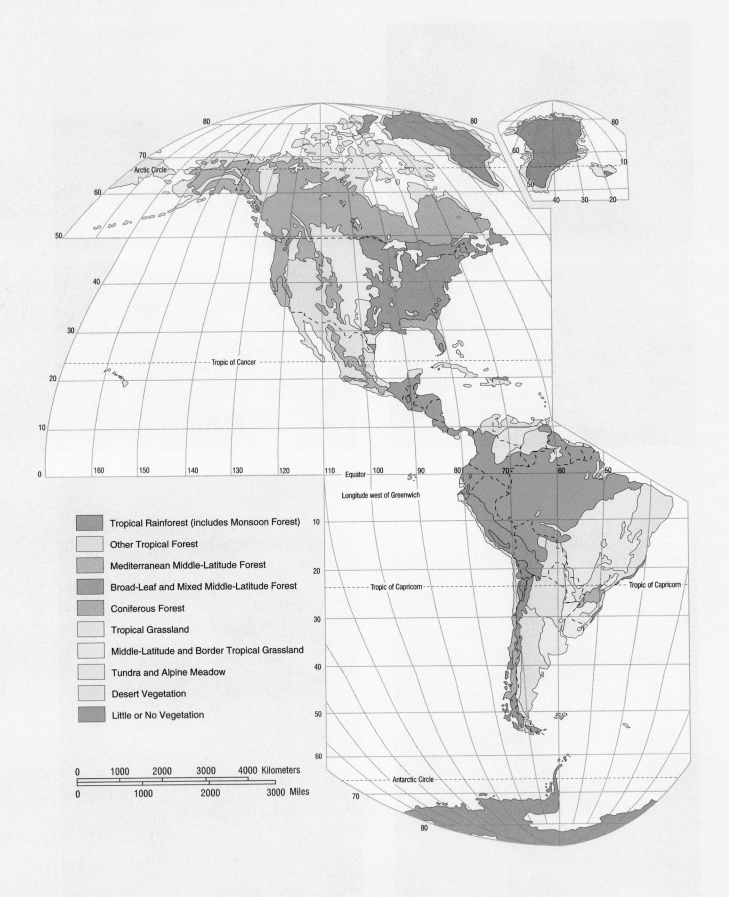

Figure 11.21 World map of natural vegetation

A Western Paragraphic Projection
developed at Western Illinois University

canopy of leaves overhead. They vary enormously in density and physical appearance. Some are evergreen and either needle- or broad-leaf; others are deciduous, dropping their leaves to reduce moisture losses during dry seasons or when soil water is frozen. Forests are found only where the annual moisture balance is positive—where moisture availability considerably exceeds potential evapotranspiration in the growing season. Thus, they occur in the tropics, where either the ITCZ or the monsoonal circulation brings plentiful rainfall, and in the middle latitudes, where precipitation is associated with cyclonic depressions along the polar front, with summer convectional rainfall, or with orographic uplift.

Tropical and middle-latitude forests have evolved different characteristics in response to the nature of the physical limitations in each area. In general, tropical forests have developed in less restrictive forest environments. Temperatures are always high, though not extreme, in the humid tropics, encouraging rapid and luxuriant growth. Middle-latitude forests, on the other hand, must adapt to combat either seasonal cold (ranging from occasional frosts to subzero temperatures) or seasonal drought (which may occur at the worst possible time for vegetative processes).

Tropical Forests

The forests of the tropics are far from uniform in appearance and composition. They grade poleward from the equatorial rainforests, which support Earth's greatest biomass, to the last scattering of low trees that overlook seemingly endless expanses of tall grass or desert shrubs on the tropical margins. We have subdivided the tropical forests into three distinct biomes: the tropical rainforest, the monsoon rainforest, and other tropical forest types, primarily thornbush and scrub. Of course, there are gradations (ecotones) between the different types, as well as distinctive variations that are found in individual localities only.

Figure 11.22 Schematic diagram showing distribution of Earth's major biomes as they are related to temperature (latitude) and the availability of moisture. Within the tropics and middle latitudes, there are distinctly different biomes as total biomass decreases with decreasing precipitation. ■ **What major biome dominates the wetter margins of all latitudes but the Arctic?**

Tropical Rainforest In the equatorial lowlands dominated by Köppen's tropical rainforest climate, the only physical limitation for vegetation growth is competition between adjacent species. The competition is for light. Temperatures are high enough to promote constant growth, and water is always sufficient. Thus, we find forests consisting of an amazing number of broad-leaf evergreen tree species of rather similar appearance, since special adaptations are not required (Fig. 11.23). A cross section of the forest often reveals concentrations of leaf canopies at several different levels. The trees composing the distinctive individual tiers have similar light requirements—lower than those of the higher tiers but higher than those of the lower tiers. Little or no sunlight reaches the forest floor, which may support ferns but is often rather sparsely vegetated. The forest is literally bound together by vines, **lianas,** that climb the trunks of the forest trees and intertwine in the canopy in their own search for light. Aerial plants may cover the limbs of the forest giants, deriving nutrients from the water and the plant debris that falls from higher levels. Light and variable wind conditions in the rainforest (associated with the latitudinal belt of the doldrums) preclude wind from being an effective agent of seed and pollen dispersal, so large colorful fruits and flowers, designed to attract animals who will unwittingly carry out seed and pollen dispersal, prevail.

The forest trees commonly depend on grotesquely flared or buttressed bases for support, since their root systems are shallow. This is a consequence of the richness of the surface soil and the poverty of its lower levels. The

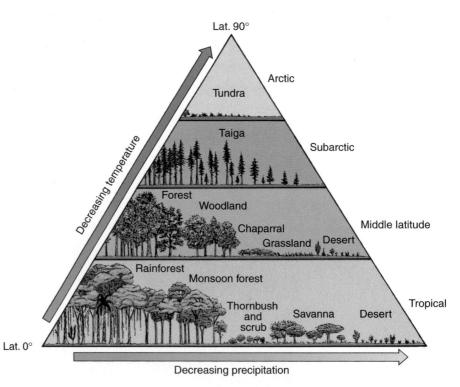

Influence of latitude and moisture on distribution of biomes

rainforest vegetation and soil are intimately associated. The forest litter is quickly decomposed, its nutrients released and almost immediately reabsorbed by the forest root systems, which consequently remain near the surface. Tropical soils that maintain the amazing biomass of the rainforest are fertile only as long as the forest remains undisturbed. Clearing the forest interrupts the crucial cycling of nutrients between the vegetation and the soil; the copious amounts of water percolating through the soil leach away its soluble constituents, leaving behind only inert iron and aluminum oxides that cannot support forest growth. The present rate of clearing threatens to wipe out the worldwide tropical rainforests within the foreseeable future. The largest remaining areas of unmodified rainforests are in the upper Amazon Basin, where they cover hundreds of thousands of square kilometers.

Because of the darkness and extensive root systems present on the forest floor, animals of the tropical rainforest are primarily arboreal. A wide variety of species of tree-dwelling monkeys and lemurs, snakes, tree frogs, birds, and insects characterize the rainforest. Even the large herbivorous and carnivorous mammals—such as sloths and jaguars, respectively—are primarily arboreal.

Within the areas of rainforest and extending outward beyond its limits along streams are patches or strips of jungle. Jungle consists of an almost impenetrable tangle of vegetation contrasting strongly with the relatively open nature (at ground level) of the true rainforest. It is often composed of secondary growth that quickly invades the rainforest where a clearing has allowed light to penetrate to the forest floor. Jungle commonly extends into the drier areas beyond the forest margins along the courses of streams. There it forms a gallery of vegetative growth closing over the watercourse and hence has been called **galeria forest.**

Monsoon Rainforest In areas of monsoonal circulation, there is an alternation between the dry monsoon season, when the dominant flow of air is from the land to the sea, and the wet monsoon season, when the atmospheric circulation reverses, bringing moist air onshore along tropical coasts. The wet monsoon season rainfall may be very high, even hundreds of centimeters where air is forced upward by topographic barriers. In any case, it is sufficient to produce a forest that, once established, remains despite the dry monsoon season. Monsoon forests may have discernible tiers of vegetation related to the varying light demands of different species, and they are included with the tropical rainforest in Figure 11.21. However, the number of species is less than in the true rainforest, and the overall height and density of vegetation are also somewhat less. Some of the species are evergreen, but many are deciduous.

Other Tropical Forests, Thornbush, and Scrub Where seasonal drought has precluded the development of true rainforest, or where soil characteristics prevent the growth of such vegetation, variant types of tropical forests have

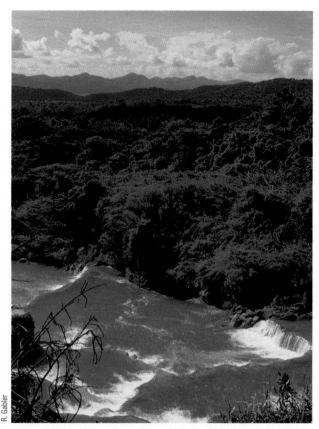

Figure 11.23 A rainforest near El Salto Falls in eastern Mexico. The dense nature of the forest canopy effectively conceals the vast number of different evergreen tree species and relatively open forest floor. ■ **How might this rainforest differ from the rainforests of the Pacific Northwest of the United States?**

developed. These tend to be found on the subtropical margins of the rainforests and on old plateau surfaces where soils are especially poor in nutrients. The vegetation included in this category varies enormously but is generally low-growing in comparison to rainforest, without any semblance of a tiered structure, and is denser at ground level. It is commonly thorny, indicating defensive adaptation against browsing animals, and it shows resistance to drought in that it is generally deciduous, dropping its leaves to conserve moisture during the dry winter season. Ordinarily, grass is present beneath the trees and shrubs. As we move away from the equatorial zone, we find the trees more widely spaced and the grassy areas becoming dominant. Along tropical coastlines, a specially adapted plant community, known as mangrove, thrives (see again Fig. 11.13). Here trees are able to grow in salt water.

Middle-Latitude Forests

The forest biomes of the middle latitudes differ from those of the tropics because the dominant trees have evolved mechanisms to withstand periods of water deprivation due to low temperatures and annual variations in precipitation. Evergreen and deciduous plants are present, equipped to cope with seasonal extremes not encountered in tropical latitudes.

Mediterranean Sclerophyllous Woodland   Surrounding the Mediterranean Sea and on the southwest coasts of the continents between approximately 30° and 40°N and S latitude, we have seen that a distinctive climate exists—Köppen's mesothermal hot and dry summer type (Mediterranean). Here annual temperature variations are moderate, and freezing temperatures are rare. However, there is little or no rainfall during the warmest months, and plants must be drought resistant. This requirement has resulted in the evolution of a distinctive vegetation that is relatively low-growing, with small, hard-surfaced leaves and roots that probe deeply for water. The leaves must be capable of photosynthesis with minimum transpiration of moisture. The general look of the vegetation is a thick scrub plant community, called chaparral in the western United States and *maquis* in the Mediterranean region (see again Fig. 10.3). Wherever moisture is concentrated in depressions or on the cooler north-facing hill slopes, deciduous and evergreen oaks occur in groves (Fig. 11.24). Drought-resistant needle-leaf trees, especially pines, are also part of the overall vegetation association. Thus, the vegetation is a mosaic related to site characteristics and microclimate. Nevertheless, the similarity of the natural vegetative cover in such widely separated areas as Spain, Turkey, and California is astonishing. (Note the location of Mediterranean middle-latitude forest, Fig. 11.21.)

Broad-Leaf Deciduous Forest   The humid regions of the middle latitudes experience a seasonal rhythm dominated by warm air in the summer and invasions of cold polar air in the winter. To avoid frost damage during the colder winters and to survive periods of total moisture deprivation when the ground is frozen, trees whose leaves have large transpiring surfaces drop these leaves and become dormant, coming to life and producing new leaves only when

the danger period is past (Fig. 11.25). A large variety of trees have evolved this mechanism; certain oaks, hickory, chestnut, beech, and maples are common examples. The seasonal rhythms produce some beautiful scenes, particularly during the periods of transition between dormancy and activity, with the sprouting of new leaves in the spring and the brilliant coloration of the fall as chemical substances draw back into the plant for winter storage.

The trees of the deciduous forest may be almost as tall as those of the tropical rainforests and, like them, produce a closed canopy of leaves overhead or, in the cold season, an interlaced network of bare branches. However, lacking a multistoried structure and having lower density as a whole, the middle-latitude deciduous forests allow much more light to reach ground level. Forests of this type are the natural vegetation in much of western Europe, eastern Asia, and eastern North America. To the north and south, they merge with mixed forests composed of broad-leaf deciduous trees and conifers. (Broad-leaf deciduous forests and mixed forests are combined in Figure 11.21.) Both the broad-leaf deciduous and mixed forests have been largely logged off or cleared for agricultural land, and the original vegetation of these regions is rarely seen.

Broad-Leaf Evergreen Forest   Beyond the tropics, broad-leaf trees remain evergreen (active throughout the year) in significant areas only in certain Southern Hemisphere locations. Here the mild maritime influence is strong enough to prevent either dangerous seasonal droughts or severely low winter temperatures. Southeastern Australia and portions of New Zealand, South Africa, and southern Chile are the principal areas of this type. In the Northern Hemisphere, broad-leaf evergreen forest may once have been significant in eastern Asia,

Figure 11.24   The distinctive sclerophyllous evergreen vegetation type encountered wherever hot, dry summers alternate with rainy winters. Oaks commonly occupy relatively damp sites, such as the gullies and windward slopes seen here in southern California.   ■   **What are the general characteristics of sclerophyllous vegetation?**

R. Gabler

R. Gabler

(a)

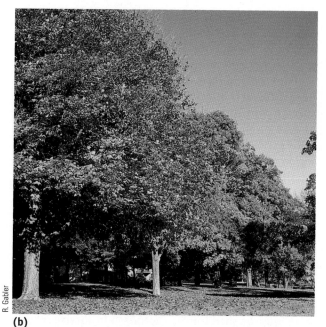

R. Gabler

(b)

Figure 11.25 The appearance of hardwood forests in middle-latitude regions with cold winters changes dramatically with the seasons. The green leaves of summer (a) change to reds, golds, and browns in fall (b) and drop to the forest floor in winter (c). Leaf dropping in areas of cold winters, such as this example in western Illinois, is a means of minimizing transpiration and moisture loss when the soil water is frozen. ■ **What length of growing season (frost-free period) is associated with the climate of western Illinois?**

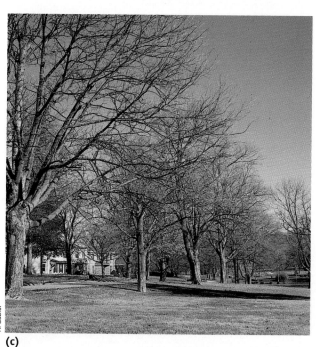

R. Gabler

(c)

but it has long since been cleared for cultivation. Limited areas occur in the United States in Florida and along the Gulf Coast as a belt of evergreen oak and magnolia.

Mixed Forest Poleward and equatorward, the broad-leaf deciduous forests in North America, Europe, and Asia gradually merge into mixed forests, including needle-leaf coniferous trees, normally pines. In general, where conditions permit the growth of broad-leaf deciduous trees, coniferous trees cannot compete successfully with them. Thus, in mixed forests, the conifers, which are actually more adaptable to soil and moisture deficiencies, are found in the less hospitable sites: in sandy areas,

on acid soils, or where the soil itself is thin. The northern mixed forests reflect the transition to colder climates with increasing latitude; eventually, conifers become dominant in this direction. The southern mixed forests are more problematic in origin. In the United States, they are transitional to pine forests situated on sandy soils of the coastal plain. In Eurasia, they coincide with highlands dominated by conifers during a stage in the plant succession that began with the change of climatic environments at the end of the ice ages, some 10,000 years ago.

Coniferous Forest The coniferous forests occupy the frontiers of tree growth. They survive where most of the broad-leaf species cannot endure the climatic severity and impoverished soils. The hard, narrow needles of coniferous species transpire much less moisture than do broad leaves, so that needle-leaf species can tolerate conditions of physiologic drought (unavailability of moisture because of excessive soil permeability, a dry season, or frozen soil water) without defoliation. Pines, in particular, also demand little from the soil in the form of soluble plant nutrients, especially basic elements such as calcium, magnesium, sodium, and potassium. Thus, they grow in sandy places and where the soil is acid in character. As a whole, conifers are particularly well adapted to regions having long, severe winters combined with summers warm enough for vigorous plant growth.

Because all but a few exceptions retain their leaves (needles) throughout the year, they are ready to begin photosynthesis as soon as temperatures permit without having to produce a new set of leaves to do the work.

Thus we find a great band of coniferous forests (the **boreal forests,** or **taiga**) dominated by spruce and fir species, with pines on sandy soils, sweeping the full breadth of North America and Eurasia northward of the 50th parallel of latitude, approximately occupying the region of Köppen's subarctic climate (see again Fig. 9.4). Conifers differ from other trees in that their seeds are not enclosed in a case or fruit but are carried naked on cones. All are needle-leaf and drought-resisting, but a few are not evergreen. Thus, a large portion of eastern Siberia is dominated by larch, which produces a deciduous, coniferous forest. In this area, January mean temperatures may be –35° to –51°C (–30° to –60°F). Here we encounter the most severe winter climate in which trees can maintain themselves, and even needle-leaf foliage must be shed for the vegetation to survive. Hardy broad-leaf deciduous birch trees share this extreme climate with the indomitable larches.

Extensive coniferous forests are not confined to high-latitude areas of short summers. Higher elevations in middle-latitude mountains of the Northern Hemisphere have forests of pine, hemlock, and fir (Fig. 11.26), with subalpine larch and specially adapted pine species characterizing the harshest and highest forest sites. The forests along the sandy coastal plain of the eastern United States are there in part because of the sandy soils, but they may also reflect a stage in plant succession that in time will lead to domination by broad-leaf types. Similarly, a temporary stage in the postglacial vegetation succession of the Great Lakes area included magnificent forests of white pine and hemlock that were completely logged off during the late nineteenth century.

A more maritime coniferous forest occupies the west coast of North America extending from southern Alaska to central California. It is made up of sequoias, Douglas fir, cedar, hemlock, and, farther north, Sitka spruce. Many of the California sequoias are thousands of years old and more than 100 meters (330 ft) high. The southern regions experience summer drought, and farther north, sandy, acidic, or coarse-textured soils dominate.

Grassland Biomes

Grasses, like conifers, appear in a variety of settings and are part of many diverse plant communities. They are, in fact, an initial form in most plant successions. However, there are enormous, continuous expanses of grasslands on Earth. In general, it is thought that grasses are

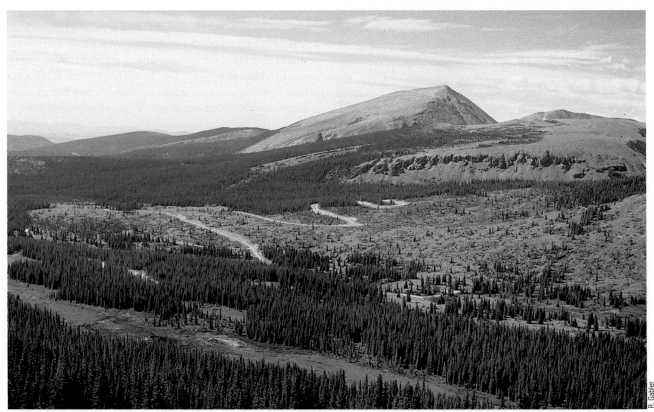

Figure 11.26 As this photograph in the Colorado Rocky Mountains indicates, evergreen coniferous forest is the characteristic vegetation of higher-elevation regions in the middle latitudes. The needle-leaf trees are an adaptation to the physiologic drought of the winter season, which is longer and more severe at higher elevations. ■ **Why are needle-leaf trees better adapted to physiologic drought than broad-leaf trees?**

Figure 11.27 The savanna biome of eastern Africa. This classic landscape of grasses and scattered trees supports dwindling herds of grazing animals as land use patterns change. ■ **What human activities have replaced much of the eastern Africa savanna?**

Donald Marshall Collection

dominant only where trees and shrubs cannot maintain themselves because of either excessive or deficient moisture in the soil. On the global scale, grassland biomes are located in continental interiors where most, if not all, of the precipitation falls in the summer. Two great geographic realms of grasslands are generally recognized: the tropical and the middle-latitude grasslands. However, it is difficult to define grasslands of either type using any specific climatic parameter, and geographers suspect that human interference with the natural vegetation has caused expansion of grasslands into forests in both the tropical and middle latitudes.

Tropical Savanna Grasslands

The tropical grassland biome differs from grassland biomes of the middle latitudes in that it ordinarily includes a scattering of trees; this is implied in the term **savanna** (Fig. 11.27). In fact, the demarcation between tropical scrub forest and savanna is seldom a clear one. The savanna grasses tend to be tall and coarse, with bare ground visible between the individual tufts. The related tree species generally are low-growing and wide-crowned forms, having both drought- and fire-resisting qualities, indicating that fires frequently sweep the savannas during the drought season. Indeed, the role of fire in maintaining savannas (suppressing tree invasions) and the role of humans in creating or expanding the savannas at the expense of the forest have been topics of continuing discussion among geographers. Savannas occur under a variety of temperature and rainfall conditions but generally fall within the limits of Köppen's tropical savanna type.

They commonly occur on red-colored soils, leached of all but iron and aluminum oxides, which become brick-like or slaggy when dried. They likewise coincide with areas in which the level of the water table (the zone below which all soil and rock pore space is saturated by water) fluctuates dramatically. The up-and-down movement of the water table may in itself inhibit forest development, and it is no doubt a factor in the peculiar chemical nature of savanna soils. Large migratory herds of grazing animals and associated predators, responding to the periodically abundant grasses followed by seasonal drought, characterized the savanna prior to widespread human disruption.

Middle-Latitude Grasslands

The zone of transition between the middle-latitude deserts and forests is occupied by the middle-latitude grasslands. On their dry margins, they pass gradually into deserts in Eurasia and are cut off westward by mountains in North America. However, on their humid side, they terminate rather abruptly against the forest margin, again raising questions as to whether their limits are natural or have been created by human activities, particularly the intentional use of fire to drive game animals. The middle-latitude grasslands of North America, like the African savannas, formerly supported enormous herds of grazing animals—in this case, antelope and bison, which were the principal means of support of the American Plains Indians.

Like the savannas, the middle-latitude grasslands were diverse in appearance. They, too, consisted of varying associations of plant species that were never uniform in composition. In North America, the grasses were as much as 3 meters (10 ft) tall in the more humid sections, as in Iowa, Indiana, and Illinois, but only 15 centimeters (6 in.) high on the dry margins from New Mexico to western Canada. Thus, the middle-latitude grassland biomes are usually divided into tall-grass and short-grass prairie, often with a zone of mixture recognized between them. Unlike growth in the tropical savannas, the germination and growth of middle-latitude grasses are attuned to the melting of winter snows, followed by summer rainfall. Whether the grasses are annuals that complete their life cycle in one growing season or perennials that grow from year to year, they are dormant in the winter season. Also, unlike in the savannas, the soils beneath these grasslands are extremely rich in organic matter and soluble nutrients. As a consequence, most of the middle-latitude grasslands have been completely transformed by agricultural activity. Their wild grasses have been replaced by domesticated varieties—wheat, corn, and barley—and they have become the "breadbaskets" of the world.

Tall-Grass Prairie   The tall-grass prairies were an impressive sight; in some better-watered areas, they made up endless seas of grass moving in the breeze, reaching higher than a horse's back. Flowering plants were

Figure 11.28 Short-grass prairie vegetation of the Nebraska Sand Hills. Although the tall-grass prairies have been completely transformed by humans, vast areas of short-grass prairie remain because of their low and unpredictable precipitation, which makes them hazardous for agriculture. ■ **What would have been obviously different if a photograph had been taken here 200 years ago?**

conspicuous, adding to the effect. Unfortunately, this **tall-grass prairie** scene, which inspired much vivid description by those first encountering it, is no longer visible anywhere. The tall-grass prairies, which once reached continuously from Alberta to Texas, have been destroyed. Their tough sod, formed by the dense grass root network, defeated the first wooden plows, which had served well enough in breaking up the forest soils. But the steel plow, invented in the 1830s, subdued the sod and was aided by the introduction of subsurface tile for draining the nearly flat uplands and by the simultaneous appearance of well-digging machinery and barbed wire. These four innovations transformed the tall-grass prairie from grazing land to cropland.

In North America, the tall-grass prairie pushed as far eastward as Lake Michigan. Why trees did not invade the prairie in this relatively humid area remains an unanswered question. Farther west, shallow-rooted grass cover is fully understandable because the lower soil levels, to which tree roots must penetrate for adequate support and sustenance, are bone dry. In such areas, trees can survive only along streams or where depressions collect water.

In Eurasia, tall-grass prairies were found on a large scale in a discontinuous belt from Hungary, through Ukraine, Russia, and central Asia, to northern China. The grasslands are known in South America as the pampas of Uruguay and Argentina and in South Africa as the Veldt. Today, all of these areas have been changed by agriculture. The factor that seems to account best for the tall-grass prairie—precipitation that is both moderate and variable in amount from year to year—is the principal hazard in the use of these regions as farmland. However, this hazard becomes much greater in the areas of short-grass prairie.

Short-Grass Prairie   West of the 100th meridian in the United States and extending across Eurasia from the Black Sea to northern China, roughly coinciding with the areas of Köppen's middle-latitude steppe climate, are vast, nearly level grasslands composed of a mixture of tall and short grass, with short grass becoming dominant in the direction of lower annual precipitation totals (Fig. 11.28). On the Great Plains between the Rocky Mountains and the tall-grass prairies, the **short-grass prairie**

Figure 11.29 Ocotillo in bloom in the Shavers Valley of the Mojave Desert, California. ■ **What drought-resistant adaptations by this ocotillo plant can be easily observed?**

zone more or less coincides with the zone in which moisture rarely penetrates more than 60 centimeters (2 ft) into the soil, so that the subsoil is permanently dry. Moving toward the drier areas, the grassland vegetation association dwindles in diversity and, more conspicuously, in height to less than 30 centimeters (1 ft). This is a consequence of reduction in numbers of tall-growing species and greater abundance of shallow-rooted and lower-growing types. The total amount of ground cover also declines toward the drier margins as the deeply rooted, sod-forming grasses of the prairie grassland give way to bunchgrass species (so called because, instead of forming a continuous grass cover, they occur in isolated clumps or bunches). In their natural conditions, the short-grass prairies of North America and Eurasia supported higher densities of grazing animals—bison and antelope in the former, wild horses in the latter—than did the tall-grass prairies. Indeed, it is suspected that the specific plant association of the short-grass regions may have been a consequence of overgrazing under natural conditions. The short-grass prairies cannot be cultivated without the use of irrigation or dry farming methods, so they remain primarily the domain of wide-ranging grazing animals; however, today's animals are domesticated cattle, not the thundering self-sufficient herds of wild species that formerly made these plains one of Earth's marvels.

Desert

Eventually, lack of precipitation can become too severe even for the hardy grasses. Where evapotranspiration demands greatly exceed available moisture throughout the year, as in Köppen's desert climates, either special forms of plant life have evolved or the surface is bare. Plants that actively combat low precipitation are equipped to probe deeply or widely for moisture, to reduce moisture losses to the minimum, or to store moisture when it is available. Other plants evade drought by merely lying dormant, perhaps for years, until enough moisture is available to assure successful growth and reproduction. The desert biome is recognized by the presence of plants that are either drought-resisting or drought-evading (Fig. 11.29). In extremely dry deserts, only a few plants can survive, and ground cover is much less common (Fig. 11.30).

Plants that have evolved mechanisms to combat drought are known as *xerophytes*. They are perennial shrubs whose root systems below ground are much more extensive than their visible parts or that have evolved tiny leaves with a waxy covering to combat transpiration. They may have leaves that are needlelike or trunks and limbs that photosynthesize like leaves or that have expandable tissues or accordionlike stems to store water when it is plentiful (the succulent cacti). They may be

Figure 11.30 The absence of ground cover in this photograph of the Namib Desert in Africa is a clear indication of the extremely low annual rainfall experienced in this region. ■ **Are there any reasons why humans might be found in such desolate regions?**

plants that can tolerate excessively saline water or shrubs that shed their leaves until sufficient moisture is available for new leaf growth. The nonxerophytic vegetation consists mainly of short-lived annuals that germinate and hurry through their complete life cycle of growth—leaf production, flowering, and seed dispersal—in a matter of weeks when triggered by moisture availability. Like other species, these ephemeral plants also require days of a certain length, so they appear only in particular months; therefore, the month-to-month and year-to-year variation in form and appearance of desert vegetation is enormous. Animals of the deserts are primarily nocturnal to avoid the searing heat of the daytime, and many have evolved long ears, noses, legs, and tails that allow for greater blood circulation and cooling. The similar life forms and habits of the different plant and animal species found in the deserts of widely separated continents are a remarkable display of repeated evolution to insure survival in similar climatic settings.

Arctic and Alpine Tundra

Proceeding upward in elevation and poleward in latitude, we finally come to regions in which the growing season is too brief to permit tree growth. Nearing the poles, we enter a vast realm dominated by subfreezing temperatures and thin snow cover much of the year, so that the ground is frozen to depths of hundreds of meters. Only the top 36 to 60 centimeters (15–25 in.) thaw during the short summer interval. Still, vegetation survives here and in fact forms a nearly complete cover over the surface. Such vegetation must be equipped to tolerate frozen subsoils (permafrost), icy winds, a low sun angle, summer frosts, and soil that is waterlogged during the short growing season. The result is **tundra**—a mixture of grasses, flowering herbs, sedges, mosses, lichens, and occasional low-growing shrubs. Most of the plants are perennials that produce buds close to or beneath the soil surface, protected from the wind. Many of the plants show xerophytic adaptation—such as small, hard leaves—in response to extreme physiologic drought resulting from wind stress. This is particularly true in areas of alpine tundra, where extended periods of waterlogging of soils are uncommon. The effect of wind is evident from the fact that the less exposed valleys within the tundra region are often occupied by coniferous woodlands.

In a band of varying width reaching across northern Alaska, Canada, Scandinavia, and northern Russia, several types of tundra are recognized: *bush tundra,* consisting of dwarf willow, birch, and alder, which grow along the edge of the coniferous forest; *grass tundra,* which is hummocky and water-soaked during the summer (Fig. 11.31); and *desert tundra,* in which expanses of bare rocks may be covered by colorful lichens. In a few ice-free valleys of Antarctica, only desert tundra occurs.

Alpine conditions are not exactly like those in the Arctic latitudes. The deeper snow cover of the high mountains prevents the development of permafrost, and the summer sun results in considerably more evaporation. However, many high areas are swept clean of snow by wind and are thus extremely exposed, supporting only desert tundra. Microclimate becomes an important control of vegetation because of the varying exposures to sun and wind. Nevertheless, the short growing season and severe wind stress produce an overall plant community similar to that in the Arctic regions (Fig. 11.32).

Animals of the tundra obviously must cope with long periods of extreme cold as well as darkness. Many animals hibernate through the long Arctic winters (or the cold and windy alpine winters); others, such as the caribou of Alaska and Canada, migrate into the boreal forest to escape the extreme cold exacerbated by Arctic winds. Year-round residents, such as the polar bear and musk ox, must have extremely large fat reserves around their large chests, in addition to extremely efficient fur. Many insects, such as the ubiquitous mosquito, cope with the extreme climate by emerging from eggs in the spring,

Figure 11.31 Tundra fireweed in bloom during the brief Arctic summer in Alaska. The photograph was taken near Atigun Pass in the Brooks Mountain Range. ■ **Why might alpine tundra be of more use to humans than Arctic tundra?**

Randy Brandon/Alaska Stock Images

maturing and laying eggs, and dying within one short Arctic summer. Enough eggs survive, buried under insulating snow, to continue the cycle in the following year.

Marine Ecosystems

The living organisms of the ocean can be divided into three groups according to where or how they live in the ocean. The first group is called **plankton.** Plankton is made up mostly of the ocean's smallest—usually microscopic—plants (**phytoplankton**) and animals (**zooplankton**). These tiny plants and animals float freely with the movements of ocean water, are true "drifters,"

Figure 11.32 Close examination of the tundra biome of the mountainous western United States reveals a mixture of grasses, sedges, and rushes. ■ **Why does all vegetation, including the occasional shrub or stunted tree, grow so close to the ground?**

and form the basis of the oceanic food chain. The second group is composed of the animals that swim in the water. This group, called **nekton,** includes fish, squid, marine reptiles, and marine mammals such as whales and seals (Fig. 11.33). The third group is comprised of the plants and animals that live on the ocean floor. This group is called the **benthos.** It includes corals, sponges, and many algae; such burrowing or crawling animals as the barnacle, crab, lobster, and oyster; and attached plants such as turtle grass and kelp.

Life in the ocean depends on the sun's energy and on the nutrients available in the water. The phytoplankton

are the most important link in the ocean food chain. They take the dissolved nutrients in the water and, through the process of photosynthesis, produce oxygen and foods needed by zooplankton and by the smallest nekton. Phytoplankton are the only food source for these animals, which in turn form the food source for larger carnivorous fish and marine mammals. These in turn are prey for still larger animals.

The marine food chain just described is part of a full food cycle in the ocean, for through the excretions of animals and the decomposition of both plants and animals in the ocean, chemical nutrients are returned to the water and are again made available for transformation by phytoplankton into usable foods. The phytoplankton also play an essential role in the production of oxygen for Earth's atmosphere.

The uneven distribution of nutrients in the oceans and the fact that sunlight can penetrate only to a depth of about 120 meters (400 ft), depending on the clarity of the water,

Figure 11.33 Elephant seals bask in the nutrient-rich and cool upwelled waters off the San Benitos Islands, Mexico. ■ **To which group of ocean organisms do these elephant seals belong?**

NASA

Figure 11.34 Distribution of chlorophyll-producing marine plankton mapped by space observations of ocean color. Purple to yellow to red indicates increasing chlorophyll concentration.

means that the distribution of marine organisms is also variable. Most organisms are concentrated in the upper layers of the ocean, where the most solar energy is available. In deep waters, where the ocean floor lies below the level to which sunlight penetrates, the benthos organisms depend on whatever nutrients and plant and animal detritus filter down to them. For this reason, benthos animals are scarce in deep and dark ocean waters. They are most common in shallow waters near coasts, such as coral reefs and tide pools, where there is sunlight and a rich supply of nutrients and where phytoplankton and zooplankton are abundant as well.

The waters of the continental shelf have the highest concentration of marine life. The supply of chemical nutrients is greater in waters near the continents, where nutrients are washed into the sea from rivers. Marine organisms are also concentrated where deeper waters rise (upwelling) to the surface layers, where sunlight is available. Such vertical exchanges are sometimes the result of variations in salinity or density. A similar situation occurs where convection causes bottom layers of water to rise and

mix with top layers, as is the case in cold polar waters and in middle-latitude waters during the colder winter months. Marine life is also abundant in areas where there is a mixing of cold and warm ocean currents, as there is off the northeastern coast of the United States (Fig. 11.34).

During the 1977 dives of the manned submersible vessel *Alvin* off the East Pacific Rise, scientists for the first time observed abundant sea life on the floor of the ocean at depths of more than 2500 meters (8100 ft). It was previously presumed that these cold (2°C/36°F), dark waters were a virtual biological desert. However, the undersea volcanic mountain range produces vents of warm, mineral-rich waters, which nourish bacteria and large colonies of crabs, clams, mussels, and giant 3-meter (10-ft) red tube worms. The discovery of this deep-ocean ecosystem, which exists without the benefit of sunlight, has caused scientists to rethink old theories about the ocean and its chemistry. This unusual "chemosynthetic" vent community has been observed more recently at several other oceanic ridge sites in the Pacific and Atlantic Oceans.

DEFINE & RECALL

abiotic	decomposer (detritivore)	primary productivity	natural vegetation
autotroph	trophic structure	secondary	plant community
heterotroph	food chain	productivity	succession
herbivore	trophic level	habitat	gap
carnivore	food web	ecological niche	climax community
omnivore	biomass	generalist	mosaic

matrix	biome	savanna	phytoplankton
patch	liana	tall-grass prairie	zooplankton
corridor	galeria forest	short-grass prairie	nekton
tolerance	boreal forest	tundra	benthos
ecotone	(taiga)	plankton	

DISCUSS & REVIEW

1. Give reasons why the study of ecosystems is important in the world today. Give several examples of natural ecosystems within easy driving distance of your own residence.

2. What are the four basic components of an ecosystem?

3. How can an organism belong to more than one trophic level? Give an example.

4. How does the biomass at one trophic level usually compare in weight with that at the next level?

5. What is the difference between primary and secondary productivity? Which is most important to the ecosystem?

6. What factors are most critical in affecting the net primary productivity of a terrestrial ecosystem?

7. Why are generalists in the majority in ecosystems?

8. In what ways has the original theory of succession been modified?

9. How do the terms *mosaic, matrix, patch, corridor,* and *ecotone* relate to each other and to a vegetation landscape?

10. What are the important environmental controls of ecosystems? Which are the most important on a worldwide basis?

11. What are the four major types of Earth biomes?

12. What important climatic characteristics are related to each of the major biomes?

13. Describe a true tropical rainforest. How does such a forest differ from jungle?

14. What are the distinctive features of chaparral vegetation? What climatic conditions are associated with chaparral?

15. What are the major differences between tropical and middle-latitude forests?

16. What conditions of climate or soil might be anticipated for each of the following in the middle latitudes: broad-leaf deciduous forest; broad-leaf evergreen forest; needle-leaf coniferous forest?

17. What factors contribute to the development and maintenance of savannas?

18. How have xerophytes adapted to desert climatic conditions?

19. What are the chief characteristics of tundra vegetation? What adaptations to climate does tundra vegetation make?

20. Which link in ocean food chains is most important?

21. The highest concentrations of marine life (see Fig. 11.34) are found in which parts of the oceans? Why?

CONSIDER & RESPOND

1. Arrange the following organisms into six logical trophic levels in order from the first to the last.

   a. owl; fungi; vulture; grass; bobcat; rabbit

   b. herbivorous zooplankton; small fish; phytoplankton; shark; bottom bacteria; large fish

2. Refer to Table 11.2. Based on the means, which is more important for terrestrial ecosystem productivity between the Equator and the polar regions: annual temperatures or the availability of water? Use examples from the table to explain your choice.

3. What broad groups might you use if you were to classify Earth's vegetation on the basis of latitudinal zones? Why, do you suppose, did your textbook authors

choose not to do this, electing instead to identify broad groups based on major terrestrial ecosystems (biomes)?

4. Refer to Figure 11.21, which shows considerable variation in the natural vegetation of the United States between 30° and 40° north latitude. Describe the broad changes in vegetation that occur within that latitudinal band as you move from the east coast to the west coast of the United States.

5. Examine the climatic variation (see Fig. 9.4) within the same latitudinal band described in the previous question. Does there appear to be a relationship between natural vegetation and climate? If so, describe this relationship.

SOILS AND PROCESSES OF SOIL DEVELOPMENT

Soil serves as an outstanding example of the integration of Earth's subsystems.

Which of the four major subsystems are involved?

Why is soil such a good example of subsystem interaction and integration?

Why should soil be considered an open system?

Soil water is essential to vegetation, partly because it is the means by which plants receive dissolved nutrients required for growth.

Why is gravitational water such an effective solution agent?

How is capillary water important during periods of drought?

Fertility of a soil depends on many factors, and a soil that is fertile for one crop or vegetation type may not be for another.

What factors determine a soil's fertility?

How are acidity and alkalinity related to soil fertility?

How is the nature of the vegetation involved?

On a global scale, climate exerts a major influence over soil formation.

How do temperature and precipitation affect the development of soils?

How does climate affect soils indirectly through vegetation?

What is the role of the moisture regime in soil development?

Soils are most productive when no single soil-forming factor is dominant.

Why is this so?

What often results when natural vegetation is removed or a single soil-forming process proceeds unchecked?

The National Resources Conservation Service (NRCS) soil classification system was devised so that soil scientists and agriculturalists could identify differences among soils.

How is the system important to physical geographers?

Why is there a general correlation among the distributions of NRCS soil orders, climatic classes, and vegetation types on a regional scale?

Soils are among the world's most important, least understood, and most widely abused natural resources.

What determines a natural resource?

Why is there so much ignorance concerning soils?

How are soils abused?

What can be done about it?

A vineyard in South Australia grows well in brightly colored soils called terra rossa (red earth).

In an urban society, few people outside the fields of science and agriculture give much thought to how important soils are to life on Earth. This is a serious oversight in an age of environmental awareness and concern. Composed primarily of decomposed rock materials and varying amounts of water, gases, and organic materials, soil covers most land surfaces with a fragile mantle that is an indispensable resource. Soil has been called the "skin of the Earth," and its condition and nature reflect the environmental factors under which it was formed and exists today. As with most environmental systems, soils reflect and respond to the influence of many natural processes, but human impacts on soils have been increasing with population growth and the development of civilization.

Soil is a dynamic natural body capable of supporting a vegetative cover. It contains chemical solutions, gases, organic refuse, and an active fauna. The complex physical, chemical, and biological processes that take place among the components of a soil are an integral part of its dynamic character. As an active body, soil responds to changes in climate (especially to temperature and moisture), to land surface configuration, to vegetative cover and composition, and to animal activity. The result is that soil develops and matures in response to its environment. Soil is an outstanding example of how Earth's subsystems interact and can overlap (Fig. 12.1). In fact, because soils integrate the major subsystems of Earth so well, they are sometimes considered a separate system called the *pedosphere* (from Greek: *pedon,* ground).

The development of a soil depends on a large number of factors. But when soils are viewed on a world regional scale, the dominant influence is climate. The relationships between climates and soil types, as well as the association of soils with climate-controlled vegetation, were considered in Chapters 9 and 10.

Principal Soil Components

What is soil actually made of? What does the scoop of a bulldozer contain when it shovels up a load of soil? What characteristics of the ground does it take to support Earth's varied vegetation? Soils contain four major components, and the processes that act on these components are many and varied, but soils operate as an *open system.* Matter and energy flow into and out of a soil, and are also held in storage within the soil (Fig. 12.2). Understanding these flows—inputs and outputs, the components and processes involved, and how they vary from soil to soil—is a key to developing an appreciation of the complex nature of this critical natural resource. The four major components of soil are inorganic materials, soil water, soil air, and organic matter.

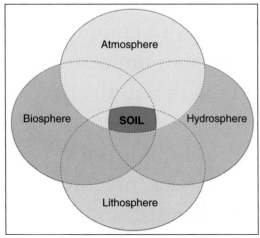

Figure 12.1 The links between soil and the major Earth subsystems. ■ **Why is soil considered to be such an integrator of Earth systems?**

Inorganic Materials

Soil is made up of insoluble mineral material—minerals that will not dissolve in water to form a solution—and also soluble minerals or chemicals in solution. Most minerals found in soils are combinations of the most common elements of Earth's crust: silicon, aluminum, oxygen, and iron. Some of these occur in chemical compounds; others are found in the air and water that are also present in, and a part of, soil. A large number of chemical elements and compounds are necessary to sustain the flora and fauna of Earth's ecosystems. Important among these, in addition to the four elements listed previously, are carbon, hydrogen, nitrogen, sodium, potassium, zinc, copper, iodine, and compounds of these elements.

The chemical components that form part of a soil come from many sources. Some are derived from the breakdown of underlying rocks or from accumulations of loose sediments. Others enter as solutions in water or as a part of the chemical structure of water itself. Still others are part of the air found in soils or are derived from organic activities, which help to disintegrate rocks, release gases, or create new chemical compounds.

Because plants need many chemical substances for growth, knowledge of a soil's mineral and chemical content is necessary to determine its productive potential. Frequently, soil deficiencies in a specific substance can be rectified through fertilization to increase the soil's productivity (Fig.12.3).

Soil Water

Soil water supplies the moisture necessary for chemical reactions that sustain life, and also provides nutrients in a form that can be extracted and used by vegetation. Plants need air, water, and minerals to function, live, and grow. They depend on the soil in which they are rooted for much of their supply of these necessities. Soil water is not pure water; rather, it is a solution bearing soluble nutrients.

The original source of soil moisture is precipitation. When precipitation falls on the land, either it is absorbed or it runs off downslope to a stream that eventually channels it into a larger body of water. The water absorbed by the soil washes over and through various soil materials, dissolving some of these materials and carrying them through the soil.

Water is found in soil in several different circumstances. The water that percolates downward through the soil, pulled by the force of gravity, is called **gravitational water.** Gravitational water moves downward through voids between individual soil particles and clumps of soil toward the *water table*—the level below which all spaces between the soil and rock particles are filled with water. As a consequence, water cannot percolate any deeper. The quantity of gravitational water in a soil is related to several conditions, including the amount of rainfall that

Figure 12.2   Soil as an open system: the relation of soil to environmental factors. Soil is an example of an open system because it receives inputs of matter and energy, stores part of these inputs, and outputs matter and energy. Note the inputs and outputs on this diagram.   ■   **What are some examples of energy and matter that flow into and out of the soil system?**

has occurred, the time since the last rainfall, evaporation rates, the ease with which water moves through the soil, and the space available for water storage.

Gravitational water functions in the soil in several ways (Fig. 12.4). First, as gravitational water moves

down through a soil, the water takes with it the finer particles from the upper layers of soil. This removal of soil components from the topsoil is called **eluviation.** Eventually, as gravitational water percolates downward, it deposits fine particles (clay and silt) that were removed

James A. Suger/CORBIS

Figure 12.3   Fertilizers increase the productivity of soils.   ■   **Why can soil fertilizer be either useful or detrimental when it is introduced into the soil system?**

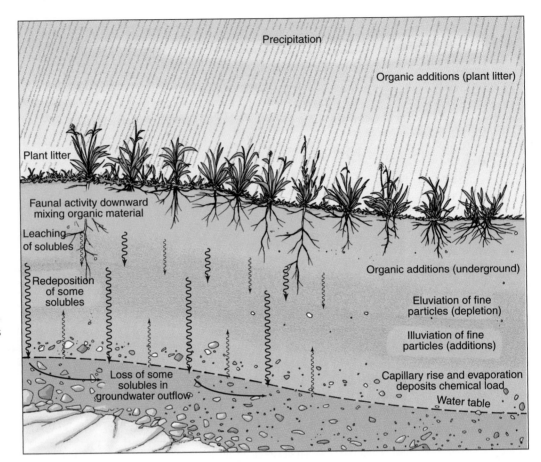

Figure 12.4 The role of water in soil development processes. Water is important in moving nutrients and particles vertically, both up and down, in the soil layer.
■ **How does deposition by capillary water differ from deposition (illuviation) by gravitational water?**

from the topsoil. This process of deposition in the subsoil is called **illuviation.** Gravitational water serves as an agent of transportation and mixing as it moves soil particles from one level to another. One result of eluviation is that the topsoil's texture becomes coarser as finer particles are removed. Consequently, the topsoil's ability to retain water is reduced, as illuviation enhances the subsoil's water retention capability. This process may be carried to such extremes that the subsoil becomes extremely dense and compact, forming a clay **hardpan.**

As gravitational water moves downward, it dissolves soluble inorganic soil components and carries them into deeper levels of the soil, perhaps to the zone where all open spaces are saturated. This process of depleting nutrients in the upper soil by the flow through of water is called **leaching.** Intensive leaching is common under conditions of heavy precipitation, and can rob a topsoil of all but the most inert and insoluble substances.

The processes of leaching and eluviation strongly influence the characteristic **stratification,** or layered changes with depth, found in soils. The upper portions are composed of coarse material and are somewhat impoverished in soluble nutrients. Fine particles and some of the substances dissolved from the upper soil are deposited in the lower portion, which becomes dense and sometimes is strongly colored by accumulated iron compounds.

Some soil water adheres to soil particles and soil clumps by surface tension (the property that causes small droplets of water to form rounded beads instead of spreading out in a thin film). This soil water, called **capillary water,** serves as a storage supply of water for plants. Capillary water can move in all directions through soil, its migration determined by a tendency to move from areas with more water to areas with less. Thus, during periods between rainfalls, when there is no gravitational water flowing through the soil, capillary water can move upward or horizontally to supply plant roots with necessary moisture and dissolved nutrients.

When capillary water migrates upward in a soil, it moves minerals from the subsoil toward the surface. If this water evaporates from the soil, the formerly dissolved minerals are left behind, generally as alkaline or saline deposits in the topsoil. Such deposits can be detrimental to plants and animals existing in the soil. Lime (calcium carbonate) deposited in this way may produce a cementlike layer, called *caliche,* that, like clay hardpan, prevents the downward percolation of water.

Soil water is also found in a very thin film, invisible to the naked eye, that is bound to the surface of soil particles by strong electrical forces. Because this water, called **hygroscopic water,** does not move through the soil, it cannot supply plants with the moisture they need.

Soil Air

A large part of soil—in some cases, as much as 50 percent—is made up of the voids between soil particles and between clumps or aggregates of soil particles. When not filled with water, these spaces are filled with air. Soil air is much like the air of the lower atmosphere, though it is likely to have less oxygen, more carbon dioxide, and a fairly high relative humidity because of the presence of capillary and hygroscopic water.

For most microorganisms and plants that live in the ground, soil air supplies oxygen and carbon dioxide necessary for life processes. Thus, the problem with water-saturated soils is not so much the excess water itself but that with all the pore spaces filled with water, there is no air supply. It is because of this lack of air that many plants and animals find it difficult to survive in water-saturated soils.

Organic Matter

In addition to various chemical compounds (both soluble and insoluble), air, other gases, and water, soil also contains organic matter. The decayed remains of plant and animal material, partially transformed by bacterial action, are called **humus.** Humus is important to soils in several ways, but it is most important as a catalyst in chemical reactions by which plants extract nutrients from the soil and for its role in restoring minerals to the soil. Humus improves soil by making it more workable and increasing its capacity to retain water. Humus also provides a food source for the enormous variety of microscopic organisms that live in soil.

The organisms that live in soil range from countless microscopic bacteria to earthworms, rodents, and other burrowers. Many animals contribute to the development and enrichment of soils. They are important in the creation of humus from inert plant litter and in mixing organic material deeper into the soil. In addition, the chemical and mechanical functions of plants and their root systems are an integral part of the soil-forming system.

Each of the soil constituents is important in determining the characteristics of a particular soil variety. Soils vary locally, regionally, and globally, as do the proportions of their constituents. The interactive links between a soil and the vegetation and climate at its location are strong. For example, soils in middle-latitude grasslands normally have a very high proportion of organic debris or humus; those in deserts are baked dry, have very little water, and are rich in soluble constituents such as lime and salt; tropical soils have a noticeably high iron and aluminum oxide content. Knowledge of a soil's water, mineral, and organic components and their proportions can help determine its potential productivity and what the best use for that particular soil might be (Fig.12.5).

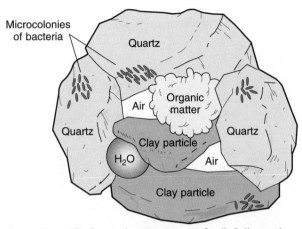

Figure 12.5 The four major components of soil. Soil contains a complex assemblage of inorganic rock and minerals, water, air, and organic matter. The interaction among these components, and the proportion in which each is present, are important factors in the kind of soil that develops.
From Purves et al. *Life: The Science of Biology,* 4/e. Used with permission of Sinauer Associates, Inc.

Characteristics of Soil

The physical properties of soils that can be readily tested or examined are used to describe and differentiate among soil types. The most important properties include color, texture, structure, acidity or alkalinity, and capacity to hold and transmit water and air.

Color

Color may not be the most important physical attribute of soil, but it is certainly the most visible. Most people are aware of the variations in soil color from place to place. For example, the well-known red clay soils of Georgia are not far from Alabama's belt of black soil. Soil colors vary from black to brown to reds, yellows, grays, and near-whites, and each color offers a clue to the physical and chemical characteristics of a particular soil. When describing soils in the field or laboratory, soil scientists use a book of standardized soil colors to clearly and precisely identify this coloration (Fig.12.6).

Humus or decomposed organic matter is black or brown, and soils that are high in humus content tend to be dark. As the humus content of soil decreases, because of either low organic activity or loss of organics through leaching, soil colors gradually fade to light brown or gray. A large proportion of humus usually indicates that a soil is highly fertile, because humus acts as a catalyst for chemical processes that allow plants to obtain nutrients from the soil. For this reason, dark brown or black soils are often referred to as *rich* and are considered superior soils. It should be noted, however, that this is not always the case, because some black or dark brown soils have little or no humus content and get their dark color from other factors.

Soils that are red or yellow typically indicate the presence of iron compounds. In moist climates, a light gray

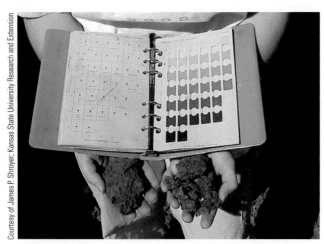

Figure 12.6 Determining soil color. A standardized color classification system is used to determine precise color by comparing the soil to the color samples found in Munsell soil color books.

or white soil indicates that iron has been removed, leaving oxides of silicon and aluminum; in dry climates, the same color typically indicates a high proportion of salts.

Soil colors provide useful clues to the physical and chemical characteristics of soils, and make the job of recognizing different soil types easier. But color alone does not answer all of the important questions about a soil's qualities or fertility.

Texture

Soil texture refers to the size (or distribution of sizes) of particles that make up the soil (Fig. 12.7). In **clayey** soils, the dominant size is **clay** particles, defined as having diameters of less than 0.002 millimeters (soil scientists universally use the metric system). In **silty** soils, the dominant **silt** particles are defined as being between 0.002 and 0.05 millimeters. **Sandy** soils have mostly **sand**-sized particles, with diameters between 0.05 and 2.0 millimeters. Individual particles with diameters larger than 2.0 millimeters are regarded as pebbles, gravel, or rock fragments and technically are not soil particles. Because no soil is made up of uniformly sized particles, the proportion of particles in various size ranges determines the soil texture.

For example, a soil composed of 50 percent silt-sized particles, 45 percent clay, and 5 percent sand would be considered a silty clay. A triangular graph (Fig. 12.8) is used to plot different classes of soil texture and the percentage ranges of each **soil grade** (as sand, silt, and clay are called) within each class. Point A within the silty clay class represents the example just given. A second soil sample (B) that is 20 percent silt, 30 percent clay, and 50 percent sand would be referred to as a sandy clay loam. **Loam** soils, which occupy the central areas of the triangular diagram, are those in which none of the three grades (sizes) of soil particles is greatly dominant. It is

interesting to note that the loam soils are generally those best suited for supporting plant life.

Soil texture is important because it helps determine the capacity of a soil to retain moisture and air that are necessary for plant growth. Soils with a greater proportion of larger particles tend to be well aerated and allow water to pass through (or **infiltrate**) the soil more quickly—sometimes so quickly that plants are unable to use the water. Clay soils present the opposite problem, because they transmit water very slowly, become waterlogged, and are deficient in air. An important process of cultivation is the aeration of soil. Plowing, disking, or shoveling a soil opens up its structure and increases its air content.

Structure

In most soils, individual particles form larger distinctive masses or clumps, known as **soil peds,** which give a soil its particular structure. The structure of a soil is an important factor in its workability. Structure also influences a soil's **permeability** (the rate at which fluids such as water pass through the soil) and its **porosity** (the amount of space in a soil that may contain fluids). Permeability, which is usually greatest in sandy soils, and porosity, which is usually greatest in clayey soils, con-

Figure 12.7 Particle sizes in soil. *Sand, silt,* and *clay* are terms that refer to the size of these particles for scientific and engineering purposes. Here, the sizes of each can be compared. Clay particles are tiny, sheetlike particles that cannot be seen.

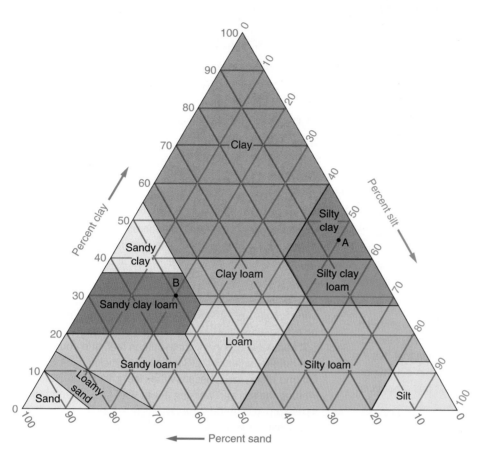

Figure 12.8 The texture of a soil can be represented by a point on this diagram. Texture is determined by sieving the soil to determine the percentage of particles falling into the size ranges for clay, silt, and sand. Note that each of the three axes of the triangle is in a different color, and the line colors also correspond (clay-red, silt-blue, sand-green). ■ **What would a soil that contains 40% sand, 40% silt, and 20% clay be classified as?**

trols soil drainage as well as the amount of moisture available to plants and organisms. Soils with similar textures may have different structures, and vice versa. Consequently, although being similar in some ways, one soil may be more productive than another.

Soil structure may be influenced by outside factors such as moisture regime and the nature of nutrient cycles by which plants and the soil interchange chemicals, keeping certain ones in the system while others are leached away. We have all seen the structural change in certain soils from when they are wet to when they have been dried by the sunshine. Soil structure can also be influenced by human activities such as cultivation, irrigation, and fertilization. Certain fertilizers, as well as lime or decayed organic debris in a soil, affect structure by encouraging clumping of soil particles and maintenance of those groups or clumps. Excess sodium and magnesium have the opposite effect, causing clay soils to be a sticky muck when wet and like concrete when dry. The absence of smaller particles hinders the development of a definite soil structure. This is one reason why sandy beaches and sand dunes, which are composed of the larger soil particles, have little or no apparent soil structure. This also explains why some soils have more structure below the layers closest to the surface, because smaller particles of topsoil have been moved to lower layers by soil water.

Scientists classify soil structures according to their form. These range from columns, prisms, and angular blocks, to nut-like spheroids, laminated plates, crumbs, and granules (Fig. 12.9). It is important to know that massive and fine structures tend to be less useful than aggregates of intermediate size and stability, which permit good drainage and aeration.

Acidity and Alkalinity

The chemical processes and exchanges that take place in soils and plant systems are what make a soil fertile or infertile. The general nature of soil chemistry is usually expressed as the degree to which a soil departs from neutrality toward either acidity or alkalinity (baseness).

The acidity or alkalinity of a soil is measured on a scale of 0 to 14, called the **pH scale.** A pH reading indicates the concentration of reactive hydrogen ions present in the soil moisture. The pH scale is logarithmic, meaning that each change in a whole number on the scale represents a tenfold change in the hydrogen present. It is also inverted, meaning that the lower the pH value, the greater the amount of hydrogen present. Low pH values indicate acid soil moisture; high values indicate alkaline conditions (Fig. 12.10). With increased rainfall, leaching increases, gradually replacing soil elements such as sodium (Na_2), potassium (K), magnesium

Basic Soil Analysis

After studying just one chapter on soils, no one would expect any introductory student to be able to do a detailed soil analysis. However, there are some basic observations anyone can perform to better understand a few properties of a local soil. No equipment is required to make analyses; only visual and hands-on examinations are necessary.

Soil Color

Soil color can hold clues to the composition and/or the formation processes of that soil. Figure 12.6 shows the Munsell color book, a standard guide for matching and recognizing precise colors of soil types. Here are some common soil colors, but each color can also appear in a wide variety of tones:

Red: Reddish soil usually indicates that oxidation has been an active process—that is, oxygen has chemically reacted with the soil minerals. Red is also an indication that there is iron in the soil. Just like rusting iron, many iron-rich minerals turn red when oxidized. The formula for this process is

$FeO + O_2 \rightarrow Fe_2O_3$ [Ferrous Oxide + Oxygen becomes Hematite]

Blue/Silver/Gray: These tones mean the soil has likely been reduced; in other words, oxygen has been removed from the soil.

$Fe_2O_3 \rightarrow FeO + O_2$ [The above formula in reverse]

White: Usually denotes calcium carbonate ($CaCO_3$) or salt (such as NaCl) may be present in the soil.

Black: A very dark color may be an indicator of the amount of organic material present in the soil. Darker soils tend to have a higher organic content.

More sophisticated field or laboratory analyses are required for absolute identification, but these examples will allow a good working hypothesis for soil characteristics represented by colors.

Soil Texture

The particle sizes in a soil determine its texture and can be estimated in the field with simple hands-on observations. Soil texture is a property that you can feel, and your fingers can help in the analysis. Sand-sized particles can be easily felt and recognized, because they feel gritty to the touch. Wetting the soil and working it with your hands can help in this process. If the sample doesn't feel gritty, but rather is smooth to the touch, then the soil contains silt or clay.

Platelike

Platy

Lenticular

Prismlike

Prismatic

Columnar

Blocklike

Spheroidal

Blocky

Nuciform

Granular

Crumb

Figure 12.9 Classification of soil structure on the basis of soil peds. ■ **How does soil structure affect a soil's usefulness or suitability for agriculture?**

(Mg), and calcium (Ca) with hydrogen. This process is enhanced by rain because falling raindrops pick up atmospheric carbon dioxide and become slightly acidic: $H_2O + CO_2 = H_2CO_3$ (carbonic acid). A result of these processes is that desert soils tend to be alkaline and soils in humid regions tend to be acidic (Fig. 12.11).

Soil acidity or alkalinity helps determine the availability of nutrients to plants and ultimately affects plant growth. Plants receive virtually all of their nutrients in solution. That is, plants are unable to absorb nutrients that are not dissolved in liquid. However, if the soil moisture lacks some degree of acidity, soil water has little ability to dissolve these nutrients. As a result, even though nutrients are in the soil, plants may not have access to

If the sample feels sticky and you can squeeze a small soil sample into a ribbon (like modeling clay), then clay-sized particles are abundant. Actual percentages of particle sizes in a soil sample are best established in a laboratory.

Soil Structure

The shape of clumps that a soil makes when it is broken apart is called structure, and can be examined by breaking up a handful of soil. The peds (or small clumps of soil) may take on some distinctive shapes. Though the peds may form a variety of shapes, some of the more common are granular (denoting a presence of sand) and platy (showing a presence of clay). Other soil structures, such as blocky, columnar, or prismatic, are shown in Figure 12.9.

Although these simple procedures will not yield a complete analysis of a soil sample, they can certainly be the first steps in the process. It is interesting to note that pedologists (soil scientists) while in the field perform many of these same procedures.

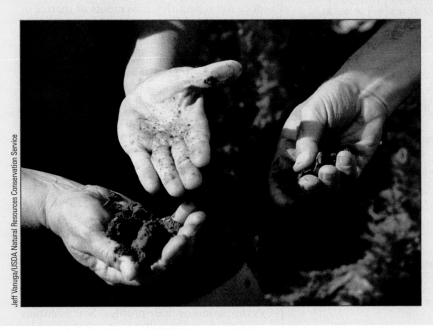

them. To correct this alkalinity, which is common in the soils of many arid regions, and to make the soil more productive, farmers can flush the soil by irrigating under conditions of good drainage.

Strongly acidic soils are also detrimental to plant growth. In soils that are acidic, soil moisture dissolves nutrients, but they may be leached away before plant roots can absorb them. This process can make the soil more acidic, accentuating the problem. This situation often occurs in soils and climates that are otherwise suitable to plant growth. Fortunately for agriculture, soil acidity can generally be corrected by adding lime to the soil.

Soil scientists have shown that most complex plants will grow only in soils with pH levels between pH 4 and pH 10. The optimum pH for plant productivity varies with the plant species. Around the world, vegetation has evolved and adapted to a variety of climates and soil environments, both of which can affect soil pH. Thus, barley can tolerate alkaline soils, but camellias and rhododendrons prefer more acid conditions. Those who bring a diversity of plants into their gardens must be prepared to produce soils of varying pH values for their specimens. This is achieved through **fertilization**—the addition to the soil of nutrients that meet the specific requirements of the plants to be grown.

In addition to affecting plant growth through the availability of nutrients, soil acidity or alkalinity also affects microorganisms living in the soil. Microorganisms are highly sensitive to a soil's pH, and each type has an optimum environmental setting.

Soil Profile Development

Soils begin to develop when rocks or deposits of loose material, the **parent material** on which soil will develop, are colonized by plant and animal life. Once organic processes begin to take place among mineral particles or disintegrated rock fragments, differences begin to develop from the surface down through the soil parent material. Initially, this vertical differentiation results from factors such as the gradual surface accumulation of organic litter and the removal of fine particles and dissolved matter from upper layers by percolating water that deposits these materials at a lower level. As climate, vegetation, animal life, and steepness of slope affect soil formation over time, this vertical differentiation becomes more and more apparent. Often, especially in middle latitudes, fully developed soils exhibit a vertical zonation with distinct layers, called **soil horizons,** that are distinguished by their different physical and chemical properties.

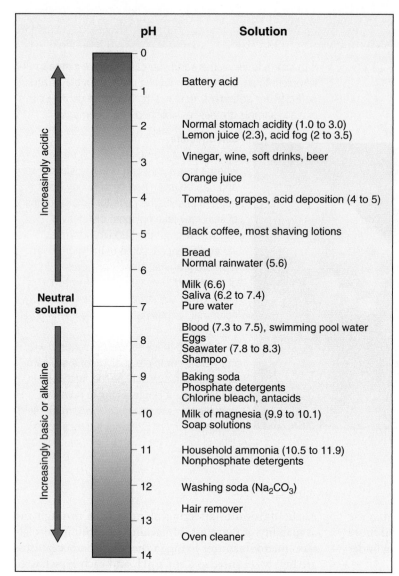

Figure 12.10 The pH scale of acidity, neutrality, and alkalinity. The degree of acidity or of alkalinity, called pH, can be better understood when numbers on the scale are linked to common substances. Low pH means acidic, and high pH means alkaline; a reading of 7 is neutral.

cept in soils that have a high organic content and in which there is vertical mixing, the *B* horizon generally has little humus. The *C horizon* is the weathered parent material from which the soil is developed—fragments of the bedrock directly beneath, or transported and deposited rock material. The *C* horizon does not reflect the movements of matter or organic activity in the higher zones. The lowest layer, sometimes called the *R horizon,* is unchanged bedrock or unmodified rock material transported to the site by water, wind, or glacial action. Certain horizons in some soils may not be as well developed as others, and some horizons may be missing altogether. Because soils and the processes that form them vary, and can be transitional between horizons, the boundaries between horizons may be either sharp or gradational. Some color and texture variation within horizons is common.

The vertical cross section of a soil from its surface down to the parent material from which it is formed is called a **soil profile** (Fig. 12.12). The differences among the infinite variety of soils that exist are apparent in an examination of their profiles. Soil scientists group and classify soils largely on the basis of differences in the soil profiles they display and in the processes that developed the horizons (Fig. 12.13).

Factors Affecting Soil Formation

Many agents, processes, and influences interact to develop a soil. Because of variations among soil components and processes of formation, no two soils are identical in all their characteristics. One important process (actually a set of processes that will be discussed in more detail in a later chapter) is weathering. Rock *weathering* refers to the many natural processes that act to break down rocks into smaller fragments. Two major categories of weathering occur: chemical reactions that cause rocks and minerals to decompose, and physical processes that cause the breakup of rocks. Just as statues, monuments, and buildings become "weather-beaten" over time, rocks exposed to the elements eventually break up and decompose.

The factors controlling the formation and distribution of different soil types are parent material, organic processes, climate, land surface configuration, and time. In the 1940s, Hans Jenny, a distinguished soil scientist, observed that soil was a function of climate, organic matter, relief, parent material, and time—factors that

Most middle-latitude soils have several recognizable layers. At the surface, in regions with sufficient vegetation and moderate rates of organic decomposition, is the *O horizon.* This is a layer of loose organic debris and raw humus. Immediately below this layer is the *A horizon.* Typically, this horizon is a dark layer because of a concentration of organic matter and decomposed humus. The *A* horizon is often referred to as "topsoil." Beneath this horizon, some soils have a lighter colored *E horizon,* named for the eluvial processes that dominate. The lighter color is a product of enhanced leaching, and can even be white in humid middle-latitude soils if the colorful soil bases are removed. Below this is a zone of accumulation, the *B horizon,* where much of the materials removed from the *A* and *E* horizons are deposited. Ex-

Figure 12.11 Alkaline and acidic soils in the United States, based on general climate. In general, soils in the East tend to be acidic and those in the West, alkaline. The dividing line corresponds fairly well with the 30-inch annual precipitation isohyet. ■ **What environmental factors might cause this east-west variation? How might you explain the places west of the 30-inch line that are acidic?**

from weathering-resistant rocks have a high level of similarity with their parent materials. Some bedrock materials are easily weathered, however, and the soils that develop from them tend to bear a greater similarity to other soils in regions of similar climate than to those of comparable parentage that formed in a different climate. It is common to find that one or more of the other soil-forming factors may have a far greater influence on the soil than does the parent material. In fact, on a world-wide basis, climate and the associated plant communities cause stronger and larger variations in soil characteristics than do parent materials. The differences among soils based on variations in parent material are visible on a local level, however, and are of more than casual interest to soil scientists and agriculturists.

As a soil develops and matures, the influence of parent material on its characteristics declines. Thus, a younger or less mature soil, provided the other agents such as climate are similar, will show more similarity to its parent material than will soil that is well developed, having been formed over a long time.

Both residual and transported parent material affects the soil that develops from it in very specific ways. First, the chemicals and nutrients available to plants and animals

are easy to remember by their initials arranged in the following order: **Cl, O, R, P, T.** Of these factors, parent material is distinctive because it is the raw material. The other factors influence the type of soil that forms from the parent material.

Parent Material

All soil is derived from the weathered fragments of rock material. If this weathered material accumulates in place—through the physical and chemical breakdown of bedrock directly beneath the soil—we refer to the fragments as **residual parent material.** If the rock fragments from which a soil is formed have been carried to the site and deposited by streams, waves, winds, or glaciers, this mass of deposits, which will eventually develop a soil, is called **transported parent material.** It is the development and action of organic matter through the life cycles of organisms and the climatic environment that are primarily responsible for differentiating soil from the fragmentary rock source or parent material that is present beneath it.

Parent material influences the characteristics of the soil derived from it in varying degrees. Some parent materials, such as sandstone that contains extremely hard and resistant sand-sized fragments, are far less subject to weathering than others. Soils that develop

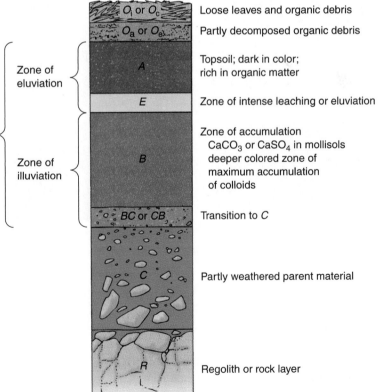

Figure 12.12 Soils are categorized by the degree of development and the physical characteristics of their horizons. *Regolith* is a generic term for broken bedrock fragments at or very near the surface. ■ **In which soil profiles of Figure 12.13 is it easiest to recognize the different layers or horizons?**

Oxisol, central Puerto Rico

Vertisol, Lajas Valley, Puerto Rico

Ultisol, North Carolina Piedmont

Alfisol, southern Michigan

Spodosol, northern New York

Inceptisol, northern Alaska

Figure 12.13   Soil profiles representing a variety of climatic and other soil-forming factors

living in a soil are derived from that soil's parent material (Fig. 12.14). For example, parent materials deficient in calcium will produce soils that are low in calcium, and its natural fauna and plant cover will be of types that require little calcium. Artificial addition of lime (calcium carbonate) can correct such a deficiency. Likewise, a parent material rich in aluminum will produce a soil that is rich in aluminum. In fact, the main source of metallic aluminum is bauxite ore found in tropical soils, where it

has been concentrated by intense leaching that removes the other bases.

The size of the particles that result from weathering of parent material is a prime factor in the determination of soil texture and structure. A rock material such as sandstone, which contains little clay and is weathered into relatively coarse fragments, will make up a soil of coarse texture. Parent materials exert an important influence on the availability of air and water to a soil's living population.

Marbut Memorial Slide Collection, Courtesy Soil Science Society of America

Mollisol, central Iowa

Marbut Memorial Slide Collection, Courtesy Soil Science Society of America

Mollisol, southeastern South Dakota

Marbut Memorial Slide Collection, Courtesy Soil Science Society of America

Mollisol, eastern New Mexico

Marbut Memorial Slide Collection, Courtesy Soil Science Society of America

Mollisol, eastern Colorado

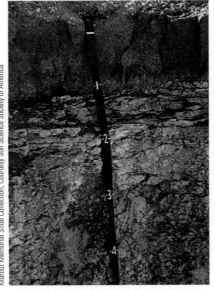

Marbut Memorial Slide Collection, Courtesy Soil Science Society of America

Aridisol, southern New Mexico

Marbut Memorial Slide Collection, Courtesy Soil Science Society of America

Aridisol, central Nevada

Organic Activity

Plants and animals affect soil formation in many ways. The life processes of the dominant plants are as important to the soil as its microorganisms—the microscopically small plants and animals that live in most soils.

Generally, a dense vegetative cover protects the soil from being removed through erosion by running water or wind. Forests provide a protective canopy and also produce a mulch of surface litter, which keeps rain from beating directly on the soil and increases the proportion of rainfall entering the soil rather than running off its surface. Vegetation can also affect the evaporation rate. A sparse vegetative cover will allow greater evaporation of soil moisture than will thick protective vegetation. This evaporation, in turn, affects the movement of capillary water toward the surface. The nature of the plant community also determines the nutrient cycles that are a

R. Gabler

Figure 12.14 Despite strong leaching under a wet tropical climate, Hawaiian soils remain high in nutrients because their parent material is of recent volcanic origin. ■ **What other parent materials provide the basis for continuously fertile soils in wet tropical climates?**

part of soil formation. Certain elements are absorbed by plants and then returned to the soil after plants die and decompose. Soluble nutrients that are not used by plants can be lost through leaching, which can impoverish the soil (Fig. 12.15). The roots of larger plants affect the soil structure as well by making it more porous and by absorbing water and plant nutrients from the soil.

All parts of vegetation (leaves, bark, branches, flowers, and root networks) contribute to the organic content of a soil when the plants die. The organic content of soil varies with the nature of its associated plant life. A grass-covered prairie supplies much more organic matter than the incomplete vegetative cover found in desert regions. There is some question, however, as to whether forests or grasslands (with their thick root network) furnish the soil with greater organic content. But there is no question that many of the world's grassland regions, like the American prairies, provide some of the richest and most fertile soils for cultivation, in part because of the high amount of organic matter that is naturally present.

The process of decay, aided by bacterial action, transforms organic matter into the jellylike mass called humus. As noted earlier, humus is important to soils in many ways. For one, it serves as the primary food supply for soil microorganisms. Humus also affects soil structure by enhancing its water retention capabilities and workability. The process of microorganisms acting on humus returns to the soil organic and inorganic materials that are necessary for other plant life. Consequently, in most soils there is a direct correlation between humus content and fertility.

Bacteria are perhaps the most important microorganisms that live in soils and contribute to soil formation. Bacteria feed on organic matter and humus and break down the debris of living things into their organic and inorganic components, allowing the formation of new organic compounds that promote plant growth. It is difficult to estimate the number of bacteria, fungi, and other microscopic plants and animals that live in a soil, though it has been suggested that there are as many as 1 billion per gram of soil. Whatever the number, it is enormous. The activities and remains of these microorganisms, minute though they are individually, add considerably to the organic content of a soil.

Earthworms, nematodes, ants, termites, wood lice, centipedes, burrowing rodents, snails, and slugs also stir up the soil, mixing mineral components from lower levels with organic components from the upper portion. Earthworms are especially important in soil formation because they take soil in, pass it through their digestive tracts, and excrete it in casts. The process not only helps to mix the soil, but it also changes the texture, structure, and chemical qualities of the soil. In the late 1800s, Charles Darwin estimated that earthworm casts produced each year would equal as much as 10 to 15 tons per acre. As for the number of earthworms themselves, sampling suggested that the total weight of earthworms beneath a pasture in New Zealand equaled the weight of the sheep grazing above them.

Climate

The discussions of climatic regions and processes in Chapters 9 and 10 clearly demonstrated that, on a world scale, climate is a major factor in soil formation. Of course, many variations among soils are apparent on a smaller-than-global basis. The differences that are apparent at a local level tend to reflect the influence of other factors such as parent material, land surface configuration, vegetation type, and time.

Temperature directly affects the activity of soil microorganisms, which in turn affects the rates of decay and decomposition of organic matter. In hot equatorial regions, the intense activity of soil microorganisms precludes any thick accumulation of organic debris or humus. Figure 12.16 shows that the amount of organic matter and humus in a soil increases toward the middle latitudes and away from the tropics. In the Köppen mesothermal *C* and microthermal *D* climates, soil microorganism activity slows enough to allow the accumulation of rich layers of decaying organic matter and humus.

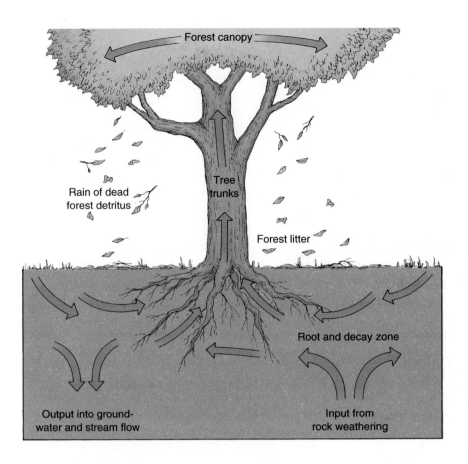

Figure 12.15   The nutrient cycle in a forest. In a wet climate, if trees do not take up soluble nutrients in the soil, they are flushed away in groundwater and lost permanently. The more demanding the forest vegetation, the richer the resulting soil. Pines are notoriously undemanding of soluble bases; consequently, in pine forests bases are lost by leaching and are not replaced by the vegetation, resulting in soil deterioration. ■ **In what way does the destruction of tropical rain-forests offer a notorious illustration of the nutrient cycle?**

However, moving poleward into colder regions, the combination of retarded microorganism activity and limited plant growth results in a thin accumulation of organic matter that may be unaffected by decomposition or only partially decomposed.

Temperature also affects the rates of chemical reactions that take place in soil, many of which make available nutrients necessary for plant growth. Chemical activity tends to increase and decrease directly with temperature.

As a result, parent materials of soils in hot equatorial regions are altered to a far greater degree by chemical means than are parent materials in colder zones.

Temperature affects soil indirectly through its influence on vegetation. We know that vegetation associations are adapted to certain temperature regimes. Soils generally reflect the character of plant cover as a result of nutrient cycles that tend to keep both vegetation and soil in chemical equilibrium.

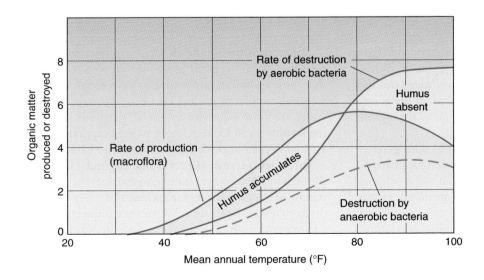

Figure 12.16   Relationship of temperature to production and destruction of organic matter in the soil. ■ **What range of mean annual temperatures is most favorable for the accumulation of humus?**

Amy Jo Steffen

Courtesy Amy Jo Steffen

University of Wisconsin, Eau Claire
BS in Geography (emphasis in Natural Resource Management) with a minor in Geology

I have always had an appreciation for the landscape, because I still remember first seeing the Rocky Mountains at age 7. I looked up and saw the most beautiful sight I had ever seen. They were grand in pictures, but not nearly as breathtaking as they were that day. I had always loved the outdoors, but didn't realize my passion for the environment and the landscape until I took a geography course my sopho-more year—that's when I declared Geography as my major.

I am a cartographer for Environmental Services, Inc., in Raleigh, North Carolina. I manage and process GPS data, print field maps, and create maps and graphics for reports, using GIS, CAD, and design programs. My company mainly deals with wetland and stream delineation and mitigation for transportation routes. We also deal with other environmental concerns such as HazMat sites, borrow pits, endangered species, and cultural resources.

As a geographer, I am an important link between my company and our clients. The maps I create display spatial data in pictures that are understood more easily by our clients.

Learning GIS and design programs has been a huge help. Knowing the basics makes it easier to jump into your work and learn the more difficult tasks as you progress. Spatial relationships have been very important in my work as well. The relationships between soils, slope, surface water, land cover, and/or property owners are important in identifying possible mitigation areas or archeology sites.

The concepts I use the most include coordinate systems and map scales. We receive data from clients in different coordinate systems and have to convert our data to fit theirs so that it can be represented properly in the final report. I also have to choose the right scale to use for each map, so that the right amount of detail is portrayed.

For what I'm doing now, courses in GPS, GIS, cartography, remote sensing, and physical and cultural landscapes are most important. However, working with biologists, archeologists, and geologists has made me realize that I can more easily understand what it is they want to portray because I've also had some courses in those disciplines. Even an art course in drawing, color, or graphic design is helpful. You can create a map with all the right information, but if its presentation isn't clear, it doesn't have the same value.

In 2001, my company was hired by the North Carolina Department of Transportation to do a mitigation site search for the Lumber River Basin. We gathered the soils, vegetation, and topographic information and used GIS to overlay them to find possible wetland mitigation areas. We chose areas that fit the criteria and were of a certain size and printed them on aerial photos to be field-checked. We are now in the process of preserving, enhancing, and reclaiming many streams and wetlands in the basin.

Almost every subject of study can be related back to geography. In biology, it could be the location and migration of certain animals. In geology, it's the spatial relationship between different rock bodies that help to tell Earth's history. In economics, when the locations of high- and low-income areas are looked at, it's geography. In English, the spread of a specific phrase or language uses geography. Geography is also an important part of everyday life, but most people don't recognize it. Every time you get into your car to go somewhere, you use geography to guide you—direction, distance, physical markers, and so on. Everyone uses geography!

Moisture conditions affect the development and character of soils more clearly than any other factor. Without precipitation—and, consequently, soil water and the chemicals dissolved therein—terrestrial plant life is impossible. An absence of plants greatly diminishes the organic content and thereby the fertility of a soil.

We have already discussed the effects of both gravitational and capillary water on soil structure, texture, color, and development. As precipitation is the original source of soil water (disregarding the minor contribution of dew), the amount of precipitation received by a soil affects the rates and degrees of leaching, eluviation, and illuviation that occur, and thereby the rates of soil formation and horizon development.

When considering the effect of precipitation on the quantity and movement of soil water, we should note that the evaporation rate is a factor as well. Salt and gypsum deposits from the upward migration of capillary water are more extensive in hot, dry regions—such as the southwestern United States, where evaporation rates are high—than in colder, dry regions (Fig. 12.17).

Just as temperature affects soil development indirectly through its effects on vegetation, so too does the moisture regime of a region. In regions where the warm season is dry, soils tend to be alkaline. Where the summers are warm and extremely wet, and the winter season warm and dry, intense chemical weathering and fluctuating water tables produce a peculiar type of soil crust known as **laterite.**

Figure 12.17 Harmful concentrations of chemical substances are frequently encountered in arid regions. Shown here is a surface salt crust resulting from the upward capillary rise of water from a water table that is saline and close to the surface. Evaporation at the surface deposits the salt. ■ **What negative soil effects can result when humans practice irrigated agriculture in arid regions?**

This crust is common in the tropics, where it is quarried for building material (Fig. 12.18).

Land Surface Configuration

The slope of the land, its relief, and its aspect (the direction it faces) influence soil development both directly and indirectly. Steep slopes are generally better drained than gentler ones. They are also subject to rapid runoff of surface water. As a consequence, there is less infiltration of water on steeper slopes, which retards the soil-forming processes. This retardation inhibits the development of mature layered soils, sometimes to the extent that no soil will develop on the parent material. In addition, rapid runoff on steep slopes can erode surfaces as fast or faster than soil can develop on them. On gentler slopes, because there is less runoff and more infiltration, more water is available for soil-forming processes and to support vegetative growth, and erosion is not as intense. In fact, erosion rates on gently rolling hills are often just enough to offset the ongoing production of soil from the parent material. It is typically gently sloping land that allows mature soils of the most ideal characteristics to develop.

Valley floors and flatlands are often very poorly drained. When this is the case and water levels in the soil are near the surface, gravitational water is unable to percolate downward, and capillary water may move harmful concentrations of salt and alkaline substances to the surface. This condition presents a constant problem in the irrigation of flatlands because irrigation tends to raise the water table. Artificial drainage ditches must be provided to lower high water tables in such instances.

Slope aspect affects microclimate. North-facing slopes in the Northern Hemisphere have microclimates that are cooler and wetter than those on south-facing exposures, which receive the sun's rays more directly and are therefore warmer and drier. Local variations in soil depth, texture, and profile development result directly from microclimatic differences.

Land surface configuration also affects soil development indirectly through its effects on vegetation. Where steep slopes prevent the development of a mature soil to support abundant vegetation, the modest plant cover diminishes the amount of organic debris available for the soil.

Time

Soils have a tendency to develop toward a state of equilibrium with their environment. A soil is mature when it has reached such a condition of equilibrium. Young soils are still in the process of alteration to achieve equilibrium with their environmental conditions. Mature soils have well-developed and stable horizons that indicate the conditions under which they developed. Young or immature soils typically have poorly developed horizons or none at all. Very old soils may have horizons that are so well developed as to present problems. Such soils frequently contain dense pans or crusts in their *B* horizons. These may consist of eluviated clays, chemically concentrated calcium carbonate (lime), silica, or oxides of iron and aluminum. Soils on ancient flat surfaces in the tropics, where soil formation is rapid, frequently present such problems. To use these soils for agriculture, farmers may

Figure 12.18 Laterite cut for building stone and stacked along a village road in the state of Orissa, India. ■ **Why is building with brick or stone rather than wood so important in heavily populated, less developed nations such as India?**

need to break up these crusts with dynamite or, less dramatically, by deep plowing.

Another effect of time is that as soils mature, they are decreasingly influenced by parent material and increasingly reflect their climatic and vegetative environments. Climate is usually the most apparent of all influences on soil development at a global scale, provided sufficient time has passed for the soil to reach maturity.

The importance of time in soil formation is especially clear in the case of soils developed on transported parent materials. Depositional surfaces are in many cases quite recent in geologic terms and have not been exposed to weathering processes long enough for a mature soil to develop. Recent deposits of transported materials have not yet been leached of their soluble nutrients, nor has their soil developed undesirable characteristics. Deposition is occurring today in a variety of settings: on the floodplains of large rivers, where the accumulating sediment is known as *alluvium;* downwind from dry areas, where dust settles out of the atmosphere to form deep blankets of wind-deposited silts, called *loess;* and in volcanic regions showered by ash and covered by lava. Ten thousand years ago, glaciers withdrew from vast areas, leaving behind jumbled and often complex deposits of rocks, sand, silt, and clay. In terms of agricultural productivity, the world's best soils are found on young alluvium, loess, volcanic ash and lava, and certain types of glacial deposits.

Somewhat older deposits of similar materials may be less productive, if they have been leached of vital nutrients or have developed unfavorable structures.

Because of the great number and variability of materials and processes involved in the formation of soils, there is no fixed amount of time that it takes for a soil to become mature. Generally, though, it takes many hundreds to thousands of years for a soil to reach maturity.

Soil-Forming Regimes

Although a nearly infinite number of factors combine and interact to produce soils, these processes can be classified into a limited number of general **soil-forming regimes.** The characteristics that differentiate major soil types result from their soil-forming regimes. The variations among these soil-forming regimes result primarily from differences in climate and vegetation. At the broadest scale of generalization, climatic differences produce three primary soil-forming regimes: laterization, podzolization, and calcification (Fig. 12.19).

Laterization

Laterization is a soil-forming regime that occurs in humid tropical and subtropical climates as a result of high temperatures and abundant precipitation. These

Figure 12.19   Profile development in the three major soil-forming regimes: (a) laterization; (b) podzolization; (c) calcification.
■ **How are these three generalized soil profiles related to Figure 12.16?**

climatic environments encourage rapid breakdown of rocks and decomposition of nearly all minerals. Despite the dense vegetation that is typical of these climates, little humus is incorporated into the soil because of rapid decomposition of plant litter and enormous numbers of soil microorganisms. Because of abundant moisture, eluviation and leaching play dominant roles in the formation of soils in the humid tropics. Lateritic soils are generally reddish in color from iron oxides; the term *laterite* means bricklike. There is no *O* horizon, and the *A* horizon loses fine soil particles as well as most minerals and bases except for iron and aluminum compounds, which are insoluble primarily because of the absence of organic acids from humification (Fig. 12.19a). As a result, the topsoil is reddish, coarse in texture, and tends to be porous. In contrast to the *A* horizon, the *B* horizon has a heavy concentration of illuviated materials. Where tropical forest vegetation remains, soluble nutrients released by weathering are quickly absorbed by vegetation, which eventually returns them to the soil where they are reabsorbed by plants. This rapid cycling of nutrients prevents the total leaching away of the bases, so the soil is only moderately acidic. Removal of vegetation permits total leaching of bases, resulting in the formation of crusts of iron and aluminum compounds (laterites), as well as accelerated erosion of the *A* horizon.

Laterization operates year-round because of the small seasonal variations in temperature or precipitation in the humid tropics. This continuous activity, and strong weathering of parent material, causes some tropical soils to develop to depths of as much as 8 meters (25 ft) or more.

Podzolization

Podzolization occurs in its purest form in the high middle latitudes, where the climate is moist with short, cool summers and long, severe winters. The typical coniferous forests of these climate regions are an integral part of the podzolization process.

Where temperatures are low much of the year, microorganism activity is reduced enough that humus does accumulate; however, because of the small number of animals living in the soil, there is little mixing of humus below the surface. Leaching and eluviation by acidic soil solutions remove the soluble bases and aluminum and iron compounds from the *A* horizon (Fig. 12.19b). The remaining silica gives a distinctive ash-gray color to the *E* horizon (*podzol* is derived from a Russian word meaning "ashy"). Because most coniferous trees require a minimum amount of bases, they return little basic material to the soil. The needles they drop are chemically acidic. This contributes to the acidic quality of the soil. Indeed, it is difficult to say whether the soil is acidic because of the vegetative cover, or whether the vegetative cover is adapted to the acidic soil.

Podzolization can take place beyond the typical cold, moist climate if the parent material is highly acidic, as

Figure 12.20 This shallow-rooted grass is growing in a thin layer of topsoil over a much thicker layer of calcium carbonate called calcrete. ■ **What precipitation characteristics are associated with the calcification soil-forming process?**

on the sands common along the east coast of the United States. The pine forests that grow in such acidic conditions return acids to the soil, promoting the process of podzolization.

Calcification

The third distinctive soil-forming regime is called **calcification.** In contrast to both laterization and podzolization, which require humid climates, calcification occurs in climates where evapotranspiration exceeds precipitation. In areas of low precipitation, the air is often loaded with alkali dusts such as calcium carbonate ($CaCO_3$). When calm conditions prevail or when it rains, the dust settles across the landscape and accumulates in the soil. The rainfall is just sufficient to translocate these materials to the *B* horizon (Fig. 12.19c). Over hundreds to thousands of years, the $CaCO_3$-enriched dust concentrates in the *B* horizons, forming hard layers of *caliche* or the much thicker *calcretes* (Fig. 12.20). These accumulations form by the upward (capillary) movement of dissolved alkaline salts in groundwater when the water table is near the surface.

Calcification is important in the climatic regions where moisture penetration is shallow. The subsoil is too dry to support tree growth, and shallow-rooted grass or shrubs are the primary forms of vegetation. Calcification is enhanced as grasses use calcium, drawing it up from lower soil layers and returning it to the soil when the grasses die. Grasses and their dense root networks provide large amounts of organic matter, which typically is mixed deep into the soil by burrowing animals. Middle-latitude grassland soils are rich both in bases and in humus, and are the world's most productive agricultural soils.

It is interesting that a soil transect across the United States from Illinois to Colorado shows a gradual westward *decrease* in soil humus content and thickness of the *A* horizon, while it shows an *increase* in the prominence of calcium carbonate at the base of the *B* horizon. The most productive soils are found about midway in this transect, where both humus and base contents are relatively high. The deserts of the American West generally have no humus, and the rise of capillary water can leave not only calcium carbonate, but also sodium chloride (salt) at the surface.

Regimes of Local Importance

Two additional soil-forming regimes merit attention. Both characterize areas with locally poor drainage, although they occur under very different climatic conditions. The first, **salinization,** or the concentration of salts in the soil, is often detrimental to plant growth. Salinization occurs in stream valleys, interior basins, and other low-lying areas, particularly in arid regions with high groundwater tables. The high groundwater can be the result of adjacent mountain ranges or stream flow originating from humid regions (Fig. 12.21), but can also be a consequence of intensive irrigation. Rapid evaporation of water leaves behind a high concentration of soluble salts in the soil, destroying its agricultural productivity. An extreme example of salinization is the fertile crescent of Mesopotamia (Iraq), where thousands of years of irrigated agriculture in the desert have led to soils too salty to cultivate today (see again Fig. 12.17 for an example). The second, **gleization,** occurs in poorly drained areas under cold and wet environmental conditions. This process is typically associated with peat bogs, where the soil is an accumulation of humus layers overlying a blue-gray layer of thick, gummy, water-saturated clay. Unreduced iron in early stages of decomposition imparts a blue-gray color to the soil. In formerly glaciated, poorly drained regions such as northern Russia, Ireland, Scotland, and Scandinavia, peat has long been harvested and used as an important source of energy (Fig. 12.22).

Soil Classification

Soils, like climates and other phenomena that vary spatially, can be classified. The agency in the United States that is responsible for soil classification (termed **soil taxonomy**) is the Soil Survey Division of the Natural Resources Conservation Service (NRCS), a branch of the Department of Agriculture. As with any classification system, the methods and categories are continually being updated.

Soil classifications are published in **soil surveys,** which outline and describe the kinds of soils in a region and include maps of the distribution of soil types, usually at the county level. These documents, available for most parts of the United States, are useful reference sources for factors such as soil fertility, irrigation, and drainage.

The NRCS Soil Classification System

The National Resources Conservation Service soil classification system is based on the nature of soil horizons. The largest classification of soils is the **soil order,** of which eleven are recognized (Fig. 12.23). To provide greater detail, soil orders can also be subdivided into suborders, great groups, subgroups, families, and series. More than 10,000 soil series have been recognized in the United States. Some soil orders reflect regional climatic

Figure 12.21 Salt deposits associated with high groundwater levels along the Amorgosa River in the area of Death Valley ■ **What causes salinization, and why is it often associated with agriculture?**

Norman Meek

R. Gabler

Figure 12.22 Peat has been harvested for fuel for hundreds of years in bogs such as this one in northern Scotland. Note the spade markings where peat has been removed and the spade-sized peat blocks (in the background) cut by hand and awaiting transport to a nearby farm, where they will be stored, dried, and used as fireplace fuel. ■ **What are the similarities and differences between peat and various grades of coal?**

conditions because they have developed during the Holocene under relatively stable climates. During this interval, the dominant soil-forming processes have produced similar types of soil horizons for regions with similar climates and vegetation cover. However, other soil orders reflect the recency or type of parent material, so the distribution of these soils does not conform to climatic regions.

The system of soil types uses names derived from root words of classical languages such as Latin, Arabic, and Greek to refer to the different soil categories. The names, like the system, are precise and consistent, and were chosen to describe the characteristics that distinguish one soil from another—recognizable features and processes that are used to classify a soil in the proper category.

When examining a soil for classification under the NRCS system, particular attention is paid to horizons and textures that characterize the soil. Some of these horizons are situated below the surface (**subsurface horizons**); others, called **epipedons,** are surface layers that usually exhibit the darker shading associated with the presence of organic material (humus). Examples of some of the more common horizons, illustrating how names were chosen to represent actual soil properties, are found in Table 12.1.

NRCS Soil Orders

Seven of the eleven soil orders largely reflect the influence of Holocene climates, those of the past 10,000 years or less, and they are discussed here in a sequence that reflects increasing moisture levels. Frequent comparison of Figure 12.23 with Figure 9.4 will illustrate the relationship between soils and climate within the United States; these relationships are also shown in Figure 12.24.

Aridisols are soils of desert regions; they develop primarily under conditions where precipitation is less than

half of potential evaporation. Consequently, most Aridisols reflect the calcification process. Where groundwater tables are high, evidence of salinization may also be present. Although ordinary horizon development is weak because of lack of water movement in the soil, there is often a subsurface accumulation of calcium carbonate (calcic horizon), salt (salic horizon), or calcium sulfate (gypsic horizon). Soil humus is minimal because vegetation is sparse in deserts; therefore, desert soils are often light in color. Aridisols are usually alkaline, but because few nutrients have been leached, they can produce bountiful harvests if irrigated to reduce the pH and salinity. Geographically, Aridisols are the most common soils on Earth because deserts cover such a large portion of the land surface.

Mollisols are most closely associated with grassland regions and are among the best soils for sustained agriculture. Because they are located in semiarid climates, Mollisols are not heavily leached, and thus they have a generous supply of bases, especially calcium. The characteristic horizon of a Mollisol is a mollic epipedon—a thick, dark-colored surface layer rich in organic matter from the decay of abundant root material. Grasslands and associated Mollisols served as the grazing lands for countless herds of antelope, bison, and horses. Before the invention of the steel plow, the thick root material made this soil nearly uncultivable in the United States, and thus led to the widespread public image of the Great Plains as a "Great American Desert." Today, Mollisols support most of the grain production from domesticated grasses (wheat, rye, oats, corn, barley, and sorghum). In regions of adequate precipitation, such as the tall-grass prairies of the American Midwest, the combination of soils and climate is unexcelled for agriculture. In areas of lesser precipitation, periodic drought is a constant threat, and the temptation of fertile soils was the

Figure 12.23   Distribution of NRCS soil orders in the United States

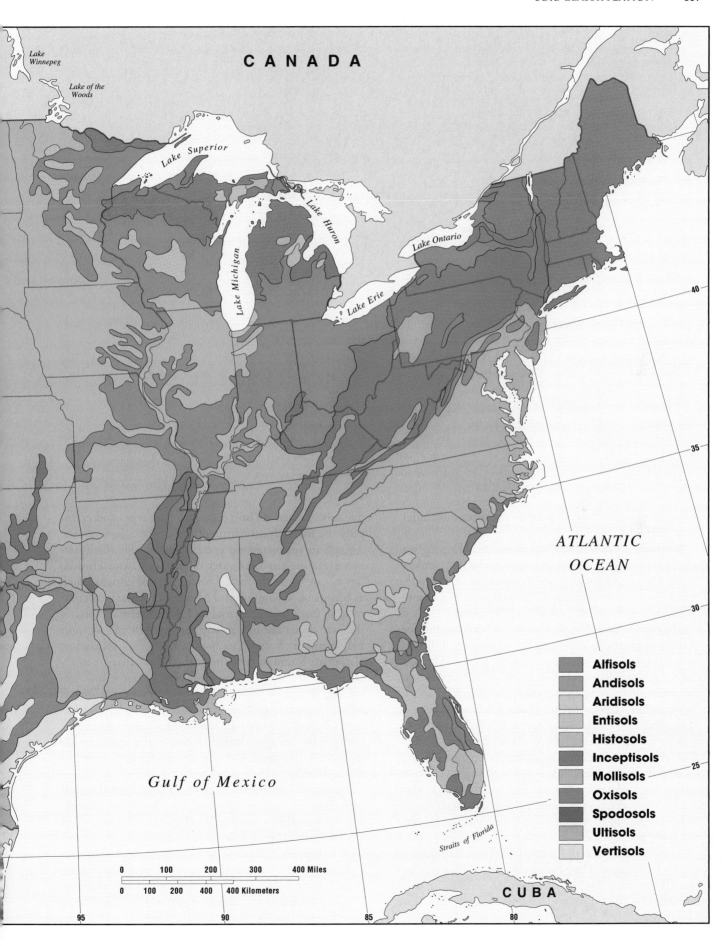

CANADA

Lake Winnepeg

Lake of the Woods

Lake Superior

Lake Huron

Lake Michigan

Lake Ontario

Lake Erie

ATLANTIC OCEAN

40

35

30

25

Gulf of Mexico

Straits of Florida

CUBA

Alfisols
Andisols
Aridisols
Entisols
Histosols
Inceptisols
Mollisols
Oxisols
Spodosols
Ultisols
Vertisols

| 0 | 100 | 200 | 300 | 400 Miles |

| 0 | 100 | 200 | 400 | 400 Kilometers |

95   90   85   80

Table 12.1   Selected Common Horizons in the NRCS Soil Classification System*

Oxic horizon (from *oxygen*)
A subsurface layer that contains hydrated oxides of iron and aluminum. Found in tropical and subtropical climates at low elevations.

Argillic horizon (from Latin: *argilla,* clay)
Usually formed beneath the *A* horizon by illuviation, this layer contains a high percentage of accumulated silicate clays.

Ochric epipedon (from Greek: *ochros,* pale)
A surface horizon that is light in color and very low in organic matter or very thin.

Albic horizon (from Latin: *albus,* white)
Commonly the *A2* horizon overlying a spodic horizon, this layer is usually sandy and very light in color due to the removal of clay and iron oxides.

Spodic horizon (from Greek: *spodos,* wood ash)
Usually underlying the *A2* horizon, the spodic horizon is dark in color because of the illuviation of humus, aluminum oxides, and often, iron oxides.

Mollic epipedon (from Latin: *mollis,* soft)
Very high in content of basic substances (calcium, magnesium, potassium), the mollic epipedon is a relatively thick and dark-colored surface layer.

Calcic horizon (from *calcium*)
A subsurface horizon rich in accumulated calcium carbonate or magnesium carbonate.

Salic horizon (from *salt*)
A layer of soil material at least 6 inches thick and containing at least 2 percent salt. Common in desert basins.

Gypsic horizon (from *gypsum*)
A subsurface soil horizon rich in accumulated calcium sulfate (gypsum).

* This table includes only some of the more common horizons.

---

downfall of many farmers prior to the advent of center-pivot irrigation.

Vertisols are typically found in regions of strong seasonal precipitation, such as tropical wet and dry climates. In the United States, they are most common where the weathering of rocks produces clay-rich soils. The combination of clayey soils in a wet and dry climate leads to the drying of the soil and consequent shrinkage and formation of deep cracks during the dry season, followed by expansion of the soil during the wet season. The constant shrink-swell process disrupts horizon formation to the point that soil scientists often describe Vertisols as "self-plowing" soils. Anything solid constructed atop or in heaving soils, such as highways, sidewalks, and basements, may be disrupted. Vertisols are dark in color, are high in bases, and contain considerable organic material derived from the grasslands or savanna vegetation with which they are normally associated. Although they harden when dry, and become sticky and difficult to cultivate when swollen with moisture, Vertisols can be agriculturally productive.

Alfisols occur in a wide variety of climatic settings in the United States. They are characterized by a subsurface clay horizon (argillic *B* horizon), a medium to high base supply, and a light-colored ochric epipedon. The five suborders of *Alf*isols reflect climate types and exemplify the hierarchical nature of the classification system: Aqu*alfs* are seasonally wet and can be found in mesothermal areas such as Louisiana, Mississippi, and Florida; Bor*alfs* are found in moist, microthermal climates such as Montana, Wyoming, and Minnesota; Ud*alfs* are common in both microthermal and mesothermal climates that are moist enough to support agriculture without irrigation, such as Wisconsin, Ohio, and Tennessee; Ust*alfs* are found in mesothermal climates that are intermittently dry, such as Texas and New Mexico; and Xer-*alfs* are found in California's Mediterranean climate, which is characterized by wet winters and long, dry summers. Because of their abundant bases, Alfisols can be very productive agriculturally if the local deficiencies are corrected: irrigation for the dry suborders, properly drained fields for the wet suborders.

Spodosols are most closely associated with the podzolization soil-forming process. They are readily identified by their strong horizon development. There is often a white or light gray *E* horizon (albic horizon) covered with a thin, black layer of partially decomposed humus and underlain by a colorful *B* horizon enriched in relocated iron and aluminum compounds (spodic horizon). Spodosols are generally low in bases and form in porous

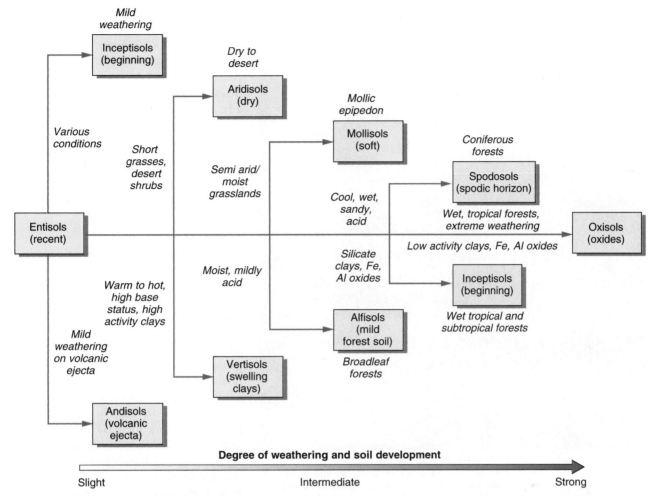

Figure 12.24 NRCS soil orders. The soil orders of the Natural Resource Conservation Service can be linked to the parent materials, climate, and vegetation of the region in which they formed. The linkages form a tree like pattern, as seen here. ■ **How is degree of weathering related to climatic characteristics?**

From *Nature and Properties of Soils*, 12/e by Brady and Weil, © 1998. Reprinted by permission of Pearson Education, Inc., Upper Saddle River, NJ.

substrates such as glacial drift or beach sands. In New England and Michigan, Spodosols are also acidic. In these regions, as well as in similar regions in northern Russia, Scandinavia, and Poland, only a few types of agricultural plants, such as cucumbers and potatoes, can tolerate the microthermal climates and sandy, acidic soils. Consequently, the cuisine of these regions directly reflects the Spodosols that dominate the areas.

Ultisols, like Spodosols, are also low in bases because they develop in moist or wet regions. Ultisols are characterized by a subsurface clay horizon (argillic horizon) and are often yellow or red in color because of the accumulation of residual iron and aluminum oxides in the *A* horizon. In North America, the Ultisols are most closely associated with the southeastern United States. When first cleared of forests, these soils can be agriculturally productive for several decades. But a combination of high rainfall and the associated runoff and erosion from the fields decreases the natural fertility of the soils. Ulti-

sols remain productive only with the continuous application of fertilizers. Today, forests cover many former cotton and tobacco fields of the southeastern United States because of a reduction in soil fertility and extensive soil erosion (see again Fig. 10.9).

Oxisols have developed over long periods of time in regions with high temperatures and heavy annual rainfall. They are almost entirely leached of soluble bases and are characterized by a thick horizon of iron and aluminum oxides (oxic horizon). The soil consists mainly of minerals that resist weathering (for example, quartz, kaolin, hydrated oxides). Oxisols are most closely associated with the humid tropics, but they also extend into savanna and tropical thorn forest regions. In the United States, Oxisols are present only in Hawaii. Oxisols are dominated by laterization and retain their natural fertility only as long as the soils and forest cover maintain their delicate equilibrium. The bases in the tropical rainforests are stored mainly in the vegetation. When a tree

Figure 12.25 Gully erosion is one of the more spectacular examples of poor agricultural practices. It produces permanent alteration of the landscape and guarantees that the original productivity of the land cannot be regained. ■ **What could have been done to prevent this example of land mismanagement?**

beginning, usually because of a very cold climate, repeated flood-related deposition, or a high rate of erosion. In the United States, Inceptisols are most common in Alaska, the lower Mississippi River floodplain, and the western Appalachians. Globally, Inceptisols are especially important along the lower portions of the great river systems of South Asia, such as the Ganges-Brahmaputra, the Irrawaddy, the Chao Phraya, and the Mekong. In these areas, the sediments associated with periodic flooding constantly enrich Inceptisols, and they form the basis for agriculture that supports millions of people.

Histosols develop in poorly drained areas, such as swamps, meadows, or bogs, as a product of gleization. They are largely composed of slowly decomposing plant material. The waterlogged conditions of the soil deprive bacteria of the oxygen necessary to prevent accumulation of organic matter. Although Histosols may be found in low areas with poor drainage at all latitudes, they are most common in tundra areas or in recently glaciated, high-latitude locations such as Canada, Ireland, and Scotland. Histosols in the subpolar latitudes are commonly acidic and only suitable for special bog crops such as cranberries. Histosols are important as the primary source of peat, which is a fuel source in regions where Histosols are common.

Andisols are soils that develop on volcanic parent materials, usually volcanic ash (dust particles emitted by volcanoes). They often have a low bulk density as well as substantial proportions of glassy minerals and the weathering products of volcanic rocks. Because many of these soils are replenished by eruptions, they are often fertile compared to soils in surrounding regions. Intensive agriculture atop Andisols supports the dense populations of the Philippines and West Indies. However, in tropical climates, intensive cultivation without proper terracing of the fields can lead to severe erosion and dissection of the landscape. In the United States, Andisols are most common on the slopes of, and downwind from, the Cascade volcanoes and, to a lesser extent, in Hawaii and Alaska.

Regardless of their composition, origin, or state of development, Earth's soils remain one of our most important and vulnerable resources. The word *fertility,* so often associated with soils, has a meaning that takes into consideration the usefulness of these soils to humans. Soils are fertile in respect to their effectiveness in producing specific vegetation. Some soils may be fertile for corn

dies, epiphytes and insects must recycle the bases rapidly before the heavy rainfall leaches them from the system. The burning of vegetation associated with slash-and-burn agriculture in rainforests releases the nutrients necessary for crop growth but quickly results in their loss from the ecosystem. Many tropical Oxisols that once supported lush forests are now heavily dissected and only support a combination of weeds, shrubs, and grasses.

The remaining four soil orders are limited in extent, and are not closely associated with climatic patterns.

Entisols are soils that lack horizons, usually because of their recent development. They are often associated with the continuing erosion of sloping land in mountainous regions or with the frequent deposition of alluvium by flooding. They can also occur in areas of heavy sand accumulation (Nebraska Sand Hills), where the porous and mobile substrate greatly decreases the rate of horizon development.

Inceptisols are young soils with weak horizon development. The processes of *A* horizon depletion (eluviation) and *B* horizon deposition (illuviation) are just

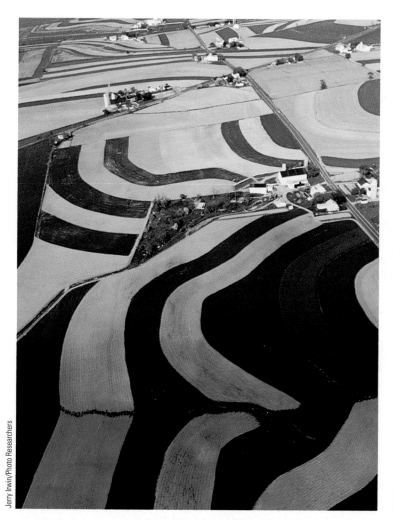

and others for potatoes. Other soils retain their fertility only as long as they remain in delicate equilibrium with their vegetative cover.

It is clearly the responsibility of all of us who enjoy the agricultural end products of farm, ranch, and orchard, or who simply appreciate the beauty of forest and field, to recognize and help protect our valuable soils. Although space does not permit a thorough examination of the problems of soil erosion, soil depletion, and land mismanagement, we should be conscious of their existence in the world today (Fig. 12.25). We should also be aware that each of these problems has reasonable solutions (Fig. 12.26). Maintaining soil fertility and usefulness is a critical challenge to humanity and one of the essentials in our continuing struggle to protect the resources of our natural environment.

Jerry Irwin/Photo Researchers

Figure 12.26 The contour farming techniques used on this farm are excellent examples of conservation methods designed to preserve the soil resource. ■ **What other soil conservation practices are often used to preserve the soil resource?**

DEFINE & RECALL

soil	silt	Cl, O, R, P, T	epipedon
gravitational water	sandy	residual parent material	Aridisol
eluviation	sand	transported parent material	Mollisol
illuviation	soil grade	laterite	Vertisol
hardpan	loam	soil-forming regimes	Alfisol
leaching	infiltrate	laterization	Spodosol
stratification	soil ped	podzolization	Ultisol
capillary water	permeability	calcification	Oxisol
hygroscopic water	porosity	salinization	Entisol
humus	pH scale	gleization	Inceptisol
soil texture	fertilization	soil taxonomy	Histosol
clayey	parent material	soil survey	Andisol
clay	soil horizon	soil order	
silty	soil profile	subsurface horizon	

DISCUSS & REVIEW

1. Why is soil an outstanding example of integration and interaction among Earth's subsystems?

2. Describe the different circumstances in which water is found in soil.

3. What is the effect of eluviation if carried to an extreme? What is the effect of illuviation if carried to an extreme?

4. Under what conditions does leaching take place? What is the effect of leaching on the soil and, consequently, on the vegetation it supports?

5. How can capillary water contribute to the formation of caliche? What is the effect of caliche on drainage?

6. How might soil air differ from air in the atmosphere? What is the effect on life when air is excluded from water-saturated soils?

7. How is humus formed? What relation does humus have to soil fertility?

8. What conclusions can you draw from the color of the soil in your area? How might color relate to fertility?

9. How is texture used to classify soils? Describe the ways scientists have classified soil structure.

10. What pH range indicates soil suitable for most complex plants?

11. What are the general characteristics of each horizon in a soil profile? How are soil profiles important to scientists?

12. What factors are involved in the formation of soils? Which is most important on a global scale?

13. How does transported parent material differ from residual parent material? List those factors that help to determine how much effect the parent material will have on the soil.

14. What are the most important effects of parent material on soil?

15. List a number of ways in which humus is important to soils.

16. How does the presence of earthworms alter soil?

17. Describe the various ways in which temperature and precipitation are related to soil formation.

18. The Bonneville Salt Flats in Utah are well known as a natural soil formation that provides a perfect surface for auto racing. How do you suppose these salt flats were formed?

19. Describe the three major soil-forming regimes.

20. Which soil orders of the NRCS soil classification system have the most agricultural potential? Why?

CONSIDER & RESPOND

1. Refer to Figure 12.8. Using the texture triangle, determine the textures of the following soil samples.

	Sand	Silt	Clay
a.	35%	45%	20%
b.	75%	15%	10%
c.	10%	60%	30%
d.	5%	45%	50%

   What are the percentages of sand, silt, and clay of the following soil textures? (Note: Answers may vary, but they should total 100 percent.)

   e. Sandy clay   f. Silty loam

2. Refer to Figure 12.12 and associated pages in the text.

   a. What horizons make up the zone of eluviation?

   b. What are two processes that occur in the zone of eluviation?

   c. The various *B* horizons are in what zone?

   d. Weathered parent material is the major constituent of what horizon?

   e. Partly decomposed organic debris makes up which horizon?

3. Refer to Table 12.1.

   a. What materials accumulate in an argillic horizon?

   b. Which would generally be better suited for agriculture—a soil with an ochric epipedon or a mollic epipedon? Why?

   c. What name would be given to a 7-inch-thick horizon that contained at least 2 percent salt?

4. Use Figure 12.23, the map of world climatic regions (Figure 9.4), and an atlas to respond to the following:

   What soil order or orders and associated climate or climates are most often found together in the Corn Belt, located principally in Illinois, Indiana, Iowa, and Ohio?

In the northern Great Lakes Region (Lake Huron, Lake Michigan, and Lake Superior)? In the grasslands astride the 50th parallel of latitude in North America? In most of Nevada, southeastern California, and southern Arizona? In the southeastern United States between the Appalachians and the Coastal Plain?

5. Where would you rank soils in terms of importance among a nation's environmental resources? Give your opinion of the overall value of soils in the United States and the extent to which these soils are preserved and protected.

EARTH MATERIALS, THE LITHOSPHERE, AND PLATE TECTONICS

Knowledge of the origin, nature, and structure of rocks is important to physical geographers because these three factors influence the character and appearance of the surface environment.

What surface characteristics does bedrock influence? How can topography be influenced by bedrock? How are humans affected by bedrock properties?

The rock cycle provides a good conceptual model for understanding the relationships among internal and external processes and materials that create, alter, and recycle Earth materials that make up the surface of our planet.

What are these processes and materials, and how do they interact to produce the major types of rocks? Could the rock cycle be considered an Earth subsystem? Why or why not? How could human activities affect parts of the rock cycle, or vice versa?

Only eight chemical elements account for almost 99 percent of Earth's crust by weight, and the most common minerals are combinations of these same elements.

What does this suggest about mineral classification? What does it suggest concerning the most common rocks in Earth's crust?

Earth's surface configuration and the locations of continents, oceans, and major landforms have changed through geologic time and continue to change today, "with no clear beginning and no definite end in sight."

What does this mean? What is its significance?

Plate tectonics theory has an all-encompassing relationship to the Earth sciences, similar to that which evolution has to the life sciences.

What is the importance of such a theory? Why is science so concerned with theory?

For a scientific theory to be accepted, it is not sufficient merely to describe what happens; the theory must also explain *how* it happens and be supported by adequate evidence.

How is this statement related to the concept of continental drift? What processes explain plate tectonics?

Earth Systems Today CD

Chapter 4: Rocks & the Rock Cycle, Rock Cycle
Chapter 4: Rocks & the Rock Cycle, Rock Laboratory: Igneous
Chapter 4: Rocks & the Rock Cycle, Rock Laboratory: Sedimentary
Chapter 4: Rocks & the Rock Cycle, Rock Laboratory: Metamorphic
Chapter 2: Earth's Layers, Core Studies
Chapter 9: Plate Tectonics, Plate Location
Chapter 9: Plate Tectonics, Plate Boundaries
Chapter 3: Geologic Time

The cooling of molten lava into rock is a fundamental process in shaping Earth's surface.

If we could travel back in time to a few hundred million years ago, we probably would not recognize our planet. A map of the world drawn to represent that time period would show a very different physical geography. The most basic geographic features that a world map or globe shows—the shapes of oceans and continents, their locations, and their orientations relative to the poles and equator—would be very different from those of today. Where extensive plains or major mountain ranges exist today, seas existed in the past—fossil fish have been found on peaks thousands of feet above sea level. Fossilized remains of tropical plants are located in the polar regions, and ancient tropical soils exist where it is not hot or humid today. These dramatic environmental changes, along with the current shapes, sizes, and distribution of our planet's major surface features, require scientific explanations. They also demonstrate the interrelatedness of the four major subsystems of

Earth: atmosphere, hydrosphere, biosphere, and lithosphere. The remainder of this textbook will focus mainly on the lithosphere and hydrosphere, but will not neglect the importance of the biosphere and atmosphere as they interact with and affect Earth's surface environments. Understanding the geography of the past and the processes that shaped environmental conditions then helps us to think about, and consider in scientific terms, the geography of the future as well as the interactions of processes and Earth materials that affect our planet's surface today.

Earth's Crustal Composition

Earth's rocky exterior is called the **crust.** It is composed of many different rocks and minerals that respond in diverse ways and at varying rates to Earth-shaping processes. To understand how these processes work, physical geographers must know the different types of rocks and their primary characteristics, especially how they respond to tectonic and gradational processes. **Tectonic processes** are those involved in breaking and deforming the Earth's exterior, often creating elevational differences in the surface. **Gradation processes** are the many surficial processes that act to reduce these differences by wearing away high places and filling in low areas, mainly through erosion and deposition.

Minerals are naturally occurring inorganic substances that are the building blocks of rocks. They are well-defined combinations of atomic elements with distinct chemical formulas and recognizable physical characteristics. A **rock** is an aggregate (a whole made up of parts) of mineral particles. The most common elements of Earth's crust (and, therefore, the minerals and rocks that make up the crust) are oxygen and silicon, followed by aluminum and iron, and the bases: calcium, sodium, potassium, and magnesium. As you can see in Table 13.1, these eight chemical elements—out of the more than 100 known—account for almost 99 percent of the Earth's crust by weight. The most common minerals are combinations of these eight elements.

(a)

(b)

Figure 13.1 The geometric arrangement of atoms determines the crystal form of a mineral: **(a)** pyrite crystals, also known as "fool's gold"; **(b)** quartz crystals. ■ **What is a mineral?**

Minerals

Every mineral has distinctive and recognizable physical qualities that aid in its identification—for example, a particular color, luster, hardness, tendency to fracture, and specific gravity. The characteristic properties of some of the most common rock-forming minerals are presented in Appendix D. Minerals are crystalline in nature, although crystals may be evident only when viewed through a microscope. Mineral crystals also have consistent geometric forms that express their molecular structure (Fig. 13.1). Electrical bonds hold the atomic elements composing a mineral together; to be stable, a mineral must have a balance between positive and negative charges. Accordingly, minerals are solid crystalline structures held together either by covalent atomic bonding or by electrostatic attraction between oppositely charged ions. Chemical bonding affects the resistance and strength of minerals, and thereby of the rocks they form. Minerals with weak internal bonds undergo chemical alteration most easily. Ions may leave or be traded within their structure, producing physical changes that are the chemical basis of rock weathering.

Table 13.1	Most Common Elements in Earth's Crust
Element	**Percentage of the Earth's Crust by Weight**
Oxygen (O)	46.60
Silicon (Si)	27.72
Aluminum (Al)	8.13
Iron (Fe)	5.00
Calcium (Ca)	3.63
Sodium (Na)	2.83
Potassium (K)	2.70
Magnesium (Mg)	2.09
Total	98.70

Source: J. Green, "Geochemical Table of the Elements for 1953," *Bulletin of the Geological Society of America* 64 (1953).

Because of the ease with which certain elements combine with a variety of others, there are discrete families of minerals. The most active of these elements are silicon, oxygen, and carbon. Consequently, the most common mineral groups are silicates, oxides, and carbonates. The **silicates** are by far the largest and most important mineral group, constituting 92 percent of Earth's crust. They are generally created by the cooling of **magma** (a subsurface molten mass containing the elements from which minerals can form), which causes the crystallization of different minerals at successively lower temperatures (Fig. 13.2). Silicates are compounds of oxygen and silica with one or more metals and/or bases. Dark, heavy, iron-rich silicate minerals are the first to crystallize (at high temperatures); light-colored, lighter-weight, silica-rich minerals crystallize later at cooler temperatures. Quartz is one of the last silicates to crystallize. The order of formation of minerals parallels their relative stability in a rock, as silicate minerals that crystallize later tend to be more resistant. In the weathering of granite (a rock that includes a variety of silicates), the dark, heavy minerals are the first to decompose, and quartz is one of the last.

Figure 13.2   The crystallization of silicate minerals from magma. The order of mineral formation is a function of temperature and pressure conditions. In general, dark, iron-rich minerals form solid crystals first and tend to sink in the magma. Lighter minerals, richer in silica, tend to crystallize later at lower temperatures and pressures as the body of magma cools into rock.   ■ **How is the crystallization process related to the freezing of ice in water?**

Rocks

Most rocks consist of several minerals (Fig. 13.3). Each mineral in a rock remains separate and retains its own distinctive properties, which together determine the properties of the rock. Though most rocks are made of a variety of different minerals, a few kinds of rocks are each composed of a single mineral. Rocks are the building materials of the lithosphere. They are lifted, pushed down, and deformed by tectonic processes; they are weathered and eroded by gradational forces, to be deposited as sediment elsewhere.

Solid rock that underlies loose surface material is called **bedrock.** Where bedrock is not exposed at the surface, a cover of broken and decomposed rock fragments called **regolith** overlies it. Soil may or may not have formed on the regolith (see again Fig. 12.12). Trenches cut into the land surface do not always reveal the presence of all three layers. On a mountain slope, for example, running water may remove weathered material as fast as it forms, so that the bedrock is left exposed. Such exposed bedrock is called an **outcrop** (Fig. 13.4).

Although the number of common rock-forming minerals is limited, they combine in a multitude of ways to produce an enormous variety of rock types. Nevertheless, rocks can be categorized as one of three major types, based on their origin. These rock types are igneous, sedimentary, and metamorphic.

The Rock Cycle   Like landforms, many rocks do not remain in their original form indefinitely but instead, in the long term, tend to undergo processes of transformation. The **rock cycle** is a conceptual model for understanding processes that generate, alter, and transform mineral material into different kinds of rocks and the components that make up rocks. The term *cycle* is used because existing rocks supply the material to make new and sometimes very different rocks, so that existing rocks can be "recycled" to form new rocks. The geologic age of a rock is the time when it assumed its current state; metamorphism, or melting, and other rock-forming processes reset the age of origin (Fig. 13.5). Although a complete cycle is shown in Figure. 13.5, many rocks do not go through all steps of the rock cycle,

Figure 13.3   Most rocks consist of several minerals. This piece of cut and polished granite displays intergrown crystals of differing composition, color, and size that give the rock a distinctive appearance.

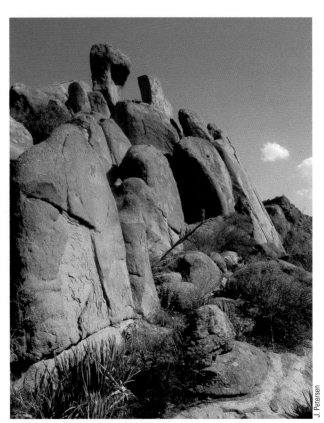

Figure 13.4 Outcrops are masses of solid rock exposed at the surface. ■ **What physical characteristics of these rock outcrops has caused them to protrude above the general land surface?**

as shown by the arrows that cut across the diagram. **Igneous rocks** form by cooling and crystallizing from a molten state. Igneous rocks can revert to their original state by remelting; they can be changed into metamorphic rocks through heat, pressure, and/or chemical action; or they may be weathered into fragments that are eroded, transported, and deposited to form sedimentary rocks. **Sedimentary rocks** form from particles and deposits derived from any of the three basic rock types. **Metamorphic rocks** are rocks that have been changed through heat and pressure; they can be created out of igneous or sedimentary rock, or metamorphosed further into a new rock type. In addition, metamorphic rocks can be heated sufficiently to become magma and cool as igneous rocks.

Igneous Rocks When molten rock material cools and solidifies, igneous rocks are formed. Below Earth's surface, this melt is called magma. The igneous material with which we are most familiar is **lava,** the surface form of magma (Fig. 13.6). Lava is spewed forth by volcanoes at temperatures as high as 1090°C (2000°F).

Molten material that emerges at Earth's surface and solidifies forms **extrusive igneous rock,** or volcanic rock (after Vulcan, the Roman god of fire), such as solidified lavas. If rising magma does not break through to the surface but solidifies within Earth, the resulting internally

cooled and crystallized rock is known as **intrusive igneous rock.** When intrusive masses solidify deep beneath the surface, they are sometimes referred to as **plutonic rocks** (after Pluto, the Roman god of the underworld). Igneous rocks vary in chemical composition, tendency to fracture, texture, crystalline structure, and presence or absence of layering. They may be grouped or classified in terms of their crystal size or texture as well as their mineral composition.

Extrusive rocks, as well as some intrusive rocks that have pushed close to the surface, undergo rapid cooling and solidification. Rapid cooling forms fine-grained igneous rocks, typical of lavas, such as *basalt* or *rhyolite* (Fig.13.7). Only small crystals are produced under conditions of rapid cooling because there is little time for crystal growth prior to solidification. In some instances, cooling is so rapid that the resulting rock has a glassy texture, as in *obsidian* (volcanic glass).

Intrusive rocks cool very slowly because surrounding masses of rock retard the loss of heat from molten magma. Slow cooling allows more time for larger crystal formation prior to solidification. Rocks formed in this manner are coarse-grained and have crystal grains that are easily visible without magnification. *Granite* and *gabbro* illustrate this coarse texture (Fig. 13.8).

The chemical composition of igneous rocks varies from acidic (rich in light minerals, especially silica) to basic (low in silica and rich in heavy minerals, such as compounds of iron and magnesium—the *ferromagnesian minerals*). Granite, an acidic, coarse-grained, intrusive rock, has the same chemical and mineral composition as rhyolite, a fine-grained extrusive rock. Likewise, gabbro is the coarse-grained, intrusive counterpart of fine-grained basalt, both having been formed from chemically basic magma, but one at the surface and the other at depth.

Many igneous rocks are fractured and broken along cracks that may be spaced or arranged in regular geometric patterns. Earth scientists call such simple fractures or cracks in rock **joints.** Although joints are not limited to rocks that formed from a molten state, one way they develop is by molten rocks' shrinking in volume and cracking as they cool and solidify. Basalt is famous for its distinctive jointing, which commonly forms hexagonal columns of solidified lava. Devil's Postpile in California and Devil's Tower in Wyoming are landforms consisting of hexagonal columns of lava formed by **columnar joints** (Fig. 13.9). Jointing in rocks of all types may also be caused by regional stresses in the crust.

Sedimentary Rocks As their name implies, sedimentary rocks are derived from accumulated sediments—unconsolidated mineral materials that have been eroded, transported, and deposited as a result of gradational processes. After the materials have accumulated in horizontal layers, pressure from the material above compacts the sediment, expelling water and

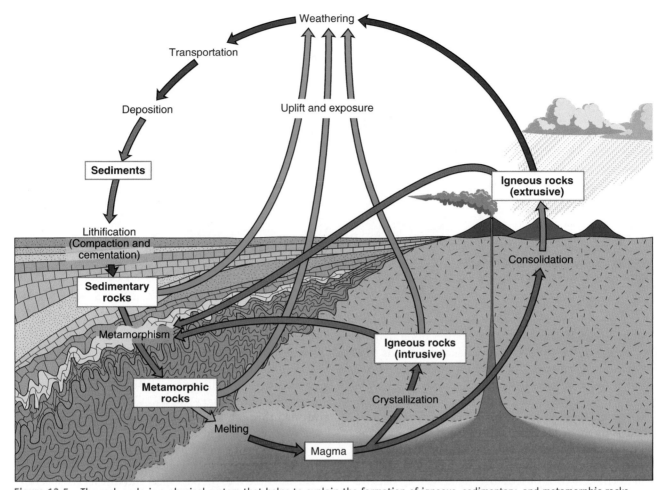

Figure 13.5 The rock cycle is a physical system that helps to explain the formation of igneous, sedimentary, and metamorphic rocks. Internal and surface processes interact to form the many kinds of rocks on our planet. Note the links that bypass some of the parts of the cycle. ■ **What conditions are necessary to change igneous rock to metamorphic rock, and metamorphic rock to igneous rock?**

Figure 13.6 Basalt, a fine-grained extrusive igneous rock, forms these fresh lava flows on the island of Hawaii. ■ **What visible characteristics in this picture suggest that the lava flow is recent?**

reducing pore space. Cementation occurs when silicon dioxide, calcium carbonate, or iron oxide accumulates and solidifies between particles of sediment. The processes of compaction and cementation transform (lithify) sediments into a solid, coherent layer of rock. Most sedimentary materials are fragments (cobbles, pebbles, sand, silt, or clay; see Appendix D) eroded from previously existing rocks that were transported and deposited in a riverbed, on a beach or sand dune, on a lake bottom, on the ocean floor, or in other environments where deposition occurs. Sedimentary rocks that form from fragments of preexisting rocks are called **clastic** (from Latin: *clastus,* broken).

Clastic sedimentary rocks include *conglomerate, sandstone, siltstone,* and *shale* (Fig. 13.10). Conglomerate is a solid mass of cemented boulders, cobbles, pebbles, or gravel; it may have sand filling in spaces between large particles. It is a hard rock and relatively resistant to weathering. Most sandstone is formed by the cementation of sand-sized quartz grains. Sandstone is usually granular, porous, and resistant to weathering. However,

Rapid cooling

Slow cooling

Figure 13.7 The rate of cooling for igneous rocks is reflected in their crystalline texture. Igneous rocks that cooled rapidly (typically at or near the surface) have fine crystals. Rocks that cooled slowly have a coarser crystalline structure. ■ **Where do coarsely crystalline igneous rocks tend to cool?**

the cementing material largely determines its strength and hardness. If cemented by substances other than silica (such as calcite or iron oxide), sandstone tends to be more easily weathered. Siltstone is similar to sandstone, though much finer textured, being formed of smaller silt-sized particles. Shale is produced from the compaction of very fine-grained sediments, especially clay. Shale is finely bedded, smooth-textured, and impermeable. It is also brittle and easily cracked, broken, or flaked apart. Clastic sedimentary rocks may be further classified as either marine or terrestrial (continental), depending on their origin. Marine sandstones have typically formed originally in near-shore coastal zones; terrestrial sandstones generally originate in desert or floodplain environments on land. The nature and arrangement of sediments in a sedimentary rock provide a great deal of

evidence for the kind of environment under which it was deposited and the processes of deposition, whether in a streambed, on a beach, or on the deep ocean floor.

Sedimentary rocks also form from the remains of organisms, both plants and animals. Such rocks are **organic sedimentary rocks.** *Coal,* for example, was formed through the accumulation and compaction of decayed vegetation in acid, swampy environments where water-saturated ground prevented oxidation. The initial transformation of such material produces peat, which, when subjected to deeper burial and further compaction, is lithified to produce coal. Most of the world's greatest coal deposits originated during a geologic time about 300 million years ago known as the Mississippian and Pennsylvanian Periods, called the Carboniferous Period in Europe.

Figure 13.8 Granite, a coarse-grained intrusive igneous rock, forms the mountainous terrain of the Sierra Nevada in California. ■ **What is the difference between granite and basalt?**

R. Sager

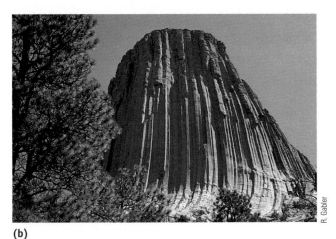

(a)

(b)

Figure 13.9   Hexagonal columnar jointing forms basalt columns: (a) Devil's Postpile National Monument, California. (b) Devil's Tower National Monument   ■   **Why is the cliff so steep here?**

Other organic sedimentary rocks developed from the remains of organisms in lakes and seas. The remains of shellfish, corals, and microscopic drifting organisms called plankton sank to the bottom of such water bodies, where they were cemented and compacted together to form *limestones,* which typically contain fossils such as shells and

(a)   **(b)**

Figure 13.10   Clastic sedimentary rocks are classified by the size and/or shape of the sediment particles they contain. Conglomerate (a) is made of cemented pebbles, and sandstone (b) of cemented sand grains   ■   **Why do the shapes and sizes of sediments in sedimentary rocks differ?**

Figure 13.11   The White Cliffs of Dover. These striking, steep white cliffs along the English Channel are made of chalky limestone from the skeletal remains of microscopic marine organisms.   ■   **What special type of limestone is this very white variety?**

coral fragments. At certain times in Earth history, however, calcium carbonate (lime) dissolved and accumulated in the water until the sea became saturated with this mineral, and the lime was precipitated out to form a deposit on the bottom. Many fine-grained limestones formed in this manner, including the chalk seen in the White Cliffs of Dover along the English Channel (Fig. 13.11).

Limestone, therefore, may vary from a jagged and cemented complex of visible shells or fossil skeletal material to a smooth-textured rock. Where magnesium is an important constituent along with calcium carbonate, the rock is called *dolomite.* Because the calcium carbonate that makes limestone can slowly dissolve in water, limestone in arid climates tends to be resistant, but in humid environments it tends to be weak.

Mineral salts that have reached saturation in evaporating seas or lakes will precipitate out to form a variety of sedimentary deposits that are useful to humans. These include *gypsum* (used in wallboard), *halite* (common salt), and *borates,* which are important in hundreds of products from fertilizer, fiberglass, and pharmaceuticals to detergents and photographic chemicals.

Because sediments accumulate in layers, most sedimentary rocks display distinctive layering, or **stratification.** Each layer represents the depositional record of a past environment and the layers, or **strata,** at a site provide evidence of the local geologic history. For example, a layer of sandstone representing an ancient beach may lie directly beneath shale layers that represent an offshore

R. Sager

Figure 13.12 Bedding planes and an unconformity. Different types of sedimentary rocks may be deposited at various time periods. Each sequence of rocks consists of strata (layers) separated by bedding planes, such as these at Grand Canyon National Park, Arizona. ■ **Where would the youngest strata be located in this photo?**

environment, suggesting that first this was a beach and later the sea covered the site. At other times, organic deposits may dominate. The varieties of sediment and rocks that develop from these deposits produce differentiated strata, or beds. The **bedding planes,** or boundaries between sedimentary strata, indicate changes in the nature of the deposits but no real break in the sequence of deposition (Fig. 13.12). Where a marked mismatch and an irregular, eroded surface occur between beds, the contact between the rocks is called an **unconformity.** This indicates an interruption in the sequence of deposition during which erosion dominated and removed some of the older deposits before sedimentation was renewed. Within some sedimentary rocks, especially sandstones and shales, fine "micro-bedding" may occur. Thus, we may find **lamination planes** parallel to major bedding planes in sandstones or shales (from Latin: *lamina,* thin leaf). Lamination planes provide evidence of minute variations in deposition with successive tides on a shore, for example, or represent seasonal changes in deposition. Another form of micro-bedding, called **cross-bedding,** is characterized by a pattern of thin layers at an angle to the main layering, often reflecting shifts of direction by waves along a coast or winds over sand dunes (Fig. 13.13).

Many sedimentary rocks become jointed, or fractured, as they lithify or when they are subjected to crustal stresses. Limestone especially may be massively jointed. The impressive slabs of rock at Arches National Park, Utah, owe their vertical, tabular shape to joints in great beds of sandstone (Fig. 13.14).

Structures such as bedding planes, lamination planes, and joints are important in the development of physical landscapes, because these structures are weak points in the rock where weathering and erosion can attack. Joints

allow water to penetrate deeply into some rock masses, causing them to be removed at a faster rate than the surrounding rock.

Metamorphic Rocks The word *metamorphic* means "changed form," and that is just what metamorphic rocks are. Enormous heat and pressure deep in Earth's crust, often associated with tectonic activity, can totally reconstitute rock, changing it to a new product. Usually the resulting rock is harder and more compact, has a foliated crystalline structure, and is more resistant to weathering than it was before being metamorphosed.

Metamorphism occurs most commonly where crustal rocks are subjected to great pressures by tectonic processes, or deep burial, or where rising magma generates heat, pressure, solutions, and gases that modify the nearby rock. Metamorphism either wholly or partially fuses the affected rock, so that it can be deformed, or flow slightly, without becoming molten. This process causes minerals to recrystallize with an orientation at right angles to the direction of pressure or stress (Fig. 13.15). Such metamorphism produces rocks whose minerals are segregated in wavy bands, or platy surfaces known as **foliations.** Where the foliations are very fine, individual minerals have a flattened but wavy, "platy" structure, and the

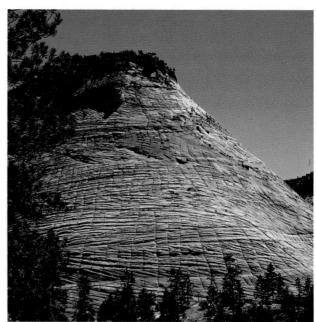

R. Sager

Figure 13.13 Cross-bedding in sandstone at Zion National Park, Utah. ■ **This zigzag pattern usually indicates what type of condition during deposition?**

Courtesy Sheila Brazier

Figure 13.14   Vertical jointing of sandstone strata at Arches National Park, Utah.
■ **What would cause joints such as the one in the upper strata in this photo?**

Earth's Planetary Structure

Scientists still know relatively little about the Earth's interior. Increased knowledge of its structure, its composition, and the processes going on within Earth will help answer questions about crustal motion, earthquakes, volcanic eruptions, the formation of mineral deposits, and the origins of continents and of our planet itself. As a prelude to understanding Earth's surface and the processes that modify it, we need to examine the planet's internal structure, as well as the materials that make up its interior and its solid lithospheric skin, which forms the surface on which we live.

Earth has a radius of about 6400 kilometers (4000 mi), but scientists have been able to penetrate, and examine directly, only its thin outer skin. Through direct means such as mining and drilling, we have gained a modest knowledge of Earth's interior. The lure of gold has taken miners to a depth of 3.5 kilometers (2.2 mi) in South Africa, and drilling for oil and gas has penetrated to several times that distance. These explorations have been helpful in providing information about Earth's uppermost layers, but they are really only scratches in the planet's surface.

Most of what we have learned about Earth's interior structure and composition has been deduced through indirect means. The most important evidence that scientists have used to gain such indirect knowledge is the behavior of earthquake waves and other shock waves (generated by controlled explosions) as they pass through Earth. A sensitive instrument called a **seismograph** (Fig. 13.17) can record these seismic waves, even when the earthquake is centered thousands of kilometers away or even on the opposite side of the planet.

There are two major types of seismic waves, which travel at various speeds in materials of different densities and states. These are P (primary) waves, which travel fastest and arrive first at the seismograph recording a quake, and S (secondary) waves, which travel more slowly. Repeated patterns of P and S waves recorded by seismographs suggest that these waves are refracted, or bent, as they pass through marked changes in the density of materials within Earth's interior. The waves speed up in denser material and slow down in material that is less dense. P waves also pass through molten rock, but S waves do not (Fig. 13.18). By analyzing worldwide patterns of many earthquake waves for many years, scientists have developed a general model of Earth's interior. Such information, supplemented by studies of Earth's magnetism and gravitational pull, suggests a series of layers, or zones, in Earth's internal structure. These zones, from the innermost to the surface, are known as the core, the mantle, and the crust (Fig. 13.18).

rocks tend to flake along these bands. A common metamorphic rock with thin foliations is called *schist.*

Where the foliations develop into broad mineral bands, the rock is extremely hard and is known as *gneiss* (pronounced "nice"). Coarse-grained rocks such as granite generally metamorphose into gneiss, whereas finer-grained rocks may produce schists. Some shales produce a hard metamorphic rock known as *slate,* which exhibits a tendency to break apart or cleave along smooth, flat surfaces.

Rocks that originally were composed of one dominant mineral are not foliated by metamorphism. Limestone is metamorphosed into much denser *marble,* and impurities in the rock can produce a beautiful variety of colors (Fig. 13.16). Silica-rich sandstones fuse into solid layers of quartz, known as *quartzite.* Quartzite is brittle, but harder than steel, and almost inert chemically. Thus, it is virtually immune to chemical weathering and commonly forms cliffs or rugged mountain peaks after the surrounding, less resistant rocks have been removed by erosion.

Minnesota Geological Survey

(a)   **(b)**

Minnesota Geological Survey

Figure 13.15   Metamorphic rocks and foliations. (a) Gneiss and (b) schist illustrate differences in degree of metamorphism. Metamorphic foliations (platy or wavy layers) are not bedding planes, but rather tend at form at right angles to bedding planes when layered rocks are subjected to the heat and pressure of metamorphism.

R. Sager

Figure 13.16   High marble cliffs in Little Cottonwood Canyon in Alta, Utah.
■ **What is the origin of marble?**

 Earth's Core

The **core** forms one-third of Earth's mass and has a radius of about 3360 kilometers (2100 mi). Earth's core is under enormous pressure—several million times atmospheric pressure at sea level. It is believed to be composed primarily of iron and nickel and has two distinct sections. The **outer core** is 2400 kilometers (1500 mi) thick and is made of iron and nickel. Because it blocks the passage of seismic S waves that will not travel through fluids, Earth scientists assume that the outer core is molten.

The **inner core** of Earth, however, appears to be solid iron. Scientists have an explanation for why it is solid. The melting point of mineral material depends not only on temperature but on pressure as well. The pressure on this innermost part of Earth is so great that the inner core remains solid; that is, its melting point has been raised to a temperature above even the high temperatures found there. The outer core, though its temperatures

are lower, is under less pressure and can exist in a molten state. Estimates of internal temperatures are 4800°C (8643°F) at the core-mantle boundary and increase to 6900°C (12,423°F) at the very center of Earth.

Density at the core-mantle boundary is estimated at 10 grams per cubic centimeter, increasing to 13 g/cm³ at Earth's center. These high core densities contrast with the lower densities of Earth's mantle (3.3 to 5.5 g/cm³) and crust (2.7 to 3.0 g/cm³) and give our planet an average density of 5.5 g/cm³. The high density of Earth's core is one reason that scientists believe that iron is its primary component; the density and composition of meteorites also support this theory.

Earth's Mantle

The **mantle** is about 2885 kilometers (1800 mi) thick and constitutes about two-thirds of Earth's mass. Earthquake waves that pass through the mantle indicate that this zone of Earth's interior is composed of dense rocky material that is solid, in contrast to the molten outer core that lies beneath it. Although most of the mantle is solid, part of the upper mantle material has the characteristics of a **plastic solid,** meaning that it can easily deform and can "flow" a few centimeters per year (an inch or two per year). Scientists agree that the mantle consists of dark, heavy silicate rocks that are high in iron and magnesium.

The mantle contains several layers, or zones, of differing strength and rigidity. The uppermost layer of the mantle is solid and together with the crust forms the **lithosphere.** The term *lithosphere* has traditionally been used to describe the entire solid Earth (as discussed in Chapter 1). In recent decades, however, scientists have used this term in a more precise way to describe the solid outer shell of the Earth, including the crust and the rigid upper mantle down to the plastic layer (Fig. 13.19). Beneath the lithosphere at a depth of about 100 to 700 kilometers (62–435 mi), in the upper mantle, is the **asthenosphere** (from Greek: *asthenos,* without strength), a thick layer of plastic mantle material. The material in the asthenosphere flows both vertically and horizontally, dragging segments of the overlying, rigid lithosphere with it. Many Earth scientists now believe that the energy for tectonic forces that break and deform the crust comes from movement within the asthenosphere produced by thermal convection currents originating within the mantle and heated by decaying radioactive materials in the planet's interior.

The interface between the crust and upper mantle is marked by a significant change of density, or a discontinuity, as indicated by an abrupt increase in the velocity of earthquake waves descending through this internal

USGS Volcano Hazards Program

Figure 13.17   Seismographs record earthquake waves for scientific study. This seismograph trace shows multiple earthquakes associated with volcanic eruptions.

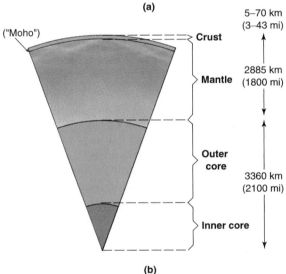

Figure 13.18   The internal structure of Earth as revealed by seismic waves. (a) The presence of internal changes is revealed by the refraction of P (primary) seismic waves and the inability of S (secondary) waves to pass through Earth's liquid outer core. (b) Cross section through Earth's internal structural zones.   ■ **How does the thickness of the crust compare with the thicknesses of the core and mantle?**

boundary. Scientists have labeled this zone the **Mohorovicic discontinuity,** or **Moho** for short, after the Croatian geophysicist who first detected it in 1909. The Moho does not lie at a constant depth around Earth, but generally mirrors the surface topography, being deepest under mountain ranges where the crust is thick, and rising to within 8 kilometers (5 mi) of the ocean floor (Fig. 13.19). During the 1980s, the former Soviet Union drilled to a depth of more than 13 kilometers (8 mi) in unsuccessful attempts to penetrate Earth's crust and recover samples of the mantle.

Earth's Crust

The only portion of the lithosphere of which Earth scientists have direct knowledge is the **crust** and its related surface materials, which form only about 1 percent of Earth's planetary mass. Earth's crust forms the exterior of the lithosphere and is of primary importance in understanding surface processes and landforms. Earth's deep interior components, the core and mantle, are of concern to physical geography because they are responsible for, and can help to explain, changes in the lithosphere, particularly the crust that forms the ocean floors and continents. It is the rocks and debris of the crust from which soils are formed and which we penetrate in search of energy and mineral resources.

Earth's crust is less dense than either the core or the mantle. It is also thin in comparison to the size of the planet. Crustal thickness varies from 5 kilometers (3 mi) in the ocean basins to as much as 70 kilometers (43 mi) under some continental mountain systems. The average thickness of continental crust is about 32 to 40 kilometers (20–25 mi). The crust is relatively cold, rigid, and brittle compared to the mantle. It responds to stress by fracturing, wrinkling, and raising or lowering rocks into upwarps and downwarps.

Two kinds of Earth crust can be distinguished by their thickness, location, and composition. One is **oceanic crust,** which is composed of material similar to basalt, a solidified, dark-colored, and fine-grained lava. In comparison to the crust that forms the continents, oceanic crust is more basic chemically, is thinner, and is

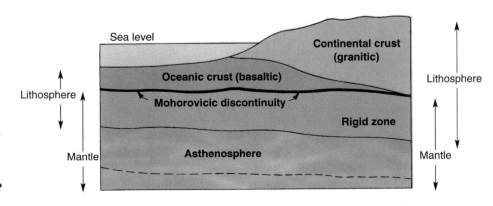

Figure 13.19   Earth has two distinct types of crust: oceanic and continental. Together with the rigid upper zone of the mantle, these form the lithosphere. Below the lithosphere is the plastic zone in the mantle called the asthenosphere.   ■ **What two zones are separated by the Mohorovicic discontinuity (Moho)?**

composed of darker, heavier rocks and thus has a higher density (3.0 g/cm^3). Forming the deep ocean floors, its most common minerals are compounds of silica (Si) that are relatively high in iron (Fe), and magnesium (Mg). Basalt is the most common rock on Earth and appears in great lava outflows on all of the continents as well as on the seafloor.

Continental crust forms the major landmasses; it is much thicker, less dense (2.7 g/cm^3), more acidic, and lighter in color than oceanic crust. Although rocks of every type and of all geologic ages may be found on the continents, the average density of continental crust is similar to that of granite, a common coarse-grained intrusive igneous rock. Thus, continental crust is regarded as granitic, in contrast to the basaltic oceanic crust (see again Fig. 13.19). Granite, like basalt, forms from a molten state, but it forms deep underground as opposed to basalt, which cools and solidifies at the surface. Where the surface elevation is high, such as where there are mountains or plateaus, the crust is thicker; at lower elevations, continental crust is thinner, but not as thin as the crust of the seafloor.

Continents in Motion: The Search for a Unifying Theory

Scientists in all disciplines constantly search for broad explanations that shed light on the detailed facts, recurring patterns, and interrelated processes that they observe and analyze. The theories of evolution and natural selection proposed by Charles Darwin and his scientific colleagues in the mid-nineteenth century comprised just such an explanation. Those theories ultimately revised the thinking of scientists in many fields and remain an important model for biologists and other life scientists today.

What are some of the fundamental theories or concepts in the Earth sciences? At one time, scientists believed that Earth's landscapes were created in great cataclysms—that the Grand Canyon, for example, split open one violent day and has remained that way ever since, or that the Rocky Mountains appeared overnight. This theory, called **catastrophism,** is now rejected. For almost two centuries, geographers, geologists, and other Earth scientists have accepted the theory of **uniformitarianism**—the idea that processes, both internal and external to Earth's surface, are operating today in the same manner as they have for millions of years. Uniformitarianism, however, does not mean that these processes operate at the same rates, either all the time or equally everywhere on Earth. In fact, the character of our planet's surface features has resulted from variations in the intensity of internal and external processes influenced by their geographic location. These processes have also varied in intensity and location throughout Earth's history. The accumulation of regular or periodic changes in the Earth

system that seem relatively small to us can dramatically change a landscape or our planet's surface after progressing, even on an irregular basis, for millions of years.

Are there hypotheses or ideas similar to evolution that explain how and why Earth processes work? Is there one broad theory that can help to explain such diverse phenomena as the growth of continents, the movement of solid rock beneath Los Angeles, the location of great mountain ranges, differing patterns of temperature in the rocks of the seafloor, and the violent volcanic eruptions on the island of Montserrat in the West Indies? The answer is yes: the theory of plate tectonics. Over the past few decades, this theoretical framework has had a major impact on the Earth sciences just as the theory of evolution affected the life sciences more than a century ago.

Continental Drift

Most of us have noted on a world map that the Atlantic coasts of South America and Africa look as if they fit together. In fact, if a globe were made into a spherical jigsaw puzzle, several widely separated landmasses could fit alongside each other without large gaps or overlaps (Fig. 13.20). Is there a scientific explanation?

In the early 1900s, Alfred Wegener, a German climatologist, proposed the theory of **continental drift.** He hypothesized that all the continents had once been part of one supercontinent, which he called Pangaea, that later divided into two large landmasses, one in the Southern Hemisphere (Gondwana), and one in the Northern Hemisphere (Laurasia). These two supercontinents also broke apart, and their fragments (the present continents) drifted to their existing positions. Wegener's evidence for continental drift included similar geographic patterns of plant fossils found on different continents that could not result from chance. The continents must once have been joined so as to allow the spread of those early land plants. Furthermore, evidence of climatic changes (for instance, ancient evidence of glaciation where the Sahara Desert is today and tropical fossils found in Antarctica) could be explained best by large landmasses moving from one climatic zone to another.

The reaction of the scientific community to Wegener's proposal ranged from skepticism to ridicule. A major objection to his hypothesis was that neither he nor anyone else could provide an acceptable explanation for the energy needed to break up huge landmasses and move them over the rigid crust and across vast oceans.

Wegener did propose that perhaps the crust was not as rigid as had been believed. If instead it were plastic under certain conditions, and if the lighter material of the continents "floated" on the heavier rocks in which they were imbedded, then movement of the continents away from each other might be possible. Nevertheless, Wegener was still unable to find a scientifically acceptable mechanism for the movement responsible for continental drift.

Figure 13.20 The geographic basis for Wegener's continental drift hypothesis. Note the close correlation of the edges of the continents that face one another across the width of the Atlantic Ocean. The actual fit is even closer if the continental slopes are matched. ■ **About how long ago is it now assumed that the continents fit together as one landmass?**

Supporting Evidence for Continental Drift

About a half-century later, Earth scientists began giving serious consideration to Wegener's notion of moving continents. Beginning in the late 1950s and early 1960s, new information appeared from research in oceanography, geophysics, and other Earth sciences, aided by sonar and radioactive dating, the improvement of magnetometers, and the coming of the computer age. Earth scientists were flooded with new evidence indicating that portions of the lithosphere (including the continents) had indeed been on the move.

As one example, scientists were originally unable to explain the unusual orientations of magnetic fields in igneous rocks that had cooled millions of years before. Minute iron minerals within these rocks displayed magnetic fields that were oriented in directions that pointed far away from today's magnetic north pole, and sometimes in a completely opposite direction, toward the south magnetic pole. The orientation of this magnetism is locked into rocks as they solidify, recording the direction to magnetic north at the time of cooling, and is known as **paleomagnetism.** Rocks of different ages, or on different continents, show magnetic orientations at a variety of angles to Earth's magnetic field as it is today. At first it was assumed that Earth's magnetic poles had wandered, which they do, but not enough to explain this much variation, or the varying paleomagnetism in rocks of the same age. It was also discovered that many times in the geologic past, the polarity of Earth's magnetic field had reversed. A historical record of polarity reversals was recorded in the iron-rich basaltic rocks of the seafloor.

A model of the continents was projected backward in time so that the magnetic orientations of the rocks coincided with Earth's magnetic field during past periods of geologic history. Paleomagnetism indicated an almost perfect fit of the continental jigsaw puzzle about 200 million years ago, when the two supercontinents split apart and the Atlantic Ocean began to form. Laurasia in the Northern Hemisphere consisted of North America, Europe, and Asia. Gondwana in the Southern Hemisphere was made up of South America, Africa, Australia, Antarctica, and India (Fig. 13.21).

Supporting evidence for crustal movement came from a variety of other sources. Widely separated patterns of similar fossil reptiles and plants found in Australia, India, South Africa, South America, and Antarctica had also previously been a puzzle. The fossils represented organisms that in each instance were so similar and specialized that they could not have developed without land connections between their now distant locations. Once the continents were reassembled on a paleomap representing the time when the fossil organisms were living, the locations where these fossils were found fit together in one place, just as certain organisms today are found in limited locations on Earth. Other environmental aspects also fit together in logical spatial patterns that helped reconstruct ancient geographies in map form.

The fit of landmasses was discovered to be even better a few hundred meters below sea level along the true continental edges, the undersea continental slopes. Mountain ranges on opposite sides of oceans, carefully matched by geographic trend, rock ages, and rock types, were shown to be continuous when continents were joined. Our knowledge of the geographic distribution of Earth's environments relative to climatic zones also provided a key. Evidence of ancient glaciation in India and South Africa, or of tropical forest climates (coal deposits) in Alaska and Antarctica, could only be explained by crustal movement, and their locational regions fit together well with reconstructions of ancient latitudes.

Paleomagnetism and Pole Reversal

As discussed earlier in this text, Earth acts as a gigantic magnet. As with all magnets, there is a magnetic field around our globe, and these field lines converge at two opposite poles (a north pole and a south pole). These lines and poles can be located and measured. For millennia, humans have used Earth's magnetic force when defining cardinal directions (compasses always point to the magnetic north pole). As we know from previous discussions, true north and magnetic north are not the same, but are located fairly close to each other. Thus, for navigational purposes, one must take into account the magnetic declination (or the angular distance between the two) to keep the cardinal directions accurate.

The exact cause of Earth's magnetic field is not entirely understood, but it is certainly related to unstable currents flowing through layers of molten rock (magma) found at different depths beneath Earth's surface. Because of the great pressures found near the center of the planet, Earth's core, which consists mainly of iron, exists as a solid despite the intense heat at the core. This solid iron core, which contains magnetite (iron's magnetic ore), seems to be a key feature in the formation and maintenance of the magnetic field. Imagine, if you will, an iron core surrounded by swirling masses of molten rock and metal, all nested within a planet that is spinning on its axis. All of these minerals and ores (some of them with magnetic properties) are flowing and circulating around each other. In addition to all of this subsurface movement, one must also consider magnetic waves and storms that Earth receives from the sun. These blasts of magnetic energy have also influenced Earth's magnetic properties. If we take all these factors into

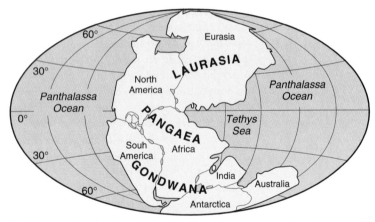

Figure 13.21 The development of continents in Earth history. The supercontinent Pangaea included all of today's major landmasses joined together. Pangaea later split to form Laurasia in the Northern Hemisphere and Gondwana in the Southern Hemisphere. Further plate motion has produced the continents as they are today. ■ **How has continental movement affected the climates of landmasses?**

Plate Tectonics

Plate tectonics, the modern version of continental drift, suggests that the lithosphere (crust and solid upper mantle) consists of large segments called **lithospheric plates** that "float" on the plastic asthenosphere (Fig. 13.22). These plates move as distinct units—in some places separating from each other (diverging), in other places sliding alongside each other (moving laterally), and elsewhere coming together (converging). Seven major plates have proportions as large as or larger than continents or ocean basins. Five other plates are of minor size, although they have maintained their own identity and direction of movement for some time. Several other plates are quite small and exist in active zones at the boundaries between

major plates. All major plates consist of both continental and oceanic crust, although the largest, the Pacific plate, is primarily oceanic. To understand how plate tectonics operates and why plates move, we must consider the scientific evidence that was gathered to test this theory. We should also evaluate how well this theory holds up under rigorous examination. The supporting evidence, however, is overwhelming.

Convection Currents and Seafloor Spreading

In the 1960s, the key to plate tectonics theory was found while studying and mapping the ocean floors. First, detailed undersea mapping revealed a system of midocean ridges (also called oceanic ridges or rises) with configurations remarkably similar to the continental coastlines (Fig. 13.23). Second, it was discovered that parallel bands of basaltic seafloor displayed similar magnetic properties in identical patterns on both sides of midocean ridges in the Atlantic and Pacific Oceans. Third, scientists made the surprising discovery that although some continental rocks are nearly 4 billion years old, rocks on the ocean floor are all geologically young—they have been in existence less than 250 million years. Fourth, the oldest rocks of the seafloor are under the deepest ocean waters or nearest to the continents, whereas the youngest ocean floor rocks are nearest to the midocean ridges. Finally, temperatures of rocks on the ocean floor vary significantly, being hottest near the ridges and becoming progressively cooler farther away.

account, the concept of a magnetic field around a dynamic Earth begins to take shape. The field of study called *geomagnetism* seeks to find the exact causes of this force and what makes it change over time.

Through the 4.5 billion years of Earth history, Earth's magnetic field has been constantly changing, and so have the positions of the poles. There has been so much change that every few million years, the poles have actually reversed! *Paleomagnetism* (the term *paleo* means ancient) is the field of study that deals with changes in Earth's magnetic field through time and attempts to establish the previous positions of the magnetic poles.

We know that pole reversals have occurred through time by studying the orientation of magnetic mineral crystals in rocks of varying ages. Cooled lava flows provide the best rock to use for this research. Before lava cools to form solid rock, its semiliquid state allows magnetic minerals to align themselves to magnetic north, much like tiny compass needles. When the rock solidifies, the alignment to magnetic north is then frozen in time. If volcanic rocks are radiometrically dated, and their position on the ground has not moved since their formation, then they will point to the magnetic north in existence during their formation. This method of research has shown that throughout Earth history, pole reversals have occurred many times. During the history of human navigation by compass, the needles have always pointed north, but since it is likely that there will be another pole reversal in Earth's future, there may come a day when all compass needles point south.

Only one logical explanation emerged to fit all of the new evidence. It became apparent that new oceanic crust is being formed at the midocean ridges while older oceanic crust is being destroyed along certain margins of ocean basins (Fig. 13.24). The emergence of this new oceanic crust is associated with the movement of great sections or plates of lithosphere away from the midocean ridges. Earth scientists label this phenomenon **seafloor spreading.** The plates move laterally and separate at an average rate of 2 to 5 centimeters (1–2 in.) per year

Figure 13.22 Earth's major tectonic plates and their general directions of movement. Earth's crust is broken into giant slabs, known as plates. Most tectonic and volcanic activity occurs along plate boundaries, where plates separate, collide, or slide past one another.
■ **Do the edges of continents and plate boundaries correlate closely? Do all the plates have a continent located on them?**

Figure 13.23 A computer-generated relief map of Earth's surface. This global view shows major landform features of both the continental and oceanic crusts and helps to confirm plate tectonics. ■ **Can you see the outline of any of the major plate boundaries (compare this relief map to Figure 13.22)? How would this type of modern data have helped Alfred Wegener when he first proposed his theory of continental drift?**

above the flowing plastic asthenosphere in the mantle. Ocean floors are not permanent. The young age of oceanic crust results from the creation of new rock at undersea ridges and the movement of the seafloor with lithospheric plates toward remelting and destruction of older rock at the margins of ocean basins.

Plate tectonics includes a plausible explanation of the mechanism involved in seafloor spreading and lithospheric motion. The mechanism is **convection**—the transfer of hot mantle material upward toward Earth's surface and cooler material downward as part of huge subcrustal convection cells. Mantle material flowing in the asthenosphere rises to the base of the lithosphere and spreads laterally when it hits that boundary. Mantle material flowing in opposite directions drags the lithospheric plates with it, breaking the brittle midocean ridge open. Molten basalt wells up into the fractures, cooling and sealing them to form new seafloor. In this process, the ocean becomes wider by the width of the now sealed fracture. The convective motion continues as solidified crustal material moves away from the ridges. Apparently, in a time frame of 250 million years or less, older oceanic crust is consumed in the deep trenches near the boundaries where some sections of the lithosphere meet (see Fig. 13.24).

 Tectonic Plate Movement

The shifting of tectonic plates relative to one another provides an explanation for many of Earth's surface features. Plate tectonics theory enables physical geographers to better understand not only our planet's ancient geography but the modern global distributions and spatial relationships among such diverse but often related phenomena as earthquakes, volcanic activity, zones of crustal movement, and major landform features (see again Fig. 13.23). Let us briefly examine the three ways in which lithospheric plates relate to one another along their boundaries as a result of tectonic movement.

Plate Divergence Tectonic **plate divergence,** the pulling apart of plates, is directly related to seafloor spreading (Fig. 13.25). Tension-producing forces cause the crust to thin and weaken. Shallow earthquakes are often associated with crustal stretching, and asthenospheric magma wells up between crustal fractures, forming new crustal ridges and ocean floor as plates move away from each other. The formation of new crust in these spreading centers gives the label *constructive plate margins* to these zones. Occasional "oceanic" volcanoes like those of Iceland, the Azores, and Tristan da Cunha mark such boundaries.

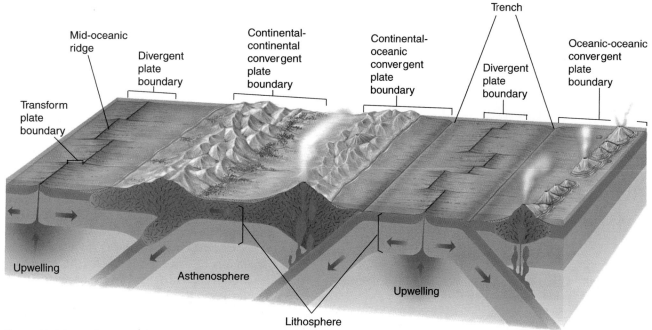

Figure 13.24   Environmental Systems: Plate Tectonic Movement. Unlike the other major Earth systems discussed, the plate tectonics system does not derive its energy from the sun. Instead, the movements of Earth's lithosphere are a result of energy derived from Earth's interior. Exactly how deep the convection cells of heat energy move within the mantle and outer core of Earth is still debated. The most common theory is that convection cells flow within the mantle's asthenosphere, and the rigid upper mantle and crust above are carried along with these currents at a rate of movement that approximates human fingernail growth.

As these plates move, they interact with adjoining plates, forming different boundary types, each having distinct landform features. This diagram shows three major plate boundary types: spreading centers, subduction zones, and collision zones. Spreading centers, at divergent plate boundaries, are shown on the left and to the right of the center of the diagram, are constructive boundaries as they form new crustal material at the surface along active rift zones. Over time, newer material pushes older rock progressively away from the active rift zone in both directions. Earth's oceanic divergent plate boundaries form the midocean ridge system, which runs through all the major oceans.

Subduction zones, as shown on the right side of the diagram, occur where two plates are converging. This is a destructive boundary where crustal material returns to Earth's interior. The denser oceanic plate is forced by gravity and plate movement to subduct beneath the thicker but less dense continental plate. Surface features common to subduction zones are deep ocean trenches and volcanic mountain ranges or island arcs. The best examples of subduction are found around the Pacific Ring of Fire, in volcanically active and earthquake-prone sites such as those in Japan, Chile, New Zealand, and the northwest coast of the United States.

Continental collision zones, as shown on the left, are found where two continental plates collide. Massive mountain building occurs as the crust thickens because of compression, and volcanoes tend to be absent in these regions. The world's highest mountains, the Himalayas, formed when the Indian plate collided with Eurasia. The Alps were formed in a similar manner in a collision between the African and Eurasian plates.

Though most plate divergence occurs along oceanic ridges, this process can also break up continents (Fig. 13.26). The Atlantic Ocean floor formed as the continent that included South America and Africa broke up and moved apart 2 to 4 centimeters (1–2 in.) per year over millions of years. The Atlantic Ocean continues to grow today at about the same rate. The best modern example of divergence on a continent is the rift valley system of East Africa, stretching from the Red Sea to Lake Malawi. Crustal blocks that have moved downward with respect to the crust on either side, with lakes occupying many of the depressions, characterize the entire system, including the Sinai Peninsula and the Dead Sea. Measurable widening of the Red Sea suggests that it may be a "proto-ocean" between Africa and the Arabian Peninsula, similar to the young Atlantic between Africa and South America about 200 million years ago (Fig. 13.27).

Figure 13.25   Seafloor spreading at an oceanic divergent plate boundary.

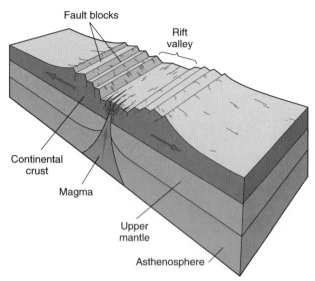

Figure 13.26 A continental divergent plate boundary breaks continents into smaller landmasses.

Plate Convergence Tectonic **plate convergence** causes a maximum of crustal activity. Despite the relatively slow rates of plate movement (in terms of human perception), the incredible energy involved in convergence causes the brittle crust to crumple as one plate overrides another. The denser plate is forced deep below the surface in a process called **subduction.** Subduction is most common where denser oceanic crust collides with and descends below lighter continental crust (Fig. 13.28). This is the situation along South America's Pacific coast, where the Nazca plate subducts beneath the South American plate, and in Japan, where the Pacific plate dips under the Eurasian plate. As oceanic crust, and the lithospheric plate of which it forms a part, is subducted, it descends into the asthenosphere to be melted and recycled into Earth's interior.

Deep oceanic trenches, such as the Peru-Chile trench and the Japanese trench, form where the crust is dragged downward into the mantle. Frequently, hundreds of meters of sediments that are deposited on the seafloor or continental margins are carried into these trenches to later form sedimentary rocks. At convergent boundaries such as subduction zones, rocks can be squeezed and con-

torted between colliding plates, becoming uplifted and greatly deformed or metamorphosed. These processes at convergent plate margins have produced many great mountain ranges, such as the Andes. A subducting plate is heated as it plunges downward into the mantle. Its rocks are melted, and the resulting hot magmas migrate upward into the overriding plate. Where molten rock reaches the surface, it forms a series of volcanic peaks in mountain regions, as in the Cascades Range of the northwestern United States. Where two oceanic plates meet, the older, denser one will subduct below the younger, less dense oceanic plate, and volcanoes may develop, forming major **island arcs** on the overriding plate between the continents and the ocean trenches. The Aleutians, the Kuriles, and the Marianas are all examples of island arcs near oceanic trenches that border the Pacific plate.

As the subducting plate grinds downward, enormous friction is produced, which explains the occurrence of major earthquakes in these regions. Subduction zones are sometimes referred to as Benioff zones, after the seismologist Hugo Benioff, who first plotted the existence of earthquakes extending downward at a steep angle on the leading edge of a subducting plate (see again Fig. 13.24).

Where two continental plates collide, the result is massive folding and crustal block movement rather than volcanic activity. This crustal thickening generally produces major mountain ranges at sites of **continental collision.** The Himalayas, the Tibetan Plateau, and other high Eurasian ranges were formed in this way as the continental Indian plate collided with Eurasia some 40 million years ago. The Indian plate is still pushing into Asia today to produce the highest mountains in the world. The Alps were formed as the African plate was thrust

Figure 13.27 The Sinai Peninsula viewed from the space shuttle, looking along the Mediterranean coast of Egypt. North is to the right corner in this image. To the east of the Sinai is the rift system that forms the Gulf of Aqaba. To the west is the Red Sea rift. The irrigated valley of the Nile River is clearly visible as it flows northward across the desert into the Mediterranean Sea. ■ **What two major landmasses are separated by the Red Sea?**

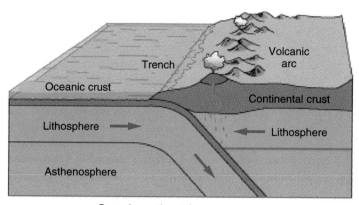

Oceanic-continental convergence

Figure 13.28   An oceanic-continental convergent plate boundary, where continent and sea floor collide.

against the Eurasian plate. Continental collision causes two continents or landmasses to fuse or join together, creating a new larger land mass (Fig. 13.29). This process, which closes an ocean basin that once separated the colliding landmasses, has been called *continental suturing* (suturing is a medical term for closing a wound).

Thus, zones where plates are converging mark locations of major and more tectonically active landforms on our planet: huge mountain ranges, chains of volcanoes, and deep ocean trenches. The distinctive spatial arrangement of these features worldwide can be understood within the framework of plate tectonics.

Transform Movement   **Transform movement** along plate boundaries occurs where plates neither pull apart nor converge but instead slide laterally past each other as they move in opposite directions. Such a boundary exists along the San Andreas Fault zone in California (Fig. 13.30). Mexico's Baja peninsula and southern California are west of the fault on the Pacific plate. San Francisco and other parts of California east of the fault zone are on the North American plate. In the fault zone, the Pacific plate is moving laterally northwestward in relation to the North American plate at a rate of about 8 centimeters (3 in.) a year (80 km or about 50 mi per mil-

lion years). If this movement continues at this rate, Los Angeles will move alongside San Francisco (450 miles northwest) in about 10 million years and eventually pass that city on its way to finally colliding with the Aleutian Islands at a subduction zone.

Another type of lateral plate movement occurs on ocean floors in areas of plate divergence. As plates pull apart, they usually do so along a series of fracture zones that tend to form at right angles to the major zone of plate contact. These crosshatched plate boundaries along which lateral movement takes place are called *transform faults*. Transform faults, or fracture zones, are common along midocean ridges, but examples can also be seen elsewhere, as on the seafloor offshore from the Pacific Northwest coast between the Pacific and Juan de Fuca plates (see again Fig. 13.30). Transform faults are caused as adjacent plates move at variable rates, causing lateral movement of one plate relative to the other. The most rapid plate motion is on the East Pacific rise, where the rate of movement is more than 17 centimeters (5 in.) per year.

Growth of Continents

The origin of continents remains a mystery. It is clear that the continents tend to have a core area of very old igneous and metamorphic rocks that may represent the deeply eroded roots of ancient mountains. These core regions have been worn down by hundreds of millions of years of erosion to form areas of relatively low relief that are located far from active plate boundaries and have a history of tectonic stability over an immense period of time. These ancient crystalline rock areas are called **continental shields.** The Canadian, Scandinavian, and Siberian shields are outstanding examples. Around the peripheries of the exposed shields, flat-lying, younger sedimentary rocks indicate the presence of a stable and rigid mass beneath, as in the American Midwest, western Siberia, and much of Africa. Most continents appear to grow outward by mountain building and uplift around the sediment-covered margins of ancient continental shields. This process is clearly related to the plate tectonic concept, for the descent of lithosphere in an ocean trench along a continent generates new molten material that forms deep plutonic masses and volcanoes on the continental edge. Simultaneously, the continental edges tend to be buckled and uplifted by plate convergence and associated

Mountain range

High plateau

Continental crust

Continental crust

Lithosphere

Lithosphere

Asthenosphere

Ancient oceanic crust

Continental-continental convergence

Figure 13.29   Continental collision along a convergent plate boundary fuses two land masses together.

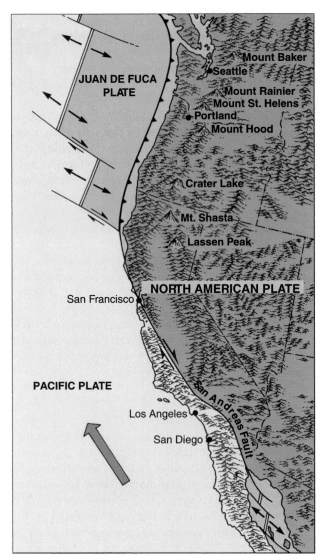

Figure 13.30 A lateral plate boundary. A major plate boundary of this type exists where the Pacific plate moves laterally along the North American plate in western North America. The San Andreas Fault system is formed along this boundary. The area west of the fault is moving northwestward in relation to the plate on the opposite side. Sliding movements occasionally produce major earthquakes as the plates slip past each other. ■ **Note that north of San Francisco the boundary type changes. What type of boundary is located there, and what type of surface features indicates this change?**

subduction processes. Oceanic rocks and marine sediments peel off and pile against continental margins to form ranges such as the Coast Ranges of California and Oregon and the Olympic Mountains of Washington. Except where two continents have collided to become one, mountain building is generally restricted to those plate boundaries where igneous plutons, surface volcanism, and the squeezing up of marine sediments all add new crustal material to the continental mass.

Earth scientists have also suggested that continents grow by **accretion**—that is, by adding numerous chunks of crust to the main continent by collision. North America

may have grown in this manner over the past 200 million years by adding segments of crust, known as **microplate terranes** (a term that should not be confused with the term *terrain*), as it moved westward over the Pacific and former oceanic plates. Though their exact origin is still speculative, parts of western North America from Alaska to California may have originated south of the equator. Terranes, which have their own distinct geology from that of the continent to which they are now joined, may have originally been offshore island arcs, undersea volcanoes, or islands made of continental fragments, such as New Zealand or Madagascar.

Paleogeography

The study of past geographic environments is known as **paleogeography.** The goal of paleogeography is to try to reconstruct the past environment of a geographic region based on geologic and climatic evidence. For students of physical geography, it generally seems that the present is complex enough without trying to know what the geography of ancient times was like. However, peering into the past helps us to forecast and prepare for changes in the future.

The immensity of geologic time over which major Earth events or processes (such as plate tectonics, ice ages, or the formation and erosion of mountain ranges) have taken place is difficult to picture in the human time frame of days, months, and years. The geologic timescale is a calendar of Earth history (Table 13.2). It is divided into *Eras,* the largest units of time, such as the Mesozoic Era (which means "middle life"), and shorter *Periods,* such as the Jurassic Period. *Epochs,* such as the Pleistocene Epoch (recent ice age), are the smallest time units and are used within the Cenozoic Era ("recent life"), for which geologic evidence is more abundant.

If we took a 24-hour day to represent the approximately 4.6-billion-year history of Earth, the Precambrian would consume the first 21 hours, an era of which we know very little. The most recent Period, the Quaternary, which has lasted about 2 million years, would take less than 35 seconds, and human beginnings, over about the last 4 million years, about a minute.

If we look at evidence for the paleogeography of the Mesozoic Era (245 million to 65 million years ago), for instance, we would find a much different physical geography than exists today. This was a time when the supercontinents Gondwana and Laurasia gradually split as new ocean floors widened, creating the continents that are familiar to us today. A paleogeographic map of North America during the Jurassic Period (144 million to 208 million years ago) shows that the ancestral Appalachian Mountains were in existence, as were the ancestral Rockies (Fig. 13.31). Most of the Great Plains and the present Pacific Coast region were beneath shallow seas, while the Midwest of North America was a lowland region.

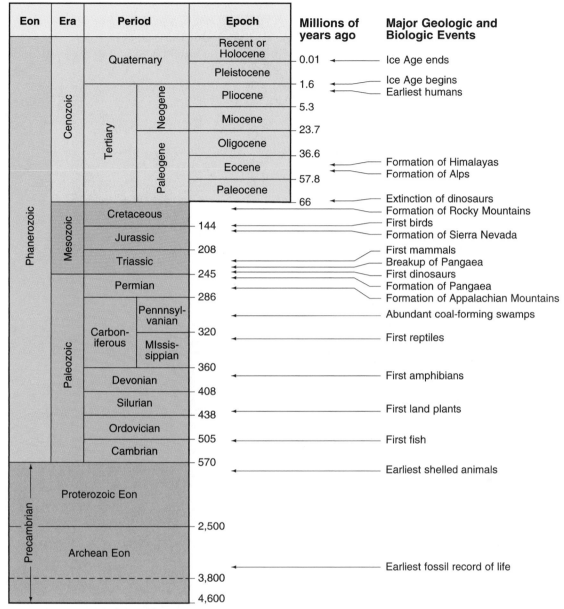

Eon	Era	Period		Epoch	Millions of years ago	Major Geologic and Biologic Events
Phanerozoic	Cenozoic	Quaternary		Recent or Holocene	0.01	Ice Age ends
				Pleistocene	1.6	Ice Age begins / Earliest humans
		Tertiary	Neogene	Pliocene	5.3	
				Miocene	23.7	
			Paleogene	Oligocene	36.6	
				Eocene	57.8	Formation of Himalayas / Formation of Alps
				Paleocene	66	Extinction of dinosaurs
	Mesozoic	Cretaceous				Formation of Rocky Mountains / First birds
		Jurassic			144	Formation of Sierra Nevada
		Triassic			208	First mammals / Breakup of Pangaea
		Permian			245	First dinosaurs / Formation of Pangaea
	Paleozoic				286	Formation of Appalachian Mountains
		Carboniferous	Pennnsylvanian		320	Abundant coal-forming swamps
			Mississippian		360	First reptiles
		Devonian			408	First amphibians
		Silurian			438	First land plants
		Ordovician			505	First fish
		Cambrian			570	Earliest shelled animals
Precambrian		Proterozoic Eon			2,500	
		Archean Eon			3,800	Earliest fossil record of life
					4,600	

Table 13.2   Geologic Timescale

Mesozoic climates were much different as well, but were changing as North America drifted northwest. By the Jurassic Period, much of the present United States was centered at about 30°N latitude and experienced warmer climates than today. Fern and conifer forests were common. The Mesozoic was the "age of the dinosaurs," large reptiles that ruled the land and the sea. Other life also thrived, including marine plants and invertebrates, insects, and the earliest birds and mammals.

The Mesozoic Era ended with an episode of great extinctions, including the end of the dinosaurs. Geologists, paleontologists, and paleogeographers are not in agreement as to what caused these great extinctions. Some of the strongest evidence points to a large meteorite striking Earth 65 million years ago, disrupting global climate. Other evidence points to plate tectonic changes in the distribution of oceans and continents, or increased volcanic activity, either of which could cause rapid climatic changes that might possibly trigger mass extinctions.

Each era and period of Earth's geologic history had a unique paleogeography with its own distribution of land and sea, climate regions, plants, and animal life. Early humans lived under different environmental conditions than we do today. How did these changing conditions influence the present distribution of human populations? (See the World Map of Population Density, Fig. 1.17.)

Our maps of ancient geologic time show only approximations of generalized patterns of mountains, plains, coasts, and oceans, with some environmental characteristics. Much of the evidence, and the rocks that bear this information, have been lost through metamorphism or erosion, buried under younger sediments or lava flows,

Figure 13.31 Paleogeographic map of North America during the Jurassic Period of the Mesozoic Era. The continent had a very different geography 180 million years ago. Its location, size, landforms, and coastlines were completely different from today. ■ **What was the environment like during the Jurassic at the location where you now live?**

or recycled into the Earth's interior. The farther back in time and the older the time period, the sketchier is the paleoenvironmental information presented on the map. Paleomaps, like other maps, are simplified models of the regions they represent.

As time passes and additional evidence is collected, paleogeographers may be able to fill in more of the empty spaces on those maps of the past that are so unfamiliar to us. These paleogeographic studies aim not only at understanding the past but also at understanding

today's environments and physical landscapes, how they have developed, and how processes act to change them today. We can also study the far more abundant evidence of Earth processes that are ongoing today or that operated in the recent geologic past. Applying uniformitarianism and the theory of plate tectonics to our knowledge of how the Earth system and its subsystems function will give us a better understanding of our planet's past, as well as its present, and will facilitate better forecasts of its potential future.

DEFINE & RECALL

crust	intrusive igneous rock	inner core	lithospheric plates
tectonic processes	plutonic rocks	mantle	seafloor spreading
gradational processes	joints	plastic solid	convection
mineral	columnar joints	lithosphere (crust and	plate divergence
rock	clastic	upper mantle)	plate convergence
silicates	organic sedimentary rocks	asthenosphere	subduction
magma	stratification	Mohorovicic discontinuity	island arcs
bedrock	strata	(Moho)	continental collision
regolith	bedding planes	crust	transform movement
outcrop	unconformity	oceanic crust	continental shields
rock cycle	lamination planes	continental crust	accretion
igneous rocks	cross-bedding	catastrophism	microplate terranes
sedimentary rocks	foliations	uniformitarianism	paleogeography
metamorphic rocks	seismograph	continental drift	
lava	core	paleomagnetism	
extrusive igneous rock	outer core	plate tectonics	

DISCUSS & REVIEW

1. What are the differences between tectonic and gradational processes?

2. List the most common elements in Earth's crust. What is a mineral? What is a rock?

3. Describe the three major classifications of rock and the means by which they are formed. Give an example of each.

4. Describe the rock cycle. What is magma, and how is it related to the rock cycle?

5. Identify the major zones of Earth from the center to the surface. How do these zones differ from one another? What is the special significance of the asthenosphere?

6. Define the differences between continental crust, oceanic crust, lithosphere, and asthenosphere.

7. What evidence did Wegener rely on in the formulation of his theory of continental drift? What evidence did he lack? What evidence has since been found to support his theory?

8. What type of plate boundary is found near the Andes, along the San Andreas Fault, in Iceland, and near the Himalayas?

9. Define paleogeography. Why are geographers interested in this topic?

CONSIDER & RESPOND

1. Explain why the eastern United States has relatively little tectonic activity in comparison to the western United States.

2. How is the concept of uniformitarianism related to plate tectonics theory and the arrangement of the continents today? How is it related to the geographic distribution of tectonically active regions?

3. List the four major plate boundary types. Name a country that is located along each one of the plate boundary types.

VOLCANOES, EARTHQUAKES, AND TECTONIC LANDFORMS

CHAPTER PREVIEW

Many natural processes that build and shape Earth's surface operate in a manner that is not predictable, steady, and continuous, but rather unpredictable, variable, and episodic.

Can humans expect to control these natural forces?

How should humans deal with these forces and the landforms they create?

Intense natural processes such as volcanic eruptions and earthquakes can lead to great human suffering and economic loss, but each occurrence has a direct cause and a rational explanation well understood by Earth scientists.

How is this linked to disaster prediction? Why do humans often avoid clear warning signs?

Tectonic uplift and volcanism can produce mountains; in doing so, these processes also affect other parts of the Earth system.

What other environmental changes in the Earth system may occur because of mountain building?

How is this important to human beings?

The locations of tectonic plate boundaries have a direct correlation with the spatial distribution of some of Earth's great "hazard zones."

In what ways are volcanism and faulting involved in this relationship?

Why do humans persist in living in or near areas subject to volcanic and earthquake activity?

Earth System Today CD

Chapter 6: Volcanism, Magma Chemistry & Explosivity
Chapter 6: Volcanism, Volcanic Landforms
Chapter 6: Volcanism, Volcano Watch, USA–Kilauea, HI
Chapter 6: Volcanism, Distribution of Volcanism
Chapter 8: Earthquakes & Tsunamis, Earthquakes in Space & Time
Chapter 8: Earthquakes & Tsunamis, Seismic Case Study, Alaska, 1964-Measurement
Chapter 8: Earthquakes & Tsunamis, Seismic Risk, USA

This landscape in Iceland was created by a combination of faulting and volcanism.

Some of the world's largest and most spectacular landforms, particularly mountains and mountain ranges, have been created primarily by tectonic or volcanic activity. Plate tectonics and related processes exert a tremendous influence on the physical landscapes of Earth, affecting the shapes of continents and coastlines as well as the character and locations of mountain ranges, individual mountains, and many less imposing landforms. The geographic distributions of such features are not random. Volcanoes and associated igneous forms are most common along plate boundaries, as are earthquake fault zones and the major mountain ranges on Earth. Although tectonic and volcanic forces originate from within Earth, they are major controls of topography—affecting the appearance, shape, height, and size of Earth's surface features. These forces can build large mountain systems, but they also produce a great variety of other landforms. Areas of the crust can be uplifted by deep-seated processes or built up by materials that are ejected from the interior onto Earth's surface. Volcanic eruptions and earthquake activity can produce scenic landscapes, but they also present serious natural hazards to humans and their property. This chapter and those that follow will focus on understanding how certain landforms and landscapes develop, and how potential hazards or other environmental concerns are related to the natural processes that cause them. An awareness that we exist on a dynamic planet and an understanding of how different processes interact to shape landforms and physical landscapes are important to human welfare and to the preservation of the environments that humans occupy.

Landforms and Geomorphology

Landscapes and landforms are often described by their **relief**—a term for the difference in elevation between the highest and lowest points within a specific area, or for a particular surface feature that is being described. Earth's **landforms**—its topography or terrain features—result primarily from interactions among those processes that operate to increase relief by raising or lowering land surfaces and those processes, collectively called *gradation,* that work to reduce relief through erosion of high places and the filling in of lower areas (Fig. 14.1). The study of landforms is called **geomorphology.** Without variations in relief, our planet would be a smooth, featureless sphere and certainly a much less interesting place, especially when one considers that its surface would be completely covered by oceans. It would be hard to imagine Earth without dramatic terrain as seen in the Himalayas, Alps, Andes, Rockies, and Appalachians, or in the Grand Canyon. Mountains and other high relief landforms typically are made of resistant rocks that undergo slow rates of **degradation** (erosional removal), or they have formed so recently that degradation has not yet had time to wear them away. In some cases, both of these factors are involved. Fundamentally, where uplift operates (or has operated) faster than degradation, positive relief features such as mountains and hills are formed, and where degradation operates (or has operated) faster than uplift, valleys and low relief features develop. Relief is also reduced by **aggradation,** the filling in of low places with sediments that have been eroded from higher landforms.

Natural processes on Earth are often referred to as operating at a certain rate—for example, a certain number of centimeters or millimeters per year. However, these are average rates, and most surface processes of uplift or gradation tend to operate in an episodic manner. Periods of relative calm continue until some threshold condition is reached, resulting in short "bursts" of rapid activity that cause significant environmental modification. Earthquakes and volcanoes provide good examples

Figure 14.1 The Teton range of Wyoming stands high above the valley floor because rates of uplift exceed the forces of weathering and erosion, which operate to reduce their height.

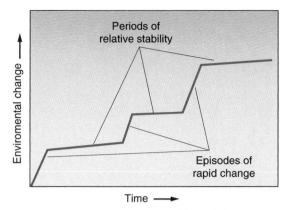

Figure 14.2 Punctuated Equilibrium. Many Earth processes operate slowly most of the time, but are punctuated by events that cause relatively major change in a landscape or natural system. For example, many small earthquakes (or volcanic eruptions) may occur over time that result in little or modest change. Occasionally large events (earthquakes or eruptions) occur that cause major changes in a short period of time. A great variety of processes that affect the Earth system operate in this episodic manner.

of this type of Earth change. They tend to function in a manner that is not continuous, steady, and easily predictable, but rather in a way that is usually quiescent but occasionally chaotic, with rapid activity. Unfortunately, the timing of these periods of increased activity is often difficult or impossible to predict with certainty. Many physical and life processes in the Earth system operate in this manner, termed **punctuated equilibrium** (Fig. 14.2). After an event of major change, the landscape adjusts to the new conditions produced by that occurrence and stays in relative equilibrium, experiencing only modest change and adjustment, until the next major event again disrupts the system. For example, major earthquake faulting or volcanic eruption can dramatically change a landscape that previously was undergoing slow or modest change by gradational processes.

In this chapter, we will first study the processes related to the uplift or building up of land surfaces through volcanic and tectonic activity, and then examine the landforms and rock structures associated with these activities. Later chapters will focus on landforms, landscapes, and processes associated with various types of gradation.

Igneous Processes and Landforms

Volcanism refers to the rise of magma and its cooling at Earth's surface. It also includes the extrusive rocks and landforms created by this surface activity, such as **volcanoes**—mountains or hills formed by the eruptive expulsion of molten or solid rock material from beneath the surface. **Plutonism** refers to igneous processes that occur below Earth's surface. It also includes the intrusive rocks and rock masses formed from magma that exists or cools in the subsurface.

Volcanoes

No spectacle in nature is as awesome as an explosive volcanic eruption (Fig. 14.3). The most violent eruptions are so cataclysmic that thousands of human lives may be lost and the previous landscape completely destroyed.

Volcanic eruptions vary greatly in character, and the volcanic landforms that result are extremely diverse. Variations in eruptive styles and in the landforms produced by eruptions result mainly from chemical differences in the magma feeding the eruption. Some magmas cool enough before eruption so that heavy minerals have crystallized while light minerals, those high in silica, remain molten. Magmas also contain a great deal of gases that remain dissolved when the magma is under high pressure at great depths. As molten rock rises closer to the surface, pressure decreases, which may release the gases. If the gases cannot be readily vented to the atmosphere, or do

Figure 14.3 The explosive eruption of Mt. Pinatubo in the Phillipines was a clear demonstration of the hazards associated with living near an active volcano.

not remain dissolved by particularly hot magma, explosive expansion of these gases can produce a violent eruption. If the magma is extremely hot, crystal formation will not yet have begun, and the gases will still be dissolved. If such molten material leaks or is forced to the surface, the resulting eruption will not be highly explosive, although enormous amounts of highly fluid lava may be produced. **Lava flows** are layers of molten rock that flow over the surface, later cooling and solidifying. Fluid layers become **pahoehoe,** or ropy lava flows; the more viscous, pasty lavas are called **aa,** or blocky lava flows. Both terms are of Hawaiian origin (Fig. 14.4).

Figure 14.4   The names of the two major types of lava flow surfaces use Hawaiian terminology. Hotter, more fluid lava forms pahoehoe, or ropy flow surfaces (lower portion of the photo). Thicker, viscous lava forms block flows, called aa (upper portion of the photo).

In addition to emitting lava flows and gases, most volcanic eruptions hurl into the air molten material that hardens at the surface and also expel solid lava fragments of various sizes. These **pyroclastic materials** (from Greek: *pyros,* fire; *clastus,* broken), also referred to as **tephra,** vary from huge volcanic "bombs" to cinders to **volcanic ash,** which ranges in size from sand-sized particles to extremely fine volcanic dust. In the most explosive eruptions, dust-sized volcanic ash may be hurled into the atmosphere (Fig. 14.5) to an altitude of 10,000 meters (32,808 ft) or

Figure 14.5   An eruption of Rabaul volcano near Java, blew a dense cloud of fine volcanic ash high into the atmosphere.

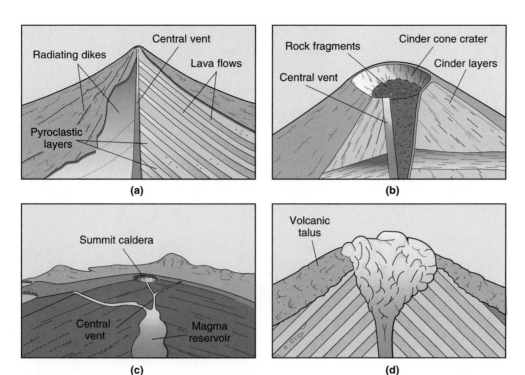

Figure 14.6 The four basic types of volcanoes: (a) composite cone or stratovolcano; (b) cinder cone; (c) shield volcano; (d) plug dome.

more. The 1991 eruptions of Mount Pinatubo in the Philippines ejected a volcanic aerosol cloud that circled the globe. The suspended material caused spectacular red-orange sunsets and may have lowered global temperatures slightly by reflecting solar energy back to space.

Types of Volcanoes and Volcanic Hazards

There are four basic types of volcanoes: composite cones, cinder cones, shield volcanoes, and plug domes (Fig. 14.6). Many volcanoes are formed by eruptions that vary between a mix of lava flows and pyroclastic materials and therefore are known as **composite cones** (Fig. 14.6a).

Most of the well-known, classically shaped volcanoes of the world are of this type: Fujiyama in Japan, Cotopaxi in Ecuador, Vesuvius in Italy, Mount Rainier in Washington, and Mount Shasta in California (Fig. 14.7). Composed of great volumes of lava—particularly andesite lava—as well as ash and larger tephra, composite cones, also known as *stratovolcanoes* (consisting of stratified pyroclastics and lava), are potentially dangerous. The viscous andesite lava of composite cones can trap gases until the pressure from expanding gases produces an explosive eruption. Andesite is named for the Andes of South America, which include the highest volcano on Earth, Nevados Ojos del Salado, which rises to 6,887 meters (22,595 ft) on the border between Chile and Argentina.

Figure 14.7 Composite cones are formed from layers of lava and pyroclastic material. (a) Cotopaxi in Ecuador is a classic composite cone formed in the Andes. (b) Mt. Shasta is a composite cone in the Cascade Range of northern California. Shastina is a secondary crater to the right of the main summit cone. ■ **Along what type of tectonic plate boundary are both these volcanoes located?**

On May 18, 1980, residents of the Pacific Northwest—and, soon after, people throughout the United States—were shocked by the realization that every age in history is a volcanic age. Mount St. Helens, a stratovolcano in southwestern Washington that had been smoldering and venting steam and ash for several weeks, exploded with incredible force. A menacing bulge had been growing on the side of the mountain, and Earth scientists warned of a possible major eruption, but no one could forecast the magnitude or the exact timing of the blast. Within minutes, nearly 400 meters (1300 ft) of the mountain's north summit had disappeared by being blasted into the sky and down the mountainside (Fig. 14.8). Unlike most volcanic eruptions, in which the eruptive force is directed vertically, much of the explosion blew pyroclastic debris laterally outward from where the bulge had been. An incredible storm cloud of intensely hot steam, searing noxious gases, and volcanic ash, blasted at a speed of more than 320 kilometers (200 mi) per hour, obliterated forests, lakes, streams, and camping sites for a distance of nearly 32 kilometers (20 mi). Huge mudflows of ash and melted ice choked streams and valleys and added to the devastation by engulfing everything in their paths. More than 500 square kilometers (193 sq mi) of magnificent forest and recreation land were destroyed. Hundreds of homes were buried or badly damaged, choking ash several centimeters thick covered nearby cities, untold numbers of wildlife were killed, and more than 60 people lost their lives in the eruption. It was a minor event in Earth's history, but a sharp reminder of the awesome power of natural forces.

Some of the worst natural disasters in history have occurred in the shadows of composite cones. Mount Vesuvius, in Italy, killed more than 20,000 people in the cities

(b)

Austin Post/ USGS

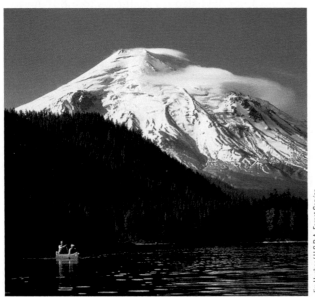

(a)

Jim Hughes/ U.S.D.A. Forest Service

(c)

C. Crisafulli

Figure 14.8   Mount St. Helens, Washington, in the Cascade Range of the Pacific Northwest. (a) A view of Mount St. Helens and Spirit Lake prior to the eruption. (b) On May 18, 1980, at 8:32 A.M. Mount St. Helens erupted violently. (c) A view of Mount St. Helens after the eruption. The massive landslide and blast removed more than 4.2 cubic kilometers (1 cu mi) of material from the mountain's north slope, leaving a crater more than 400 meters (1300 ft) deep. The blast cloud and monstrous mudflows destroyed the surrounding forests and lakes and took 60 human lives. ■ **What type of volcano is Mount St. Helens?**

Figure 14.9 Volcanic destruction on Montserrat. The Caribbean island of Montserrat was struck by a series of volcanic eruptions, beginning in 1995 that devastated much of the island. The town of Plymouth, shown here, has been completely abandoned because of the destruction and threat of future eruptions.

of Pompeii and Herculaneum in A.D. 79. Mount Etna, on the Italian island of Sicily, destroyed 14 cities in 1669, killing more than 20,000 people. Today, Mount Etna is active much of the time and remains Europe's most dangerous volcano. The greatest volcanic eruption in recent history was the explosion of Krakatoa in the Dutch East Indies (now Indonesia) in 1883. The violent eruption killed more than 36,000 persons, many as a result of the subsequent **tsunamis** (seismic sea waves, discussed in Chapter 20) that swept the coasts of Java and Sumatra. In 1985, the Andean composite cone Nevado del Ruiz, in the center of Colombia's coffee-growing region, erupted and melted its snowcap, sending torrents of mud and debris down its slopes to bury cities and villages, resulting in a death toll in excess of 23,000. The 1991 eruption of Mt. Pinatubo in the Philippines killed more than 300 people and caused suspected climatic effects for years following the eruption. In 1997, a series of violent eruptions from the Soufriere Hills volcano destroyed more than half of the Caribbean island of Montserrat with volcanic ash and pyroclastic flows (Fig. 14.9). In 2002, Mexico City, perhaps the world's most populous urban area, was being threatened by Popocatepetl, a large, active composite cone volcano that is 80 kilome-

ters (50 miles) away. At this distance, ash falls from a major eruption would be the most severe hazard to be expected. Volcanic ash can cause breathing problems and stall vehicles by choking the air intakes of combustion engines. The heavy weight of significant ash accumulations on building roofs can cause them to collapse.

The smallest type of volcano, typically only a few hundred to more than a thousand feet high, is known as a **cinder cone** and consists largely of tephra (see Fig. 14.6b). Gas-charged eruptions "spray" molten lava and solid pyroclastic fragments into the air. These particles fall, under the influence of gravity, to build a layered pile of cinders in the form of a volcanic cone. The cinders that make up these volcanoes are typically fragments of basaltic lava. Cinder cones are illustrated by Craters of the Moon and Menan Buttes in Idaho (see Map Interpretation: Volcanic Landforms), Mt. Capulin in New Mexico (Fig. 14.10), and Sunset Crater, Arizona. In 1943, a remarkable cinder cone called Paricutín grew from a crack in a Mexican cornfield to a height of 92 meters (300 ft) in five days, and to more than 360 meters (1200 ft) in a year. Eventually, the volcano began erupting basaltic lava flows, which buried a nearby village except for the top of a church steeple.

Shield volcanoes are built by the accumulation of many lava flows and relatively little ash or cinders (see Fig. 14.6c). The gently sloping, dome-shaped cones of Hawaii best illustrate this type of volcano (Fig. 14.11). They erupt extremely hot basaltic lava with temperatures of more than 1090°C (2000°F). Escape of gases and steam may hurl molten lava a few hundred meters into the air, with some buildup of cinders (lava clots that congeal in the air), but the major feature is the outpouring of very fluid basaltic lava flows. Compared to other volcano types, these eruptions are mild, although still potentially damaging and dangerous. The accumulation of flow layers develops broad, dome-shaped volcanoes with very gentle slopes. On the "Big Island" of Hawaii, active shield volcanoes also erupt lava from fissures on their flanks, so that living on the island's edges, away from the summit craters, does not guarantee safety from volcanic hazards. Neighborhoods in Hawaii have been destroyed or threatened by lava flows in recent years.

By contrast, **plug dome** volcanoes extrude extremely stiff, taffy-like lava that fills the initial pyroclastic cone without flowing beyond it (see Fig. 14.6d). Plug domes are characteristically steep-sided volcanoes with summits composed of solid lava. The vents of these volcanoes become plugged by thick lava,

Figure 14.10 Mt. Capulin, New Mexico, is an excellent example of a cinder cone. ■ **What are pyroclastics (also called tephra)?**

VOLCANIC LANDFORMS

The Map

The Menan Buttes map area is located on the upper Snake River Plain in eastern Idaho. The Snake River Plain is a region of recent lava flows that extends across southern Idaho. It is part of the vast volcanic Columbia Plateau, which covers more than 520,000 square kilometers (200,000 sq mi) of the northwestern United States. The lava originated from fissure eruptions that spread vast amounts of fluid basaltic lava across the landscape, accumulating to a thickness of several thousand feet. Most of the Snake River Plain has an elevation between 900 and 1500 meters (3000 and 5000 ft). Rising above the basalt plain are numerous volcanic peaks, including Menan Buttes and Craters of the Moon. The Snake River flows westward across the region.

The Snake River Plain has a semiarid or steppe climate. Because of the moderately high elevation of the plain, temperatures are cooler than in nearby lowlands, with an average annual temperature of about 10°C (50°F). Precipitation is between 25 and 50 centimeters (10 and 20 in.). It is unpredictable from year to year, and there are thunderstorms during the summer months. The low rainfall total is mainly a result of a rainshadow effect. Moisture-producing storms from the Pacific Ocean are blocked from reaching the region by the Cascade Range and the Idaho Batholith section of the northern Rockies. The upper Snake River Plain and the Menan Buttes area are left with dry, adiabatically warmed air descending from the lee slopes of the mountains. The vegetation cover in the area is sparse, mainly characterized by sagebrush and bunch grasses.

Interpreting the Map

1. What type of volcano are Menan Buttes? On what landform characteristics is your decision based?

2. Define a butte. Are these landforms really buttes?

3. What is the local relief of the northern Menan Butte? What is the depth of the crater of each butte?

4. Do you think that these volcanoes are active at present? What evidence from the map and aerial photograph indicates activity or a period of inactivity?

5. Sketch an east-west profile across the center of northern Menan Butte from the railroad tracks to the channel of Henry's Fork. Is this profile typical of a volcanic summit and crater? What is the horizontal distance of the profile?

6. Slope ratio can be calculated by dividing the relief by the horizontal distance. For example, a 3000-foot-high mountain slope with a horizontal distance of 9000 feet would have a slope ratio of 1:3. What is the slope ratio for the western slope of northern Menan Butte from the crater ridge down to the railroad tracks at the foot of the butte?

7. This is a shaded relief topographic map and differs from most of the other topographic maps in this book. What is the major advantage of this shaded relief mapping technique? Are there any disadvantages, compared with the regular contour topographic maps?

8. Compare the southern Menan Butte on the map with the vertical aerial photograph on this page. Why would it be useful to have both a map and an aerial photograph when studying landforms? What is the chief advantage of each?

Vertical aerial photograph of one of the Menan Buttes, Idaho

**Opposite:
Menan Buttes, Idaho
Scale: 1:24,000
Contour interval = 10 ft
U.S. Geological Survey**

(a)

(b)

USGS

Figure 14.11 Shield volcanoes of Hawaii Volcanoes National Park. (a) The gentle slopes surrounding the active Kilauea Crater. (b) The active Hawaiian craters emit an extremely hot, fluid, basaltic lava. As the lava cools, it changes from its molten orange color to a black pahoehoe basalt. ■ **Why are Hawaiian volcanoes less explosive than those of the Cascades or Andes?**

which allows great pressures to build, so plug domes, like composite cones, have the potential for very explosive eruptions. In 1903, Mount Pelée, a plug dome on the French West Indies island of Martinique, destroyed in a single blast all but one person from a town of 30,000. Lassen Peak in California is a large plug dome that has been active in this century (Fig. 14.12). Other plug domes exist in Japan, Guatemala, and the Aleutian Islands.

Occasionally a volcano may expel so much material in eruptions that its summit collapses into its emptied magma chamber, forming a large crater known as a **caldera.** The best-known caldera in North America is Crater Lake in Oregon, a circular body of water 10 kilometers (6 mi) across and almost 610 meters (2000 ft) deep, surrounded by near-vertical cliffs as much as 610 meters (2000 ft) high. The caldera that contains Crater Lake formed by the prehistoric collapse of a composite cone volcano. The rim of the crater crests at about 2440 meters (8000 ft). A cinder cone, Wizard Island, has built up from the floor of the caldera and rises above the lake's surface (Fig. 14.13). The area of Yellowstone National Park is the site of three ancient calderas, and Valles

Caldera in New Mexico is another excellent example. Krakatoa in Indonesia and Santorini (Thera) in the Greek Islands are also calderas. Calderas are also found in the Philippines, the Azores, Japan, Nicaragua, Tanzania, and Italy, many of them occupied by deep lakes.

 Distribution of Volcanism

The majority of Earth's volcanoes are concentrated in several well-defined zones along plate boundaries, but a few active volcanic regions have a more sporadic distribution. The midoceanic ridge system where crustal plates are diverging is volcanic, as are the oceanic islands on the mid-Atlantic ridge, such as the Azores and Iceland. Volcanoes also tend to occur where continental plates are spreading; examples along the East African rift valleys include Mt. Kilimanjaro and Mt. Kenya. Plate divergence, whether it breaks continents or seafloor, produces deep fractures in the crust, providing avenues for molten rock to reach the surface. A rim of active volcanoes, known as the "Pacific Ring of Fire," almost completely circles the Pacific Ocean (Fig. 14.14). Many of these volcanoes erupt andesite lava and can be dangerously explosive. When oceanic crust descends below continental crust and into oceanic trench systems, it melts and is reconstituted into magma. The magma moves up to the surface under the continental borders to produce plutons (see following section) and volcanism. In fact, wherever crustal material is subducted into a deep-ocean

Alan L. Mayo/ Geophoto Publishing Co.

Figure 14.12 Plug dome volcanoes extrude stiff acidic flows, which form very steep slopes. Lassen Peak, located in northern California, is a plug dome and the southernmost volcano in the Cascade Range. It was last active between 1914 and 1921. ■ **Why are plug dome volcanoes considered dangerous?**

(a)

Raven Maps and Images

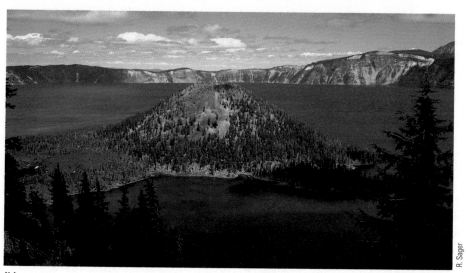

(b)

R. Sager

Figure 14.13 Crater Lake National Park, Oregon. (a) Crater Lake is the best-known caldera in North America. It formed when a violent eruption of Mt. Mazama about 6000 years ago caused the summit to collapse, forming a deep crater. (b) Later the crater filled with water, forming the 610-meter-deep (2000 ft) Crater Lake. Wizard Island is a secondary volcanic peak formed in the caldera. ■ **Do you think other Cascade volcanoes could repeat this kind of event?**

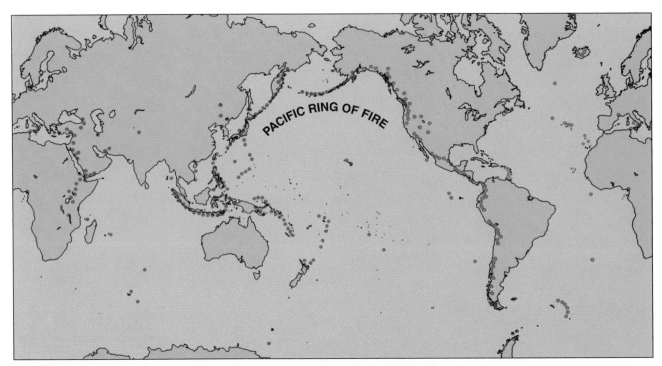

Figure 14.14   Map of major volcanic regions of the world. Note the strong correlation with the plate boundaries in Figure 13.22. Intraplate "hot spots," such as Hawaii, are the only volcanic areas not near plate boundaries.   ■   **Which plate has the greatest number of volcanoes around it?**

trench system, volcanoes occur in the vicinity, typically along the edge of the overriding plate. This subduction process produces the Pacific ring of volcanoes that includes parts of the Andes, the Cascades Range, and the Aleutians; the Kuril Islands and the Kamchatka Peninsula in Russia; and Japan, the Philippines, New Guinea, Tonga, and New Zealand.

Another volcanic belt marks the line of collision between northward-moving Southern Hemisphere plates and the Eurasian plate. The volcanoes of the Mediterranean region, Turkey, Iran, and Indonesia are located along this line.

The movement of Earth's major lithospheric plates influences the distribution of most volcanism, as it does the tectonic processes of folding and faulting, which we will discuss next. However, some active volcanic areas do not occur near plate boundaries but exist within the central parts of plates, far from the more active edges. The Hawaiian Islands, the Galapagos Islands, and the area of Yellowstone National Park are examples of these intraplate "hot spots." Some may be associated with a plume of magma rising from the mantle.

Not all volcanism involves erupting volcanoes. Some continental areas are covered with enormous accumulations of lava, called **lava plateaus** or **flood basalts,** which consist of hundreds of overlapping lava flows. In past geologic periods, basaltic magmas welled up quietly through many separate **fissures** and flowed across the existing landscape, often engulfing it in lava to depths of thousands of meters. The Columbia Plateau in Washington, Oregon, and Idaho, covering 520,000 square kilometers (200,000 sq mi), is a major example of such a lava plateau (Fig. 14.15), as is most of India's Deccan Plateau.

Figure 14.15   Columbia Plateau flood basalts in southwestern Idaho. River erosion has cut a deep canyon to expose the uppermost layers of basalt underlying the Columbia Plateau.

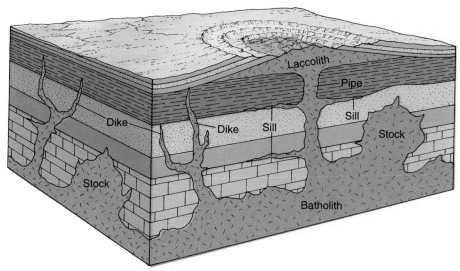

Figure 14.16 The major plutonic features. These intrusive igneous features crystallize below the surface. Later these forms may be exposed by erosion. ■ **Of what type of rock are many of these plutons composed?**

Plutons

A great variety of shapes, sizes, and forms of magma bodies can result from plutonic activity, but when first formed most have no effect on the shape of Earth's surface terrain. **Plutons** are intrusive igneous bodies that eventually cool and solidify in the subsurface, often at great depth, so it is only after millions of years of uplift and erosion that some of these forms become exposed at the surface to become part of the landscape. What usually happens is that a plutonic body, sometimes called an *intrusion,* is composed of granite or other intrusive igneous rocks that are eventually exposed by erosion and either stand higher or are reduced lower than the landscape around them, depending on their resistance relative to surrounding rocks (Fig. 14.16).

A relatively small, irregularly shaped pluton is called a **stock.** A stock is usually limited in area to less than 100 square kilometers (40 sq mi). The largest plutons, termed **batholiths,** are larger than 100 square kilometers and are complex masses of solidified magma, usually granite. Batholiths melt, metamorphose, or push aside other rocks as they grow within Earth's crust. Batholiths vary in size; some are as much as several hundred kilometers across and thousands of meters thick, and they form kilometers beneath Earth's surface. Batholiths form the core of many major mountain ranges, primarily because uplift has caused older covering rocks to be eroded away, and granite tends to be resistant to degradation processes. The Sierra Nevada, Idaho, Rocky Mountain, Coast, and Baja California batholiths cover areas of hundreds of thousands of square kilometers of granite landscapes in western North America.

When magma forces its way into cracks and between layers, disturbing but not melting the surrounding rock and coming near to the surface, it forms other kinds of **igneous intrusions.** A **laccolith** develops when molten magma flows horizontally between rock layers, bulging the layers nearer to the surface upward to form a solidified mushroom-shaped structure. A laccolith resembles a mushroom because it is usually connected with a magma source by a pipe or stem. The resulting uplift on Earth's surface is often compared to a giant blister, with the magma beneath the surface layers comparable to the fluid beneath the skin of a blister. Like batholiths, laccoliths, although typically much smaller, often form the core of mountains or hills after millions of years of erosion have worn away the surface cover of sedimentary rocks. The LaSal, Abajo, and Henry Mountains in southern Utah have been formed from laccoliths, as have many other mountains in the American West.

Smaller but no less interesting landforms created by intrusive activity may be exposed at the surface through erosion. Magma under pressure can intrude into a vertical fracture in the crust that runs across the general trend of the surrounding rocks. As it solidifies, it forms a wall-like sheet of igneous rock known as a **dike** (Fig. 14.17).

Figure 14.17 An exposed dike in central Colorado. This resistant rock was intruded into softer sandstone. Later the sandstone was eroded away, leaving the resistant dike exposed.

Figure 14.18 Sills form where magma intrudes between parallel layers of surrounding rocks. This photo in Yellowstone Park, Wyoming, shows two sills intruded within layers of volcanic ash. ■ **What is the major difference between a dike and a sill?**

When exposed by erosion, dikes appear as flat vertical walls of resistant rock rising above the countryside. At Shiprock, in New Mexico, resistant dikes many kilometers in length rise vertically to more than 90 meters (300 ft) above the plateau. Magma can also intrude between and parallel to layers of rock, solidifying in a horizontal sheet of igneous rock called a **sill** (Fig. 14.18). The Palisades, along New York's Hudson River, provide an example. Shiprock is a **volcanic neck,** a spire or mountain that formed from the exposed (formerly subsurface) pipe that fed a long-extinct volcano situated above it about 30 million years ago (Fig. 14.19).

Tectonic Processes, Rock Structure, and Landforms

Tectonic processes include the bending, folding, warping, and fracturing of Earth's crust that are largely a product of the conflicting motions of Earth's lithospheric plates. Most of our information about such deformation comes from observing aspects of the rock **structure**— the nature, configuration, and arrangement of rocks as they extend into the subsurface. Sedimentary rocks are especially useful to study because we know that when they were formed the layers in these rocks were nearly horizontal and the youngest layers are always on top of older ones. If layers of sedimentary rocks are tilted, bent, or displaced, then we can assume some kind of deformation has taken place.

Earth scientists describe the tilting of rock layers or inclination of other rock surfaces as the **dip.** Dip is measured in degrees of angle from the horizontal. The **strike** of such beds is their compass direction (in map view), taken at right angles to the dip. Thus, we might say that certain layers of outcropping rocks have a dip of 40° (to the) southeast and a northeast-southwest strike (Fig. 14.20).

Earth's crust has been subjected to tectonic pressures and tensions throughout its history, though the stresses seem to have been greater during some geologic periods than others. Crustal rocks have responded to these stresses by wrinkling, warping, and fracturing, as well as by being pushed up or by sinking down. Most of these changes have occurred slowly over hundreds of thousands or millions of years, but others have been rapid and cataclysmic. The varied responses of rocks to these stresses may yield a variety of complex configurations, as evidenced by the nature of deformation in the structure of the rocks. Gradational processes act on these structures as soon as they begin to appear at Earth's surface, working to wear them down, fill them in, and level the landscape.

Tectonic movements may involve uplift or depression of large crustal sections, and may result from either **compression,** which tends to shorten and thicken the crust, or **tension,** which tends to stretch and fracture the rocks

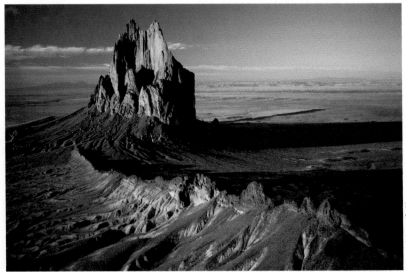

Figure 14.19 Shiprock, New Mexico, is a volcanic neck. Volcanic necks form when the resistant intrusive pipe of the volcano is exposed by erosion. ■ **Why do you think this feature is called Shiprock?**

Figure 14.20 Dip and strike in sedimentary rocks. Dip is the angle and direction that rock layers tilt, measured from the horizontal. Strike is the compass direction of the line marking the intersection of the rock layers with Earth's surface. The strike is always at right angles to the dip.
■ **What are the dip and strike of the upper strata of sandstone in this diagram? (Assume that north is toward the top of the diagram.)**

(Fig. 14.21). Generally speaking, compression usually results in "wrinkles" in the crustal rocks, which may eventually shear or slide over themselves horizontally if the forces are great enough. The crust tends to fracture, and crustal blocks may move downward because of the loss of support related to tensional stresses.

Warping of Earth's Crust

The broad, gentle deformation of Earth's crust over large areas is called **warping.** The state of Florida and Mexico's Yucatán Peninsula were raised to their present levels from below the ocean by crustal warping. Evidence for this warping in Florida is found in the number of marine fossils in the sedimentary rocks there. The extensive

Colorado Plateau has been uplifted thousands of feet in similar fashion. Slow downward warping of the crust and consequent inundation of low plains by the ocean have created the North Sea. The crust beneath Hudson Bay is rising slowly, so that eventually the bay will disappear. Another example occurs around Lake Superior, where the north shore is rising while the south shore is slowly sinking. The last two examples are a consequence of the disappearance of continental ice sheets, whose sheer weight depressed these areas hundreds of meters. With the disappearance of the ice, the areas are rebounding to their former levels.

The rebounding of Earth's crust is related to **isostasy,** the concept that the crust is floating in hydrostatic equilibrium on denser material in the asthenosphere. The effect of isostasy is that adding weight (in the form of sediment deposits, glacial ice, or a large body of water) causes the crust to sink, whereas the removal of weight (by erosion, deglaciation, or disappearance of large bodies of water) allows the crust to rise (isostatic rebound) or to float a bit higher. Although much of the broad crustal warping occurring today is related to isostasy, the kind of warping that is raising the Colorado Plateau may be due to deep-seated upwelling.

Crustal warping is an important process of landform change that affects broad parts of Earth's surface. Although it is a tectonic process, it is often not directly related to widespread tension or compression in Earth's crust;

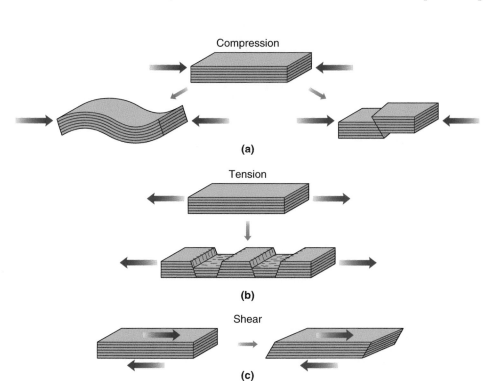

Figure 14.21 Three types of stress and the deformation of rock layers. (a) Compression can cause rocks to bend and fold or break and shift along a fault. (b) Tension, or pulling apart, generally causes rocks to break and shift along a fault. (c) Shear stress pulls or pushes rocks in opposite directions and may cause blocks of rock to break and slide alongside each other.

Mapping the Distribution of Earthquake Intensity

When an earthquake strikes a populated area, one of the first reports of geoscience information released is the magnitude of the tremor. Magnitude is a numerical expression of an earthquake's size at its source in terms of energy released. In this sense earthquakes can be compared to explosions: for example, a magnitude 4.0 earthquake releases energy equivalent to 1000 tons of TNT exploding. Because the scale is logarithmic, a 6.9 magnitude is the equivalent of 22.7 million tons of TNT.

Magnitude is an index of an earthquake's power. Larger earthquakes have the potential to cause more damage and human suffering than smaller ones do, but the situation is far more complex than that. A moderate earthquake in a densely populated area may cause great injury and dam-

age, yet a very large earthquake in an isolated region may not affect humans at all. Many factors relating to physical geography can influence the impact of an earthquake on people and their built environment. In general, the farther a location is from the earthquake epicenter, the less the effect of the shaking, but this generalization does not always apply. An earthquake in 1985 caused great damage in Mexico City, including the complete collapse of buildings, yet it was centered 385 kms (240 mi) away.

The Mercalli Scale of earthquake intensity (I–XII) was devised to measure the impact of a tremor on people, their homes, buildings, bridges, and other elements of human habitation. Although every earthquake has only one magnitude, the intensity can vary greatly from place to place, so a range of intensities will be encountered. The impact of a

(a)

Folding

The wrinkling of Earth's crust, known as **folding,** usually occurs in response to slow lateral compression, or the effect of compressive stress on rocks that are deeply buried in the subsurface. If you spread a cloth on a table

rather, it proceeds from vertical movements probably related to lateral flow of material in the asthenosphere.

and push it from one side, the folds that result are much like the folds in Earth's crust that result from various kinds of pressure. Placing both hands some distance apart on the cloth and moving one hand toward you and the other away produces an even larger number of folds. Thus, if underlying rocks move laterally past one another, compressional folding can also occur.

Folds produced naturally in rock may be very small, covering an area of a few centimeters, or they may be single earthquake on a region vaires spatially, and the patterns of Mercalli intensity can be mapped. Earthquake intensity maps use lines called isoseismals, which are lines of equal shaking and earthquake damage, expressed in Mercalli intensity levels. Patterns of isoseismals are useful in assessing what local conditions contributed to spatial variations in areas of greater or lesser shaking and impact. Earthquake intensity factors that vary geographically include conditions of soil or bedrock, population density, construction type and quality, and topography. Areas on unconsolidated earth materials, or with poor construction, or having dense populations tend to suffer more from shaking and tend to experience greater damage. Mexico City's problem was that it was built in a topographic basin filled with deep alluvial deposits. These unconsolidated sedimentary deposits shook violently, even though the earthquake epicenter was distant from the urban area.

Geographic patterns of isoseismals can be analyzed to explain why certain areas suffered more than others did. Analyzing areas of intensities that are higher or lower than expected, given their locations, helps us to understand the reasons for spatial variations in earthquake hazard. The geographic patterns of Mercalli intensity that are generated by small tremors help in planning for the possibility of a larger earthquake in the same area. The patterns of shaking and isoseismals are likely to be similar for a larger earthquake with the same epicenter. The area affected, however, would be more extensive and levels of intensity would increase for each isoseismal.

(b)

enormous, with vertical distances between the crests and troughs measured in kilometers. Folds can be tight or broad, symmetrical or asymmetrical (Fig. 14.22). Much of the Appalachian Mountain system is an example of fairly simple folding on a large scale (see also Fig. 15.20). In contrast, highly complex folding formed the Alps, where folds are overturned, sheared off, and piled on top of one another. Almost all mountain systems exhibit some degree of folding, though the folding is usually found in connection with some faulting (fracturing of the crust).

In structural terms, upfolds are called **anticlines** and downfolds are called **synclines** (see Fig. 14.22). The flanks of folds that form anticlinal crests and synclinal troughs are the fold limbs. Folds may be symmetrical where the compressional forces are somewhat even from both sides. As these forces become stronger from one side, the folds become more and more asymmetrical,

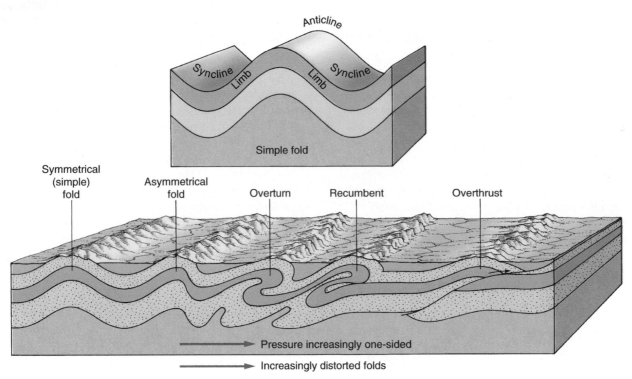

Figure 14.22 Diagram of the major types of folds

until they are overturned (Fig 14.23). Some may be so lopsided that the fold lies horizontally; these are known as **recumbent folds.** Ultimately bending rocks may not yield further, so the upper part of the fold shears or breaks and slides over the lower. This process and the rock structure produced are each referred to as an **overthrust** (thrust fault). Major overthrusts occur along the northern Rocky Mountains (Fig.14.24) and in the southern Appalachians. Recumbent folds and overthrusts are important in the formation of complex mountains such as the Andes, Alps, and Himalayas.

Faulting

Some rocks are too rigid to bend into folds, and they rupture instead. Both tension and compression can cause rock masses to fracture and move differentially with respect to one another. When slippage or displacement of the crust occurs along a fracture plane, the fracture is called a **fault** (Fig. 14.25). The instantaneous movement along a fault during an earthquake varies from fractions of a centimeter to several meters. Faulting can move rocks vertically (up and down), laterally, or both. The maximum horizontal displacement along the San Andreas Fault in California during the 1906 San Francisco earthquake was more than 6 meters (21 ft). A vertical displacement of more than 10 meters (33 ft) occurred during the Alaskan earthquake of 1964. The cumulative displacement along a major fault over millions of years may be tens of kilometers vertically or hundreds of kilo-

meters horizontally, though the vast majority of faults have offsets that are much smaller.

Vertical displacement along a fracture occurs when one side of the fault moves up or drops down in relation to the other side. This kind of movement, up or down along the *dip* of the fracture plane extending into Earth, is known as **dip slip.** Faults that have vertical displacement and have resulted from tension are called **normal faults;** those resulting from compression are called **reverse faults.** When compression pushes one crustal slab up and over another, the displacement is called a **thrust fault** (overthrust).

Figure 14.23 Folded sedimentary strata at Ecola Beach, Oregon.
■ **Identify the anticlines and synclines in this photo.**

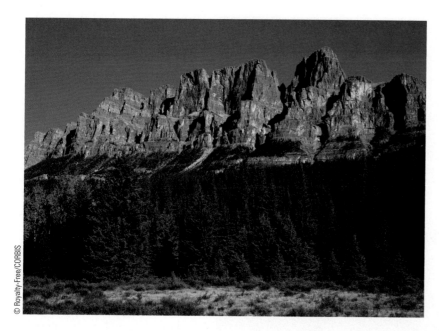

Figure 14.24 The Canadian Rockies are complex mountains where folds have been broken and sheared, forming spectacular overthrusts.

Figure 14.25 Diagrams of the major types of faults

Figure 14.26   The east front of the Sierra Nevada in California is a steep fault scarp.

in slope. Precipitation on the windward side of the range is increased because of orographic lifting. The steep lee side of the tilted fault block becomes more arid than before, because it now lies in the rain shadow of the range. Increased precipitation and lower temperatures at higher elevations combine to change the climate of the raised land significantly, and the change in climate then changes vegetation, soils, and animal life. Soils are also affected by increases in surface runoff and erosion. The process of uplift extended over several million years, so these changes have been gradual, but episodic. The Sierra Nevada are continuing to rise rapidly, geologically speaking— perhaps an average of a centimeter per year. In addition, it should be remembered that the gradational agents (water, ice, wind, and gravity, alone or in combination) begin working to level the land as uplift begins. The Sierra Nevada, for instance, are much different today than they once were, for they have been carved and etched by thousands of years of glaciation, stream erosion, and gravitational movement of weathered material. These processes have

The steep topographic escarpment of the uplifted block where vertical displacement has offset the land surface is called a **fault scarp** or **fault escarpment.** Such fault scarps can account for spectacular mountain walls, especially in the western United States. The east face of the 645-kilometer-long (405 mi) Sierra Nevada of California is a classic example of a fault scarp as it rises steeply 3350 meters (11,000 ft) above the desert. In contrast, the west side of the Sierra Nevada (the "back slope") descends very gently over a distance of 100 kilometers (60 mi) through rolling foothills. The Sierra Nevada range is a great **tilted fault block** that presents its steep fault scarp to the east (Fig. 14.26). The equally dramatic Teton Range of Wyoming rises in the same way, with its fault scarp facing to the east as well. The fault block that forms the walls of Santa Elena Canyon in Big Bend National Park, Texas, is also an excellent example of a fault scarp, modified so little by degradation that it preserves much of its block shape (Fig. 14.27). The Colorado Plateau steps down to the Great Basin by a series of fault scarps that face westward in southern Utah and northern Arizona.

Let us consider for a moment the effects on other physical systems as uplifting occurred in the Sierra Nevada. Stream erosion is accelerated by the increase

Figure 14.27   The steep fault scarp at Santa Elena Canyon, Texas-Mexico. The Grande has cut a canyon into the uplifted block. The left wall of the canyon is Mexico and the right wall is in the United States.

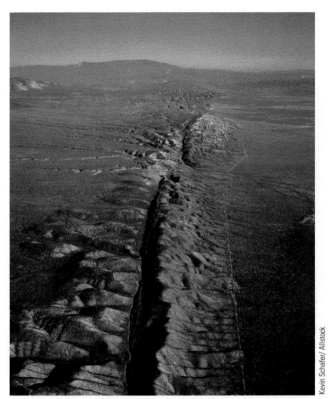

Figure 14.28 The San Andreas Fault along the Carrizo Plain in central California. The area west of the fault is moving northwestward, in relation to the area on the eastern side. Note that erosion has formed a long, narrow valley in the faulting-weakened rock along the fault zone. ■ **What type of fault is the San Andreas?**

carved and shaped valleys in the fault block, leaving spectacular canyons and mountain peaks.

The displacement along some faults is horizontal rather than vertical. All such faults descend steeply into Earth's crust, but the slippage is parallel to the surface trace, or *strike,* of the fault. This lateral movement creates a **strike-slip fault.** Active strike-slip faults cause horizontal displacement of roads, railroad tracks, fences, streambeds, and other features that are astride the fault. The San Andreas Fault, which runs through much of California, is a strike-slip fault (Fig. 14.28). Faults frequently develop in parallel sets. Where the block between two parallel faults is elevated above the land to either side, because it has been pushed up or because the land to either side has dropped down, the raised portion

is called a **horst** (Fig. 14.29). The central European Highlands north of the Alps contain many horsts, as does the Basin and Range region of the southwestern United States. The Basin and Range region extends eastward from California to Utah and southward from Oregon to New Mexico, including all of Nevada. The great Ruwenzori Range of East Africa is a horst, as is the Sinai Peninsula between the fault troughs in the Gulfs of Suez and Aqaba (see again Fig. 13.27).

The opposite of a horst is a **graben,** a depression between two facing fault scarps (see Fig. 14.29) that may occur when the block between two faults drops down or when the land to either side of the faults is uplifted. Classic examples of grabens are Death Valley, California (Fig. 14.30), and the middle Rhine Valley in Germany. Lowlands that are much longer than they are wide and that are related to faulting are called **rift valleys.** Examples of rift valleys are the Rio Grande rift of New Mexico and Colorado; the Great Rift Valley of East Africa, in which are located Lakes Tanganyika and Malawi; and the Dead Sea rift valley, where that body of water lies at an elevation some 390 meters (1280 ft) below the Mediterranean Sea, which is only 64 kilometers (40 mi) away. Rift valleys also run along the centers of oceanic ridges and along much of the San Andreas Fault zone.

Immense regions may be affected by crustal shattering along faults, producing a mosaic of horsts, grabens, and tilted fault blocks. The outstanding example is the Basin and Range region of the Western United States, named for the many north-south trending downdropped basins and uplifted mountain blocks (ranges).

Faulting and folding are both important processes of landform development, especially mountain building. In fact, the two processes—faulting and folding—often occur in the same areas, and sometimes in the vicinity of volcanism, the other important mountain-building process. These processes are active today, as attested to by recent earthquake activity.

Earthquakes

Earthquakes, the most prominent evidence of present-day tectonic movement, are vibrations of Earth that occur when the accumulating strain of crustal deformation is suddenly released by displacement along a fault. An earthquake does not cause displacement. Rather, the shock waves of an earthquake signify the release of energy caused by the sudden, lurching movement of crustal blocks past one another. It is the earthquake's

Figure 14.29 Horsts and grabens are blocks of earth material that are bounded by roughly parallel vertical faults. An uplifted block between two faults is a horst and a graben is a down-dropped block.

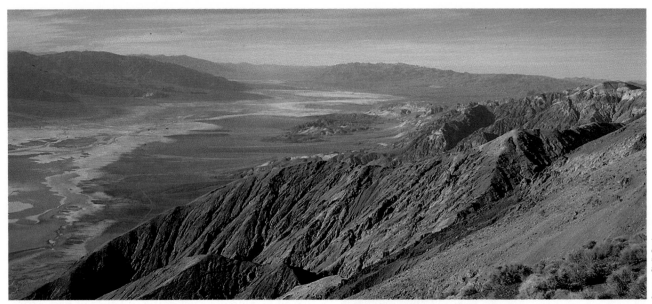

Figure 14.30 Death Valley, California, is a classic example of a topographic graben. The valley floor is 86 meters (282 ft) below sea level—the lowest elevation in North America. ■ **What is the difference between a graben and a horst?**

surface waves, which pass along the crustal exterior or emerge on the surface from below, that cause the damage and subsequent loss of life that we associate with major earth tremors. The point at which an earthquake originates is the **focus,** which may occur anywhere from near the surface to a depth of 700 kilometers (435 mi). The earthquake **epicenter** is the point on Earth's surface that lies directly above the focus and is where the strongest shock is normally felt (Fig. 14.31).

Most earthquakes are so slight that we cannot feel them, and they produce no visible damage. Usually they occur deep within Earth, and no displacement is visible at the surface. Others are strong enough to rattle a few dishes, while a few are so strong as to topple buildings, break power and gas lines and water pipes, and trigger rockslides or landslides. Aftershocks may follow the main quake as crustal movement settles. Geophysicists are currently investigating the possibility that foreshocks may alert us to major earthquakes, though evidence is at present inconclusive.

Measuring the Size of Earthquakes   The severity of an earthquake can be expressed in two ways: (1) the size of the event as a physical Earth process and (2) the degree of its impact on humans. These two methods may be related; that is, all other factors being equal, large earthquakes should have a greater impact on humans than smaller ones would. However, large tremors may strike in distant, uninhabited places, thus having limited impact, and smaller quakes may strike densely populated areas, causing much damage and suffering. The two systems, which together measure both the size and the human impact of an earthquake, help scientists and planners understand hazard potential for a locality.

The scale of **earthquake magnitude,** originally developed by Charles F. Richter in 1935, is based on the energy released by an earthquake as recorded on seismographs. Earthquake magnitude is expressed in a number, generally with one decimal place. Every increase of one whole number in magnitude (for example, from 6 to 7) means the earthquake wave motion is 10 times greater, but the actual energy released may be 30 times greater. The extremely destructive 1906 San Francisco earthquake occurred before the magnitude scale was devised but has been estimated to have been the equivalent of a magnitude 8.3. The strongest earthquake in North

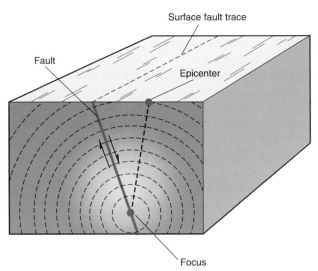

Figure 14.31 The center of energy release for an earthquake is the focus—the location where the fault broke—which is typically at some depth beneath the surface. The epicenter is the location on the surface directly above the focus. ■ **Why is the epicenter not located where the fault crosses the Earth's surface?**

America (8.6) occurred in Alaska in 1964. The 1989 Loma Prieta earthquake near San Francisco measured 7.1, but it was still responsible for the deaths of more than 60 people. Most were crushed in their automobiles when an elevated freeway collapsed. Many factors other than magnitude affect the severity of damage and loss of human life resulting from an earthquake. Each earthquake has only one magnitude reading that represents the earthquake's size in terms of energy released.

A very different earthquake scale, called **earthquake intensity,** is used to record and understand patterns of damage caused by the earthquake and the degree of its impact on people and their property. Called the **modified Mercalli scale** of earthquake intensity, it uses a system of categories numbered from I to XII (Table 14.1) to describe the effects of the earthquake on people and the spatial variation in impacts from the tremor. Roman numerals are used to avoid confusion with the magnitude scales. Although a maximum intensity is determined, an earthquake produces a variety of intensity levels, depending on local conditions. These conditions include distance to the epicenter, how long the shaking lasted, the severity of shaking resulting from the local surface materials affected, population density, and building construction in the affected area. After an earthquake, observers gather information about Mercalli intensity levels by noting the damage and talking to residents about their experiences. Locations and the intensity of responses are mapped, so that geographic patterns of damage and intensity of shaking can be analyzed. Understanding the spatial patterns of damage and ground response helps us to plan and prepare for future tremors.

Earthquake Hazards The location of an earthquake epicenter in relation to population density, construction type, and the stability of earth materials on which buildings are constructed greatly influences its impact on humans. A strong tremor (magnitude 7.5) that struck the sparsely settled Mojave Desert communities of California in 1992 caused little loss of life. Yet a magnitude 6.9 quake in 1988 in Armenia (in the former Soviet Union) destroyed two cities, with a death toll of 25,000. The effects of an 8.1 magnitude earthquake with an epicenter 240 miles from Mexico City in 1985 caused a death toll of more than 9000 and extensive damage to many buildings, including high-rise sections of the city. Loosely deposited earth materials tend to shake violently in comparison to areas of solid bedrock. Mexico's densely populated capital city is built on an ancient lake bed, made of soft sediments that shook greatly causing much destruction, even though the earthquake was centered in a distant mountain region.

A Los Angeles area earthquake in 1994 killed about 60 people, collapsed freeways, and caused commuter traffic jams for months afterward (Fig. 14.32a). This earthquake had a magnitude of 6.7, with the epicenter at Northridge in the San Fernando Valley, and was not considered the "big one" that southern Californians have

Table 14.1 Modified Mercalli Intensity Scale

I.	Not felt except by a very few under especially favorable conditions.
II.	Felt only by a few persons at rest, especially on upper floors of buildings.
III.	Felt quite noticeably by persons indoors, especially on upper floors of buildings. Many people do not recognize it as an earthquake. Standing motor cars may rock slightly. Vibrations similar to the passing of a truck. Duration estimated.
IV.	Felt indoors by many, outdoors by few during the day. At night, some awakened. Dishes, windows, doors disturbed; walls make cracking sound. Sensation like heavy truck striking building. Standing motor cars rocked noticeably.
V.	Felt by nearly everyone; many awakened. Some dishes, windows broken. Unstable objects overturned. Pendulum clocks may stop.
VI.	Felt by all, many frightened. Some heavy furniture moved; a few instances of fallen plaster. Damage slight.
VII.	Damage negligible in buildings of good design and construction; slight to moderate in well-built ordinary structures; considerable damage in poorly built or badly designed structures; some chimneys broken.
VIII.	Damage slight in specially designed structures; considerable damage in ordinary substantial buildings with partial collapse. Damage great in poorly built structures. Fall of chimneys, factory stacks, columns, monuments, walls. Heavy furniture overturned.
IX.	Damage considerable in specially designed structures; well-designed frame structures thrown out of plumb. Damage great in substantial buildings, with partial collapse. Buildings shifted off foundations.
X.	Some well-built wooden structures destroyed; most masonry and frame structures destroyed with foundations. Rails bent.
XI.	Few, if any (masonry) structures remain standing. Bridges destroyed. Rails bent greatly.
XII.	Damage total. Lines of sight and level are distorted. Objects thrown into the air.

Source: Abridged from *The Severity of an Earthquake: A U.S. Geological Survey General Interest Publication.* U.S. Government Printing Office: 1989–288–913 been warned is coming. The $30 billion costs from this earthquake were the greatest in U.S. history as 24,000 buildings were damaged. Residents of the Pacific Northwest have also been warned about an impending "big one" as the North American plate slips over the subducting Juan de Fuca plate. In February of 2001, a 6.8 magnitude earthquake struck near Olympia, Washington, damaging buildings, roads, and bridges in a region that includes Seattle and many nearby communities.

Other relatively recent earthquake tragedies have occurred in Peru in 1970 (magnitude 7.8), where 65,000 lives were lost when enormous avalanches obliterated entire mountain villages; and in Iran in 1990 (magnitude 7.7),

where at least 50,000 died in remote areas of the country. In January 1995, a 7.2 magnitude quake struck the densely populated port city of Kobe, Japan (Fig. 14.32b). It killed more than 5000 people and destroyed more than 50,000 buildings, with property damage estimated at about $100 billion. In terms of economic value, this was the most expensive natural disaster in history.

Although most major earthquakes are related to known faults and occur in mountainous regions, the one most widely felt in North America was not. It was one of a series of tremors during 1811 and 1812, centered near New Madrid, Missouri, and was felt from Canada to the Gulf of Mexico and from the Rocky Mountains to the Atlantic Ocean. Fortunately, the area was not densely settled at that stage in history. Because of that huge earthquake, the area near St. Louis is still considered to be at risk from a major earthquake, as shown on the seismic risk map for the conterminous United States (Fig. 14.33). Though they are not common in these regions, recent earthquakes have occurred in New England, New York, and the Mississippi Valley. Probably no area on Earth is what could be called entirely "earthquake safe." The map in Figure 14.34 shows world regions where earthquakes are most common.

(a)

(b)

Figure 14.32 Earthquake damage along the Pacific rim. (a) The San Fernando Valley earthquake occurred on January 17, 1994. The epicenter was below Northridge, California, and the quake killed about 60 people in the Los Angeles area. (b) Exactly one year later, on January 17, 1995, an earthquake struck Kobe, Japan. The deadly tremor killed more than 5000 people and destroyed more than 50,000 buildings.

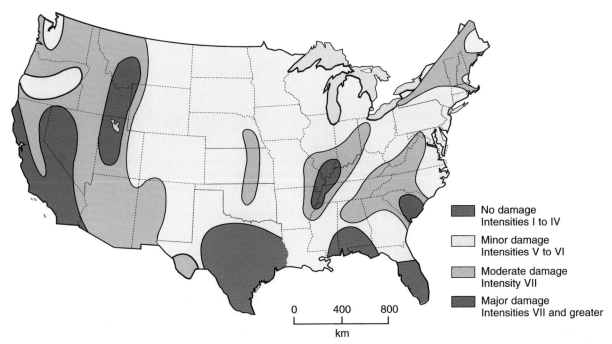

Figure 14.33   Earthquake Hazard Potential in the conterminous United States. This map shows a very generalized determination of the greatest expectable earthquake intensities for the 48 states shown.   ■   **What is the earthquake hazard potential for where you live, and what does that level of intensity mean according to the Mercalli scale in Table 14.1?**

Figure 14.34   Map of the locations of earthquake epicenters. Compare this map with Figure 13.22, which shows the major tectonic plates and demonstrates the strong relationship of earthquakes to plate boundaries. Also compare this map with Figure 14.14, the map of world volcanic regions.   ■   **Is there a strong correlation between volcanic and earthquake locations? What parts of the world are susceptible to both tectonic hazards?**

Figure 14.35 Relationships between structure and topography. Not all upfolds form mountains, nor do all downfolds form valleys. The bedrock structure and its configuration may or may not be directly reflected in the terrain. (a) The structure is an anticline (left) and a syncline (right), but the surface is only one of low relief. (b) Here, the center of a downfold, a syncline, has resisted erosion to form a mountain peak. (c) A valley has been eroded into the crest of an anticline.

Note the especially strong geographic correlation between earthquakes and volcanic activity (see Fig. 14.14 for comparison). Volcanic and tectonic processes are a natural functioning part of the Earth system. People who live in areas affected by those processes, and the governments responsible for regions where volcanic, tectonic or other potentially hazardous natural processes are active, must make detailed plans for coping with disasters before they occur.

Relationships between Rock Structure and Topography

The landform features that result from tectonic activity vary from microscopic fractures to major mountain ranges. However, few of these remain in original form because of the leveling action of agents of gradation. [It is important to note the difference between surface topography (terrain features) and the structural arrangement of the rocks, as shown in Figure 14.35.] For instance, the limb of an anticline may form a surface ridge, but it can also form a valley because of erosional activity on weak or broken rocks. Some mountains in the Alps are the erosional remnants of synclines, which are structural downfolds. As another example, Nashville, Tennessee, occupies a topographic valley, yet it is sited in the remains of a structural dome. An important distinction is that the terms *mountain* and *ridge* refer to surface landforms (the topography), whereas the terms *anticline* and *normal fault* signify rock structures, which may or may not be visible at the surface. It is also important to remember that the topographic variation on Earth's surface has resulted mostly from three major factors: the interaction among forces that act to create relief, the gradational processes that operate to shape landforms and reduce relief, and the relative strength or resistance of different rock types to degradational processes.

DEFINE & RECALL

relief	tsunami	structure	thrust fault
landforms	cinder cone	dip	fault scarp (fault
geomorphology	shield volcano	strike	escarpment)
degradation	plug dome	compression	tilted fault block
aggradation	caldera	tension	strike-slip fault
punctuated equilibrium	lava plateau	warping	horst
volcanism	(flood basalts)	isostasy	graben
volcanoes	fissure	folding	rift valley
plutonism	pluton	anticline	earthquake
lava flows	stock	syncline	focus
pahoehoe	batholith	recumbent fold	epicenter
aa	igneous intrusion	overthrust	earthquake magnitude
pyroclastic materials	laccolith	fault	earthquake intensity
(tephra)	dike	dip-slip fault	modified Mercalli scale
volcanic ash	sill	normal fault	
composite cone	volcanic neck	reverse fault	

DISCUSS & REVIEW

1. What are the four basic types of volcanoes? Give an example of each.

2. Which is the most dangerous type of volcano? Explain why.

3. What factors affect the geographic distribution of volcanoes?

4. How do plutonism and volcanism differ from each other?

5. What is a batholith? What is the difference between a dike and a sill?

6. What are the major differences between folding and faulting? What causes this type of crustal activity?

7. Draw a diagram of folding, showing anticlines, synclines, and thrusts.

8. What causes an earthquake? What is the relationship between the focus and the epicenter of an earthquake?

9. What is the major difference between volcanism and solid tectonic activity? What structural and landform features are the results of each of these two types of tectonic activity?

10. How do earthquake magnitude and earthquake intensity differ? Why are there two systems for evaluating earthquakes?

CONSIDER & RESPOND

1. Name the tectonic activity that would be most responsible for the following:

   Sierra Nevada

   Cascades

   Basin and Range region

   Ridge and Valley section of the Appalachians

2. Name several areas in the United States that are highly susceptible to natural hazards from tectonic activity. Give examples of populated locations that may someday face potential disasters from earthquakes and/or volcanic activity.

3. List five countries that lie along the "Pacific Rim of Fire." Name a few countries that also face tectonic hazards but that are not in the Pacific region.

4. Can you recall any recent earthquakes or volcanic eruptions in the news? If so, where did these occur?

5. Assume you are a regional planner for an urban area in the western United States. What hazards must you plan for if the region has active fault zones and volcanoes? What recommendations would you make as far as land use and settlement patterns are concerned to lessen the danger from these tectonic hazards?

GRADATION, WEATHERING, AND MASS WASTING

Active tectonism and volcanism cause uplift and create new landforms that are then shaped and modified by gradational processes.

What evidence in Earth's crust and on its surface supports this statement?

What does the statement suggest concerning the appearance and distribution of past and future landforms?

The theoretical end product of gradation is a land surface reduced to base level (sea level).

Why has this not happened?

What are the sources of the energy that produce gradation?

Weathering and mass wasting are initial steps in the overall process of gradation.

How are they important in shaping Earth's surface features?

Why are they initial steps?

How are they related to erosion and deposition?

Sedimentation, volcanism, faulting, and folding produce rocks and structures that vary in their resistance and susceptibility to gradational processes.

Why is this so?

What factors cause rocks to vary in this manner?

How are these processes related to joints and fractures?

The rates at which rocks weather are not equal because different rocks have varying resistance and different climatic conditions influence the type and intensity of weathering.

In what ways does climate influence weathering?

How does differential weathering influence the shape and development of landforms?

Despite spectacular mass wasting events, the imperceptible downhill motion of soil and regolith, called creep, is probably the gravity process with the greatest worldwide impact.

How can this be so?

What factors facilitate creep?

Although significant personal risks and potential economic losses may be involved, people still live in many of the world's greatest "hazard zones."

Why do they continue to live in such areas?

What can be done to reduce the risks?

Little Rock as viewed from Enchanted Rock, Texas.

The previous two chapters have summarized theories, materials, processes, and structures associated with the creation of topographic irregularities on Earth's surface. Now we will begin to examine processes that simultaneously work to modify and perhaps eventually remove these irregularities. *Degradation* wears down areas of high relief, and eroded materials are transported and deposited in areas of lower elevation. *Aggradation* processes also construct landform features by building depositional landforms such as deltas (Chapter 17), alluvial fans and sand dunes (Chapter 18), and glacial moraines (Chapter 19). Gradational forces originate on and act upon Earth's surface, in contrast to volcanic and tectonic forces that originate in Earth's interior.

Ultimately, gradation acts to level and smooth Earth's surface, through erosion and deposition in association with rock weathering and gravitational processes. Degradational processes, however, can also produce rugged landscapes. For example, the surfaces of a plateau or shield volcano may be relatively smooth in their original state, although their elevations have been raised thousands of meters. But degradation, working through its primary agent—running water—gradually begins to erode valleys separated by higher land, creating an area of mountains or hills as an intermediate stage in the gradation process. Eventually, as degradation by running water wears away more and more of the uplifted land surfaces, the irregularities originally created by water erosion may also be reduced.

Gradation and Tectonism

As noted previously, the landforms and elevational differences on Earth's surface develop as a function of two sets of opposing forces and processes—those that tend to raise the landscape (Fig. 15.1a) and those that operate to wear away landforms (Fig. 15.1b). Ultimately, if uplift ceases, the land surface may be reduced to a level and form beyond which little or no modification by the gradational processes is possible (Fig. 15.1c). This is, however, a conceptual model of what may actually occur.

In reality, degradation and aggradation are ongoing processes, and their agents are constantly at work everywhere there are differences in elevation. Also, the tectonic and volcanic processes that build up Earth's surface reoccur periodically, creating new landforms and increasing relief to be further modified by gradation. Topographic relief on Earth is a function of recency and amount of uplift, rates of degradation, and how well the Earth materials that make up rocks and landforms resist processes that work to break them down and remove them. Areas of hard rocks and recent uplift tend to be mountainous, with high elevation and high relief. Areas with soft rocks, and regions that have been tectonically inactive for a long time in terms of Earth history, tend to have low relief and low elevation.

Gradation includes the removal of loose material, its transportation to another location, and its deposition. **Erosion**—the wearing down of landscapes—and the **transportation** of rock materials act together to lower the land elevation. **Deposition** fills in depressions or raises elevation where sediments or other rock materials accumulate. The combined result of erosion and deposition is a gradual reduction of irregularities in elevation, particularly if uplift ceases or operates at a rate that is low compared to the rates of gradation (see Fig. 15.1c). Theoretically, the ultimate effect of gradation is the reduction of the land surface to **base level**—a surface so flat that erosional forces can no longer affect it. *Ultimate base level* is sea level, below which, under normal conditions, gradational forces will not lower the land surface.

The removal of rock material from one location and its deposition elsewhere are accomplished by transporting agents, the most important of which are water, glaciers, wind, and gravity. The following two chapters discuss how water underground and on the surface, particularly in streams and rivers, acts as a gradational agent. Chapter 18 covers differences in the development of landforms in arid

(a) Tectonic disturbance

Uplift

Deposition · Erosion · Deposition

(b) Gradation begins

Depositional plain · Erosional remnants · Depositional plain

(c) Nearly complete gradation

Figure 15.1 The opposing processes of uplift and gradation: (a) tectonic uplift; (b) gradation begins; (c) nearly complete gradation. ■ **What is the ultimate effect of the gradation process?**

environments and considers the wind as a factor in erosion and deposition. Chapter 19 examines glacial ice as an agent of gradation, and Chapter 21 discusses landform processes in coastal zones. Throughout the study of landform modification, the contributions of minor gradational agents, such as humans, other animals, and plants, will also be noted.

Gravity and solar radiation are the ultimate sources of energy for gradation. Gravity is the major force that moves rock and weathered debris downslope to lower elevations. Such removal and deposition is called **mass wasting** or **mass movement.** Toward the end of this chapter, we will consider some common processes of mass movement, how they shape the landscape, and their potential impacts on humans.

Weathering, mass wasting, and the various erosional agents usually do not work alone to shape and develop a particular landform. More often they combine, simultaneously or sequentially, to modify and level the landscape. For example, the northern Rockies were produced by uplift, but much of the spectacular scenery seen there today was produced by glacial activity, running water, and mass wasting that have sculpted mountains and valleys and, to varying degrees, continue to modify the landscape (Figs. 15.2 and 15.3). However, to simplify the explanation of how gradational forces and processes operate and to examine their resultant landforms, each of these processes and their associated land features will be discussed separately.

Weathering

Weathering includes various processes by which rocks at or near Earth's surface are disintegrated or decomposed in preparation for removal and transportation. Eventually, weathering can break down and loosen massive rocks into fragments small enough to be moved downslope by erosional or mass wasting processes.

Some Earth scientists include weathering in the concept of gradation. However, rock weathering can occur without gradation. Weathering does not by itself level the land; it only breaks rocks into smaller fragments so they can be transported, thus allowing gradation to take place. The definition of *weathering* is the breakdown of surface rock material *in place*, which involves no movement of the resulting fragments by weathering processes (Fig. 15.4). It is important to note, however, that practically no erosion or mass wasting will occur without initial preparation by weathering.

Weathering is not a single process, but many processes that fall into two basic categories. **Physical** or **mechanical weathering** *disintegrates* rocks without altering their chemical composition. **Chemical weathering** *decomposes* rock through a variety of chemical reactions. Both types of rock weathering take place at or near Earth's surface, though evidence of weathering has been found as deep as 185 meters (600 ft) underground. Variations in the depths to which weathering occurs, as well as differences in the types and rates of weathering, depend primarily on four factors: (1) the structure and composition of the rocks, particularly rock strength and resistance to weathering processes; (2) climate (thus the term weathering); (3) the land surface configuration; and (4) the vegetative cover. Relationships between climatic variables, especially temperature and moisture conditions, and rates and types of weathering processes are of particular interest to geographers (Fig. 15.5). In most environments, either physical or chemical weathering processes may dominate, but disintegration and decomposition generally work together.

Physical Weathering

The mechanical disintegration of rocks by physical weathering is important to gradation processes in two ways. First, smaller rock particles are more easily removed

Figure 15.2   The scenery of most high mountain regions has been formed by at least three separate phases of gradational development: stream cutting and valley formation during tectonic uplift; glacial enlargement of former stream valleys, with intense frost weathering above the level of the ice; and postglacial weathering and mass wasting with recent stream cutting. ■ **How does the profile of a valley change at each phase?**

Figure 15.3 An aerial view of mountain summits in the northern Rockies. The mountains in the foreground are the White Cloud Peaks in the Sawtooth Range of Idaho. ■ **Can you identify evidence of the three phases shown in Figure 15.2?**

and transported by one of the gradational agents. Second, the fragmentation of a large rock into smaller ones encourages chemical weathering because it increases a rock's surface area, thus exposing more of the rock to possible chemical decomposition.

Joints and fractures that develop in igneous, sedimentary, and metamorphic rocks are areas of weakness that expose more area of the rock and encourage physical weathering by making fragmentation easier. Chemical weathering also proceeds faster along joint planes or in zones of stress where the rocks are broken. Jointing can be found in any rock that has been subjected to crustal stress (Fig. 15.6). The processes of volcanism, mountain building, and rock formation often produce fractures in the rocks that can be exploited by agents of gradation, weathering, and mass wasting to shape landforms. Landscapes and landforms that develop by weathering in fractures, combined with erosion, are strongly influenced by the spatial pattern of joints in the rock.

Plants, animals, and humans also contribute to the disintegration of rocks. Plant roots wedge into cracks, and as their roots grow, they exert pressure on the rocks, eventually fragmenting them (Fig. 15.7). Burrowing animals, such as gophers and ground squirrels, weaken rocks by fragmentation and disintegration. Certain human activities

Figure 15.4 A greatly weathered boulder in Death Valley. This once hard boulder has been completely broken down by weathering into clays and mineral fragments.
■ **Why haven't the fragments been washed away to form sediments?**

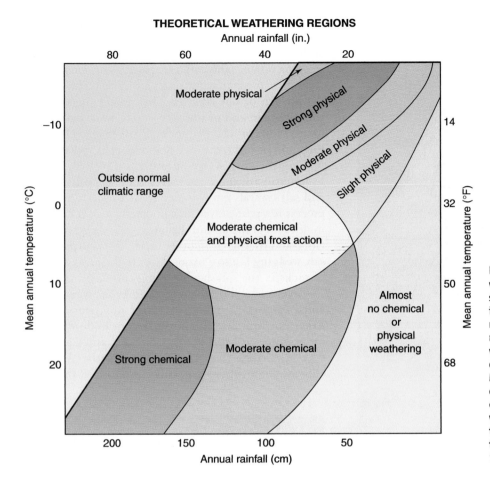

THEORETICAL WEATHERING REGIONS

Annual rainfall (in.)

Moderate physical

Strong physical

Moderate physical

Slight physical

Outside normal climatic range

Moderate chemical and physical frost action

Almost no chemical or physical weathering

Strong chemical

Moderate chemical

Mean annual temperature (°C)

Mean annual temperature (°F)

Annual rainfall (cm)

Figure 15.5 This diagram of weathering regions shows relationships between climate and weathering processes. Physical weathering is most active where temperature and rainfall are both low. Chemical weathering is most active in regions of high temperature and rainfall. Most world regions experience a combination of both physical and chemical weathering. ■ **In what weathering region would an area that had an annual mean temperature of 5°C and annual rainfall of 100 centimeters be located?**

also encourage physical weathering. To greatly varying degrees, bulldozers, off-road vehicles, bombs, mining, quarrying, and even hikers all shatter rocks into smaller fragments.

The freezing of water in fractures and joints is one of the most important means of physical weathering. When water freezes, it expands in volume by 9 percent, exerting tremendous outward force. Drivers who fail to fill their radiators with antifreeze in preparation for winter or for a ski trip to the mountains find this an expensive lesson. In cold winter areas, water pipes in homes may burst if they are not sufficiently insulated to protect them from freezing.

Similarly, when water seeps into cracks in exposed rock and freezes, the expanding ice forms a wedge that can split the rock apart (Fig. 15.8). This process, called **frost wedging,** is most intense where cycles of daytime thawing and nighttime freezes are frequent and intense—for example, in middle- to high-latitude regions and at high elevations. This weathering process is attacking even the "ageless" 60-foot-high granite faces of presidents on Mount Rushmore. Freeze-thaw cycles have enlarged fractures in the granite since the sculptured faces were completed in 1941 (Fig. 15.9).

Frost wedging is an important weathering process in most high-latitude regions, but it is most active in high mountains at or near the tree line. Frost wedging breaks

up the exposed bedrock at these higher elevations (Fig. 15.10), and if fragments break loose on a steep slope, gravity causes them to fall or tumble downslope. Broken rock fragments that accumulate at the base of steep slopes or cliffs, called **talus,** often produce a landform called a

Figure 15.6 Jointed and weathered granite near the summit of Mt. Whitney in the Sierra Nevada of California. The closer the joint spacing in a rock mass, the more rapid the weathering, both physical and chemical. ■ **Why do high mountain regions such as this one have such severe weathering?**

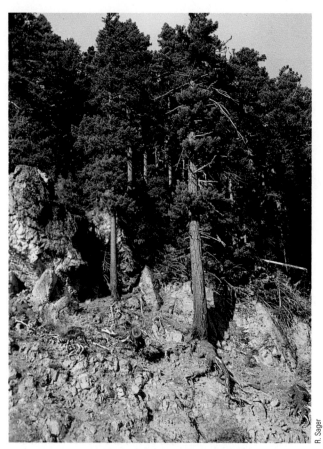

Figure 15.7   Physical weathering in Olympic National Park, Washington. As tree roots grow, the rock is wedged apart.   ■ **Besides trees, what other organisms cause organic weathering?**

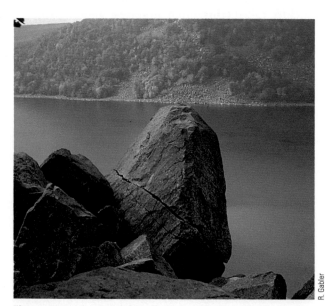

Figure 15.8   Frost wedging of a rock at Devil's Lake, Wisconsin. As water freezes in rock joints, it expands, wedging further into the rock.   ■ **What is the reason that frost wedging can split such a large rock?**

talus cone, named after its distinctive shape (Fig. 15.11). Any process that causes rocks to break, fall, and accumulate at the base of a slope can produce talus, but where frost wedging occurs, talus cones are particularly conspicuous.

Similar mechanical disintegration also results from the growth of salt crystals. This process occurs most commonly in arid or rocky coastal regions, particularly in sandstone or fractured rocks. Between periods of wetting (rain or surf), the rock dries out and capillary moisture moves salty water outward to the rock surface. Evaporation of the water leaves saline deposits behind, and salt crystals grow, wedging the rock apart. As this process is repeated over time in desert areas, it can create multitudes of niches and shallow caves in large sandstone cliffs, and overhangs in other rocks. **Salt weathering** or **salt wedging** is also a major factor in the breaking down of rocks in the splash zone along cliff-lined seacoasts.

Caves and cliff overhangs created by salt wedging provided important living spaces for prehistoric Native American communities in the American Southwest—for example, at Mesa Verde, Colorado, and Canyon de Chelly, Arizona (Fig. 15.12). These Native American cliff-dwelling villages generally faced toward the south,

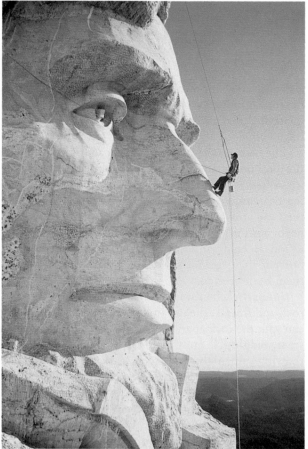

Figure 15.9   Even the granite face of President Lincoln at Mount Rushmore is aging as a result of freeze-thaw cycles and frost wedging.   ■ **Will he need a facelift someday?**

Figure 15.10 Water expands when it freezes into ice. The freezing of water in rock fractures creates a force of expansion great enough to break up rocks. ■ **How could this process of frost wedging affect the public in areas where freezes are common in winter?**

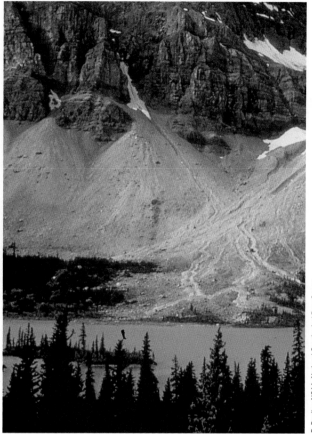

Figure 15.11 Talus cones in the Canadian Rockies formed by weathering and rockfalls at the base of steep slopes. ■ **Why does this rock debris form a cone shape?**

so they were shaded when the sun was high in summer and warmed by the sun's lower-angle rays in winter.

Granular disintegration, another type of physical weathering, can result from a variety of processes that break rock down into its constituent grains or particles, especially when the rocks consist of coarse-grained sediments or mineral crystals. Granite, composed of quartz, feldspar, and mica, is one example. These minerals expand with heat and contract with cold at different rates, producing stresses that may cause the rock to disintegrate. Granular disintegration is also a common product of chemical weathering of the weaker minerals in granite and other coarse-grained rocks.

It was once believed that extreme diurnal temperature changes in desert regions were a primary cause of rock disintegration there. Laboratory studies now refute this. Fire, however, can be an effective weathering agent. Because rock is a poor conductor of heat, fire expands only the rock's outer shell. In addition, fracturing is compounded by steam pressure formed from water in the rocks. Thus, bedrock and boulders exposed to forest and brush fires often undergo cracking and accelerated granular disintegration.

Exfoliation refers to the *spalling*, or breaking off, of curved rock shells parallel or concentric to the surface. Exfoliation is especially common in granite and was once thought to be caused by temperature changes. However, this process may actually be related to both chemical and physical weathering processes. Shallow exfoliation, which produces thin sheets or spalls a few millimeters or so thick, is associated with hydration, a chemical weath-

Figure 15.12 Ruins of prehistoric Native American dwellings built into the cliffs at Montezuma Castle National Monument, Arizona. ■ **Can you think of reasons why people chose these cliffs for their dwellings?**

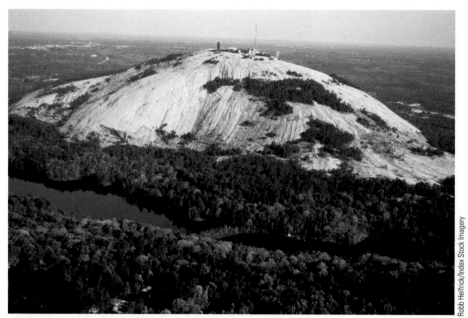

Figure 15.13 Stone Mountain, in Georgia, is a huge granite exfoliation dome. ■ **Why is granite so susceptible to exfoliation?**

ering process. The unloading of pressure on rocks that were once deeply buried can also result in exfoliation. When uplift and subsequent erosion expose at the surface rocks that were once deep underground, the pressure on the rocks is lessened, and they can minutely expand. As they do so, fractures develop, creating concentric shells of rock that resemble the layers of an onion. This characteristic fracturing is especially apparent in massive intrusive rocks, such as granite, that crystallized under very high pressures. As rock layers are peeled away from large, solid rock masses, immense dome-shaped hills or mountains, called **exfoliation domes,** are formed. Some of the world's best-known exfoliation domes are Stone Mountain in Georgia (Fig. 15.13), Half Dome in Yosemite National Park, and Sugar Loaf Mountain, which overlooks Rio de Janeiro, Brazil. Enchanted Rock and other nearby granitic hills in central Texas are excellent examples of granite exfoliation domes (see the chapter opening photograph).

Chemical Weathering

Chemical weathering, or **decomposition,** prepares rocks for removal and transportation by gradational agents in three ways. First, chemical alteration forms new minerals that are softer, finer, and less resistant to erosion. Second, chemical weathering may create new substances—through the addition of water or through other chemical change—that are greater in volume than the original mineral material. This expansion fractures rocks, accelerating the rates of both physical and chemical weathering by increasing the surface area exposed to those processes, and by weakening the rock mass. Third, chemical weathering

may dissolve minerals in a chemically active water solution (weak acid), making them easy to remove and transport. As more and more minerals are removed, the number and size of pore spaces and other openings in the rock are increased, allowing the rate of weathering to increase.

Almost without exception, chemical weathering can take place only in the presence of water. Rates of chemical weathering are always increased when water is present, so chemical weathering is most rapid in humid climates. Arid climates have enough moisture to allow some chemical weathering to take place, but the rounded boulders that show exfoliation and granular disintegration, common in arid regions where granite is present, may have developed much of their present form in earlier times of wetter climates (Fig. 15.14). The tendency of weathering to remove edges and corners of rocks, rounding them off to form spheroidally weathered boulders, is called *spheroidal weathering.* Although spheroidal weathering is common, the term does not refer to a specific weathering process but is typically the product of a combination of weathering processes.

Chemical reactions are also more rapid at high temperatures. Consequently, humid tropical regions are subject to more intense chemical weathering than are places with lower temperatures. Subarctic and polar climates

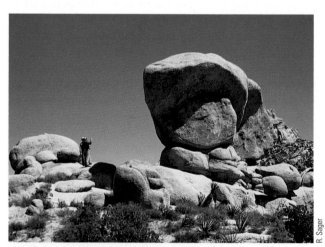

Figure 15.14 Block disintegration of granite rocks, Mojave Desert. The blocks of rock became rounded or "spheroidal" as angles were attacked by both physical and chemical weathering. ■ **Can you see where the original joints were located?**

are also subject to chemical weathering, but the effects of physical weathering are much more visible in these regions. Thus, in hot, humid regions, such as the tropical rainforest, savanna, and monsoon climates, chemical weathering is more significant in affecting the shape of landforms than is physical weathering. The landforms and rocks of warm, humid climates display the importance of chemical weathering in their rounded shapes. In contrast, the landforms and rocks of drier and cooler regions, where physical weathering dominates, are sharp, angular, and jagged (see again Fig. 15.5).

The principal processes of chemical weathering are hydration and hydrolysis, oxidation, carbonation, and solution. The most important agents performing these processes are water, oxygen, and carbon dioxide, all of which are common in soil, rock, precipitation, groundwater, and air.

Hydration and Hydrolysis   Although both processes involve the addition of water to a chemical substance, hydration and hydrolysis are distinctly different. In **hydration,** water molecules are attached to the crystalline structure of a mineral without chemical change. Hydration causes a mineral to expand, which can result in either granular disintegration or exfoliation of thin sheets. Hydration also weakens minerals, making them more susceptible to other chemical weathering processes—particularly oxidation and hydrolysis—as well as to physical weathering.

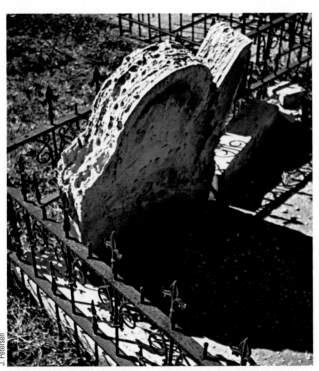

Figure 15.15   Weathering of limestone tombstones and an iron fence. This tombstone was not made of rock that was very resistant to weathering processes. ■ **The reddish fence rails result from what kind of chemical weathering?**

Hydrolysis involves a chemical change as the hydrogen and oxygen in water (H_2O) disassociate and combine with another substance. Many common rock-forming minerals are susceptible to hydrolysis, particularly the silicate minerals that form igneous rocks. Hydrolysis, like hydration, results in expansion that can create minute fractures in the rock. Hydrolysis is not limited to rock directly exposed at the surface; in warm, humid climates, it can decay rock to a depth of 30 meters (100 ft) or more.

Oxidation   The chemical union of oxygen with another substance to form a new product is called **oxidation.** The chemical compounds formed by oxidation are usually softer and finer than the original substances and have a greater volume. One of the most common oxides, as these mineral compounds are called, is iron rust, derived from the combination of iron and oxygen (Fig. 15.15). Rust is composed of two iron oxides: the minerals hematite (red) and limonite (yellow). The rusty stains visible on many rocks are often a combination of hematite and limonite, which are softer and more easily removed from rocks than the original iron-bearing minerals from which they were formed.

Solution   Though it does not actually involve chemical change, **solution** is an important process of chemical weathering. Once minerals dissolve in water, they are easily removed and transported by groundwater, soil water, or surface water flow.

Some minerals, such as calcium carbonate (calcite), sodium chloride (salt), and calcium sulfate (gypsum), form rocks that are soluble in neutral or acidic waters. Mineral salts that are immediately water-soluble are called **evaporites,** because they precipitate out of solution when water becomes saturated with them, as water loss through evaporation concentrates the minerals left behind. Many of the mineral compounds produced by chemical weathering are also soluble. Even silica becomes soluble in hot, wet climates. Removal of silica is a major feature of the laterization process (see Chapter 12).

Many minerals that are insoluble or only slightly soluble in pure water are more easily dissolved in slightly acidic water. Rain and soil water, by absorbing carbon dioxide from the atmosphere and from decaying vegetation in the soil, form a weak solution of carbonic acid. This acidic water is capable of dissolving a variety of minerals, notably calcite or calcium carbonate, the ingredient of common limestone (Fig. 15.16). When acted on by carbonic acid, calcium carbonate forms calcium bicarbonate, a salt that is soluble in water. Thus, the solution of limestone involves both **carbonation** (creation of calcium bicarbonate) and solution. In humid regions, the leaching (or removal) of salts such as calcium bicarbonate can severely weaken rocks by greatly increasing the size of pore openings.

A form of weathering that spans both chemical and physical processes is the concept of *shrink and swell*. This process refers to soils that expand and contract with the addition and removal of water.

Soils consist of both organic and inorganic constituents. The inorganic components are of two types. First are silts and sands, derived from parent materials that have not been appreciably altered by chemical weathering. Second are the clay colloids, derived through alteration of the original material. These clay colloids (particles with very small mass but with a high surface area) are tiny, platelike crystals. They are important because they influence the cation–exchange capacity (CEC) of the soil. Cations are positively charged atoms, such as potassium, calcium, and magnesium (K^+, Ca^{++}, and Mg^{++}). These are termed *metallic cations* to differentiate them from the nonmetallic cations, such as hydrogen (H^+). A high CEC means that clay colloids are well supplied with cations that can be exchanged for different cations in a soil solution. It is from this soil solution that plants acquire cations and nutrients for healthy growth. In a high-CEC situation, the H^+ in water (H_2O) can also enter and exit the crystal structure of clay minerals.

The capacity of the colloids to exchange cations is a function of their chemical composition and their crystal structure. Clay minerals have layered, "sandwich-like"

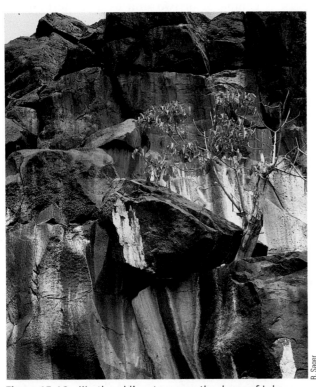

Figure 15.16 Weathered limestone near the shores of Lake Boringo, Kenya. Exposed limestone quickly discolors and becomes pitted by solution in wet topical climates. ■ **Why is limestone so intensely weathered under these climatic conditions?**

R. Sager

Acid rain and acid fogs, caused by air pollutants dissolved into atmospheric moisture, have accelerated chemical weathering in recent decades. Because many of the world's great monuments and sculptures are made of limestone (calcium carbonate) or marble (calcite), there is a growing concern about weathering damage to these treasures. The Parthenon in Greece, the Taj Mahal in India, and the Great Sphinx in Egypt are examples where

chemical solution and salt buildup are damaging and rotting away monument surfaces (Fig. 15.17).

Carbonic acid is not the only acid active in chemical weathering by solution. Other acids, derived primarily from decaying organic matter and present in groundwater, are also capable of dissolving rock minerals. Although most organic weathering takes place below the soil surface, it can also affect exposed rock. Lichens and mosses that grow on rock surfaces secrete acid substances that assist rock disintegration and thus produce nutrients for these simple life forms as they colonize bare rock.

Differential Weathering

Different rock types are affected by different weathering processes, and are removed at different rates. Granite is susceptible to being altered by oxidation, hydration, and hydrolysis, particularly under warm, humid conditions. Accordingly, it is often covered by a deep weathered regolith if it has been exposed to a warm, humid environment. Shale is chemically inert but mechanically weak, susceptible to hydration, which converts it back to clay, its original material. Sandstone is only as strong as its cement, which varies from soluble calcite to inert silica.

Wherever a variety of rock types occurs in a landscape, some will be relatively strong, or resistant to weathering, and others will be weak, or easily altered and removed. The relative resistance of a rock to weathering is a function of rock type and nature, along with the climatic conditions that it is exposed to over time. A rock that is strong compared to other rocks in one region may be weak relative to the same kinds of rocks in another area. In humid regions, limestone is removed quite rapidly by carbonation and solution, but in arid regions, limestones tend to be resistant. Likewise, rocks that are resistant in a climate dominated by chemical weathering may be weak structures. Their CEC depends on the clay mineral type and how much that type has been weathered. *Illite* and *vermiculite* are two clay minerals associated with slightly weathered soils; *montmorillonite* is an important component of moderately weathered soils; and *kaolinite* is characteristic of highly weathered soils. Of these four basic clay minerals, it is montmorillonite that expands and contracts the most. Water molecules can actually enter and exit the layered crystalline structure and cause the soil to expand when wet and contract when dry. This contraction and expansion (or shrinking and swelling) of these clay minerals can apply physical pressure on surrounding soil particles and rock surfaces, causing them to physically weather at a faster rate. According to

some sources, the expansion in volume can vary from a very small percentage to more than 100 percent. In most cases, however, the expansion in volume is less than 50 percent.

We can now return to our opening statement. The addition and removal of water from clay mineral structures is a chemical weathering process known as *hydration*. However when the clay minerals expand and contract in volume, they can then exert the forces associated with physical processes on surrounding earth materials. The end result of chemical and physical processes working in combination is a new process, shrink and swell—a weathering agent of considerable significance where water is present in clay-bearing soils.

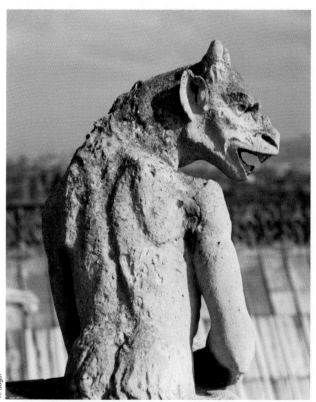

R. Sager

Figure 15.17 Sculptures in solid rock that are hundreds of years old, such as this one in Paris, have experienced accelerated erosion, particularly in recent years, because they exist in an urban environment. ■ **Why have the last several decades been so damaging to this statue?**

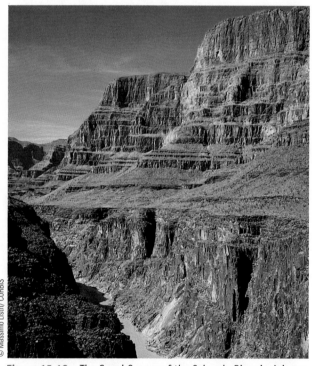

© Massimo Listri/ CORBIS

Figure 15.18 The Grand Canyon of the Colorado River in Arizona is a classic example of differential weathering and erosion in an arid climate. This cross section shows the relationship between rock type and surface form in the canyon. ■ **What surface features do the more resistant rocks, such as sandstone, form?**

where physical weathering processes dominate. In general, the more massive the rock (the fewer the joints and bedding planes), the more resistant it is to all types of weathering. The process by which certain rock types weather more rapidly than others, resulting in an uneven landscape, is known as **differential weathering.**

Relationships between Rock Structure and Topography

The variations in rock resistance to weathering exert a strong and often highly visible control on the appearance of landforms and landscapes. Differential weathering and erosion operate to make layers in rock structure more detectable unless a deep mantle of weathered regolith covers the area, as is common in warm, humid regions.

An outstanding example of how differential weathering and erosion can expose rock structure is the scenery at Arizona's Grand Canyon (Fig. 15.18). In the dry climate of that region, limestone is resistant, as are sandstones and conglomerates, while shale is relatively weak. Thus, stair-stepped cliffs wall the Grand Canyon with ledges composed of limestone, sandstone, or conglomerate, separated by gentler slopes of shale. At the canyon

Figure 15.19 Weathering, mass wasting and erosion work together to make differences in rock structure and strength visible in the landscape. Layers of varying thickness and resistance result in a distinctive array of cliffs and slopes.

Figure 15.20 A satellite image of Pennsylvania's Ridge and Valley section of the Appalachians. The image shows the effects of weathering and erosion on folded rock layers of different resistance. Resistant rocks form ridges, while weaker rocks form valleys. The image also shows the pattern of anticlines and synclines formed by an ancient continental collision that occurred before the formation of the Atlantic Ocean. The Susquehanna River cuts across the ridges at the northern (top) part of the image. This river flowed across the region prior to the development of the present ridge and valley landscape. ■ **Can you see how the topography of the Ridge and Valley section influences human settlement patterns?**

base, resistant and ancient metamorphic rocks have produced a steep-walled inner gorge.

Rock structures such as varying strata, joints, folds, and fault scarps tend to be more prominent and obvious in arid and semiarid environments or in other landscapes where chemical weathering is minimal. Resistant rocks stand out in the topography as cliffs, ridges, or mountains, while weaker rocks undergo greater weathering and erosion to form gentler slopes, valleys, and subdued hills (Fig. 15.19).

An excellent regional example of differential weathering and erosion can be seen in the Appalachian Ridge and Valley region of the eastern United States (Fig. 15.20). The rock structure here consists of sandstone, conglomerate, shale, and limestone folded into anticlines and synclines. These folds have been eroded so that the edges of the steeply dipping rock layers are exposed at the surface. In this region of humid climate, forested ridges composed of resistant sandstone and conglomerate stand up to 700 meters (2000 ft) above agricultural lowlands excavated out of weaker shales and soluble limestones.

Differential weathering and erosion is often used as a single expression because as weathering breaks up rocks at differential rates, it combines with erosion—the removal of weathered fragments—to produce the landforms we see. Weathering and erosion (as well as the deposition that follows) are closely linked processes in the development of landforms. Erosion removes loose regolith, thus exposing the less-weathered rock underneath to accelerate weathering processes at the surface.

The Importance of Weathering

The fragmented materials prepared by chemical and physical weathering are not only available to be transported—either by direct gravitational transfer or as part of the sediments carried by rivers, glaciers, or wind—but they also provide raw material for soil development. Regolith formed by weathering is the major source of inorganic soil constituents, without which most vegetation could not grow. Weathered rock material is also a major source of nutrients in the oceans as rivers carry sediments to deposit on or near the coastlines.

Chemical weathering plays an important role in breaking down rocks into their mineral components and in creating new compounds. In this way, valuable ores of iron, aluminum, tin, manganese, and uranium may be concentrated over long periods of time. This effect usually occurs under humid, tropical conditions as part of the laterization process (see Chapter 12). Such secondary ores are to be distinguished from primary metallic ores, which are concentrated by magmatic processes.

Figure 15.21 The scar and debris pile from the rockslide at Frank, Alberta, in 1903. In less than two minutes, millions of tons of rock slid down the steep slopes of Turtle Mountain and buried the town of Frank. ■ **What might trigger such a catastrophic event?**

Mass Wasting

Mass wasting, also called *mass movement,* is a collective term for all varieties of downslope transport of surface materials in direct response to gravitational force. Everywhere on the planet's surface, gravity pulls objects toward Earth's center. On sloping surfaces, this force encourages downward movement of loose or weak material. Mass movement occurs in a wide variety of ways. A single rock rolling and tumbling downhill is a form of gravitational transfer, as is an entire hillside sliding hundreds or thousands of feet downslope, burying homes, cars, and trees. Some mass movement is catastrophic in scale and violence (Fig. 15.21). However, many types of mass wasting operate so slowly that they are imperceptible, although tilted telephone poles, gravestones, and fence posts indirectly reveal the process.

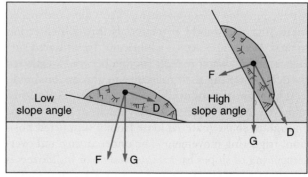

Figure 15.22 Rates of mass wasting are strongly related to the slope angle and the strength of the earth materials that make up the slope. Other factors such as amount and type of vegetation cover and amount of soil moisture or groundwater also influence downslope movement of materials by mass wasting.
■ **How might vegetative cover or moisture content affect the potential for downslope movement of soil?**

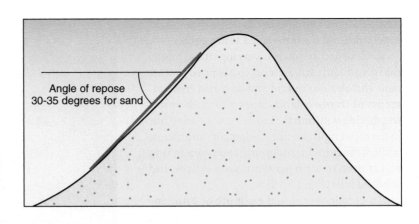

Figure 15.23 The angle of repose results from friction between loose fragments that fall or are dumped under the influence of gravity. The same size and shape of particles and the same size dump truck loads formed these nearly identical piles, all with the same height and slope angle. ■ **Would angular particles form steeper or gentler slopes in comparison to rounded particles if dumped in this manner?**

The combination of all forms of mass wasting rivals the gradational work of water as a modifier of physical landscapes, because gravitational force is always present. Wherever there is loose regolith or soil on a slope, gravity will be able to cause some movement downhill. The steeper the slope, the stronger the friction or rock strength must be to resist downslope motion (Fig. 15.22). Because gravitational force overcomes rock friction or strength more effectively on steeper slopes, it pulls down the regolith faster there. Consequently, regolith on steep slopes is apt to be thinner than it is on gentler slopes, and bedrock may be exposed.

Any surface materials that do not have the strength or stability to resist the force of gravity will respond by rolling, falling, sliding, or flowing downslope, stopping at the bottom of the slope or wherever there is enough friction to resist gravitational pull. The **angle of repose** is the steepest slope angle that loose fragments will form as they pile up after falling. Loose rocks and gravel that have been dropped by dump trucks accumulate at the angle of repose to form distinctive, cone-shaped piles (Fig. 15.23).

Gravity is obviously the main cause of mass movement. However, other factors are also important, especially in providing the initial instability leading toward movement. Water is the most important additional factor encouraging mass movement. Water fills pore spaces, increasing the weight of materials into which it has soaked. Water also acts as a lubricant. In saturated surface materials, water reduces friction between weathered particles, lowering their resistance to gravitational pull.

The shaking produced by earthquakes, explosions, and even the movements of heavy trucks or trains can be enough to shake material loose from a supported position, triggering movement. The undercutting and oversteepening of slopes by streams, waves, or bulldozers is an especially conspicuous factor in mass movement. Groundwater, meltwater, and alternate cycles of freezing and thawing or wetting and drying can contribute to mass movement. Even plant root systems and burrowing animals promote mass movement.

Classification of Mass Wasting Processes

Mass wasting processes are classified according to three major factors, each of which can be stated as a question (see Table 15.1).

1. Was the movement fast or slow?
2. What kind of earth materials moved?
3. What kind of motion took place?

Speed of Movement A simplified way to consider speed of movement is to classify mass wasting processes that can be directly witnessed while they occur as *rapid*. Processes that are ongoing but only evident when the land surface and features on it are observed or measured over time can be termed *slow* mass wasting. It is important to note, however, that designations *fast* and *slow* are generalities; the

Table 15.1			
Motion	**Material**	**Speed**	**Effect**
Creep	Soil	Slow	
Slump	Soil or debris	Slow or fast	
Slide	Rock or debris	Fast or slow	
Flow	Debris or mud	Fast or slow	
Avalanche	Ice and snow or debris or rock	Fast	
Fall	Rock	Fast	

speed of movement can vary greatly according to the slope, the nature of materials, and the triggering factor involved. Also, mass movement that is initially slow may be a precursor to more immediately destructive rapid motion, and most materials that undergo rapid mass wasting continue to be unstable with slow movements.

Type of Earth Materials Anything on Earth's surface that exists in or on an unstable (or potentially unstable) land surface is susceptible to gravity-induced movement, and can be transported downslope as a result of mass wasting. However, it is usually the movement of surface materials that causes problems for humans and the constructed environments in which they live. Mass wasting involves many kinds of surface materials, some of which do not require definition here, such as rock, soil, snow, and ice. Other kinds of materials include **debris,** a mix of rock and soil, and **mud,** a mix of soil or fine rock material and coarser rocks with water.

Kinds of Motion The ways that masses of surface materials move in response to gravity can vary greatly, and are key determinants of the mass wasting processes involved. The major kinds of motions are fall, avalanche, slide, flow, creep, and slump. Each of these motions is defined in Table 15.1 and discussed in the following sections.

Most mass wasting processes are named for a combination of the kind of surface materials transported and the kind of motion. Thus, *rockfall* specifically refers to rocks falling through the air until they come to rest, *rockslide* refers to a rock mass sliding (moving more or less as a unit) downslope; *debris flow* means a blend of rock and soil mixing and tumbling (flowing) as it moves downslope. *Creep* is a slow process, generally involving

either soil or regolith. The term is typically used alone, without mentioning the materials moved, as is *slump,* which can affect a wide variety of surface materials.

Rapid Mass Movement

Varieties of rapid mass movement include landslides, rockslides, rockfalls, slumps, earthflows, and mudflows. They are distinguished from the slower types in that their movement can be directly witnessed and their effects on the land surface are more dramatic. The speed of movement varies by the particular situation, depending on the quantity and composition of the material being moved, steepness of slope, amount of water involved, the nature of vegetative cover, and the triggering factor. Rapid mass movements usually leave a visible scar, where material has been removed upslope, and a deposit below, consisting of debris that has slid, fallen, flowed, or rolled downslope.

Landslide Though the term is often used to refer to any form of rapid mass movement, Earth scientists give the name **landslide** to the rapid downslope movement of a mass of material that moves as a unit and carries with it loose material that sometimes includes great masses of bedrock. The movement is sudden, and the moving mass can attain very high velocities. Most landslides involve the downslope sliding or flowing of some combination of rock and soil and may involve a variety of downslope motions. Large landslides are relatively rare but are often newsworthy because of their destructive qualities. If a more specific term for the kind of mass wasting that occurred is difficult to ascertain, any rapid downslope movement of surface materials may be called

Figure 15.24 Landslides not only modify the landscape, they also can cause much destruction when buildings are constructed in areas susceptible to mass movements ■ **What kinds of environments might be most susceptible to landsliding?**

© AFP/CORBIS

Figure 15.25 Looking down the valley from the site of the 1959 earthquake-caused slide on the Madison River in Montana. This rapid slide of rock, soil, and trees completely blocked the river valley, creating a new body of water, Earthquake Lake, seen in the right foreground. The massive slide killed 28 people in a valley campground. ■ **Why can earthquakes trigger landslides?**

a landslide, using the term in a generic sense. Landslides of all kinds can threaten the lives of people who live in areas that are unstable and potentially active in terms of mass wasting processes (Fig. 15.24).

Landslides occur most frequently on steep slopes during or after periods of heavy rain because of the additional weight of water and its lubricating qualities. They are a particular hazard in the clay-rich soil areas of the northern Appalachians and New England. Earthquakes have also triggered many landslides, because shaking adds additional forces and generally reduces friction. During the 1994 Northridge earthquake, thousands of landslides were triggered in the local Santa Susanna Mountains. They are also common on slopes that are undercut by streams or wave activity, or any natural or human-induced process that produces oversteepened slopes. More than 25 fatalities occur each year in the United States as a result of landslides.

Rockslide Whereas some landslides involve only the surface regolith, larger movements may detach huge masses of rock. These **rockslides** are truly enormous, with volumes measured in cubic kilometers. Anything in their path is obliterated. More important, they may dam up valleys, which soon become filled with lakes. When the lakes become deep enough, they may wash out the dams, producing catastrophic sudden floods in the valleys downstream. Thus, immediately after such a major rockslide, workers clear out the resulting dam and control the outlet of water trapped behind it. This was done successfully after the Hebgen Lake slide in south-

western Montana in 1959 (Fig. 15.25). This slide, one of the largest in North American history, was triggered by an earthquake and killed 28 people camped along the Madison River.

Huge rockslides have resulted from rock mass instability related to rock structures and the undercutting of slopes by streams, glaciers, or waves. Like all landslides, major rockslides usually occur during exceptionally wet periods when the rock mass, or a sliding plane at its base, is well lubricated. Earthquakes have been associated with many, but not all, large historic rockslides. Today, there are many locations in mountain regions where enormous slabs of rock supported by weak materials are poised on the brink of detachment, waiting only for an unusually wet year or a jarring earthquake to set them in motion.

Rockfall When an individual rock or several rocks fall downslope, clattering over other loose rock and debris, the movement is called a **rockfall**. The rocks or rock

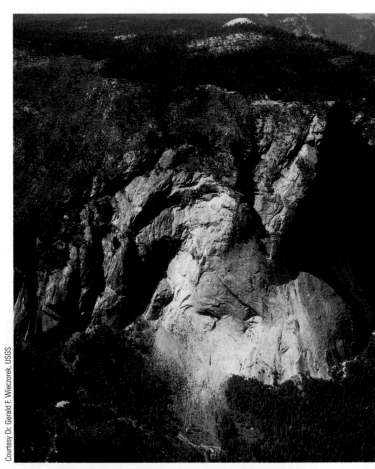

Figure 15.26 The scar from the Happy Isles rockfall that occurred in Yosemite Valley on July 10, 1996. The rockslide killed one hiker and injured numerous others. The slide was so rapid (estimated at more than 250 kilometers per hour, or 160 miles per hour) and large that it created an air blast that set off a giant dust cloud that rose from the valley floor.

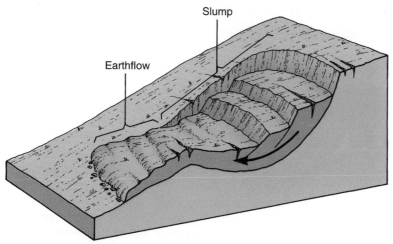

Figure 15.27   Diagram of a slump and earthflow.   ■ **What is the difference between these two types of mass movement?**

fragments involved in rockfalls can vary greatly. A rock-fall may consist of tiny granular particles skittering downslope or a huge boulder bounding downhill and fragmenting along the way. In steep mountainous areas where rockfalls are common, cone-shaped accumulations of rock fragments build up at the base of the cliffs, forming the talus slopes or cones mentioned earlier (see again Fig. 15.11). Rockfalls are particularly common in spring, when a traveler can encounter rocks scattered on mountain roads. Meltwaters and alternate periods of freezing and thawing at that time disturb precariously balanced rock masses, loosening them from their previously secure positions.

The steep granite cliffs of Yosemite Valley, California, have experienced some very large rockfalls. In July of 1996, one hiker was killed and numerous others injured by such an event. A huge, 200-ton mass of rock broke away from a cliff, slid 200 meters (650 ft) down a steep slope, and then fell airborne for another 550 meters (1800 ft) before impacting with great force. The rockfall was estimated to have moved downslope at more than 250 kilometers per hour (160 miles per hour). The moving rock produced a destructive blast of compressed

air that destroyed trees hundreds of meters from the cliff as the rockfall crashed to the valley floor (Fig. 15.26).

Slump and Earthflow   Sometimes a mass of soil on a slope slips or collapses in a backward rotation, down at the top and up at the base, so that what was once a portion of the hill slope ends up tilting backward (Fig. 15.27). The curved backward rotation of such a **slump** distinguishes it from a landslide. Slumps are likely to occur where moisture is concentrated at the base of a water-soaked mass of clay-rich soil. Slumps are conspicuous phenomena during exceptionally wet winters in California, where they frequently damage expensive hillside homes. Slumps often change into **earthflows** in a downslope direction. Earthflows are slow-moving masses of soil and/or rock fragments.

Debris Flow and Mudflow   The movement of water and rock debris mixed in various proportions is called a **debris flow** or **mudflow.** Debris flows and mudflows are fast-moving masses that combine tumbling and mixing movements of rock and soil. The main difference in composition is the water content. Mudflows tend to be water-saturated and generally move faster and farther than debris flows. In general, debris flows and mudflows follow valleys, although they can also flow as sheets downslope.

Mudflows and debris flows are the most fluid of all mass movements. They result from torrential rainfalls on steep, poorly vegetated slopes, typically in regions with distinct wet and dry seasons, such as coastal California, Oregon, and Washington (Fig. 15.28). The rains flush weathered debris into canyons, where it is mixed

Figure 15.28   A debris flow in Laguna Beach, California. During torrential rains from El Niño storms in 1998, many debris flows of mud, soil, vegetation, and human structures occurred in the mountains of southern California.

Doug Morton/USGS

Hazard Potential in the Cascades

With the violent eruption of Mount St. Helens in 1980 the majestic volcanic mountains located in northwest North America known as the Cascades served notice that they contain natural hazards of great significance. One of these hazards is a massive volcanic mudflow known as a *lahar*—the Indonesian word for such a traumatic event. These rapid flows are a slurry mix of rock debris, mud, glacial ice, and water. In 1985, such a volcanic mudflow killed 23,000 people in the Colombian town of Armero, located in a valley 50 kilometers (30 mi) from the Andean volcanic summit of Nevado del Ruiz. The lahar destroyed and then buried the town. A major eruption is not necessary to set off such a deadly event. It can be triggered by a minor eruption, which heats the ice and snow at the summit, or an earthquake, which destabilizes the steep, volcanic rock-, ash-, and ice-covered slopes.

Partly due to its lahar potential, Mt. Rainier in the Cascades was targeted by scientists as one of several high-risk world volcanoes for intensive research under the International Decade for Natural Disaster Reduction (IDNDR). Mt. Rainier was chosen because it is potentially more dangerous than any other Cascade volcano for several reasons. First, Mt. Rainier is the highest and steepest of the peaks. Second, due to corrosive volcanic gases seeping from its active summit area, the rocks have been weakened and the

with floodwaters. The result is a torrent of mud, gravel, boulders, plant debris, and water that can knock out bridges and destroy buildings and homes. Such torrents frequently occur in wet winters following dry summers in which fires have destroyed hillside vegetation and left the soil vulnerable to removal by water runoff.

Serious mudflow hazards exist in many regions with active volcanoes. Here, steep slopes may be covered with hundreds of feet of loose volcanic ash. During eruptions, emitted steam, cooling and falling as rain, saturates the ash, sending down dangerous and fast-moving volcanic mudflows, known as **lahars.** Of particular concern are high volcanic peaks capped with glaciers and snowfields. Should an eruption melt the ice, rapid and catastrophic lahars rush down the mountains with little warning and bury entire valleys and towns. In the United States, there is a concern over some of the high Cascade volcanoes in the Pacific Northwest. Mounts Rainier, Baker, Hood, and Shasta all have the conditions in place, including nearby populated areas, for potentially disastrous mudflows to occur.

Slow Mass Movement

Despite spectacular examples of rapid mass movement's force and visual impact on the land, slow mass movement, especially creep, has a far greater cumulative effect on Earth's surface configuration. **Creep** is the slow downhill movement of soil and regolith. Creep is so gradual that it is visually imperceptible to the observer. Yet creep is the most widespread and persistent form of mass wasting, as it affects nearly all slopes where there are weathered materials and soil available for movement.

For the most part, creep does not produce distinctive landforms. Instead, it gradually moves surface materials

downward to the base of sloping land, where it accumulates and can be carried away by one of the erosional agents, usually running water. Sometimes the material that creeps downslope is not removed but serves instead to fill in low areas and level the bottom of slopes.

Several factors facilitate creep; in most locations, these factors act in combination to move regolith and soil slowly downhill. When soil water freezes, it expands with great force, pushing soil particles upward; when thawing occurs, the soil contracts. Soil particles sink back down, but gravity causes them to be displaced slightly down-

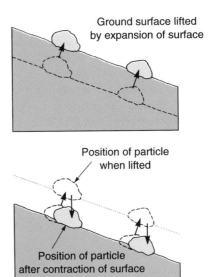

Ground surface lifted by expansion of surface

Position of particle when lifted

Position of particle after contraction of surface

Figure 15.29 The relationship between creep and the expansion and contraction of soil on a slope. This diagram shows how freeze-thaw or wet-dry cycles produce a mechanism that moves particles downslope. ■ **Are there places near where you live that show evidence of creep?**

unstable clay-rich debris could slip. Third, Rainier is second only to Mount St. Helens in earthquake activity, and one of these could jar loose massive amounts of material. Fourth, there is more water stored in Rainier's summit ice cap, its dozen major glaciers, and deep winter snow cover than in the ice and snow of all the other Cascade volcanoes combined. Last, adding to the danger is the rapid growth of the urban area at the base of its western slopes.

Mt. Rainier also has a history of past lahar activity. About 5700 years ago a lahar known as the Osceola mudflow removed the summit of Rainier. Scientists have noted that the summit was probably 620 meters (2000 ft) higher prior to the top sliding off the steep volcano. This lahar changed river courses and buried river deltas and valleys with as much

as 31 meters (100 ft) of material. The flow reached as far as 100 kilometers (62 mi) from the summit and covered an area estimated at 505 square kilometers (195 sq mi).

During the 10-year IDNDR program, scientists used remote sensing, field mapping, and volcanic gas sampling to learn more about Mt. Rainier. They hoped to improve their ability to monitor the mountain, predict eruptions, and give ample warning of the potential for a castastrophic lahar roaring downslope toward the populated Puget Sound lowlands. The volcano's hazardous history should play a role in future land use planning and in disaster preparation. As one scientist commented, "I think a lot of people don't realize the hazard potential that Mt. Rainier has. They just look up and see a beautiful mountain."

slope of their original position, as shown in Figure 15.29. Cycles of freezing and thawing cause repeated lifting and downslope sinking of soil and rock particles, resulting in downslope creep of the surface materials. The rate of movement is very slow, usually less than a few centimeters per year.

Alternating cycles of wetting and drying can also cause expansion and contraction of soil or regolith to produce a net downslope movement. This effect occurs because expansion tends to be greater on the downslope side, in response to the ever-present pull of gravity. Likewise, there is less contraction on the downslope side.

Small burrowing animals such as gophers and ground squirrels are effective soil movers. The tunnels they dig help push soil downhill, and the excavated material they bring to the surface also tends to fall downslope. Plant roots, too, push soils outward. Even the trampling of soil and regolith by humans or grazing animals tends to push surface material downhill. Every step up or down a steep slope, or even across it, pushes some surface material downslope to a slightly lower position. Because the factors that encourage soil creep act more strongly near the surface, and friction increases with depth, rates of creep are greater at the surface and diminish with depth. Therefore, telephone poles, fence posts, and other human structures (Fig. 15.30)—even trees (Fig. 15.31), all of which are anchored at a level below the surface, exhibit a downslope tilt when affected by creep.

Solifluction is the relatively slow downslope movement of soil and/or regolith that is saturated with water. Solifluction, which means "soil flow," is most common at high latitudes or high elevations, particularly where there is a layer of permafrost beneath the surface. During summer, when temperatures are higher, only the top few meters of the ground thaw. Because the frozen permafrost

layer beneath the surface prevents downward percolation of meltwater, the upper, unfrozen soil layer becomes water saturated, creating a soggy mass that sags slowly downslope in response to gravity until the next surface freeze arrives. The annual movement of soil may amount to only several centimeters. Solifluction is encouraged by the sparseness of vegetation found in tundra regions. Plant roots on gentle slopes act to retard mass movements, but where permafrost exists, roots are very shallow. The upper portion of the soil that freezes and thaws annually and that moves when saturated is called the *active layer*.

Evidence of solifluction can be seen in many tundra landscapes, where irregular lobes of soil produce hummocky terrain or mounds (Fig. 15.32). Slopes affected by solifluction typically exhibit fractures and wrinkles formed by compression or tension during downslope movement. The general effect of both creep and solifluction is to produce rounded hillcrests and a landscape free of sharp angular features—the subdued landscapes usually associated with humid climates.

Weathering, Mass Wasting, and the Landscape

In this chapter, we have concentrated on only two phenomena associated with gradation: weathering and mass movement. Although neither weathering nor the slower forms of mass movement attract much human attention, they vitally affect the landscape. They can create distinctive landforms without help from other, more familiar processes, and every slope reflects the local nature of weathering and mass wasting processes. The influence that rock structure has on the landscape is controlled by the relative resistance of the structure's rocks and minerals,

Figure 15.30 Diagram of common visible landscape effects of creep on natural and human features. ■ **What other human structures might be damaged by creep?**

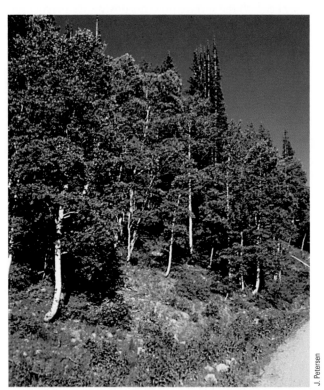

Figure 15.31 Trees attempt to grow vertically, but their trunks become bent if surface creep is occurring.

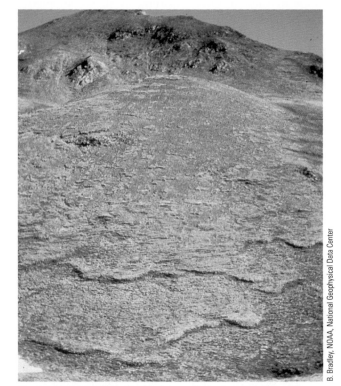

Figure 15.32 Solifluction flow near Suslositna, Alaska. ■ **Solifluction is commonly found in what climate region?**

the nature of the weathering processes involved, and the way that weathered regolith is affected by gravity. Steep slopes reflect resistant rocks and slow weathering, and gentle slopes result from rapid weathering or weak rocks (which can also develop from rapid weathering). Rounded slope forms suggest chemical weathering and slow but active mass movement (creep or solifluction);

angular slopes are associated with physical weathering and mass movements in the form of rockfalls, rockslides, and debris flows. There are, however, exceptions to these general relationships. In the following chapters, we will consider landforms produced by other dynamic, but more generally visible agents of gradation: running water, groundwater, wind, glacial ice, and coastal waves.

DEFINE & RECALL

erosion
transportation
deposition
base level
mass wasting
  (mass movement)
weathering
physical (mechanical)
  weathering
chemical weathering

frost wedging
talus
talus cone
salt weathering (salt
  wedging)
exfoliation
exfoliation dome
decomposition
hydration
hydrolysis

oxidation
solution
evaporites
carbonation
differential weathering
angle of repose
debris
mud
landslide
rockslide

rockfall
slump
earthflow
debris flow
mudflow
lahar
creep
solifluction

DISCUSS & REVIEW

1. How does weathering differ from the transporting agents of gradation?

2. How are the joints and fractures in a rock related to the rate at which weathering takes place? How does physical weathering encourage chemical weathering in rock?

3. What are several ways in which expansion and contraction can affect the weathering of rock?

4. Why is chemical weathering more rapid in humid climates than in more arid climates?

5. In what ways does chemical decomposition prepare rock for gradation? Give examples of oxidation, hydration, and solution.

6. Compare the ways in which hydrolysis and carbonation work below Earth's surface.

7. What types of rocks best resist all types of weathering? How visible are they in the landscape, compared to less resistant rocks?

8. What function does weathering perform in shaping landforms?

9. What two factors encourage mass movement the most? How do they work together?

10. What factors facilitate creep?

11. Describe the general effects of solifluction on the landscape. With what climates is solifluction usually associated?

12. What conditions encourage landslides and rockslides?

13. State the primary differences between a lahar and a debris flow. What causes the development of each?

14. Suggest ways in which weathering and mass movement affect human lives.

CONSIDER & RESPOND

1. Based on Figure 15.5, in what weathering region would your local area be classified?

2. Based on Figure 15.5, in what weathering regions would the following geographic regions be located?

  Brazil's Amazon Basin

  The North Slope of Alaska

  The summit of Pike's Peak, Colorado

  The Mojave Desert of southern California

  The Appalachian Mountains of Pennsylvania

3. What would you recommend as a solution to prevent the loss of valuable historical monuments to weathering processes?

4. If you were an urban planner in a city with numerous steep slopes, what major hazards would you have to plan for? What recommendations would you make to lessen these dangers to the community?

CHAPTER 16

UNDERGROUND WATER AND KARST LANDFORMS

Because of its usefulness and unique properties, water is an essential natural resource for life on Earth and the most distinctive feature of our planet's environment.

What are the peculiar properties of water that make it so important? What role does water play in making Earth a unique planet?

Groundwater and surface water are closely interrelated components of the hydrosphere, a major Earth subsystem.

How are they related? What is the movement of water through the hydrosphere called? What is the source of power that drives the system?

The depth (location) of the water table and water quality as well as quantity are the most important factors in the availability of groundwater.

How does the water table reflect precipitation? Surface landforms? What human activities may affect the water table and groundwater quality?

The dissolution, transportation, and deposition of soluble bedrock by groundwater produce distinctive features both above and below Earth's surface that characterize karst topography.

How are these processes involved in the development of karst topography? What unusual surface and subsurface features result from these processes?

Karst landscapes are found primarily in regions where limestone bedrock lies at or near the surface.

Why is this so? What areas of the United States and of the rest of the world have karst landscapes?

Earth Systems Today CD

Chapter 7: Groundwater, *Groundwater Basics*
Chapter 7: Groundwater, *Tapping the Ground—Permeability*
Chapter 7: Groundwater, *Tapping the Ground—Core of Depression and Multiple Wells*

The work of groundwater can produce limestone caverns of great beauty.

Fresh water is a precious, but limited, natural resource. Seventy percent of the fresh water on Earth is frozen as ice and snow in the polar regions and thus generally unavailable for human use. When most people think of fresh water, they envision rivers and lakes, both of which are important sources, but they represent only about another 1 percent of this resource. The remaining fresh water, nearly 30 percent, is found and stored beneath the surface as **groundwater.** In this chapter, we will look at the significance of groundwater for human life, especially as it affects our domestic, agricultural, and industrial water supplies. Also, in keeping with our continuing examination of processes that shape the land, we will consider the gradational effects of groundwater. Although of minor gradational importance when compared with streams on a worldwide basis, groundwater has a significant impact on landforms in certain areas.

Occurrence and Supply of Groundwater

By far the largest proportion of groundwater is derived from atmospheric sources and represents Earth's primary supply of fresh water (Fig. 16.1). Groundwater is brought to the surface by springs and wells, and it contributes significantly to standing water bodies, such as lakes, and to running water in streams.

A small portion of groundwater is so deep below Earth's surface that it has probably never been part of the hydrologic cycle. Other groundwater was once involved in the hydrologic cycle but, because of changes in Earth's surface, has been locked out of that cycle for a long period of time. This water is trapped in layers of sediment laid down by ancient rivers or seas. Future changes in the lithosphere could release these trapped waters and return them to the hydrologic cycle. Through volcanic activity, these waters could be released in the form of *geothermal energy* (steam and hot water). The most obvious evidence of this activity occurs in hot springs and geysers.

Groundwater Zones and the Water Table

Groundwater includes water found within the soil as well as in the loose rock and bedrock in the subsurface. Under conditions of modest precipitation and good drainage, as water infiltrates into the ground, it first passes through a level where both air and water fill spaces and pores within soil and rock (see Fig. 16.1). This level is called the **zone of aeration.** Further movement downward brings the water to a second level, called the **zone of saturation,** where all the openings are filled with water. Between the two levels, and marking the upper limit of the zone of saturation, is a surface called the **water table.** The water table does not remain fixed in

Figure 16.1 Environmental Systems: The Groundwater System. The groundwater system is actually a subsystem of Earth's hydrologic cycle. The major input for groundwater is precipitation, as rainwater or snowmelt percolates into the ground. Some water may also become groundwater via surface runoff when streams and lakes contribute to the groundwater system. The major output, or natural loss, of groundwater occurs when it is released at the ground surface through wells or in springs that flow directly out of bedrock or surface material.

The groundwater system consists of three major parts: the zone of aeration, the zone of saturation, and, between them, a fluctuating water table. Water percolates downward through the zone of aeration, where air fills most of the openings between soil and rock particles. Eventually a depth is reached at which all openings are saturated with water—the zone of saturation. The water table is the depth-surface where the zones of aeration and saturation meet. Water table depths vary in response to changes in precipitation and outflow. During dry seasons or years, the water table drops; during wet seasons or years, the water table rises.

More than 50 percent of the U.S. population receives their drinking water from groundwater sources. In the arid western United States, groundwater is also the major source of irrigation water, and overpumping of wells has caused the water table to drop drastically in this region. Groundwater is generally more difficult to pollute than surface water, but because it moves and recharges so slowly, it is also harder to clean up. Some major sources of groundwater pollution are leaking septic systems, animal feed lots, leaking underground storage tanks, and seepage related to disposal of toxic materials.

position, but fluctuates with the quantity of recent precipitation and the amounts of outflow to the surface or removal by pumping. After heavy precipitation or snowmelt, the water table will rise. Since the depth to the water table generally reflects the precipitation amount, it generally lies closer to the surface in humid regions than in arid regions.

In humid regions, there are actually three groundwater zones (see Fig. 16.1). The lowest zone is always saturated. The upper zone is almost never saturated. Between these two is a zone that is saturated under conditions of ample precipitation, but not saturated under conditions of low precipitation. It is through this middle zone that the water table fluctuates. Obviously, a well or spring originating within the permanently saturated zone will always have water, but one originating in the intermediate zone of fluctuation will run dry when the water table falls below it. In many desert regions, there is no saturated zone at all. Soon after rainstorms, water evaporates at or just below the surface. If considerable groundwater is present, it may be very old, having accumulated during a past period of greater precipitation. When such water is extracted in wells, it is not replaced from the atmosphere under current climatic aridity, and the water table falls rapidly. Pumping ancient water from the subsurface has been called **water mining,** because this resource has a limited supply that will not last indefinitely.

The water table is typically not level but tends to follow the general contours of the land surface, being higher under hills or other high topography and lower beneath valleys or depressions. Because water tends to seek its own level, and because it is affected by gravitational force, the groundwater under higher land surfaces tends to flow downslope to a lower elevation, as would a stream on the surface. Thus, the water table is usually closer to the surface under low places than under high places.

In humid regions of low relief, the water table is often so high that it intersects the ground surface, producing lakes, ponds, or marshes such as those common in New England and along the Gulf Coast from Louisiana to Florida. Where the landscape is one of hills and narrow valleys, the lowest points on the water table are controlled by the position of valley floors. As a stream deepens its valley, it may cut below the elevation of the water table. Groundwater then flows downslope along the gradient of the water table and into the stream. This *effluent* condition keeps the stream flowing between rains.

Many streams in semiarid and arid regions flow only seasonally or immediately after rains. In semiarid regions, the water table typically lies below the streambed during dry periods and rises to intersect the streambed during wet periods. These streams are fed by groundwater only during wet seasons, but they lose water by seepage into the subsurface during dry *influent* periods. In most desert regions, this influent condition is continuous because the groundwater table is deep below the surface, so that no groundwater is available to feed streams. Under these conditions, surface water flows only during and immediately after rains. Much of this downstream flow is eventually lost by seepage into the dry ground under the stream channel (Fig. 16.2).

Humid:
permanent
stream flow

High water table (wet season)
Low water table (dry season)

(a) Effluent condition

Semiarid:
intermittent
stream flow

High water table (wet season)
Low water table (dry season)

(b) Influent condition

Arid:
episodic flow

No water table except
by seepage from stream

(c) Influent condition

Figure 16.2 The influence of the water table on stream flow. (a) In humid regions, groundwater flows into major stream channels year-round, providing streams with continuous flow. This is known as a gaining stream (effluent condition). (b) In semiarid regions, with a dry season, the water table may drop below the streambed, so that the stream dries up until the next wet period. (c) In desert regions, the absence of a water table can mean that stream runoff occurs only immediately after a rainfall, and flow diminishes downstream because of losses by seepage. This is known as a losing stream (influent) condition.

Carmen Yelle, B.S., Environmental Science, emphasizing Physical Geography, The University of Alabama

Courtesy Carmen Yelle

My early interests were in meteorology, which is under the general field of physical geography. I decided to major in geography when I took my first physical geography course. The processes and natural cycles of Earth sparked my interest. I continued taking physical geography courses, but in my junior year, when an environmental science degree was offered, I switched. Most of the required classes for this degree were also in physical geography.

I work for the Alabama Department of Environmental Management (ADEM). This is the state agency that administers major federal environmental laws, including the Clean Water Act, the Clean Air Act, and the Safe Drinking Water Act, as well as federal laws concerning hazardous and solid wastes. I am in the Office of Education and Outreach in the Nonpoint Source (NPS) Unit. Our department manages Environmental Protection Agency grants. We operate under the Clean Water Act to help protect listed streams from further degradation and to improve their water quality. I also participate in educational activities to inform the general public about nonpoint source pollution (NPS) and its effect on water quality.

In college, I learned about the dynamics of streams and rivers, gaining knowledge that is very important in my career. Knowing how to read topographic maps is a geographic skill that I apply to my work. When dealing with nonpoint source pollution, you have to understand drainage patterns in a watershed and be able to identify them on a topographic map in order to figure out where the pollution is coming from. I have to understand the concepts of physical geography to determine how pollutants enter the stream and what can be done to alleviate the problem.

For students interested in a career like mine, I recommend being involved in an internship. I interned for the Natural Resource Conservation Service, a part of the U.S. Department of Agriculture. They have a volunteer program called the "Earth Team." This experience helped me to directly apply what I was learning in class. Other students interned for the United States Geological Survey. There are many other agencies that are interested in taking on geography interns.

Physical geography illustrates how Earth operates as a system of interconnected parts and helps you to understand relationships between water, air, and land as well as how each is affected by the others. Physical geography is an important course, not only for geography majors but also for those in other science fields, because it focuses on the big picture of Earth's environments.

Factors Affecting the Distribution of Groundwater

The quantity, quality, and availability of groundwater in an area depend on a variety of factors. Most fundamental is the amount of precipitation that falls in a given location and in the areas that drain into it. Second is the rate of evaporation. A third factor is the amount and type of vegetation cover. Although dense vegetation transpires great amounts of moisture back to the atmosphere, it also inhibits rapid runoff of rainfall, encourages infiltration of water into the ground, and lowers evaporation rates by providing shade. Thus, the overall effect of forests in a humid region is to increase the supply of groundwater.

Two additional factors that affect the distribution of groundwater are the porosity and permeability of the soil and rocks. **Porosity** refers to the amount or proportion of space between the particles that make up the soil or rock. Thus, loose gravels and sands, consisting of coarse particles that do not completely interlock, are exceedingly porous and can hold large amounts of water. Dense, interlocking crystalline rocks such as granite, with virtually no pore space, can hold little water within the rock itself.

Granite and other crystalline rocks, however, may hold water within joints, which allow the passage of this water rather freely. The **permeability** of a material—its ability to allow passage of water through it—is related to the number and size of spaces within the rock, which may be large pore spaces or joints, bedding planes, or other fractures (Fig. 16.3). Thus, jointed granite can be described as permeable, even though the rock itself is not porous. Lavas such as basalt may contain much pore space from gas bubbles frozen into the rock, but typically these holes are not interconnected, so the rock itself is porous but not permeable. In volcanic areas, groundwater permeability also tends to be provided by fractures. *Porosity* and *permeability* are not synonymous terms. A coarse-grained sandstone, large-pored and jointed, will typically be porous and permeable; a finely porous and unjointed clay may contain significant amounts of water but at the same time be impermeable, because the water clings to the tiny particles by surface tension. Porosity therefore affects the storage of groundwater (if the spaces are available to the water), and permeability affects the rates and volumes of groundwater movement. Both of these factors affect the availability of water for wells and springs.

A sequence of porous and permeable layers of rock or sediments that acts as a container and transmitter of water is called an **aquifer** (from Latin: *aqua,* water; *ferre,* to carry). Although any type of rock material that is porous and permeable can serve as an aquifer, most aquifers that supply water for human use are sandstones, limestones, or deposits of loose, coarse sediments such as sand and gravel. A rock layer that is relatively impermeable, such as slate or shale, restricts the passage of water and limits its storage, and therefore is called an **aquiclude** (from Latin: *aqua,* water; *claudere,* to close off).

**Jointed sandstone
porous and permeable**

(a)

**Unjointed clay
nonporous and impermeable**

(b)

Figure 16.3 The relationship between porosity and permeability in two different sedimentary rocks. (a) Jointed sandstone is porous and permeable. (b) Unjointed clay is nonporous and impermeable.
■ **Which of these two rock types is the ideal aquifer?**

Sometimes an aquifer will be found between two aquicludes. In this case, water flows in the aquifer much as it would in a water pipe or hose. Water can pass through the aquifer and cannot escape outward through the aquicludes. Furthermore, soil water percolating downward may be prevented from reaching the zone of saturation by an aquiclude. An accumulation of such water above an aquiclude is called a **perched water table** (Fig. 16.4).

Careless drilling can puncture the aquiclude supporting a perched water table so that the water drains out. The well must then be deepened to reach the true water table.

Springs are outflows of groundwater to the surface. They are related to many causes—landform configuration, geologic structure, level of the water table, and the relative position of various types of aquicludes. Springs may occur along a valley wall where a stream or river has cut through the land to a level lower than a perched water table. There an impermeable layer of rock prevents further downward penetration of groundwater and thus forces the water to move horizontally. Spring flows are perennial if the water table always remains at a level above the outlet of the spring; otherwise the spring is intermittent, flowing only when the water table is at a high enough level to feed water to the outlet.

Availability of Groundwater

Groundwater is an extremely vital resource to many areas of the United States and most of the world. In fact, half the U.S. population derives its drinking water from groundwater. In some states, such as Florida, more than 90 percent of the drinking water is from groundwater. Today, more than two-thirds of the groundwater used in the United States is for irrigation. One of the largest aquifers is the Ogallala aquifer that underlies the Great Plains from west Texas to South Dakota and supplies more than 30 percent of the irrigation groundwater in the United States.

Figure 16.4 A perched water table can develop where impermeable rock strata exist above and below an aquifer. In this example, the perched water table is underlain by a dry zone, between impermeable beds of shale and clay. Below the dry zone is the regional water table and the zone of saturation through which water flows toward the nearby river. ■ **Is a perched water table a reliable source of groundwater?**

Following the Flow of Underground Water

It is simple enough to trace the flow of a stream as it runs along the surface. However, when water sinks below ground, as often is the case in karst regions, the actual magnitude and direction of the flow is anyone's guess unless the underground flow is traced with dyes. Dye tracing can have a variety of uses, including: cave mapping, contaminated groundwater tracing, locating sewer leaks, mapping pipelines, and a multitude of others. Needless to say, dye tracing has become a widespread practice.

Fluorescent dyes are the most successful water tracers ever developed. Fluorescein, a green fluorescent dye, was first created in 1871 and a few years later was used to trace sinking portions of the Danube River in parts of Germany. Shortly thereafter, a more water soluble, disodium salt form of the dye, was produced under the trade name "Uranine." Today there are many fluorescent dyes, but only about 10 of them are considered safe.

Initially, dye tracing was strictly a qualitative science. Dye was injected into a location and then people would sit around and wait for it to appear at another location. Later it was discovered that some fluorescent dyes could be absorbed by charcoal grains. The dye can then be removed from the charcoal with an alcohol solvent. This major discovery allowed for mesh packets of charcoal (sometimes called 'bugs') to be left, unattended, in springs to wait for the dye cloud to pass by. Having people stand by at numerous emergence points was no longer an issue.

There was still one problem, however: both methods relied upon the human eye to detect the presence of the dye. This problem was addressed by the introduction of a device called a fluorometer. This device uses some of the principles of fluorescence in order to identify and quantify the fluorescent molecules (dyes). Fluorescent dyes absorb light at one energy level (or wavelength) and emit light at a lower energy level (or longer wavelength). The difference

Groundwater also plays a major role in supporting many wetlands and forming shallow lakes in land depressions. These bodies of water, fed by groundwater, are extremely important for thousands of resident and migratory birds. The flow of groundwater is vital for the survival of the Everglades in southern Florida. This "river of grass" and its great variety of birds and other animals are totally dependent on the continued southward movement of groundwater flow through the region.

Wells

Wells are artificial openings in the land surface dug or drilled below the water table. Water is extracted from wells by lifting devices ranging from simple rope-drawn water buckets to pumps powered by gasoline or electricity. In shallow wells, the supply of water depends on fluctuations in the water table. Deeper wells that penetrate into lower aquifers provide more reliable sources of water and are less affected by seasonal periods of drought.

In an area where there are many wells, or a limited groundwater supply, the removal of groundwater may exceed the intake of water that replenishes the supply (through a process called *groundwater recharge*). In many areas that are irrigated from wells, water tables have fallen below the depth of the original wells (Fig. 16.5). Progressively deeper wells must be dug (or the old ones extended) in order to reach the lower supply of water. In the Ganges valley of northern India, the development of deep modern wells to replace shallow hand-dug wells has resulted in significantly lower water tables. In the High Plains of Texas, Oklahoma, and Nebraska, the drawdown of the Ogallala aquifer, mainly for the irrigation of crops, is of serious concern.

In certain environments, particularly where high groundwater demand has led to heavy pumping, sinking of the land, called *subsidence,* can occur as the water pressure is reduced. The Central Valley of California, Mexico City, and Venice, Italy, all have subsidence problems. In southern California, groundwater is artificially replaced by diverting rivers over permeable deposits. This process is known as *artificial recharge.*

Because groundwater filters down through many layers of soil and rock before it reaches an aquifer, it is free of sediment but often carries a large chemical load dissolved from the materials through which it passes. Groundwater is thus said to be *hard* in comparison with *soft* rainwater. Moreover, just as increases in population, urbanization, and industrialization have resulted in the pollution of some of our surface waters, they have also resulted in the pollution of some of our groundwater supplies. The most recent danger to groundwater supply has been from toxic waste seepage. In coastal regions, where groundwater pumping has lowered the water table, salt water from the sea seeps in and replaces the fresh water because the pressure is no longer sufficient to hold back the salt water. This saltwater replacement has occurred in many localities, notably in southern Florida, Long Island (New York), and Israel.

Artesian Wells

In an *artesian system,* water under pressure rises to a level above the local water table. In **artesian wells,** water rises to the surface and flows out under its own pressure, without pumping. Certain conditions are prerequisite to an artesian structure (Fig. 16.6). First, a permeable aquifer such as sandstone or limestone must be exposed between the absorbed wavelength and the emitted wavelength is referred to as the Stoke's Shift (or $\Delta\lambda$). Depending on how much energy is lost in the fluorescence process, the Stoke's Shift can be a very small or a very large number. At the time, this was a great breakthrough in fluorescence analysis, allowing quantification down to 1 part per billion (1 ppb).

Since the mid-1980s, the synchronously scanning spectrofluorophotometer (SSS) has been available and has revolutionized fluorescence analysis and dye tracing technologies. While still exploiting the same principles of fluorescence as a fluorometer does, this new instrument uses a series of motorized mirrors and lenses to more accurately identify a variety of different fluorescent dyes in their various states of excitation and emission. Identification and magnitude of the dye is now accomplished with great speed and accuracy. With the help of these well-established dye techniques and new ultra-sensitive instrumentation, the underground movement of any liquid can be accurately traced.

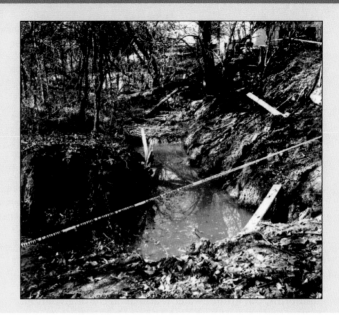

at the surface in an area of high precipitation or infiltration. This aquifer must absorb water at the surface, must incline downward hundreds of meters below Earth's surface, and must be confined between impermeable layers that prevent escape of the water. These conditions cause the aquifer to act as a pipe that conducts water through the subsurface. Provided with no exit, the water in the "pipe" is under pressure from the water above it (closer to the area of intake at the surface). As a consequence of this pressure, water will move toward any available outlet. If that outlet happens to be a well drilled through the impervious layer and into the aquifer, the water will rise in the well, sometimes gushing out at the surface. The height to which the water rises depends on the amount of pressure exerted on the water. Pressure, in turn, depends on the quantity of water in the aquifer (more water, more pressure), on the angle of incline (steeper slope, more pressure), and on the number of other outlets, usually wells, available to the water (more outlets, less pressure). Sandstone exposed at the surface in Colorado and South Dakota transmits artesian water eastward to wells as far as 320 kilometers (200 mi) away. Other well-known artesian systems are found in Olympia, Washington, the western Sahara, and eastern Australia's Great Artesian Basin, which is the largest artesian formation in the world. The word *artesian* is derived from the Artois region of France, where such systems were once common.

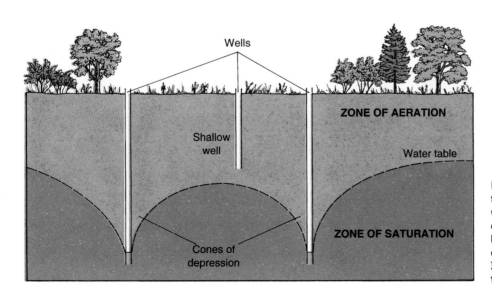

Figure 16.5 Cones of depression in the water table caused by pumping wells. In areas with many wells, adjacent cones of depression intersect, lowering the regional water table and causing shallow wells to go dry. ■ **In extreme cases, what other condition may occur?**

Figure 16.6 Conditions producing an artesian system. The Dakota sandstone, an aquifer that averages 30 meters in thickness, transmits water from the Black Hills to more than 320 kilometers eastward beneath South Dakota. ■ **What is unique about artesian wells?**

Land Sculpture by Solution

The runoff of surface water is the major process that shapes landforms, but groundwater is also an agent of gradation. It is, of course, vital in the subsurface chemical weathering process, and, like surface water, it dissolves, removes, transports, and deposits material prepared by weathering.

The principal mechanical effect of groundwater is to encourage mass movement by lubricating weathered material and soil, producing slumps, debris flows, mudflows, and landslides. But it is by chemical action that groundwater contributes most to gradational processes. Through the removal of rock materials by solution and the deposition of those materials elsewhere, groundwater is an effective land-shaping agent, especially in areas where limestone is present. Because limestone can be dissolved by acidic groundwater or surface water, distinctive landscapes develop wherever these waters can act on limestone or other rock that is significantly soluble in water (Fig. 16.7).

Karst Landscapes and Landforms

Overwhelmingly the most common soluble rock is limestone, and distinctive landscape features occur where limestone strata exist at or near the surface. The eastern Mediterranean region in particular exhibits features of limestone solution on a large scale. These are most clearly developed on the Karst Plateau along Croatia's scenic Dalmatian Coast. Landforms developed by solution, particularly in limestone, are called **karst** landforms after this classic locality. Other karst regions are located in many areas of the United States, Mexico's Yucatán Peninsula, the larger Caribbean islands, central France, southern China, and Laos.

Because limestone is a common sedimentary rock type, features created by the action of surface water or groundwater are found in many parts of the world (Fig. 16.8). However, a true karst landscape, in which

Figure 16.7 The formation of sinkholes. (a) Solution sinkholes develop where surface water funneling into the subsurface dissolves a closed depression into the landscape. (b) Collapse sinkholes form when an underground cavern system has grown through solution until the cave ceiling is no longer able to support itself and falls into the void below. ■ **Which type of sinkhole poses the greatest potential threat to houses and roads?**

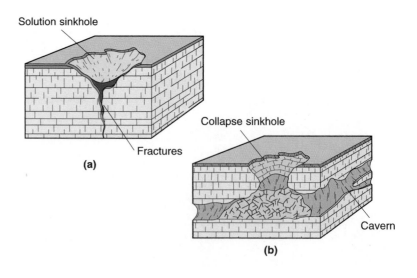

solution of limestone has been the dominant agent of land formation, is rather uncommon because of the special circumstances required.

A humid climate with plenty of precipitation is most conducive to karst development. Karst features are typically absent or are not well developed in arid climates.

However, some arid regions have karst features that originated during a period when the climate was much wetter than it is today.

Another important factor in the development of karst landforms is active movement of groundwater, because water saturated with dissolved calcium carbonate flows

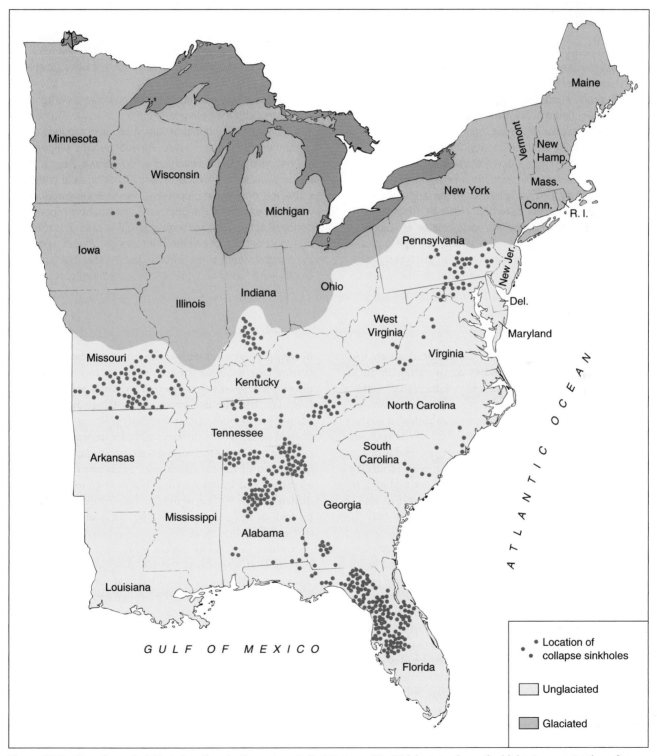

Figure 16.8   Map of recent sinkhole locations in the eastern United States.   ■ **Which states have the highest concentration of recently formed sinkholes?**

Sinkhole Formation

Sinkholes may not be a problem in many areas of the United States but they certainly can create havoc when they occur in some regions with karst landforms, as evidenced by the accompanying photograph of the Dishman Lane road collapse near Bowling Green, Kentucky. Most sinkholes develop slowly through long-term solution of underlying limestone bedrock where the water table is above the contact zone between bedrock and overlying loose rock and soil (regolith). As solution proceeds, openings are formed in the upper layers of the bedrock and regolith slowly slumps into the openings, creating sinkholes at the surface. These slow-forming sinkholes do not constitute an environmental hazard because they are clearly evident in the landscape and human activities are not disturbed. However, when surface material suddenly collapses into voids created by solution in underlying bedrock, there can be serious damage to property in the area. Earth scientists recognize two types of sinkhole-forming collapse: regolith collapse and bedrock collapse.

Sudden regolith collapse can occur, especially in high water table regions, during drought periods or at times of lowered water table due to excessive pumping from wells. These collapses are caused by the loss of buoyant groundwater support for the regolith arches covering the openings in the limestone. Adding additional weight to regolith through the construction of roads and buildings or due to the impounding of water above openings in the bedrock increases the likelihood that regolith arches will suddenly collapse.

The creation of a sinkhole by the collapse of bedrock associated with the enlargement of an underground cave or cavern is rare but when it occurs, it can also produce a significant environmental hazard. As underground passages are enlarged in the limestone, the roofs above the passages become weakened and can eventually collapse to create a sinkhole of sizeable proportions. Physical geographers who have examined the Dishman Lane sinkhole report evidence of actual cave roof collapse in this case. It is likely that most of the roof collapsed thousands of years ago leaving a regolith arch at the surface, but the fact that limestone blocks were found in the rubble indicated that part of the

away to be quickly replaced by unsaturated water. Vigorous movement of groundwater occurs when an outlet at a lower level is available, such as a deeply cut stream valley or a tectonic depression.

Limestone landscapes may have an absence of surface streams, because the high permeability of the rock encourages infiltration of surface water underground. Frequently, the surface junction between impermeable clays and shales and permeable limestones is marked by a series of *swallow holes* down which streams and rivers are "lost," or disappear. These seepage points may be widened and deepened by solution into great vertical *potholes,* very often tens or even hundreds of meters deep.

Surface outcrops of limestone are pitted and pockmarked by chemical solution, especially along joints, forming large, flat, furrowed exposures of limestone pavements. Because seepage of water into the subsurface tends to be concentrated where joints intersect, these intersections are subject to accelerated solution, which develops circular depressions, called **sinkholes**—more specifically, **solution sinkholes** if solution was the dominant process of formation (Fig. 16.7a). Solution sinkholes develop as rock is dissolved and removed by water infiltrating downward into the subsurface.

Slowly circulating water below the surface, flowing along joints and between bedding planes, also dissolves limestone, creating **caves** and **caverns** as a system of connected passageways within the limestone bedrock. Collapsing cave roofs are the second major cause of sinkhole development—in this case, called **collapse sinkholes** (Fig. 16.7b). The two processes, collapse and solution, cooperate to make most sinkholes. Whether they are termed collapse or solution sinkholes depends on which of these two processes were dominant in their formation. Collapse and solution sinkholes often occur together in a region, and based on their form, they may be either difficult or easy to recognize. Solution sinkholes tend to be funnel-shaped, and collapse sinkholes tend to be steep-walled, but these shapes vary greatly.

Sudden collapse of sinkholes is a significant natural hazard that can cause severe property damage and human injury (Fig. 16.8). These rapidly forming sinkholes are usually caused by excessive groundwater withdrawal for human use, or they occur during drought periods. Either of these conditions will lower the water table, causing a loss of buoyant support for the ground above, followed by collapse. Rapid sinkhole collapse has damaged roads and railroads and has even swallowed buildings (Fig. 16.9).

Sinkholes, also called **dolines** in some regions, may enlarge and merge over time to form larger karst depressions (Fig. 16.10). As sinkholes, potholes, and dolines coalesce, the larger depressions that develop are often linearly arranged along former underground water-cave roof collapsed in this event. The resulting sinkhole affected over an acre of surface area and caused considerable property damage.

courses, forming **valley sinks** or **uvalas.** (The terms *doline* and *uvala* are derived from Slavic languages used in the former Yugoslavia.)

In many well-developed karst landscapes, a complex underground drainage system all but replaces the normal surface patterns of water flow. The landscape may consist of large valleys that contain no streams. The valleys were originally excavated by surface streams until the river exposed and broke through into a cavern system. River water was then diverted to underground paths as the stream disappeared into a hole in the cave roof called a **swallow hole** (Fig. 16.11). This process is

Figure 16.9 This large sinkhole formed in Winter Park, Florida, when a dropping water table caused the surface to collapse into an underground cavern system during a severe drought in May 1981.
■ **What human activities might contribute to such hazards?**

449

KARST TOPOGRAPHY

The Map

The Interlachen area is in northern central Florida. Florida's peninsula is the emerged portion of a gentle anticline called the Peninsular Arch. The region is underlain by thousands of feet of marine limestones and shales. This great thickness of marine sediments originated in the Mesozoic Era, when Florida was a marine basin. As the arch rose, Florida became a shallow shelf and eventually became elevated above sea level.

Although outsiders think of Florida mainly as a state with magnificent beaches and warm winter weather from its humid subtropical climate, it is also a state with hundreds of lakes dotting its center. The lake region is formed on the Ocala Uplift, a gentle arch of limestone that reaches to 46 m (150 ft) above sea level. Lake Okeechobee is the largest of these lakes and has an average depth of less than 4.5 m (15 ft). Most of the lakes, such as those in the Interlachen map area, are much smaller.

Florida's central lake region is an ideal area for studying karst topography. Both the surface and the subsurface features express the geomorphic effects of groundwater. Extensive cavern systems exist beneath the surface. Much of the state's runoff is channeled through huge aquifers, and groundwater springs are quite common.

Interpreting the Map

1. This area would be classified as what type of major landform surface? What is its average elevation?

2. What contour interval is used on this map? Why do you think the cartographers chose this interval?

3. On what type of bedrock is the map area situated? Do you think the climate has any influence on the landforms in the area? Explain.

4. What landform features on the map indicate that this is a karst region?

5. What are the round, steep depressions called? Why are some of the depressions occupied by lakes?

6. Locate Clubhouse Lake on the full-page map (scale 1:62,500) and the smaller map (1: 24,000). What is the elevation of Clubhouse Lake? What is its maximum width?

7. What is the area north of Lake Grandin?

8. What is the approximate elevation of the water table? (*Note:* You can determine this from the elevation of the lakes' water surface.)

9. Underground, the water flows through an aquifer. Define an "aquifer" and list the characteristics an aquifer must have. What is the general direction of groundwater flow in the aquifer underlying the Interlachen area?

10. Since much of central Florida is rapidly urbanizing, what problems and hazards do you anticipate in this karst area?

Putnam Hall, Florida
Scale 1:24,000
Contour interval = 10 ft
U.S. Geological Survey

Opposite:
Interlachen, Florida
Scale 1:62,500
Contour interval = 10 ft
U.S. Geological Survey

Figure 16.10 Idealized development of karst topography. (a) Surface solution at joint intersections in limestone encourages the development of sinkholes. Caverns form by groundwater solution along fracture patterns and between bedding planes. Cavern ceilings may collapse, causing larger and deeper sinkholes. Surface streams may disappear into sinkholes to join the groundwater flow. (b) Eventually, the limestone landscape may be divided by merging sinkholes and longer karst valleys, called uvalas. (c) Finally, limestone remnants, called haystack hills or hums, are isolated above an exposed surface of insoluble rocks. ■ **Overall, what has been the general effect on the landscape of groundwater solution?**

Figure 16.11 A swallow hole and a disappearing stream. In areas of karst topography with humid climates, streams may disappear into a hole in the ground, flow underground in a cave system, and perhaps emerge elsewhere as a major spring.

Courtesy Parris Lyew-Ayee

Figure 16.12 Tropical karst photo with hums and/or intergrown sinkholes. Solution and karst weathering can be intense in wet tropical environments to form a maze of intergrown sinkholes.

characteristic of the Mammoth Cave area in Kentucky. Some of the "lost rivers" may reemerge as springs where they have cut down to impervious beds below the limestone.

After intense and long-term karst development, especially in the wet tropics, only limestone remnants are left standing above insoluble rock below. These remnants are usually in the form of small, steep-sided, and cave-riddled karst hills called **haystack hills** or **hums** (Fig. 16.12). Examples of this landscape are found in Puerto Rico, Cuba, and Jamaica. They have been described as "eggbox" landscapes, because an aerial view of the numerous sinkholes and hums resembles the surface of an egg box. If the limestone hums are particularly high and steep sided, the landscape is called *tower karst*. Spectacular examples of tower karst landscapes are found in southern China (Fig. 16.13).

Limestone Caverns and Cave Features

Underground caverns are the most spectacular forms created by the solution of limestone. Groundwater, sometimes flowing much like an underground stream, carves out networks of caverns that later become filled with air and decorated with depositional forms. Examples of these in the United States are Carlsbad Caverns in New Mexico, Mammoth and Colossal Caves in Kentucky, and Luray and Shenandoah Caverns in Virginia. In fact, 34 states have caverns open to the public. Some are quite extensive, with rooms more than 30 meters (100 ft) high and with kilometers of connecting passageways.

One factor necessary for cavern development in karst landscapes is the fracturing of limestone strata. Groundwater solution works along and widens joints, faults, and bedding planes, and the fracture pattern is often evident in the rectangular patterns of cave passageways. Cave maps illustrate well this relationship between caverns and fracture distributions (Fig. 16.14). Streams do not run through all caverns, but dry caverns often show evidence of previous stream activity, such as sedimentation of clay and silt on the cavern floor.

Jeffrery Alford/Asia Access

Figure 16.13 Karst towers (hums) of Guilin, a limestone region in southern China.

Figure 16.14 Cave map showing structural control by fractures. The pattern that most karst caverns assume is strongly controlled by zones of weakness in the limestone, so they tend to have straight passageways that form a regular pattern, reflecting the spacing of fractures in the rock. Groundwater flowing in the fractures widens them to develop the cavern system.

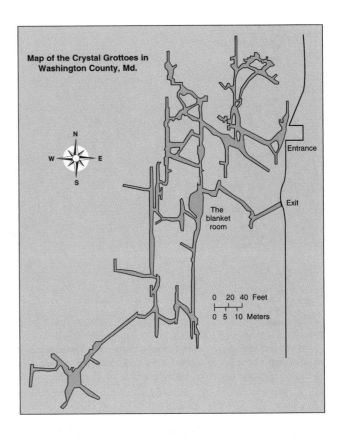

Depositional features in limestone caverns, which develop when previously dissolved minerals, particularly calcite, precipitate out of groundwater, produce some of the most beautiful forms found in nature. Where water drips from the ceiling, carbon dioxide gas is vented from the water into the cave air. This makes the water less acidic and thus less able to hold the lime in solution. The dripping water leaves behind a deposit of calcium carbonate called *travertine* or *dripstone*. As these travertine deposits grow downward, they form icicle-like spikes called **stalactites** that hang from the ceiling. Calcium-saturated water dripping onto the floor of a cavern builds up similar but more massive structures called **stalagmites.** Stalactites and stalagmites often meet and continue growing to form columns or pillars (Fig. 16.15). Depositional forms in caves (**speleothems**) develop in a great variety of textures and shapes, and many are both delicate and ornate (Fig. 16.16).

Limestone caverns vary greatly in size, shape, and interior character. Some are well decorated with speleothems, and others are not. Many have several levels and are almost spongelike in structure, whereas others are linear in pattern. The variation in cavern forms often indicates variations in mode of origin. Most large caverns appear to have developed either at the water table, where the rate of solution is most rapid, or below the water table in the zone of saturation. Subsequently, a decline in the water table level, caused by the incision of surface streams, climatic change, or tectonic uplift, replaced the cavern water with air, allowing speleothem formation to begin. Underground rivers deepen some air-filled caverns, and collapse of their ceilings enlarges them upward. Many small caves appear to have been formed entirely

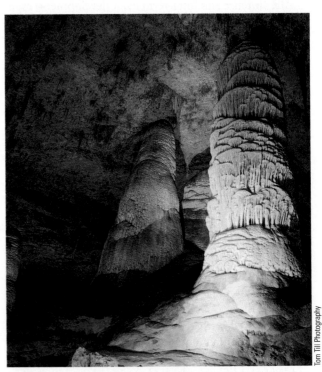

Figure 16.15 Cavern formations (speleothems) of dripstone in Carlsbad Caverns National Park, New Mexico. ■ **How do you explain the presence of such huge caverns in such a dry climate?**

Tom Bean/Getty Images

Figure 16.16   Stalactites form very slowly in places where water drips from the cave ceiling and deposits some of its dissolved minerals, typically lime. These stalactites in the Lehman Caves, Great Basin National Park, Nevada, are hollow, and may grow into long, narrow and very delicate tubes up to several feet in length, called soda straws.

above the water table by water percolating downward through the zone of aeration.

Cavern development is a complex process, involving such variables as rock structure, groundwater chemistry, and hydrology, as well as the regional tectonic and erosional history. The science of **speleology** (cavern studies) is particularly challenging. Our knowledge of caves has come from explorations as deep as hundreds of meters underground, by "spelunkers" who have made scientific observations while crawling through mud and water or even over bat droppings in dark passages (Fig. 16.17). Recent cave exploration and mapping of the water-filled caves below the surface in Florida and other karst regions have involved scuba diving. Cave diving in such fully submerged, totally dark, and confined passageways, which may contain dangerous currents, is an extremely risky operation.

Geothermal Water

Hot springs and geysers are both intriguing groundwater-related phenomena. Water heated at its source is referred to as **geothermal water.** It has probably been heated by contact with hot solidified rocks below the surface and is often accompanied by steam. Geothermal waters flowing or bubbling out fairly continuously form a **hot spring.** Water flow that is intermittent and somewhat eruptive, expelling water and steam vertically out of the ground, as at Old Faithful in Yellowstone National

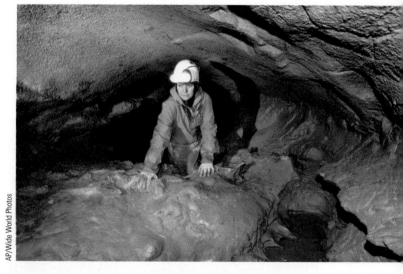

AP/Wide World Photos

Figure 16.17   Spelunker exploring the Lost River Cave System near Orleans, Indiana. Caving, also called spelunking, can be exciting but is potentially hazardous, so it requires skill, good equipment, and proper judgement of conditions. ■ **What are some of the potential hazards of caving?**

Park, produces a **geyser** (Fig. 16.18). The word *geyser* is an Icelandic term for the steam eruptions that are so common on that volcanic island. Geysers appear to erupt when the steam pressure below reaches a critical level and forces the column of superheated water and steam out of the fissure in an explosive manner.

Most hot springs and geysers contain significant amounts of minerals in solution, which become deposited in various forms, often as terraces or cones around the vent. These calcareous tufa and siliceous geyserite deposits are often colorful and impressive forms (Fig. 16.19).

Geothermal activity is usually associated with areas of tectonic instability, especially along plate boundaries. This geothermal energy has been used to produce electricity in areas such as California, Mexico, New Zealand, Italy, and Iceland. The best geothermal water for harnessing energy is not only very hot for generating steam, but also "clean"—relatively free of dissolved minerals that can clog pipes and generating equipment.

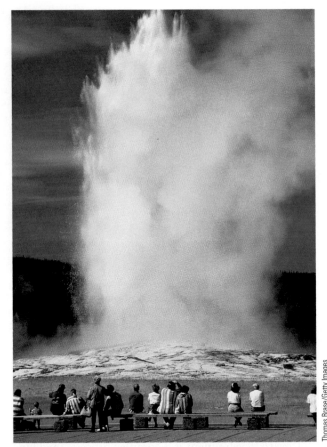

Thomas Brase/Getty Images

Figure 16.18   One of the world's most famous geysers is "Old Faithful" in Yellowstone National Park, Wyoming.

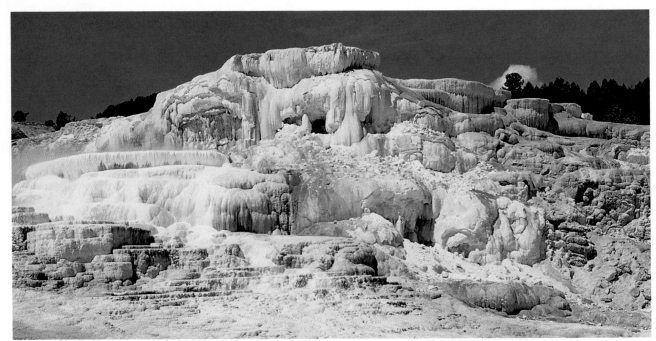

R. Gabler

Figure 16.19   Hot spring calcareous deposits in Yellowstone National Park, Wyoming.   ■   **Explain how these unique deposits are formed.**

DEFINE & RECALL

groundwater
zone of aeration
zone of saturation
water table
water mining
porosity
permeability
aquifer

aquiclude
perched water table
spring
well
artesian well
karst
sinkhole
solution sinkhole

caves (cavern)
collapse sinkhole
doline
valley sink (uvala)
swallow hole
haystack hill (hum)
stalactite
stalagmite

speleothem
speleology
geothermal water
hot spring
geyser

DISCUSS & REVIEW

1. What is the water table? How is it related to climate?

2. What is porosity? What is permeability? How are these two characteristics related to groundwater?

3. Define and describe the conditions necessary for an aquifer.

4. Explain how an artesian well works.

5. Define karst. What conditions are necessary for a region to have karst landforms?

6. Describe a sinkhole. What is the formation process involved?

7. Explain how haystack hills are formed. What type of climate would be most conducive to their formation?

8. Describe how caverns form, and name some common cavern features.

9. What is a geyser? What is the energy source for a geyser? Where are geysers most often located?

CONSIDER & RESPOND

1. If you were a water resources geographer and had to plan for the development of groundwater resources for a community, what would be your major considerations?

2. Describe the major landform features in a region of karst topography. Over time, how will the region be changed?

3. What are some human impacts and environmental problems related to the use of groundwater?

Running water, through sheet wash and stream flow, is the most important single agent of erosion and does more work than all of the other gradational agents combined.

What does this tell us about Earth's landforms? How is this significant to humans?

Moderate increases in the volume, velocity, and turbulence of stream flow can cause major increases in the load-carrying capacity of a stream.

What does this tell you about stream erosion and deposition? What does it suggest about a stream at flood stage?

What are the major characteristics of stream erosion, transportation, and deposition processes?

The upper, middle, and lower courses of streams differ significantly in terms of gradational processes and the landform features that result from these processes.

What features generally characterize the upper, middle, and lower courses of streams? What gradational processes are most important in the three different stream courses?

A stream system provides an excellent example for illustrating the concepts of energy budget, water budget, and equilibrium as applied to natural systems.

What is meant by a budget in a natural system? Why study these budgets?

Why is equilibrium in natural systems an important concept?

The Grand Canyon, here viewed from space, is a spectacular example of the work of streams. North is to the right on this image.

Flowing water exerts a greater influence in shaping the landforms of our planet than any other gradational process. Through both erosion and deposition, running water modifies existing landforms, and creates others, as it flows over land surfaces or in confined channels. With few exceptions, nearly every world region has been shaped by the power of water flowing downslope. This is true for both humid and arid regions, although a major exception would be the perennially frozen polar regions. The long-term effects of flowing water, whether dominated by erosion or deposition, can be quite dramatic. Two prime examples in the United States illustrate the effectiveness of water in creating landforms: the Black Canyon of the Gunnison River (Fig. 17.1a), carved by long-term river erosion into the Rocky Mountains, and the Mississippi River delta (Fig. 17.1b), where deposition is building new land into the Gulf of Mexico.

The Stream System

Streams are bodies of water that flow in well-defined channels (except during times of flood). In the Earth sciences, the term *stream* means water flowing in a natural channel of any size, although in general usage we describe large streams as rivers and use many local terms for smaller streams (creek, brook, run, draw, bayou). Processes associated with the work of streams are known as **fluvial processes** (from Latin: *fluvius,* river).

Most of the water in streams comes from one or more of three main sources that deliver water to the land: precipitation, meltwater from ice and snow, and groundwater from springs. Each of these sources can contribute to the flow of water on a land surface, called **runoff.** For short distances, runoff may occur in sheets (as **sheet wash,** or unconcentrated flow), or it may form tiny channels called rills. Eventually, overland flow develops a pattern of well-defined channels, which join to form larger channels and eventually a stream system. Surface runoff and stream flow are essential components in the hydrologic cycle that return water to the oceans. By distributing water over the land, these processes also provide (or contribute to) the soil moisture necessary for supporting Earth's rich variety of biotic environments.

A river or stream and its **tributaries** (the smaller streams that feed into a larger one), along with the area it drains, make up a **stream system.** A stream system has three major subsystems: (1) a *catchment system* that collects and directs runoff into (2) a *transport system,* consisting of a network of channels that move water and sediment, and (3) a *depositional* or *dispersal system*—the locations to which water and sediment are eventually transported. The land surface drained by a stream (or one or more of its tributaries) is called a **drainage basin, watershed,** or **catchment basin** (Fig. 17.2). Drainage basins are the surface areas that feed runoff into a stream, and they exist for every channel. Large river systems drain extensive watersheds that consist of many sub-basins, one for each tributary. Drainage basins are open systems that involve both inputs and outputs of water, materials, and energy. Knowing the boundaries of a drainage basin and its sub-basins is critical in properly managing the water resources of a watershed. For example, pollution discovered in a river generally comes from a location in its drainage basin, either at the point where the pollutant was detected or from somewhere upstream. This knowledge helps us to track, detect, and correct sources of pollution.

(a)

(b)

Figure 17.1 (a) The Black Canyon of the Gunnison has been carved through a part of the Rocky Mountains in Colorado through long-term fluvial erosion by the Gunnison River (assisted by weathering and mass movement). The Gunnison flows westward into the Colorado River and the Gulf of California. The Canyon is 829 m (2,722 ft) deep, 80 km (50 miles) long, and very narrow in width, with steep walls formed by resistant rocks. (b) Through fluvial deposition, the Mississippi River has been building its delta outward into the Gulf of Mexico for thousands of years. The river slows as it enters the Gulf, depositing much of the muddy sediments that have been transported from its drainage basin. This is a false-color composite, a digital image. Muddy water is light blue, and clearer, deeper water is dark blue.

Vertical erosion
large bed load

Drainage divide of
upland drainage basin

Interfluves

Increased
suspended and
dissolved load

Drainage
divide of
upland
drainage
basin

Major load
capacity

Figure 17.2 **Environmental Systems: The Hydrologic System—Streams.** The stream or surface runoff system is a subsystem of the hydrologic cycle. Its major water input is from precipitation. However, groundwater may also contribute to the stream system in regions with a high water table. The major water output for most stream systems returns water into the ocean. Output or loss of water also occurs by evaporation back to the atmosphere and by infiltration into the groundwater system. Stream systems are divided into natural regions known as drainage basins (also called watersheds), separated from each other by divides; interfluves are areas that separate streams flowing within the same drainage basin. Drainage basins are fundamental natural regions of critical importance to life in both humid and arid regions (refer to top of pages 496–497 for more information).

Streams are complex systems of moving water that involve energy transfers and the transport of a variety of surface materials. Energy enters the system with precipitation. The runoff flows downslope increasing the amount of energy available to the stream for cutting and eroding channels.

Materials transported by streams, known as load, enter the stream system by erosion and mass movement, particularly in the headwaters of a drainage basin. Much of the surface material eroded here consists of large particles, including boulders. Coarse material is carried along the channel bottom as bed load. As the number and size of tributaries increases downstream, the amount of load carried by the stream generally increases dramatically. This is especially true for finer materials suspended in water (suspended load) and dissolved minerals (dissolved load). Load leaves the stream system when carried and deposited in the sea at the river mouth. Streams also deposit sediment along the channel as they overflow their banks during floods. Human activities can change the amount of load available for stream systems by building dams, altering land with construction projects, overgrazing, and clearing forests. These activities may also affect water quality downstream, where communities may depend upon the stream system for their water supply.

The higher land between two tributary valleys of a drainage basin is referred to as an **interfluve** (from Latin: *inter*, between; *fluvius*, river). On an interfluve that separates two drainage basins, there is an imaginary line called a **divide.** Some divides are ridges, but in many cases the higher land is not necessarily ridge-shaped. For example, drainage divides can exist on high plains, on tablelands, or along the crests of rounded hills. Surface runoff flows into one stream system (or sub-basin) on one side of the divide, and toward another stream system (or another sub-basin) on the opposite side (Fig. 17.3). The *Continental Divide* separates North America into a western region where most runoff flows to the Pacific Ocean,

and an eastern region where runoff flows to the Atlantic Ocean. Generally, the Continental Divide follows high ridges in the Rocky Mountains, but in some locations it also lies on gently sloping high plains.

Runoff and Stream Flow

When precipitation falls on the ground, several factors interact to determine whether or not surface flow will occur. Basically, runoff is generated when the amount, duration, and/or rate of precipitation exceed the ability of the ground and its environment to soak up the moisture. The process of water seeping into the surface is

Figure. 17.3   An aerial photograph of a small stream channel network, basin, and divide. Every arm of the stream has its own sub-basin (separated by divides) that combines with others to form the basin of the main stream channel.   ■   **Can you mentally trace the outline of the main drainage basin shown here?**

Texas Natural Resources Information System

snowmelt, the size and nature of drainage basins, and regional climatic patterns. Not all streams flow during the entire year. In arid and semiarid regions, streams often flow only after a heavy rain or during a rainy season, if there is one. These are called **ephemeral streams.** In contrast, **perennial streams** flow all year, though not always with the same volume or at the same velocity. In times of high flow, many streams overflow their banks, flooding the flat lowlands on either side called **floodplains.**

Stream flow, or **stream discharge,** is the volume of water flowing past a cross section of the channel in a given unit of time (stated in either cubic meters or cubic feet per second). Stream discharge data and information are important to collect and analyze for several reasons. First, discharge can be used to indicate and compare the size of stream systems. Also, because a stream's discharge changes in response to runoff—during storms, floods, or droughts—a stream system continually makes adjustments to these changes in flow. Discharge data from a stream is one index of the severity of a flood.

Discharge is calculated by a simple equation: $Q = AV$. Discharge (Q) is equal to the stream cross-sectional area (A) times the average stream velocity (V). This equation can also be stated as: $Q = wdV$ because the cross-sectional

called **infiltration,** and ability of a ground surface to hold water is called the **infiltration capacity.** Factors that tend to increase infiltration and infiltration capacity include high permeability of surface materials, deep soils, dense vegetative cover, gentle land slopes, and dry conditions prior to the precipitation event. The amount and intensity of runoff also depend on these factors, because any condition or process that increases infiltration will reduce runoff. As part of a natural system, these factors are interrelated. The amount of runoff is equal to the precipitation amount minus water "lost" to the environment through evaporation, infiltration, and interception by vegetation (Fig. 17.4). For example, for a given rainfall amount, less vegetative cover will mean more runoff. Any factor that impedes infiltration will increase runoff, and vice versa. Human activities can alter many of these variables, and in some places the natural runoff has been greatly modified by urbanization, dam building, mining, logging, and agriculture.

Stream flow varies considerably around the world, responding to short-term storms, seasonality of precipitation,

Figure. 17.4   The amount of runoff that occurs is a function of several factors, including the intensity and duration of a rainstorm. Surface features, however, are also important. Any factor that increases infiltration and evapotranspiration will reduce the amount of water available to run off, and vice versa. Deep soil, dense vegetation, fractured bedrock, and gentle slopes tend to reduce runoff. Thin or absent soils, thin vegetation, and steep slopes tend to increase runoff.

Figure. 17.5 These university students are measuring a stream's velocity, depth, and width, in order to find discharge, the volume flowing in a stream per unit time.

area (*A*) is equivalent to channel width (*w*) times channel depth (*d*), two factors that are relatively easy to measure in the field (Fig. 17.5). Understanding this equation is easy, yet also very important. Because it is an equation, both sides must always be equal, so a change on one side must be accompanied by a change on the other. If dis-

charge increases, the velocity and/or the size of the channel must also increase, and this is what happens during a flood. The discharge equation ($Q = AV$) illustrates the natural balance of water flow in a stream system. Discharge is also important in understanding changes in a stream's energy, its ability to erode its channel, and the amount and/or size of sediment that a stream can transport.

The United States Geological Survey maintains more than 6000 gaging stations to measure stream discharge in the United States (Fig. 17.6). In the past, data from these stream gages had to be collected at the site, but today many gaging stations operate on solar energy and beam their data to a satellite that relays them to a receiving station. With this system, stream flow changes can be monitored at the time they occur, which is very beneficial in issuing flood warnings, and in understanding how streams change in response to changes in discharge.

The record of changes in discharge in a stream over time is a **stream hydrograph** (Fig. 17.7). These hydrographs can indicate how high and fast the water level rises in response to a precipitation event. During (and just after) a rainfall that produces runoff, the level of the stream will rise in response to the increased discharge.

(a)
Satellite radio antenna
Recorder
Shelf
Floor
Float to record water level
Water surface
Water surface
Valves   Intakes

(b)

Figure 17.6 (a) The U.S. Geological Survey (USGS) monitors the flow of streams and rivers using gaging stations like the one illustrated here. When river level rises or falls, so does the water in the lower part of the station, connected to the channel by intake pipes. A gaging float moves up and down with the water level, and this motion is measured and recorded. Until recent years, a person had to come to the station to retrieve the data or to see how much water flowed in the channel during a flood. Today, stream gaging stations electronically beam flow data to a satellite that transfers the information to a receiving station. Now it is possible to obtain data from many stations at once and to instantly monitor how a stream responds to a storm at the same time that a flood event is occurring. (b) USGS stream gaging station. These are commonly located where highway bridges cross rivers or streams. Note the antenna for sending stream flow data through a satellite link to the USGS receiving station. ■ **What are the advantages of receiving flow data automatically and instantly at a receiving station distant from the stream channel?**

Andy Wohlsperger

Courtesy Andy Wohlsperger

University Education: After 2 years of coursework at the University of Texas at El Paso, Andy transferred to the University of Texas at Austin, where he matriculated.

Early Interests: Growing up in Germany, where a classroom focus on geography is present even in the early grades, helped Andy develop a deep curiosity for geography. However, Andy's subsequent interest in geography, and particularly GIS, was fueled by a physical geography class he took at UTEP.

Work Experience: While at the University of Texas, he worked for a professor in the School of Business Administration, creating layouts for the display of marketing data for several research studies. After graduation, Andy worked for an aerial photography/topographic mapping company as a stereo-plotter operator. Here he gained most of his aerial photo interpretation skills and knowledge of DEM and DTM creation. About seven months after graduation, he was hired by Watershed Concepts, a major engineering firm, where he works as GIS staff. He helps produce FEMA (Federal Emergency Management Agency) certified Digital Flood Insurance Rate Maps and Flood Recovery Maps. He also does general GIS work: gathering, projecting, storing, and displaying geographic data and creating map layouts. The firm uses advanced GIS methodologies and customized software solutions to solve their clients' water resources needs. Andy fits into his company's mission because of his knowledge and training in GIS and remote sensing, and his general understanding of geographic concepts.

I decided to major in geography mainly because I've always been very interested in my surrounding world, especially nature, politics, economics, and world affairs. Geography is such a broad field that whatever you are interested in you will find a niche in this field to integrate all of those interests. I'm also very interested in new technologies; therefore, GIS and Remote Sensing attracted me right away.

What did you learn in your physical geography education (particularly GIS) that is now important to you in your career?

Everything! In the past six months, I've probably used every concept I learned in my GIS, remote sensing, and cartography courses (projections, layouts, resolution, buffers, databases, intersecting, plotting, data display, the list goes on . . .).

How do you apply your knowledge of geography and geographic concepts to your work?

I mainly apply my knowledge with the use of computer applications and software programs. However, if these concepts are not understood outside of the computer world, GIS work can become a very confusing field to work in.

What sort of academic preparation would you recommend to students who are interested in a career similar to yours?

Take all the techniques courses you can. GIS, remote sensing, basic cartography, digital cartography, physical geography, and so on. Make friends with the computer and the software applications; you will need to be able to use them to succeed.

Do you have some interesting examples about how you've used your training in physical geography and GIS in your career?

I was working in FEMA's Disaster Field Office in San Antonio during the 2002 July Central Texas floods for data gathering and onsite GIS support. I was put on the spot during a net meeting Web conference with our main office in North Carolina using ArcView and other applications. We needed some reservoir and stream files right away to be integrated into the conference. I located the data on the Web, remembering the data site from my GIS classes, and downloaded the data, which were not, however, in ArcViews shapefile format. I remembered a lab exercise in which we had translated data using an ArcView script, and I was able to import the data into ArcView.

Why do you think that physical geography is an important course for those who are majoring in other fields?

A physical geography course answers a lot of questions. Why is the sky blue? Why are deserts where they are? How would global warming affect us? El Niño? La Niña? These are all issues that we are exposed to every day in our natural surrounding or that are addressed in the news. Therefore, we should all have a basic understanding of these issues. My wife just recently told me that she never knew how important geography was until she took a physical geography class. She also said that this class was the most beneficial class in her college career when it comes to day-to-day events and understanding our world.

The water level of the stream will peak at the time the flow is greatest, and then fall as the river eventually returns to its "normal" level of flow, called **base flow.** The discharge of a stream is recorded by a gaging station in the form of a hydrograph curve plotted on a graph that represents this rise, peak, and fall of stream flow. The shape of the curve can be used to understand a great deal about how a watershed and a stream channel respond to an increase in runoff, particularly during and after floods. The higher the peak, the greater the discharge when a flood is at its greatest level, called the *flood crest.* How high a river rises and how fast it reaches peak flow

Figure 17.7 A stream hydrograph measures changes in flow and water depth, recorded over a period of time by a gaging station. This hydrograph example shows the rise in a river in response to flood runoff, the flood peak, and the recession of flow waters following the end of a storm. Note the lag between the time that precipitation starts and the rise in the river. ■ **Why would such a time lag occur between the rainfall and rise in the river?**

in response to a certain amount of precipitation are both important in making preparations for future floods.

The hydrograph curve recorded during a flood event on a stream consists of three major parts: (1) the *rising limb*, (2) the *peak flow* or *flood crest*, and (3) the *receding limb*. Any conditions in the watershed that contribute to high rates of runoff will make the stream rise faster; the rising limb will be steep, representing a rapid rise in discharge to the peak flow in a short amount of time. Rising limbs that are represented by very steep curves indicate flash flood conditions. After the flood crest passes, the stream discharge will decline, but typically the return to base flow takes longer (a more gentle curve on the receding limb) because water continues to seep into the channel from the precipitation-saturated ground of the watershed. Studies have shown that urbanization of a watershed (particularly small drainage basins) tends to make the flood peak rise higher and faster than was the case before human development in the drainage basin. It is ironic that urbanization in such watersheds increases the risk of flooding for local residents, for reasons other than the fact that more people are living there. Reduction of vegetation and its replacement with impermeable surfaces like roads, roofs, and other paved areas contribute to higher rates of runoff and increased runoff, two conditions that directly affect the flow of the stream (Fig. 17.8). When a drainage basin becomes more highly developed and populated, the potential flood size, the rapidness of flood onset, and the flood hazard all tend to become greater.

In general, as the size of the drainage basin increases, so do the length of the river and its discharge. Other factors that affect precipitation and runoff amounts, such as climate and the nature of the land surface, can also have an impact on these relationships. Among the major rivers of the world, the Amazon has by far the greatest discharge and the largest drainage basin; the Mississippi–

Missouri system is ranked fourth in terms of discharge (Table 17.1). Looking at Table 17.1, why would the Nile, the world's longest river, have relatively low discharge for its length and size of drainage basin?

Erosion by Streams

The force of gravity affects water flowing in a stream in two main ways. First, gravitational force causes water to flow downslope, and second, gravity causes water to exert pressure on the streambed. The strength of these forces, translated into flow velocity and the pressure exerted on the streambed, depends on (1) the steepness of the stream gradient and (2) the depth of water in the channel. The volume of water in the stream is directly related to the depth, provided the channel width remains constant. When the volume of flow in a stream increases, the amount of energy available to shape the land also increases.

About 95 percent of a stream's energy is consumed in overcoming external friction against channel walls and internal friction among water molecules, eddies, and currents. The remaining energy, probably less than 5 percent, is used for vertical or lateral erosion of the streambed, channel, and banks and for moving the **stream load**— the sediment that is transported downstream.

Streams continually undergo changes in seeking to sustain a balance among these energy uses. Dynamic

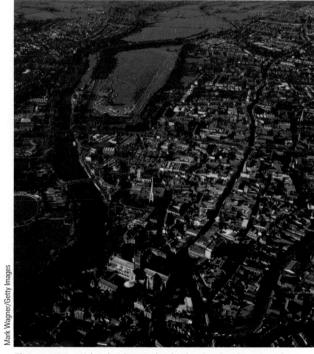

Mark Wagner/Getty Images

Figure 17.8 Urbanization and suburbanization increase the impermeable cover in the drainage basins that these settlements in Worcester, England, occupy. As a result, both the amount and the rate of runoff from those areas increase compared to their pre-settlement state. ■ **How would these factors affect the flood potential of urban and suburban areas?**

Table 17.1   Ten Largest Rivers of the World

	Length		Area of Drainage Basin		Discharge	
	km	mi	sq km (×1000)	sq mi	1000 m³/s	1000 cfs
Amazon	6276	3900	6133	2368	112–140	4000–5000
Congo (Zaire)	4666	2900	4014	1550	39.2	1400
Chang Jiang (Yangtze)	5793	3600	1942	750	21.5	770
Mississippi–Missouri	6260	3890	3222	1244	17.4	620
Yenisei	4506	2800	2590	1000	17.2	615
Lena	4280	2660	2424	936	15.3	547
Paraná	2414	1500	2305	890	14.7	526
Ob	5150	3200	2484	959	12.3	441
Amur	4666	2900	1844	712	9.5	338
Nile	6695	4160	2978	1150	2.8	100

Source: Adapted from Morisawa, *Streams: Their Dynamics and Morphology.* New York, McGraw-Hill Book Company, 1968

equilibrium is maintained by adjustments among channel slope, shape, and roughness; amount of load eroded, carried, or deposited; and the velocity and discharge of stream flow. These last two factors can change very rapidly. This is one reason why velocity and discharge (which is also partly a function of velocity) are such significant aspects of a stream's behavior, regulating its ability to erode and to transport sediment (Fig. 17.9). Friction along the bottom and sides of a channel slows

down the flow, so the maximum velocity in a stream is usually found in the center of the channel slightly below the surface. Anyone who boats, rafts, canoes, or floats on inner tubes on rivers or streams knows that not only does the flow vary seasonally, but the velocity also varies with its position in the channel.

The ability of a stream to pick up and carry materials is largely determined by the volume of water, its depth, its friction with the streambed, the strength and size of

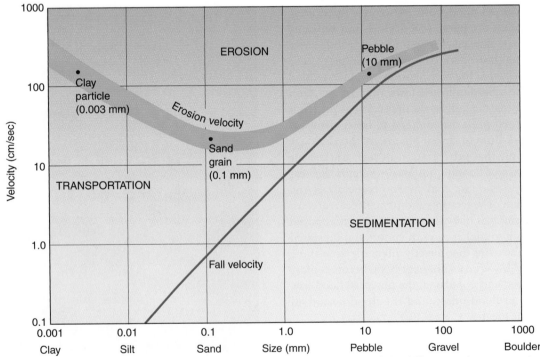

Figure 17.9 This graph shows the relationship between stream flow velocity and the ability to erode or transport material of varying sizes (inability to erode or transport particles of a particular size or larger will result in deposition). Note that small pebbles (particles with a diameter of 10 mm, for example) need a high stream flow velocity to be moved because of their size and weight. The fine silts and clays (smaller than 0.05 mm) also need high velocities for movement because they stick together cohesively. Sand-sized particles (between 0.05 and 2.0 mm) are relatively easily eroded and transported, compared to clays or gravel (a mix of any particles larger than 2.0 mm).

the rocks over which it flows, and the degree of *stream turbulence.* Turbulence is chaotic flow that mixes and churns the water; it is controlled by channel roughness and the gradient on which the stream is flowing. A rough channel bottom increases the intensity of turbulent flow. A small increase in velocity can result in a significant increase in turbulence, which in turn greatly increases the rate of erosion as well as the load-carrying capacity of the stream (as shown in Fig. 17.9).

Volume of stream flow and resistance of the streambed also affect erosion rates. Where rocks are highly resistant, erosion proceeds very slowly. Because erosion rates are affected by so many variables (flow energy, turbulence, volume of stream flow, resistance of rocks, and friction with the streambed), erosion in a stream channel varies greatly from place to place. All streams, however, have a **base level,** an elevation beneath which they cannot erode. At the mouth of a stream—its outlet into an ocean or lake—either sea level or the lake level forms the stream's base level. The ultimate base level for virtually all stream gradation is sea level (Fig. 17.10). Channel and valley deepening (downward erosion) can occur as long as there is enough slope and water depth to allow streams to flow toward their outlet.

A stream flowing on land that is significantly above its base level will gradually deepen, widen, and lengthen its valley through **degradation,** erosion that is accomplished in several ways. First, **hydraulic action** occurs as turbulent river currents wedge under rock slabs on the bed and pound away at the riverbanks or below waterfalls. **Plunge pools** at the base of waterfalls and rapids reveal the power of stream erosion where it is directed toward a localized point (Fig. 17.11). Hydraulic action also displaces loose particles from the streambed or channel walls and carries the material along in the current. The sediments that a stream may transport range in size from clay to silt, sand, pebbles, cobbles, and boulders. Another term used for coarse sediment is gravel, which is a general term for rock particle sizes that are larger than sand.

As rock particles bounce, scrape, and drag along the bottom and sides of a stream channel, they break off additional rock fragments. This form of erosion is called **stream abrasion.** Under certain conditions, stream abrasion makes round depressions called **potholes** in the rock of the streambed (Fig. 17.12). Potholes are formed only in special circumstances, such as below waterfalls or swirling rapids, or at points of structural weakness, such as joint intersections in the stream's bed. They can range in diameter and depth from a few centimeters to many meters. If you peer into a pothole, you can often see one or more round stones at the bottom. These are the *grinders.* Swirling whirlpool movements of the stream water cause such stones to grind the bedrock and enlarge the pothole, while finer sediments are carried away. As potholes widen and join, the streambed gradually deepens.

Some fluvial erosion occurs as a result of stream water's chemically dissolving the bedrock. This process, called *solution* (or *dissolution*), has a limited effect on many rocks but may be significant in limestone areas.

As rock particles are transported downstream, they are gradually reduced in size, and their shape changes from angular to rounded. This process is called **attrition**—the wear and tear on sediments as they tumble and bounce against one another and against the stream channel. Attrition explains why rounded pebbles are found in streambeds and why the material carried in the

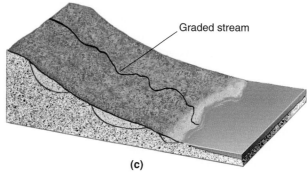

Figure 17.10 Idealized representation of a stream's base level. (a) A stream with many temporary base levels. (b) The stream profile becomes smoother as the stream is graded toward ultimate base level through erosion and deposition. Temporary base levels, formed by resistant bedrock obstructions, control the levels of the river upstream. (c) Eventually streams that flow to the ocean tend to develop a smooth profile graded to sea level (ultimate base level) that flattens downstream and is steeper in the headwaters.
■ **What would account for the rapids and waterfalls in the initial gradient?**

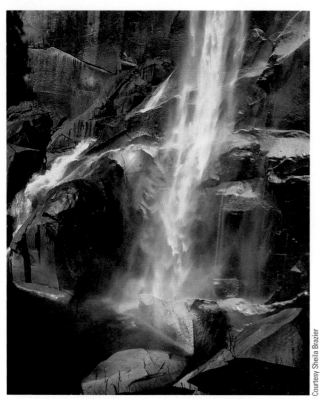

Courtesy Sheila Brazier

Figure 17.11 Plunge pool at the base of Vernal Falls in Yosemite National Park, California. ■ **Why is a deep plunge pool found at the base of most waterfalls?**

Figure 17.12 These potholes were cut into a streambed of solid rock during times of high flow. Potholes result from stream abrasion and the swirling currents in a fast-flowing river. ■ **Explain how a pothole is carved into bedrock on the stream bottom.**

lower reaches of most large rivers is composed primarily of silt- or clay-sized particles and dissolved minerals.

The slope of a streambed is its **gradient**—the vertical drop over a horizontal distance (meters per kilometer, or feet per mile). In general, steeper gradients produce greater flow energy and increased turbulence in the stream, which can result in increased erosion. Stream gradients are usually steepest at the headwaters and in new tributaries, and diminish in a down-stream direction (Fig. 17.13). However, the greater volume and depth of water farther downstream accelerates flow velocity even more than the higher gradient upstream, so the ability to pick up and carry materials is equivalent or even increases downstream.

As erosion continues, an idealized gradient is reached in which the frictional loss of energy is counterbalanced by the energy produced by the stream's flow. A stream in this equilibrium state is said to be in a graded condition. A **graded stream** has just the velocity and

discharge necessary to transport the load eroded from the drainage basin. The determination of a graded stream is based on a theoretically balanced state averaged over a period of years. Near the mouth, stream erosion develops a graded condition much sooner than at the headwaters, where the discharge is lower. As the stream gradient is progressively reduced, a graded condition is reached far-ther and farther upstream, eventually producing a smooth stream gradient profile (see again Fig. 17.13).

Stream erosion widens and lengthens stream channels and the valleys they occupy. Lengthening occurs primar-ily at the source through **headward erosion,** accom-plished partly by surface runoff flowing into a stream

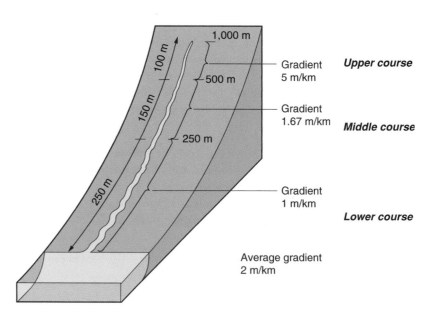

Figure 17.13 An idealized gradient profile for a graded stream showing the upper, middle, and lower courses of a stream.

Figure 17.14 Valley widening is accomplished by meandering, as shown here by the Tazlina River, Alaska. This is due to the increased stream velocity on the outside turn of each meander, eroding and undercutting the valley side. ■ **What occurs along the inside turn of a meander?**

and partly by springs undermining the slope. The lengthening of a river's course in an upstream direction is particularly important where erosional gullies are rapidly dissecting agricultural land. Such gullying may be counteracted by soil conservation practices to reduce erosional soil loss. Channel lengthening also occurs as the stream becomes more sinuous and thus decreases its gradient. Streams flowing on low-gradient floodplains do not generally flow in straight lines, but *meander* from side to side in the valley. **Meanders** are broad, sweeping bends in a channel that develop on most streams where they flow in low gradients on floodplains.

Valley widening is accomplished by lateral erosion on the outside of bends and commonly accompanies the outward expansion of meanders (Fig. 17.14). Lateral undercutting encourages slumping of the banks into the channel. Channel widening is greatly encouraged by mass movement of material down the side slopes of a stream valley. Tributaries flowing into a larger stream also help to widen the valley through which the trunk stream flows. Streams also widen their valleys by lateral shifts in their channel position from one side of the valley to the other.

Stream Transportation

The stream directly erodes some fluvially transported sediment, but a far greater proportion of the sediment is brought to the stream by surface runoff and mass movement. As previously noted, these materials transported by fluvial processes are called the stream load. Streams transport their load in several ways (Fig. 17.15). Some

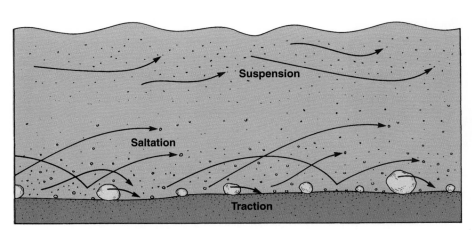

Figure 17.15 Transport of solid load in a stream. Clay and silt particles are carried in suspension; sand travels by suspension and saltation; and larger material, from pebbles to boulders, moves by traction. ■ **What is the difference between traction and saltation?**

minerals are dissolved in the water and thus carried in *solution.* The finest solid particles are carried in *suspension,* buoyed by vertical turbulence. Such particles can remain suspended as long as the force of upward turbulence is stronger than the downward settling tendency of the particles. Sediment particles too large to be carried in suspension slide and roll along by **traction.** When large particles hop and bounce along the channel bottom, the process is called **saltation** (from French: *sauter,* to jump).

The load of a stream is measured by the weight of the material it is transporting. There are three main types of stream load. In most streams, the largest portion of sediment load is **suspended load**—the materials being carried in suspension. **Bed load** consists of particles that roll or saltate along the streambed. **Dissolved load** is that portion that is held in solution. The relative proportion of each type of load varies with flow rates, water volume, and the nature of the drainage basin (vegetative cover, slope, rock and soil types). Some rivers transport huge suspended loads; the highest is that of the Huang He in northern China, known as the "Yellow River" because of the color of its suspended load. The Huang He carries a silty suspended load of more than 1 million tons per year. Com-

Figure 17.16   Some rivers carry a tremendous load of sediment in suspension, and show it in their muddy appearance, as in this aerial view of the Mississippi River in Louisiana. Much of the load carried by the Mississippi River is brought to it by its major tributaries.   ■   **Check an atlas to discover which major tributaries enter the Mississippi below Rock Island, Illinois. What states do those rivers help to drain?**

pared to the Mississippi River, the Huang He carries five times the suspended sediment load with only one-fifth the discharge. Rivers that are carrying a high suspended load look characteristically muddy (Fig. 17.16).

Two terms are used to convey the relationship between load and a stream's ability to transport this material: **Stream capacity** refers to how much load a stream can carry; **stream competence** is determined by the diameter of the largest particle it can transport as bed load. Both capacity and competence can increase in response

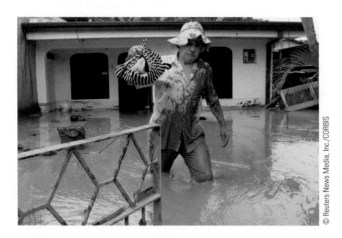

Figure 17.17   It is not just the power of flowing water that causes damage in a flood. A flooding stream typically has a high suspended load, can transport large heavy materials, and may carry floating debris that can damage homes and other structures, as in this image from the village of Orosi de Cartago, Costa Rica, 35 km (22 mi) southeast of the capital, San Jose.

NASA/Earth Sciences and Image Analysis Laboratory at Johnson Space Center

Figure 17.18 The braided stream channel of the Brahmaputra River in Tibet, viewed from the International Space Station. Stream braiding results from an abundant bed load of coarse sediment that obstructs flow and separates the main stream into numerous strands. Braided channels typically occur when a stream has low discharge compared to the amount of bed load. ■ **Under what climatic conditions might a braided stream develop?**

to a relatively small increase in velocity, discharge, and gradient. A stream that doubles its velocity during a flood may increase its sediment load six to eight times. The boulders seen in many mountain streams arrived there during some past flood that greatly increased stream competence; they will be moved again when a flow of similar magnitude occurs. Rapid increases in discharge can temporarily upset the equilibrium in a stream, causing it to pick up more load through erosion in order to reestablish a balance. In fact, rivers do most of their heavy earth-moving work during short periods of flood (Fig. 17.17).

Stream Deposition

Because a stream's capacity to carry material depends primarily on velocity and discharge, a reduction in either of these elements will cause a stream to reduce its load through deposition. Deposition, also called **aggradation,** occurs in locations where the velocity is slowed, such as on the insides of bends in the channel, on floodplains (or valley floors), at river mouths (deltas), and where the stream gradient abruptly flattens. **Alluvium** is the general name given to fluvial deposits, regardless of the type or size of material. Alluvium is usually recognized by the characteristic sorting and/or rounding of sediments that streams perform. A stream sorts particles by size, transporting the sizes that it can and depositing larger ones. As velocity fluctuates, particle sizes that can be picked up, transported, and deposited vary accordingly. The velocity of a stream generally varies with the changes in discharge that occur between wet and dry seasons or in response to storm events. The alluvium deposited by a stream with fluctuating discharge will exhibit changes in stream flow by depositing alternating layers of coarse and fine material.

If a stream has a heavy bed load in relation to its discharge, it deposits much of its load as sand and gravel bars in the streambed. These obstructions break the stream into strands that interweave, separate, and rejoin, giving a braided appearance to the channel. Such a pattern is called a **braided stream** (Fig. 17.18). This pattern may develop wherever the coarse sediment input into a stream is extremely high owing to banks of loose sand and gravel or the presence of a nearly unlimited coarse bed load, such as that found downstream from glaciers and also in many desert areas. Braided streams are common on the Great Plains (for example, the Platte River), in the desert Southwest, in Alaska, and in Canada's Yukon.

Land Sculpture by Streams

One way to understand the variety of landform features resulting from fluvial processes is to examine a typical river course from its headwaters to its mouth. In the following discussion of upper, middle, and lower river courses, you should note that erosion tends to be more significant in the upper course, whereas deposition is more important in the lower course. Upper, middle, and lower course in this sense refers to the gradient of the stream and how it changes in an upstream or downstream direction. For most stream channels, the gradient diminishes continually downstream as the stream flows toward base level (see again Fig. 17.13). There are exceptions, however, as some streams begin on, and continually flow over, a low gradient (Fig 17.19a), and other streams, particularly small ones on mountainous coasts, flow down a steep gradient all the way to the sea (Fig. 17.19b). A stream may also have several changes in gradient (increasing or decreasing) along its course.

(a)

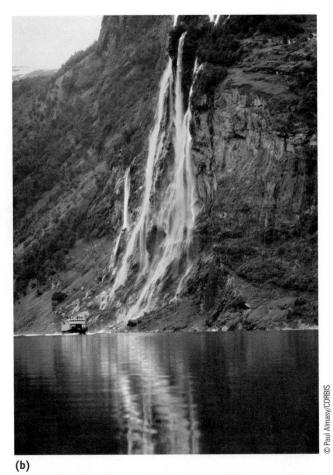

(b)

Figure 17.19 Not all streams fit the generalized pattern of characteristics for upper, middle, and lower stream segments. (a) The Mississippi River, here a relatively small stream, meanders on a low gradient near its headwaters in Minnesota, far upstream from its mouth. (b) This stream ends by cascading down an extremely steep gradient into the ocean in the Geiranger Fjord, located in the Sogn Og Romsdal region of central Norway.

Features of the Upper Course

At the headwaters, in the upper course of a river where the gradient is steepest, vertical erosion, particularly abrasion, operates strongly as the stream seeks base level through downcutting. Erosion in the upper course generally creates a steep-sided valley, a gorge, or a ravine as the stream channel in the bottom of the valley cuts deeply into the land. No floodplain is present, as the valley walls typically slope directly to the edge of the stream channel. Steep valley sides encourage mass movement of rock material directly into the flowing stream. Valleys of this type, dominated by the downcutting activity of the stream, are often called **V-shaped valleys** because of their shape (Fig. 17.20).

Because a stream in its upper course is generally in contact with bedrock, the effects of *differential erosion* are significant. Differential erosion occurs where rivers

Figure 17.20 Near the headwaters of a stream, valleys typically have a characteristic "V" shape. The stream flows in a steep-walled valley, with rapids and waterfalls, as shown here in Yellowstone Canyon, Wyoming. ■ **Describe the gradient of the Yellowstone River (compare with Fig. 17.19a).**

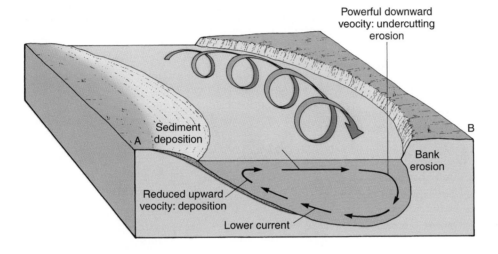

Figure 17.21   Characteristics of a meandering river channel. Note that water flowing in a channel has a tendency to flow downstream in a "corkscrew" fashion, which moves water against one side of the channel and then to the opposite side. The up-and-down motion of the water contributes to the processes of erosion and deposition.
■ **Why do the meanders become more sinuous as time passes?**

cut through rock layers of varying resistance. Typically, rivers flowing over resistant rock have a steeper gradient than where they encounter weaker rock. A steep gradient gives the stream flow more energy. Rapids and waterfalls indicate the resistant materials and steep gradients most common in the upper course of streams and their tributaries (see again Fig. 17.20). Where rocks are particularly resistant to weathering and erosion, valleys will be narrow, steep-sided gorges or canyons; where rocks are less resistant, valleys tend to be more spacious.

Many streams spill from lake to lake in their upper courses, either over open land (like the Niagara River at Niagara Falls, between Lake Erie and Lake Ontario) or through gorges. In either case, the lakes are eventually eliminated as stream erosion lowers their outlets or deposition fills the lake by building deltas at the inlet point.

Features of the Middle Course

In places where a stream has reduced its gradient, smoothed out its channel bed, and begun to approach its base level, vertical erosion becomes less significant and lateral erosion of the channel sides assumes a more important role. Combined with mass movement, this effect displaces the channel laterally, eroding valley walls to produce a wider valley floor with a smoother stream gradient. Lateral erosion on the outside of river bends eventually produces a narrow floodplain along the banks. Vertical erosion is minimized because the stream is flowing over a gentler gradient, so it has little or no excess energy for eroding toward base level. The locations of the highest velocity flow in the stream, deeper channel depths, and the direction of centrifugal force against the banks shift from one side of the stream channel to the other. These factors cause lateral erosion of one bank and deposition on the other side of the channel.

Here the river valley contains a floodplain but still maintains definite valley walls. The river becomes sinuous, forming wide bends or meanders on the valley floor. Such a stream erodes a **cut bank** on the outside of a meander loop and deposits material on a **point bar** on the inside of the meander (Fig. 17.21). Centrifugal force accelerates stream velocity, deepens the channel and forces water against the outside of these bends, allowing sediment to be picked up, and decelerates velocity on the inside, causing sediment to be deposited. Over time, this activity increases the size of the stream meanders, further widening the floodplain.

The *meander core,* the land between the curves of a stream's meanders, is relatively flat and covered with alluvium left by subsiding floodwaters. As lateral erosion continues and the meander size increases, the area of flat land covered with rich alluvium increases. Though flooding of such land is always a potential hazard, the richness of a floodplain's alluvial soils offers an irresistible lure for farmers.

Features of the Lower Course

In the lower course of a river, where the gradient is minimal, deposition becomes more significant in shaping the landscape. The lower valley of a major river is wider than the width of the meander belt and typically shows evidence of many shifts in the river's course (Fig. 17.22). In most cases, this great valley width results from the lower course flowing on a vast alluvial floodplain rather than an erosional plain (see Map Interpretation: Fluvial Landforms). In other words, the alluvium filling the valley is much deeper than the stream channel. These giant lower valleys are typically sediment-filled marine **estuaries** (river mouths drowned by the sea where salt and fresh water mix)—as in the case of the Mississippi River—or subsiding tectonic depressions, as in the Amazon and Congo Basins.

During floods, floodplains of a river's lower course are inundated with sediment-laden water that deposits alluvium on the valley floor. These plains accordingly are called **alluvial plains.**

A common landform of the lower course, an **oxbow lake,** provides evidence of a meandering river's shifts in position over time. Especially during floods, oxbow lakes form when a portion of the channel is cut off as the stream seeks a shorter, steeper, and straighter path. The cut-off section of the channel, now isolated from the river, forms the oxbow lake (Fig. 17.23).

During flooding, a sharp reduction in a stream's velocity caused by friction at the channel's edge results in heavy sedimentation along the riverbanks. Over time, the banks are built up into **natural levees** through successive deposition that occurs most often on the immediate banks of the river. Natural levees along the Mississippi River rise up to 5 meters (16 ft) above the rest of the floodplain.

When a river has less energy during times of reduced discharge, deposition will also occur in the channel. Thus, in its lower course, a river can sometimes raise the level of its channel bed. In some instances, as in China's Huang He and the Yuba River in northern California, the stream channel may be raised above the surrounding floodplains, although the natural levees tend to hold the water in the channel. Flooding presents a very serious danger in this situation because the so-called "bottomlands" of the floodplain actually lie below the level of the river. Sometimes humans attempt to control streams by building up levees artificially in order to keep the river in

Figure 17.22 A colorized radar image shows part of the Mississippi River floodplain along the Arkansas-Louisiana-Mississippi state lines. Images like this help us to assess flood potential and to learn much about the geomorphic history of the river and its floodplain. The colors are used to enhance landscape features such as water bodies (dark), field patterns, and forested areas (green). Note how the river has changed its channel position many times, leaving oxbow lakes and meander scars on the floodplain.

its channel. Unfortunately, artificial levees tend to make the river higher and deeper, so that when floodwaters go over the levees, they may be even more extensive and destructive. This situation happened during the 1993 Mississippi River floods in the Midwest (Fig. 17.24).

The presence of levees—both natural and artificial—can prevent tributaries in the lower course from joining the main stream. Smaller streams are forced to flow parallel to the main river until a convenient junction can be found. These parallel tributaries are called **yazoo streams,** named after the Yazoo River, which parallels the Mississippi River for more than 160 kilometers (100 mi) until it finally joins the larger river near Vicksburg, Mississippi.

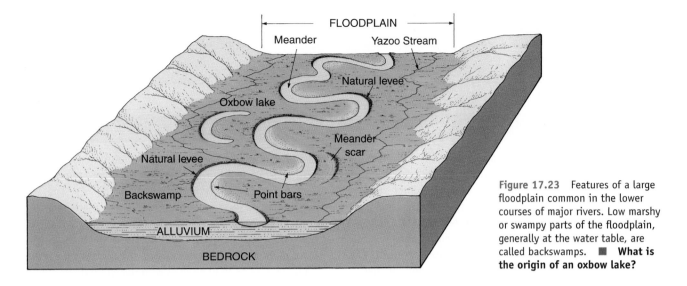

Figure 17.23 Features of a large floodplain common in the lower courses of major rivers. Low marshy or swampy parts of the floodplain, generally at the water table, are called backswamps. ■ **What is the origin of an oxbow lake?**

When a stream flows into a body of water such as a lake or an ocean, both flow velocity and load capacity are reduced. Consequently, the stream deposits its load—larger particles first, and then finer ones. The resulting landform is a **delta,** named after the triangular shape of the Greek letter (Δ), which some deltas, like that of the Nile River, resemble when seen on a map or from space (Fig. 17.25a). Most deltas, such as the Niger and Nile deltas, have a triangular arc shape.

Multiple channels flowing away from the main stream are called **distributaries,** a common channel form on deltas. Natural levees build up along the banks of these distributary channels, even developing outward from the coast into a body of water. Continued deposition and delta formation can build new land far out from the original shoreline. Rich alluvial deposits and the abundance of moisture allow vegetation to quickly become established on this fertile new land and further

secure the delta's position. Delta lands, such as those of the Mekong, Indus, and Ganges Rivers, form important agricultural areas that feed the dense populations of many parts of Asia.

In general, deltas can attain great size only where the sediment supply is high, where the continental shelf does not drop off sharply, and where waves, currents, and tides are not strong. Waves, currents, and tides tend to transport alluvial deposits that form the delta away as they are deposited by the river, so some rivers have no delta at all.

The *bird's-foot delta* of the Mississippi is a special type that results from land subsidence during delta formation, leaving only the natural levees of the distributary channels above sea level (looking like a bird's foot). The Mississippi delta (Fig. 17.25b) has changed rapidly over the years as new distributaries deposit alluvium that subsequently sinks, leaving only the levee crests extending seaward along the edge of the channels.

Another type of delta forms at the heads of some marine estuaries. As previously noted, estuaries are river valleys that have been drowned by the sea, either because the sea level has risen or because the land has subsided (Fig. 17.25c). Deltas form slowly in estuaries because of the counteraction of tides and sometimes because of low sediment input caused by a gentle gradient.

Figure 17.24 During floods, low areas adjacent to the river are inundated with sediment-laden water that flows over the banks to deposit alluvium, mainly silts and clays, on the floodplain. This is the Missouri River floodplain at Jefferson City, Missouri, during the 1993 Midwest flood. ■ **What would the river floodwaters leave behind in flooded homes after the water recedes?**

FLUVIAL LANDFORMS

The Map

Campti is in northwestern Louisiana on the Gulf Coastal Plain. This coastal region stretches from northern Florida to the Texas-Mexico border, and extends inland for more than 200 miles in some areas. Elevations on the Gulf Coastal Plain gradually increase from sea level at the shoreline to several hundred feet far inland. The region is underlain by gently dipping sedimentary rock layers. The surface material includes marine sediments and alluvial deposits from rivers that cross the region, especially those of the Mississippi drainage system. This is a land of meanders, natural levees, and bayous.

The Red River's headwaters are in the semiarid plains of the Texas Panhandle, but it flows eastward toward an increasingly more humid climate. About midcourse, the Red River enters a humid subtropical climate region, which supports rich farmland and dense forests. The river flows into the Mississippi in southern Louisiana, about 161 kilometers (100 mi) downstream of the map area. The Red River is the southernmost major tributary of the Mississippi.

Louisiana has mild winters with hot, humid summers. Annual rainfall totals for Campti average about 127 centimeters (50 in.). The warm waters of the Gulf of Mexico supply vast amounts of atmospheric energy and moisture, producing a high frequency of thunderstorms, tornadoes, and, on occasion, hurricanes that strike the Gulf Coast.

Interpreting the Map

1. How would you describe the general topography of the Campti map area? What is the local relief? What is the elevation of the banks of the Red River at Campti?

2. What kind of landform is the low relief surface that the river is flowing on? Is this mainly an erosional or depositional landform?

3. In what general direction does the Red River flow? Is the direction of flow easy or hard to determine just from the map area? Why?

4. Does the Red River have a gentle or steep gradient? Why is it difficult to determine the river gradient from the map area?

5. What is the origin of Smith Island? Adjacent to Smith Island is Old River; what is this feature called?

6. Explain the stippled brown areas in the meanders south of Campti. Are these areas on the inside or the outside of the meander bend? Explain why.

7. How would you describe the features labeled a "bayou"?

8. Is this map area more typical of the upper, middle, or lower course of a river?

9. Although this is not a tectonically active region, what do you think would happen to this area if it were uplifted? What would be the major geomorphic process change?

U.S. Department of Agriculture

Campti, Louisiana
Scale 1:62,500
Contour interval = 20 ft
U.S. Geological Survey

(a)

(b)

(c)

Figure 17.25 Satellite views of river deltas. (a) The Nile delta, the classic triangular arc river delta, as it enters the Mediterranean Sea. (b) A false-color satellite view of upper Chesapeake Bay. Chesapeake Bay was formed by submergence of the lower valley of the Susquehanna River as sea level rose at the end of the last glaciation. The Potomac River enters the main estuary at the bottom left of the image, and numerous minor tidal estuaries are found along the margins of the bay.(c) The unusual shape of the Mississippi River delta resembles a bird's foot. ■ **What change in a river's equilibrium causes the development of a delta?**

Changes in Base Level

Changes in the base level of a stream can occur as a result of either tectonic processes or climatic change. Landmasses may rise in response to faulting or other tectonic uplift, or perhaps sea level may drop in response to the growth of glaciers. In either case, the result will be similar. Streams flowing in that region (or, in the case of sea level change, worldwide) will be above their base level and will now have the energy to regrade (downcut) their channels to the new base level. Valleys with waterfalls or rapids may develop as these channels are deepened by

erosion. The landscape and its stream are then said to be **rejuvenated,** because the stream's gradient has steepened, increasing its energy for erosion and its competence for carrying load, and these changes also affect the landscape.

If new uplift occurs gradually in an area where stream meanders have formed, these meanders may become **entrenched** as the stream deepens its valley (Fig. 17.26). Now, instead of eroding the land laterally, with meanders migrating across an alluvial plain, the rejuvenated stream's primary activity is vertical incision.

It is important to note that virtually all rivers reaching the sea downcut their valleys during the Pleistocene in response to the lowering of sea level associated with continental glaciation. The accumulation of water on the land, as glacier ice, caused sea level to drop as much as 120 meters (400 ft), and lowered the base level for coastal streams. Consequently, the streams eroded deep valleys for their channels. Subsequent melting of the glaciers again elevated base level. This base level rise caused the streams to deposit their sediment loads, fill-

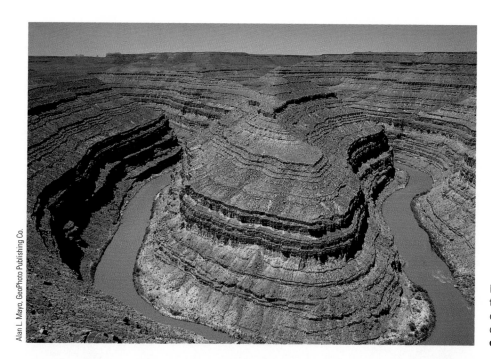

Alan L. Mayo, GeoPhoto Publishing Co.

Figure 17.26 The "Goosenecks" of the San Juan River in Utah provide classic examples of entrenched meanders. ■ **Explain how rejuvenation could form such spectacular features.**

ing valleys with sediment, as the streams adjusted their channels to a second new base level. These consecutive changes in base level produced broad, flat, alluvial floodplains above buried valleys cut far below today's sea level.

While a drop in base level causes downcutting and a rise causes deposition, an upset of stream equilibrium resulting from sizable increases or decreases in discharge or load can have similar effects on the landscape. Research has shown that variations in base level, tectonic movements, and changes in stream equilibrium can each cause down-cutting in stream valleys, so that the valley is slightly deepened and remnants of the older, higher valley floor are preserved in "stair-stepped" banks along the walls of the valley. These remnants of previous valley floors are **stream terraces.** Multiple terraces are a consequence of successive periods of down-cutting and deposition (Fig. 17.27). Stream terraces provide a great deal of evidence about the geomorphic history of the river and its surrounding region.

Stream Patterns

When viewed from the air or on maps, the channel networks of stream systems form distinct **drainage patterns** (also called **stream patterns**). Two primary factors that influence stream patterns are the bedrock structure and the nature of the land surface. A **dendritic** (from Greek: *dendros,* tree) stream pattern (Fig. 17.28a) is an irregular branching pattern with tributaries joining larger streams at acute angles (less than 90°). A dendritic stream pattern is the most common type, in part because water flow in this pattern of channel network is highly efficient. Dendritic patterns form where the underlying rock structure does not strongly control the position of

stream channels. Hence dendritic patterns tend to develop in areas where the rocks have a roughly equal resistance to weathering and erosion. In contrast, a **trellis** stream pattern consists of long, parallel streams linked by short, right-angled segments (Fig. 17.28b). Trellis drainage is usually evidence of folding, where parallel outcrops of erodible rocks form valleys between more resistant ridges, as in the Ridge and Valley region of the Appalachians. A **radial** pattern develops where streams flow away from a common high point on cone- or dome-shaped geologic structures, such as volcanoes and domal uplifts (Fig. 17.28c). The opposite pattern is **centripetal,** with the streams converging on a central area, as in a basin of *interior drainage* (Fig. 17.28d). **Rectangular** patterns occur where streams follow sets of intersecting fractures to produce a blocky network of straight channels with right-angle bends (Fig. 17.28e).

Most stream patterns that drain a large region require a considerable amount of Earth history to become well established (varying from tens of thousands to millions of years). In regions that were covered by extensive glacier ice until about 10,000 years ago, streams flow on low gradients left by receding glaciers, wandering between marshes and small lakes in a chaotic pattern called **deranged** drainage (Fig. 17.28f). These streams have yet to establish a "normal" drainage pattern because, given the low gradient, there has not been enough time since deglaciation for a stream network to become well developed.

Many streams follow the "grain" of the topography or bedrock structure, eroding valleys in weaker rocks and flowing between divides formed on resistant rocks. Many examples also exist, however, of streams that flow across (transverse to) the structure, cutting a gap or canyon through mountains or ridges. These **transverse**

(a)

Figure 17.27. (a) Diagram of stream terraces formed by three phases of change in the stream that may have developed through rejuvenation (changes in base level through rise or fall of sea level, or uplift/subsidence of the land surface). Major changes in the equilibrium of streams (from erosion-dominated to deposition-dominated, or vice versa) can also form terraces. The terrace development process has produced three successively lower valley floors along the river in this diagram. The oldest terrace is the highest level (1); the youngest is the lowest level (3). (b) River terraces in the Tien Shan Mountains of China. ■ **How many terraces can be clearly identified in this photo?**

(b)

Courtesy Sheila Brazier

streams can be puzzling, giving rise to questions such as how a stream can cut a gorge through a mountain range, and how a stream can move from one side of a mountain range to the other. Such streams are probably either *antecedent* or *superimposed*. Antecedent streams existed before the formation of the mountains that they flow through, maintaining their courses by cutting an ever-deepening canyon as gradual mountain building took place across their paths. The Columbia River Gorge through the Cascade Range in Washington and Oregon and many of the great canyons in the Rocky Mountain region (Royal Gorge and the Black Canyon of the Gunnison, both in Colorado, and Flaming Gorge in Utah/Wyoming) probably originated as antecedent streams (Fig. 17.29a). Other rivers, such as those of the central Appalachians, have cut gaps through mountains in a very different manner. These streams originated on earlier rock strata, since stripped away by erosion, so that

the streams have been superimposed on the rocks beneath (Fig. 17.29b). This sequence would explain why, in many instances, the rivers flow *across* folded rock structure, creating water gaps through mountain ridges. Examples include the Cumberland Gap and the gap formed by the Susquehanna River in Pennsylvania, both of which were important travel routes for the first American settlers crossing the Appalachians.

Quantitative Fluvial Geomorphology

Many studies in geomorphology are based on statistical analyses of data measured through instrumentation in the field, from maps, or from remotely sensed imagery. Measurements and quantitative data allow geomorphologists to understand and scientifically test relation-

Figure 17.28 Drainage patterns often reflect bedrock structure. (a) A dendritic pattern is found on homogeneous rocks. (b) A trellised pattern indicates parallel valleys of weak rock between ridges of resistant rock. (c) A radial pattern indicates a domed upland or volcano. (d) A centripetal pattern indicates a structural lowland or basin. (e) A rectangular pattern indicates linear joint patterns in the bedrock structure. (f) A deranged pattern typically results following the retreat of continental ice sheets; it is characterized by a chaotic arrangement of channels connecting small lakes and marshes.

ships concerning how landforms develop and how the processes responsible for their formation operate. Quantitative methods are important in studying virtually all aspects of the Earth system and are used by climatologists, biogeographers, geomorphologists, soil scientists, and hydrologists. A favorite topic for quantitative research by geomorphologists is the stream system. Because of the importance of streams to human existence, extensive efforts have been undertaken to gather the data necessary to help us understand how streams work, to predict their behavior, to compare streams objectively, and to describe their nature. Streams are dynamic systems in a constant state of change, and many types of quantitative data must be gathered, tested, and applied for descriptive and analytical purposes. An understanding of fluid mechanics and energy flow may also be integrated with that of geomorphic processes and landform descriptions. An example of this quantitative approach was discussed earlier in the section on equilibrium in streams ($Q = AV$).

The value of quantitative data can be illustrated by questions such as the following: How can we objectively compare streams in terms of their size or importance as a stream system? Do stream systems develop in a logical

manner, or is the process completely random? If a channel network is a part of a stream system, and a system is a set of interacting parts, how do parts of the stream system interact with one another? A first step to finding the answers to these questions is to objectively compare streams within a river basin, or to compare similar river basins. The problem, though, is how to determine what "similar" means in an objective and quantitative manner. A solution was suggested in the 1940s by a hydrologist, Robert Horton, who proposed a method (since modified by others) for determining **stream order.** First-order streams are the uppermost headwater channels, formed without tributaries, which can be seen in the field or mapped on large-scale topographic maps. When two or more first-order streams join, the channel becomes larger, and the stream is classified as a second-order stream. When two or more second-order streams intersect, the stream is classified as a third-order stream, and so on. Generally, higher order means larger and longer channels, but fewer channel segments. Drainage basins, in turn, can be classified according to the highest-order stream in the basin, called the *trunk stream* (Fig. 17.30). To answer our questions at the start of the paragraph, the size or importance of a stream or basin

(a)

(b)

Figure 17.29 How transverse streams form valleys and canyons across mountains. (a) An antecedent stream and its valley. A stream maintains its course by cutting a valley through a mountain as it is uplifted gradually over much time. (b) A superimposed stream and its valley. This stream has uncovered and excavated ancient structures that were buried beneath the surface. As the stream erodes the landscape downward, it cuts across and through the ancient structures.

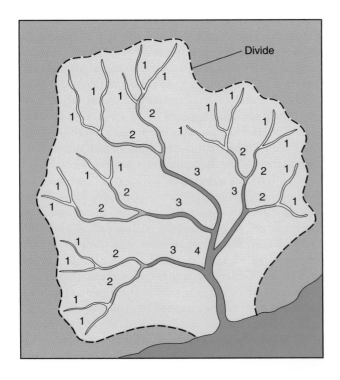

can be objectively compared using the stream order system: The Mississippi River would be classed as a tenth-order stream, and the Amazon would probably be a thirteenth-order stream. Further, stream systems do seem to develop in a logical manner, and their parts (tributaries, sub-basins) are generally well adjusted to one another. Higher-order streams, for example, tend to have larger basins, longer channels, larger channels, and greater discharge than lower-order streams (and vice versa). These quantitative relationships help us to understand how stream systems operate.

Figure 17.30 The hierarchy of stream ordering is illustrated by this fourth-order watershed. Note that when two streams of a lower order flow together, the downstream segment will become the next higher order (note also that lower number tributaries flowing into a higher one do not change the order of the downstream segment). Each segment of the stream has only one order following this system. ■ **What is the highest-order stream called?**

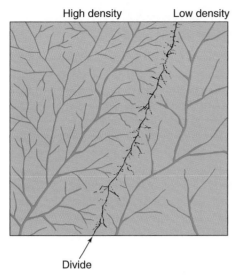

High density   Low density

Divide

Figure 17.31 Drainage density—the length of channels per unit area—varies according to several environmental factors. For example, other factors being equal, soft, highly erodible, and impermeable rocks tend to have higher drainage density than areas dominated by strong, resistant, and or permeable rocks. Slope and vegetation cover can also affect drainage density.
■ **What kind of drainage density would you expect in an area of steep slopes and sparse vegetation cover?**

Drainage density studies provide information on the degree of erosional dissection in parts of a basin area. The formula used is $D = L/A$: Drainage density (D) is equal to the total length of all stream channels in the basin (L) divided by the area of the drainage basin (A). The easily eroded Dakota Badlands have an extremely high drainage density of over 125 (125 km of channel per 1 sq km of land), whereas very resistant granite hills may have a drainage density of only 5 (5 km of stream channel for every square kilometer of basin area). High drainage density is associated with steep slopes, sparse vegetation, weak rocks, high erosion rates, and high runoff. To understand this concept, think about what would happen on a hillside if the vegetation were burned off in a fire. Erosion would rapidly cut gullies, creating more channels where fewer existed before. In other words, the drainage

density would increase. We could use the quantitative measure of $D = L/A$ to determine precisely the immediate change and to monitor the change over time. Figure 17.31 illustrates drainage density.

Future quantitative studies not only will help us better understand the origins and formational processes of landforms and landscapes, but will also allow us to apply quantitative and scientific methods to help predict water supplies and flood hazards, estimate soil erosion, and trace sources of pollution.

The Importance of Surface Waters

Streams

Historically, people have used streams and rivers for a variety of purposes. The settlement and growth of the United States would have been very different without the Mississippi River and its far-reaching tributary system that drains most of the United States between the Appalachians and the Rockies. The Mississippi, like many other rivers, has been used for exploration, migration, and settlement, and the number of major cities along it (Minneapolis/St. Paul, St. Louis, Memphis, and New Orleans, to name a few) provides evidence of people's tendency to settle along rivers. The Mississippi also provides inexpensive transportation of bulk cargo. Today, our navigable rivers still compete successfully with railroads and trucks as carriers of grain, lumber, and mineral fuels.

Rivers and streams have generated power for mills and, more recently, for hydroelectricity. Rivers provide irrigation water, and the alluvial soils of floodplains and river banks are often productive agricultural lands.

Figure 17.32 Living on an active floodplain has its risks, as seen in this photo of the Coast Guard rescuing a man stranded on a rooftop in Olivehurst, California. A levee along the Feather River ruptured, sending acres of water into the Sutter County community.
■ **What can be done to prepare for or to avoid flood problems?**

Restoring the Everglades

In Chapter 1, human alteration of the waterflow system leading to the Everglades of south Florida was cited as an example of flood control programs gone awry. Although straightening and ditching of the Kissimmee River, the building of a 38-foot-high dike to control outflow from Lake Okeechobee, and the construction of discharge canals across southeast Florida controlled the periodic flooding of the region's tropical savanna climate, these human modifications of the environment created a host of new problems. Today, as floodwaters rush to the sea, underground water supplies are not replaced and water quality diminishes. Perhaps most significantly for those scientists concerned about the environment has been the progressive deterioration and disappearance of the Everglades.

The Everglades can never be restored to its historical size and condition. Nearly half of the area originally covered by the Everglades is now devoted to buildings, roads, sugar plantations, and vegetable farms. When water filled with excess nutrients runs off agricultural fields and reaches the Everglades during flood periods, native species die and invading plants flourish. However, there are now clear indications that what humans damage unwittingly, humans, with enough resolve, can attempt to repair.

Programs to restore the Kissimmee-Okeechobee-Everglades ecosystem to some of its original natural state are already underway. A demonstration project that returned waterflow to a portion of the Kissimmee River floodplain proved that the floodplain and its ecosystems could be restored if large sections of the river were dechannelized. Florida and the U.S. federal government are currently supporting the program at an estimated cost of $400 million to be shared equally between them.

Now it is the Everglades' turn. As a result of painstaking research and years of negotiation, the state and federal governments have agreed to fund the $8 billion cost of restoring the Everglades as nearly as possible to its original state. Some levees will be removed, water will flow again through the Everglades in natural channels, new surface reservoirs for water storage will be constructed, and large tracts of agricultural land will become marshes for the filtering out of ex-

---

We also use streams as sources of food and water, for recreation (boating, fishing, water skiing, swimming), and as a depository for wastes.

There are many benefits to living near streams and rivers. However, settlement along a river has its risks, particularly in the form of floods. Variability of stream flow constitutes the greatest problem for life along rivers and is also an impediment to their use. Stream channels can generally carry (and contain) the maximum flows that are estimated to occur once every year or two. The maximum flows that are probable over longer periods of 5, 10, 100, or 1000 years overflow the channel and inundate the surrounding land, sometimes with disastrous results (17.32). Similarly, exceptionally low flows may produce crises in water supply and bring river transportation to a halt.

Because stream flow can be so variable, and in some cases unreliable, humans today regulate most rivers in some way. Many river systems now consist of a series of **reservoirs**—artificial lakes impounded

behind dams. The dams contain potentially devastating floodwaters and store the discharge of wet periods to make the water available during dry seasons or drought years (Fig. 17.33). An example of such human river-basin management is the Tennessee River Valley, which was "tamed" during the 1930s. A look at a hydrographic map will show that most of our great rivers, such as the Missouri, Columbia, and Colorado, have been transformed by dam construction into a string of reservoirs. Unfortunately, the life of these reservoirs, like that of any lake, will only be a few centuries at most, because they will

Figure 17.33 The multipurpose Lookout Point Dam on the Middle Fork of the Willamette River, Oregon. ■ **Name several functions that multipurpose dams serve.**

U.S. Army Corps of Engineers, Portland District

cess nutrients. However, the real key to the ecological health of the Everglades is water control: protection against over-flooding during the rainy season, and adequate water supplies during the dry season. Project engineers expect to provide this control by developing the largest underground water-storage system ever constructed. During flood periods, fresh water from Lake Okeechobee will be forced downward through a series of nearly 200 new wells. The water will be stored in the naturally porous rock that underlies south Florida and pumped to the surface again during the dry season. The lake holds so much water that engineers estimate that it will take five months for the lake level to drop a foot with the well pumps running at their full capacity of 1.6 billion gallons a day. The real victory for biogeographers and other ecologists occurs because all parties to the project have agreed that the first 80 percent of all that water will be sent directly to the Everglades. Even though the total cost of the water storage plan alone is estimated at nearly $2 billion, humans have finally placed the environment ahead of agriculture, urban water supplies, and other human needs.

Restored wetlands along the original course of the Kissimmee River.

R. Gabler

gradually fill with sediment carried by the inflowing streams. To protect the natural flow and environments of the few remaining undeveloped streams and rivers in the United States, Congress enacted the Wild and Scenic Rivers Act in 1968.

Lakes

Lakes are standing bodies of inland water, and the water in most lakes is runoff that is held in temporary storage. Most of the world's lakes, such as Lake Superior, Lake Tahoe, and Lake Victoria, contain fresh water. However, some lakes, such as the Caspian Sea, Dead Sea, and Great Salt Lake in Utah, are salty, because they exist in closed basins where evaporation recycles water to the atmosphere, leaving dissolved minerals to accumulate in these lakes.

Natural lakes form wherever the water supply is adequate and geomorphic or topographic processes have created depressions on the land surface. The majority of the world's lakes, such as North America's five Great Lakes and Minnesota's "10,000 lakes," are products of glaciation. Rivers, groundwater, tectonic activity, volcanism, and human activities also produce lakes. Lake Baikal in Russia is the world's deepest lake. This long, narrow Siberian lake is more than 1525 meters (5000 ft) deep and occupies a fault depression. Crater Lake in Oregon, North America's deepest lake, is in a caldera formed by the collapse of a volcano.

Lakes and ponds (small, shallow lakes) are temporary features on Earth's surface. Few have been in existence for more than 10,000 years, and most are very recent in terms of Earth history (Fig. 17.34). As

R. Sager

Figure 17.34   The majority of lakes in the world are of glacial origin; shown here is Stanley Lake in the Sawtooth Range of Idaho. ■ **Why are lakes considered temporary features?**

soon as lakes are formed, their destruction begins through several processes, including sedimentation, biological activity, and perhaps down-cutting of an outlet by a stream.

Streams that flow into lakes deliver sediment—sand, silt, and gravel—forming deltas that will eventually fill the entire lake. Plants and animals, using nutrients brought by streams, occupy the shore and build communities on the bottom, helping the shoreline advance farther into the lake area. As the lake becomes smaller and shallower, plants take root on its bed, and it becomes a marsh. Eventually the marsh is filled and becomes a meadow. This inevitable process is accompanied by progressive changes in the lake's flora and fauna. The first occupants are life forms adapted to cold, deep water. As the process continues, there is a shift toward organisms adapted to progressively shallower and warmer water. Eventually, toads hop where pike and trout formerly swam and humans fished.

Lakes are important to humans for more than their scenic appeal and their value for fishing or recreational activities. Like oceans, lakes affect the nearby climates, particularly by reducing daily and seasonal temperature ranges and by increasing humidity. Major fruit-producing areas exist near lakes in Florida, New York, Michigan, and Wisconsin because of the moderating temperature effects of lakes. Lakes can also cause a downwind increase in precipitation—snow or rain generated by the "lake effect," in which storms either pick up more moisture from the lake or undergo uplift as they move over a relatively warm body of water.

The benefits of lakes are such that humans have produced tens of thousands of artificial lakes through the construction of dams. Reservoirs are some of humankind's most ambitious construction projects. However, because lake water tends to stratify by temperature, it does not mix as well as rivers or the oceans. Poor circulation makes lakes easily susceptible to destruction by the chemical, thermal, and biological pollution often resulting from human activities. The Great Lakes—Lake Erie in particular—provide instructive examples of the damage that can be done to a large, complex natural system by human misuse over a short period of time.

DEFINE & RECALL

stream
fluvial processes
runoff
sheet wash
tributary
stream system
drainage basin (watershed, catchment basin)
interfluve
divide
infiltration
infiltration capacity
ephemeral stream
perennial stream
floodplain
stream discharge
stream hydrograph

base flow
stream load
base level
degradation
hydraulic action
plunge pool
stream abrasion
pothole (stream)
attrition
gradient (stream)
graded stream
headward erosion
meander
traction
saltation
suspended load
bed load

dissolved load
stream capacity
stream competence
aggradation
alluvium
braided stream
V-shaped valley
cut bank
point bar
estuary
alluvial plain
oxbow lake
natural levee
yazoo stream
delta
distributary
rejuvenated stream

entrenched meander
stream terraces
drainage pattern
    (stream pattern)
dendritic stream pattern
trellis stream pattern
radial stream pattern
centripetal stream pattern
rectangular stream pattern
deranged stream pattern
transverse streams
stream order
drainage density
reservoirs

DISCUSS & REVIEW

1. On a worldwide basis, what is the most important gradational process operating to shape the landscapes of the continents?

2. What dominant stream process formed the Black Canyon of the Gunnison? What was the dominant stream process that formed the delta of the Mississippi River?

3. What factors affect the discharge of a stream?

4. Why is it important to monitor stream gaging stations and the stream flow data that they provide?

5. Which variables affect the erosion rate of a streambed? What determines the base level to which a streambed will erode?

6. What constitutes the largest portion of a stream's load? State the ways a stream can transport its load.

7. How does a stream sort alluvium? Explain the relationship of this sorting to velocity changes in a stream.

8. Where in the course of a stream is the gradient steepest? How do rapids and waterfalls in streambeds relate to rock resistance and steepness of gradient?

9. Describe the development of natural levees.

10. What would cause rejuvenation of a stream? What unique features would distinguish a rejuvenated stream?

11. What are stream terraces? How are they formed? How did changes in sea level during the Pleistocene cause stream terraces to form on land?

12. Name the two primary variables that influence a stream system's pattern. How do the various patterns reflect these variables?

13. Discuss the positive and negative aspects of building dams.

14. What kinds of processes help form lakes? Why are lakes such short-lived features of Earth's surface?

CONSIDER & RESPOND

1. One aspect of the concept of equilibrium in streams is that a stream adjusts several factors to match the discharge that it is carrying (velocity, gradient, load capacity, load competence, channel width and depth). How would each of these factors change in response to a flood event during each of the three flood phases—rising, cresting, and receding? What effects would these changes have on the landscape?

2. In the study of a drainage basin, what types of geographic observations and quantitative data might prove useful in planning for flood control and water supply?

3. How does urbanization affect runoff, discharge, and flood potential in a watershed?

ARID LANDFORMS AND EOLIAN PROCESSES

In most deserts, running water only operates occasionally, so it may seem odd that water rather than wind is the chief agent of gradation in dry regions.

How can this be? What features in arid regions were produced predominantly by water action?

A variety of landforms in many deserts today could not have been created under present conditions of aridity; they are evidence of landscapes formed under earlier, wetter climatic conditions.

What is the evidence of climatic change? How might the most recent cooler, wetter period be associated with glaciation?

Landforms of fluvial deposition in valleys and along the base of mountains in desert areas are attractive locations for human settlement.

What combination of factors causes this to be true? What regions of the United States are excellent examples? What are the most common landforms of fluvial erosion and deposition in arid environments?

Most people associate deserts with sand dunes, but the land surface in arid regions is more typically covered by sparse vegetation and bare exposures of regolith, alluvium, or bedrock.

What type of landscape is found in the majority of the world's deserts?

Under what conditions are the various types of sand dunes formed?

Over most land surfaces, wind exerts little influence on landform development, but under certain conditions, it can be an important agent of gradation.

Why is wind so limited as a major agent in landform development?

Under what conditions is the wind an important agent of gradation?

Deposition of loess by the wind has had a positive impact on humans in a number of major world regions.

Why is this so? What regions are involved, and what were the sources of the loess?

R. Sager

The red rocks form striking cliffs at Sedona, Arizona.

Relatively barren of vegetation, arid regions display distinctive scenery that is unlike that of most other environments. Hill slopes in humid regions tend to be rounded and mantled in soil. Desert mountains and hill slopes are generally angular, with extensive bedrock exposures, separated by low relief areas either filled in by sedimentation or controlled by resistant rock layers. Many desert regions in North America hold a mythical fascination for the people who live there as well as for ecotourists, two groups of people who appreciate arid landscapes. A great number of movies have used striking desert scenery of all kinds as a location for filming. Desert landscapes often display, in stark beauty, the colors, characteristics, and structure of the rocks that make up the area. The desert's barrenness reveals much evidence about landforms and geomorphic processes that would be much more difficult to observe in humid environments, under their cover of soil and vegetation. Much of our understanding of how landscapes and landforms develop in a wide variety of environments has come from important studies and scientific explorations conducted in desert regions.

It is a fallacy that most desert landforms are produced by wind erosion. The wind does play an important role in arid regions, but most deserts of the world are rocky rather than sandy. The effects of wind erosion are mainly confined to moving loose rock fragments of sand-size materials and to removing fine, dust-sized particles from desert regions. Sparse vegetation permits wind gradation to operate on a grand scale in deserts. We should also understand that eolian processes and landforms are not confined to arid regions; in fact, they are also very conspicuous in coastal areas, and anywhere that loose sediments are exposed to winds strong enough to move them. We will examine eolian, or wind-produced, landforms in detail in the second half of this chapter.

Water in the Desert

Landforms, rather than vegetation, typically dominate desert scenery. Where little or no vegetation exists, running water, when it is available, is extremely effective in shaping the land. Evaporation and precipitation are important factors in determining the amount of water available for surface runoff, stream flow, and gradational work. In addition, these same climatic factors affect the density of vegetative cover. The absence of a continuous vegetative and soil cover, a direct result of arid climates, gives desert landforms their unique character. Without sufficient vegetation to break the force of raindrop impacts, and without the binding effects of root networks, a blanket of moisture-retentive soil cannot accumulate on slopes. Grains of rock loosened by weathering may be swept away by the next rainfall. Although desert landscapes strongly reflect a deficiency of water, the effects of running water are widely evident, on slopes as well as in valley bottoms (Fig. 18.1). Much of the rain that falls, sparse as it is, encounters soil-free surfaces and runs off immediately, eroding the land.

Desert climates characteristically receive small amounts of precipitation and are subjected to high rates of potential evaporation. Sometimes years may pass without any rain in certain desert areas, though such situations are exceptional. Most desert locations receive some precipitation each year, although the frequency and amounts are highly unpredictable. Rains that do fall are often brief and limited in their coverage, although they can also be intense. Times of rainfall are short, unreliable, and difficult to predict, yet high potential evaporation continues throughout much of the year. The most important impact of rain on landform development in deserts is that when rainfall does occur, much of it falls on impermeable surfaces, producing intense runoff, generating flash floods, and operating as a powerful agent of erosion.

Although running water is a highly effective agent of landform development in deserts today, it only operates occasionally. In most desert regions, running water is active only during and shortly after rainstorms. The importance of water as a gradational agent in arid regions is related not only to the climate of those areas today, but also to past climates.

Paleogeographic evidence indicates that most deserts have not always had the arid climates that exist today. Geomorphologists studying arid regions have found certain landforms that are incompatible with the present climate, and attribute these to the work of water under earlier, wetter climates. A great majority of desert areas were wetter in the past, most recently during the Pleistocene Epoch. While glaciers were advancing in high latitudes and in mountain regions, precipitation increased in the middle and subtropical latitudes where deserts are found today. Cooler climates also meant lower evaporation rates for these regions. Evidence of pluvial (wetter) periods in many of today's deserts include lake deposits, wave-cut shorelines (Fig. 18.2), and immense canyons occupied by streams that today are too small to have eroded such large valleys.

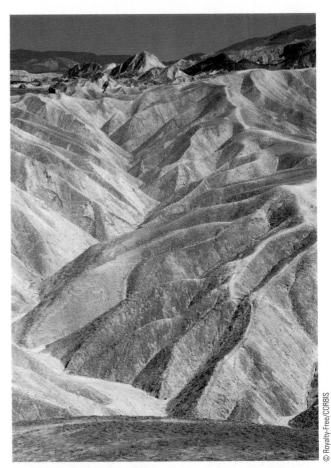

© Royalty-Free/CORBIS

Figure 18.1 In almost all deserts, even in the most arid locations, the effects of erosion and deposition by running water are prominent in the landscape. This is Death Valley, California.
■ **Why do you think there is a high drainage density here?**

Figure 18.2 Many desert areas in Nevada, Utah, and California, have remnant shorelines that were created by wave action in lakes during the Pleistocene. The Great Basin region of the Western United States contained many such lakes at that time.

■ **What can we learn about climatic change from studying these relict lake features?**

Most desert streams are *ephemeral;* that is, they flow only during and shortly after a particularly rainy period. The rest of the time, the streambed is exposed and dry. There is no groundwater input to sustain a *base flow* between periods of surface runoff. Ephemeral streams generally lose water volume by seepage into the ground rather than gaining volume from groundwater inflow, as do most streams in humid regions.

Many desert streams do not receive enough discharge to flow to the sea before evaporating or infiltrating into their dry beds. The mountains that have contributed to aridity through the rain shadow effect can also block desert streams from flowing to the ocean, and the streams terminate in interior depressions to form temporary lakes. These, too, eventually evaporate and disappear, then reappear when rains provide adequate inflow. In arid regions, however, many closed basins were filled with water to form lakes during the cooler and wetter times of the Pleistocene Epoch.

Where surface runoff drains into closed desert basins, sea level does not govern erosional base level, as it does in streams that flow into the ocean. Low areas characterized by this pattern of ephemeral streams that are controlled by a local base level are known as **interior drainage basins** (Fig. 18.3). When deposition raises the surface of the basin, local base level also rises. If tectonic activity lowers the basin floor, local base level is also lowered, and stream erosion is rejuvenated. Some basins of interior drainage have floors below sea level, as in Death Valley, California, the Dead Sea Basin, the Turfan Basin in western China, and Australia's Lake Eyre (see Map Interpretation: Desert Landforms).

Figure 18.3 The Sierra Nevada (top of photo) posed a topographic barrier to streams, even during the wetter times of the Pleistocene so that only a few flowed to the sea. The others filled basins to form lakes, most of which are completely dry today. This image shows the now dry bed of Owens Lake, which shrank because of climatic change and was desiccated as its waters were diverted to urban areas in Southern California.

DESERT LANDFORMS

The Map

The map shows a section of Death Valley National Park, California. This is part of the Basin and Range region, characterized by fault-block mountains (ranges) separated by down-faulted valleys (basins). As the rugged

A digital terrain model shows the topography of Death Valley. Figure 18.18 presents an alternate view with a topographic profile (black foreground) at about 3× vertical exaggeration.

NASA/JPL/CACR

ranges erode, the basins fill with debris carried by infrequent flash floods.

The mountain block that forms the western slope of Death Valley (in the map area) is the Panamint Range. The highest summit in that range, Telescope Peak, reaches an elevation of 3368 meters (11,049 ft). The Amargosa Range forms the eastern boundary. Death Valley's lowest elevation is −86 meters below mean sea level (−282 ft).

The present climate of the Basin and Range is arid, except for high-elevation mountains that receive more precipitation. With an average annual rainfall of 3.5 centimeters (1.7 in.) and a potential evaporation that may exceed 380 centimeters (150 in.), Death Valley has the most extreme aridity in the region. Summer temperatures commonly exceed 40°C (104°F), and the record maximum is 57°C (134°F). In winter, much of the Basin and Range region has freezing nighttime temperatures.

During the Pleistocene, lakes several hundred feet deep filled numerous basins in the region. Death Valley was occupied by Lake Manly, and traces of this 183 meter (~550 ft) deep lake's shoreline can be seen from the valley floor. This is a classic site for studying desert landforms.

Interpreting the Map

Note: In answering these questions it will be helpful to refer to both the figures and text on pages 498–503.

1. Based on the general location of Death Valley, why is the area so arid?

2. Describe the general topography of the map area.

3. What is the lowest elevation on the map? What is significant about the elevation of Death Valley?

4. What specific type of arid landform is the blue striped feature located in the depression? Why are the edges of this feature shown with blue dashed lines?

5. What types of surface materials compose the feature in Question 4? Explain how this feature was formed.

6. What specific type of landform is indicated on the map by the curved contours at the base of the mountains? What do the blue ------------- lines crossing the contours represent?

7. What types of surface materials compose the feature in Question 6? Explain how this feature was formed.

8. Note that the large, curved landforms at the mountain base coalesce. What is the specific landform name for such broad alluvial features?

9. What evidence from the map indicates that this is an interior drainage basin?

10. Sketch a general east-west profile from the benchmark (elevation −270 ft, at Devil's Speedway to the 2389-foot benchmark on the mountain straight to the west. Label the following landforms: mountain front, pediment, alluvial fan, and basin.

Opposite:
Furnace Creek, California
Scale 1:62,500
Contour interval = 80 ft
U.S. Geological Survey

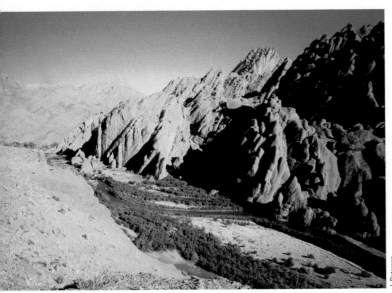

Figure 18.4   A stream flows through a deep gorge in the Atlas Mountains of Morocco. This is the arid, rain-shadow side of the mountains, facing the Sahara to the east. This stream loses water by infiltration and evaporation and disappears into the Sahara. Note the steeply dipping, folded rocks of the Atlas and the thin line of vegetation along the stream channel. ■ **Was the gorge eroded by the stream with this amount of flow? If not, what factors might have produced more discharge to erode the deep canyon?**

Most streams found in deserts originate in nearby humid regions or in cooler, wetter mountain areas. Many streams in arid regions, however, have insufficient discharge to sustain flow across a large desert (Fig. 18.4). Without inflow provided by tributaries or groundwater to replenish water losses caused by evaporation and underground seepage, the stream flow dwindles and finally disappears. The Humboldt River in Nevada is an outstanding example; after flowing 465 kilometers (290 mi), the river disappears into the Humboldt Basin, a closed depression. Only a few large rivers that originate in humid uplands have sufficient volume to survive the long journey across hundreds of kilometers of desert to the sea (Fig. 18.5). Called **exotic streams,** these rivers erode toward a base level governed by sea level and provide drainage that is external to the arid region. Classic examples of exotic streams include the Nile (Egypt and Sudan), Tigris-Euphrates (Iraq), Indus (Pakistan), Murray (Australia), and Colorado (United States and Mexico) Rivers.

Water as a Gradational Agent in Arid Lands

When rain falls in the desert, sheets of water run down unprotected slopes, picking up sediment. Channels become filled with flooding streams of muddy water. The material removed by runoff and surface streams is transported, just as in humid lands, until velocity and/or vol-

ume of stream flow decreases sufficiently for deposition to occur. Eventually these streams disappear because seepage and evaporation losses exceed their flow. Huge amounts of sediment can be deposited along the way as the stream loses volume and velocity. The processes of erosion, transportation, and deposition by running water are essentially the same in both arid and humid lands. However, the resulting landforms differ because of the intermittent nature of desert runoff, the presence of interior drainage that does not flow to the sea, and the lack of vegetation to protect surface materials against rapid erosion.

Fluvial Landforms of Arid Environments

Among the most common desert landforms created by surface runoff and erosion are the channels of ephemeral streams. Created by rushing surface waters, these vertical-

Figure 18.5   A false-color satellite image of the Nile River meandering across the Sahara in Egypt. The dark-red irrigated croplands contrast with the barren desert terrain. The Nile is an exotic stream. Its headwaters are in the wetter climates of the Ethiopian Highlands and lakes in the east African rift zone, which support its northward flow across the Sahara to the Mediterranean Sea. ■ **Can you name three other exotic streams?**

R. Sagler

Figure 18.6 Dry streambeds such as this one in southern Utah are caused by runoff channeled after rains in arid regions. These stream-cut channels are called arroyos in the southwestern United States. ■ **Why do desert stream channels have a high flash-flood risk?**

walled stream channels, usually cut into unconsolidated alluvium, are called **arroyos** or **washes** in the southwestern United States, **barrancas** in Latin America, and **wadis** in North Africa and southwest Asia (Fig. 18.6). Arroyos are prone to flash floods that make them potentially dangerous sites for desert camping. Though it may sound bizarre, many people have drowned in the desert during flash floods.

Where weak, easily erodible clays and shales underlie steep slopes, rapid runoff produces a dense network of barren ridges dissected by a maze of V-shaped gullies and dry ravines. Early fur trappers in the Dakotas called such areas "bad lands to cross" (Fig. 18.7). The phrase stuck, and those regions are still called the Badlands. Similar areas with rugged, barren, and highly dissected **badlands** topography with extremely high drainage density can be

R. Gabler

Figure 18.7 The Badlands of South Dakota. Impermeable clays that lack a soil cover produce rapid runoff, leading to intensive gully erosion and a high drainage density. ■ **Was this rugged terrain named appropriately?**

Watersheds As Critical Natural Regions

Perhaps the most environmentally logical division of Earth's land surface is the watershed, the drainage basin for a stream or river system. Humankind has a vital responsibility to monitor, maintain, and protect the quality of freshwater resources. Surface water from watershed sources provides much of the potable water resources for the world's population. As the population grows, the need for fresh water increases, as does the human impact on these resources.

In recent years many government agencies have designated watersheds as critical regions for environmental management, and have worked to map the drainage divides that form the boundaries for these areas. There are several sound reasons for this spatial strategy. Watersheds are not only natural regions, but they also function as spatially coherent environmental systems.

Virtually all of Earth's land surface forms part of a watershed. Along with the drainage network, stream, or river that occupies these basins, watersheds form well-integrated natural systems that involve water quality, soil, rock, terrain, vegetation, wildlife, domesticated animals, and humans. The environmental processes and physical components of a watershed are strongly interrelated. Problems in one part of the system are likely to cause problems elsewhere.

Complex biotic habitats exist in these basins and they are greatly affected by the environmental quality of the watershed. Increasing human populations and land use intensities in previously natural drainage basins place pressures on these habitats and the water quality in the stream

seen in Death Valley National Park, California, and Big Bend National Park, Texas. Badlands generally do not form naturally in humid climates because the vegetation there slows runoff and erosion, decreasing drainage density. Removing the vegetation from clay or shale areas by overgrazing, mining, or logging, however, can cause badlands topography to develop even in humid areas.

Plateaus are regions of low relief and high elevation, generally dominated by a structure of horizontal rock layers. Many striking plateaus exist in the deserts and semiarid regions of the world. An excellent example in the United States is the Colorado Plateau, centered on the Four Corners area of Utah, Arizona, New Mexico, and Colorado. In tectonically uplifted desert plateau regions such as this, streams and their tributaries respond to uplift by cutting steep-sided canyons. Where the canyon walls consist of horizontal layers of alternating resistant rocks and erodible rocks, differential weathering and erosion exert strong influence on the canyon walls. Canyons in these areas tend to have stair-stepped walls, with near-vertical cliffs marking the resistant layers (ordinarily sandstone, limestone, or basalt) and weaker rocks (often shales) forming the slopes. The distinctive walls of the Grand Canyon have this appearance, which exposes the structure of horizontal rock layers of varying thickness and resistance (Fig. 18.8). The rim of the Grand Canyon is a flat-topped cliff formed by a **caprock,** a term that refers to a resistant horizontal layer that forms (caps) the top of a landform. Caprocks not only form the rims of canyons; they also form the summits of other kinds of flat-topped landforms that, although they are found in many climatic regions, are more characteristic of deserts.

Figure 18.8 The earliest explorers of the Grand Canyon took along an artist to record the geomorphology of the canyon, here beautifully shown in great detail. Aridity creates an environment where the bare rocks are exposed to our view; differential weathering and erosion form the stair step quality to the walls of the Grand Canyon.

U.S. Geological Survey

system. Monitoring and managing watersheds requires an interdisciplinary effort that considers aspects of all four of the world's major spheres.

Watersheds share the hierarchy of stream numbers and can be subdivided into smaller sub-basins for local studies and management. Watersheds of large river systems can also be studied in a broader regional scale. Water quality can be monitored for the whole system for any sub-basin. Streams form a network of flow in one direction (downstream), and problems with pollution can be located by tracing them upstream.

Watersheds are clearly defined, well-integrated, natural regions of critical importance to life on Earth, and they make a logical spatial division for environmental management. Unfortunately, rarely do political and administrative regions coincide with the divide that defines the edge of a watershed. A stream system may flow through many counties, cities, states, and governmentally administered lands, and perhaps through more than one country. Each of these jurisdictions may have very different needs, goals, and strategies for using and managing their part of a watershed and often a variety of strategies results in direct conflict. Yet cooperative management of the watershed system as a whole is the best approach. The United States Environmental Protection Agency of the government advocates and urges this management approach, and many cooperative river basin authorities have been established to encourage a united effort to protect a shared watershed. This watershed-oriented management strategy, based on a fundamental natural region, is an important step in protecting our freshwater resources.

Weathering and erosion will eventually reduce the extent of a caprock until only flat-topped, steep-sided **mesas** remain (*mesa* means "table" in Spanish). Mesas are relatively common landscape features in the Colorado Plateau region. Through additional erosion of the caprock from all sides (Fig. 18.9), a mesa may be reduced to a smaller flat-topped hill, an erosion remnant, called a **butte.** Variations in the hill slope form of buttes, mesas, and plateaus are related to the height of the cliff at the top, which is controlled by the thickness of the caprock in comparison to the size of landform feature. Mesas and buttes in a landscape are generally evidence that uplift occurred in the past and that weathering and erosion has been extensive since that time. Monument Valley, in the Navajo Tribal Reservation on the Utah-Arizona state line, is an exquisite example of such a landscape formed with a caprock that is particularly thick, contributing to the distinctive scenery (Fig. 18.10). Many famous western movies have been filmed in Monument Valley and in nearby areas of the Colorado Plateau because of the striking, colorful, and photogenic desert landscape.

Erosion of slopes fringing a desert basin or plain generally results from sheetwash and gully development. As these slopes gradually undergo erosional retreat (aided by weathering), a gently sloping surface eroded into bedrock, called a **pediment,** is formed (Fig. 18.11). Characteristically in desert areas, there tends to be a sharp break in slope at the base of steep hills or mountains, which rise at angles of 20 to 30 degrees or steeper, and the gentle pediment, whose slope is usually only 2 to 7 degrees.

Geomorphologists do not agree on how pediments are formed, perhaps because different processes may be responsible for their formation in various world regions. However, there is general agreement that most pediments are erosional surfaces created by the action of running water. In some areas, however, weathering may also have played a strong role in the development of pediments.

Plateau

Mesa

Butte

Figure 18.9 Diagram of landforms developed through weathering and erosion in an area of horizontal rock layers with a resistant caprock (such as the Colorado Plateau of Arizona, Colorado, New Mexico, and Utah)

Figure 18.10 Monument Valley, Arizona, with prominent buttes and mesas. The caprock here is particularly thick, and represents a rock layer that once covered the entire region. The buttes and mesas are erosional remnants of that layer. ■ **Compare this photo to the diagram in Figure 18.9. How were these landforms produced?**

R. Sager

Landforms of Arid Fluvial Deposition

Deposition is as important as erosion in creating landform features in arid regions, and in many areas the deposition by water does as much to level the land as does erosion. Many desert areas have wide expanses of alluvium deposited either in closed basins or at the base of mountains by streams as they lose water in the arid environment. As the flow of a stream diminishes, so does its capacity—its ability to transport load. Most landforms developed by fluvial deposition in arid lands are not exclusive to desert regions, but are common and visible in any dry environment that is barren of soil and sparse in vegetative cover.

Alluvial Fans   Where streams, particularly ephemeral ones, flow out of narrow canyons and onto open plains, their channels may flare out to become wide and shallow. The greater external friction of the shallow channel reduces stream velocity and carrying power. Discharge also decreases

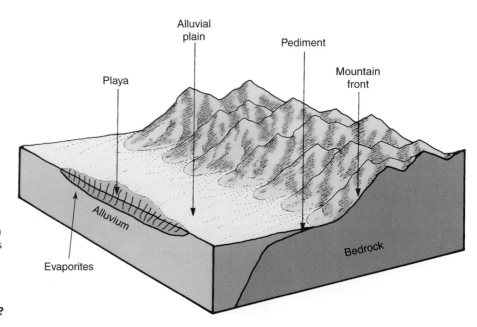

Figure 18.11 Pediments are erosion surfaces cut into bedrock at the bases of mountain slopes in arid regions. At many locations, pediment surfaces are covered with a veneer of alluvial deposits. ■ **How can a pediment be distinguished from an alluvial fan?**

Alan L. Mayo, GeoPhoto Publishing Co.

Figure 18.12 Alluvial fans form at the base of a mountain range where streams generally lose their carrying power. The apex of a fan forms at the mouth of the canyon or wash where the stream encounters more gently sloping land. Fans are particularly common landforms in the arid Basin and Range region of the western United States. For scale, note that a road runs across the lower portion of this alluvial fan in Death Valley, California. ■ **Where would the coarser deposits be found on an alluvial fan?**

as water seeps from the channel into coarse alluvium below. As a result, most of the sediment load carried by such streams is deposited along the base of the highland.

Upstream, the channel is constrained by bedrock valley walls, but as it flows out of the canyon, the channel is free to shift its position. As sediments are deposited on the plain area, the channel, acting much like a braided stream, shifts, subdivides, and continues to deposit alluvium. The location where a stream exits the canyon mouth serves as a pivotal point as the channels shift their location more downslope and away from the canyon. Deposition takes place radially away from that point to create a depositional landform with the shape of an Oriental fan (or sometimes a cone), called an **alluvial fan** (Fig. 18.12).

An important characteristic of an alluvial fan is the sorting of sediment that typically occurs on its surface. Coarse sediments like boulders and pebbles are deposited near the fan's apex, where the slope is steepest. The alluvium becomes increasingly finer downslope on the fan surface. Alluvial fans tend to develop in arid regions because of intermittent flow and heavy sediment loads carried during brief periods of runoff. In some regions, alluvial fans are also composed largely of debris-flow or mudflow deposits, which are not sorted like the alluvium deposited by running water.

Alluvial fans are particularly common where ephemeral streams laden with debris flow out of a mountainous region onto arid plains or into interior basins. In the western United States, alluvial fans are a major landform feature in landscapes consisting of fault-block mountains and basins, as in the Great Basin of California, Nevada, and Utah (Fig. 18.13). Here streams periodically rush from

canyons cut into uplifted fault-block mountains and deposit their load of alluvial debris in the adjacent basins.

Alluvial fans develop in such environmental settings for several reasons. First, highland areas in desert regions are subject to intense erosion, primarily because of their thin vegetative cover, steep slopes, and the orographically intensified downpours that can occur over mountains. In addition, streams in arid regions typically carry a larger sediment load (in comparison to the discharge) than comparable streams in more humid regions. As the streams flow from confined mountain canyons into desert basins, they deposit most of their coarse sediment near the canyon mouth. Flowing into the desert basin, width increases; their depth decreases, and their volumes are significantly reduced through infiltration. Not far from the canyon, the stream itself may disappear. (Note: In humid areas, streams generally have enough discharge to continue flowing in restricted valleys after leaving highland areas. Perennially flowing streams also tend to transport sediment to a delta. As a result, most highland streams that flow into the lower lands of humid regions do not create large alluvial fans.)

Where a veneer of alluvium has been deposited on a pediment, the land surface may closely resemble a water-deposited alluvial fan. In some situations, it may not be possible to determine the existence of an underlying pediment without either excavating through the surface alluvium or finding the erosional pediment surface exposed in the walls of washes or gullies. In locations where no extensive pediment exists, alluvial deposits beneath the fans can be tens or even hundreds of meters thick. In contrast, the layer of alluvium overlying a pediment is

Fig. 18.13. Map of the Great Basin of the western United States. More than 100 lakes formed in the fault block basins of this region during the Pleistocene Epoch. ■ **Why are there only a few remnant lakes in this region today?**

only a relatively thin layer, no more than a few meters deep and sometimes much less. Along the bases of mountains in arid regions, adjacent alluvial fans may become so large that they join together to form an undulating, ramplike surface called a **bajada** (Fig. 18.14). A bajada consists of adjacent alluvial fans that have coalesced to form an "apron" of alluvium around the moun-

tain base. Where the associated streams are large, fans tend to be less steep, less permeable, and larger in area. Where extensive fans coalesce over wide areas, they form a **piedmont alluvial plain,** like the area surrounding Phoenix, Arizona (Fig. 18.15).

Piedmont alluvial plains generally have rich soils and the potential to be transformed into productive agricul-

R. Sager

Figure 18.14 A bajada is formed when a series of alluvial fans join, forming a continuous alluvial slope along the front of an eroding mountain range. This example is from the Mojave Desert of California. ■ **Why would a series of alluvial fans have a tendency to eventually join to form a bajada?**

tural lands. The major obstacle is inadequate water supply to grow crops in an arid environment. In many world regions, arid alluvial plains are irrigated with water diverted from mountain areas or obtained from reservoirs on exotic streams. The alluvial farmlands near Phoenix are a good example, producing citrus fruits, dates, cotton, alfalfa, and vegetables.

Playas Desert basins of interior drainage surrounded by mountains are sometimes called **bolsons.** Most bolsons were formed as faulting uplifted the fringing highlands to create a basin in between the up-faulted mountains. A normally dry lake bed, known as a **playa,** commonly occupies the lowest part of a bolson. Floods occasionally transform a playa into a shallow pond, called a **playa lake,** in the space of a few hours. The lake may be gone the next day, or it may persist for weeks (Fig. 18.16). Playa lakes lose most of their water to evaporation in the desert air, although seepage may play a role. Repeated cycles of inflow and evaporation can leave behind thick mineral deposits crystallized from evaporating brines: chlorides, sulfates, and carbonates (for example, rock salt, gypsum, and lime, respectively). When the playa lake dries out, its bed may be encrusted with sparkling white salt deposits, although some playas form a salt crust from groundwater loss through surface evaporation. Saline playas are known as **salt flats** or **salinas** (Fig. 18.17). Most playas have a clay surface, baked hard by the desert sun when it is dry but extremely sticky and

slippery when it is wet. Playas are among the flattest landforms on Earth.

Playas are useful in several ways. For one, companies mine the rich deposits of evaporite minerals, including such important industrial chemicals as potash, salt, borates, and sodium nitrate, that have been deposited in playa beds. Also, the extensive, flat surfaces of some playas make them suitable as racetracks and airstrips. Utah's famous Bonneville Salt Flats are the bed of an

NASA/Goddard Space Flight Center Scientific Visualization Studio

Fig. 18.15 Extensive desert alluvial plain in Arizona

Figure 18.16   Badwater, in Death Valley, California. The playa surface, seen just behind the edge of the water, is the lowest elevation in North America at 86 meters (282 ft) below sea level. Badwater, the small lake, is fed by groundwater that flows down from the surrounding mountains into alluvial deposits under the surface of the basin. The water quickly evaporates in this extremely arid environment.   ■ **Why was this small lake called Badwater?**

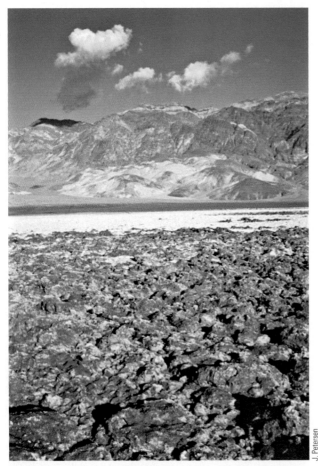

Fig. 18.17   This view of Death Valley shows deposits of salt, left behind by evaporation of both groundwater and runoff from the surrounding mountains. As the water evaporates in this arid environment, it leaves minerals behind to accumulate on the floor of the basin.

extinct Pleistocene lake. The western portion of the Salt Flats, where world land speed records are set, still floods to a depth of one or two feet in the cool, wet, winter season. The hard, flat playa surface at Edwards Air Force Base, in California's Mojave Desert, serves as a landing site for the space shuttle. Occasionally this playa also floods, and the shuttle must be diverted to an alternate landing location.

Landscape Development in Deserts

Landscape development in arid climates is comparable in many ways to that in humid climates. Similar processes may be operating on similar materials, but these processes operate at much different rates in humid environments compared to arid regions. Relationships between weathering and these two widely differing climatic regimes have already been discussed; similar relationships exist between gradation processes, resulting landforms, and the two climatic extremes. The major differences in the results of gradation in arid climates, as compared to humid climates, are caused by the great expanses of exposed bedrock, a lack of continuous water flow, and a more active role of the wind in arid regions.

An excellent example of desert landscape development based on uplift and associated gradation is found in the Basin and Range region of the western United States. The region extends from west Texas to eastern Oregon; it includes all of Nevada and large portions of New Mexico, Arizona, Utah, and eastern California. Here, more than 200 mountain ranges, with basins be-

Figure 18.18   A false-color satellite image of Death Valley, California, which is a desert bolson in the Basin and Range region. Death Valley lies between the Amargosa Range to the northeast and the Panamint Range to the southwest. Alluvial fans dominate the edges of the valley.
■ **What do you think the white areas are in the center of the valley?**

tween them, dominate the topography. Much of this area is part of the Great Basin subregion, characterized by interior drainage and centered on Nevada (Fig. 18.18). No rivers flow to the sea in the Great Basin of the western United States. However, other areas also have interior drainage, such as the Coachella Valley in southern California.

Uplifted fault-block mountains rise thousands of feet above the desert basins, and many form continuous ranges. These high ranges, such as the Guadalupe Mountains, Sandia Mountains, Warner Mountains, and the Panamint Range, encourage orographic rainfall. Fluvial erosion dissects the mountain blocks to form peaks and interfluves between arroyos and canyons. Pediments develop along the base of the mountains (Fig. 18.19a). Fluvial deposition builds alluvial fans at the base of the range. As time passes, if uplift is slow or ceases, erosion and deposition subdue relief, resulting in lower mountains, broader pediments, and increased valley filling as alluvial fans coalesce into bajadas (18.19b).

Many areas of the Basin and Range region, and its spectacular counterparts in western China and Iran, are so tectonically active that uplift exceeds erosion. The landscapes of the Mojave Desert in California and parts of the Sonoran Desert in Arizona, however, have localities where uplift along faults has been inactive long enough for erosion and deposition to dominate the landscape. Under these conditions, relief is reduced, and the landscape becomes dominated by extensive desert plains and wide bajadas interrupted by a few isolated, resistant erosion remnants called **inselbergs** (island mountains) as reminders of earlier, tectonically active, mountainous landscapes (18.19c). The ancient and geologically stable deserts of Australia and the Sahara also have areas of inselbergs surrounded by extensive desert plains (Fig. 18.20).

Although fault-block mountains and fault-block basins dominate the geologic structure of immense regions of the American West and other arid locations around the world, it is important to know that desert landscapes are as varied as those of other climatic environments. Deserts exist at localities where the landscape has developed in nearly every imaginable geologic setting, including volcanoes, ash deposits and lava flows, folded rocks forming ridges and valleys, horizontal strata, and exposures of massive intrusive rocks. The geologic structures and geomorphic processes found in desert areas are, for the most part, the same as those found in humid regions. It is variations in the effects and rates of these processes that make the desert landscape distinctive.

Wind As a Gradational Agent

On a worldwide basis, wind is a less effective force than running water, waves, groundwater, moving ice, or mass movement in leveling the land. Under certain conditions, however, wind can be a significant agent in the modification of topography. Landforms—whether in the desert or elsewhere—that are created by wind (Fig. 18.21) are called **eolian** landforms (after Aeolus, the god of winds in classical Greek mythology). Note that outside of North America, *eolian* is usually spelled *aeolian*.

The three primary conditions necessary for wind to become an effective gradational agent are a sparse vegetative cover, the presence of dry, loose materials at the surface, and a wind velocity that is high enough to pick up and move those surface materials. These three conditions are most widespread in arid regions and on beaches, though they are also found on, or adjacent to, dry lake beds, areas of recent alluvial or glacial deposition, newly plowed fields, and overgrazed lands.

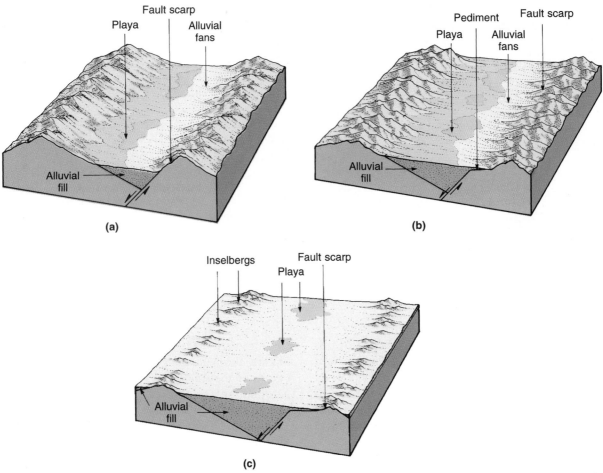

Figure 18.19 An idealized sequence of desert landscape evolution in an arid region of fault-block mountains and intervening basins, such as the Basin and Range region. (a) Alluvial fans develop at the base of uplifted fault-blocks as stream downcut and erode the ranges. (b) After uplift ceases, the range fronts erode back, producing pediments, and the basins continue to fill with eroded material, forming bajada and playa surfaces. (c) Eventually, over millions of years, erosion may reduce the mountains to isolated remnants called inselbergs, which rise above flat desert plains.   ■ **Can you locate some inselbergs on the image of the Phoenix area in figure 8.15?**

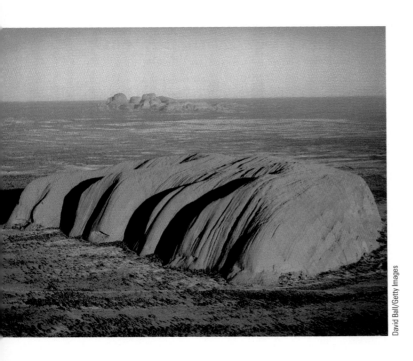

A dense vegetative cover reduces wind velocity near the surface, absorbs the wind's force, prevents wind from being directed against the land surface, and holds materials in place with its root network. Without such a protective cover, fine-grained and sufficiently dry surface materials are subject to removal by strong gusts of wind. If surface particles are damp, they tend to adhere together in wind-resistant aggregates, but the arid conditions of deserts make these regions most susceptible to wind erosion.

Figure 18.20 Ayers Rock is a striking sandstone inselberg, an erosional remnant, rising above the arid, flat interior of the Australian Outback. The Olgas, in the background, formed as another erosional remnant divided into many bedrock hills by the widening of joints.   ■ **Explain how inselbergs form.**

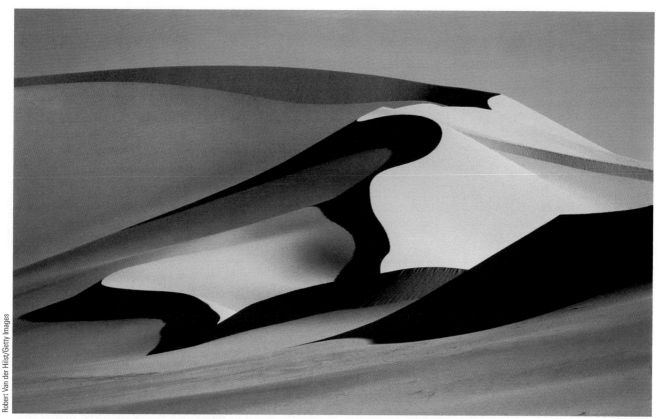

Robert Van der Hilst/Getty Images

Figure 18.21   This view of sand dunes in Africa's Namib desert shows how eolian processess can create a stunningly beautiful landscape.
■ **How does the process that formed these dunes differ from the process that forms an alluvial fan?**

Many eolian processes operate in ways that are similar to those of running water. Running water and wind detach and remove materials by comparable means. Both gradational agents transport material by traction, saltation, and suspension. However, unlike water, the wind cannot transport material in solution. Furthermore, wind has hardly any lateral or vertical limitations on movement. The result is that the dissemination of material by the wind can be far more widespread and unpredictable than that by streams.

Another similarity between the gradational properties of wind and running water is that flow velocity controls the size of particles each can pick up and carry (their competence). However, because air is a much less dense fluid medium than water, its competence is generally limited to only sand-size or finer material. Wind erosion selectively transports finer particles and leaves behind coarser or heavier particles that the wind is incapable of lifting. Also, similar to water deposits, wind deposits are stratified according to changes in velocity.

Wind Erosion

Strong winds frequently blow through arid regions, whipping up loose surface materials and transporting them within turbulent air currents. Wind removes or erodes surface materials by two main processes (Fig. 18.22). The first of these is **deflation,** which is similar to the hydraulic force of running water. As wind velocity increases, the first loose particles to be affected are the finest ones present at the surface, essentially fine dust. The finest particles transported by winds are carried in suspension, buoyed by vertical currents. Such particles will remain in suspension as long as the strength of upward air currents exceeds the tendency of particles to settle out to the ground. If the wind velocity surpasses 16 kilometers (10 mi) per hour, surface sand grains will be put into motion. Particles that are too large to be carried in suspension are bounced along the ground by the transportation process of saltation. As these particles bounce along, they dislodge other particles that are then

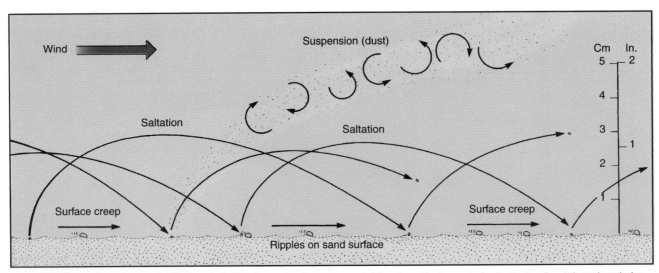

Figure 18.22 The movement of sand by wind saltation and surface creep resulting from the impact of saltating grains. When the wind is strong, the forward ripple movement of sand is rapid enough to be seen. ■ **Why are only sand-sized or smaller particles moved by the wind?**

added to the wind's load and driven forward on the ground as surface sand creep. The particles carried in suspension by the wind make up its suspended load; the particles that bump and bounce along the ground make up the wind's bed load. The velocity required to pick up and start to move a particle is greater than that required to keep it moving, partly because of inertia and partly because impacts with particles that are already moving will set other particles in motion.

The second way that the wind erodes is by **abrasion.** This process is analogous to abrasion produced by the transported load of streams, breaking waves, and glacial ice, but it operates on a much more limited scale. In order for the wind to remove surface materials by abra-

sion, it must carry sediment that is harder than the surface rocks. As these wind-blown sediments hit a mass of weaker material, they erode the surface. Quartz sand is a very effective and common abrasive agent in eolian processes. Yet sand grains are relatively large and heavy for the wind to move and rarely are lifted higher than 1 meter (3 ft) above the surface. Consequently, the effect of natural sandblast is limited to a zone close to ground level.

Where loose dust particles exist on the land surface, they will be picked up and carried in suspension by strong winds. The result is a thick, dark, swiftly moving dust cloud swirling over the land. **Dust storms** (Fig. 18.23) can be so severe that visibility drops to nearly zero and almost all sunlight is blocked. They can also be highly

Figure 18.23 Dust storms occur when strong winds mobilize, erode, and suspend silt particles, picked up in areas of barren soil or alluvium. This is a dust storm in Texas during the "Dust Bowl" times of the 1930s.

Figure 18.24 Desert pavement. Removal of fine materials by wind and unchanneled running water leaves a surface of pebbles and rock fragments, forming a desert pavement. ■ **Is desert pavement a surface indestructible to human activities? Why?**

destructive, removing layers of surface materials and depositing them elsewhere, sometimes in a thick, choking layer, all within a matter of a few hours. **Sandstorms** occur in areas where sand is abundant at the surface. Because sand grains are larger and heavier than dust particles, most sandstorms are confined to a low level near the surface. Evidence of the restricted height of desert sandstorms can be seen on vehicles that have traveled through the desert, as well as on fence posts, utility poles, and other structures. After a sandstorm, the pitting, gouging, and abrading effects of natural sandblast are more damaging to the lower portions of the vehicle, or whatever object the sand was abrading.

Erosion by deflation can produce depressions in a barren surface of unconsolidated materials. These depressions, which can vary in size from a few centimeters to a few kilometers in diameter, are called **deflation hollows** or **blowouts.** In the Kalahari Desert, deflation hollows collect rainwater, attracting animals and their hunters, the Bushmen. Often blowouts form where there was already a slight depression in the surface, such as in silty playa deposits.

Deflation is a factor in producing a close-fitting surface mosaic of rock fragments called **desert pavement** (**reg** in North Africa, **gibber** in Australia), which is common in many desert regions. Because erosion by deflation is selective, smaller clay and silt particles from a surface of mixed particle sizes are selectively eroded and transported to another location, and sometimes even the coarser, heavier sand grains are removed as well. This selective removal leaves behind the largest particles, pebbles and rock fragments, which together form the cobbled surface of desert pavement (Fig. 18.24). This erosional feature is widespread in parts of the Sahara, interior Australia, the Gobi in central Asia, and the American Southwest. Some recent research indicates that desert pavement may also be a product of sheet flow, unchanneled running water. Regardless of its origin, desert

pavement is important for the protection it affords the finer material below the surface layer of coarse pebbles and rocks. Pavement formation stabilizes desert surfaces by preventing continuous erosion. Unfortunately, off-road recreational vehicles may disturb this stability, thus damaging desert ecological systems.

Where abrasion or sandblasting is at work on rocks of varying resistance, differential erosion etches away softer rocks while the more resistant rocks remain. Eolian abrasion also produces **ventifacts** (wind-fashioned rocks). A ventifact is a rock fragment that has been trimmed flat on one or more sides by sandblast. Flat faces called facets, which meet along sharp edges, are produced. This highly distinctive rock form is an indication of sandblast. Although not extremely common, ventifacts are plentiful where wind and surface rock conditions are ideal for their formation.

Another feature often attributed to wind abrasion is the pedestal or balancing rock, commonly and incorrectly thought to form where sandblast attacks the base of an individual rock so that the larger top part appears balanced on a thinner pedestal below. Actually, such forms result from various weathering processes in the damper environment at the base of an exposed rock and are not typically related to sandblast (Fig. 18.25).

Where the land surface is exposed bedrock, wind abrasion can polish and groove the rock surface but will not significantly erode it. Rates of eolian erosion in arid regions depend primarily on the strength of the materials exposed. Only where rocks are soft or weak can abrasion and deflation be very effective.

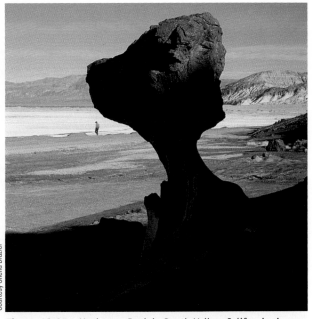

Figure 18.25 Mushroom Rock in Death Valley, California, is a pedestal rock caused by desert weathering processes. ■ **What would account for such an unusual shape?**

Wind Deposition

All materials transported by wind are deposited some-where, generally in a distinctive manner that is related to characteristics of the wind as well as nature and size of the deposits. Coarser, sand-sized material is often de-posited in drifts in the shape of hills or ridges, called **dunes.** Finer silty materials, called **loess,** can be transported long distances from their source area, blanketing and sometimes modifying the existing topography (Fig. 18.26).

Sand Dunes   To many people, the word *desert* evokes the image of endless sand dunes, blinding sandstorms, a blazing sun, mirages, and an occasional palm oasis. Al-though such deserts do exist, particularly in Arabia and North Africa, most deserts have rocky or gravelly sur-faces, scrubby vegetation, and few or no sand dunes. Nevertheless, sand dunes are certainly the most spectac-ular features of wind deposition, whether they occur as seemingly endless dune regions called "sand seas," or *ergs,* in the Sahara or as hills behind a beach in Oregon or on Cape Cod (Fig. 18.27).

Dune topography is almost infinite in its variation. For instance, dunes in the great ergs of the Sahara and Arabia look like rolling sea waves. Others have aero-dynamic crescent forms. Eolian sand deposits can also form "sand sheets," with no dune formation at all. Laboratory experiments indicate that the different dune forms result from the amount of sand available, the strength and di-rection of dominant winds, and the amount of vegeta-tive cover. As wind-carrying sand encounters surface obstacles or topographic obstructions that decrease its velocity, sand is deposited and piles up in drifts. These sand piles further decrease wind velocity, so that the dunes grow

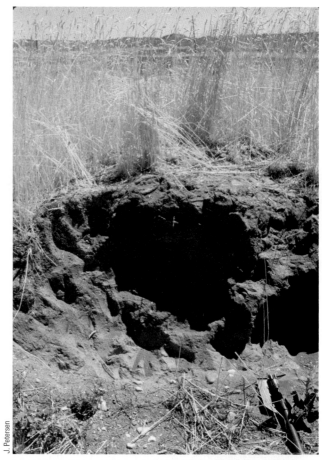

J. Petersen

Figure 18.26   Loess, seen here near Ogden, Utah, is wind-deposited silt.

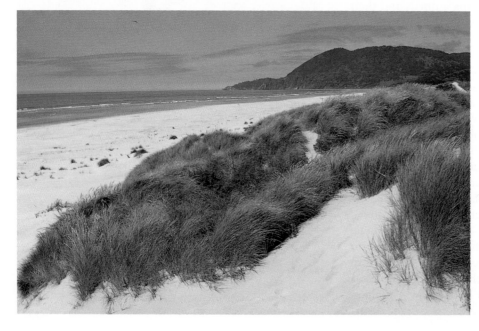

Figure 18.27   Dunes along the Oregon coast. ■ **Why are coast-lines such good locations for dune formation?**

larger until equilibrium is reached between dune size and the ability of winds to feed sand to the dune.

Sand dunes may be classified as either *active* or *stabilized* (Fig. 18.28). Active dunes change their shape or advance downwind as a result of wind action. Dunes may change their shape with variations in wind direction and/or wind strength. Dunes move forward as the wind erodes sand off of their windward slope. This causes sand ripples to migrate up to the dune crest, and sand is deposited on the steep leeward slope, or **slip face,** which is at the angle of repose for dry sand (about 35 degrees, the steepest slope dry sand can retain without slipping or falling). When wind direction and velocity are relatively constant, a dune can move forward while maintaining its form (Fig. 18.29). The speed at which active dunes move downwind varies greatly, but as with many processes, the movement is episodic; the dune advances only when the wind is strong enough to move sand from the windward to the leeward side. Large dunes travel extremely slowly, but smaller ones may move up to 40 meters (130 ft) a year. During sandstorms, a dune may migrate more than 1 meter (3 ft) in a single day. Some dunes are affected by seasonal wind reversals so that they do not advance, but the crest at the top moves back and forth annually under the influence of seasonally opposing winds.

Stabilized dunes maintain their shape and position over time. Vegetative cover normally stabilizes dunes.

R. Sager

(a)

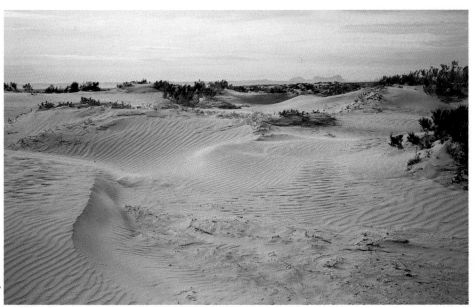

R. Sager

(b)

Figure 18.28 Active and stabilized dunes. (a) Active dunes usually have sharp crests. The gentle back slopes are erosional and face windward, while the leeward advancing slip faces are steep and depositional. The slip face is at the angle of repose for dry sand, as shown by these active dunes. (b) If plants can establish themselves in a dune area, they bring up moisture beneath the dunes, which stabilizes the dune movement. Stabilized dunes tend to have more rounded crests.

■ **Why are active dunes sharper, and stabilized dunes more rounded in form?**

Invasions by Sand Dunes

This chapter discusses the various types and movements of sand dunes within arid regions of Earth. Because sand dunes can play a role in preserving wildlife species and protecting human development, protecting dunes against environmental degradation is important. However, there is also a counterargument to the view of dunes as beneficial. Sand dunes can and do invade delicate environments and human communities, and this invasion of massive amounts of sand can be devastating.

For sand dunes to actively move, three conditions are necessary: a source of sand, winds that are strong enough to move the sand, and a lack of stabilizing vegetation. Depending on the relative balance of these three factors, different types of dunes may result. For example, parabolic dunes can form with a moderate amount of sand but with some vegetation present; longitudinal dunes can exist with minimal sand but strong winds; transverse dunes need abundant sand and little or no vegetation. If sand supply is minimal, winds are weak, and/or there is abundant vegetation, no dunes will form.

In trying to predict if sand dunes will be active or stabilized, research in the Colorado Plateau has shown that comparing precipitation (P) and potential evapotranspiration (PE) rates in a region often holds the answer. As discussed in Chapter 9, arid regions are deficient in moisture overall; that is, the ratio of P to PE is less than 1. The ques-

In places where plants, including trees, lie in the path of an advancing dune, the sand cover may move over and smother the vegetation. Where invading dunes and blowing sands are a problem, attempts are frequently made to plant grasses or other vegetation to stabilize the dunes, halting their advance. Vegetation is able to stabilize a sand dune if plants can gain a foothold and send roots down to moisture within the dune (Fig. 18.30). This task is difficult for most plants, because the sand itself offers little in the way of nutrients or moisture.

One large region of stabilized dunes is the Sand Hills of Nebraska. This area features giant dunes that formed during a drier interglacial period and have been stabilized by a cover of grasses (Fig. 18.31). Similar stabilized dunes are found along the southern edge of the Sahara, where the desert has clearly extended farther toward the equator in the recent geologic past. Both instances involve changes in climate that affected sand supply, wind patterns, and moisture availability (see Map Interpretation: Eolian Landforms).

Types of Sand Dunes Sand dunes are classified according to their shape and their relationship to the wind direction. The different types are also related to the available supply of sand, which affects not only the dune size but also its shape.

Barchans are crescent-shaped dunes (Fig. 18.32a). Their windward slope is the convex curve of the crescent, a gentle slope up which sand is moved. The steeper concave (in map view) leeward slope, or slip face, is at the angle of repose. The crescent's two horns point downwind (Fig. 18.33). Barchans form in areas of low sand supply where winds are strong enough to move sand downwind in a single prevailing direction. Although they form as isolated dunes, barchans often appear in swarms.

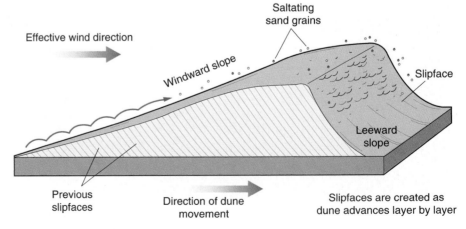

Figure 18.29 Active dunes move downwind. The wind moves sand up the back slope (windward) of the dune toward the sharp crest. The sand then slides down the steeper slip face (leeward) of the dune, causing the dune to advance. Sand supply and wind speed and duration are major factors in the speed of dune advance. ■ **Which way (to the left or right) is this dune moving?**

Effective wind direction

Saltating sand grains

Windward slope

Slipface

Leeward slope

Previous slipfaces

Direction of dune movement

Slipfaces are created as dune advances layer by layer

tion then becomes, is there enough moisture to support stabilizing vegetation? In regions where the P/PE is between 0.3 and 0.5, there is still enough moisture to support sagebrush and grasses at lower elevations and piñon pine and juniper at higher elevations. Thus, at this time, the Colorado Plateau Region, despite its abundance of sand and wind capable of moving sand, has few actively migrating dunes.

Dune migration is more fully quantified by comparing the P/PE ratio with W the percent of time wind is capable of moving sand. The ratio of W, to the P/PE value is referred to as the *dune mobility index.* Between the two factors of this index, W is not as critical as P/PE; thus, a decrease in moisture is the factor that may activate dune mobility in the future. Dunes are therefore more mobile in extremely dry regions and arid regions undergoing severe drought.

There are multiple examples of areas where sand dune migration threatens human development. In areas of Saudi Arabia, teams of workers have dug into shifting dunes in attempts to alter their movement as they invade precious desert oases. These invaluable water sources are vital to desert survival. In southern Peru, dunes from the northern reaches of the Atacama Desert invade cities that border the region. In the Colorado Plateau region, the Navajo and Hopi Indian Reservations lie downwind from large areas of sand dunes. Activation of these sand dunes in the future may disrupt their housing, grazing for their sheep and cattle, and their farming practices. Only the continual monitoring of moisture conditions to identify regions of severe drought can provide warning that massive deposits of sand may again be on the move.

Parabolic dunes are somewhat similar to barchans but have a reverse orientation (Fig. 18.32b). Here, the long points of the curving dune trail behind, pointing upwind, and the slip face has a convex shape when viewed from above. Such dunes often resemble hairpins that are open in the upwind direction. Often their "tails" are stabilized by vegetation. Parabolic dunes most commonly occur along beaches.

Transverse dunes form where winds blow from a constant direction and there is an abundant supply of sand covering the landscape (Fig. 18.32c). Transverse dunes take the shape of a series of troughs and ridge crests whose peaks run perpendicular to the direction of prevailing winds (hence the name *transverse*). These

Robert J. Sager

Figure 18.30 A sand dune invades a vegetated area on Padre Island, Texas. In coastal dune regions, there is almost always constant change as dunes move inland and plants invade the dune area. ■ **Explain how plants can stabilize dunes.**

David B. Loope, University of Nebraska

Figure 18.31 The rolling grazing lands of the Sand Hills in central Nebraska were once a major region of active sand dunes. ■ **Why are these dunes no longer active today?**

EOLIAN LANDFORMS

The Map

Eolian processes formed the Sand Hills region, the largest expanse of sand in North America. The region covers over 52,000 square kilometers (20,000 sq mi) of central and western Nebraska.

The Sand Hills region was part of an extensive North American desert some 5000 years ago. The sand dunes here reached more than 120 meters (400 ft) high and inundated postglacial peat bogs and rivers. As the climate became wetter, vegetation growth invaded the dunes, greatly reducing eolian erosion and transport. The vegetation anchored the sand and the stabilized dunes developed a more rounded form. Underlying the Sand Hills is the Ogallala aquifer. The high water table of the aquifer supports the many lakes nestled between the dunes.

The Sand Hills area has a middle-latitude steppe environment (BSk) and receives about 50 centimeters (20 in.) of precipitation annually. Temperatures have a great annual range, from freezing winters to very hot summers. During summer months, the region is often pelted by hailstorms and thunderstorms; during the winter months, the area is subject to blinding blizzards.

Vegetation is mainly bunch grasses that can survive on the dry, sandy, and hilly slopes. Some species of bunch grasses have extensive root systems that may extend more than 1 meter (3 ft) into the sandy soil. The lakes and marshes in the interdunal valleys support a marsh plant community that, in turn, supports thousands of migratory and local birds. Currently, the main land use in the region is cattle grazing. Some scientists are predicting that the Sand Hills area will lose its protective grass cover if global warming continues and will revert to an active area of migrating desert sand dunes.

Interpreting the Map

1. What is the approximate relief between the dune crests and the interdunal valleys?

2. What is the general linear direction of the dunes and valleys?

3. Which side of the dunes has the steepest slopes?

4. If the slip face is the steepest slope of the dunes, what was the prevailing wind direction when the dunes were active?

5. Based on your answers to the previous three questions, determine what type of sand dune formed the Sand Hills.

6. What is the general direction of groundwater flow in the aquifer beneath the Sand Hills? (*Note:* Use the elevation of the lakes to determine the water table elevation.)

7. Sketch a north-south profile across the middle of the map from School No. 94 to the eastern end of School Section Lake (in section 16). Label the following landform features: dune crests, dune slip faces, interdunal valleys, and lakes.

8. What cultural features on the map indicate the dominant land use for the region?

A location map of the Nebraska Sand Hills, which cover almost one-third of the state.

Opposite:
Steverson Lake, Nebraska
Scale 1:62,500
Contour interval = 20 ft
U.S. Geological Survey

Figure 18.32   Five principal sand dune types: (a) barchan, (b) parabolic, (c) transverse, (d) longitudinal, and (e) star. The wind direction is shown in all figures, as indicated by arrows.   ■   **What factors play a role in which type of dune will be found in a region?**

dunes look like sea waves. The windward slope of transverse dunes is gentle like that of the barchans, while the steeper leeward slip face is at the angle of repose.

Longitudinal dunes are long, narrow dunes that are aligned parallel to the prevailing wind direction (Fig. 18.32d). A small sand supply and strong winds are important factors contributing to the formation of longitudinal dunes. There is no consistent distinction between the back slopes and slip faces of these dunes, and their summits may be either rounded or sharp. Longitudinal dunes cross vast areas of interior Australia, where they are known as *sand ridges*. A type of longitudinal dune called a **seif** (from the Arabic for "sword") is found in

the deserts of Arabia and North Africa. Seifs are huge, sharp-crested dunes, sometimes hundreds of kilometers long, whose troughs are almost free of sand (Fig. 18.34). They may reach 180 meters (600 ft) in height.

Star dunes are pyramidal shaped dunes in which ridges of sand radiate out from a peaklike center to resemble a star in map view (Fig. 18.32e). These dunes are most common in areas where there is a great quantity of sand and the climate is extremely hot and dry. Star dunes do not move their position, but ridges and slip faces shift orientation with wind variations. It is thought that convectional uplift of air currents shapes the sand into a star-shaped, pyramidal dune.

Gary J. James/Biological Photo Service

Figure 18.33 Barchan sand dunes at Bahia Magdalena, Baja California, Mexico. ■ **What is the dominant wind direction in this photo (see Fig. 18.32)?**

NASA

Figure 18.34 A satellite view of longitudinal dunes in the Sahara. The area in the photograph is approximately 160 kilometers (100 mi) across. ■ **Estimate the length of the dunes in this satellite image.**

Dune Protection To many who visit the desert or the beaches, dunes are one of nature's most beautiful landforms. However, dune areas are also very attractive sites for recreation, and they are particularly exciting for drivers of ORVs (off-road vehicles). Although dunes appear to be indestructible and rapidly changing environments that do not damage easily, this is far from the truth. Dunes are actually fragile environments with easily impacted ecologies. Because dune regions are the result of an environmental balance between moving dunes and the plants trying to stabilize them, the environmental

equilibrium is easily upset. Many of the most spectacular dune areas in the United States have special protection in national parks, national monuments, or national seashores, such as White Sands, New Mexico; Great Sand Dunes, Colorado; Indiana Dunes, Indiana; and Cape Cod, Massachusetts. Many dune areas, however, do not have special protection, and environmental degradation is a constant threat.

There are many practical reasons for dune preservation. In coastal zones, dunes play an important role in coastal protection and are sometimes the last defense of

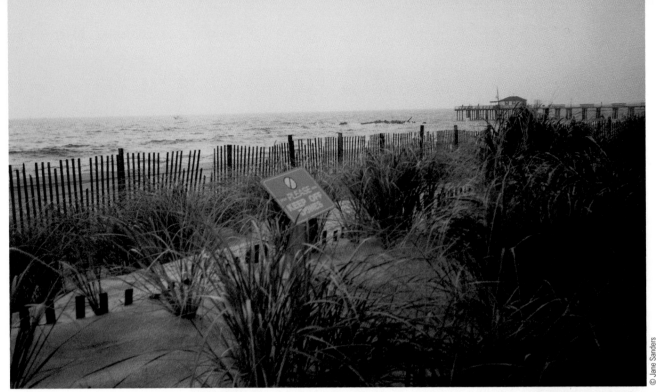

Figure 18.35 A sign indicating protected dune area along the New Jersey coast. ■ **Why do you think some dune areas need to be protected from human activities such as "dune buggies" and recreational vehicles?**

coastal communities from storm waves (Fig. 18.35). They are particularly important along the low-lying Gulf and Atlantic coasts of the United States, where occasional hurricanes or "northeasters" batter coastlines and erode beaches in front of the dunes. In nations such as the Netherlands, coastal dunes are extremely important because the land behind them is below sea level and a breach through the dunes could mean disaster. Coastal dune regions also are critical wildlife habitats, especially for many bird species.

Loess Deposits   The wind can carry dust-sized particles of clay and silt, removed by deflation, for hundreds or thousands of kilometers before depositing them. Eventually these particles settle out to form a tan or grayish blanket of loess that may cover or bury the existing topography over widespread areas. These deposits vary in thickness from a few centimeters or less to more than 100 meters (330 ft). In northern China downwind from the Gobi Desert, the loess is 30 to 90 meters (100–300 ft) thick (Fig. 18.36). Loess may originate from deserts, river floodplains, or other unvegetated surfaces. The extensive loess deposits of the American Midwest and

Figure 18.36 A steep gully eroded into the thick loess deposits of northern China. ■ **Where is the origin of these loess deposits?**

Europe were derived from glacial and meltwater deposits of retreating glaciers. As winds blew across the barren glaciated areas or meltwater floodplains, they picked up a large load of fine sediment and deposited it as loess in downwind regions.

Certain interesting characteristics of loess affect the landscape where it forms the surface material. For example, though fine and dusty to the touch, loess maintains vertical walls when cut through naturally by a stream or artificially by a road. Sometimes slumping will occur on these steep faces. Slumping gives a steplike profile to many loess bluffs. In addition, loess is easily eroded by either the wind or running water because of its fine texture and unconsolidated character. As a result, loess-covered regions that are unprotected by vegetation often become gullied. Where loess covers hills, gully erosion and slumping are conspicuous processes. A particularly severe erosion problem is responsible for the recent collapse of high loess bluffs along the Mississippi River at Vicksburg, Mississippi (Fig. 18.37).

Because of its high calcium carbonate content and young unleached characteristics, loess is the parent material for many of Earth's most fertile agricultural soils. Extensive loess deposits are found in northern China, the Pampas of Argentina, the North European Plain, Ukraine, and Kazakhstan. In the United States, the Midwestern Plains, the Mississippi Valley, and the Palouse region of eastern Washington are underlain by rich loess soils (Fig. 18.38). All of these areas are extremely productive grain-farming regions.

Vicksburg Convention and Visitors Bureau

Figure 18.37 The steep and unstable loess bluffs of Vicksburg, Mississippi. ■ **Why might the instability of loess cliffs be a problem?**

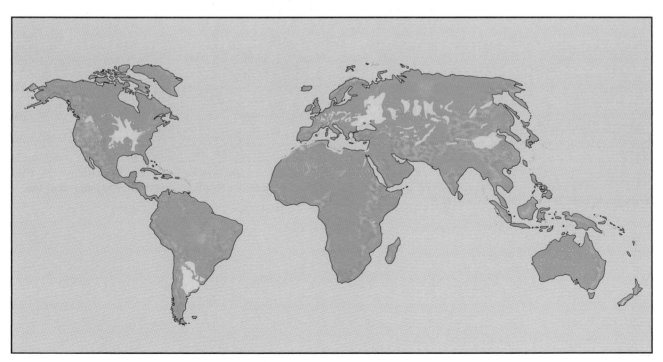

Figure 18.38 Map of the major loess regions of the world. Most loess deposits are peripheral to deserts and recently glaciated regions.

Arid landscapes and eolian landscapes in any climatic environment can be beautiful and stark, but those who are unfamiliar with these environments too often misunderstand them. Deserts are not wastelands, and dunes are not merely piles of sand. It is true that most desert and eolian areas have little to offer in terms of directly exploitable economic value, but their unique characteristics and scenic beauty fully qualify them for preservation and protection. The austere, angular character of their landforms, the fragility of their environments, and

the opportunities they provide for learning about how certain Earth systems operate continue to attract those who seek to understand and experience arid and eolian environments (Fig. 18.39). One only needs to count the number of national parks, national monuments, and other scenic attractions in the arid southwestern United States and in regions of sand dunes to find ample support for their survival. Deserts and dune localities are places with a beauty all their own and are areas worthy of appreciation and appropriate environmental protection. *Vive la difference!*

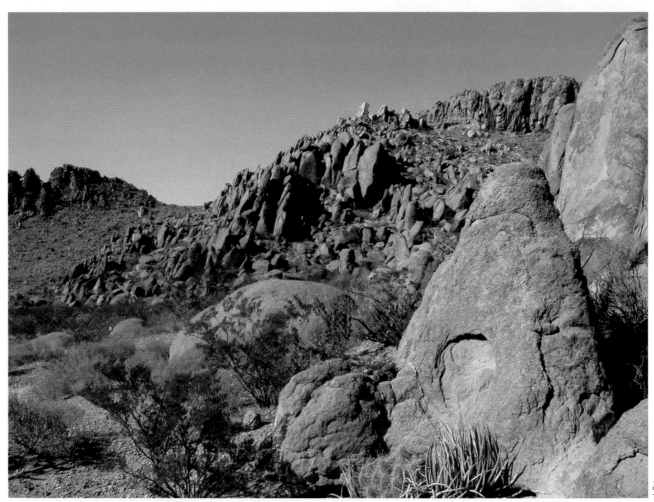

Figure 18.39   This image of the Grapevine Hills in Big Bend National Park, Texas, illustrates the dramatic beauty of desert landscapes. The arid climate of the Chihuahuan Desert has produced an intriguing region featuring fractured granitic hills and boulders.   ■   **What aspects of this environment make this an attractive landscape?**

J. Petersen

DEFINE & RECALL

base flow	pediment	deflation	slip face
interior drainage basin	alluvial fan	abrasion	barchan
exotic stream	bajada	dust storm	parabolic dune
arroyo (wash, barranca, wadi)	piedmont alluvial plain	sandstorm	transverse dune
badlands	bolson	deflation hollow (blowout)	longitudinal dune
plateau	playa	desert pavement (reg, gibber)	seif
caprock	playa lake		star dune
mesa	salt flat (salina)	ventifact	
butte	inselberg	dune	
	eolian	loess	

DISCUSS & REVIEW

1. What are eolian landforms? What gradational agent most significantly affects the desert landscape?

2. How do climate and vegetation affect landforms in the desert?

3. How does an exotic stream differ from an intermittent stream? Give three examples of exotic streams.

4. Why do landforms produced by erosion in the desert differ from erosional landforms in more humid regions?

5. How does the formation of a mesa or butte differ from the formation of a pediment?

6. How is a bajada formed?

7. Why do temporary lakes form in bolsons? How are such lakes related to salinas?

8. Why are most sandstorms confined to a low level? How do they differ from dust storms?

9. How does deflation affect the desert landscape?

10. Describe the formation of a ventifact.

11. What is the difference between a barchan and a transverse dune?

12. Name several regions in the world where loess deposits are located. What was the probable origin of the loess for each of these regions? What is the major type of economic activity for these loess regions today?

CONSIDER & RESPOND

1. How does stream gradation in arid regions differ from stream gradation in humid regions? What are the most common erosional and depositional landforms associated with stream gradation in arid regions?

2. What forms of evidence exist today to indicate that many desert regions had much different climates in the past?

3. Name several types of arid landforms commonly found in young, tectonically active desert regions, such as the southwestern United States. Name several types of arid landforms commonly found in deserts, such as those in central Australia, that have had much more time to become gradational in nature.

4. What type of sand dunes might be expected to develop under the following conditions:

    a. Moderate winds from a constant direction and abundant sand over a large desert area?

    b. Low sand supply and winds from a constant direction?

    c. Abundant sand, moderate winds, and sparse vegetation along beaches?

5. Give some reasons why regions of loess are more common in the Northern than in the Southern Hemisphere.

6. Why are there so many national parks and monuments in the arid parts of the American Southwest? What landform characteristics and other factors of desert landscapes draw people to these locations?

GLACIAL SYSTEMS AND LANDFORMS

A glacier serves as an excellent example of an open system.

Why is a glacier considered an open system? What materials enter and leave the system? How does energy move through the system?

The two major forms of glaciation, though similar in some respects, have their own impact on the landscape and originate in various areas in different ways.

What are the two forms, and how do they originate?

Much of the stark grandeur and scenic beauty of mountain regions is a result of alpine glaciation.

What characteristics of most high mountains are products of alpine glaciation? How might human use of mountains be influenced by alpine glaciation?

The ice budget of a glacial system is determined by two major factors: the amount of snow (ice) that accumulates each year at the source of the glacier and the amount of ablation (melting, evaporation, and sublimation) near the snout of the glacier.

What happens to the glacier when accumulation exceeds ablation? When ablation exceeds accumulation? What happens when dynamic equilibrium exists?

The most visible effects of continental glaciation in North America are from the Wisconsinan, the last stage of Pleistocene glaciation.

Why are the effects of Wisconsinan glaciation so prominent in the physical landscape? How did Pleistocene glaciation change the physical appearance of North America?

Although both are produced by continental glaciation, erosional and depositional glacial landforms exhibit major differences.

What are the major differences? How did they develop through glaciation? How has human use been affected by these landscapes?

Susitna Glacier and Mt. Hayes, Alaska

One can hardly imagine greater natural beauty than in the Swiss Alps, the Canadian Rockies, or the coastal fjord country of Norway and Alaska. Rugged mountain peaks rise high above deep, lake-filled valleys or narrow, deep-sea lanes, creating the ultimate in scenic appeal for many people. Glaciers, masses of moving ice, have transformed the appearance of high mountains, as well as large portions of continents, into distinctive landscapes. The flowing ice of glaciers is the most effective and spectacular geomorphic agent on Earth's surface.

Glacier Formation and the Hydrologic Cycle

Glaciers are flowing masses of ice that have accumulated on land in areas where more snow falls during a year than melts (Fig. 19.1). Snow falls as flakes composed of hexagonal ice crystals of intricate beauty and variety. Once on the land surface, however, snow is soon transformed into a more compacted mass of smaller, rounded grains. As the air space around them is lessened by compaction and melting, the grains become denser. With further melting, refreezing, and increased weight from newer snowfall above, the snow reaches a granular recrystallized stage between flakes and ice known as **firn.** With additional time, pressure, and refrozen meltwater from above, the firn will be recrystallized into ice. The small firn granules will become larger, interlocked ice crystals. When the ice is thick enough, usually more than 30 meters (100 ft), the weight of the snow and firn above will cause the crystals to become plastic and flow outward or downward from the area of snow accumulation (see pages 528–530, "How Does a Glacier Flow?").

Glaciers are open systems, with snow entering the system and mainly meltwater leaving the system in a constant cycle. The glacial system is controlled by two basic climatic conditions: precipitation in the form of snow, and freezing temperatures. First, there must be sufficient snowfall to exceed the annual loss through melting, evaporation, sublimation, and **calving** (Fig. 19.2), which occurs when the glacier loses chunks as icebergs to the ocean or other water body. Mountains along middle-latitude coastlines and even at high elevations near the equator can support glaciers because of heavy orographic snowfall, despite intense sunshine and warm surrounding climates. Yet some very cold polar regions in Subarctic Alaska and Siberia, and a few valleys in Antarctica, have no glaciers because the climate is too dry.

A second climatic condition is temperature. Summer temperatures must not be high for too long, or all the snowfall from the previous winter will melt. Surplus snowfall is essential because it allows the pressure of accumulated snow over the years to transform older buried snow into firn and glacial ice and creates depths great enough for the ice to flow. Glaciers are sometimes classi-

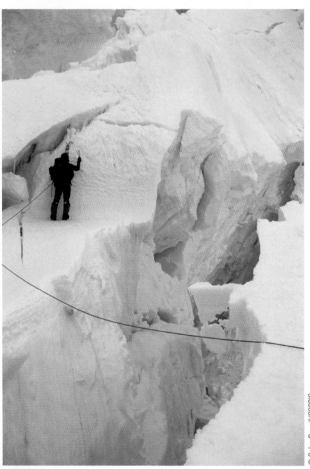

Figure 19.1 Glaciers form in environments that recieve more snowfall in a year than will melt. This mountaineer considers negotiating a crevasse on a snow-covered glacier on the mountain of Huascaran in the Peruvian Cordillera Blanca range.

Figure 19.2 Glaciers that flow into the sea or a deep lake have icebergs break off in a process called calving. This glacier is in Alaska.

fied by temperature, either as more active *temperate glaciers,* or as colder, slower-flowing *polar glaciers.*

Glaciers are an important part of Earth's hydrologic cycle and are second only to the oceans in the total amount of water they contain. About 2 percent of Earth's water is currently frozen as ice. Two percent may be a deceiving figure, however, because more than 80 percent of the world's *fresh* water is locked up as ice in glaciers, the vast majority of it in Antarctica and Greenland. The total amount of ice is even more impressive if we estimate the water that would be released upon the hypothetical melting of the world's glaciers. Sea level would rise about 60 meters (200 ft). This would change the geography of the planet considerably. In contrast, should another ice age occur, sea level would drop drastically (see Geography as the Spatial Science, Climatic Change and Its Impact on Coastlines in Chapter 8). During the last major glacial advance, sea level dropped about 120 meters (400 ft).

When snow falls on high mountains or in polar regions, it may become part of the glacial system. Unlike the water in stream systems, which returns rapidly to the sea or atmosphere (unless it is stored in lakes or as groundwater), the snow that becomes part of a glacier is stored as ice in a much more slowly moving system. Here water may be stored as ice for hundreds or even hundreds of thousands of years before being released as meltwater into the liquid part of the hydrologic system. In the meantime, however, glacial ice is not stagnant. It moves slowly across the land with tremendous energy, carving into even the hardest rock formations. The glacier reshapes the landscape as it engulfs, pushes, drags, and finally deposits rock debris, often in places far from its original location. Long after the land is released from its icy entombment, a tremendous variety of glacial landforms remains as a reminder of the energy of the glacial system.

During most of Earth history, glaciers did not exist on the planet. However, when a period of time occurs during which significant areas of the middle latitudes are covered by glaciers, we call it an *ice age.* At the present time, glaciers cover about 10 percent of Earth's land surface.

Present-day glaciers are found in Antarctica, in Greenland, and at high elevations on all the continents except Australia. During recent Earth history, from about 2.4 million years ago to about 10,000 years before the present, nearly a third of Earth's land area was periodically covered by ice thousands of meters thick. In the much more distant past, other ice ages have occurred.

Types of Glaciers

There are two major types of glaciers, alpine glaciers and continental ice sheets. **Alpine glaciers** are fed by ice and snow in mountain areas, usually accumulating in depressions initiated by stream erosion. Those that are confined by the rock walls of the valley they occupy may also be called **valley glaciers** (Fig. 19.3). Another variety

Figure 19.3 A valley glacier in Alaska's Kenai Peninsula. A clearly defined equilibrium line separates the white upper accumulation zone from the blue-toned lower ablation zone. ■ **Explain what firn is.**

Austin Post, U.S. Geological Survey

of alpine glacier, the **piedmont glacier,** forms where valley glaciers coalesce and spread out over flatter land at the base of a mountainous region because the valley walls no longer confine them (Fig. 19.4). Whereas piedmont glaciers result from glacial flow beyond the limits of confining valleys, some alpine glaciers do not reach the valleys below the zone of high peaks. Instead, they occupy distinctive high amphitheater-like depressions, called **cirques,** and are known as **cirque glaciers** (Fig. 19.5).

Alpine glaciers create the characteristic rugged scenery of high mountains in most world regions. They can currently be found in the Rockies, Sierra Nevada, Cascades, the Olympic Mountains, the Coast Ranges, and in numerous Alaskan ranges of North America. They also exist in the Andes, in New Zealand, in the Alps, and in the Himalayas, Pamirs, and other high Asian mountain ranges. Small glaciers are even found on tropical mountains in New Guinea and in East Africa on Mounts Kenya and Kilimanjaro and in the Ruwenzori Range. The largest alpine glaciers in existence today are found in Alaska and the Himalayas, where some reach lengths of more than 100 kilometers (62 mi).

The second and larger type of glacier, far more extensive than the valley glacier, is the **continental ice sheet,** (Fig. 19.6). At one time, continental ice sheets covered as much as 30 percent of the Earth's land. They still blanket Greenland and Antarctica to a depth of 3 kilometers (2 mi) or more, and smaller glaciers called **ice caps** are present on some Arctic islands and on Iceland. In contrast to alpine glaciers, continental ice sheets are unconfined, flowing over even higher portions of the land. Thus, they are not elongated but flow radially outward in all directions from their source area.

Features of an Alpine Glacier

Alpine glaciers can be divided into two parts: a **zone of accumulation** and a **zone of ablation** (Fig. 19.7). The upper portion of the glacier, where snowfall exceeds **ablation** (losses through melting, evaporation, and sublimation), is the zone of accumulation. The lower portion of the glacier, where ablation exceeds snowfall, is the zone of ablation. This lower zone develops because, under the stress of gravity or pressure from ice thickness, glacial ice flows constantly from the zone of accumulation down to the zone of ablation. Downslope movement is seen in the development of a great crack or crevasse, known as the **bergschrund,** around the head of the glacier (see again Fig. 19.5). It shows that the ice mass is pulling away from the confining rock walls of the

Figure 19.4 The Malaspina Glacier, Alaska. This large piedmont glacier lies at the flat base of the St. Elias Mountains. ■ **How do you think this type of glacier forms?**

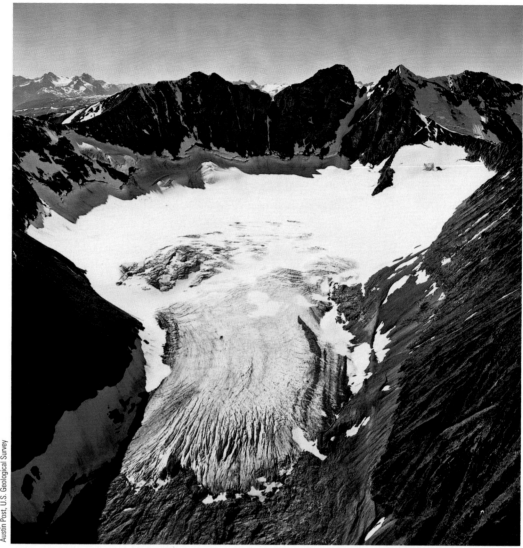

Austin Post, U.S. Geological Survey

Figure 19.5 A cirque glacier in Alaska. A bergschrund, or great crevasse, can be seen along the upper margin of the glacier. ■ **What does such a large crevasse indicate about this glacier?**

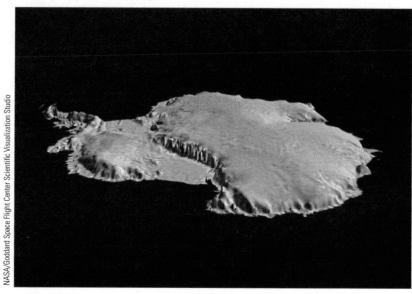

NASA/Goddard Space Flight Center Scientific Visualization Studio

Figure 19.6 Continental ice sheet. This terrain model of Antarctica shows the dome-like form of an ice sheet. Ice flows away from the center of the continent, near the South Pole, where the pressure is greatest, because the ice is thickest there. ■ **These glaciers flow in a roughly radial pattern. Remembering radial drainage of streams, what does this mean?**

Glacial Ice Is Blue!

When we make ice in our freezers, clear colorless water turns to relatively clear ice cubes. The ice cubes may contain some white crystalline forms, but for all intents and purposes the ice is clear. In nature, however, the process of making ice is very different from that of an ice-maker in a freezer unit. As snow falls in colder, higher latitudes and altitudes, it forms a layer of snow (known as snow pack) on the surface. Each successive snowfall forms another layer as it piles onto the previous snow pack. The weight of the successive (newer) layers of snow helps to compact the older layers beneath. Through time, the layers of snow become layers of solid ice. While compaction is occurring, there may be periods of partial melting and refreezing, which may speed the process along.

Snow is white in color because the hexagonal crystalline structure of the snowflakes reflects all the wavelengths of

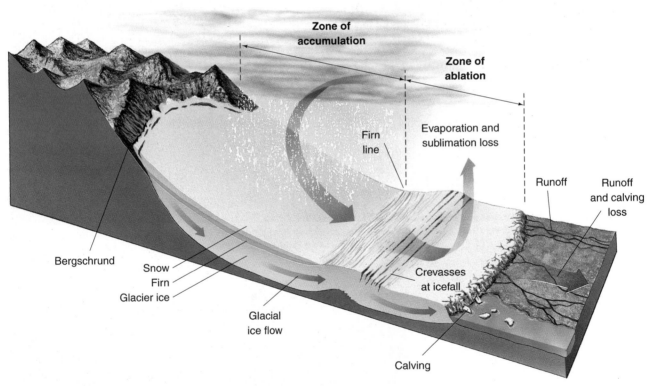

Figure 19.7 Environmental Systems: The Hydrologic System—Glaciers. Glacial systems are controlled by two major climatic factors: the input of snowfall (accumulation) and the loss of snow and ice (ablation) resulting from evaporation or melting. Most of the glacial growth is in the zone of accumulation, where temperatures permit freezing much of the year and annual snowfall exceeds melting. Over the years in the accumulation zone, snow becomes buried under newer snowfall, changes into granular firn, and much later transforms into glacial ice. When the ice reaches a threshold depth, the deeply buried ice will deform plastically and flow downslope.

Most loss from a glacier is in the zone of ablation—the area below the equilibrium line where annual melting exceeds snowfall. Here the glacier loses its mass from direct evaporation and sublimation from the glacier's surface. Losses also occur from runoff produced by melting, and by calving icebergs at the terminus if the glacier flows into a body of water deep enough to float them.

A glacial system is in equilibrium if a balance is achieved between inputs of snow and ice and outputs by ablation. If accumulation exceeds ablation for several years, the glacier's snout will advance; if ablation exceeds accumulation, the glacier will lose mass, and the glacier will recede. At present, most of the world's alpine glaciers are receding, possibly because of a global warming trend. Many scientists are concerned about the potential effects of global warming on glacial systems, especially on the continental ice sheets of Antarctica and Greenland. Increasing world temperatures can shift glacial systems toward a new equilibrium, at a smaller size and mass, with more rapid ice loss through melting and calving. These changes, in turn, could affect world sea level and could cause local increases in flooding hazards.

visible light with equal intensity, thus appearing white to the human eye. However, when the snow strata in a glacier are compacted through the years (perhaps hundreds or thousands of years), the ice becomes denser; that is, more ice crystals are squeezed into the same volume. The denser ice reflects more of the shorter wavelengths of visible light (the blue end of the color spectrum). Therefore, the denser the ice, the more blue it appears. However, one must be careful not to assume that deeper blue layers in a glacier

are the oldest layers. Again, melting and refreezing may affect the density of the ice layer, and the packing of a wet snow as opposed to drier snow layers can affect the density of a particular layer. Nevertheless, what is certain when looking at massive ice accumulations in nature, such as glaciers, ice caps (smaller and more localized), and ice sheets (greater and regional in extent), is that shades of the color blue will appear.

cirque and is moving outward and downward. The end of a glacier—its **terminus** or **snout**—marks the farthest advance of the glacier. If the terminus reaches the sea, it will calve icebergs.

The **equilibrium line,** which separates the two zones of a glacier, represents an equilibrium point between snowfall and ablation. The equilibrium line tends to coincide with the late summer snow line. Several factors influence the level of the snow line (see Fig. 19.3). Latitude and elevation, which affect temperature, are important factors. Equally important is the amount of snowfall received during the winter. In general, with colder temperatures and greater snowfalls, the snow line will go down in elevation. Other factors causing variation in the elevation of the snow line include the amount of insolation. A shady mountain slope, or one where there is a high percentage of cloud cover, will have a lower snow line than one that receives more insolation. Wind is another factor, because it produces snowdrifts on the lee side of mountain ranges. Thus, in the middle latitudes of the Northern Hemisphere, the snow line is lower on the north (shaded) and east (leeward) slopes of mountains. Consequently, the most significant glacier development is on these slopes (Fig. 19.8).

Some alpine glaciers are caused primarily not by snowfall on the glacier itself, but by the accumulation of snow blowing, drifting, and avalanching onto the glacier's surface. The Colorado Rockies and the Ural Mountains in Russia provide good examples of alpine glaciers formed by snowdrift.

Equilibrium and the Glacial Budget

When the snout of a glacier does not move, the glacier is said to be in a state of equilibrium—that is, a balance in the system has been achieved between accumulation and

Figure 19.8 This topographic map of part of Rocky Mountain National Park in Colorado, illustrates the impact of slope aspect on glaciation. In the northern hemisphere at mid to high latitudes, slopes that face north or northeast tend to be shaded from the sun. Thus, more snow and ice can accumulate there and glaciation tends to be more extensive. More snow melts from south-facing slopes because they receive more direct sunlight. (see Chapter 3.)

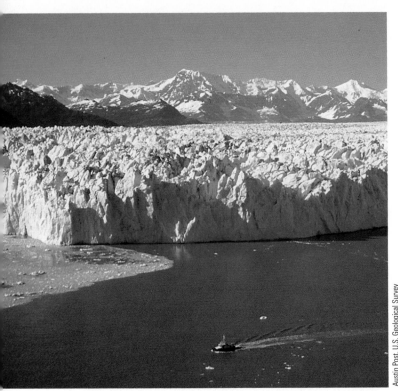

Austin Post, U.S. Geological Survey

Figure 19.9 The receding Columbia Glacier, Alaska. As this large glacier retreats, it calves icebergs into Prince William Sound. ■
Why is the rapid receding of this glacier of particular concern to scientists?

ablation of ice and snow. As long as this condition is maintained, the glacier's snout will remain in the same location, although the glacial ice continues to flow forward, like the stairs of an escalator.

Let us assume that exceedingly heavy amounts of snow are received for several years, and the glacier's equilibrium is upset. Under the pressure of this increased snow accumulation, more ice will be produced, and the snout will advance until it reaches a new point of equilibrium where receipt of ice and snow equals wastage. A deficit of snow in the glacial budget will cause the snout to recede, by melting back faster than it is flowing, until equilibrium between receipt and wastage is again achieved. An increase or decrease in temperature will also cause an advance or a retreat of the glacier.

However, even in a retreating glacier, the ice is constantly flowing forward. Forward movement stops only if the ice becomes too thin to flow at all. Glaciers can fluctuate from year to year, and most changes in the position of the terminus are seen by studying glaciers over the longer term (Fig. 19.9). From about 1890 to the present, most Northern Hemisphere glaciers have been receding. This overall retreat may be an indication of global warming, yet each individual glacier has its own balance. For example, in 1986, Alaska's Hubbard Glacier advanced so rapidly that it cut off Russell Fjord from the ocean, trapping many seals and porpoises. Yet just a few hundred miles away, the giant Columbia Glacier is rapidly receding and also calving increased numbers of icebergs into Prince William Sound. Scientists and the Coast Guard are concerned because further retreat will increase the hazard from icebergs to nearby oil tanker lanes from the Trans-Alaska Pipeline.

How Does a Glacier Flow?

The mechanics by which an alpine glacier flows are complex, and some explanations are still theoretical. Most of the movement, however, is caused by gravity of valley glaciers. It is believed that a glacier flows by a combination of processes, one of which is *basal slip* over its rock floor (Fig. 19.10). Steep slopes, which allow gravity to work, and meltwater, which reduces friction by lubricating, must both be involved. Basal slip is particularly important in middle-latitude glaciers and during summer, when much of the glacier is near its melting point and meltwater is available. During winter and in polar-climate glaciers, with little meltwater available, the glacier may freeze onto the bedrock and basal slip may be prevented.

Internal plastic deformation is another means by which ice flows. At depth, the

Figure 19.10 Most of the ice movement in a glacier is caused by ice that deforms under pressure and flows as a plastic solid. A lesser part of glacial motion is the result of the ice sliding on the bedrock at its base. This diagram shows that the upper part of the ice moves faster than the lower ice that is in friction with the bed. Plastic flow is also cumulative upward, so the velocity is greatest on the glacier's surface. The brittle ice above the plastic layer cracks to form crevasses as it is transported by the flowing ice below.

R. Sager

(a)

R. Sager

(b)

Figure 19.11 Crevasses form in the upper surface of glaciers. (a) A large crevasse on the Aleyska Glacier, Alaska. (b) Crevasses are particularly numerous where the ice flow speeds up, especially at icefalls, such as this one on the Chilkat Glacier, Alaska.

tremendous weight of the ice causes individual ice crystals to arrange themselves in parallel layers and then slide over each other, much like a deck of cards. This type of flow seems to occur only below about 30 meters (100 ft) of ice, in an area known as the *zone of plastic flow* .

The upper surface of the glacier is brittle and does not have plastic deformation. Motion of the ice in this upper *zone of brittle flow* occurs by fracturing and faulting. Here the ice breaks and cracks as a solid material. These cracks, called **crevasses,** are common where the ice mass is stretched, particularly along the margins and snouts of alpine glaciers (Fig. 19.11). Where glaciers flow over steep slopes, they form an **icefall.** Here, intersecting crevasses break the ice into a morass of unstable ice blocks. Icefalls are extremely dangerous areas for mountain climbers and scientists who venture onto the ice (Fig. 19.12). Locations where the snout of the glacier meets seawater can also be dangerous, as large blocks of ice calve off and topple into the water, creating large waves.

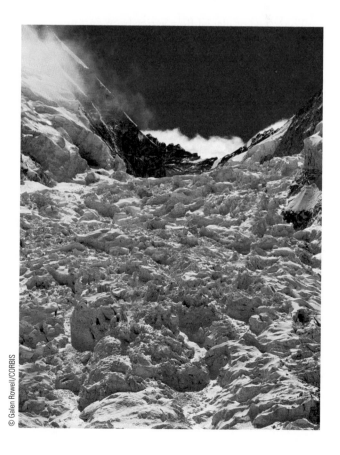

Figure 19.12 An icefall is the glacial equivalent of rapids or waterfalls in a river. Although glacial ice flows much more slowly than running water, these are still the most rapidly moving and changing part of a glacier. The Khumbu Icefall, shown here, is one of the most treacherous parts of a Mt. Everest climb because it is made of huge ice blocks that can shift at any time.

© Galen Rowell/CORBIS

Austin Post, U.S. Geological Survey

Figure 19.13 Glacial abrasion produces smooth rock surfaces with scratches and grooves parallel to the direction of ice movement. ■ **Can you see evidence of the direction of ice flow from this photograph in Mount Rainier National Park, Washington?**

The speed of glacial flow varies from an imperceptible fraction of a centimeter per day to as much as 30 meters (100 ft) per day. In addition, the flow of an individual glacier varies from time to time because of changes in the dynamic equilibrium, and from place to place because of variation in the gradient over which it flows or in the amount of friction encountered with adjacent rock. Glaciers flow more rapidly where there is a steep slope, where the ice is thickest, and where temperatures are warmest. For example, the Nisqually Glacier, on the steep slopes of Washington's Mt. Rainier, moves 38 centimeters (16 in.) per day in summer. As a general rule, temperate alpine glaciers flow much faster than the cold polar Antarctic or Greenland Ice Sheets.

The velocity of ice flow varies, depending on its location in the glacier. In the middle of the upper surface, where friction is least, speed of flow is greatest. On the sides and bottom surfaces of the glacier, friction slows the rate of flow.

Sometimes a glacier's velocity will increase by many times its normal rate, causing the glacier to travel hundreds of meters per year. The reasons for such enormous *glacial surges,* as these velocity increases are called, are not completely clear, although lubrication of the glacial bed by pockets of meltwater explains some of them.

Glaciers As Agents of Gradation

As a glacier moves over the land, whether as ice sheet or an alpine glacier, it scrapes along the land surface, picking up and carrying along with it boulders and rock fragments. This erosive process of lifting and incorporating rock and soil into the glacial ice is called **glacial plucking.** Weathering processes, particularly the freezing of water in bedrock joints and fractures, which breaks rock fragments loose, encourages plucking.

Glaciers also drag rock fragments along their undersides. The abrasive rock fragments traveling on the bed of a glacier become tools of erosion themselves as they scrape and gouge stationary surface rocks. The **striations** (gouges, grooves, and scratches) produced by such **glacial abrasion,** as this erosion process is called, can be seen as evidence of previous glaciation in areas devoid of glaciers today (Fig. 19.13). Striations in the bedrock indicate the direction of flow long after a glacier has disappeared.

Rock surfaces subjected to intense glacial abrasion are typically smoother and more rounded than those produced mainly by plucking. The glacially produced landform of such rock surfaces is the **roche moutonnée**—a bedrock hill that is smoothly rounded on the upstream side most subject to abrasion, with plucking evident on the downstream side (Fig. 19.14).

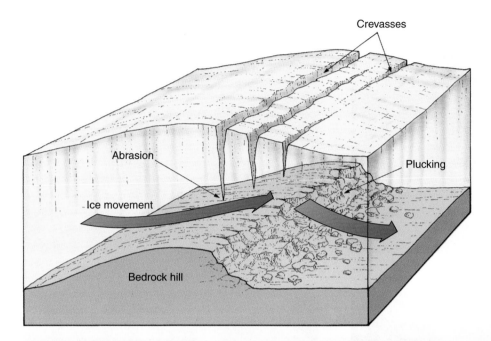

Figure 19.14 The characteristic asymmetric form produced by glacial erosion of a bedrock hill. Arrows indicate the direction of ice flow. This glacial landform is known as a roche moutonnée, or sheep-backed rock. ■ **Why would crevasses form above this feature?**

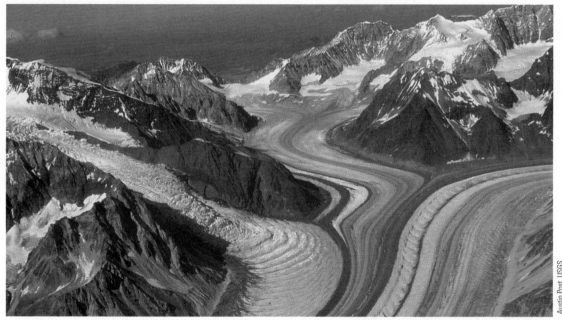

Austin Post, USGS

Figure 19.15 Where valley glaciers join, the eroded materials from glacial erosion, weathering, and mass movement form characteristic stripes as shown here in the Alaska Range. ■ **Explain how these stripes can be found in the middle of a valley glacier.**

Plucking and abrasion provide glaciers with a load of rock fragments of all sizes, from the finest-ground rock flour to giant boulders. Much of the load is concentrated near the surfaces of the glacier that are in contact with land surfaces—the source of most of a glacier's load. Additional material, particularly on alpine glaciers, is transported on the upper glacial surface, where it has been deposited by processes associated with erosion or mass wasting. Ice wedging on the high unglaciated slopes above the glacier and rockfall produce large quantities of rock debris along the valley sides. Where valley glaciers join, these materials cause the characteristic stripes on the surface of a glacier (Fig. 19.15). The volume and ve-

locity of a glacier do not determine the particle sizes it can erode and transport, as is the case with running water. As a consequence, the sorting of sediment by size that occurs as running water changes speed does not occur with flowing ice.

Glacial erosion is especially aggressive in removing nonresistant rock, often producing depressions. The bottom ice currents (but not the ice surface) may move obstructions even if the ice has to move slightly uphill to drag material out of the depression. Because ice currents can move upward, after the ice has melted a glaciated surface may have many depressions. After the retreat of a glacier, these depressions often fill with water to form lakes.

(a)

(b)

R. Gaber

R. Sager

(c)

R. Sager

(d)

R. Sager

Figure 19.16   Erosional landforms produced by alpine glaciation. (a) A cirque and tarn (lake) formed by headward erosion of a glacier in the Canadian Rockies. (b) Jagged sawtooth spines of rock, such as these in the French Alps, are called arêtes. (c) Mt. Assiniboine in the Canadian Rockies is a classic example of a horn. Horns are formed when several glaciers cut headward into a mountain peak. (d) Glaciers carve steep-sided U-shaped valleys called glacial troughs. Yosemite Valley, California, is a classic example of a glacial trough.

Erosional Landforms Produced by Alpine Glaciation

Ice thickness and erosion are greatest at the head of alpine glaciers. Glacial erosion works headward even as the glacier flows downslope. The headward erosion of a glacier produces a valley head shaped like a steep-sided bowl or amphitheater, called a *cirque headwall*. Glacial

undercutting of rock walls above the ice level, combined with mechanical weathering, causes mass movement and increases the steepness of the bowl's walls, forming a cirque. If a lake forms in the cirque depression after the ice melts, it is called a **tarn**, or cirque lake (Fig. 19.16a).

Often two or more cirques will form near the top of the same mountain. As the cirques of two ridge glaciers are enlarged, the wall of rock between them will be shaped into a jagged, sawtooth spine of rock, called an **arête** (Fig. 19.16b). Where three or more cirques meet at

Figure 19.17 The alpine glacial topography of the Sierra Nevada at Yosemite National Park in California were beautifully and accurately painted for the Park Service by a landscape artist. ■ **What features of glaciation do you recognize in this painting?**

a mountain summit, they form a characteristic pyramid-like peak called a **horn** (Fig. 19.16c). The Matterhorn, in the Swiss Alps, is the world's classic example. A **col** is a pass formed by the headward erosion of two cirques that have intersected to produce a low saddle in a high mountain ridge or arête.

Unlike streams, which initially erode V-shaped valleys, glaciers erode characteristically steep-sided U-shaped valleys called **glacial troughs** (Fig. 19.16d). In addition, a glacier's tendency to move straight ahead rather than to meander causes it to straighten out the original valley.

By eroding weaker rocks on the valley floor, the glacier often creates a sequence of rock steps and excavated basins. During glaciation, ice falls will be present at the steps. The result is a "glacial stairway." When the ice recedes, rock-bound lakes may fill the basins, often looking like beads connected by a glacial stream flowing down the glacial trough. Such lake chains are called **paternoster lakes.**

A large glacier often has tributaries that merge with it. These tributary glaciers, like the main ice stream, carve U-shaped channels. However, because they have less volume than the main glacier, the rate of erosion in

these tributaries is less rapid. As a result, their troughs are smaller and not as deep as those of the main glacier. Nevertheless, during peak glacial phases, the surface of the ice that flows from the smaller glacier is at the same level as the ice in the larger glacier. Not until the two glaciers begin to wane does the difference in height between their trough floors become apparent. The higher trough of the old tributary glacier is called a **hanging valley.** A stream that flows down such a channel will drop down to the lower channel by a high waterfall or a series of steep rapids. Yosemite Falls and Bridalveil Falls in Yosemite National Park are excellent examples of hanging valley waterfalls. Yosemite Valley itself is a classic example of a glacial trough.

Landscapes eroded by alpine glaciers show a sharp contrast between the glacial troughs scoured smooth by ice flow and the jagged peaks above the former ice levels. The rugged quality of these upper surfaces is caused primarily by mechanical weathering above the ice surface and by undercutting at the head of the glacier. California's Sierra Nevada (Fig. 19.17) and Glacier National Park, Montana, are excellent sites of rugged alpine glacial

(a) Preglacial fluvial topography

(b) Maximum glaciation

(c) Postglacial landforms

Figure 19.18   The development of glacial landforms in an alpine region. (a) Mountain topography prior to glaciation. (b) Maximum valley glaciation. (c) Major postglacial landforms.   ■   **How have the valley profiles changed from preglacial to postglacial times?**

terrain (see Map Interpretation: Alpine Glaciation). A hypothetical sequence in the development of alpine glacial landforms is illustrated in Figure 19.18.

In coastal areas where glaciers flow to the sea, they calve icebergs. Today, this process is ongoing along the coasts of British Columbia, southern Alaska, Chile, Greenland, and Antarctica, and it was formerly characteristic of Scotland, Norway, Iceland, and New Zealand. If these glaciers recede landward, the sea invades the former glacial trough, creating a deep, inland finger of the sea called a **fjord** (Fig. 19.19). This invasion by the sea is possible primarily because—unlike streams, which can erode only to base level—glaciers can erode below sea level. Because of its lower density, compared to liquid water, ice must be nine-tenths submerged before it will float. Thus, the sea may enter deep glacial channels once the ice has melted. In addition, many fjords formed dur-

ing periods of large-scale glaciation when sea level was lower and were later submerged as sea level rose with melting of the glaciers. Most of the deep, narrow channels of Washington's Puget Sound were carved into bedrock by glacial erosion and later invaded by the sea.

Depositional Landforms Produced by Alpine Glaciation

Alpine glaciers carry debris on their surfaces, frozen in their interiors, and dragged along at the bottom. As mentioned previously, the load they carry includes boulders, rocks, and fragments plucked by glaciers from their channel sides and floor, as well as smaller fragments and particles produced by abrasion. Rockfall from steep trough walls may also supply a glacier with debris. Glaciers can deposit huge chunks of bedrock, fine rock flour, layers of pollen, dead trees and other plants, soil, and volcanic dust.

All glacial deposits are included within the general term **drift,** whether they are unsorted and unstratified ice deposits or orderly deposits of meltwater streams issuing from the glacier. To differentiate these two types of deposits, the term **till** is applied to unsorted drift laid down by ice. **Glaciofluvial deposits** refer to better sorted and stratified meltwater deposits.

Glaciers deposit a portion of their load when their capacity is reduced. Landforms constructed from glacial deposits, called **moraines,** occur along the margins of glaciers. The ridgelike deposits laid down along the side margins are called **lateral moraines** (Fig. 19.20a). When two glaciers flow together, their interior lateral moraines form a **medial moraine** in the center of the new glacier. At the snout of the glacier, all debris carried forward by the "conveyor belt" of ice and pushed ahead of the glacier is deposited in a jumbled heap of rocks and fine material, called an **end moraine** (Fig. 19.20b). End moraines marking the farthest advance of the snout are **terminal moraines.** End moraines deposited as a consequence of a halt in a snout that is receding, followed by a stabilization of the ice front prior to further receding, are called **recessional moraines.** Glaciers also deposit a great deal of till along the floor of their channel as they retreat, particularly near the snout, where melting is greatest, where the ice becomes thinner, and where the load capacity is consequently reduced. This glacial till deposit is called **ground moraine.**

Braided streams of meltwater, laden with sediment, commonly issue from the glacier terminus. The sediment, called **glacial outwash,** is deposited beyond the terminal moraine, with larger rocks and debris deposited first and then the progressively finer particles. Often re-

— 4.9 *miles* —

Figure 19.19   Fjords, like those shown in this digital image, are glacial valleys that were drowned by the sea during the Pleistocene epoch, as the glaciers receded and sea levels rose.

sembling an alluvial fan confined by valley walls, this depositional form is called a **valley train.** Valleys in glaciated regions may be filled to depths of a few hundred meters by outwash, producing extremely flat valley floors (Fig. 19.21).

(a)

(b)

Figure 19.20   Depositional landform features due to alpine glaciation. (a) Moraines on the west slope of the Wind River Range, Wyoming. (b) The glacial deposits of an end moraine (left side of photo) mark an earlier position of this glacier's terminus, from which it has retreated in recent times. A moraine-dammed lake has formed by the ponding of meltwater between the glacier and the moraine. This glacier is part of the MacBeth Icefields in Canada's Kootenay Lake Forest District. ■ **Can you identify the lateral and end moraines in this photo?**

ALPINE GLACIATION

The Map

The Chief Mountain, Montana, map area is located in Glacier National Park, which adjoins Canada's Waterton Lakes National Park located across the international border. Alpine glaciation has produced much of the spectacular

An oblique aerial photo of Glacier National Park.

Robert Sager

scenery of this region of the Northern Rocky Mountains. Most of the glaciated landscape was produced during the Pleistocene. Today, glaciers still exist in Glacier National Park, but because of rapid melting in recent decades, it is estimated that they may be gone within 70 years. Prior to glaciation, this region was severely faulted and folded during the formation of the Northern Rocky Mountains.

The oblique aerial photo clearly shows the rugged terrain of this map area. The steep slopes, glaciers, lakes, and sharp arête ridges are obvious landform characteristics of alpine glaciated regions.

Temperature is a primary control of the highland climate of this region. Elevation is another important factor. As you would expect, the rapid decrease in temperature with increasing elevation results in a variety of microclimates within alpine regions. Exposure is also an important highland climate control. West-facing slopes receive the warm afternoon sun, whereas east-facing slopes are sunlit only in the cool of the early morning.

Interpreting the Map

1. What is the approximate local relief depicted by this map? How does the scale of this topographic map compare to those previously shown in the book (is this map of larger or smaller scale)?

2. Examining the topographic map and the aerial photograph of this mountain region, do you think most of the glaciated landscape was produced by erosion or deposition?

3. Locate Grinnell, Swiftcurrent, and Sperry glaciers. What type of glaciers are they?

4. What evidence indicates that the glaciers were once larger and extended farther down the valleys?

5. Note that most of the existing glaciers are located to the northeast of the mountain summits. Explain this orientation. At what elevation are the glaciers found?

6. What types of glacial landform are the following features?

a. The features occupied by Gunsite, Iceberg , and Ipasha Lakes

b. The feature occupied by McDonald Lake in the southwest corner of the map) and Lake Josephine

c. Mt. Gould and Mt. Wilbur

d. The series of lakes that occupy the valley of Swiftcurrent Creek

7. Along the high ridges runs a dashed line labeled "Continental Divide." What is its significance?

8. If you were to hike southeast from Auto Camp on McDonald Creek up Avalanche Creek to the base of Sperry Glacier how far would you travel, and how much elevation change would you encounter?

**Opposite:
Chief Mountain, Montana
Scale 1:125,000
Contour interval = 100 ft
U.S. Geological Survey**

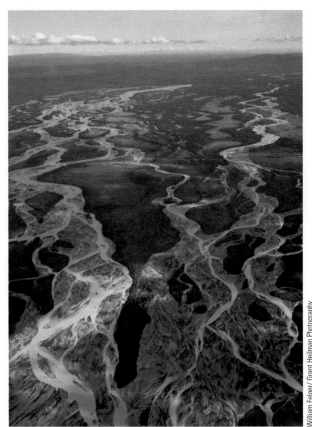

Figure 19.21   Meltwaters from glaciers in the Chugach Mountains of Alaska formed these braided streams from deposits of glacial outwash.   ■   **What is the source of all this material?**

Continental Ice Sheets

Continental ice sheets differ from alpine glaciers, especially in size and shape (Fig. 19.22). However, as agents of landform gradation, ice sheets are similar in many ways to alpine glaciers, and much of what we have discussed about alpine glaciers also applies to continental ice sheets. The differences in the gradational activities of the two types of glaciers are primarily differences in scale, attributable to the enormous disparity in size between the two.

Existing Ice Sheets

Glacial ice currently covers about 10 percent of Earth's land area, and alpine glaciers can be found in mountain regions on most continents. However, in area and mass,

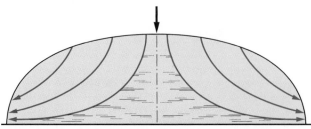

Figure 19.22   This diagram shows how ice flows in an ice sheet, downward from the center where the glacier is thickest, and outward.

alpine glaciers are insignificant compared to the ice sheets of Greenland and Antarctica, which account for 96 percent of the area covered by glaciers today. Glaciers resembling those of Greenland and Antarctica but on a much smaller scale, called ice caps, are also present in Iceland, in the arctic islands of Canada and Russia, in Alaska, and in the Canadian Rockies.

The Greenland ice sheet covers the world's largest island with a layer of ice that is more than 3000 meters (2 mi) thick in the center. The only land exposed in Greenland is a narrow, mountainous coastal strip (Fig. 19.23). Where the ice reaches the sea, it usually does so through fjords. These ice flows to the sea resemble alpine glaciers and are called **outlet glaciers.** When glacial ice reaches the sea, melting and the action of waves and tides break off huge chunks that float away. The resulting **icebergs** are a hazard to vessels in the North Atlantic shipping lanes south of Greenland. Tragic maritime disasters, such as the sinking of the *Titanic,* have been caused by collisions with these huge irregular chunks of ice, which are nine-tenths submerged and thus mostly invisible to ships. Today, with iceberg tracking by radar, satellites, and the ships and aircraft of the International Ice Patrol, these sea disasters are minimized.

The Antarctic ice sheet covers some 13 million square kilometers (5 million sq mi), an area almost seven-and-one-half times the size of the Greenland ice sheet, which covers 1.7 million square kilometers (650,000 sq mi). Little is known about the land beneath the thick layer of ice in Antarctica (Fig. 19.24). As in Greenland, little land is exposed in Antarctica, and the weight of the 4500 meter (nearly 3 mi) thick ice in some interior areas has depressed the land well below sea level. Where the ice reaches the sea, it floats in enormous flat-topped plates called **ice shelves.** These ice shelves are the source of icebergs in Antarctic waters, which do not have the irregular shape of Greenland's icebergs. Because they do not float into heavily used shipping lanes, these huge tabular-shaped Antarctic icebergs are not as much of a hazard to navigation (Fig. 19.25). They do, however, add to the problem of access to Antarctica for scientists. The huge wall of the ice shelf itself, the broken-up and melting sea ice, and the extreme climate combine to make Antarctica inaccessible to all but the hardiest individuals and equipment. This icy continent serves, however, as a natural laboratory for scientists from many countries to study Antarctic glaciology, climatology, and ecology.

Pleistocene Ice Sheets

The Antarctic and Greenland ice sheets are remnants of the Pleistocene. The Pleistocene Epoch, or ice ages, was the time of recent geologic history that experienced numerous major advances and retreats of continental ice sheets over large portions of the world's land. There were also a great number of minor fluctuations in the glaciers on

(a)

(b)

Figure 19.23 The Greenland ice sheet. (a) Except for the mountainous edges, the Greenland ice sheet almost completely covers the world's largest island. Ice thickness is more than 3000 meters (10,000 ft) and depresses the bedrock below sea level. In this satellite view, several outlet glaciers flow seaward from the ice sheet to the east coast of the island. (b) Map showing the extent of the Greenland ice sheet.

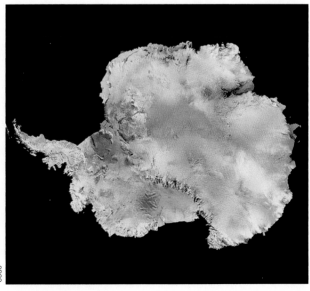

Figure 19.24 The Antarctic ice sheet. The world's largest ice sheet covers an area larger than the United States and Mexico and has a thickness more than 4500 meters (14,000 ft). This infrared satellite mosaic image covers the whole south polar continent. Notice that most of Antarctica is ice-covered (white and blue on the image). The only rocky areas (darker areas on the image) are the Antarctic Peninsula and the Transantarctic Mountains. Large ice shelves flow to the coastline, the largest being the Ross Ice Shelf. The image is oriented with the Greenwich Meridian at the top.

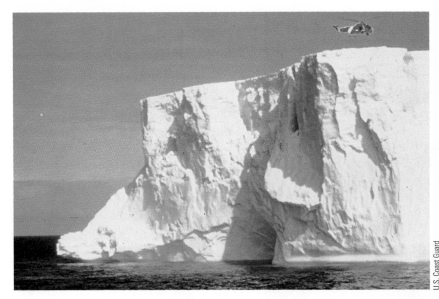

Figure 19.25 The Antarctic ice sheet calves enormous flat-topped tabular icebergs. ■ **What portion of the iceberg is hidden below the sea surface?**

U.S. Coast Guard

Earth during the Pleistocene, which began about 2.4 million years ago and ended around 10,000 years before the present. During the times when Pleistocene glaciers advanced, ice sheets spread outward over the land from centers in Canada, Scandinavia, and eastern Siberia, and also from high mountain ranges, to cover nearly a third of Earth's land surface (Fig. 19.26). At the same time, sea ice expanded equatorward. In the Northern Hemisphere, sea ice was present along coasts as far south as Delaware in North America and Spain in Europe. Between each glacial advance, a warmer *interglacial* occurred, during which the enormous continental ice sheets and sea ice retreated and almost completely disappeared. An examination of glacial deposits has determined that within each major glacial advance, many minor retreats and advances reflected smaller changes in global temperature and precipitation.

The gradational effects of the last major glacial advance, known as the Wisconsinan stage, which ended about 10,000 years ago, are the most visible in landscapes today. The glacial landforms created during the Wisconsinan stage are relatively new and have not been destroyed to any great extent by other gradational agents. Consequently, we can derive a fairly clear picture of the extent and actions of the ice sheets at that time.

The land area covered during the major glacial advances in North America and Eurasia is depicted in Figure 19.26. Continental ice sheets in North America extended as far south as the Missouri and Ohio Rivers and covered nearly all of Canada and much of the northern Great Plains, the Midwest, and the northeastern part of the United States. In New England, the ice was thick enough to overrun the highest mountains, including Mount Washington, with an elevation of 2063 meters

Fig. 19.26 The maximum extent of glacial ice coverage in the Northern Hemisphere during the Pleistocene. Much of North America and Eurasia was covered by ice up to several thousand meters thick. ■ **Why were some very cold areas, such as portions of interior Alaska and Siberia, ice-free during this time?**

(6288 ft). The ice was more than 2000 meters (6500 ft) thick in the Great Lakes area. In Europe, glaciers spread over what is now most of Great Britain, Ireland, Scandinavia, northern Germany, Poland, and western Russia. In much of Siberia and interior Alaska, it was too cold and dry (as it is today) to generate such massive snow and ice accumulations as occurred further south in North America and northwestern Europe.

During each advance of the ice sheets, alpine glaciers were much more extensive and massive in highland areas than they are today. In fact, it was the examination of landforms in the Swiss Alps that led Louis Agassiz in 1840 to publish the theory of past glaciations. Though at first considered radical, the theory today is acknowledged as fact and is supported by countless detailed studies of glacial deposits and erosional forms in various parts of the world.

Where did the water locked up in all the ice and snow come from? Its original source was the oceans. During the periods of glacial advance, there was a general lowering of sea level, exposing large portions of the continental shelf and forming land bridges across the present-day North, Bering, and Java Seas. The most recent melting and glacial retreat raised the oceans a similar amount—about 120 meters (400 ft). Evidence for this rise in sea level can be seen along many submerging coastlines around the world.

Ice Sheets and Erosional Landforms

Ice sheets erode the land in ways similar to alpine glaciers but on a much larger scale. As a result, landforms created by ice sheet erosion are far more extensive than those formed by alpine glaciation, stretching over millions of square kilometers of North America, Scandinavia, and Russia. As ice sheets flowed over the land, they gouged Earth's surface, enlarging valleys that already existed, scouring out rock basins, and smoothing off existing hills. The eroding ice sheets removed most of the soil and then attacked the bedrock itself.

Today, these **ice-scoured plains** are areas of low, rounded hills, lake-filled depressions, and wide exposures of bedrock (Fig. 19.27). The ice sheets covered and totally disrupted the former stream patterns. Because the last glaciation was so recent, new stream drainage systems have not had time to form a well-integrated system of stream channels. Ice-scoured plains are characterized by thousands of lakes, marshes, and areas of *muskeg* (poorly drained areas, grown over with vegetation, that form in cold climates). The major characteristics of glacially eroded landscapes, such as those found in Canada and Finland, are the great expanses of exposed gouged bedrock and standing water.

Movement of Continental Ice Sheets

It is a popular misconception that all continental ice sheets originate at the poles and spread toward the equator. Actually, the great centers of Pleistocene glacial accumulation (aside from highland areas, Antarctica, and Greenland) were in the upper middle latitudes, in the vicinity of Hudson Bay in Canada, on the Scandinavian Shield, and in eastern Siberia. These huge ice sheets also covered some fairly rugged terrain, such as New York's Adirondacks, New England's White Mountains, and the highlands of Scandinavia and Scotland (see again Fig. 19.26).

Continental ice sheets began to flow outward in all directions from a central zone of accumulation when the ice was thick enough to allow for plastic flow. As with valley glaciers, the initial flow direction of advancing ice sheets was determined largely by the path of least resistance, found in preexisting valleys and belts of softer rock. Thus, the radial expansion is analogous to the spreading of pancake batter in a frying pan. If enough batter is poured into the center of the pan, it will eventually spread outward to cover the entire bottom of the pan. Like an alpine glacier, an ice sheet flows outward from a zone of accumulation to a zone of ablation. Like alpine glaciers, ice sheets advance and recede with small changes in temperature and snowfall, and will melt and disappear when ablation exceeds accumulation.

NASA

Figure 19.27 This section of the Canadian Shield, deeply eroded and scoured by ice sheets, exposes the remains of an ancient meteor crater, gouged out to make a circular lake (reservoir). Although the overall relief is low, note how erosion has exploited zones of weakness in fractures to make linear valleys and lakes.

Ice Sheets and Depositional Landforms

Again, scale makes ice sheet deposition different from that of alpine glaciers. Though terminal and recessional moraines, ground moraines, and glaciofluvial deposition are found, these deposits form significantly larger features in the landscapes caused by ice sheets (see Fig. 19.28 and Map Interpretation: Continental Glaciation).

Terminal and Recessional Moraines   Terminal and moraines form belts of low hills and ridges deposited on the land in glaciated areas. These features are rarely more than 60 meters (200 ft) high (Fig. 19.29). The last major

Pleistocene glacial advance through New England left a terminal moraine running the length of New York's Long Island and formed the offshore islands of Martha's Vineyard and Nantucket. Recessional moraines formed both Cape Cod and Lake Michigan's rounded southern end. End moraines are usually arc-shaped and convex toward the direction of ice flow. Their pattern and placement indicate that the ice sheets did not maintain an even front but spread out in tongue-shaped lobes channeled by the previous terrain (Fig. 19.30). The positions of terminal and recessional moraines give evidence of more than simply the direction of ice flow. In addition, the character of deposited materials can be examined to detect the sequence of advances and retreats of each successive ice sheet.

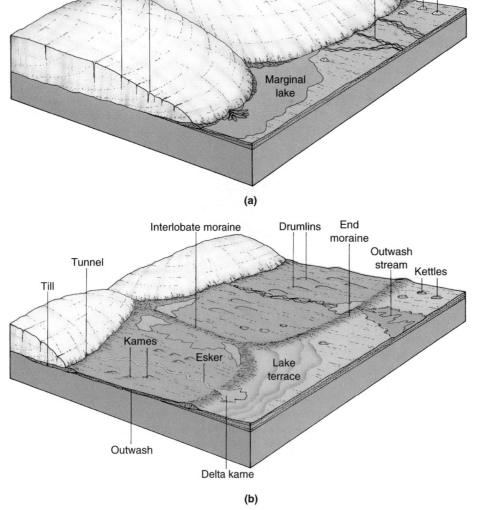

(a)

(b)

Figure 19.28 Landscape alteration by ice sheet deposition. (a) Features associated with ice stagnation at the edge of the ice sheet. (b) Landforms resulting from further modification by glacial meltwater as the ice sheet retreats.   ■   **What is the direct cause of most of these landscape features?**

John S. Shelton

Figure 19.29   The hilly topography of an end moraine deposited by a continental ice sheet. This aerial photo is of an end moraine on the Waterville Plateau in eastern Washington.   ■   **How do you account for the bumpier terrain at the left of the photo compared to the smoother surface of the plain at the right? Refer to Figure 19.28(b) to help explain your answer.**

Figure 19.30   Glacial deposits in the Great Lakes region.   ■   **Why do the many end moraines have such a curved pattern?**

CONTINENTAL GLACIATION

The Map

The Jackson, Michigan, map area is located in the Great Lakes section of the Central Lowlands. Massive continental ice sheets covered this region during the Pleistocene. As the glaciers melted and the ice sheets retreated, a totally new terrain, vastly different from that of preglacial times,

Michigan Department of Natural Resources

Landsat image of the region near Jackson, Michigan

was exposed. Today, evidence of glaciation is found throughout the region. The most obvious glacially produced landforms are the thousands of lakes (including the Great Lakes), the knobby terrain, and moraine ridges. Moraines left by advancing and retreating tongue-shaped ice lobes that extended generally southward from the main continental ice sheet also influenced the shapes of the Great Lakes. The ice sheet, its deposits of sediment, and its meltwaters also created other, smaller-scale features. The glaciers left a jumbled mosaic of deposits from boulders to sand, silt, and clay that has produced a hilly and hummocky terrain. The overall relief is low, in part because of glacial erosion. Many landform features of continental glaciation are well illustrated on the Jackson, Michigan, 7.5 minute quadrangle map.

This region has a humid continental climate. The summers are mild and pleasant. Excessively warm and humid air seldom reaches this area for more than a few days at a time. Instead, cool but pleasant evening temperatures tend to be the rule in summer. Winters, however, are long and often harsh. Snow can be abundant and on the ground continuously for many weeks or even months at a time. The annual temperature range is significant and precipitation occurs year round, but is provided primarily by storms moving along polar fronts.

Interpreting the Map

1. Describe the general topography of the map area.

2. What is the local relief? Why is it so difficult to find the exact highest and lowest points on this map?

3. Does the topography of this region indicate glacial erosion or glacial deposition?

4. Does the area appear to be well drained? What are the three main hydrographic features that indicate the drainage conditions?

5. What type of glacial landform is Blue Ridge? What are its dimensions (length and average height)?

6. How is a feature such as Blue Ridge formed? What economic value might it have?

7. What is the majority of the surface material in the map area? What is the term for this type of surface cover?

8. What probably caused the many small depressions and rounded lakes? What are these depressions called?

9. Describe how the topography of the Jackson, Michigan, area appears on the satellite image.

10. What is the advantage of having satellite image and topographic map coverage of an area of study?

**Opposite:
Jackson, Michigan
Scale 1:62,500
Contour interval = 10 ft
U.S. Geological Survey**

The Driftless Area—A natural region

The Driftless Area, a region in the midwestern United States, is located mainly in Wisconsin, but also occupies sections of the adjacent states of Iowa, Minnesota, and Illinois. The unifying geographic characteristic of this region is a landscape that is free of deposits left by the Pleistocene ice sheets that covered most of Canada, and extended over the northern Midwest and northeastern United States. As discussed in this chapter, "drift" is a term for glacial deposits, so the regional name is appropriate and linked to its geomorphic history. There is some debate about whether earlier Pleistocene glaciations covered the

region, but the most recent glacial advance did not override this locality. The landscape of this region is very different from the muted terrain and low rounded hills of adjoining glaciated areas. Narrow stream valleys, steep bluffs, caves, sinkholes, and loess-covered hills produce a scenic landscape with many landforms that would not have survived erosion by a massive glacier.

In some parts of the United States, lobes of the ice sheet extended as much as 300 miles farther south than the latitude of the Driftless Area. Why this region was glacier-free has to do with the topography directly to the north. A highland of resistant rock, called the Superior Upland, caused

Till Plains In the zone of ice sheet deposition, massive amounts of unsorted glacial till accumulated, often to depths of 30 meters (100 ft) or more. Because of the uneven nature of deposition, the configuration of land covered by till varies from place to place. In some areas, the till is too thin to hide the original contours of the land; in other regions, thick deposits of till form broad, rolling plains of low relief. Small hills and slight depressions, some filled with water, characterize most **till plains,** reflecting the uneven glacial deposition. Some of the best agricultural land of the United States is found on the gently rolling till plains of Illinois and Iowa. The young, dark-colored, grassland soils (mostly mollisols) that developed on the till are extremely fertile.

Outwash Plains Beyond the belts of hills formed by the terminal and recessional moraines lie the **outwash plains.** These are extensive, relatively smooth plains covered with sorted deposits carried forward by the meltwater from the ice sheets. Outwash plains, which may cover hundreds of square kilometers, are analogous to the valley trains of alpine glaciers.

Pits, called **kettles,** formed when blocks of ice were buried in the glacial deposits, mark some outwash plains, till plains, and moraines. Eventually, where the blocks of ice melted, holes remained. Today, countless kettles form lakes in the glaciated landscapes. For example, most of Minnesota's famous 10,000 lakes are **kettle lakes** (Fig. 19.31).

Drumlins A **drumlin** is a streamlined hill, often about 0.5 kilometer (0.3 mi) in length and less than 50 meters (160 ft) high, molded in glacial drift on the till plains (Fig. 19.32a). Drumlins are oddly shaped, resembling half an egg or the convex side of a teaspoon, and are usually found in swarms, with as many as a hundred or more clustered together. Glaciologists do not yet understand

exactly how drumlins were formed. The most conspicuous feature is their elongated shape, which follows the direction of ice flow. Their broad, steep noses face the direction from which the ice advanced, while their gently sloping, narrow tails point in the direction of ice flow. Thus, their geometry is the reverse of that of roches moutonnées. They are well developed in Canada, Ireland, and in the states of New York and Wisconsin. Boston's Bunker Hill, a drumlin, is one of America's best-known historical sites.

Eskers An **esker** is a narrow, winding ridge composed of glaciofluvial gravels (Fig. 19.32b). Some eskers are as long as 200 kilometers (130 mi), but usually they do not exceed several kilometers. It is believed that most eskers were formed by streams of meltwater flowing in ice tunnels at the base of ice sheets. Eskers are prime sources of gravel and sand for the construction industry. Being natural embankments, they are frequently used in marshy, glaciated landscapes as highway and railroad beds. Eskers are especially well developed in Finland, Sweden, and Russia.

Kames Roughly conical hills composed of sorted glaciofluvial deposits are called **kames.** They are presumed to have formed in contact with glacial ice when sediments accumulated in ice pits, in crevasses, and among jumbles of detached ice blocks. Like eskers, kames are excellent sources for mining sand and gravel and are especially common in New England. **Kame terraces** are landforms resulting from accumulations of glaciofluvial sand and gravel along the margins of ice lobes that melt in valleys in areas of hilly relief. Examples of kame terraces can be seen in New England and New York.

Erratics Scattered in and on the surface of glacial deposits may be boulders called **erratics** (Fig. 19.33) because the rock they consist of differs from the local the front of the glacier to split into two lobes, diverting the southward-flowing ice around this topographic obstruction. The diverging lobes flowed into two valleys that today form Lake Superior and Lake Michigan. These two lowlands were oriented in directions that channeled the glacial lobes away from the Driftless Area. Although not overridden by these glaciers, the region did receive outwash deposits and a cover of wind-deposited loess, both derived from the surrounding glaciers and their sediments.

The Driftless Area has a unique landscape because of its isolation, an unglaciated "peninsula" almost completely surrounded by extensive glaciated regions. The rocks, soils, terrain, relief, vegetation and habitats contrast strongly with the nearby drift-covered areas, and provide a glimpse of what the preglacial landscapes of this midwestern region may have been like. The region also provides distinctive habitats for flora and fauna that do not exist in the glaciated terrain. Unique and unusual aspects of the topography and ecology in this region attract ecotourists and offer countless opportunities for scientific study. For these reasons there are many protected natural lands in the Driftless Area, an excellent example of a natural region defined by its physical geographic characteristics.

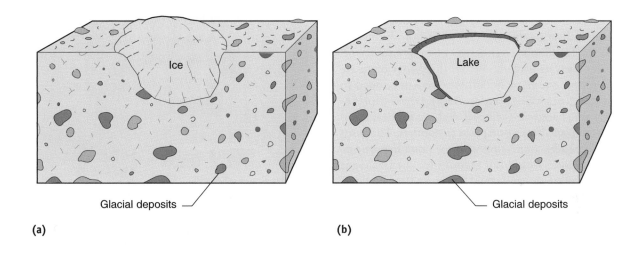

Glacial deposits

(a)

Glacial deposits

(b)

© Galen Rowell/CORBIS

(c)

Figure 19.31 Kettle lakes form (a) when huge blocks of ice melt away after burial or partial burial in glacial deposits. (c) This kettle, near the headwaters of the Thelon River in the Northwest Territories, Canada, formed in an area of ground moraine deposits, as can be seen by the hummocky topography.

(a)   **(b)**

Figure 19.32 Drumlins and eskers. (a) Porcupine Islands, a series of submerged drumlins at Acadia National Park, Maine. Drumlins are streamlined hills elongated in the direction of ice flow. (b) An esker near Albert Lea, Minnesota. Eskers are depositional ridges formed by meltwater deposits in a tunnel under the ice. ■ **What economic importance do eskers have?**

bedrock. Often the source regions of erratics can be identified, indicating the direction of ice flow during the Pleistocene. In Illinois, erratics may be from a source region as far away as Canada. Before the theory of an ice age was proposed, many hypotheses were developed to explain the existence of erratics. Among these explanations was one based on the belief that the Biblical flood transported rocks from one place to another. The term *drift* originated in connection with the flood hypothesis. However, a flood would not account for the striations present on the erratics and could not move large boulders hundreds of kilometers. Also, such boulders would be found worldwide, not just in the middle and high latitudes.

Glacial Lakes and Periglacial Landscapes

The ability of ice sheets to form lake basins has already been mentioned. Millions of glacially created lakes are still in existence, formed by the continental ice sheets that once covered much of North America and Eurasia. The ice sheets created some lakes by scooping out deep elongated basins along former stream valleys. If the glacier deposited a moraine dam at one end of such a basin, a meltwater lake would remain after the retreat of the ice sheet. New York's beautiful Finger Lakes are a classic example of moraine-dammed lakes in ice-deepened elongated basins (Fig. 19.34). Alpine glaciers can produce similar results, as in the case of Washington's Lake Chelan and Lakes Maggiore, Como, and Garda in the Italian Alps. Although many of the lakes formed by ice sheets no longer exist, the **glaciolacustrine** (from *glacial,* ice; *lacustrine,* lake) deposits that they laid down prove their former existence and size.

Some lakes formed where the disruption of drainage by glacial deposition, or the ice itself, prevented depressions from being drained of meltwater. This situation

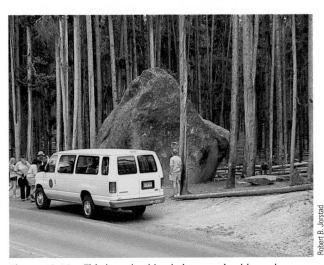

Figure 19.33 This huge boulder, being examined by a class on a field trip to Yellowstone National Park is a glacial erratic, transported and deposited by a glacier. ■ **What does this indicate about the ability of flowing ice to modify the terrian?**

usually developed where water was trapped between a large end moraine and the ice front, or where the land sloped toward, instead of away from, the ice front. In both situations, **ice-marginal lakes** were formed from meltwater. They were drained and ceased to exist when the retreat of the ice front uncovered an outlet route.

During their existence, the floors of these ice-marginal lakes accumulated layers of fine sediment. The result of this sedimentation can be seen in the extremely flat surfaces that characterize most glaciolacustrine plains. The outstanding example of such a plain is the valley of the Red River in North Dakota, Minnesota, and Manitoba. This plain, one of the flattest landscapes in the world, is of great agricultural significance because it is well suited to growing wheat. The Red River flows northward and eventually flows into Lake Winnipeg. The plain is the result of deposition in a vast Pleistocene

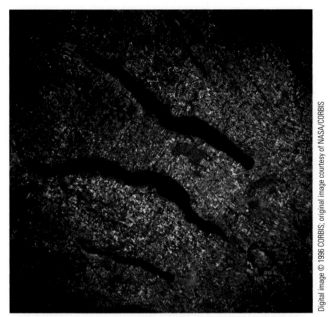

Figure 19.34 The Finger Lakes of New York were excavated by glacial erosion.

Digital image © 1996 CORBIS; original image courtesy of NASA/CORBIS

lake held between the front of the receding continental ice sheet on the north and moraine dams and higher topography to the south. This lake has been called Lake Agassiz, after the Swiss scientist who espoused the theory of an ice age. Lake Winnipeg is the last remnant of Lake Agassiz, lying in the deepest part of the ice-scoured and sediment-filled lowland.

Another ice-marginal lake produced much more spectacular landscape features, but not in the area of the lake itself. In northern Idaho, a glacial lobe moving southward from Canada blocked the valley of a major tributary of the Columbia River and caused the formation of

an enormous ice-dammed lake known as Glacial Lake Missoula. This lake covered almost 7800 square kilometers (3000 sq mi) and was 610 meters (2000 ft) deep at the ice dam. On occasions when the ice dam failed, Lake Missoula emptied in stupendous floods that engulfed much of eastern Washington. The racing floodwaters scoured the basaltic terrain, producing Washington's channeled scablands consisting of intertwining steep-sided troughs (**coulees**), dry waterfalls, scoured-out basins, and other features quite unlike those associated with normal stream erosion, particularly in their gigantic size (Fig. 19.35).

The Great Lakes of the eastern United States and Canada make up the world's largest lake system. Lakes Superior, Michigan, Huron, Erie, and Ontario occupy former river valleys that were vastly enlarged and deepened by glacial erosion. All of the lake basins except that of Lake Erie have been gouged out to depths below sea level and have irregular bedrock floors lying beneath thick blankets of glacial till.

The history of the Great Lakes is very complex, resulting from the back-and-forth movement of the ice front that produced many changes of lake levels and overflow in varying directions at different times (Fig. 19.36). The earliest lakes appear to have emerged near the southern tip of the Lake Michigan Basin (Pleistocene Lake Chicago) and the western end of the Erie Basin (Pleistocene Lake Maumee). These lakes drained westward to the Mississippi through the Illinois and Wabash Rivers.

Ice retreat exposed the southern fringe of the Ontario Basin (Pleistocene Lake Iroquois), which was occupied by a lake with an eastern outlet through New York's Mohawk and Hudson Valleys. By this time, the basin of Lake Huron was emerging, with an outlet westward across Michigan through the Grand River to Lake Michigan. This outlet also channeled the overflow from the Erie Basin as the Wabash route ceased to function.

Jeff Schmaltz, MODIS Rapid Response Team, NASA/GSFC

Figure 19.35 This area of eastern Washington State (tan colored) is called the channeled scablands because of the tremendous river channels that criss cross the region (dark grey). The failure of glacier formed dams during the Pleistocene released huge amounts of water, perhaps ten times the flow of all the rivers of the world today that scoured the landscape.

(a) Glacial retreat

(b) Port Huron glacial advance

(c) Glacial retreat (post-Valderan)

(d) Postglacial Great Lakes

Figure 19.36 Stages in the formation of the Great Lakes of North America as the ice sheet receded at the close of the Pleistocene Epoch.
■ **Name and locate the five Great Lakes.**

Further recession of the ice exposed the western portion of the Superior Basin (Pleistocene Lake Duluth), which overflowed westward to the Mississippi through the St. Croix River in Minnesota. Lakes Michigan, Huron, Erie, and Ontario were now linked to overflow eastward to the Mohawk and Hudson Valleys, with Lakes Michigan and Huron spilling along the ice front into Lake Iroquois rather than following their present route through Lakes St. Clair and Erie.

About 9000 years ago, the St. Lawrence outlet was exposed after the ice retreated northward, and the Great Lakes formed a single system emptying to the east—the upper lakes entering the St. Lawrence by way of the Ottawa River. This low outlet permitted the lakes to diminish well below their present levels and areas. However, complete deglaciation caused Earth's crust to slowly rise and raised the outlet. This process so enlarged the lakes that the old Illinois River outlet of Lake Michigan began to function again, and a new connection formed between Lakes Huron and Erie through Lake St. Clair. Eventually, the St. Lawrence outlet was lowered, the Illinois River link was abandoned, and crustal rise terminated the Ottawa River link between the upper and

lower lakes. With these developments, the modern lake system was finally established.

In addition to alterations in the terrain caused directly by glaciation, other landscape changes occurred. The cool Pleistocene middle-latitude arctic and subarctic conditions created **periglacial landscapes** (*periglacial* means "near the ice"). Permafrost conditions created landforms, such as ice wedges, patterned ground, and smoothed hill slopes due to solifluction, that are peculiar to tundra climates and areas where frost action dominates (see Chapter 15).

The weight of the ice depressed the land surface several hundred meters. As the glacier receded, the land rose slowly in a process known as **isostatic rebound.** This process helped form today's Great Lakes, and areas such as Hudson Bay and the Baltic Sea may someday emerge above sea level. Measurable isostatic rebound still raises elevations of areas such as Sweden, Canada, and some of eastern Siberia by up to 2 centimeters (1 in.) per year. Should Greenland and Antarctica lose their ice sheets someday, their depressed central land areas would also rise to reach isostatic balance.

DEFINE & RECALL

glacier	equilibrium line	drift	ice-scoured plain
firn	crevasse	till	till plain
calving	icefall	glaciofluvial deposit	outwash plain
alpine glacier	glacial plucking	moraine	kettle
valley glacier	striation	lateral moraine	kettle lake
piedmont glacier	glacial abrasion	medial moraine	drumlin
cirque	roche moutonnée	end moraine	esker
cirque glacier	tarn	terminal moraine	kame
continental ice sheet	arête	recessional moraine	kame terrace
ice cap	horn	ground moraine	erratic
zone of accumulation	col	glacial outwash	glaciolacustrine
zone of ablation	glacial trough	valley train	ice-marginal lake
ablation	paternoster lakes	outlet glacier	coulee
bergschrund	hanging valley	iceberg	periglacial landscape
terminus (snout)	fjord	ice shelf	isostatic rebound

DISCUSS & REVIEW

1. Describe the two major types of glaciers, and explain how each is formed.

2. Diagram and label the characteristic parts of an alpine glacier.

3. How does a glacier maintain its budget in a state of equilibrium?

4. Explain how a glacier moves.

5. How does a glacier accumulate its load? Cite examples of evidence of glacial erosion and movement.

6. What is the difference between *till* and *outwash*?

7. Explain how one can distinguish glacial valleys from stream valleys in mountain areas.

8. How are hanging valleys formed?

9. Describe the major types of deposition by alpine glaciers.

10. Where are the two major existing ice sheets located? How do they compare in area and thickness?

11. How has ice sheet erosion altered the landscape? What kinds of landscapes are produced after continental glaciers deposit their loads and recede?

12. What are the differences between eskers, drumlins, and kames?

13. Explain the relationship of glaciation to the origin and history of the Great Lakes.

14. How does the present extent of continental ice sheet coverage compare with the maximum extent of the Pleistocene ice sheet coverage?

CONSIDER & RESPOND

1. Using your textbook and lecture notes, give the correct glacial landform term for each of the following definitions.

    a. Amphitheater-shaped head of a glacial valley

    b. Peaked mountain summit formed by erosion of valley glaciers on several sides

    c. Glacial deposits located along the sides of glaciated valleys

    d. Steep-sided "sawtoothed ridge" between two glacial valleys

    e. Material deposited by meltwater beyond the leading edge of the glacier

2. Name five coastal areas of the world where fjords can be found. What type of climate do most of these coastal regions have?

3. Assume that the maximum advance of the Pleistocene ice sheet were occurring today. Describe what the following present-day locations would be like in terms of climate and terrain.

    a. Upper Michigan

    b. Coastal New Jersey

    c. Eastern Washington

    d. Central interior Alaska

4. Describe the ice budget of a glacial system. What two major factors control this budget? How is this related to the movement of glaciers?

5. Glaciation is truly an interdisciplinary topic. In addition to physical geographers, what other scientists are involved in the study of glaciers? Why are they so interested in glacial ice?

THE GLOBAL OCEAN

Oceans cover more than 70 percent of Earth's surface, yet oceanography is among the youngest branches of science and we have just begun to investigate its associated content, concepts, and unanswered questions.

Why is oceanography such a young science? What does the global ocean mean to you personally? What are the four major divisions of the global ocean?

There are significant surface and depth differences in the salinity, temperature, density, and pressure of seawater.

What are the most important differences? Why does salinity vary in different geographic locations? What effects and limitations does water pressure have on humans?

Contrary to early scientific opinion, there are topographic irregularities on the ocean floor comparable to those that exist on the continents.

What did early scientists believe the ocean floor looked like? Why? What are the major features associated with the continental margins and the deep-ocean floor?

Islands are landmasses that are smaller than continents and surrounded by water.

What are the three major types of islands? How are they formed? What are coral reefs?

The surface movements of the global ocean have a major impact on humans and their activities.

What are the three most important ocean movements? How do they impact humans?

Earth Sciences CD

Chapter 9: Plate Tectonics, *Triple Junctions and Seafloor Study*
Chapter 5: Waves, Tides, & Currents, *Tidal Forces and Tidal Shifts*
Chapter 5: Waves, Tides, & Currents, *Wave Properties,*
Wave Progression,
Wind-Wave Relationships
Chapter 8: Earthquakes & Tsunamis, *Tsunamis,* and
Seismic Case Study,
Alaska, 1964-Tsunami

Jacques Descloitres, MODIS Land Rapid Response Team, NASA/FSFC

The Atlantic Ocean, the Caribbean Sea, and the Gulf of Mexico merge together in this scene. The bright blue-green colors in the water indicate dense colonies of plankton.

Knowledge of the oceans and seas is of major importance to an understanding of physical geography. Oceans cover most of Earth's surface and play a major role in the physical systems of the planet. For instance, the oceans are the world's major collector of solar radiation, and they absorb, store, and distribute much of Earth's energy. The oceans play a significant role in Earth's hydrologic cycle and in the global circulation system, and they are a major control of weather and climate. The oceans are a vital producer of the world's oxygen and an important absorber of carbon dioxide. The floors of the global ocean contain two-thirds of Earth's surface topography, and its submarine geological processes are still being discovered. Finally, the oceans are a source for food, minerals, and energy resources that, hopefully, will serve humankind for centuries to come.

Introduction to the Oceans

The oceans contain more than 97 percent of Earth's water and cover 71 percent of Earth's surface. In fact, the name for our planet, Earth, is a misnomer. It should probably be called Oceanus or Hydro, because water dominates its surface and it is the only planet in the solar system with liquid water at the surface. Yet our knowledge of the oceans is not very extensive because they are difficult to explore. The oceans are vast, dark at depth, and the pressures existing at ocean depths present a hostile environment to humans. Only since the British expedition aboard HMS *Challenger,* from 1872 to 1876, have scientists seriously begun to explore the ocean's complexity. Oceanography was not generally recognized as an important science until the second half of the twentieth century. Most of our knowledge of the ocean has been developed with very recent technologies, such as sonar (using sound waves to determine ocean floor topography), deep-diving submersibles, underwater photography and videography, deep-ocean drilling ships, environmental satellites, and the global positioning system (GPS).

The oceans actually form a single, large, continuous body of water. The "global ocean" is divided into four individual oceans that are distinguished largely on the basis of their geographic locations, although they do differ somewhat in certain marine characteristics as well. The four oceans are the Pacific, Atlantic, Indian, and Arctic Oceans. The approximate area of the Pacific Ocean (Fig. 20.1) is 166 million square kilometers (64 million sq mi). The Atlantic is about half that size, or 83 million square kilometers (32 million sq mi), and the Indian Ocean is about 73 million square kilometers (28 million sq mi). The Arctic Ocean is much smaller, at 13 million square kilometers (5 million sq mi). In comparison, the continent of North America is 23 million square kilometers (9 million sq mi) in area. In fact, the total land area of Earth—149.5 million square kilometers (57.5 million sq mi)—is smaller than the area of the Pacific Ocean, which is the largest geographic feature on Earth's surface.

The oceans vary greatly in depth as well as in size. The Pacific Ocean has an average depth of 4200 meters (14,000 ft), and the Atlantic and Indian Oceans have average depths of 3900 meters (13,000 ft). The Arctic Ocean is the shallowest, averaging only 930 meters (3250 ft). However, these figures are somewhat misleading, as the ocean floors are not just flat plains, but have mountains, trenches, and basins that vary considerably in depth. The Pacific reaches a maximum depth of more than 11,000 meters (36,000 ft).

Seas are saltwater bodies that are smaller than oceans and somewhat enclosed by land. Unlike lakes, which can be of either fresh or salt water, seas interchange water with the oceans. Examples of major seas include the Mediterranean, Baltic, Bering, Arabian, Caribbean, Coral, North, Black, Yellow, and Red Seas. Some large semi-enclosed areas of the oceans are also called gulfs or bays, but are actually seas as well. Examples of these include the Gulf of Mexico, Gulf of Alaska, Gulf of California, and Bay of Bengal. Large salty lakes, such as the Aral, Caspian, and Dead "Seas," are incorrectly named, as they are landlocked and are not at global mean sea level.

Characteristics of Ocean Waters

The world ocean is 96.5 percent water by weight, but seawater is a dilute solution of many dissolved solids, dominated by salts. On average, 3.5 percent, or 35 parts per thousand (ppt) is dissolved solids. Of these dissolved minerals, the most common substance by far is sodium chloride, or common table salt (Fig. 20.2). Other important constituents of seawater are magnesium, sulfur, calcium, and potassium. The oceans also contain the dissolved gases of the atmosphere, especially oxygen and great amounts of carbon dioxide.

Ocean waters contain virtually every element found on Earth. Most, however, are trace elements found only in concentrations of a few parts per billion (ppb). For example, there are about 40 pounds of gold and 200 pounds of lead per cubic mile of ocean water. We now extract salt (sodium chloride) and magnesium from the sea, and the future extraction of other substances, including hydrogen for energy, may also prove worthwhile.

NASA/Goddard Space Flight Center, The SeaWiFS Project and ORBIMAGE Science Visualization Studio

Figure 20.1   This view shows the vast extent of the Pacific Ocean.
■ **Why has Earth been referred to as the "ocean planet"?**

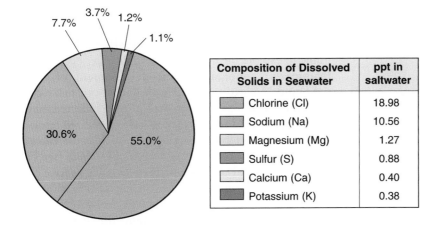

Composition of Dissolved Solids in Seawater	ppt in saltwater
Chlorine (Cl)	18.98
Sodium (Na)	10.56
Magnesium (Mg)	1.27
Sulfur (S)	0.88
Calcium (Ca)	0.40
Potassium (K)	0.38

The measurement of all dissolved solids in seawater is referred to as **salinity,** and it varies throughout Earth's oceans and seas. Although, as already mentioned, the average salinity is 3.5 percent (35 ppt), this can vary in the open ocean from 3.2 to 3.8 percent (32–38 ppt). The salinity is even greater in enclosed or partially enclosed seas.

Several factors affect salinity—primarily, amounts of precipitation and rates of evaporation. In humid regions, where precipitation is high, fresh water tends to dilute the seawater and reduce salinity. The flow of major rivers into the sea also lowers the concentration of dissolved salts. Arid and semiarid regions have a higher salinity, because precipitation is low, few rivers exist, and evaporation is high.

Ocean temperatures also vary and, together with salinity, affect density. Density is the mass per unit volume of a substance and is measured in kilograms per cubic meter (or pounds per cubic foot). The maximum density of fresh water occurs at about 4°C (39°F) and is 1000 kilograms/cubic meter (62.4 lb/cu ft). Salinity and density are directly proportional—the greater the salinity, the denser the water. The salinity of ocean water raises water density to about 1025 kilograms/cubic meter (64 lb/cu ft). Above 4°C/39°F in fresh water (slightly lower in salty water), an increase in temperature results in a decrease in density. The densities of ocean waters affect circulation because density differences cause gravitational displacement of water. Because warm water is less dense than cold, it tends to float above colder, denser water. Cold or highly saline water near the surface sinks and is replaced by warmer or less saline water. This effect produces patterns of surface-water movement that vary with the seasons. These density differences also cause deep-ocean currents (*thermohaline* currents) that circulate even the ocean's deepest waters.

Surface Variations

The ocean surface has significant regional variations in salinity. Salinity is highest in subtropical high pressure regions near 30°N and S, because of low precipitation and high evaporation rates. Salinity decreases toward the equator because of abundant rainfall, heavy stream flow, and lower evaporation rate due to increased cloud cover and humidity. For example, salinity is only 2 percent (20 ppt) near the equatorial coast of Brazil because of the diluting influence of the Amazon River as well as the abundant rainfall and reduced evaporation rate.

Low salinity is also found in the polar regions, because of the extremely low evaporation rate and the summer freshwater inflow from melting snow and glacial ice. Moving equatorward, salinity increases in the middle latitudes with their more temperate ocean waters and greater evaporation, especially during the summer.

The highest marine salinities are found in semi-enclosed seas in hot, arid regions, where the evaporation rates are very high and there is no stream flow. For example, the Red Sea, between the Sahara and the Arabian Desert, has a salinity of more than 4 percent (40 ppt). Desert basin lakes have the world's highest salinity; the Dead Sea is so saline— 23.8 percent (238 ppt)—that most aquatic organisms cannot survive (hence its name). The Great Salt Lake in Utah has a maximum salinity of 22 percent (220 ppt).

Variations with Depth

The vertical temperature distribution in the oceans results from solar heating that is concentrated in the surface waters. Some of this heat is mixed into deeper water by conduction and currents. Oceans tend to have three vertical temperature zones: (1) a surface layer of generally warmer water, (2) a transition zone of rapidly cooling waters with increasing depth, and (3) the uniformly cold (average 3.5°C/37°F) waters of the deep ocean. The change from the heated surface layer to the cold deep water occurs in a well-defined transition zone called the **thermocline.** The thermocline is most apparent in the tropics and during summer in middle latitudes, when surface waters are most strongly heated. In polar waters, there may be no thermocline as seawater temperatures are often as cold at the surface as they are at great depth.

Pressure increases rapidly with depth (Fig. 20.3). The effect is comparable to the increase in atmospheric pressure as we descend down a mountain slope to a lower elevation. However, water is more than 800 times denser than air, and water pressure increases by 1 atmosphere (14.7 pounds per square inch) for every 10 meters (33 ft) of water depth. At a depth of 300 meters (1000 ft), water pressure is 445 pounds per square inch. At 600 meters (2000 ft), it is 892 pounds per square inch, and at 1800 meters (6000 ft), it is 2685 pounds per square inch. At the greatest ocean depths in the Pacific's Mariana

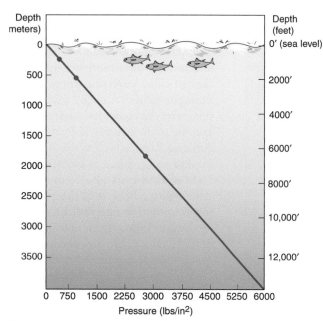

Figure 20.3 Water pressure in the ocean increases directly with increasing depth.

Trench, pressure exceeds 16,000 pounds per square inch. Dangerous water pressure will strain air-filled human organs such as eardrums and lungs, even during shallow descents into the ocean. For this reason, the deepest scuba dives have been only to about 100 meters (330 ft). However, deep-diving submarines and research submersibles can descend thousands of meters while maintaining a normal atmospheric pressure for humans inside (Fig. 20.4).

 # The Ocean Floor

It was not until the twentieth century that scientists were able to explore the ocean floor and discover details of its submarine topography. The development of sonar (SOund Navigation And Ranging) that can make continuous recordings of ocean depths using reflected sound waves has helped enormously, as have specialized research ships designed for deep-sea drilling, such as the *Glomar Challenger,* and deep-diving vehicles like *Alvin* (see Fig. 20.4). Recent advances in marine technologies have produced materials and engineering designs able to withstand the enormous pressures far below the ocean surface.

At present, we know more about the surface of the moon than we do about our own seabed. However, recent innovations are rapidly increasing our knowledge and ability to explore deep-sea terrain. Undersea research laboratories allow scuba-diving scientists to stay in shallower reaches of the ocean for days at a time. New side-scanning sonar can record swaths of seafloor several kilometers wide with great accuracy and can produce three-dimensional bathymetric contour maps. Camera and video systems can image the deepest sections of the ocean floor. Though submersibles such as *Alvin* will continue to take scientists to the ocean floor, robotics are the way of the future for ocean exploration. Remote-controlled submersibles such as *Argo-Jason,* which helped locate HMS *Titanic* and other famous sunken vessels, are the safest and most efficient deep-ocean exploration vehicles.

Before the most recent explorations of the sea, it was believed that most of the ocean floor consisted of flat

Figure 20.4 The deep-diving submersible Alvin has taken scientists to the ocean bottom to study the oceanic ridges and make numerous scientific discoveries. Inside Alvin, crew members are under normal atmospheric pressure, while outside, the submersible is under tremendous pressure from the ocean depths.
■ **Does all undersea exploration require manned submersibles? Why or why not?**

Rod Catanach, Woods Hole Oceanographic Institution

NOAA/NGDC

Figure 20.5 The sea floor topography in this image was obtained from satellite data. Radar altimeters can measure the seawater surface accurately to within a few centimeters. Variations in gravity cause surface to bulge slightly over seamounts and oceanic ridges creating a small mound of water at the sea surface. Sea level dips slightly over trenches. Thus the ocean bottom can be "seen" from space.
■ **Identify the Mid-Atlantic Ridge and the trenches around the Pacific Rim from this image. What colors represent these features?**

plains. We now know that this is far from the truth. Beneath the ocean waters are landforms greater in size and extent than those found on the continents (Fig. 20.5). Mountains, basins, plains, volcanoes, escarpments, canyons, and trenches are all present beneath the ocean (Fig. 20.6).

In the 1960s, Bruce Heezen and Marie Tharp produced the first global map of the ocean floor (Fig. 20.7, pages 558–559). Since then, mapping of the ocean floor's topography, known as **bathymetry,** has advanced rapidly, with thousands of echo soundings by ships and the use of computer-imaging techniques to produce detailed bathymetric maps. Amazingly, the sea bottom

can also be mapped from satellites, based on the measurement of sea level. Variations in gravity cause sea level to bulge slightly over higher areas such as submarine mountains and to dip over deep marine trenches (see again Fig. 20.5).

The ocean waters spill over the rim of their basins and onto the continental edges. Thus, much of the shallow seafloor is geologically a part of the continents. As a result, the ocean basins have two major topographic subdivisions: (1) the continental margins, consisting of the continental shelf and continental slope, which are part of the continental crust; and (2) the deep-ocean floor, which is formed by oceanic crust (see again Fig. 20.5).

Figure 20.6 The major features of the ocean floor are shown in this profile of the North Atlantic Ocean from the east coast of the United States to the Mid-Atlantic Ridge. ■ **What is the deepest part of this section of the North Atlantic Ocean?**

Figure 20.7 The "Floor of the Ocean" map by Heezen and Tharp. This map was produced in the 1960s and was the first detailed submarine relief map of the world ocean floor. It still serves well for a general view of the major ocean floor features. ■ **Locate the deepest point in the global ocean.**

The Floor of the Oceans map, © Marie Tharp

Features of the Continental Margins

Continental Shelf   The **continental shelf** consists of the edges of the continents flooded by ocean water. This part of the seafloor slopes gently from sea level to depths of about 200 meters (650 ft) below sea level. The continental shelf varies a great deal in width. In some places, it is almost nonexistent; in others, it is as wide as 1300 kilometers (800 mi). Generally, where broad continental plains slope to meet the ocean (passive margins), the continental shelf is quite wide. Where mountains are found along the continental edge (active margins), the shelf is narrow. These extremes are found on the two coasts of the United States. Beyond the "passive" plains of the Atlantic and Gulf coasts, the continental shelf extends outward as much as 480 kilometers (300 mi). On the Pacific Coast, with its "active" mountainous margins, the continental shelf is narrow. It is not surprising that the character of the continental shelf is related to the character of the nearby land, especially when we remember that the shelf is an extension of the continents and geologically is not a part of the ocean basins.

The continental shelf is the "treasure chest of the sea." The amount of sunlight it receives and the supply of nutrients washed into these waters from the continent's rivers support vast populations of marine plants and animals. Thus, 90 percent of the fish we eat are caught in the shallow waters above the continental shelf. Virtually all the lobster, crab, shrimp, and shellfish live in these rich waters. Vast offshore reserves of gas and oil are stored in the shelf sediments, as are gravels (for construction) and some minerals, including phosphates (for fertilizer) and diamonds. Presently, about 25 percent of the world's petroleum is produced from offshore oil rigs.

The topography of the continental shelf is not as smooth as we might expect from the effects of constant wave motion, settling of sediments (sand, silt, and clay) brought by rivers from the adjacent land, and global sea-level changes. Although it is usually a relatively smooth, gently sloping plain, the shelf often has ridges, depressions, hills, valleys, and canyons. Some of the higher features break the water surface and appear as islands, such as New York's Long Island and Massachusetts' Martha's Vineyard. During periods of lower sea level during the Pleistocene, when much water was stored on the land as glacial ice, the continental shelves formed land bridges connecting Alaska to Siberia and Great Britain to continental Europe. The teeth and bones of Pleistocene land mammals are occasionally brought up in fishing nets from continental shelves previously exposed as land areas. The balance between water in the oceans and glacial ice on land is important in determining the location of the coastline. If global warming accelerates and causes rapid glacial melting, many low-lying coastal areas will become submerged parts of the continental shelves.

Submarine canyons are probably the most striking features of the continental margin. These canyons are steep-sided, erosional valleys cut into the sediment and rock of the shelf and slope. They resemble canyons cut by rivers on land (Fig. 20.8), and some are larger than the Grand Canyon. V-shaped in cross section, these submarine canyons are often found opposite the mouths of major rivers, such as the Congo or Hudson. This distri-

Figure 20.8   Monterey Canyon off the central California coast is a submarine canyon that is deeper than the Grand Canyon of the Colorado River. This image was created from high-resolution satellite imagery and high-tech digital lithography. The region is now designated the Monterey Bay National Marine Sanctuary.
■ **What is the probable origin of submarine canyons such as this one?**

bution has led to the hypothesis that rivers helped in forming the canyons. Yet some submarine canyons are not located off present-day or ancient river mouths. Many hypotheses have been proposed as to how these submarine canyons were formed. The most widely accepted explanation is that they were carved by **turbidity currents**—periodic submarine flows of sediment and water (a slurry) that move from the continental shelves down to the deeper ocean floor. These gravity flows of dense turbid water have a high capacity for eroding into the seabed. The upper portions of the canyons may have formed by surface erosion during the Pleistocene, when lower sea levels exposed much of the present-day continental shelves. The lower portions of some submarine canyons are cut thousands of meters below sea level and have fans of sediment at their mouths on the ocean floor.

Continental Slope Marking the outer edge of the continental shelf is a relatively steep drop, usually 3000 to 3600 meters (10,000–12,000 ft), to the ocean floor (Fig. 20.9). This drop, called the **continental slope,** forms the true boundary between the continents and the ocean basins. The continental slope begins at depths of 120 to 180 meters (400–600 ft).

The continental slope is actually not an extremely steep incline, but descends at an angle of about 15 degrees. However, this is much steeper than the gentle surface of the adjacent continental shelf, which slopes at an average of only 1 or 2 degrees. What is most characteristic of the continental slope is its great descent, usually some 3600 meters (12,000 ft), but sometimes as much as 9000 meters (30,000 ft), to the ocean floor or into depths of the trenches. There is evidence of massive submarine mudslides or slump flows off the edges of some continental slopes.

Some ocean floor areas at the base of the continental slope form a gently sloping surface, known as the **conti-**

nental rise (see Fig. 20.9). These continental rises are well developed along the passive margins of the Indian and Atlantic Oceans, especially off major river mouths, such as in the Bay of Bengal. However, they are almost nonexistent along the active Pacific margins, where trenches are found at the edges of the continental margins. The continental rise consists of muddy sediments from turbidity currents and landslide deposits from the shelf and slope. Deep-ocean currents may distribute these materials along the base of the continental slopes and out onto the deep-ocean floor.

Features of the Deep-Ocean Floor

The deep-ocean floor lies at an average depth of about 4,000 meters (12,000 ft) below sea level. Until recently it was believed to be a relatively smooth submarine plain. Now, with sophisticated bathymetric maps, we have discovered that the topography of the deep-ocean floor is as irregular as that of the continents. The ridges and depressions of the ocean floor often rival and surpass the relief of continental landforms. In contrast, large areas of the deep ocean are blanketed by thousands of meters of marine sediments forming virtually featureless plains.

Erosional and depositional agents are important in shaping the continents (see Chapters 15–19). Running water, glaciers, wind, weathering, and gravity all work to smooth and level the land. They do this by wearing down higher landforms and by filling in depressions. Under the sea, these leveling agents are either missing entirely or much less active than they are on land. Consequently, many submarine landforms created by volcanic action and crustal tectonics remain essentially in their original form. The two most impressive submarine features of the deep-ocean floor are oceanic ridges and trenches. Other significant terrain features of the deep-ocean are abyssal plains, seamounts, and guyots.

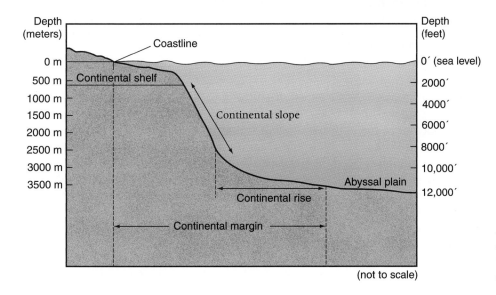

Figure 20.9 A general profile of the major features of the continental margin. ■ **Where is the true boundary between the continents and the deep-ocean basins?**

Figure 20.10   Relief map of the floor of the North Atlantic Ocean, showing such features as the Mid-Atlantic Ridge, the Puerto Rico Trench, abyssal plains, numerous seamounts, and the continental shelves and slopes of surrounding continents.   ■ **What is the most obvious submarine topographic feature in the North Atlantic Ocean?**

Oceanic Ridges   The **oceanic ridges** (or **midocean ridges**) form an interconnected network of submarine volcanic mountains that extends through all four oceans (see again Figs. 20.5 and 20.7). The best known of these, the Mid-Atlantic Ridge (Fig. 20.10), extends southward from the Arctic Ocean to Iceland. It then trends southward, midway between the Americas and Africa, almost to Antarctica before it turns eastward south of Africa and extends into the Indian Ocean. The Mid-Indian Ridge forms an inverted "Y" with one arm extending into the Red Sea (see again Fig. 20.5). Its other arm links the Mid-Atlantic Ridge to the Pacific Ridge and on to the

East Pacific Rise that extends across the South Pacific and then northward into the Gulf of California, separating the Baja California peninsula from mainland Mexico. This continuous global network of submarine mountains is 64,000 kilometers (40,000 mi) long and averages about 1600 kilometers (1000 mi) wide. The oceanic ridges rise an average of 1500 to 3000 meters (5000–10,000 ft) above the ocean floor.

In some areas, the highest peaks of the midocean ridges rise above sea level to form islands. Iceland, the Azores, and Ascension Island are all high volcanic peaks of the Mid-Atlantic Ridge. The Azores rise more than

8100 meters (27,000 ft) above the ocean floor. From the results of oceanographic research, scientists now know that new material is being added to Earth's oceanic crust by undersea volcanic activity along the oceanic ridges.

The topography of oceanic ridges is very rugged, and the whole range consists of volcanic rocks, chiefly basalts forming "pillow lavas." Pillow lavas are globular formations formed by lava cooling rapidly under seawater. Numerous fractures and faults add to the complexity of the oceanic ridges. Running through the middle of the oceanic ridges is a central rift valley that may reach several kilometers in width. It is volcanically active and the center of earthquake activity. Most dramatic are the undersea volcanoes called "black smokers" (Fig. 20.11) that spew out hydrothermal fluids (hot water mixed with fine grains of iron and zinc sulfides). Fracture zones that cut perpendicularly across the oceanic ridges offset the central rift valleys. These fracture zones are actually transform faults, where sections of oceanic plates slide past each other. They cause the rift valleys to have a zigzag pattern and usually form long fault scarps that may extend several thousand kilometers.

The Deep Sea Drilling Project (DSDP) operated the drill ship *Glomar Challenger* between 1968 and 1983.

Figure 20.12   The drill ship JOIDES Resolution explores bottom sediments under the Ocean Drilling Program (ODP). A computer-controlled positioning system supported by two main shafts and 12 side thrusters maintains the ship's position over a specific sea bottom site while it drills in water depths up to 8000 meters.
■ **Why are drill cores of ocean sediments so important to scientists?**

Glomar Challenger drilled more than 1000 cores in all the major oceans. The most remarkable discovery was the confirmation of seafloor spreading. The addition of new oceanic crust rock along the midocean ridges is pushing older parts of the crust apart. Deep-ocean drill cores are vital to earth scientists seeking to explain further detail on the theory of plate tectonics. The Ocean Drilling Program (ODP) continues to explore the ocean bottom with the newer drill ship *JOIDES Resolution* (Fig. 20.12).

Trenches   **Trenches** are the deepest parts of the ocean. They are long, narrow, arc-shaped, steep-sided, and aptly named. Most trenches are not located in the middle of ocean basins, but rather close to and along active oceanic margins. Trenches are found adjacent to zones that have a concentration of volcanic and earthquake activity and are most common on the seaward (convex) side of curving **island arcs** formed by volcanoes, such as the Aleutian Islands.

Figure 20.11   Hydrothermal vents on the mid-ocean ridges are called "black smokers."

Figure 20.13 A satellite image of the western Pacific Ocean, created from gravity data, shows the major trenches off Asia. Note the great depths of the Japan, Mariana, and Philippine trenches. ■ **What caused the formation of these trenches?**

Most trenches, including the deepest ones, are found around the Pacific "Ring of Fire." Challenger Deep, in the North Pacific's Mariana Trench, is the deepest known part of the ocean, reaching 10,915 meters (35,810 ft) below sea level. In 1960, Jacques Piccard and Donald Walsh descended in the bathyscaphe *Trieste* to the bottom of the Mariana Trench. Placed in that trench, Mount Everest would still have a mile of seawater above its summit. Other major trenches in the Pacific are the Aleutian, Kuril, Japan, Philippine, Tonga, and Peru-Chile Trenches (Fig. 20.13). The deepest part of the Atlantic is the Puerto Rico Trench, at 8648 meters (28,374 ft), and the Java Trench, at 7125 meters (23,377 ft), is the Indian Ocean's deepest point (see again Figs. 20.5 and 20.6).

As with the oceanic ridges, the trenches play a major role in Earth's geologic evolution. As noted in Chapter 13, earth scientists believe that oceanic plates are descending below continental plates and are being recycled to Earth's interior by way of the trenches. This process, known as *subduction,* is a key concept in the theory of plate tectonics. Ocean-floor drilling has shown that no ocean-floor rock is older than about 200 million years. The youngest oceanic crust is being created at oceanic ridges, while the oldest oceanic crustal rock is being destroyed in the trenches by subduction. Thus, oceanic crust is completely recycled over a period of a few hundred million years.

Abyssal Plains **Abyssal plains** are vast submarine plains of very low relief. They lie at depths of 3000 to 6000 meters (10,000–20,000 ft) and cover about 40 percent of the ocean floor. Most abyssal plains are covered by layers of sediments that bury the relief of the oceanic crust. Much of this sedimentary blanket consists of fine brown and red clays, contributed by turbidity flows from the continental slopes, wind deposition, and volcanic eruptions. A significant portion also consists of the remains of microscopic marine organisms. Such organic marine deposits are known as **ooze.** Britain's famous white chalk cliffs of Dover are uplifted ancient ocean-floor ooze deposits (Fig. 20.14). In regions of slow deposition, manganese nodules form on the abyssal plains. In the future, these potato-sized concentrations of manganese, iron, copper, and nickel may be mined.

Seamounts and Guyots One of the most common features of submarine topography is the **seamount,** a submarine volcanic mountain. Unlike the continuous system of volcanic mountains that forms the oceanic

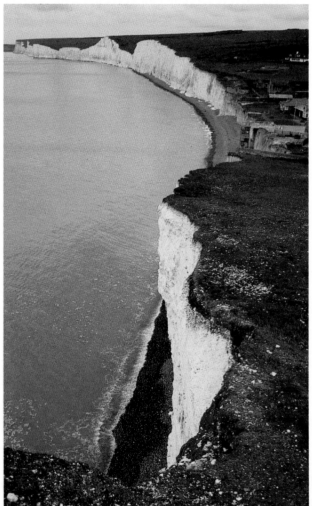

Figure 20.14 The Seven Sisters section of the chalk-white cliffs of Dover along the English Channel. ■ **What is the origin of the cliffs of Dover?**

Rod Catanach, Woods Hole Oceanographic Institution

Figure 20.15   The Hawaiian Island chain. Hawaii, the largest and youngest island, is located over an active "hot spot" in the Pacific plate. The other islands and seamounts of the Hawaiian chain become inactive and sink into the oceanic crust as the Pacific plate drifts to the northwest.   ■   **What is the difference between an oceanic island and a seamount?**

ridges, seamounts are relatively isolated volcanic mountains or groups of mountains, usually with an elevation above the ocean floor of 900 meters (3000 ft) or more. Often steep-sided with small summits, seamounts are volcanic peaks that grow from the deep-ocean floor. Some are high enough to break the ocean surface and appear as oceanic islands, such as the Canary Islands, Tahiti, and Hawaii (Fig. 20.15).

Guyots are flat-topped seamounts, found at depths of a few thousand meters below sea level. The origin of their flat tops at such depths is not certain. Research indicates that they are volcanoes whose summits have been planed off by wave erosion. Later they subsided to their present depths, possibly during lateral movement of the seafloor away from the oceanic ridges and "hot spots."

Islands and Coral Reefs

There are three basic types of islands: continental, oceanic, and coral. **Continental islands** are usually found on the continental shelves. They are geologically part of the continent but are separated from it because of global sea-level changes or regional tectonic activity. The world's largest islands, such as Greenland, New Guinea, Borneo, and Great Britain, are continental. Smaller continental islands include New York's Long Island, California's Channel Islands, and Vancouver Island off the west coast of Canada. The hundreds of barrier islands along the Gulf and Atlantic coasts of the United States are also continental. A few large continental islands, such as New Zealand and Madagascar, are isolated

"continental fragments" that separated from continents millions of years ago.

Oceanic islands are volcanoes that rise from the deep-ocean floor. They are not geologically related to the continents. Most of the oceanic islands, such as the Aleutians, Tonga, and the Marianas, occur in island arcs along the edges of the trenches. Others, like the Azores, are peaks of the oceanic ridges rising above sea level. Many oceanic islands occur in chains, such as the Hawaiian Islands. The oceanic crust sliding over a stationary "hot spot" in the mantle causes these island chains. The exact cause of these hot spots is not known. In the future, the Hawaiian Islands will move northwestward with the Pacific plate, and the islands will slowly sink deeper into the thin oceanic crust. Hence, the islands to the northwest will submerge to become seamounts. A new volcanic island, named Loihi, will form to the southeast (Fig. 20.16). Evidence of the plate motion is indicated by the fact that the youngest islands of the Hawaiian chain, Hawaii and Maui, are to the southeast, while the older islands, such as Kauai and Midway, are located to the northwest.

Coral reefs are shallow, wave-resistant structures formed by an accumulation of skeletal remains of tiny sea animals that secrete a limy skeleton of calcium carbonate. Many other organisms, including algae, sponges, and mollusks, add material to the reef structure. Reef corals need special conditions to grow—clear and well-aerated water, water temperatures above 20°C (68°F), plenty of sunlight, and normal salinity. These conditions can be found in the shallow waters of tropical regions

Mauna Loa Mauna Kea Kilauea Hilo Bay

N

20° 00'N

156° 00'W

155° 00'W

19° 00'N

156° 00'W

155° 00'W

Loihi

University of Hawaii, School of Ocean and Earth Science and Technology Transfer and Economic Development

(a)

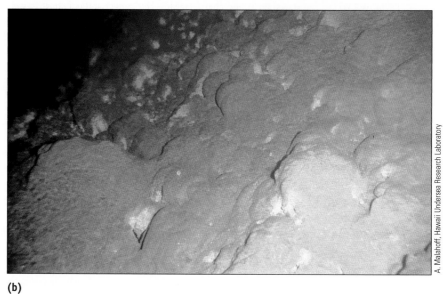

Figure 20.16 The island of Hawaii was formed over a "hot spot." (a) Two large shield volcanoes, Mauna Loa and Mauna Kea, dominate the island. Mauna Loa and Kilauea on its eastern slope are still very active. To the south of the island is Loihi seamount, an active submarine volcano, which may reach above sea level in 50,000 years. ■ **Do all islands form in this manner? Why or why not?** (b) A submarine view of the steep slopes near the summit of Loihi at a depth of 980 meters (3200 ft). The yellow areas are iron-rich minerals spewed from the volcanic vents; the white areas are bacterial mats, which grow around the active volcanic vents. The bacteria chemosynthetically feed on the chemicals spewed from the hydrothermal vents.

(b)

A. Malahoff, Hawaii Undersea Research Laboratory

such as Hawaii, the West Indies, Indonesia, the Red Sea, and the coast of Queensland in Australia. Today, coastal water pollution, dredging, souvenir coral collecting, and possibly global sea temperature warming stresses threaten the survival of many coral reefs.

A **fringing reef** is a coral reef attached to the coast (Fig. 20.17a). Fringing reefs tend to be wider where there is more wave action to bring a continuous supply of well-aerated water and additional nutrients for increased coral growth. They are usually absent where there is a river mouth because the corals cannot grow where the waters are laden with sediment or where river water lowers the salinity of the marine environment.

Sometimes the coral forms a **barrier reef,** which lies offshore, separated from the land by a shallow lagoon

(Fig. 20.17b). Most barrier reefs form in association with slowly sinking oceanic islands, growing at a pace that keeps them above sea level. Other barrier reefs, such as Australia's Great Barrier Reef, the Florida Keys, the Bahamas, and Belize were formed on continental shelves and grew upward as sea level rose after the Pleistocene ice age waned. The world's largest organic structure, the Great Barrier Reef of Australia, is more than 1930 kilometers (120 mi) long.

An **atoll** is a ring of coral reefs encircling a central lagoon that has no inner volcanic island (Fig. 20.17c). Figure 20.18 illustrates the manner in which atolls develop. In response to the subsidence of volcanic islands, the fringing reef grows upward as fast as seafloor subsidence occurs, becoming a barrier reef and finally an atoll.

(a)

(b)

(c)

Figure 20.17 The major types of coral reefs are evident in the Society Island chain of French Polynesia. (a) Moorea is a rugged young oceanic island with a fringing reef. (b) The older island of Bora Bora is subsiding and now has a fringing reef around it. (c) Tetiaroa Atoll has no surface evidence of its former volcanic core. The individual island portions of an atoll are known as "motu" in the South Pacific.

Charles Darwin proposed this explanation of atoll formation in the 1830s. It was proved correct when the U.S. Navy drilled into several Pacific atolls in the 1950s. Drilling evidence indicates that there has been as much as 1200 meters (4000 ft) of subsidence and an equal amount of reef development in the past 60 million years.

Sometimes small islands are formed on atolls from storm waves that erode the reef and heap coral debris above sea level. Such atoll islands are most common in the Pacific. Atolls pose severe challenges as environments for people. First, they have a low elevation above sea level and provide no defense against huge storm waves that can inundate the entire atoll, drowning all its inhabitants. Possible future sea-level rise from global warming may also threaten these low islands. Second, there is little fresh water available on the porous coral limestone surface. Third, little vegetation can survive in the lime-rich rock and soil of which the atoll islands are composed. The coconut palm is an exception, and coconuts were vital to the survival of early inhabitants of the atoll islands in Polynesia and Micronesia.

Fringing reef

(a)

Barrier reef

(b)

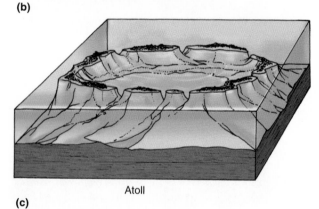

Atoll

(c)

Figure 20.18 The theory of coral reef development around oceanic islands is based on an explanation by Charles Darwin. Three successive reef forms develop from island subsidence and coral reef building. (a) First, a fringing reef forms along the shore. (b) Later, as the island erodes and subsides, a barrier reef forms. (c) Further subsidence causes the coral to build upward while the volcanic core of the island is completely submerged below a central lagoon, forming an atoll. ■ **Explain why the island subsides while the coral grows upward.**

Tides and Waves

Oceans exhibit three major forms of water movement: currents, tides, and waves. First are the major ocean currents—the global flows of surface water pushed by the major wind belts. Because ocean currents and winds play such a significant role in the global circulation system and its effect on climates, they were discussed in Chapter 5. Second are the tides—the periodic rise and fall of sea level caused by the gravitational forces of the moon, sun, and Earth. Third are waves—oscillatory motions on the ocean surface that are generated mainly by the force of winds. Coastlines (to be discussed in Chapter 21) are primarily a product of the work of waves, especially storm waves. The tides allow the waves to break closer or farther from shore and produce local coastal currents that move sediments and circulate water in bays and estuaries.

 ## Tides

The periodic rise and fall of ocean waters in response to gravitational forces acting on them are called the **tides.** For the waters to rise in one place as a high tide, they must be pulled away from another part of Earth, resulting in a low tide there. The gravitational pull of the moon and sun, and the force produced by rotation of the Earth-moon system, are the major causes of the tides (Fig. 20.19). Tides are complex phenomena; in order to explain them, we must first consider the relationship between Earth and the moon.

Because the moon revolves around Earth every 29.5 days (one "moonth," or month), one might assume that the center of its revolution would be at the center of Earth. However, both the moon and Earth actually revolve around a common center of gravity that lies within Earth and that is always on the side of Earth that faces the moon. As the Earth-moon system rotates around this axis, centrifugal force causes Earth and the moon to tend to fly away from each other. However, this tendency is blocked because the centrifugal force is exactly balanced by the gravitational attraction between Earth and the moon. Centrifugal force causes Earth's fluid hydrosphere to bulge outward on the side opposite the moon. This

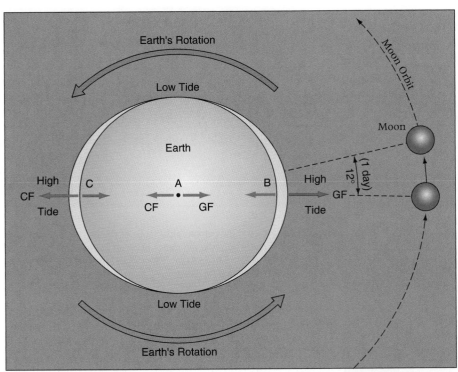

A. Gravitational force (GF) and centrifugal force (CF) are equal. Thus separation between earth and moon remains constant.

B. Gravitational force exceeds centrifugal force, causing ocean water to be pulled toward moon.

C. Centrifugal force exceeds gravitational force, causing ocean water to be forced outward away from moon.

Figure 20.19   The tides are a response to the moon's gravitational attraction (periodically reinforced or opposed by the sun), which pulls a bulge of water toward it while the centrifugal force of rotation of the Earth-moon system forces an opposing mass of water to be flung outward on the opposite side of Earth. Earth rotates through these two bulges each day. Actually, a "tidal" day is 24 hours and 50 minutes long. ■ **Explain why a tidal day is slightly longer than an Earth day. How many high tides and low tides are there during each tidal day?**

effect is one of the external forces that raises the ocean surface enough to produce a high tide. The ocean waters on the side of Earth facing the moon respond to the moon's gravitational force by bulging toward the moon. This bulge of water is directly opposite the bulge produced by centrifugal force, and Earth rotates through two high tides daily. Because they are 180° of longitude apart, they are separated by about 12 hours (half a day, because 180° is half of Earth's circumference). Between these two tidal bulges, the waters recede as they are pulled toward the areas of high tide. Accordingly, there are two low tides midway (90° of longitude) between the high tides.

Earth's 24-hour rotation, together with the moon's daily movement of 12° eastward along its monthly 360° orbit around Earth, means that theoretically coastlines will experience two high tides and two low tides every 24 hours, 50 minutes (the length of a tidal day). The time between the two high tides, called the **tidal interval,** thus averages 12 hours, 25 minutes. However, this ideal tidal pattern does not occur everywhere, although the most common tidal pattern does approach the ideal model of two high tides and two lows in a tidal day. This *semidiurnal* tidal regime is characteristic along the Atlantic coast of the United States. In a few seas that have restricted access to the open ocean, such as the Gulf of Mexico, tidal patterns of only one high and one low tide occur during a tidal day. This type of tide, called *diur-*

nal, is not very common. A third type of tidal pattern consists of two high tides of unequal height and two low tides, one lower than the other. The waters of the Pacific coast of the United States exhibit this *mixed pattern.*

Tidal regimes are extremely complex and variable, even over short distances. We have described the tides as though they were bulges of water through which Earth rotates. Actually, this "wave of equilibrium" theory does not account for many variations in tidal regimes. A somewhat different theory explains the tides as an oscillatory movement of the hydrosphere set up as a consequence of the forces we have discussed. This "stationary-wave theory" describes the tides as a back-and-forth sloshing of water, like the waves produced if a tray of water is tipped first one way and then the other. A circular motion, caused by Earth's rotation, is superimposed on this up-and-down or rocking movement.

The difference in sea level between high tide and low tide is called the **tidal range.** Tidal range varies from place to place in response to factors such as the shape of the coastline, water depth, access to the open ocean, and ocean floor topography. The average tidal range along open-ocean coastlines like the Pacific coast of the United States is 2 to 5 meters (6–15 ft). In restricted or partially enclosed seas, such as the Baltic or Mediterranean Sea, the tidal range is usually 0.7 meter (2 ft) or less. Funnel-shaped bays off major oceans, such as the Bay of Fundy

(a)

(b)

Figure 20.20 The extreme tidal range of the Bay of Fundy in Nova Scotia, Canada. Note the difference between the low tide (a) and the high tide (c), 6 hours later. ■ **Why does the Bay of Fundy have such a great tidal range?**

(c)

on Canada's east coast, produce extremely high tidal ranges. The Bay of Fundy is famous for its enormous tidal range, which averages 15 meters (50 ft) and may reach as much as 21 meters (70 ft) (Fig. 20.20). Other narrow, elongated coastal inlets that exhibit great tidal ranges are Cook Inlet, Alaska; Puget Sound, Washington; and the Gulf of California in Mexico.

Spring and Neap Tides The sun acts as a secondary tidal influence on the ocean waters, but because it is so much farther away, its tidal effect is less than half (46 per-

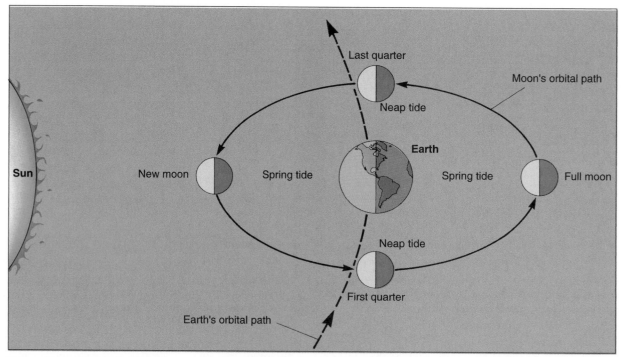

Figure 20.21 The maximum tidal ranges of spring tides occur when the moon and sun are aligned on the same side of Earth, or on opposite sides of Earth, at new moon or full moon respectively. The minimum tidal ranges of neap tides occur when the gravitational forces of the moon and sun are acting at right angles to each other, at first- and last-quarter moons. ■ **How many spring tides and neap tides occur each month?**

cent) that of the moon. When the sun, moon, and Earth are lined up, as they are when there is a new or full moon, the additional influence of the sun on ocean waters causes abnormally high and low tides, increasing the tidal range. This situation occurs every two weeks and is called **spring tide** (*spring* here does not refer to the season). A week after a spring tide, when the moon has revolved a quarter of the way around Earth, its gravitational pull on Earth is exerted at an angle of 90 degrees to that of the sun. At this time, the forces of the sun and moon tend to counteract one another. At the time of the first-quarter and last-quarter moon, the counteracting force of the sun's gravitational pull diminishes the moon's attraction. Consequently, the high tides are not as high, and the low tides are not as low. This moderated situation, which also occurs every two weeks, is called **neap tide** (Fig. 20.21).

Using astronomical information, long periods of observation of tidal patterns, and up to 40 local factors, tide tables can be prepared years in advance. Tide-predicting machines, ingenious devices of gears and pinions, were used beginning in the late 1800s. Today, computers produce tide tables of local coastal areas; these are essential to safe ship navigation and useful to sailors, fishers, scuba divers, surfers, beach joggers, clam diggers, and sea shell collectors.

Tidal Currents   The tidal movement that we have been describing is a vertical rise and fall of ocean waters. There are also horizontal currents, especially in bays or sounds, called **tidal currents.** The *flood tide* is the incoming current that accompanies rising tide; the outgoing *ebb tide* accompanies falling tide. The time period between flood

Figure 20.22   A tidal bore is the advance of high tide waters in the form of an incoming wave. This is the tidal bore at the Bay of Fundy.

and ebb, with no current, is called *slack water*. The velocity of tidal currents may be high enough to affect shipping, erode the shoreline, and move fine sediments. The tidal current through the Golden Gate, the entrance to San Francisco Bay, is 4 knots (4.6 mph) during both flood and ebb tides. In some of the channels between islands of the Inside Passage of the Alaska and British Columbia coasts, there are tidal currents of up to 10 knots (11.5 mph). These currents can be hazardous to ships trying to navigate narrow channels between the islands.

When the strong outgoing currents of rivers and estuaries oppose incoming flood tides, the tide may develop a steep, wavelike front known as a *tidal bore* (Fig. 20.22). A few rivers famous for their tidal bores are the Amazon, Yangtze, and Seine. In exceptional instances, as at the mouth of the Amazon, the tidal bore may be more than 5 meters (16 ft) in height.

The tides offer a potential for generating electrical energy, but necessary tidal conditions limit this energy source to only a few areas. The only large-scale operating tidal energy facility is located on the Rance estuary in western France. It has been operating successfully since 1966. Proposals have been made for a tidal energy facility in the Bay of Fundy between Maine and New Brunswick. However, economic forecasts have not supported the plan. Although tidal energy is low cost and pollution free, there are some disadvantages. These include the initial construction costs, the limited number of sites with adequate tidal range, and ecological concerns.

Figure 20.23   Orbital paths of water particles cause oscillatory wave motion in deep water. Wave height is the diameter of the surface orbit, which is the vertical distance from trough to crest.   ■ **Wavelength is the measurement of what distance?**

Tsunami Forecasts and Warnings

The phenomenon that is termed a *tsunami* is often referred to as a "tidal wave." This is not an accurate description. Though the level of the tides can greatly affect the magnitude of destruction when tsunamis strike, tsunamis are not related to the tides. Some scientists call tsunamis "seismic sea waves." This term is misleading as well. "Seismic" implies an earthquake generating mechanism, but tsunamis can also be caused by nonseismic events. The direct translation of the Japanese term is "harbor wave" (*tsu*, harbor; and *nami*, wave).

Tsunamis are unlike wind-generated waves (those we might see at a lake or seashore) in both their *period* (the time interval between successive waves) and their *wavelength* (the distance between two successive waves). On our West Coast, for example, spawned by a storm out in the Pacific, the waves that come in rhythmically one after another might have a period of about

(a)

NOAA, National Geophysical Data Center

(b)

NOAA, National Geophysical Data Center

Tsunami damage aftermath: (a) Coastal cliff formed by erosion from the 1992 tsunami that stuck Flores Island, Indonesia. (b) Tsunami destruction on Okushiri Island, Japan, in 1993.

 Waves

Waves are undulations at the surface of bodies of water. Contrary to appearance, waves do not transport water horizontally from one place to another, except where they break as surf along a coastline. Rather, the form or shape of waves and their energy are transmitted *through* water. The movement of waves is similar to the movement of stalks of wheat as wind blows across a wheat field, causing wavelike ripples to roll across its surface. The wheat returns to its original position after the passage of each wave. Water, too, returns to its original position (or close to it) after transmitting a wave.

Most natural water waves are initiated by wind; the few that are not are produced by seismic or volcanic activity (tsunamis) or are an effect of tidal activity (tidal bores). When wind blows across water, friction increases. Some of the wind's energy is thus transmitted to the water. Waves are the result of this energy transfer.

In deep, open bodies of water, waves appear as undulations on the surface. Figure 20.23 illustrates what happens to parts of the surface water during the passage of a

10 seconds and a wavelength of 150 meters (500 ft). A tsunami, on the other hand, may have a period of about an hour and a wavelength of more than 100 kilometers (60 mi).

As a result of their long wavelengths, tsunamis behave like shallow-water waves, where the ratio between the water depth (*D*) and its wavelength (*W*), or *D/W*, is very small. The speed of a shallow-water wave is related to the acceleration due to gravity (*g*), 9.8 meters/second/second, multiplied by *D*. In the Pacific Ocean, where typical depth is 4000 meters (13,000 ft), a tsunami travels at about 700 kilometers per hour (435 mph). Tsunamis not only move at high speed, but they can also travel great distances. In 1960, a tsunami originating off the coast of Chile traveled more than 17,000 kilometers (10,600 mi) to Japan, where it killed 200 people!

Tsunamis usually form when large areas of the ocean floor elevate or subside as a result of earthquake activity. This deformation vertically displaces the overlying water from its equilibrium position. Waves are formed as the displaced water mass acts under the influence of gravity to regain its equilibrium. Besides earthquakes, submarine landslides, collapse of submarine volcanic structures, and eruption of submarine volcanoes are all capable of creating tsunamis. Supermarine (above the ocean) causes such as coastal landslides and meteor impact can also form tsunamis, but this type dissipates quickly and rarely affects distant coastlines.

The energy of a tsunami depends on a balance between its wave speed and its wave height. As it moves into shallow coastal water, the speed decreases, and the wave height must grow to maintain the balance. Thus, a tsunami that is 1 meter high in the open ocean can grow to 30 meters (100 ft) high when it reaches the coast. A tsunami consists of a series of waves, and the first wave is often not the largest. The danger from a tsunami can last for several hours after the arrival of the first wave.

Unfortunately, the mechanism of tsunami generation is not well understood. Seismologists (those who study earth-

The PTWC locates the epicenter of the earthquake and provides estimated times of arrival for potential tsunamis in the Pacific region.

quakes) are studying the relationships between earthquake strength and the power of tsunamis. Subsurface pressure sensors, able to measure tsunamis in the open ocean, are providing valuable data as well. These data are combined to enable the Pacific Tsunami Warning Center (PTWC) in Ewa Beach, Hawaii, and its 26-nation group to share warnings throughout the Pacific Basin, and the Alaska Tsunami Warning Center (ATWC) in Palmer, Alaska, to issue appropriate warnings for the West Coast of North America. When a warning is issued, it is wise for ships to leave the shallow harbor areas and go out to sea, where the tsunamis may only be a meter high in deep water. Coastal residents are warned to evacuate the area and move quickly to higher ground. Tsunamis can come with little or no lead time, and when warnings are issued they need to be taken seriously, for this monstrous and still mysterious phenomenon can have devastating effects.

wave. First, the upward movement of water produces a **wave crest,** and then subsequent sinking of the water surface produces a **wave trough.** Water particles rise and fall, producing an endless series of waves passing along. The actual movement of water particles is circular, so that there is a small amount of forward movement during each rise. This circular or oscillatory pattern of movement is the reason why deepwater ocean waves are called *waves of oscillation* (see Fig. 20.23). The height of a wave is measured vertically from trough to crest. The horizontal distance between two wave crests is the **wavelength.** A **wave**

period, measured in seconds, is the time it takes for successive crests to pass a fixed point. A *sea state* report would read something like this: "Three- to four-foot waves from the northwest, at 10-second intervals." These reports are useful to sailors, fishers, lifeguards, and surfers.

Three factors that determine wave size are (1) wind velocity, (2) duration of the wind, and (3) fetch. **Fetch** is the distance of open water across which the wind can blow without interruption. An increase in any of these three factors produces an increase in the size (height and wavelength) of waves.

Figure 20.24 A U.S. Coast Guard vessel in storm waters in the Pacific off the Oregon coast. Within a storm, chaotic waves called "sea" occur. They become whitecaps when the wind is strong enough to blow the tops off the waves and become huge breakers when they enter the shallow water near shore. ■ **What factors would determine the height of such storm waves?**

Gentle breezes form *ripples* on the sea surface, providing roughness necessary for the wind to "grip" the water. If the wind increases, the ripples are soon transformed into larger waves. In a storm, steep, choppy, chaotic waves called *sea* occur. They become *whitecaps* if the wind is strong enough to break the tops off the waves (Fig. 20.24).

When the waves travel out of the storm area or the wind dies down, the chaotic sea is transformed into gentler, longer-period waves called *swells*. Swells can travel thousands of kilometers and completely cross an ocean. For instance, large southwest swells arriving on the California coast in summer have been generated by winter Antarctic storms occurring south of New Zealand. The regular rhythm of the swells is the type of wave motion usually associated with seasickness.

Catastrophic Waves

Tsunamis are waves caused by undersea earthquakes, volcanic eruptions, or undersea landslides. In deep-ocean water, they travel at speeds of up to 725 kilometers (450 mi) per hour and yet pass beneath a ship unnoticed. However, they are extremely dangerous as they move onshore. Tsunamis may rise more than 30 meters (100 ft) along the shore, engulfing and devastating entire coastal settlements and often causing great loss of life and property.

In 1946, an Alaskan earthquake caused a tsunami in Hilo, Hawaii, that attained a height of more than 10 meters (33 ft) and killed more than 150 people. When Krakatoa erupted in 1883, it generated a tsunami that reached more than 40 meters (130 ft) and killed more than 37,000 people in the nearby Indonesian islands.

Tsunamis are sometimes called "tidal waves," even though they are not related to tidal activity.

Similarly, **storm surges,** which are sometimes called "storm tides," are not related to tides, although they may coincide with high tide. *Storm surge* is the name given to

Figure 20.25 The city of Galveston, located on a barrier island on the Texas Gulf Coast, was devastated by a hurricane-generated storm surge in 1900. In terms of lives lost, this was the greatest natural disaster in the history of the United States. Although the exact number is not known, perhaps as many as 9000 people perished when the surge swept over the island.

the rise in sea level beneath a severe storm (low air pressure) and to the waves driven onshore by the strong winds accompanying hurricanes or typhoons. During Hurricane Camille in 1969, the sea level was more than 8 meters (25 ft) above normal along the Gulf Coast.

Like tsunamis, these high seas can be enormously destructive. In 1900, Galveston, Texas, was destroyed and more than 6,000 lives were lost as a result of a hurricane-produced storm surge (Fig. 20.25). It remains the most deadly natural disaster in U.S. history. In 1970, a storm surge produced by a tropical cyclone in the Indian Ocean drowned more than 200,000 people in the Ganges Delta area of Bangladesh, and in 1985, thousands more lives were lost in a similar storm there.

DEFINE & RECALL

salinity	trench (ocean)	barrier reef	wave trough
thermocline	island arc	atoll	wavelength
bathymetry	abyssal plain	tide	wave period
continental shelf	ooze	tidal interval	fetch
submarine canyon	seamount	tidal range	tsunami
turbidity current	guyot	spring tide	storm surge
continental slope	continental island	neap tide	
continental rise	oceanic island	tidal current	
oceanic ridge	coral reef	wave	
(midocean ridge)	fringing reef	wave crest	

DISCUSS & REVIEW

1. How do the oceans play a major role in the Earth system?

2. List Earth's four oceans. Which is the largest? What is the difference between an ocean and a sea?

3. List the major constituents of seawater. Which ones are economically recoverable for human use?

4. How do salinity and temperature affect the density of water? How is density related to oceanic circulation?

5. At which latitudes does seawater have the highest salinity? What are some factors that cause this high salinity?

6. Describe the major topographic features of the deep-ocean floor. Why have these submarine terrain features remained relatively unchanged over long periods of Earth history?

7. How are the features of the continental shelf related to the adjacent continents? What are the differences between the continental shelf and the continental slope?

8. What is the difference between a submarine canyon and a trench? How is each formed? Where are they generally located?

9. What is the relationship between the oceanic ridges and the trenches? Relate this relationship to the theory of plate tectonics.

10. What is a coral reef, and how does it form? What sea conditions are necessary for coral reef formation?

11. Describe where most of the world's coral reefs are located. Name several islands or island groups that have coral reefs. Name two continental coastlines that have coral reefs.

12. Describe the major factors that produce tides. What are some variations in tidal patterns?

13. What force causes most ocean waves? What characteristics of that force affect wave height?

CONSIDER & RESPOND

1. Many scientists consider the oceans to be Earth's last true frontier and a great resource base for the future. Why is the ocean the least explored part of our planet?

2. Refer to Figures 20.5 and 20.6. Describe how the major topographic features of the deep-ocean floor differ from the major continental features. Why is there such a difference, and what geomorphic forces are dominant in each case?

3. Differentiate between oceanic and continental islands. List several examples of each.

4. Assume you were asked to plan for the construction of a major tourist resort on a beautiful tropical atoll. What would be the attractions of such a resort location? What limitations and concerns would you have to evaluate before construction of the resort?

CHAPTER

21

COASTAL LANDFORMS

CHAPTER PREVIEW

The ever-changing coastal regions of Earth are greatly affected by human activities and should be protected from destructive uses and increasing pollution.

Why are the coasts such dynamic environments?

Why are the coastal zones the most polluted ocean environments?

Waves are the dominant agent of erosion and deposition in the narrow coastal zones where the land and water meet.

What are the major erosional and depositional coastal landforms produced by waves?

What would be the ultimate result of wave action on shorelines if only wave processes continued without interruption?

Beaches are the most visible evidence of coastal deposition, and they reflect a balance between input and removal of material by waves and currents.

Of what types of materials are beaches made?

How do coastal currents affect the formation of major coastal landforms?

Coasts are dynamic and complex systems, and they are hard to classify because of their great variety and changeable nature.

Based on global plate tectonics, what are the two major types of coastlines?

What are the differences between coastlines of emergence and coastlines of submergence?

576

The coastal features both above and below sea level are well illustrated in this digital terrain model of the Los Angeles area.

The most changeable environment in the world is the coastal zone. This is the zone where the hydrosphere, lithosphere, and atmosphere meet. It is a zone of continuous interaction between Earth's physical spheres. The coastlines of the world offer a tremendous variety of landforms, weather patterns, and marine and coastal organisms. The coastal waters of the oceans are not only dynamic from a geomorphic viewpoint; they are also biologically dynamic, with a great variety of niches for organisms to occupy.

To many people, the coasts are the most scenic and interesting landscapes in the world. The coastal regions of the world attract more tourism than any other natural environment. Places such as the Mediterranean Riviera, Hawaii, Florida, southern California, and Australia compete for the hundreds of millions of tourists who head for the seashore on their vacations each year. The coastal zones, however, include some of our most polluted waters. They are heavily affected by a multitude of human activities, including rapid urban development, port operations, offshore oil production, tourism, and agricultural runoff. The coasts need protection from these human impacts, and by understanding the natural processes that operate there, we can work toward solving future problems. In this chapter, we will examine both the dynamic forces that shape our coastal zones and the magnificent coastal landforms produced by these forces.

Coastal Landforms and Processes

Waves and currents are the major geomorphic processes associated with large bodies of water—the oceans, seas, and major lakes. Their effects are felt in a narrow, dynamic zone where land, air, and water meet, called the **shoreline.** Landform changes are continuous both landward and seaward of the shoreline. The position of the shoreline fluctuates with tides, storms, and long-term rises and falls in sea (or lake) level, as well as with tectonic movements. The **coastal zone** includes the dynamic region on land as well as areas currently submerged under water, through which the shoreline boundary fluctuates.

Waves, like streams, glaciers, and wind, erode, move, and deposit materials. By removing, transporting, and depositing material, waves continually work the narrow strip of land with which they are in contact. Although large storm waves generate the most effective erosive force, all waves that reach the coastal zone do some work in shaping the land (Fig. 21.1). Changing tidal levels allow the waves to extend their effects over a wider zone, and global sea-level changes do the same in a much longer time frame.

As waves pound the coast, their general effect is to straighten and smooth the shoreline. Peninsulas or headlands, which extend farther into the water than other parts of the land, are gradually cut back by wave action. Bays and inlets, on the other hand, are gradually filled in.

R. Sager

Figure 21.1   Large breakers pounding the shore at Point Lobos, California.   ■ **What is the origin of such large breakers?**

The Breaking of Waves

As long as the water is deep, waves of oscillation can roll along without disturbing the bottom, and the ocean (or lake or sea) floor will remain unaffected. However, when the water depth is about half the wavelength, the waves begin to "feel" the bottom and are slowed down by friction with sand and rock (Fig. 21.2). As these shallow water waves (waves of translation) slow down, their wavelength decreases, and both wave height and steepness increase. Eventually wave height reaches a point of instability, and the waves curl and collapse, producing *surf,* or *breakers*. If the waters out from shore are shallow, the surf will break some distance offshore. If, however, water remains deep adjacent to the coast, the surf is forced to break against the land. The tidal range also influences how far from shore the waves will break.

If a wave crashes directly against solid material, the impact can be sufficient to break the material apart, whether it is rock or a built structure. The direct impact of storm waves erodes coastal cliffs and headlands. More com-

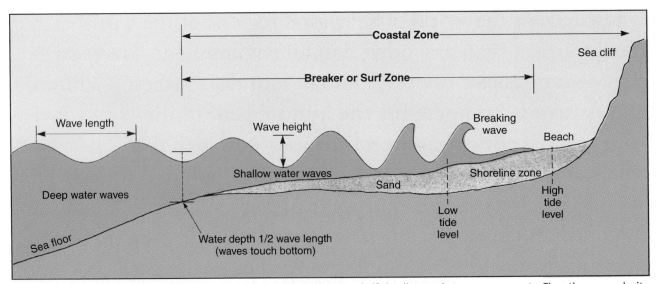

Figure 21.2   Waves begin to "feel" bottom when the water depth becomes half the distance between wave crests. Then the wave velocity and wavelength decrease while the wave height and steepness increase until breaking occurs.   ■ **Why don't the waves break in deeper water?**

monly, the waves break before striking the land directly; then the water surges forward as foaming **swash,** which picks up sand and carries it onto a beach. Gravity eventually overcomes the waning momentum of the water, which finally reverses direction and drains back down the beach slope as **backwash,** taking some sand with it. The material returned to the water in the backwash is flung back onto the shore during the breaking of the next wave.

Waves are a powerful and relentless erosive force. When waves are large, chunks of rock and sand serve as abrasive tools similar to those of a river's bed load. Thus, abrasion is an important factor in wave erosion, just as it is in fluvial erosion. The power of waves, combined with the buoyancy of water, enables large storm waves to carry rocks and even move boulders. During storms, large pieces of rock or coral may be thrown against cliffs. The force of impact has a "cannonball effect" that breaks the rock surface. Wave erosion of the coastal zone is also assisted by solution, as is the case with surface water and groundwater. There is also the power of hydraulic action from the sheer physical force of the pounding water and the explosive effect of the air compressed between breaking waves and cliff faces. The growth of salt crystals from evaporating ocean spray also helps detach mineral grains from cliffs and rocks, helping to push back the land.

Wave Refraction

Wave refraction is the bending of waves as they approach a shore. An important consequence of wave refraction is that wave energy becomes concentrated along certain parts of the shoreline and is greatly reduced in others. To see how this happens, imagine an irregular

shoreline of bays and headlands (Fig. 21.3). Offshore waves are essentially parallel to each other, and may approach the shore either directly or obliquely. However, as they approach land, the waves reach shallow water in some places sooner than in others. The shallow waters off headlands will be reached by a wave before the shallow waters in a bay. As a result, the part of a wave that approaches a headland will be slowed down and forced to break before the part that is approaching the bay. Because one part of the wave is slowed down before another part, the wave is bent, or refracted.

Consequently, wave energy is directed more toward headlands and less toward bays and coastal indentations. Because wave energy is concentrated at headlands, wave erosion is more intense there. The debris produced by wave erosion at coastal headlands and cliffs joins with fluvial sediment brought to the coast by streams and is transported to areas where there is less wave energy, accumulating in the bays between headlands. This effect is the reason why waves tend to reduce shoreline irregularity: The concentration of erosion on headlands wears them back, while beach deposition in bays fills them seaward.

Erosional Coastal Landforms

As waves erode the coastal zone, they create distinctive landforms (Fig. 21.4; also see Map Interpretation: Active-Margin Coastlines). As is the case with differential weathering and erosion by water inland, rock resistance to wave erosion is a major factor in the formation of coastal landforms. **Sea cliffs** are created where waves pound directly against steeply sloping land or resistant headlands. The erosive power of waves gradually undercuts

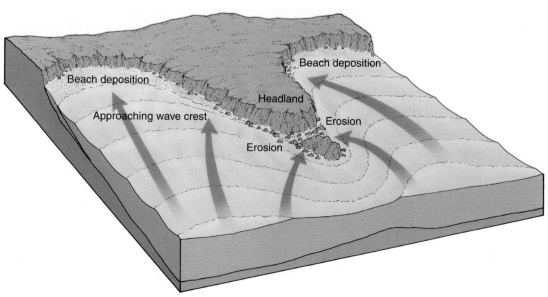

Figure 21.3 Wave refraction causes wave energy to be concentrated on headlands, eroding them back, while in bays, wave deposition causes beaches to grow seaward. ■ **How will this coastline change over a long period of time?**

ACTIVE-MARGIN COASTLINES

The Map

Point Reyes National Seashore is north of San Francisco on the rugged California coast. Point Reyes consists of resistant bedrock that is being eroded by the forces of the sea. Marine life is abundant, including many marine birds and marine mammals (note Sea Lion Cove below

High-altitude oblique aerial photo of Point Reyes.

NASA

Point Reyes). A hilly area, Punta de Los Reyes, separates Drakes Bay and Point Reyes from Tomales Bay (visible on the aerial photograph as a ribbon of water across the top, but barely showing in the northeast corner of the map).

The oblique aerial photo of Point Reyes was taken from a high-altitude NASA aircraft. Color infrared film makes vegetation appear red. The view is looking northeast. Trending across the upper portion of the photograph is the San Andreas Fault, which forms the linear Tomales Bay. The San Andreas Fault also separates the Pacific plate, on which Point Reyes is located, from the North American plate, which underlies the area east of the fault. Point Reyes is moving northwest along this fault. Emergent coastal features and recent tectonic activity characterize active-margin coastlines.

Point Reyes has a Mediterranean climate, influenced by the cool offshore California current. Here, the sea is not hospitable for swimming because of the uncomfortably cool seawater, large surf, and dangerous rip currents. The coastal location and onshore westerly winds create a truly temperate climate. Although very hot and very cold temperatures are rarely experienced, this is one of the windiest and foggiest coastlines in the United States. It is an area of rugged natural beauty with wind-sculpted trees, grasslands, rocky sea cliffs, and long sandy beaches.

Interpreting the Map

1. Which area of the coast is most exposed to wave erosion? What features indicate this type of high-energy activity?

2. Which area of the map is under the influence of strong longshore currents? What is the general direction of flow, and what coastal feature would indicate this flow?

3. Looking into the future, what may happen to Drakes Estero? What would Limantour Spit become?

4. Which area of the map appears to have strong wind activity? What would indicate this? What do you think the prevailing wind direction is?

5. Locate examples of the following coastal erosional landforms:

   a.   Headland.
   b.   Sea stack.
   c.   Sea cliffs.

6. Note that the bays and esteros (Spanish for estuary) have mud bottoms. Because the creeks and streams are very small in the region, what is the probable source and cause for the movement of the mud?

7. Note the offshore bathymetric contours (blue isolines). Which has a steeper gradient, the Pacific coast (west side) or Drakes Bay (east side)? Why do you think there is such a great difference?

**Opposite:
Point Reyes, California
Scale: 1:62,500
Contour interval = 80 ft
U.S. Geological Survey**

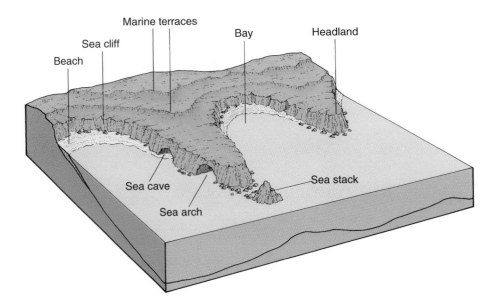

Figure 21.4 Diagram of the major coastal erosional landforms associated with wave activity.

the land, creating steep coastal cliffs (Fig. 21.5a). Where the rocks are well jointed but cohesive, wave erosion may create **sea caves** along lines of weakness (Fig. 21.5b). **Sea arches** result where two caves meet from either side of a headland (Fig. 21.5c). When the top of an arch collapses, or a sea cliff retreats and a resistant pillar is left standing, it is called a **sea stack** (Fig. 21.5d).

Waves frequently erode notches into a sea cliff. Notches are particularly common where the sea cliffs are made of limestone. Seawater dissolves the limestone just as groundwater does on land, creating notches or lateral indentations at the water level. Notches are also cut into other rock types by abrasion and are an important factor in the erosional retreat of sea cliffs. Rates of coastal ero-

(a)

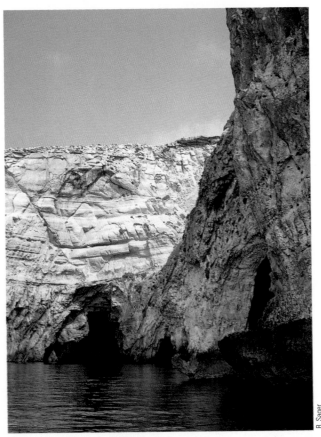

(b)

Figure 21.5 Examples of the major coastal landforms shown in Figure 21.4. (a) The rugged sea cliffs along the wave-eroded and uplifted Washington coastline. (b) Sea caves in the steep limestone sea cliffs on the Mediterranean island of Malta. (c) Sea arches, such as this one at Cabo San Lucas at the tip of Mexico's Baja California peninsula, form as sea cliffs are eroded completely through a headland. (d) A sea stack such as this one off Santa Barbara Island in Channel Islands National Park, California, forms when sea cliffs retreat, leaving a resistant pillar of rock standing above the waves.

sion are controlled by both wave action and rock type, and are accelerated during severe storms.

The presence of a wave-cut cliff implies erosion and removal of a large mass of material. The cliff is the vertical portion of a notch cut into the land. The horizontal portion of the notch is below water level and consists of a broad abrasion platform or **wave-cut bench** (Fig. 21.6). Rock material eroded from sea cliffs and wave-cut

benches come to rest in the beaches that occupy the coastal indentations. Backwash and coastal currents then carry some of the eroded material seaward. Because turbulence, important in transporting debris, is lower seaward, coarse debris accumulates and is battered into smaller particles close to the land. Finer particles move out into the sea. Thus marine sedimentation, like fluvial deposition, is organized by size and weight, with the finest

R. Sager

(c)

R. Sager

(d)

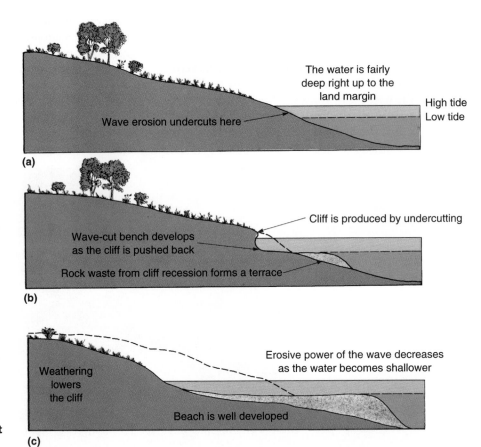

The water is fairly deep right up to the land margin

Wave erosion undercuts here

High tide
Low tide

(a)

Cliff is produced by undercutting

Wave-cut bench develops as the cliff is pushed back

Rock waste from cliff recession forms a terrace

(b)

Weathering lowers the cliff

Erosive power of the wave decreases as the water becomes shallower

Beach is well developed

(c)

Figure 21.6 Diagram of the origin of sea cliffs, wave-cut benches, and wave-built terraces. ■ **What is the difference between a wave-cut bench and a wave-built terrace?**

particles being deposited farthest from shore. As these deposits accumulate, a **wave-built terrace** is created just seaward of the wave-cut bench. Should tectonic activity uplift these wave-cut benches and wave-built terraces above sea level, they are then called **marine terraces** (Fig. 21.7). Successive periods of uplift may create a coastal topography that resembles a series of steps. Each step represents a period of time that a terrace was at sea level. Some areas, such as the Palos Verdes peninsula near Los Angeles, may have more than ten marine terraces, each representing an episode of uplift.

Beaches and Coastal Deposition

Beaches are the most visible evidence of coastal deposition. They reflect an equilibrium in the dynamic coastal system between input of material by swash and removal by backwash. Though sand-sized sediments are the most common beach material, not all beaches are made of sand. Actually, beach

deposits may consist of a wide range of sediments, from boulders to large cobbles to fine silt (Fig. 21.8). Granite, basalt, shale, conglomerate, and coral all result in beaches that differ in type, color, and texture. Where wave energy is high, particles are larger and beaches will be steeper than where only fine material is present and wave energy is low. Because the wave height generated by storms is greater in winter than in summer in the middle latitudes, winter beaches are generally narrower, steeper, and composed of coarser material than are sum-

Figure 21.7 A marine terrace on the Palos Verdes Peninsula on the southern California coast.
■ **How does a marine terrace form?**

R. Sager

mer beaches. Winter storm waves are more erosive and destructive, and smaller summer waves tend to be more depositional and constructive. On the Pacific coast of the United States, beach deposits may be removed entirely by destructive winter storm waves. Summer beaches are generally temporary accumulations of sand deposited over winter beach materials. On the Atlantic and Gulf coasts of the United States, the late summer to early fall hurricane season is a time when beach erosion can be most severe.

Coastal Currents

A **longshore current** is the parallel movement of water in the surf zone generated by obliquely approaching waves. The movement of beach sediment carried in long-shore currents is known as *longshore drift.* When waves break onto a beach at an angle to the shoreline, the swash pushes material ahead of it toward shore. The waves and consequent swash strike the shore at an angle, but the backwash, responding to gravity, moves directly down-slope perpendicular to the shoreline. As a result, beach material is pushed diagonally up the beach by swash, but not returned to its original position by backwash. Re-peated over time, this zigzag movement will cause the mass transport of tons of sediment along the shore and beach.

Rip currents (Fig. 21.9) are strong, seaward-moving currents found near the shore and are caused by the chan-neled return of water and sediments from large waves that have broken against the land. Rip currents (also known as *rip tides*) can be dangerous, because the rapidly moving water can pull swimmers out to sea. Rip currents

(b)

(d)

Figure 21.8   Beaches are the most visible evidence of wave deposition and may be made of any material deposited by waves. (a) Sand beach at Laguna Beach, California. (b) Gravel beach at Kopachuck State Park, on Washington's Puget Sound. (c) Fine-grained, light-colored sand beach, common on tropical islands with coral reefs, such as Huahine in French Polynesia. (d) Black Sand Beach on the island of Hawaii, formed from wave deposits of volcanic material.

Beach Building

Some people cite Miami Beach, Florida, as the classic example of how human effort and modern engineering can protect a coastline from the ravages of a menacing sea. During the second half of the twentieth century, Miami Beach had become the chosen holiday destination of millions of Americans from the eastern and central United States. The star attractions were the magnificent beaches. However, as natural wave erosion occurred over the years, it became obvious that the beaches were in jeopardy. The building of breakwaters or the construction of rock walls, called groins, perpendicular to the shorelines had done little to protect beaches or trap drifting sand all along Florida's eastern coast.

By 1970, the waves at Miami Beach had relentlessly eroded away so much sand that the Atlantic Ocean pounded against the sea wall and the beaches had mostly disappeared. With the beaches gone, the shoreline condominiums were threatened with serious damage from storm waves, and hotels were half empty as tourists traveled elsewhere. The prospects were bleak for Miami Beach until the U.S. Army Corps of Engineers (the federal agency responsible for the development of inland and coastal waterways) chose that resort community to experiment with a new technology for protecting the nation's shorelines. Often referred to as beach building, the technology involves the transfer of millions of cubic yards of sand from offshore sources to nearby coastlines to replenish existing beaches or build

new ones where beaches have eroded away. In remarkable feats of engineering, huge dredges mounted on barges dig sand from the sea bottom near the shore so that the sand can be pumped as a slurry through massive movable tubes to be deposited along the coast as a beach.

The Miami Beach project was funded by $72 million from a combination of federal, state, and local tax money. This was a major investment, but it was so successful that within two years Miami Beach again had a protecting coastline of sand 300 feet wide and 10 miles long, and the number of visitors to the city had increased to 21 million annually, triple the number prior to beach building.

Since the Miami Beach project, beach building has become the accepted way to respond to beach erosion. One recent newspaper analysis indicated that nearly $1.7 billion was spent on beach restoration during the decade of the 1990s alone. According to the Corps of Engineers, these projects are not meant primarily to provide beaches for recreational purposes, but to protect buildings on or near the beach. In the Corps's opinion, the true value of beach building can best be measured in savings from storm damage. For example, it has been estimated that beach renourishment along the coast of Miami–Dade County prevented property losses of $24 million during Hurricane Andrew in 1992.

When beaches are rebuilt or widened to protect against storms, the Corps of Engineers takes into consideration that erosion is inevitable. In deciding how much sand must be pumped from offshore sources for beach renewal, they first determine the width of the permanent beach necessary to protect shoreline property from storms. This is usually about

---

are often visible as streaks of foamy, turbid water flowing perpendicular to the shore and directly offshore.

Beach systems are in equilibrium when income and outgo of sand are in balance. An increase in the size of a beach may be accomplished by building an obstruction to the longshore current. This will prevent sand removal while sand input remains the same. The obstruction is accomplished by constructing a **jetty,** or **groin,** usually a concrete or rock wall built perpendicular to the beach. Of course, any obstruction starves the adjacent downcurrent beach area, which now has no beach material input but still has the usual rate of removal (Fig. 21.10). Beach deposition is frequently engineered to keep harbors free of sediment or to encourage recreational beach growth. When human activity upsets the natural sand supply, by damming rivers or building jetties to slow beach migration, the beach's temporary nature may be accelerated. In Florida and New Jersey, hundreds of millions of dollars have been spent to replenish the sand beaches. The beaches not only serve obvious recreational needs but also offer coastal settlements some protection from storm waves.

Depositional Coastal Landforms

Wave erosion of nonresistant rock materials produces a great deal of sediment and can construct a wide variety of coastal depositional landforms (Fig. 21.11a). Rivers

Figure 21.9 Rip currents move water seaward from a beach. Here the current can be seen moving offshore, opposite to the wave direction. These currents can be a great hazard to swimmers.

30 meters (100 ft). Then an amount of additional, "sacrificial" sand is added to the project to produce a beach double the size of the permanent beach. It is anticipated that in about seven years the sacrificial sand will be lost to wave erosion, but if rebuilding is repeated each time the sacrificial sand is removed, the permanent beach will remain in place and the coastal zone will be protected. It has been calculated that almost $5.5 billion has already been committed to the continuous rebuilding of existing projects over the next 50 years.

Is the battle with the environment at the beachfront worth the price we are paying? Many people are not so sure, but most environmental scientists have made up their minds and respond to the question with a resounding no, for they are concerned with a price that is not measured solely in appropriated dollars. This price is paid in destroyed beach environments, reduced offshore water quality, eliminated or displaced species, and repeated damage to food chains for coastal wildlife each time a beach is rebuilt. Clearly, beach building is a mixed blessing.

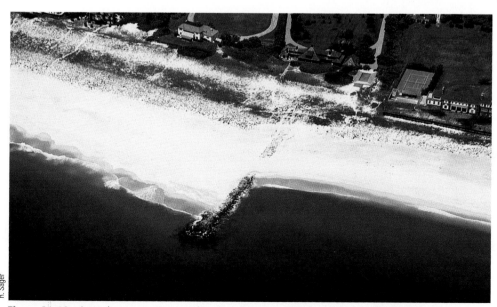

Figure 21.10 Longshore currents carry beach sediments parallel along the coast. The building of a jetty traps sand on one side, while it starves the beach on the other side, as evidenced by this jetty at East Hampton, New York. ■ **What direction does the longshore current move in this aerial photo?**

Beach Erosion and Protection

If you go to the same coastal area year after year, you may have noticed that some beaches undergo significant changes through time. At times the loose material is eroded away and expanses of beach have simply disappeared. At other times there may be an enlargement of the beach by *accretion* as the natural processes of wind, waves, and currents deposit new materials. Beaches are among the most dynamic landscapes on Earth, as their form and area are altered, sometimes changing at an alarming rate. As discussed in this chapter and Chapter 20, several natural forces affect beaches daily, seasonally, and in response to storm events. The most important processes that cause beach change are *waves, tides/tidal currents,* and *ocean currents.* In most locations, waves are the dominant force affecting beaches because their actions are continual as long as the winds generate them. Tides, with their corresponding currents of ebb and flow, also affect most oceanic coastlines. The offshore flow of ocean currents also can have a significant impact on coasts and the beaches along them.

In addition to these three major processes, there are also periodic and drastic changes caused by hurricanes and other sea storms whose *surges* and wind-generated waves can bring about rapid change in a coastline. *Tsunamis* may be large enough to devastate a shoreline. As the coastal population continues to grow, so have the efforts to protect beachfront property with a variety of methods. Some are discussed in this chapter; others are discussed below.

Protecting the sediments that make a beach can be accomplished using a number of structures, in addition to the groins or jetties described in this chapter. There are also *seawalls* (massive structures built along the shore to protect the beach against erosion) and *offshore breakwaters* (structures built out in the water but parallel to the shore, which break the energy of the incoming waves). In addition, *rock revetments* (layers of flat stones laid upon the shore to act as armor against erosive wave energy) may also protect the beach sands.

In many parts of the world, annual change in beaches is a normal and natural process that occurs seasonally, with

flowing into the sea bring additional sediment that is added to the coastal system. Material not carried out into deeper waters or deposited on the wave-built terrace is transported along the shore from the area of intense wave action to a more protected area, where deposition can take place. Where a shoreline is relatively straight except for a bay, the material carried by longshore currents may be deposited as a **spit** that continues along the shoreline into the bay mouth (Fig. 21.11b). If two spits grow together from opposite sides of a bay, or if a single spit grows completely across a bay, the result is called a **baymouth bar** (Fig. 21.11c).

The formation of a baymouth bar changes the bay into a protected lagoon. The salinity of such a lagoon

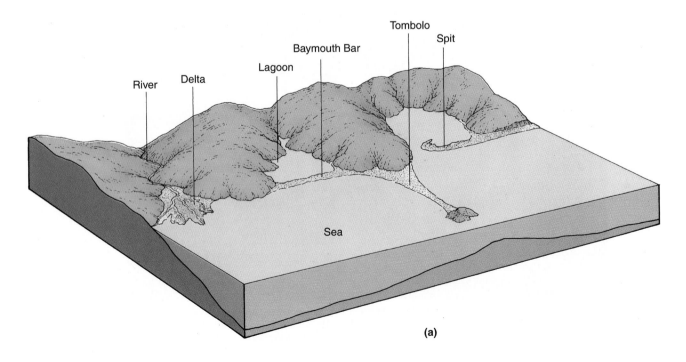

(a)

erosive, high-energy waves in winter and deposition during summer. Every beach has a distinctive character that results from the interaction of coastal processes, sediment availability, and the local environment. Measuring and monitoring the shoreline processes that affect a beach help us to understand and identify potential problem areas. Being aware of the natural processes and materials involved in a beach system is the best strategy for deciding if artificial beach protection would be beneficial and, if so, what approach would be most effective.

Wood groins along the North Sea coast of England, near Cromer.

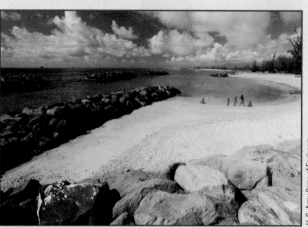

Boulders were brought in and offshore breakwaters were built to protect the beach at Sand Island, Oahu, Hawaii.

will now vary from that of the open sea, depending on river inflow and evaporation. Salinity levels will affect marine organisms living in the lagoon. A river that flows into the waters of a bay protected from wave erosion by a spit or bar will be able to build a delta. Eventually, the entire lagoon may be transformed into a salt marsh or coastal wetland by river deposits on one side trapped

(c)

(b)

Figure 21.11 (a) Diagram of the major coastal landforms formed from deposition by waves and longshore currents. (b) A spit is a beach connected at one end to the coast, as illustrated by this example on New York's Long Island. (c) A baymouth bar is connected to the coast at each end of a bay, as this one is on Long Island. (d) A tombolo is formed when wave-deposited materials connect a nearby island with the mainland, as shown by this example in Baja California, Mexico.

(d)

Aaron Young, BA, University of Florida

When I was in the Army as an artillery scout, I would occupy a hilltop for three to four days, looking for people and drawing the terrain out. It was an amazing experience to sit in the same spot where you can't see anyone else for two to three days. While observing the sun's apparent movement, I knew that I wanted to be involved in a career that involved observing aspects of the environment. Many other physical geographers have the same experience. You can look at a landscape and see different things than other people would see. My natural abilities took me into this area of geography.

Today, I work creating virtual three-dimensional harbors, computer visualizations for boating and shipping. We use satellite imagery, digital terrain elevation data (DTAT), bathymetry, and other ancillary data. We geoposition the imagery, getting it lined up to a coordinate system, and map data to the outer bounds of the harbor. We also decide what amount of detail should be presented in each area of a harbor. If you're coming into a harbor on a ship, you want a high level of detail in the outer areas, but may not need such a high level of detail farther inland.

These three-dimensional models must be very accurate and precise. We include data about buildings and cultural features. Image-to-image change detection is used to identify shoreline movement, new man-made features, and determine areas of potential change. We then create a virtual 3-d environmental model that can simulate the movements of a ship that is very similar to being at the port. Through these computer models mariners can "sail" into port a hundred times before they actually get there.

We have also created digital elevations models using NASA imagery of Mars. A great aspect of that project was that in analyzing the digital terrain, we might have been the first humans to physically see certain terrain features at such a high resolution on the Martian surface.

A good foundation would be to first learn the scientific method. Take an object or an issue or a concept and look at it from every single angle, and try to root out the core truth. What has also been helpful on a daily basis was to learn the physical characteristics of landforms. A solid background in photogrammetry is very helpful. A solid base in core GIS concepts is critical for a continued successful career in this field. As a physical geographer you try to understand the physical landscape of an area and you can apply the slowly accumulated knowledge based on experience.

Another concern is the observation that our physical landscape is going through huge anthropogenic transformations. It's somewhat disturbing to be told about, in the third person, the horrible atrocities to our environment; it's quite another the see the destruction firsthand through imagery.

Everything has a connection with everything else. It's hard to talk about anything without putting a component about how it relates to something else. Every theme, whether it is political, social, or environmental, has a geographic component to it. If one is not a physical geographer and has no plans on being one, he or she would still benefit from the physical geographer's perspective when becoming aware that we all live and have an impact on our Earth.

behind wave deposits on the other. A **tombolo** is formed if wave deposition builds up enough beach material to connect an offshore island or a sea stack to the mainland (Fig. 21.11d).

On gently sloping coasts or along coastlines where waves are forced to break at considerable distance from shore, churning motion of the water builds submerged **offshore bars** parallel to the coast. As these bars grow and emerge above sea level, they form barrier beaches and possibly the larger **barrier islands** (Fig. 21.12; also see Map Interpretation: Passive-Margin Coastlines). The origin of barrier islands is controversial, but most coastal experts agree that rising sea level since the Pleistocene has played a major role in their formation. They also agree that barrier islands are truly temporary landforms that migrate over long periods of time (Fig. 21.13) and may change drastically during severe storms, especially hurricanes. Most barrier islands have three distinct zones: (1) a sandy beach, (2) a sand dune zone, and (3) a shallow lagoon between the barrier island and the mainland (Fig. 21.14). Barrier islands are the most common type of beach deposit on low-relief coastlines. Barrier islands dominate the Atlantic and Gulf coasts from New York to Texas. Some excellent examples of long barrier islands are Fire Island (New York), Cape Hatteras (North Carolina), Cape Canaveral and Miami Beach (Florida), and Padre Island (Texas).

Types of Coasts

Because coasts are such dynamic and complex systems, influenced by tectonics, global sea level changes, storms, and marine and continental geomorphic processes, there is no universally accepted classification system. However, there are a few different coastal classifications that do contribute to an understanding of such complex systems.

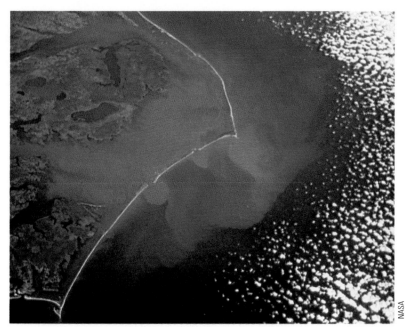

Figure 21.12   Barrier islands, such as this one near Pamlico Sound on the North Carolina coast, form from wave deposition along coasts with gentle slopes and an adequate sediment supply.   ■   **Name several other barrier islands along the U.S. Atlantic and Gulf coasts.**

Ocean (Fig 21.15). Most major tectonic activity is in the center of the ocean along the Mid-Atlantic Ridge, whereas the coasts are tectonically passive, with little mountain-building or volcanic activity. Generally, passive-margin coasts have low relief with broad coastal plains and wide continental shelves. Most have been modified by marine deposition and some subsidence. The East Coast of the United States is a good example of a passive-margin coast (see Map Interpretation: Passive-Margin Coastlines). There are exceptions among some of the younger passive-margin coasts, such as those of the Red Sea and the Gulf of California, that do not show this low-relief pattern.

Active-margin coasts are best represented by coastal regions along the Pacific Ocean (Fig. 21.16). Here, most tectonic activity is around the margins of the ocean, because the "Pacific Ring of Fire" is formed along active subduction and transform plate boundaries. Active-margin coasts are usually characterized by high relief with narrow coastal plains, narrow continental shelves, and earthquake and volcanic activity. These coasts tend to be erosional and spectacular in nature, with much less time within Earth history for the development of marine or continental depositional features.

On a global scale, coastal classification is based on plate tectonic relationships. Here, there are two major coastal types: passive-margin coasts and active-margin coasts. **Passive-margin coasts** are well represented by the coastal regions of continents along the Atlantic

Figure 21.13   Historic maps show us that barrier islands shift their shapes and positions over time, with major changes coming during storm events. This pair of photos shows hurricane generated damage to homes built along the shore on a barrier island. Compare the before and after photos.   ■   **What is the solution to problems caused by this natural process?**

PASSIVE-MARGIN COASTLINES

The Map

Eastport, New York, is located on the south shore of Long Island, 70 miles east of New York City. Long Island is part of the Atlantic Coastal Plain, which extends from Cape Cod, Massachusetts, to Florida. Water bodies such as Delaware Bay, Chesapeake Bay, and Long Island Sound embay much of the Atlantic Coastal Plain in the

Satellite image of Long Island

NASA/GSFC/SVS

eastern United States. The south shore of Long Island has low relief, and its coastal location moderates its humid continental climate.

Although this is a coastal region, its recent glacial history has played an important role in the landforms that exist today. Two east-west trending glacial terminal moraines deposited during the Pleistocene glacial advances formed Long Island. Between the coast and the south moraine is a sandy glacial outwash plain that forms the higher elevations at the northern part of the map area. As the glaciers melted, sea level rose, submerging a lowland that now forms Long Island Sound. This water body separates Long Island from the mainland of the Atlantic Coastal Plain.

The Atlantic and Gulf coasts of the United States have nearly 300 barrier islands with a combined length of over 2,500 kilometers (1600 mi). New York, especially the coastal zone of Long Island's south shore, has 15 barrier islands with a total length of over 240 kilometers (150 mi). Coastal barrier islands protect the mainland from storm waves and they form coastal wetlands on their landward side, which are a critical habitat for fish, shellfish, and birds.

Interpreting the Map

1. What is the long, narrow feature called that forms the straight shoreline labeled Westhampton Beach?

2. What is the highest elevation on the linear coastal feature? What do you think forms the highest portions of this feature?

3. What forms the outer seaward portion of the landform in Question 1?

4. Behind the linear feature is a water body. Is it deep or shallow? Is it a high-energy or low-energy environment? What is a water body such as this called?

5. Is the Eastport coastline erosional or depositional? What features support your answer? Is this a coastline of submergence or emergence?

6. Note Beaverdam Creek in the upper middle of the map area. Does it have a steep or gentle gradient?

7. Based on its gradient, does Beaverdam Creek flow seaward continuously? If not, what would influence its flow?

8. What kind of topographic feature exists as indicated by the map symbols in the area surrounding Oneck?

9. How do you think geomorphic processes will likely modify this map area in the future? Which area will probably be modified the most? Explain.

10. What type of natural hazard is this coastal area of Long Island most susceptible to? Why?

Opposite:
Eastport, New York
Scale: 1:24,000
Contour interval = 10 ft
U.S. Geological Survey

Figure 21.14   A typical barrier island consists of three major sections: a beach, dunes, and salt marsh.

Figure 21.15   A rocky passive-margin coast on the Atlantic Ocean in Acadia National Park, Maine.

The West Coast of the United States is an excellent example of an active-margin coast (see Map Interpretation: Active-Margin Coastlines).

On a regional scale, coastlines may be classified as coastlines of emergence or coastlines of submergence. **Coastlines of emergence** (Fig. 21.17) occur where the water level has been lowered or the land has risen in the coastal zone. In either case, land emerges from an area that was once below sea level. Coastal landforms dominated by wave action, such as marine terraces, sea cliffs, sea stacks, and beaches, are found above the level of the present shoreline. The position above the present shoreline of wave-created landforms serves as evidence of emergence. Coastlines of emergence were probably common during the Pleistocene ice age, prior to 12,000 years ago, because the formation of glaciers lowered sea level about 120 meters (400 ft). Features of emergence are at present best developed along active-margin coasts such as those of California, Oregon, and Washington, where marine terraces are found as much as 370 meters (1200 ft) above sea level. Less spectacular emergent coastlines are located in areas where isostatic rebound has raised the land after the retreat of the continental ice sheets, such as around the Baltic Sea and Hudson Bay.

Over the last 12,000 years, the Pleistocene ice sheets have retreated and global sea level has been gradually rising, creating **coastlines of submergence** where many features of the coastal zone are being submerged. Coastlines of submergence may also occur where tectonic forces have lowered the level of the land, as in San Francisco Bay. Great thickness of river deposits and alluvial

Figure 21.16   This digital terrain model shows an active-margin coast south of Santa Barbara, California, that has experienced much tectonic unrest via faulting and mountain building. The low coastal area between the mountains is the Oxnard Plain. There is 1.8 times vertical exaggeration in this model.

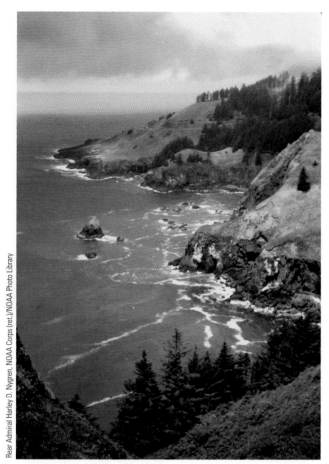

Figure 21.17 Otter Crest, Oregon. An emergent coast with mountains rising directly from the Pacific Ocean. There is little or no beach development, in this type of coastal environment.

walls that may be several thousand feet high. Fjord coastlines are found in Norway (where the term originates), Chile, New Zealand, Greenland, and Alaska. Canada, however, has more fjords than any other coastal nation. In many fjords, the glaciers have retreated far inland, but some, especially in Greenland and Alaska, have "tidewater" glaciers that calve icebergs into the cold fjord waters.

Some coastlines, such as those formed by coral reefs and river deltas, cannot be classified as either submerging or emerging. Actually, most shorelines show evidence of more than one type of development, largely because the elevation of the land and the level of the seas have changed many times during geologic history. For this reason, features of both submerged and emerged shorelines characterize many coastlines.

Since coastlines are really shaped by both continental and marine geomorphic processes, another regional classification system recognizes two types of coasts: primary and secondary coastlines. **Primary coastlines** are formed mainly by land erosion and deposition processes. Such formation is due to tectonic activity or sea level changes that cause the shoreline to change its level too rapidly for the marine processes to shape the coast. The major types of primary coastlines (with examples) are drowned river valleys (Delaware and Chesapeake Bays); glacial erosion coasts (southeastern Alaska, British Columbia,

sediment compaction, as along the Louisiana coast, may also cause coastal submergence. The features of a new coastline of submergence are related to the character of the coastal lands prior to submergence. Plains, for instance, will produce a far more regular shoreline than will a mountainous region. When areas of low relief with soft sedimentary rocks are submerged, barrier islands form with shallow bays and lagoons behind them. The classic example of this type of submerging coastline is the Atlantic and Gulf coasts of the United States.

Two special types of submerged coastlines are ria and fjord coasts. **Rias** are created where river valleys are "drowned" by a relative rise in sea level or a sinking of the coastal area (Fig. 21.18). These irregular coastlines are formed when valleys become narrow bays and the ridges form peninsulas. The Aegean coast of Greece and Turkey is an outstanding example of a ria coastline. *Fjords*, which are drowned glacial valleys, form scenically spectacular shorelines (Fig. 21.19). A fjord shoreline is highly irregular, with deep, steep-sided arms of the sea penetrating far inland in the troughs originally deepened by glaciers. Tributary streams cascade down the fjord

Figure 21.18 Submergent coasts like the Chesapeake Bay region are characterized by rias, or drowned river valleys, that developed as sea level rose at the end of the Pleistocene.

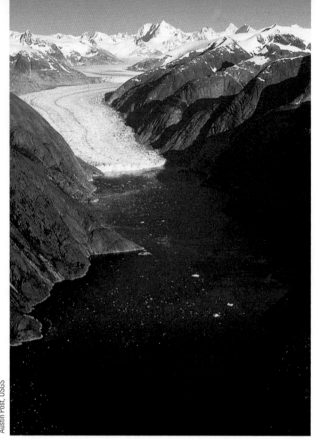

Figure 21.19 An aerial photo of an Alaskan fjord with an active "tidewater" glacier. ■ **How did this fjord form?**

Figure 21.20 The shorelines of the world are complex environments where many elements of the hydrosphere, lithosphere, atmosphere, and biosphere interact.

and Puget Sound in Washington, and from Maine to Newfoundland); glacial deposition coasts (Cape Cod, Massachusetts, and the north shore of New York's Long Island); river deltas (Mississippi Delta, Louisiana); volcanic coasts (Hawaii); and faulted coasts (California).

Secondary coastlines are those formed mainly by marine geomorphic agents, especially waves, and by marine organisms. Such features as sea cliffs, arches, stacks,

and sea caves dominate marine erosional coasts (Oregon). Marine depositional coasts have such features as barrier islands, spits, and bars (North Carolina). One example of coasts built by marine organisms is the coral reef (Florida Keys). Mangrove trees and salt-marsh grasses also trap sediments to build new land areas in shallow coastal waters.

Regardless of the classification type, coastlines are one of the most spectacular and dynamic regions on Earth (Fig. 21.20). They are the true meeting place of all Earth's spheres—hydrosphere, lithosphere, atmosphere, and biosphere.

DEFINE & RECALL

shoreline	sea arch	rip current	passive-margin coast
coastal zone	sea stack	jetty (groin)	active-margin coast
swash	wave-cut bench	spit	coastline of emergence
backwash	wave-built terrace	baymouth bar	coastline of submergence
wave refraction	marine terrace	tombolo	ria
sea cliff	beach	offshore bar	primary coastline
sea cave	longshore current	barrier island	secondary coastline

DISCUSS & REVIEW

1. What is the difference between a shoreline and the coastal zone?

2. How do ocean waves change when they enter shallow coastal waters? What is the main factor causing this change in the waves?

3. What is wave refraction? How is wave refraction related to the shape of the coastline?

4. Describe how sea cliffs form. Name three other coastal erosional landforms that form in sea cliff areas.

5. Describe a marine terrace. What does it indicate?

6. What are the differences between longshore currents and rip currents?

7. Explain the relationship between the sorting of coastal marine sediment and the development of beaches. Why do beaches change seasonally?

8. Why do baymouth bars develop? What effect do they have on bays?

9. What is the difference between an offshore bar and a barrier island? Name several examples of barrier islands. Where are most barrier islands located in the United States?

10. What are the major differences between active-margin coastlines and passive-margin coastlines? Why do they look so different?

11. What would be the major changes in the world's coastal zones if sea level were to rise?

CONSIDER & RESPOND

1. Explain why the coastal regions of the world are considered the most dynamic systems of Earth's environments.

2. What are some of the major ways humans change coastlines? Explain how this human activity may interfere with natural coastal processes.

3. Assume you are the geographer in charge of coastal zone management for a rapidly growing coastal community on the Atlantic or Gulf coast of the United States. Describe the major coastal landforms. What natural hazards would you have to plan for to protect the community?

4. Assume you are a geographer in charge of coastal zone management for a rapidly growing coastal community on the Pacific coast of the United States. Describe the major coastal landforms. What natural hazards would you have to plan for to protect the community?

DYNAMIC PHYSICAL GEOGRAPHY

Physical geography uses three scientific perspectives for analyzing Earth as a dynamic system: spatial science, physical science, and environmental science.

Why is Earth described as a dynamic system? How does geography contribute to our understanding of the Earth system?

Earth is an oasis in space, a life-support system of great environmental diversity. An amazing combination of physical characteristics makes Earth hospitable for life.

What basic characteristics of Earth combine to provide a habitat suitable for life? How does exploration of space contribute to our insights about our planet?

Only a relatively small percentage of Earth's surface can directly support large human populations.

Why does Earth's surface vary in its ability to sustain a human population? What factors set limits to the amount of life that Earth can sustain?

Planet Earth is a complex system composed of many major and minor subsystems.

What major subsystems interact to generate diversity on our planet's surface? How do thresholds relate to the operation of Earth systems?

Geography is a modern discipline that involves problem solving through various means, including the scientific method. Today, most geographic studies are supported by computer technologies.

What computer technologies are particularly applicable to geographic problem solving? How can computers help us "dissect" complex problems and "reassemble" them?

Physical geography as a science typically investigates topics related to some combination of six major geographic themes. These themes help us to develop and appreciate a geographer's perspective on the world.

What are these themes, and how do they relate to the spatial, physical, and environmental science approaches used by geographers? How does physical geography relate to you and your existence?

NASA's lunar orbiter over the lifeless surface of the moon, with a rising Earth above the horizon.

Viewed from the distance of the moon, the astonishing thing about the Earth, catching the breath, is that it is alive. The photographs show the dry, pounded surface of the moon in the foreground, dead as an old bone. Aloft, floating green beneath the moist, gleaming membrane of bright blue sky, is the rising Earth, the only exuberant thing in this part of the cosmos. If you could look long enough, you would see the swirling of the great drifts of white clouds, covering and uncovering the half-hidden masses of land. If you had been looking for a very long, geologic time, you could have seen the continents themselves in motion, drifting apart on their crustal plates, held aloft by the fire beneath.

Lewis Thomas
The Lives of a Cell

Vivid descriptions of Earth by astronauts and remarkable pictures from space have reinforced existing perceptions of our planet, provided new insights, and presented a critical learning experience to humankind. The space program yielded the first real images of Earth in space, views of a full hemisphere, rich with detail of our planet's workings and its environmental setting—a spherical island surrounded by the vast, dark emptiness of space. Previously we could only view artists' conceptions, with their surmised interpretations of this scene. Today we can examine, study, and appreciate authentic images of the genuine article, Earth, in all its complexity and beauty. We now have the ability to see the actual physical geography of our planetary home.

Physical geography involves studying spatial variations in terrestrial, aquatic, biotic, and atmospheric environments, the factors responsible for this variability, and the dynamic processes that operate the Earth system. These factors, along with geography's integrative nature, provide a perspective that contributes greatly to our knowledge of Earth. This text has illustrated how physical geographers apply and integrate spatial science, physical science, and environmental science to solve problems and answer questions, to increase our knowledge, and to deepen our understanding of Earth. Having an understanding of physical geography sharpens people's observational skills as they examine our world, its features, and its environments. This knowledge also fosters a stronger awareness of our planet's complexity and beauty. These are a few of the reasons why astronauts receive training in physical geography, helping them to be skilled observers of Earth and its features from space as they gather data and imagery (Fig. 22.1).

Looking outward to space has fostered a fresh, inward look at Earth and a greater appreciation for the intricacies of its environmental systems. Acquiring new knowledge about our own planet is, in fact, a primary justification for studying the solar system and the universe beyond. The space program has focused human efforts on examining and understanding our planet, how it operates, how it functions as a habitat for life, and its place in the universe. These aspects of Earth have long been subjects of interest to physical geographers, and as new images, tools, and techniques become available our knowledge of its complex systems increases.

As you near the conclusion of your physical geography course, consider how much of this knowledge you have acquired. Are your attitudes toward Earth and its environments different than they were before? In the years to come, what should you remember about physical geography? What knowledge, ideas, theories, and skills should you develop further? This chapter should help you think about the role of physical geography in your future.

A Special Planet

Is Earth unique? It has been argued that, in its ability to support life, Earth may not be unique in the expanse of space. This opinion suggests that somewhere, with so many billions of stars in the universe, other planets in distant solar systems may support living organisms. Further, this reasoning contends that, given the vastness of space, the existence of life as complex as that on Earth is likely. There are two problems with these contentions, however. First, the mere fact that space is vast does not prove that life exists elsewhere. The scientific method (see again the box in Chapter 1) requires that the burden of proof be on those who propose an idea, or hypothesis; it is not the responsibility of others to prove that the idea is invalid. Second, even our most distant space probes have yet to discover other planets capable of supporting life as we know it. Yet human and scientific curiosity continue to drive quests to answer the intriguing and compelling question of whether life exists elsewhere in the universe.

A consideration in determining whether or not our planet is unique, at least among features of the known universe, is the nature of the Earth system. Other than receiving inputs of solar energy, Earth is a self-sustaining life support system. The environmental richness and the

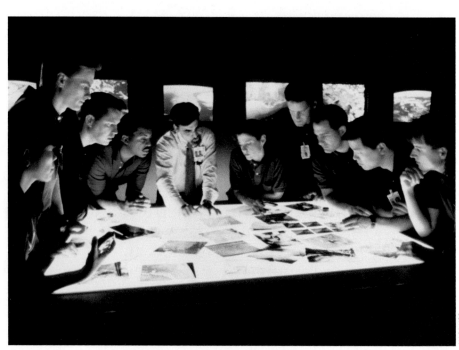

Figure 22.1 Space shuttle astronauts receive training in physical geography before their missions. Here a NASA geographer explains some of the Earth features they might study and photograph while they are in orbit.

NASA/JSC

nurturing characteristics that make our planet suitable for life are a consequence of an amazing combination of factors. Earth orbits at just the right distance from the sun, a star that radiates the right amount and wavelengths of energy to provide appropriate light for photosynthesis and vision. Of course, organisms have also developed mechanisms for using the energy that the sun provides and had to adapt to this environment or they would perish. Our planet absorbs sunlight, converting it into thermal energy that heats the land surface, atmosphere, and hydrosphere within a temperature range suitable for life to exist. Rotating on its axis, most of Earth receives a daily dose of direct solar energy throughout the year. The planet's rotation causes daytime and nighttime temperatures to be within a range that can support life, at least on most of Earth's surface.

Atmospheric gases allow insolation to penetrate while absorbing and storing heat energy and slowing its loss to space (Fig. 22.2a). Without this aspect of the greenhouse effect, our oceans would freeze over, shutting down the hydrologic cycle. Our planet's surface consists of water and terrestrial materials that store heat energy, received from the sun as light. The inclination of Earth's axis, combined with our annual revolution, cause the Northern and Southern Hemispheres to face both toward and away from the sun, so that the direct solar rays track over a wide tropical belt each year. These factors give us the seasons, and also act to moderate air, land, and water temperatures in extreme latitudes. Currents in the oceans and the atmosphere transfer heat energy from warmer regions to cooler ones, also moderating our global climates.

NASA/Goddard Space Flight Center/ The SeaWiFS Project and ORBIMAGE Science Visualization Studio

Figure 22.2 Spatial patterns on satellite images of hemispheres reveal features of the Earth system and interrelatedness of Earth's subsystems. (a) Atmosphere: Variations in water vapor levels in the atmosphere; (b) Hydrosphere: The vastness of the Pacific Ocean; (c) Lithosphere: The crust of Earth seen without the oceans; (d) Biosphere: The relative density of plant life on land and in the oceans. ■ **What are some of the geographic similarities seen in the patterns displayed in these four images?**

Our planet's mass results in enough gravitational force to hold an atmosphere in place. Earth also possesses abundant, life-sustaining surface and groundwater (Fig. 22.2b). Yet landmasses project above sea level because of plate movement, tectonic activity, and isostasy. Without these crustal dynamics, Earth would be completely covered by oceans. Crustal activity and lithospheric materials combine with processes of the hydrosphere, atmosphere, and biosphere to produce an intriguing variety of landforms and landscapes (Fig. 22.2c). Minerals in the crust undergo weathering to provide soil and nutrients for plant life. The surface relief caused by uplift or subsidence, and by erosion or deposition, interacts with the atmosphere and hydrosphere to provide a variety of elevational environments for the biosphere.

The biosphere is exceedingly complex, with organisms occupying diverse niches in virtually every environment in the planet's hydrosphere and near-surface lithosphere (Fig. 22.2d). Earth is a complex system of interrelated features, processes, organisms, and regions. In our vast universe, the existence of this complex suite of environmental variables occurring together on even one solitary planet is indeed a miracle. Any discovery of other life in the universe would still not diminish the miraculous nature of Earth.

So, Earth—the life support system for humanity and the rest of the biosphere—is an exceptional place. The diversity of environmental conditions that support life on Earth is tremendous. Every ecosystem offers different challenges and advantages to the success or survival of indigenous life forms, including humans. Geographic transitions and regional boundaries exist in a nearly unlimited variety, and their expression in the landscape ranges from obvious to subtle—food for thought for geographers in the millennia to come.

Only a relatively small part of Earth's surface, however, can directly sustain a human population (Fig. 22.3). Seventy-one percent of Earth is covered by water. The remaining 29 percent is land, but a considerable portion is in areas unsuitable for human habitation because of extreme environmental conditions. These locations include climatically extreme desert, glacial, and tundra regions, as well as places with very high elevations. Although these areas may be unsuitable for human occupation, they are not "wastelands" (an unfortunate term) because they support other life and play a role in Earth's environmental systems. Other lands are also excluded as potential living space for humans because they are conserved as biosphere reserves (Fig. 22.4), parks, and wilderness areas or are necessary for the natural resources they provide, such as forests.

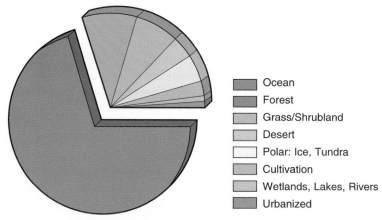

Ocean
Forest
Grass/Shrubland
Desert
Polar: Ice, Tundra
Cultivation
Wetlands, Lakes, Rivers
Urbanized

Figure 22.3 The percentages of land and water areas on Earth. Habitable land is a limited resource on our planet.

Thus, Earth's sustainable living space and natural resources are finite, while the human population continues its exponential growth. Dealing with population growth is difficult and complex, but no matter what the

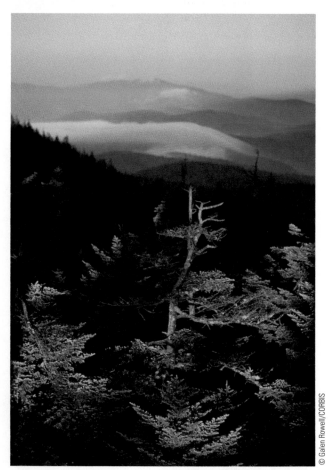

Figure 22.4 The environment of the Great Smoky Mountains, Tennessee-North Carolina, a Biosphere Reserve. An ongoing global effort seeks to conserve a wide range of threatened environments around the world by establishing Biosphere Reserves to protect these sensitive areas and preserve them for future generations.

© Galen Rowell/CORBIS

answer to that challenge may be, gathering information about our planet and its environments will always play a key role in sustaining human existence.

Integrating the Complexities of Planet Earth

How should we think about our planet—as a collection of environmental features and regions, as spatial bits and pieces; or as a unified whole, the Earth system? Actually, geographers and other earth scientists often have to do both, as each approach offers advantages. The decision is based partly on the nature, scope, and spatial scale of the problem being addressed.

In *Operating Manual for Spaceship Earth,* Buckminster Fuller noted how strange it is that people seem to be moving toward increasingly narrow academic and career specializations. He contended that the desire "to understand all and put everything together" is a basic human trait. Categories and classification schemes are human inventions. We use them to divide our seamless, complex world into meaningful segments, to organize these parts in logical ways, and to reduce complicated features to a level of simplicity with which we can mentally deal. While we classify and subdivide the Earth system to increase our understanding, we also learn from features that apparently do not fit our schemes and from overlaps that exist between categories.

Responding to our original question—whether to think about Earth as a whole or as a collection of parts—let us reexamine Earth's four major subsystems: the atmosphere, lithosphere, hydrosphere, and biosphere. To a certain extent they operate individually, but they are also strongly interrelated—one might even say separable but also inseparable. Imagine Earth without one of these major subsystems, each of which is a critical and essential part of the whole (Fig. 22.5). Studies in physical geography tend to involve aspects of each of the four major subsystems of Earth. Physical geography integrates organisms, water, air, and earth materials, as well as the potential impact of human activities, in order to understand Earth's environments, features, and regions.

Countless instances of overlap exist between these physical domains, but consider a few examples. The lithosphere incorporates part of the hydrosphere as soil moisture and groundwater. It interacts with the biosphere through organic activity in soil, and with the atmosphere as rocks weather and soil develops. The atmosphere contains water vapor—part of the hydrosphere—and volcanic gases, volcanic ash, and soil particles, all of lithospheric origin. Pollen, spores, bacteria, and smoke from burning vegetation, products of the biosphere, are in the atmosphere. The biosphere's abundance and variety would not exist without land, air, and water. Water plays a critical role in processes that support the Earth system.

Earth is further complicated by its ever-changing nature; it is a dynamic system. Fortunately, it is also robust and resilient, as negative feedback acts to maintain balance. Studies indicate that most natural systems can accommodate much change without impacts that might be catastrophic to a living population. Yet Earth systems also have limits (thresholds) to how much alteration can occur without significant environmental change. If these threshold conditions are either exceeded or not met, major problems can develop.

NASA

Figure 22.5 The Earth system. Images of a full hemisphere from space display global and regional information about the Earth system and our planet's physical geography. This image shows aspects of all four major subsystems and spatial patterns help us to know more about how they operate, interact, and integrate to produce a unified planetary system. Understanding Earth systems benefits from considering our planet and its features at many spatial scales, from global to local. ■ **What recognizable spatial and environmental patterns do you see in the physical geography shown in this image?**

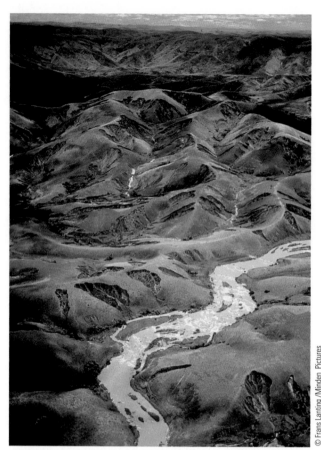

© Frans Lanting /Minden Pictures

Figure 22.6   Severe Environmental Destruction Related to Human Activities. Madagascar has suffered severe environmental degradation as a result of deforestation associated with a growing population and the cutting and burning of forests to clear lands for agriculture. Most of the island was covered with trees at the beginning of the twentieth century, but today only about 10 percent of the forest remains. Soil erosion on the barren slopes is intense as heavy tropical rains carve gullies into hillsides and choke rivers with sediment.   ■   **What are possible answers to dealing with this problem in order to conserve the land?**

Human activities are exerting ever-increasing pressures on the "normal" functions of environmental systems. These activities may be intentional, such as building flood control structures, or they may be the inadvertent effects of human activities, such as increasing releases of greenhouse gases. A valid question concerns the limits of human impact that can be accommodated by the natural processes that regulate an environmental system (Fig. 22.6). It is through stewardship, care, and maintenance of our planet's environments that humans will function as the crew of Spaceship Earth, rather than as passengers.

Physical Geography and Modern Technology

Although technology has contributed to many problems that our environment faces, technology is also critical to addressing these problems. Providing "power tools" for studying Earth, technology makes tasks easier and allows skilled users to accomplish them in a shorter time. A knowledge of geography and the expertise to apply appropriate techniques are useful, not only in academics, but also in a variety of applied professional fields.

Physical geographers today use many different techniques to gather data and information, conduct analyses, and graphically display the results of their studies. These methods are not exclusive to geography, but when they are applied using one or more of the three science perspectives of geographers, studying Earth with these techniques becomes uniquely geographical.

Computer applications are increasingly important to geographers studying complex, multifaceted problems. Through modeling, simulation, and prediction, computers allow geographers to "dissect" the Earth system (or one of its subsystems), breaking it down into its component variables. The goal is to analyze how these variables function, individually or in conjunction with other variables, and then reassemble them, subsystem by subsystem, until Earth is "whole" again. Geographic problems in the spatial, physical, or environmental sciences typically involve so many factors that several approaches to information gathering and data analysis are used (Fig. 22.7). Supported by a variety of data, information, and imagery, one form of evidence can be checked against another until a reasonable conclusion is reached.

Geography-related technology is also becoming ever more important in our daily lives. GPS receivers are replacing magnetic compasses for direction finding and

Courtesy of Environmental Research Systems, Inc.

Figure 22.7   Geographic Information Systems (GIS) facilitate storing, retrieving, and displaying information and data for solving complex spatial problems. Like any tool, however, GIS requires knowledgeable and skilled human input to operate effectively.   ■   **How would a physical geography background be helpful in applying a GIS to understand spatial relationships and solve environmental problems?**

Figure 22.8 A farmer in Bozeman, Montana, uses a GPS and a four-wheel off-road vehicle to measure his acreage. Originally designed for the military, the global positioning system has many civilian applications. GPS is revolutionizing surveying, mapping, location finding, and navigation, by deriving the precise location of the receiver and tracking its movements on a computer map display. ■ **What other kinds of civilian applications could be assisted by GPS technology?**

navigation. Mapping systems can display GPS coordinate data instantly on a computerized map as locations are surveyed in the field. GPS systems support navigation and mapping on land, on the sea, and in the air. These systems are also coming to replace traditional road maps (Fig. 22.8).

The computer has revolutionized graphic and cartographic methods used in physical geography. Maps, graphs, and charts that once took days or weeks to draw manually can now be completed in hours or minutes (Fig. 22.9). Understanding geography cartographic principles is still essential to producing effective maps, but computers help us generate maps that are economical and easily revised or customized for solving a specific problem.

Geographic information systems are now essential to many applications in business, industrial, and public sectors (see again Fig. 22.7). Countless decisions affecting your life will be based on analyses facilitated by a GIS. Remotely sensed imagery improves our ability to monitor and protect our environment, and supports earlier and more accurate assessments of environmental change. Spatial information, digital maps, and imagery are all readily available on the World Wide Web. The "information explosion" of recent years has had a dramatic impact on geographic studies.

Topics in Physical Geography—Revisited

Chapter 1 outlined five topics in physical geography that provide a guide to asking spatial questions about the Earth system. It is useful to briefly reexamine these topics and consider how they can be used to develop a geographer's view of the world.

Location

Location is essential in spatial communication. Having a reference system or a "global address" to describe a location or to find features is basic to geography. Location is the language of geography, whether it is absolute location in terms of a grid system, or relative location in terms of proximity to other features. Recall how important location has been in discussions of weather elements,

Figure 22.9 Computer technology can be used to process digital data and imagery and generate complex maps that would have been impossible, or exceedingly difficult, to produce without this technology.

Figure 22.10 The way that an environment looks is often geographically distinctive—a function of its physical characteristics, its location on Earth, and a variety of other factors we have learned in this course. This photograph shows a tropical savanna area of the East African rift valley in Kenya—a tropical-wet and dry region of faulting and volcanic activity. ■ **What factors of location and characteristics of place have combined to give this area its distinctive appearance?**

climates, environmental hazards, tectonic plate boundaries, catastrophic natural events, and major landforms. Locational information provides the spatial context that is relevant in every aspect of geography, and it inevitably leads the geographer to question *why* things are located where they are.

Characteristics of Places

Describing the physical characteristics of a place was once the goal of many geographic studies, but today that task usually represents only the beginning of our efforts to understand a site, locale, or region. Environments on Earth vary greatly according to location, elevation, and continental or oceanic position. Descriptions based on detailed observation and inventories of environmental features are still important, but explaining the factors that have contributed to the nature of a place or a feature is the intent of most studies in physical geography (Fig. 22.10).

Geographers often group places with similar characteristics into regions. Physical regions may be based on any environmental criteria. The physical and conceptual boundaries that divide natural regions are generally fuzzy rather than sharp, with edges that tend to be zones of transition rather than readily defined lines. Physical characteristics are important because environmental concerns are linked to the nature of places and regions.

Spatial Distributions and Spatial Patterns

Physical geographers are especially interested in the extent of, and factors that influence, distributions of natural features and areas. Whatever a map of distribution shows—for example, earthquake epicenters, vegetation, precipitation, storms, deserts, ocean currents, loess deposits, atmospheric pressure, glaciation—physical geographers consider these important questions: What processes explain the distribution, and what can be learned from the spatial pattern that is revealed?

Applying these questions as we encounter maps and remotely sensed images enriches our awareness and comprehension of spatial patterns. Examples can be seen in a satellite image of North America at night (Fig. 22.11) that approximates human population distributions on the continent. *Random* means that there is no pattern; *clustered* means that people exist close together, as seen in urban areas; and *uniform* means that there is a regular spacing, as seen in the lights of rural areas that use the public lands survey system (West and Midwest regions of the United States).

Analyzing distributions helps us explain locational relationships between different phenomena. The distribution of one feature may be related to the distribution of another feature, region, process, or environmental condition. An example is the global distribution of volcanism (see Fig. 14.14). Volcanic regions reveal a remarkable spatial correlation with global patterns of earthquake epicenters (see Fig. 14.34), and both distributions are geographically related to tectonic plate boundaries (see Fig. 13.22).

Spatial Interactions

Humans once thought that places existed in virtual isolation from environmental events in distant locations. We now understand how certain remote occurrences can affect our local environments. We have learned about feedback operations and systems analysis, concepts that help us understand complex environments, and how

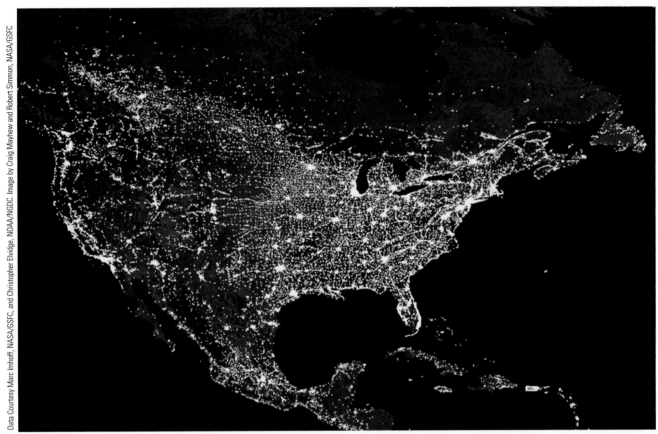

Figure 22.11 An approximation of the population distribution of North America is shown by a nighttime satellite image. Lights indicate population patterns and densities, although "light pollution" exaggerates the population density, and certain regions and cultures may use less electric lighting than others. ■ **Why do certain large regions have such a low density of lights, representing a sparse population?**

variables interact with one another. Interactions between places and regions, and the interconnections that link them, are nearly boundless.

Volcanic eruptions that put ash and dust into the atmosphere can affect weather over wide parts of Earth. El Niño, once thought to be a local condition affecting South America's west coast, is now known to be associated with global weather patterns. Deforestation of tropical rainforests is thought to have an impact on world climates. It has recently been discovered that soil nutrients are supplied to Amazon rainforests through dust carried by winds over the Atlantic from the Sahara (Fig. 22.12). Nature knows few boundaries in its environmental interactions and interrelationships.

Ever-Changing Earth

We should be aware that changes taking place in the Earth system operate at different rates and over widely varying time spans. Certain changes—such as volcanic eruptions, storms, seasonal transitions, wildfires, floods, coastal erosion, and many others—can be directly witnessed. However, environmental modification can also operate at a slow rate that requires long-term scientific study to monitor. This type of change includes gradual

slippage along a fault, shifts in climatic boundaries, vegetation succession, global sea level change, glacial movement, and revegetation of an area devastated by a volcano (Fig. 22.13).

Human modifications complicate the processes involved in Earth systems. Most scientists agree that global warming is occurring, and that human impacts on the atmosphere probably play a role in this climatic shift. As atmospheric levels of greenhouse gases rise because of human activity, so does the global average temperature. But the problem is not as simple as that, because many factors influence climate change. Climate change can occur naturally, or be a result of human activities, or result from some combination of the two.

Change in one part of the Earth system typically produces changes in other parts. If global warming were to continue until the polar glaciers melted, Earth would undergo myriad other changes. The temperature and salinity of the oceans would be altered. Sea level would rise, flooding coastal lowlands and shrinking continental landmasses. There would also be shifts in Earth's vegetation zones, and animals would migrate with these shifts. Rivers would regrade their channels to a new base level.

Determining to what degree these environmental changes would be "good" or "bad" would depend to a great extent on geography, as the impact would differ by location and region. If you lived on a lowland coast, a rising ocean would threaten your home. However, if you lived in Russia or Canada, the growing season could become longer, something that most residents in those countries would consider advantageous.

Human Interactions with the Environment

Humans are becoming an increasingly significant factor in changing the Earth, for two major reasons. First, our ability to affect Earth environments has increased with technological advances and with population growth. Second, much environmental change is cumulative over time and/or is not easily reversible. Few, if any, parts of the Earth system have not been affected in some way by human-induced modification.

Today we understand that humans have an impact on the environment, and the environment also has an impact on us. Humans have some ability, although not an unlimited ability, to adapt environments to fit their needs (Fig. 22.14), but humans will never completely control their environment. As we have learned about the Earth system, we understand that humans must work within the "rules" that govern how natural systems operate or risk creating or aggravating environmental problems.

Scientists do not seek to impose their needs and wants on Nature, but instead humbly interrogate Nature and take seriously what they find.

Carl Sagan

Physical Geography and You—Revisited

You might ask, How does knowledge of physical geography relate to me? The answer lies in you as an individual. "Geography for Life," the title of the Geography National Standards, conveys the value of a lifelong appreciation for geography. Physical geography is a part of our everyday life. Understanding how the Earth system operates contributes to our appreciation of the planet and helps us to become responsible citizens. Geographic awareness is a broadening experience in a liberal education, and it can also grow into a related career.

Virtually any professional activity that deals with maps, spatial information, and environmental data can benefit from the services of a well-trained geographer. The Career Vision series, featuring a selection of physical geographers in a variety of career fields, illustrates the relevance that expertise in physical geography, with its tools and methods, has in today's world. Many geographers are employed by federal, state, and local government agencies, corporations, businesses, and consulting firms.

At the end of Chapter 1, you were challenged to "see Earth differently" after taking

NASA, Johnson Space Center, Houston, TX

Figure 22.12 A dust storm blows a cloud of sediment and soil particles from a desert region over the ocean. Some of the particles will be deposited on the ocean and contribute to bottom sediments. Fine dust can be carried thousands of kilometers from its source, until it settles out of the air or is washed out by precipitation. Research suggests that such storms can provide nutrients to the rainforests of the Amazon Basin across the Atlantic from the Sahara. This dust storm is about 112 km (70 miles) in size. ■ **What do transoceanic dust storms suggest about the Earth as a system?**

(a)

(b)

(c)

Figure 22.13   Environmental changes in the area of Mt. St. Helens. These three images show the volcano and environs (a) before the 1980 eruption, when much of the region was covered with forest, (b) just after the eruption, with a radically changed volcanic cone, and a large area devastated by the eruption and associated events, and (c) twenty years later, as the land is becoming re-vegetated.

a course in physical geography. Perhaps it would be better to call this process "reading" instead of "seeing," because reading means having an awareness and comprehension of the Earth system, its elements, and their spatial, or geographic, nature. For many people, "reading" an environment and the elements of a landscape in a geographic context (Fig. 22.15) is an enriching experience, whether their involvement is career-oriented or avocational.

The *Titanic* and Physical Geography

The RMS *Titanic* offers an excellent case study for understanding the relevance of physical geography. The Global Positioning System (GPS) played a key role in pinpointing the wreck's location by a discovery team sup-

ported by the National Geographic Society (Fig. 22.16). One problem was finding the ship's exact location, given the featureless (in terms of landmarks) surface of the ocean floor. Dr. Robert Ballard, the scientist in charge, said that locating a ship on the ocean bottom under miles of seawater using a robotic submarine was like "lowering a penny on a thread from the top of the Empire State Building to its base and getting it to land on a toothpick." Once the wreck was found, precise GPS coordinates guaranteed that the *Titanic* could be found again.

Knowing the *Titanic*'s location, we can consider the geographic reasons for why the wreck is located at this latitude and longitude on the North Atlantic seafloor. The story concerns human error—trying to cross the Atlantic as quickly as possible on the maiden voyage of a ship that was thought to be unsinkable and immune to the environmental conditions under which it would operate. This erroneous assumption led to another tragic decision—that because of this supposed "immunity to nature," adequate lifeboats were unnecessary. Important factors in that story involve a spatial science perspective, and also relate to the physical science and environmental aspects of physical geography.

Looking at a world map, we can ask why the *Titanic* met its fate in its voyage from England to New York City at this particular latitude, which seems so far north. The answer is that the ship's crew, attempting to set a record time for an Atlantic crossing, was sailing a great circle route (Fig. 22.17), the shortest distance between two

Mark Wagner/Getty Images

Figure 22.14 Humans change the environment to fit a particular need. An aqueduct (concrete-lined, artificial stream channel) transports water from the mountains to arid and semiarid regions of southern California. Without this human-induced transfer of water, many arid regions could not support either a large population or intensive agriculture. ■ **What environmental problems may be created through the transfer of water from wet regions to arid regions?**

points on Earth. Many projections create a mistaken impression of the ship's route, because they do not portray a great circle as a straight line, which we normally associate with the shortest distance. The distortion on most world maps increases in the high latitudes. These facts illustrate the importance of spatial location and how locations are displayed on maps, and gives us a part of the answer to our question, Why there?

Second, we need to consider some physical characteristics of place. The *Titanic* hit an iceberg, but why were icebergs there? Icebergs calve off of glaciers in Greenland, to the north of where the *Titanic* met its fate, and drift southward, unfortunately into a heavily used shipping lane between Europe and North America. Because ice is only slightly lighter than seawater, only a small part of an iceberg projects above the surface as it floats in the ocean. The iceberg in the *Titanic*'s path was there because it was adrift in the Labrador Current, which flows southward between Greenland and Baffin Island. Icebergs are common in these waters off Newfoundland, Canada. Since the *Titanic* sank, they have been counted and monitored; 250 per year is the average number. The icebergs eventually shrink in size, melting as they encounter warmer waters to the south.

Another geographic factor is the climate of that world region and, more specifically, the weather conditions on that fateful April night in 1912. The North Atlantic off the east coast of Canada includes a region called the Grand Banks, one of the foggiest places on Earth. The fog is advection fog, caused by cooling of warm, moist air that follows the Gulf Stream, a warm current flowing poleward along the east coast of North America. As this warm air moves over the cold Labrador Current, a dense fog frequently forms in the Grand Banks and nearby parts of the North Atlantic. So, did the *Titanic* hit an iceberg that its crew could not see because of foggy conditions? Actually, this was not the situation because that night

Paul Chelsey/Getty Images

Figure 22.15 How have physical geographic and environmental factors influenced this landscape? Look at this photograph of an area in the Colorado Rockies. What aspects of this environment do you recognize in this landscape that make it look the way it does? ■ **What can you explain about this photograph in terms of the Earth system and its four major subsystems? Can you examine this photo with the "educated eye" of the geographer that you learned about in Chapter 1?**

These atmospheric and oceanic environmental factors combined with errors of human judgment to seal the *Titanic*'s fate. If the moon's phase had been full on that clear night, icebergs would have been easier to spot, perhaps early enough to avoid a collision. Unfortunately, the icebergs were obscured by the cover of darkness in a moonless sky, a situation compounded by pressures on the crew to maintain speed. It is ironic that if the night had been foggy—as it often is near the Grand Banks—the *Titanic* would have had to reduce its speed, perhaps avoiding collision with the iceberg with enough force to sink the ship. So, not only did human error contribute to this notorious event, but the geographic factors associated with location, the physical characteristics of place, and other aspects of physical geography also played key roles in the *Titanic*'s tragic story.

Figure 22.16 Sophisticated technology and equipment along with much searching was involved in Robert Ballard's expedition to find the wreck of the *Titanic*. Global positioning systems played a key role in locating and mapping the site in detail. Ballard's expedition was supported by the National Geographic Society.

Epilogue

As a part of a general education, knowledge of how the Earth system functions can deepen our comprehension of events involving human interaction with the environment. Physical geography can also focus our awareness of the Earth system and its natural processes. In either case, comprehension of our physical environment enriches our existence on Earth and helps us to be informed, active participants in our finite life support system rather than just idle planetary passengers. Unlike those on board the *Titanic,* we know that as crewmembers of Spaceship Earth, we are not immune to the environmental conditions under which our ship operates.

the air was cold and clear. High pressure dominated, with its typically calm conditions, so the sea was smooth. Visibility should have been good, but there was no moon that evening, making the clear skies and the calm ocean particularly dark.

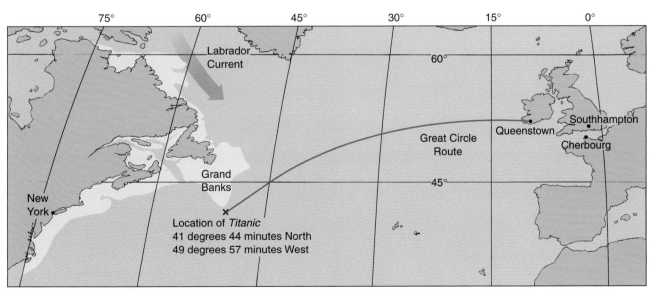

Figure 22.17 This map shows the latitude and longitude where the *Titanic* rests on the North Atlantic Ocean floor, about 3810 meters (12,500 feet) below the surface. The great circle route is shown along with a few of the physical geographic factors that contributed to the demise of the ship, such as the Labrador Current which delivers icebergs southward each spring into one of the busiest shipping lanes in the world.

International System of Units (SI), Abbreviations, and Conversions

Symbol	Multiply	By	To Find	Symbol
Length				
in	inches	25.4	millimeters	mm
ft	feet	0.305	meters	m
yd	yards	0.914	meters	m
mi	miles	1.61	kilometers	km
ft	feet	63,360	mile	mi
mm	millimeters	0.039	inches	in
m	meters	3.28	feet	ft
m	meters	1.09	yards	yd
km	kilometers	0.62	miles	mi
Area				
in^2	square inches	645.2	square millimeters	mm^2
ft^2	square feet	0.093	square meters	m^2
yd^2	square yards	0.836	square meters	m^2
ac	acres	0.405	hectares	ha
mi^2	square miles	2.59	square kilometers	km
ac	acres	43,560	square feet	ft^2
mm^2	square millimeters	0.0016	square inches	in2
m^2	square meters	10.764	square feet	ft^2
m^2	square meters	1.195	square yards	yd^2
ha	hectares	2.47	acres	ac
km^2	square kilometers	0.386	square miles	mi^2
Volume				
gal	U.S. gallons	3.785	liters	l
ft^3	cubic feet	0.028	cubic meters	m^3
yd^3	cubic yards	0.765	cubic meters	m^3
l	liters	0.264	U.S. gallons	gal
m^3	cubic meters	35.30	cubic feet	ft^3
m^3	cubic meters	1.307	cubic yards	yd^3
Velocity				
mph	miles/hour	1.61	kilometers/hour	km/h
knot	nautical miles/hour	1.85	kilometers/hour	km/h
km/h	kilometers/hour	0.62	miles/hour	mph
km/h	kilometers/hour	0.54	nautical miles/hour	knots
Mass				
oz	ounces	28.35	grams	g
lb	pounds	0.454	kilograms	kg
g	grams	0.035	ounces	oz
kg	kilograms	2.202	pounds	lb
Pressure or Stress				
mb	millibars	0.75	millimeters of mercury	mm Hg
mb	millibars	0.02953	inches of mercury	in Hg
mb	millibars	00.01450	pounds per square inch	(lb/in^2 or psi)
lbs/in^2	pounds per square inch	6.89	kilopascals	kPa
in Hg	inch of mercury	33.865	millibars	mb
kPa	kilopascals	0.145	pounds per square inch	(lb/in^2 or psi)

Standard Sea Level Pressure

29.92 in Hg
14.7 lb/in^2
1013.2 mb
760 mm Hg

Powers of Ten

nano	one-billionth	$= 10^{-9}$	=	0.000000001
micro	one-millionth	$= 10^{-6}$	=	0.000001
milli	one-thousandth	$= 10^{-3}$	=	0.001
centi	one-hundredth	$= 10^{-2}$	=	0.01
deci	one-tenth	$= 10^{-1}$	=	0.1
hecto	one hundred	$= 10^{2}$	=	100
kilo	one thousand	$= 10^{3}$	=	1000
mega	one million	$= 10^{6}$	=	1,000,000
giga	one billion	$= 10^{9}$	=	1,000,000,000

Temperature

°F	Fahrenheit	(°F − 32)/1.8	Celsius	°C
°C	Celsius	1.8 °C + 32	Fahrenheit	°F
K	Kelvins	K = C + 273	Celsius	°C

613

Topographic Maps

Mapping has changed considerably in recent years, and with the ever-increasing capabilities of computers to store, retrieve, and display graphics, this trend will continue well into the future. In the United States the U.S. Geological Survey produces the vast majority of topographic maps available. These maps have long been the tried and true tools of geographers and scientists in many other disciplines who study various aspects of the environment. Today virtually all USGS topographic maps are accessible in digital format, on computer disk, for downloading and printing, or for examining on a computer screen. This availability makes it easy to print maps or map segments at home, school, or work. Computers and the Internet have made maps much more available and accessible than they were just a few years ago. A logical question then would be, will paper maps become obsolete, given computer displays?

There are several reasons why paper maps will still be popular and useful, whether they are purchased from the source or downloaded and printed. Topographic maps are particularly important in fieldwork. Maps are highly portable, require no batteries or electrical power that could fail, and they do not suffer from technology glitches. They are reliable and easy to use and they also provide a good base for making field notes, and marking routes.

Today the USGS is working to make and maintain a seamless database of the United States, with maps, imagery, and spatial data in digital form, so areas that once were split on adjacent maps can be printed on a single sheet. This is a great change from dividing the country into quadrangles (the roughly rectangular area a map displays). Topographic quadrangles (quads), however, will continue to be in use for a long time. There are several standard quadrangles, each with a specific scale, and many other special purpose topographic maps.

7.5 minute quads—these are printed at a scale of 1:24,000, and cover 7.5 minutes of longitude and 7.5 minutes of latitude.

15 minute quads—these are printed at a scale of 1:62,500, and cover 15 minutes of longitude and fifteen minutes of latitude.

1° × 2° quads—these are printed at a scale of 1:250,000 and cover 1 degree of latitude and 2 degrees of longitude.

1:100,000 metric quads—these are printed at a scale of 1:100,000, cover 30 minutes of longitude and 60 minutes of latitude, and use metric measurements for distance and elevation.

Brown topographic contours are used to show elevation differences and the terrain.

Some rules for interpreting contours are given in Chapter 2.

Determining distances on a map, distances from a map, or an RF scale

It is important to understand representative fractions, like 1:24,000, which means that any measurement on the map will represent 24,000 of the same measurements on the ground. This knowledge is particularly significant because reproduced maps may not be printed at the original size. On a map that is reduced or enlarged the bar scale will still be accurate, but the printed RF scale will not. Enlarging or reducing a map changes the RF scale from the original.

How to find the RF of a map (or air photo, or satellite image) of unknown scale.

Here is the formula:

1/RFD = MD/GD

The numerator in an RF scale is always the number 1. The RFD is the denominator of the RF (such as 24,000). MD is map distance measured in any particular units on the map (cms, in.). GD is the true ground distance that the map distance represents (expressed in the same units used to measure the map distance). Never mix units in this calculation, but convert values into desired units afterward.

Example: How long in inches is a mile on a 1:24,000 scale map?
Important information: there are 63,360 inches in a mile. To find MD, for a known distance (mile) on a map of known scale, use this formula:

1/24,000 = MD/63,360 in.

one mile = 2.64 inches at 1:24,000

This statement has now been converted into a **stated scale** so units may be mixed.

1:24,000 Bar Scale

Here is a bar scale that can be used to make distance measurements directly from the 1:24,000 maps printed in the Map Interpretation sections. Note: check the listed RF, because not all are at a 1:24,000 scale.

BOUNDARIES

National ....
State or territorial ....
County or equivalent ....
Civil township or equivalent ....
Incorporated-city or equivalent ....
Park, reservation, or monument ....
Small park ....

LAND SURVEY SYSTEMS

U.S. Public Land Survey System:
  Township or range line ....
    Location doubtful ....
  Section line ....
    Location doubtful ....
  Found section corner; found closing corner ...
  Witness corner; meander corner ....

Other land surveys:
  Township or range line ....
  Section line ....
Land grant or mining claim; monument ....
Fence line ....

ROADS AND RELATED FEATURES

Primary highway ....
Secondary highway ....
Light duty road ....
Unimproved road ....
Trail ....
Dual highway ....
Dual highway with median strip ....
Road under construction ....
Underpass; overpass ....
Bridge ....
Drawbridge ....
Tunnel ....

BUILDINGS AND RELATED FEATURES

Dwelling or place of employment: small; large
School; church ....
Barn, warehouse, etc.: small; large ....
House omission tint ....
Racetrack ....
Airport ....
Landing strip ....
Well (other than water); windmill ....
Water tank: small; large ....
Other tank: small; large ....
Covered reservoir ....
Gaging station ....
Landmark object ....
Campground; picnic area ....
Cemetery: small; large ....

RAILROADS AND RELATED FEATURES

Standard gauge single track; station ....
Standard gauge multiple track ....
Abandoned ....
Under construction ....
Narrow gauge single track ....
Narrow gauge multiple track ....
Railroad in street ....
Juxtaposition ....
Roundhouse and turntable ....

TRANSMISSION LINES AND PIPELINES

Power transmission line; pole; tower ....
Telephone or telegraph line ....
Aboveground oil or gas pipeline ....
Underground oil or gas pipeline ....

CONTOURS

Topographic:
  Intermediate ....
  Index ....
  Supplementary ....
  Depression ....
  Cut; fill ....

Bathymetric:
  Intermediate ....
  Index ....
  Primary ....
  Index Primary ....
  Supplementary ....

MINES AND CAVES

Quarry or open pit mine ....
Gravel, sand, clay, or borrow pit ....
Mine tunnel or cave entrance ....
Prospect; mine shaft ....
Mine dump ....
Tailings ....

SURFACE FEATURES

Levee ....
Sand or mud area, dunes, or shifting sand ....
Intricate surface area ....
Gravel beach or glacial moraine ....
Tailings pond ....

VEGETATION

Woods ....
Scrub ....
Orchard ....
Vineyard ....
Mangrove ....

COASTAL FEATURES

Foreshore flat ....
Rock or coral reef ....
Rock bare or awash ....
Group of rocks bare or awash ....
Exposed wreck ....
Depth curve; sounding ....
Breakwater, pier, jetty, or wharf ....
Seawall ....

BATHYMETRIC FEATURES

Area exposed at mean low tide; sounding datum ...
Channel ....
Offshore oil or gas: well; platform ....
Sunken rock ....

RIVERS, LAKES, AND CANALS

Intermittent stream ....
Intermittent river ....
Disappearing stream ....
Perennial stream ....
Perennial river ....
Small falls; small rapids ....
Large falls; large rapids ....
Masonry dam ....
Dam with lock ....
Dam carrying road ....
Intermittent lake or pond ....
Dry lake ....
Narrow wash ....
Wide wash ....
Canal, flume, or aqueduct with lock ....
Elevated aqueduct, flume, or conduit ....
Aqueduct tunnel ....
Water well; spring or seep ....

GLACIERS AND PERMANENT SNOWFIELDS

Contours and limits ....
Form lines ....

SUBMERGED AREAS AND BOGS

Marsh or swamp ....
Submerged marsh or swamp ....
Wooded marsh or swamp ....
Submerged wooded marsh or swamp ....
Rice field ....
Land subject to inundation ....

Köppen Climate Classification

The key to understanding any system of classification is found by personally practicing use of the system. This is one reason why the Consider and Respond review sections of both Chapters 9 and 10 are based on the classification of data from sites selected throughout the world. Correctly classifying these sites in the modified Köppen system may seem complicated at first, but you will find that, after applying the system to a few of the sample locations, the determination of a correct letter symbol and associated climate name for any other site data should be routine. In addition, this will be the case whether the site data is in your textbook or provided by your instructor from some other source.

Before you begin, you should take the time to familiarize yourself with Table 1. You will note that there are precise definitions in regard to temperature or precipitation that identify a site as one of the five major climate categories in the Köppen system (*A*, Tropical; *B*, Arid; *C*, Mesothermal; *D*, Microthermal; and E, Polar). Furthermore, you will note that the additional letters, which are required to complete the letter symbol and identify the actual climate type, also have precise definitions or are determined by the use of graphs that distinguish one climate type from another. In other words, Table 1 is all you need to classify a site if monthly and annual means of precipitation and temperature are available. Table 1 should be used in a systematic fashion to determine first the major climate category and, once that is determined, the second and third letter symbols (if needed) that complete the classification. As you begin to classify, it is strongly recommended that you use the following procedure. (After a few examples you may find you can omit some steps with a glance at the statistics.)

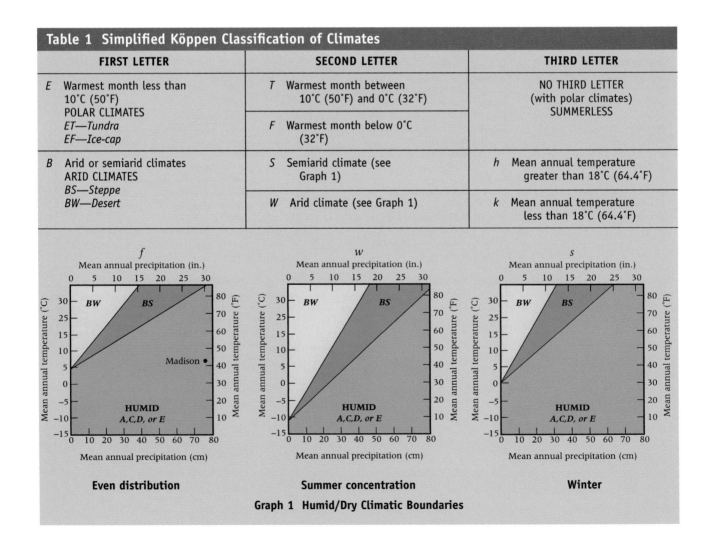

Table 1 Simplified Köppen Classification of Climates

FIRST LETTER	SECOND LETTER	THIRD LETTER
E Warmest month less than 10°C (50°F) POLAR CLIMATES *ET—Tundra* *EF—Ice-cap*	*T* Warmest month between 10°C (50°F) and 0°C (32°F)	NO THIRD LETTER (with polar climates) SUMMERLESS
	F Warmest month below 0°C (32°F)	
B Arid or semiarid climates ARID CLIMATES *BS—Steppe* *BW—Desert*	*S* Semiarid climate (see Graph 1)	*h* Mean annual temperature greater than 18°C (64.4°F)
	W Arid climate (see Graph 1)	*k* Mean annual temperature less than 18°C (64.4°F)

Even distribution **Summer concentration** **Winter**

Graph 1 Humid/Dry Climatic Boundaries

Table 1 Simplified Köppen Classification of Climates (continued)

FIRST LETTER	SECOND LETTER	THIRD LETTER
A Coolest month greater than 18°C (64.4°F) TROPICAL CLIMATES *Am—Tropical monsoon* *Aw—Tropical savanna* *Af—Tropical rainforest*	*f* Driest month has at least 6 cm (2.4 in.) of precipitation	NO THIRD LETTER (with tropical climates) WINTERLESS
	m Seasonally, excessively moist (see Graph 2)	
	w Dry winter, wet summer (see Graph 2)	

Graph 2

Graph 2   Rainfall of driest month (cm)

C Coldest month between 18°C (64.4°F) and 0°C (32°F); at least one month over 10°C (50°F) MESOTHERMAL CLIMATES *Csa, Csb — Mediterranean* *Cfa, Cwa - Humid Subtropical* *Cfb, Cfc - Marine West Coast*	*s* (DRY SUMMER) Driest month in the summer half of the year, with less than 3 cm (1.2 in.) of precipitation and less than 1/3 of the wettest winter month	*a* Warmest month above 22°C (71.6°F)
		b Warmest month below 22°C (71.6°F), with at least four months above 10°C (50°F)
D Coldest month less than 0°C (32°F); at least one month over 10°C (50°F) MICROTHERMAL CLIMATES *Dfa, Dwa — Humid Continental, Hot Summer* *Dfb, Dwb — Humid Continental, Mild Summer* *Dfc, Dwc, Dfc, Dwc — Subarctic*	*w* (DRY WINTER) Driest month in the winter half of the year, with less than 1/10 the precipitation of the wettest summer month	*c* Warmest month below 22°C (71.6°F), with one to three months above 10°C (50°F)
	f (ALWAYS MOIST) Does not meet conditions for *s* or *w* above	*d* Same as *c*, but coldest month is below {neg}38°C ({min}36.4°F)

Step 1. Ask: Is this a polar climate (*E*)? Is the warmest month less than 10°C (50°F)? If so, is the warmest month between 10°C (50°F) and 0°C (32°F) (*ET*) or below 0°C (32°F) (*EF*)? If not, move on to:

Step 2. Ask: Is there a seasonal concentration of precipitation? The answer is determined by examining the monthly precipitation data for the driest and wettest summer and winter months for the site. Take careful note of temperature data as well, because you must take into consideration whether the site is located in the Northern or Southern Hemisphere. (April to September are summer months in the Northern Hemisphere but winter months in the Southern Hemisphere. Similarly, the Northern Hemisphere winter months of October to March are summer south of the equator.) As the table indicates, a site has a dry summer (*s*) if the driest month in summer has less than 3 cm (1.2 in.) of precipitation and less than 1/3 of the precipitation of the wettest winter month. It has a dry winter (*w*) if the driest month in winter has less than 1/10 the precipitation of the wettest summer month. If the site has neither a dry summer nor a dry winter, it is classified as having an even distribution of precipitation (*f*). Move on to:

Step 3. Ask: Is this an arid climate (*B*)? You must use one of the small graphs (included in Graph 1) to decide. Based on your answer in Step 2, select one of the small graphs and compare mean annual temperature with mean annual precipitation. The graph will indicate whether the site is an arid (*B*) climate or not. If it is, the graph will indicate which one (*BW* or *BS*). You should further classify the site by adding *h* if the mean annual temperature is above 18°C (64.4°F) and *k* if it is below. If the site is neither *BW* nor *BS*, it is a humid climate (*A*, *C*, or *D*). Move on to:

Step 4a. Ask: Is this a tropical climate (*A*)? The site has a tropical climate if the temperature of the coolest month is higher than 18°C (64.4°F). If so, use Graph 2 in Table 1 to determine which tropical climate the site represents. (Note that there are no additional lowercase letters required.) If not, move on to:

Step 4b. Ask: Which major middle-latitude climate group does this site represent, mesothermal (*C*) or microthermal (*D*)? If the temperature of the coldest month is between 18°C (64.4°F) and 0°C (32°F), the site has a mesothermal climate. If it is below 0°C (32°F), it has a microthermal climate. Once you have answered the question, move on to:

Step 5. Ask: What was the distribution of precipitation? This was determined back in Step 2. Add *s, w,* or *f* for a *C* climate, or *w* or *f* for a *D* climate, to the letter symbol for the climate. Then, move on to:

Step 6. Ask: What letter should I add to my climate symbol to express the details of seasonal temperature for the site? Refer again to Table 1 and the definitions for the letter symbols. Add *a*, *b*, or *c* for the mesothermal (*C*) climates or *a*, *b*, *c*, or *d* for the microthermal (*D*) climates and you have completed the classification of your climate. However, note that you may not have come this far because you might have completed your classification at Steps 1, 3, or 4a.

We should now be ready to try out the use of Table 1, following the steps we have recommended. Data for Madison, Wisconsin, is presented below for our example.

Step 1. First, we must determine whether or not our site has an *E* climate. Since Madison has several months averaging above 10°C, it does not have an *E* climate.

Step 2. Second, we must determine if there is a seasonal concentration of precipitation. Since Madison is driest in winter, we compare the 2.5 cm of February precipitation with the precipitation of June (¹⁄₁₀ of 11.0 cm, or 1.1 cm) and conclude that Madison has neither a dry summer nor a dry winter, but instead has an even distribution of precipitation (*f*). (Note: The 2.5 cm of February precipitation is not less than ¹⁄₁₀ (1.1 cm) of June precipitation.

Step 3. Next we assess, through the use of Graphs 1(*f*), 1(*w*), or 1(*s*), whether our site is an arid climate (*BW*, *BS*) or a humid climate (*A*, *C*, or *D*). Since we have previously determined that Madison has an even distribution of precipitation, we will use Graph 1(*f*). Based upon Madison's mean annual precipitation (77.0 cm) and mean annual temperature (7°C), we conclude that Madison is a humid climate (*A*, *C,* or *D*).

Step 4. Now we must assess which humid climate type Madison falls under. Since the coldest month ({min}8°C) is below 18°C, Madison does *not* have an *A* climate. Although the warmest month (21°C) is above 10°C, the coldest month ({min}8°C) is not between 0° and 18°C, so Madison does *not* have a *C* climate. Since the warmest month (21°C) is above 10°C, and the coldest month is below 0°C, Madison *does* have a *D* climate.

Step 5. Because Madison has a *D* climate, the second letter will be *w* or *f.* Since precipitation in the driest month of winter (2.5 cm) is not less than ¹⁄₁₀ of the amount of the wettest summer month (¹⁄₁₀ {tim} 11.0 cm = 1.1 cm), Madison does *not* have a *Dw* climate. Madison, therefore, has a *Df* climate.

Step 6. Since Madison is a *Df* climate, the third letter will be *a*, *b*, *c*, or *d*. Since the average temperature of the warmest month (21°C) is not above 22°C, Madison does *not* have a *Dfa* climate. Since the average temperature of the warmest month is below 22°C, with at least 4 months above 10°C, Madison is a *Dfb* climate.

If you have not been able to follow the explanation for why Madison, Wisconsin, is classified *Dfb* (a site with a humid continental, mild summer climate), try the example again using only Table 1 before reading again the explanation at each step in the procedure. If you still are having difficulty, do not hesitate to ask your instructor for assistance. Stick with it; classification can be fun once you begin getting the correct answers. In addition, you will be better prepared for the real challenge of explaining why a site has a particular climate.

	J	F	M	A	M	J	J	A	S	O	N	D	Year
T(°C)	−8	−7	−1	7	13	19	21	21	16	10	2	−6	7
P(cm)	3.3	2.5	4.8	6.9	8.6	11.0	9.6	7.9	8.6	5.6	4.8	3.8	77.0

BASIC ROCKS

Rocks are aggregates of minerals, and although there are thousands of kinds of rocks on our planet, they can be classified into three fundamental kinds based on their origin: igneous, sedimentary, and metamorphic. The formation of rocks was outlined in Chapter 13. Having a solid knowledge of how the Rock Cycle operates (Figure 13.5), as well as its components and its processes is essential to understanding the solid Earth. Specific rock types are mentioned in several chapters of this book. Although making a positive identification of a rock type requires examining several physical properties, having a mental image of what different rocks look like will be an aid to understanding Earth processes and landforms. The following is an illustrated guide to a few common rocks to help in their identification.

Igneous Rocks

Igneous rocks are subdivided into intrusive and extrusive, depending on whether they cooled within Earth or on its exterior.

Intrusive Igneous

Intrusive igneous rocks form from a molten state by cooling and crystallizing underground, they generally cool very slowly which allows coarse crystals to form, crystals that are easy to see with the naked eye.

Granite

Granite forms deep in the crust, and has easily visible intergrown crystals of light and dark minerals, but is dominated by light colored silicate minerals. Granitic rocks are typically gray or pink, and their mineral composition is similar to that of the continental crust.

Diorite

Diorite is an intermediate intrusive rock meaning that it has a roughly equal mix of easily visible light and dark minerals, which gives it a spotted appearance. Generally diorite is dark gray in color.

Gabbro

Gabbro is a dark intrusive rock dominated by heavy, iron-rich silicate minerals. The crystals are coarse enough to be easily visible, but because of the overall dark tone, they tend to blend together. Gabbro is black and may contain some very dark green minerals.

Extrusive Igneous

Extrusive igneous rocks cool at or near the surface, and include lavas, as well as rocks made of tephra (pyroclastics), fragments blown out of a volcano. Extrusive rocks sometimes preserve gas bubble holes, and may contain visible crystals, but typically the grains are small. Cooling relatively rapidly at the surface produces fine-grained lavas.

Rhyolite

Rhyolite is a very thick lava when molten (much like melted glass), light in color and high in silica. Colors of rhyolite vary widely, but light gray, very light brown, and pink or reddish are common. Rhyolite is the extrusive equivalent of granite in terms of mineral composition.

Andesite

Andesite, named for the Andes, is an intermediate lava in terms of both mineral content and color. Associated with composite cone volcanoes, it is relatively thick when molten. Often mineral crystals are visible in a matrix of finer grains, and the color is gray to brown. Andesite is the extrusive equivalent of diorite.

Basalt

Basalt is dark, typically black, and heavier than other lavas. Associated with lava flowing from fissures and shield volcanoes, basalt is relatively low in silica, and so it has a lower viscosity than other lavas. Basaltic lava also tends to be hotter than other lavas, and because it is relatively thin when molten, it can flow for many miles before the flow cools enough to stop. Basalt is the rock of the oceanic crust, and is the extrusive equivalent of gabbro.

Tuff

Tuff is a rock made of fine tephra—volcanic ash—that was blown into the atmosphere by a volcanic eruption, and settled out in layers that blanketed the surface. Tephra (loose fragments) was converted into tuff by burial and compaction, or by being welded together from intense heat. Tuff is gray to tan.

Sedimentary Rocks

Sedimentary rocks can also be divided into two major categories, clastic and non-clastic.

Clastic Sedimentary

Clastic means made up of cemented rock fragments, such as clay, silt, sand, pebbles, cobbles, or even boulders. The sizes and shapes of clasts within a sedimentary rock provide clues about the environments under which the fragments were deposited (fluvial, eolian, glacial, coastal).

Shale

Shale is a fine-grained clastic rock that contains lithified clays, generally deposited in very thin layers. Shales represent a calm water environment, such as a sea or lake bottom. Shales vary widely in color, but most are gray or black, and it breaks up into smooth flat surfaces.

Sandstone

Sandstone is made of cemented fragments of sand size, typically made of quartz or other relatively hard mineral. Sandstones can be virtually any color, feel gritty, and may be banded or layered. Sandstone may represent ancient beaches, dunes, or fluvial deposits

Conglomerate

Conglomerate contains rounded pebbles cemented together by finer sediments. Conglomerate may represent deposits from a river's bed load, or a pebble beach.

Breccia

Breccia is similar to conglomerate, but the cemented fragments are angular. Breccia is associated with mudflows, pyroclastic flows, and fragments deposited by mass wasting.

Non-clastic Sedimentary

Non-clastic sedimentary rocks consist of materials that are not rock fragments. Examples include chemical precipitates, such as limestone, evaporites such as rock salt, and deposits of organic materials, for example coal, or limestones made of shells and coral fragments. Non-clastic rocks also represent the ancient environment under which they were deposited.

Rock Salt

Rock salt consists of sodium chloride, table salt, often with a mix of other salts. Rock salt represents the deposits left behind by the evaporation of a saline inland lake, or an arm of the sea that was cut off from the ocean by sea level change or tectonic activity. Common color is white.

Limestone

Limestone is made of deposits of calcium carbonate (lime, $CaCO_3$). Limestone can represent a wide variety of environments and vary widely in color and physical appearance. Typical colors are white or gray, and the most common environment of deposition was in shallow tropical seas, that were rich in lime. Many cave and spring deposits are also varieties of limestone.

Coal

Coal is a rock made of the carbonized remains of ancient plants. Coal deposits typically represent a low swampy environment that was invaded by sea level rise, which killed and buried dense vegetation.

Metamorphic Rocks

Metamorphic rocks can also be divided into two general categories, foliated rocks and non-foliated rocks.

Foliated

Foliated refers to metamorphic rocks that have either wavy, roughly parallel plates, or bands of light and dark minerals that formed under intense heat and pressure. The appearance of these foliations indicates the degree of metamorphism or change from the rock's original state.

Slate

Slate is metamorphosed shale, and looks much like shale, except it is harder and has very thin platy foliations. The most common color is black.

Schist

Schist has very prominent, wavy, and platy foliations generally covered with mineral crystals that formed during metamorphism. Schist represents a high degree of metamorphism and could originally have been any of a wide variety of rocks, so colors and appearance vary greatly.

Gneiss

Gneiss (pronounced "nice") is a banded metamorphic rock, with alternating bands of light and dark minerals. Gneiss represents extreme heat and pressure during metamorphism and also may originally have been any of a variety of rocks. Metamorphism of granites commonly produces gneiss.

Non-Foliated

Non-foliated metamorphic rocks do not display regular patterns of banding or platy foliations. In general, non-foliated metamorphics represent a rock that has been changed by the fusing and recrystallization of minerals in the original, often identifiable, rock.

Marble

Marble is metamorphosed limestone that has been recrystallized. Many colors and patterns of marble exist and its relative softness compared to other rocks makes it easy to cut and polish.

Quartzite

Quartzite is metamorphosed quartz sandstone in which the former sand grains have fused together to produce an extremely hard, resistant rock.

aa an angular, jagged lava flow.

abiotic natural, nonliving component of an ecosystem.

ablation loss of glacial ice and snow cover by means of melting, evaporation, and sublimation.

abrasion (corrasion) erosion of a stream by the grinding and rolling of rock particles and boulders carried by a stream, or by wave action, wind, or glacial ice.

absolute humidity mass of water vapor present per unit volume of air, expressed as grams per cubic meter, or grains per cubic foot.

absolute location location of an object on the basis of mathematical coordinates on an Earth grid.

abyssal plains deep low-relief ocean floor, generally covered with a blanket of marine sediments.

accretion the growth of continents by the adding of large pieces of their borders through collision.

acid rain rain with a pH value of less than 5.6, the pH of natural rain; often linked to the pollution associated with the burning of fossil fuels.

active-margin coast coastal region characterized by active volcanic and tectonic activity.

actual evapotranspiration the actual amount of moisture loss through evapotranspiration measured from a surface.

adiabatic heating and cooling change of temperature *within* a gas because of compression (resulting in heating) or expansion (resulting in cooling); no heat is added or subtracted from outside.

advection horizontal heat transfer within the atmosphere; air masses moved horizontally, usually by wind.

advection fog fog produced by the movement of warm moist air across a cold sea or land surface.

aggradation processes that tend to build up elevation through deposition.

air mass large portion of the atmosphere, sometimes subcontinental in size, that may move over Earth's surface as a distinct, relatively homogeneous entity.

air mass analysis explanation of weather phenomena by a study of the actions and interactions of major portions of the atmosphere.

albedo proportion of solar radiation reflected back from a surface, expressed as a percentage of radiation received on that surface.

Aleutian low center of low atmospheric pressure in the area of the Aleutian Islands, especially persistent in winter.

alluvial fan fan-shaped aggradational feature where a stream emerges from a mountain channel onto a flat plain and deposits material, generally found in arid regions.

alluvial plain a low-relief land surface formed through the deposition of sediment by running water.

alluvium fragmented Earth materials deposited by a river or stream.

alto signifies a middle-level cloud (i.e., from 2000 to 6000 meters in elevation).

alpine glacier moving glacial ice accumulated in high sheltered mountain valleys; also called *valley glacier.*

altithermal an interval of time about 7,000 years ago when the climate was hotter than it is today.

altitude heights of points above Earth's surface.

analemma a diagram that shows the declination of the sun throughout the year.

angle of inclination tilt of Earth's polar axis at an angle of 23½° from the vertical to the plane of the ecliptic.

angle of repose (rest) angle of slope created by a bed of loose sand, gravel, or rock.

annual temperature range difference between the mean daily temperatures for the warmest and coolest months of the year.

Antarctic circle parallel of latitude at 66½°S; the northern limit of the zone in the Southern Hemisphere that experiences a 24-hour period of sunlight and a 24-hour period of darkness at least once a year.

anticline upfold in a wave of crustal folding.

anticyclone an area of high atmospheric pressure, also known as a *high.*

aphelion position of Earth's orbit at farthest distance from the sun during each earth revolution.

aquiclude rock layer that restricts flow and storage of groundwater; it is impermeable and nonporous.

aquifer rock layer that is a container and transmitter of groundwater; it is both porous and permeable.

Arctic circle parallel of latitude at 66½°N; the southern limit of the zone in the Northern Hemisphere that experiences a 24-hour period of sunlight and a 24-hour period of darkness at least once a year.

arête jagged, sawtooth spine or wall of rock separating two expanding cirque basins.

arid a climate dominated by year-round moisture deficiency.

artesian well groundwater that flows toward the surface under its own pressure.

asteroid sometimes called a minor planet; any solar system body composed of rock and/or metal not exceeding 500 miles in diameter.

asthenosphere thick, plastic layer within Earth's mantle that theoretically flows in response to convection and instigates the surface movement of tectonic plates.

atmosphere blanket of air, composed of various gases, that envelops Earth.

atmospheric air pressure (barometric pressure) force per unit area that the atmosphere exerts on any surface at a particular elevation.

atmospheric controls geographic features that affect climate and weather patterns; e.g., distance from the ocean, wind direction, altitude.

atmospheric disturbance refers to variation in the secondary circulation of the atmosphere that cannot correctly be classified as a storm; e.g., front, air mass.

atmospheric effect the absorption of long-wave Earth radiation by water vapor, carbon dioxide, and dust in the atmosphere so that Earth temperatures are moderated.

atoll ring of coral reefs and islands encircling a lagoon, with no inner island.

attrition the reduction of size in sediment as it is transported downstream.

auroras colorful interaction of solar wind with ions in Earth's upper atmosphere; more commonly seen in higher latitudes. Called aurora borealis in the Northern Hemisphere (also known as the northern lights), and the aurora australis (southern lights) in the Southern Hemisphere.

autotroph organism which, because it is capable of photosynthesis, is at the foundation of a food web and is considered a basic producer.

axis an imaginary line between the geographic North Pole and South Pole, around which the planet rotates.

azimuth an angular direction of a line, point, or route measured clockwise from north 0–360°.

azimuthal map a projection that preserves the true direction from the map center to any other point on the map.

Azores high *see* Bermuda high.

backing wind shift change in wind direction counterclockwise around the compass; e.g., from east to northeast, to north, to northwest.

backwash return of water to the sea following onshore swash of waves.

badlands region of rugged, barren topography with sharp ridges and ravines; caused by gully erosion of soft materials.

bajada continuous series of alluvial fans forming a gently sloping, low-relief area along the base of a mountain range.

barchan crescent-shaped sand dune with tips that point downwind.

barometer instrument for measuring atmospheric pressure.

barrier island long, narrow, wave-built island separated from the mainland by a lagoon, formed on low-relief coastlines.

basalt a dark-colored fine-grained extrusive igneous rock generally associated with the oceanic crust and oceanic volcanoes.

base flow the "normal" level of a stream or river between events of greater discharge.

base level elevation below which a river or stream cannot erode; although sea level is the ultimate base level, basins or lakes may be local base levels.

batholith largest of the deep-seated igneous masses generally known as plutons.

bathymetry the science of mapping and measuring the ocean depths.

baymouth bar low depositional ridge extending across the mouth of a bay.

beach coastal region of unconsolidated sediments between the low tide line and the upper limit of wave action.

bearing an angular direction of a line, point, or route measured from north or from a current location to a desired location (often in 90° compass quadrants).

bedding plain boundary between different sedimentary strata marking a change in deposits.

bed load solid particles carried by a stream or river that are moved by rolling, bounding, and tumbling on the streambed.

bedrock solid rock of Earth's crust that underlies soil and other unconsolidated materials.

benthos the ocean bottom and the plants and animals that live on the seafloor.

bergschrund the large crevasse at the head of a valley glacier, beneath the cirque headwall.

Bermuda high persistent, high atmospheric pressure center located in the subtropics of the north Atlantic Ocean.

biomass amount of living material or standing crop in an ecosystem or at a particular trophic level within an ecosystem.

biome one of Earth's major terrestrial ecosystems, classified by the vegetation types that dominate the plant communities within the ecosystem.

biosphere the life forms, human, animal, or plant, of Earth that form one of the major Earth subsystems.

bolson desert basin, surrounded by mountains, with no drainage outlet.

bora cold downslope wind in the Adriatic (*see* katabatic wind).

boreal forest (taiga) coniferous forest dominated by spruce, fir, and pine found growing in subarctic conditions around the world north of the 50th parallel of latitude.

boulder a rock fragment greater than 256 mm in diameter.

braided stream stream channel with multiple subchannels that form a braided pattern flowing through alluvial deposits.

butte isolated erosional remnant of a tableland with a flat summit, often bordered by steep-sided escarpments. Buttes are usually found in arid regions of flat-lying sediments and are smaller than mesas.

calcification soil-forming process of subhumid and semiarid climates. Soil types in the mollisol order, the typical end products of the process, are characterized by little leaching or eluviation and by the accumulation of both humus and mineral bases (especially calcium carbonate, $CaCO_3$).

caldera collapsed summit area of a stratovolcano, thought to be caused by the expulsion or withdrawal of supporting magma.

calíche hardened layers of lime ($CaCO_3$) deposited at the surface of a soil by evaporating capillary water.

calorie amount of heat necessary to raise the temperature of one gram of water 1 degree Celsius.

calving the formation of icebergs by a mass of ice breaking away from the snout of a glacier at its junction with the sea or a lake.

campos region of characteristic tropical savanna vegetation in Brazil, located primarily in the Amazon Basin bordering the tropical rainforest.

Canadian high high atmospheric pressure area that tends to develop over the central North American continent in winter.

capacity the maximum amount of water vapor that can be contained in a given quantity of air at a given temperature.

capillary action the upward movement of water through tiny cracks and pore spaces.

capillary water soil water that clings to soil peds and individual soil particles as a result of surface tension. Capillary water moves in all directions through the soil from areas of surplus water to areas of deficit.

cap rock a resistant horizontal layer of rock that forms the flat top of a landform, such as a butte or a mesa.

carbonate a mineral group characterized by carbon's ability to form complex compounds of organic and inorganic origins.

carnivore animal that eats only other animals.

cartography the science of mapmaking.

catastrophism once-popular theory that all Earth's landforms developed in a relatively short time in a catastrophic fashion.

Celsius (or centigrade) scale temperature scale in which 0° is the freezing point of water and 100° its boiling point at standard sea level pressure.

centrifugal force force that pulls a rotating object away from the center of rotation.

chaparral sclerophyllous woodland vegetation found growing in the Mediterranean climate of the western United States; these seasonal, drought-resistant plants are low-growing, with small, hard-surfaced leaves and deep, water-probing roots.

chinook dry warm wind on the eastern slopes of the Rocky Mountains (see foehn wind).

cinder cone volcano formed primarily from the expulsion of cinders, ash, and other solid rock fragments.

circle of illumination line dividing the sunlit (day) hemisphere from the shaded (night) hemisphere; experienced by individuals on Earth's surface as sunrise and/or sunset.

cirque deep, sometimes steep-sided amphitheater formed at the head of an alpine valley by glacial ice erosion.

cirque glacier glacial ice limited to a cirque basin and not entering the alpine valley itself.

cirro signifies a high-level cloud (i.e., above 6000 meters in elevation).

cirrus high, detached clouds consisting of ice particles. Cirrus clouds are white and feathery or fibrous in appearance.

Cl, O, R, P, T Hans Jenny's description of the factors of soil formation: Climate, Organics, Relief, Parent material, and Time.

classification process of systematically arranging phenomena into groups, classes, or categories based on some established criteria.

clastic rock sedimentary rock formed by the compaction and cementation of pre-existing rock debris.

clay A very fine-grained mineral particle with a size less than 0.004 mm, often the product of weathering.

climate accumulated and averaged weather patterns of a locality or region; the full description is based upon long-term statistics and includes extremes or deviations from the norm.

climatology scientific study of climates of Earth and their distribution.

climax community the final step in the succession of plant communities that occupy a specific location.

climograph graphic means of giving information on mean monthly temperature and rainfall for a select location or station.

closed system system in which no substantial amount of energy and/or materials can cross its boundaries.

cloud mass of suspended water droplets (or at high altitudes, ice particles) in air above ground level.

coastal zone region of local interaction between land and a sea or ocean.

coastline of emergence coast with formerly submerged land that has risen above sea level, either by uplift or by a drop in sea level.

cobble rock fragments ranging in diameter from 64 to 256 mm.

cold front leading edge of a relatively cooler, denser air mass that advances upon a warmer, less dense air mass.

col a glacially-eroded pass between two mountain valleys.

collapse sinkhole a topographic depression formed mainly by the cave-in of a cavern or an overlying regolith arch.

color composite a digital image that combines several wavelength bands—a common color composite that is used resembles a near-infrared color photo.

columnar joints hexagonal fractures caused by lava shrinking as it cools.

comet a small body of icy and dusty matter that revolves about the sun. When a comet comes near the sun, some of its material vaporizes, forming a large head and often a tail.

composite cone (stratovolcano) volcano formed from alternating layers of lava and pyroclastic materials; generally known for violent eruptions.

compression stress resulting from convergent forces, pushing together.

conceptual model image in the mind of an Earth feature or landscape as derived from personal experiences.

condensation process by which a vapor is converted to a liquid during which energy is released in the form of latent heat.

condensation nuclei minute particles in the atmosphere (e.g., dust, smoke, pollen, sea salt) on which condensation can take place.

conduction transfer of heat within a body or between adjacent matter by means of internal molecular movement.

conformal map projection a map projection that maintains the true shape of small areas on Earth's surface.

connate water groundwater trapped in the pore spaces of sedimentary rock at the time it was first deposited; water locked out of the hydrologic cycle in sedimentary rocks.

continental collision the fusing together of landmasses as tectonic plates converge.

continental crust the less dense (av. 2.7 gm/cm^3) portion of Earth's crust that underlies all the continents.

Continental Divide line of separation dividing runoff between the Pacific and Atlantic Oceans. In North America it generally follows the crest of the Rocky Mountains.

continental drift theory proposed by Alfred Wegener stating that the continents joined, broke apart, and moved on Earth's surface; it was later replaced by the theory of plate tectonics.

continental ice sheet thick ice mass that covers a major portion of a continent and buries all but the highest mountain peaks; it usually flows from one or more areas of accumulation outward in all directions.

continental islands islands that are geologically part of a continent and are usually located on the continental shelf.

continentality the distance a particular place is located in respect to a large body of water: the greater the distance, the greater the continentality.

continental rise gently sloping depositional surface at the base of the continental slope.

continental shelf gently sloping submarine surface extending from the coast to the steep continental slope.

continental shields ancient crystalline rock cores of the continents.

continental slope steeply sloping submarine surface that is seaward of the continental shelf.

continuous data numerical or locational representations of phenomena that are present everywhere—such as air pressure, temperature, elevation.

contour interval vertical distance represented by two adjacent contour lines on a topographic map.

contour map (topographic map) map that uses contour lines to show differences in elevation (topography).

convective thunderstorm a thunderstorm produced by the convective uplift mechanism.

convection process by which a circulation is produced within an air mass or fluid body (heated material rises, cooled material sinks); also, in tectonic plate theory, the method whereby heat is transferred to Earth's surface from deep within the mantle.

convectional precipitation precipitation resulting from condensation of water vapor in an air mass that is rising convectionally as it is heated from below.

convergent wind circulation pressure-and-wind system where the airflow is inward toward the center, where pressure is lowest.

coordinate system a precise system of grid lines used to describe locations.

coral reef ridge of limestone built up by accumulation of skeletal remains of tiny sea animals.

core extremely hot and dense, innermost portion of Earth's interior; the molten outer core is 2400 km (1500 mi) thick; the solid inner core is 1120 km (700 mi) thick.

Coriolis effect effect of Earth's rotation on horizontally moving bodies, such as wind and ocean currents; such bodies tend to be deflected to the right in the Northern Hemisphere and to the left in the Southern Hemisphere.

corridor a relatively linear feature cutting across a mosaic (ecosystem supporting a particular plant community) caused by nature (a stream) or by humans (a road or powerline).

coulee snaking, steep-sided channel cut through lava formations by glacial meltwater.

creep slow downslope movement of soil and regolith caused by the pull of gravity, also refers to slow fault zone displacement.

crevasse stress crack commonly found along the margins and at the snout or terminus of a glacier.

crust relatively thin, approximately 8–64 km (5–40 mi) deep, low-density surface layer of Earth.

cumulus globular clouds, usually with a horizontal base and strong vertical development.

cut bank the bank of a stream that is on the outer side of a bend, where erosion steepens the bank.

cyclone center of low atmospheric pressure, also known as a *low.*

cyclonic precipitation precipitation formed by cyclonic uplift.

debris a mix of rock and soil.

debris flow channelled movement of Earth material mixed with water usually following drainage; a *mud-flow* is a rapidly moving debris flow with the consistency of mud.

declination the latitude on Earth at which the noon sun is directly overhead.

decomposition a term that refers to the processes of chemical weathering.

decomposer organism that promotes decay by feeding on dead plant and animal material and returns mineral nutrients to the soil or water in a form that plants can utilize.

deflation surface erosion and removal of fine Earth materials by the wind.

deflation hollow (blowout) a depression formed by wind erosion.

degradation processes that tend to reduce elevation or relief, such as erosion and weathering.

delta depositional landform where a river flows into a still body of water, such as a sea or lake.

dendritic term used to describe a drainage pattern that is treelike with tributaries joining the main stream at acute angles.

deposition accumulation of Earth materials at a new site after they have been dropped by the transporting agents: water, wind, or glacial ice.

desert pavement desert surface accumulation of pebbles and stones, the finer materials having been removed by wind and/or water erosion.

detritivore animal that feeds on dead plant and animal material.

dew tiny droplets of water on ground surfaces, grass blades, or solid objects. Dew is formed by condensation when air at the surface reaches the dew point.

dew point the temperature at which an air mass becomes saturated; any further cooling will cause condensation of water vapor in the air.

differential weathering (and erosion) process whereby different types of rock weather (and erode) at varying speeds due to differing resistance to the weathering (and erosion) processes; such differing resistance often produces distinctive landform features.

digital image an image made from computer data displayed like a mosaic of tiny squares, called pixels.

digital mapping mapmaking that employs computer techniques.

digital terrain model a computer-generated graphic representation of topography.

dike igneous intrusion that forms a vertical rock mass after molten material has been forced through a crustal fracture and cools at right angles to flat-lying rock layers.

dip the angle that a stratum of rocks or a fault makes with the horizontal plane.

dip-slip fault a vertical fault where the movement is up and down the dip of the fault surface.

discharge (stream discharge) rate of stream flow; measured as the volume of water flowing past a cross section of a stream per unit of time (cubic meters or cubic feet per second).

discrete data numerical or locational representations of phenomena that are present only at certain locations—such as earthquake epicenters, sinkholes, tornado paths.

dissolved load soluble minerals or other chemical constituents carried in water as a solution.

distributary branching stream that flows away from the main stream, common on deltas; the opposite of a tributary.

diurnal (daily) temperature range difference between the highest and lowest temperatures of the day (usually recorded hourly).

divergent wind circulation pressure-and-wind system where the airflow is outward away from the center, where pressure is highest.

divide line of separation between drainage basins; generally follows high ground or ridge lines.

doldrums zone of low pressure and calms along the equator.

doline another term for a sinkhole, generally a very large sinkhole.

Doppler radar advanced type of radar that can detect motion in storms, specifically motion toward and away from the radar signal.

drainage basin (watershed) total land surface area drained by a stream system.

drainage (stream) pattern the form of a channel network for a stream or river as viewed from a map (vertical) perspective.

drainage wind *see* katabatic wind.

drift all material deposited by glacier; includes both unsorted and unstratified material and sorted debris deposited by meltwater.

drizzle fine mist or haze of very small water droplets with a barely perceptible falling motion.

drumlin streamlined, elongated hill composed of glacial drift. Drumlins are usually found in swarms, with as many as 100 or more clustered together; their elongated shapes indicate the direction of ice flow.

dry adiabatic lapse rate rate at which a rising mass of air is cooled by expansion when no condensation is occurring (10°C/1000 m or 5.6°F/1000 ft).

dune (sand dune) mound of sand-sized materials deposited and shaped by the wind.

dust storm a moving cloud of wind-blown dust (typically silt).

dynamic equilibrium constantly changing relationship among the variables of a system, which produces a balance between the amounts of energy and/or materials that enter a system and the amounts that leave.

earthflow linear movement downslope of moist, clay-rich soil and regolith, usually exhibiting a tongue-like shape.

earthquake series of vibrations or shock waves set in motion by sudden movement along a fault.

earthquake intensity a measure of the impact of an earthquake on humans and their built environment.

earthquake magnitude a measure of the energy release of an earthquake.

Earth system set of interrelated components or variables (e.g., atmosphere, lithosphere, biosphere, hydrosphere), which interact and function together to make up Earth as it is currently constituted.

easterly wave trough-shaped, weak, low pressure cell that progresses slowly from east to west in the tradewind belt of the tropics; this type of disturbance sometimes develops into a tropical hurricane.

eccentricity cycle the change in Earth's orbit from slightly elliptical to more circular, and back to its earlier shape every 100,000 years.

ecological niche combination of role and habitat as represented by a particular species in an ecosystem.

ecology science that studies the interactions between organisms and their environment.

ecosystem community of organisms functioning together in an interdependent relationship with the environment which they occupy.

ecotone transition zone of varied natural vegetation occupying the boundary between two adjacent and differing plant communities.

effective precipitation actual precipitation available to supply plants and soil with usable moisture; does not take into consideration storm runoff or evaporation.

El Niño warm countercurrent that influences the central and eastern Pacific.

electromagnetic energy all forms of energy that share the property of moving through space (or any medium) in a wavelike pattern of electric and magnetic fields; also called radiation.

elements (weather and climate) the major elements include solar energy, temperature, pressure, winds, and precipitation.

elevation vertical distance from mean sea level to a point or object on Earth's surface.

ellipsoid of rotation a rotating, near-sphere with an elliptical (oval-shaped), rather than pure circular, plane or cross section.

eluviation removal by gravitational water of fine soil components from the surface layer (*A* horizon) of the soil.

empirical classification classification process based on statistical, physical, or observable characteristics of phenomena; it ignores the causes or theory behind their occurrence.

end moraine accumulation of rocks and fine glacial material at the terminus or snout of a glacier.

entrenched meander a meander that flows in a relatively deep and steep-sided valley.

environment surroundings, whether of man or of any other living organism; includes physical, social, and cultural conditions that affect the development of that organism.

eolian (aeolian) referring to the work of wind; associated with wind erosion, transportation, and deposition.

ephemeral stream a stream that flows only at certain times, when adequate discharge is supplied by precipitation events, ice- or snowmelt or irregular spring flow.

epicenter point on Earth's surface directly above the focus of an earthquake.

epipedon surface soil layer that possesses specific characteristics essential to the identification of soils in the National Resources Conservation Service System (Examples of epipedons may be found in Table 12.1.)

equal-area map projection a map projection on which any given areas of Earth's surface are shown in correct proportional sizes on the map.

equator great circle of Earth midway between the poles; the zero degree parallel of latitude that divides Earth into the Northern and Southern Hemispheres.

equatorial low zone of low atmospheric pressure centered more or less over the equator where heated air is rising. (*See also* doldrums.)

equidistance a property of some maps that depicts distances equally without scale variation.

equilibrium state of balance between the interconnected components of an organized whole.

equilibrium (firn) line an imaginary line along the balance point on a glacier that separates the upper area, where the annual snowfall exceeds melting and sublimation, from the lower end, where ice wastage predominates.

equinox one of two times each year (approximately March 21 and September 22) when the position of the noon sun is overhead (and its vertical rays strike) at the equator; all over Earth, day and night are of equal length.

erg desert region of active sand dunes, most common in the Sahara.

erosion removal of Earth materials by water, wind, or glacial ice.

erratic large rock or boulder transported and deposited by a glacier above bedrock of different composition.

esker narrow, winding ridge composed of glaciofluvial gravels; believed to have been formed by streams of meltwater flowing in tunnels of a stagnant ice sheet, or on a melting glacial surface.

estuary coastal waters where salt and fresh water mix.

evaporation process by which a liquid is converted to the gaseous (or vapor) state by the addition of latent heat.

evaporite mineral salts that are soluble in water and accumulate when water evaporates.

evapotranspiration combined water loss to the atmosphere from ground and water surfaces by evaporation and, from plants, by transpiration.

exfoliation progressive breaking off of concentric slabs or sheets from the exposed portions of massive rocks due to weathering.

exfoliation dome a rounded hill or mountain formed primarily by exfoliation.

exotic stream (or river) stream that originates in a humid region and has sufficient water volume to flow across a desert region.

exposure direction of mountain slopes with respect to prevailing wind direction.

extratropical disturbance *see* middle-latitude disturbance.

extrusive rock igneous rock that was erupted and solidified on Earth's surface.

Fahrenheit scale temperature scale in which 32° is the freezing point of water, and 212° its boiling point, at standard sea level pressure.

faulting movement of adjacent crustal blocks along joints, or fracture planes, in bedrock.

fault scarp (escarpment) the steep cliff or exposed face of a fault where one crustal block has been displaced vertically relative to another.

feedback sequence of changes in the elements of a system, which ultimately affects the element that was initially altered to begin the sequence.

feedback loop path of change as its effects move through the variables of a system until the effects impact the variable originally experiencing change.

fertilization adding additional nutrients to the soil.

fetch distance over open water that winds blow without interruption.

firn compact granular snow formed by partial melting and refreezing due to overlying layers of snow.

firn line boundary between the zones of ablation and accumulation on a glacier, representing the equilibrium point between net snowfall and ablation.

fissure an extensive crack or break in rocks which may allow lava to be extruded.

fjord deep, glacial trough along the coast invaded by the sea after the removal of the glacier.

floodplain the area along a stream or river that is subject to flooding.

fluvial term used to describe landform processes associated with the work of streams and rivers.

focus point within Earth's crust where an earthquake originates.

foehn wind warm, dry, downslope wind on lee of mountain range, caused by adiabatic heating of descending air.

fog mass of suspended water droplets within the atmosphere that is in contact with the ground.

folding the wrinkling of the Earth's crust due to compressional forces.

foliation process whereby metamorphic rocks tend to develop parallel banding or platy structures during formation.

food chain sequence of levels in the feeding pattern of an ecosystem.

food web feeding mosaic formed by the interrelated and overlapping food chains of an ecosystem.

friction force that acts opposite to the direction of movement or flow; for example, turbulent resistance of Earth's surface on the flow of the atmosphere.

front sloping boundary or contact surface between air masses with different properties of temperature, moisture content, density, and atmospheric pressure.

frontal lifting lifting or rising of warmer, lighter air above cooler, denser air along a frontal boundary.

frontal precipitation precipitation resulting from condensation of water vapor in an air mass that is rising over another mass along a front.

frontal thunderstorm a thunderstorm produced by the frontal uplift mechanism.

frost frozen condensation that occurs when air at ground level is cooled to a dew point of 0°C (32°F) or below; also any temperature near or below freezing that threatens sensitive plants.

frost wedging breaking apart of bedrock by the expansive power of water freezing, melting, and refreezing in joints, cracks, and crevices.

fusion (thermonuclear reaction) the fusing together of two hydrogen atoms to create one helium atom. This process releases tremendous amounts of energy.

galactic movement movement of the solar system within the Milky Way Galaxy.

galaxy a large assemblage of stars; a typical galaxy contains millions to hundreds of billions of stars.

galeria forest jungle-like vegetation extending along and over streams in tropical forest regions.

gap an area within the territory occupied by a plant community when the climax vegetation has been destroyed or damaged by some natural process, such as a hurricane, forest fire, or landslide.

General Circulation Model (GCM) complex computer simulations based on the relationships of selected variables within the Earth system that are used in attempts to predict future climates.

genetic classification classification process based on the causes, theory, or origins of phenomena; generally ignoring their statistical, physical, or observable characteristics.

generalist species that can survive on a wide range of food supplies.

geocoding the process and reference system used to tie map locations to field locations using a grid system.

Geographic Information Systems (GIS) complex computer programs that combine the features of automated (computer) cartography and database management to produce new data to solve spatial problems.

geography study of Earth phenomena; includes an analysis of distributional patterns and interrelationships among these phenomena.

geomorphology the study of the origin and development of landforms.

geostationary orbit an orbit that synchronizes a satellite's position and speed with Earth rotation so that it continually images the same location.

geostrophic winds upper-level winds in which the Coriolis effect and pressure gradient are balanced, resulting in a wind flowing parallel to the isobars.

geothermal water groundwater that has been heated by contact with hot rock in the subsurface.

geyser a fountain of hot groundwater that spews above the surface.

giant planets the four largest planets—Jupiter, Saturn, Uranus, and Neptune.

glacier a mass of ice that is flowing as a plastic solid.

glaciofluvial deposit sorted glacial drift deposited by meltwater.

glaciolacustrine deposit sorted glacial drift deposited by meltwater in lakes associated with the margins of glaciers.

glaze (freezing rain) translucent coating of ice that develops when rain strikes a freezing surface.

gleization soil-forming process of poorly drained areas in cold, wet climates. The resulting soils have a heavy surface layer of humus with a water-saturated clay horizon directly beneath.

Global Positioning System (GPS) GPS uses satellites and computers to compute positions anywhere on Earth to within a few centimeters of their true location.

gnomonic projection planar projection with greatly distorted land and water areas; valuable for navigation because all great circles on the projection appear as straight lines.

graben depressed landform or crustal trough that develops when the crust between two parallel faults is lowered relative to blocks on either side.

gradational processes processes that derive their energy indirectly from the sun and directly from Earth gravitation and serve to wear down, fill in, and level off Earth's surface.

graded stream stream where slope and channel size provide velocity just sufficient to transport the load supplied by the drainage basin; a theoretical balanced state averaged over a period of many years.

gradient a term for slope often used to describe the angle of a streambed.

granite a coarse-grained intrusive igneous rock generally associated with continental crust.

graphic (bar) scale a ruler-like device placed on maps for making direct measurements in ground distances.

gravel a general term for sediment sizes larger than sand.

gravitation the attractive force one body has for another. The force increases as the mass of the bodies increases, and the distance between them decreases.

gravitational water meteoric water that passes through the soil under the influence of gravitation.

gravity the mutual attraction of bodies or particles.

great circle any circle formed by a full circumference of the globe; the plane of a great circle passes through the center of the globe.

greenhouse effect warming of the atmosphere that occurs because short-wave solar radiation heats the planet's surface, but the loss of long-wave heat radiation is hindered by the release of gases associated with human activity (e.g., CO_2).

Greenwich mean time (GMT) time at zero degrees longitude used as the base time for Earth's 24 time zones; also called Universal Time or Zulu Time.

ground-inversion fog *see* radiation fog.

ground moraine glacial till deposited on Earth's surface beneath a melting glacier.

groundwater (underground water) all subsurface water, especially in the zone of saturation.

guyot flat-topped seamount thought to be formed by the slow subsidence of a volcanic island.

gyre broad circular patterns of major surface ocean currents produced by large subtropical high pressure systems.

habitat location within an ecosystem occupied by a particular organism.

hail form of precipitation consisting of pellets or balls of ice with a concentric layered structure usually associated with the strong convection of cumulonimbus clouds.

hamada desert plains covered with boulders or featuring large expanses of exposed bedrock.

hanging valley tributary trough that enters a main glaciated valley at a level high above the valley floor.

hardpan dense, compacted, clay-rich layer occasionally found in the subsoil (*B* horizon) that is an end product of excessive illuviation.

Haystack hill (hum) hills that remain after considerable erosion of a karst landscape.

headward erosion gullying and valley cutting that extends a stream channel in an upstream direction.

heat the total kinetic energy of all the atoms that make up a substance.

heat energy budget relationship between solar energy input, storage, and output within the Earth system.

heat island mass of warmer air overlying urban areas.

hemisphere half of a sphere; for example, the northern or southern half of Earth divided by the equator or the eastern and western half divided by a meridian.

herbivore animal that eats only living plant material.

heterotroph organism that is incapable of producing its own food and that must survive by consuming other organisms.

high *See* anticyclone.

Holocene the most recent time interval of warm, relatively stable climate that began with the retreat of major glaciers about 10,000 years ago.

horizon the visual boundary between Earth and sky; *see also* soil horizon.

horn pyramid-like peak created where three or more expanding cirques meet at a mountain summit.

horst raised landform that develops when the crust between two parallel faults is uplifted relative to blocks on either side.

human geography specialization in the systematic study of geography that focuses on the location, distribution, and spatial interaction of human (cultural) phenomena.

humidity amount of water vapor in an air mass at a given time.

humus organic matter found in the surface soil layers that is in various stages of decomposition as a result of bacterial action.

hurricane severe tropical cyclone of great size with nearly concentric isobars. Its torrential rains and high-velocity winds create unusually high seas and extensive coastal flooding; also called willy-willies, tropical cyclones, baguios, and typhoons.

hydration attachment of water molecules to molecules of other elements or compounds without chemical change.

hydrologic cycle circulation of water within the Earth system, from evaporation to condensation, precipitation, runoff, storage, and re-evaporation back into the atmosphere.

hydrolysis union of water with other substances involving chemical change and the formation of new compounds.

hydrosphere major Earth subsystem consisting of the waters of Earth, including oceans, ice, freshwater bodies, groundwater, and water within the atmosphere and biomass.

hygroscopic water water in the soil that adheres to mineral particles.

ice age period of Earth history when much of Earth's surface was covered with massive continental glaciers. The most recent ice age is referred to as the Pleistocene Epoch.

iceberg free-floating mass of glacier broken off by melting, tidal, and wave action.

ice cap small ice sheet found in highland areas that usually covers all but the highest mountain peaks.

ice fall portion of a glacier moving over and down a steep slope, creating a rigid white cascade, criss-crossed with deep crevasses.

Icelandic Low center of low atmospheric pressure located in the north Atlantic, especially persistent in winter.

ice-marginal lake temporary lake formed by the disruption of meltwater drainage by deposition along a glacial margin, usually in the area of an end moraine.

ice-scoured plain a broad area of low relief, deeply eroded by a glacier.

ice sheet mass of glacial ice thousands of feet thick that is of continental proportions and covers all but the highest points of land. The sheet usually flows from one or more areas of accumulation outward in all directions.

ice shelf large flat-topped plate of ice from the Antarctic ice cap, which overlies Antarctic waters and is a source of icebergs.

igneous rock one of the three major rock types: formed from the cooling and solidification of molten Earth material.

illuviation deposition of fine soil components in the subsoil (*B* horizon) by gravitational water.

infiltration water seeping downward into the soil or other surface materials.

infiltration capacity the greatest amount of infiltrated water that a surface material can hold.

inner core the innermost portion of the Earth's core that forms the center of the Earth and is considered to be a dense hot solid mass of iron.

inputs energy and material entering an Earth system.

inselberg remnant residual hill rising above an arid or semiarid plain; produced by stream erosion of a former mountainous area.

insolation incoming solar radiation, i.e., energy received from the sun.

instability condition of air when it is warmer than the surrounding atmosphere and is buoyant with a tendency to rise; the lapse rate of the surrounding atmosphere is greater than that of *unstable* air.

interfluve the land between two tributary streams.

interglacial warmer period between glacial advances, during which continental ice sheets and many valley glaciers retreat and disappear or are greatly reduced in size.

interior drainage pattern a closed topographic depression that water flows down into, but has no outlet, or outflowing stream.

intermittent stream stream that flows part of the time, usually only during, and shortly after, a rainy period.

International Date Line line roughly along the 180-degree meridian, where each day begins and ends; it is always a day later west of the line than east of the line.

Intertropical Convergence Zone (ITCZ) zone of low pressure and calms along the equator, where air carried by the trade winds from both sides of the equator converges and is forced to rise.

intrusive rock igneous rock that was cooled and crystallized beneath the Earth's surface.

inversion *see* temperature inversion.

isarithm line on a map that connects all points of the same numerical value, such as isotherms, isobars, and isobaths.

island arc curved row of volcanic islands along a deep oceanic trench; found near tectonic plate boundaries where subduction occurs.

isobar line drawn on a map to connect all points with the same atmospheric pressure.

isoline a line on a map that represents equal values of some numerical measurement such as lines of equal temperature or elevation contours.

isostasy theory which holds that Earth's crust *floats* in hydrostatic equilibrium in the denser plastic layer of the mantle.

isotherm line drawn on a map to connect all points with the same temperature.

jet stream high-velocity upper-air current with speeds of 120–640 kph (75–250 mph).

jetty artificial structure extending into a body of water; built to protect a harbor, inlet, or beach by modifying action of waves or currents.

joints cracks or systems of cracks revealing lines of weakness in bedrock.

jungle dense tangle of trees and vines in areas where sunlight reaches the ground surface (not a true rainforest).

kame conical hill composed of sorted glaciofluvial deposits; presumed to have formed in contact with glacial ice when sediment accumulated in ice pits, crevasses, and among jumbles of detached ice blocks.

kame terraces landform resulting from accumulation of glaciofluvial sand and gravel along the margin of a glacier occupying a valley in an area of hilly relief.

karst unique landforms developed as a result of the dissolving of limestone by groundwater.

katabatic wind downslope flow of cold, dense air that has accumulated in a high mountain valley or over an elevated plateau or ice cap.

Kelvin scale temperature scale developed by Lord Kelvin, equal to Celsius scale plus 273; no temperature can drop below absolute zero, or 0 degrees Kelvin.

kettle depression formed by the melting of an ice block buried in glacial deposits left by a retreating glacier.

kettle hole water-filled pit formed by the melting of a remnant ice block left buried in drift after the retreat of a glacier.

kettle lake a small lake or pond occupying a kettle hole.

Köppen system climate classification based on monthly and annual averages of temperature and precipitation; boundaries between climate classes are designed so that climate types coincide with vegetation regions.

laccolith massive igneous intrusion that bows overlying rock layers upwards in a domal fashion as it forces its way toward the surface.

lahar rapid form of mass movement involving mudflows from volcanic materials.

lamination planes very thin layers in rock.

La Niña cold sea-surface temperature anomaly in the Equatorial Pacific (opposite of El Niño).

land breeze air flow at night from the land toward the sea, caused by the movement of air from a zone of higher pressure associated with cooler nighttime temperatures over the land.

Landsat a family of U.S. satellites that have been returning digital images since the 1970s.

landslide mass of Earth material, including all loose debris and often portions of bedrock, moving as a unit rapidly downslope.

lapse rate *see* normal lapse rate.

latent heat of condensation energy release in the form of heat, as water is converted from the gaseous (vapor) to the liquid state.

latent heat of evaporation amount of heat absorbed by water to evaporate from a surface (i.e., 590 calories/gram of water).

lateral moraine moraine deposited along the side margin of an alpine glacier or lobe of a continental ice sheet.

laterite iron, aluminum, and manganese rich layer in the subsoil (*B* horizon) that can be an end product of laterization in the wet-dry tropics (tropical savanna climate).

laterization soil-forming process of hot, wet climates. Oxisols, the typical end product of the process, are characterized by the presence of little or no humus, the removal of soluble and most fine soil components, and the heavy accumulation of iron and aluminum compounds.

latitude angular distance (distance measured in degrees) north or south of the equator.

lava molten Earth material expelled at the surface from volcanoes or fissures. From this material extrusive igneous rock is formed.

leaching removal by gravitational water of soluble inorganic soil components from the surface layers of the soil.

leeward located on the side facing away from the wind.

levee natural raised alluvial bank along margins of a river on a floodplain; artificial levees may be constructed along river banks for flood control.

liana woody vine found in tropical forests that roots in the forest floor but uses trees for support as it grows upward toward available sunshine.

life-support system interacting and interdependent units (e.g., oxygen cycle, nitrogen cycle) that together provide an environment within which life can exist.

lightning visible electrical discharge produced within a thunderstorm.

light year the distance light travels in one year—6 trillion miles.

lithification the combined processes of compaction and cementation that transform clastic sediments into sedimentary rocks.

lithosphere solid crust of Earth that forms one of the major Earth subsystems. In a more technical definition related to tectonic plate theory, the lithosphere consists of Earth's crust and the uppermost rigid zone of the mantle, which is divided into individual plates that move independently on the plastic material of the asthenosphere.

lithospheric plates huge slab-like segments of Earth's exterior including the crust and solid part of the upper mantle.

Little Ice Age an especially cold interval of time during the early 14th century that had major impacts on civilizations in the Northern Hemisphere.

llanos region of characteristic tropical savanna vegetation in Venezuela, located primarily in the plains of the Orinoco River.

loam soil soil with a texture in which none of the three soil grades (sand, silt, or clay) predominate over the others.

loess wind-deposited silt; usually transported in dust storms and derived from arid or glaciated regions.

longitude angular distance (distance measured in degrees) east or west of the prime meridian.

longitudinal dune a linear ridge-like sand dune that is oriented parallel to the prevailing wind direction.

longshore current current flowing parallel to the shore within the surf zone, produced by waves breaking at an angle to the shore.

long-wave radiation electromagnetic radiation emitted by Earth in the form of waves more than 4.0 micrometers in amplitude, which includes heat reradiated by Earth's surface.

low *see* cyclone.

magma melt or molten Earth material, situated beneath Earth's surface, from which plutonic and intrusive igneous rock is formed.

magnetic declination horizontal angle between geographic north and magnetic north.

mantle moderately dense, relatively thick (2885 km/1800 mi) middle layer of Earth's interior that separates the crust from the outer core.

map projection any presentation of the spherical Earth on a flat surface.

maquis sclerophyllous woodland and plant community, similar to North American chaparral; can be found growing throughout the Mediterranean region.

marine terrace horizontal land surface, now above sea level, that was either wave-cut or wave-built by shoreline.

mass a measure of the total amount of matter in a body.

mass wasting (mass movement) movement of surface materials downslope as a result of Earth gravitation.

mathematical/statistical model computer-generated representation of an area or Earth system using statistical data.

matrix the dominant area of a mosaic (ecosystem supporting a particular plant community) where the major plant in the community is concentrated.

meander a broad, sweeping bend in a river or stream.

medial moraine central moraine in a large valley glacier; formed when two smaller valley glaciers come together to form the larger glacier and their interior lateral moraines merge.

mental map conceptual model of special significance in geography because it consists of spatial information.

Mercalli Scale, modified an earthquake intensity scale with Roman numerals from I–XII used to assess spatial variations in the degree of impact that a tremor generates.

Mercator projection mathematically produced, conformal map projection showing true compass bearings as straight lines.

mercury barometer instrument measuring atmospheric pressure by balancing it against a column of mercury.

meridian one half of a great circle on the globe connecting all points of equal longitude; all meridians connect the North and South Poles.

mesa flat-topped erosional remnant of a tableland characteristic of arid regions with flat-lying sediments; typically bordered by steep-sided escarpments and may cover large areas.

mesopause upper limit of mesosphere, separating it from the thermosphere.

mesosphere layer of atmosphere above the stratosphere; characterized by temperatures that decrease regularly with altitude.

mesothermal climate that displays a mild winter.

metamorphic rock one of the three major rock types; formed from other rock within the crust by change induced by heat and pressure.

meteor the luminous phenomenon observed when a small piece of solid matter enters Earth's atmosphere and burns up.

meteorite any fragment of a meteor that reaches Earth's surface.

meteorology study of the patterns and causes associated with short-term changes in the elements of the atmosphere.

microclimate climate associated with a small area at or near Earth's surface; the area may range from a few inches to one mile in size.

microthermal climate that displays a severe winter.

microplate terrane material added to continents as they collide with smaller areas of distinct geology such as volcanic islands or continent fragments.

middle-latitude disturbance convergence of cold polar and warm subtropical air masses over the middle latitudes.

millibar unit of measurement for atmospheric pressure; one millibar equals a force of 1000 dynes per square centimeter; 1013.2 millibars is standard sea level pressure.

mineral naturally occurring inorganic substance that possesses fairly definite physical characteristics and unique chemical composition.

mistral cold downslope wind in southern France (*see* katabatic wind).

Mohorovicic discontinuity (Moho) interface between Earth's crust and the more dense mantle.

monadnock erosional remnant of more resistant rock on a plain of old age; associated with a theoretical cycle of erosion in humid lands.

monsoon seasonal wind that reverses direction during the year in response to a reversal of pressure over a large landmass. The classic monsoons of Southeast Asia blow onshore in response to low pressure over Eurasia in summer and offshore in response to high pressure in winter.

moraine unsorted glacial drift deposited beneath and along the margins of a glacier.

mosaic a plant community and the ecosystem upon which it is based, viewed as a landscape of interlocking parts by ecologists.

mountain breeze air flow downslope from mountains toward valleys during the night.

multispectral scanning using a number of energy wavelength bands to create images.

mud a mix of rock and soil with ample water.

mudflow downslope movement of mud with mixing and tumbling as it moves.

muskeg poorly drained vegetation-rich marshes or swamps usually overlying permafrost areas of polar climatic regions.

natural resource any element, material, or organism existing in nature that may be useful to humans.

natural vegetation vegetation that has been allowed to develop naturally without obvious interference from or modification by humans.

navigation the science of location and finding one's way, position, or direction.

neap tide tide of less than average range; occurs at first- and third-quarter moon.

near-infrared (NIR) film photographic film that makes pictures using near-infrared light that is not visible to the human eye.

negative feedback reaction to initial change in a system that counteracts the initial change and leads to dynamic equilibrium in the system.

nekton marine organisms that swim freely in the oceans.

nimbus term used in cloud description to indicate precipitation; thus cumulonimbus is a cumulus cloud from which rain is falling.

normal fault a vertical fault with the footwall up and the hanging wall down, caused by tensional stress.

normal lapse rate decrease in temperature with altitude under normal atmospheric conditions; approximately 6.5°C/1000 m (3.6°F/1000 ft).

North Atlantic Oscillation oscillating (see-saw) pressure tendencies between the Azores High and the Icelandic Low.

northeast trades *see* trade winds.

North pole maximum north latitude (90° N), at the point marking the axis of rotation.

oblate spheroid Earth's shape—a slightly flattened sphere.

obliquity cycle the change in the tilt of the Earth's axis relative to the plane of the ecliptic over a 41,000 year period.

occluded front boundary between a rapidly advancing cold air mass and an uplifted warm air mass cut off from Earth's surface; denotes the last stage of a middle-latitude cyclone.

ocean current horizontal movement of ocean water, usually in response to major patterns of atmospheric circulation.

oceanic crust the denser (av. 3.0 g/cm³) portion of the Earth's crust that underlies the ocean basins.

oceanic islands volcanic islands that rise from the deep ocean floor.

oceanic ridge (midocean ridge) linear seismic mountain range that interconnects through all the major oceans; it is where new molten crustal material rises through the oceanic crust.

oceanic trench (trench) long, narrow depression on the sea floor usually associated with an island arc. Trenches mark the deepest portions of the oceans and are associated with subduction of oceanic crust.

omnivore animal that can feed on both plants and other animals.

ooze sedimentary deposit containing the microscopic remains of sea organisms.

open system system in which energy and/or materials can freely cross its boundaries.

organic sedimentary rocks rocks that consist of deposits of abundant organic material such as carbon from plants (coal), or the shells of sea creatures (some limestones).

orographic precipitation precipitation resulting from condensation of water vapor in an air mass that is forced to rise over a mountain range or other raised landform.

orographic thunderstorm a thunderstorm produced by the orographic uplift mechanism.

outcrop bedrock exposed at Earth's surface with no overlying regolith or soil.

outer core the upper portion of the Earth's core; considered to be composed of molten iron liquefied by the Earth's internal heat.

outputs energy and material leaving an Earth system.

outwash glacial drift deposited beyond an end moraine by glacial meltwater.

outwash plain extensive, relatively smooth plain covered with sorted deposits carried forward by the meltwater from an ice sheet.

overthrust faulting that pushes the rocks on one side a considerable distance over the opposite side of the fault; the rocks that have overridden others in this manner.

oxbow lake crescent-shaped lake or pond formed on a river floodplain in an abandoned meander channel.

oxidation chemical union of oxygen with other elements to form new chemical compounds.

oxide a mineral group composed of oxygen combining with other Earth elements, especially metallics.

oxygen-isotope analysis a dating method used to reconstruct climate history; it is based on the varying evaporation rates of different oxygen isotopes and the changing ratio between the isotopes revealed in foraminifera fossils.

ozone gas with a molecule consisting of three atoms of oxygen, (O_3); forms a layer in the upper atmosphere that serves to screen out ultraviolet radiation harmful at Earth's surface.

Pacific high persistent cell of high atmospheric pressure located in the subtropics of the North Pacific Ocean.

pahoehoe a smooth, ropy surface on a lava flow.

paleogeography the study of past geographic environments, based on climatic and geologic evidence.

paleomagnetism the historic record of changes in Earth's magnetic field.

parabolic sand dune a hairpin-shaped sand dune, most common in coastal regions.

parallel circle on the globe connecting all points of equal latitude.

parallelism tendency of Earth's polar axis to remain parallel to itself at all positions in its orbit around the sun.

passive-margin coast coastal region that is far removed from the volcanism and tectonism associated with plate boundaries.

parent material residual (derived from bedrock directly beneath) or transported (by water, wind, or ice) mineral matter from which soil is formed.

patch a gap or area within a matrix (territory occupied by a dominant plant community) where the dominant vegetation is not supported due to natural causes.

paternoster lakes chain of lakes connected by a post-glacial stream occupying the trough of a glaciated mountain valley.

patterned ground (frost polygons) polygonal shapes formed on the surface in subarctic and tundra climates, formed by repeated freezing and thawing of soils.

ped soil aggregate or mass of individual mineral particles with a distinctive shape that characterizes a soil's structure.

pediment gently sloping bedrock surface, usually covered with fluvial gravels, located at the base of a stream-eroded mountain range in an arid region.

pediplain desert plain of pediments and alluvial fans; the presumed final erosion stage in an arid region.

peneplain theoretical plain of extreme old age; the last stage in a cycle of erosion, reached when a landmass has been reduced to near base level by stream erosion in a humid region.

perched water table a minor water table that exists above the regional water table.

perennial stream a stream with regular and adequate discharge to flow all year.

periglacial landscape The environment and landforms that developed or exists today near the margins of a glacier.

perihelion position of Earth at closest distance to sun during each Earth revolution.

permafrost permanently frozen layer of subsoil and underlying rock found in subarctic and polar climates where the season is too short for summer thaw to penetrate more than a few feet below ground level.

permeability characteristic of soil or bedrock that determines the ease with which water moves through Earth material.

pH scale scale from 0 to 14 that describes the acidity or alkalinity of a substance and which is based on a measurement of hydrogen ions; pH values below 7 indicate acidic conditions; pH values above 7 indicate alkaline conditions.

photosynthesis the process by which carbohydrates (sugars and starches) are manufactured in plant cells; requires carbon dioxide, water, light, and chlorophyll (the green color in plants).

physical geography specialization in the systematic study of geography that focuses on the location, distribution, and spatial interaction of physical (environmental) phenomena.

physical model three-dimensional representation of all or a portion of Earth's surface.

pictorial/graphic model representation of a portion of Earth's surface by means of maps, photographs, graphs, or diagrams.

piedmont alluvial plain a plain created by stream deposits at the base of an upland, such as a mountain, a hilly region, or a plateau.

piedmont glacier glacier that forms where two or more valley glaciers coalesce to cover lower lands at the base of a mountainous region.

pixel the smallest area that can be resolved in a digital image. Pixels, short for "picture element," are much like pieces in a mosaic, fitted together in a grid to make an image.

plane of the ecliptic plane of Earth's orbit about the sun and the apparent annual path of the sun along the stars.

planet any of the nine largest bodies revolving about the sun, or any similar bodies that may orbit other stars.

plankton passively drifting or weakly swimming marine organisms, including both phytoplankton (plants) and zooplankton (animals).

plant community variety of individual plants living in harmony with each other and the surrounding physical environment.

plastic solid any solid material that changes its shape under stress, and retains that deformed shape after the stress is relieved.

plateau an extensive, flat-topped landform or region characterized by relatively high elevation, but low relief.

plate convergence the collision of tectonic plates that are moving toward each other.

plate divergence the separation of tectonic plates as they move away from each other.

plate tectonics theory that superseded continental drift and is based on the idea that the lithosphere is composed of a number of segments or *plates* that move independently of one another, at varying speeds, over Earth's surface.

playa dry lake bed in a desert basin.

Pleistocene the name given to the most recent "ice age" or period of Earth history experiencing cycles of continental glaciation; it commenced approximately 2.4 million years ago.

plucking *see* quarrying.

plug dome a steep-sided, explosive type of volcano with its central vent or vents plugged by the rapid congealing of its highly acidic lava.

plunge pool a depression at the base of a waterfall formed by the impact of cascading water.

plutonic rock igneous rock formed by the cooling of magma deep within the Earth's crust.

pluvial rainy time period, usually pertaining to glacial periods when deserts were wetter than at present.

podzolization soil-forming process of humid climates with long cold winter seasons. Spodosols, the typical end product of the process, are characterized by the surface accumulation of raw humus, strong acidity, and the leaching or eluviation of soluble bases and iron and aluminum compounds.

point bar a deposit of sand or gravel on the inside of a bend in a stream or river.

polar referring to the North or South Polar regions

polar easterlies easterly surface winds that move out from the polar highs toward the subpolar lows.

polar front shifting boundary between cold polar air and warm subtropical air, located within the middle latitudes and strongly influenced by the polar jet stream.

polar highs high pressure systems located near the poles where air is settling and diverging.

polar jet stream high-velocity air current within the upper air westerlies.

pollution alteration of the physical, chemical, or biological balance of the environment that has adverse effects on the normal functioning of all life forms, including humans.

porosity characteristic of soil or bedrock that relates to the amount of pore space between individual peds or soil and rock particles and which determines the water storage capacity of Earth material.

positive feedback reaction to initial change in a system that reinforces the initial change and leads to imbalance in the system.

potential evapotranspiration hypothetical rate of evapotranspiration if at all times there is a more than adequate amount of soil water for growing plants.

pothole a bedrock depression in a streambed drilled by abrasive rocks swirling in a whirlpool.

prairie grassland regions of the middle latitudes. Tall-grass prairie varied from 2 to 10 feet in height and was native to areas of moderate rainfall; short-grass prairie of lesser height remains common in subhumid and semi-arid (steppe) environments.

precession cycle changes in the time (date) of the year that perihelion occurs; the date is determined on the basis of a major period 21,000 years in length and a secondary period 19,000 years in length.

precipitation water in liquid or solid form that falls from the atmosphere and reaches Earth's surface.

pressure belts zones of high or low pressure that tend to circle Earth parallel to the equator in a theoretical model of world atmospheric pressure.

pressure gradient rate of change of atmospheric pressure horizontally with distance, measured along a line perpendicular to the isobars on a map of pressure distribution.

prevailing wind direction from which the wind for a particular location blows during the greatest proportion of the time.

primary coastline coast that has primarily developed its current form and shape from land-based processes (fluvial or glacial, erosion or deposition, tectonism), and is only moderately modified by coastal processes.

prime meridian (Greenwich meridian) half of a great circle that connects the North and South Poles and marks zero degrees longitude. By international agreement the meridian passes through the Royal Observatory at Greenwich, England.

productivity rate at which new organic material is created at a particular trophic level. Primary productivity through photosynthesis by autotrophs is at the first trophic level; secondary productivity is by heterotrophs at subsequent trophic levels.

profile a graph of changes in height over a linear distance, such as a topographic profile.

pyroclastic material solid rock material (cinders, ash, and rock fragments) thrown into the air by a volcanic eruption.

quarrying (plucking) process whereby active glaciers break away and carry forward weathered and fractured bedrock.

radiation emission of waves that transmit energy through space. (*See also* short-wave radiation and long-wave radiation.)

radiation fog fog produced by cooling of air in contact with a cold ground surface.

rain falling droplets of liquid water.

rain shadow dry, leeward side of a mountain range, resulting from the adiabatic warming of descending air.

recessional moraine end moraine deposited behind the terminal moraine, marking pauses in the retreat of a valley glacier or ice sheet.

recumbent fold a fold in rock that has been completely overturned.

reg desert surface of gravel and pebbles with finer materials removed; common to large areas in the Sahara.

regional geography specialization in the systematic study of geography that focuses on the location, distribution, and spatial interaction of phenomena organized within arbitrary areas of Earth space designated as regions.

regolith weathered surface materials that usually cover bedrock.

relative humidity ratio between the amount of water vapor in air of a given temperature and the maximum amount of water vapor that the air could hold at that temperature, if saturated; usually expressed as a percentage.

relative location location of an object in respect to its position relative to some other object or feature.

relief a measurement or expression of the difference between the highest and lowest location in a specified area.

remote sensing mechanical collection of information about the environment from a distance, usually from aircraft or spacecraft, e.g., photography, radar, infrared.

remote sensing devices variety of techniques by which information about Earth can be gathered from great heights, typically from very high-flying aircraft or spacecraft.

representative fraction (RF) scale A map scale presented as a fraction or ratio between the size of a unit on the map to the size of the same unit on the ground, as in 1/24,000 or 1:24,000.

reverse fault a vertical fault with the hanging wall up and the footwall down, caused by compressional stress.

revolution (Earth) motion of Earth along a path, or orbit, around the sun. One complete revolution requires approximately 365¼ days and determines an Earth year.

rhumb line line of true compass bearing (heading).

ria coastline with many narrow bays mainly due to submerged river valleys.

ribbon falls high, narrow waterfall dropping from a hanging glacial valley.

rift valley major lowland that forms in a graben or down-faulted crustal block.

rime ice crystals formed along the windward side of tree branches, airplane wings, etc., under conditions of supercooling.

rip current strong, narrow surface current flowing away from shore. It is produced by the return flow of water piled up near shore by incoming waves.

roche moutonnée bedrock hill subjected to intense glacial abrasion on its upstream side, with some plucking evident on the downstream side.

rockfall nearly vertical drop of individual rocks or a small rock mass caused by the pull of gravity on steep slopes.

rock flour rock fragments finely ground between the base of a glacier and the underlying bedrock surface.

rockslide rapid downslope movement of huge masses of bedrock.

Rossby waves horizontal undulations in the flow of the upper air winds of the middle and upper latitudes.

rotation (Earth) turning of Earth on its polar axis; one complete rotation requires 24 hours and determines one Earth day.

runoff flow of water from the land surface, generally in the form of streams and rivers.

salinity the amount of dissolved solids in seawater.

salinization soil-forming process of low-lying areas in desert regions; the resulting soils are characterized by a high concentration of soluble salts as a result of the evaporation of surface water.

saltation the transportation by running water or wind of particles too large to be carried in suspension; the particles are bounced along on the surface or streambed by repeated lifting and deposition.

salt flat a low-relief deposit of saline minerals, typically in desert regions.

sand sediment particles ranging in size from about .05 mm to 2.0 mm.

sandstorm strong winds blowing sand along the ground surface.

Santa Ana very dry foehn wind occurring in southern California. (*See also* foehn wind.)

satellite any body which orbits a larger primary body, for example, the moon orbiting Earth.

saturation (saturated air) point at which sufficient cooling has occurred so that an air mass contains the maximum amount of water vapor it can hold. Further cooling produces condensation of excess water vapor.

savanna tropical vegetation consisting primarily of coarse grasses, often associated with scattered low-growing trees or patches of bare ground.

scale ratio between distance as measured on Earth and the same distance as measured on a map, globe, or other representation of Earth.

sclerophyllous vegetation type commonly associated with the Mediterranean climate; characterized by tough surfaces, deep roots, and thick, shiny leaves that resist moisture loss.

sea breeze air flow by day from the sea toward the land; caused by the movement of air toward a zone of lower pressure associated with higher daytime temperatures over the land.

seafloor spreading movement of oceanic crust in opposite directions away from the midocean ridges, associated with the formation of new crust at the ridges and subduction of old crust at ocean margins.

seamount submarine volcanic peak rising from the deep ocean floor.

sea stack mass of rock that forms a small near-shore island, isolated from the shoreline by wave erosion.

secondary coastline coast that has primarily developed its current form from sea-based processes (erosion or deposition by waves and currents, reef-building).

sedimentary rock one of the three major rock types; formed by the accumulation, compaction, and cementation of fragmented Earth materials, organic remains, or chemical precipitates.

seif a term for longitudinal dune used in North Africa and the Middle East.

seismograph scientific instrument utilized to read the passage of vibratory earthquake and shock waves.

selva characteristic tropical rainforest comprised of multistoried, broad-leaf evergreen trees with significant development of lianas and relatively little undergrowth.

sextant navigation instrument used to determine latitude by star and sun positions.

sheet wash water that runs off over a surface rather than flowing in a channel.

shield volcano gentle-sloped volcano formed by the cooling and accumulation of successive fluid lava flows extruded from a central vent or system of vents.

shoreline line of intersection between a water body and a landmass.

short-wave radiation radiation energy emitted by the sun in the form of waves of less than 4.0 micrometers (1 micrometer equals one ten-thousandth of a centimeter); includes X rays, gamma rays, ultraviolet rays, and visible light waves.

Siberian high intensively developed center of high atmospheric pressure located in northern central Asia in winter.

side-looking airborne radar (SLAR) a radar system that is used for making maps of terrain features.

silicate the largest mineral group, composed of oxygen and silica and forming most of the Earth's crust.

sill igneous intrusion that forms a horizontal rock mass after molten material has been forced between rock layers and subsequently cools.

silt sediment particles with a grain size between 0.002 mm and 0.05 mm.

sinkhole circular surface depression produced by the dissolving of limestone by groundwater.

slash-and-burn (shifting) cultivation also called swidden or shifting cultivation; typical subsistence agriculture of primitive societies in the tropical rainforest. Trees are cut, the smaller residue is burned, and crops are planted between the larger trees or stumps before rapid deterioration of the soil forces a move to a new area.

sleet form of precipitation produced when raindrops freeze as they fall through a layer of cold air; may also, locally, refer to a mixture of rain and snow.

slip face the steep, downwind side of a sand dune.

slope aspect direction a mountain slope faces in respect to the sun's rays.

slump mass of soil and regolith that slips or collapses downslope with a backward rotation.

small circle any circle that is not a full circumference of the globe. The plane of a small circle does not pass through the center of the globe.

smog combination of chemical pollutants and particulate matter in the lower atmosphere, typically over urban industrial areas.

snow precipitation in the form of ice crystals.

snow line elevation in mountain regions above which summer melting is insufficient to prevent the accumulation of permanent snow or ice.

soil a dynamic, natural layer on Earth's surface that is a complex mixture of inorganic minerals, organic materials, microorganisms, water, and air.

soil grade classification of soil texture by particle size: clay (less than 0.002 mm), silt (0.002–0.05 mm), and sand (0.05–2.0 mm) are soil grades.

soil horizon distinct soil layer characteristic of vertical zonation in soils; horizons are distinguished by their general appearances and their specific chemical and physical properties.

soil profile vertical cross section of a soil that displays the various horizons or soil layers that characterize it; used for classification.

soil survey a publication of the United States Soil Survey Division of the Natural Resources Conservation Service which includes maps showing the distribution of soil within a given area, usually a county.

soil taxonomy the classification and naming of soils.

soil texture the distribution of particle sizes in a soil that give it a distinctive "feel."

solar constant rate at which insolation is received just outside Earth's atmosphere on a surface at right angles to the incoming radiation.

solar energy *see* insolation.

solar noon the time of day when the sun angle is at a maximum above the horizon (zenith).

solar system the system of the sun and the planets, their satellites, comets, meteoroids, and other objects revolving around the sun.

solid tectonic processes those processes that distort the solid Earth crust by bending, folding, warping, or fracturing (faulting).

solar wind streams of hot ions (protons and electrons) traveling outward from the sun.

solifluction slow movement or flow of water-saturated soil and regolith downslope due to gravity; causes characteristic lobes on slopes in permafrost areas where only the top few feet of Earth material thaws in summer and drainage is poor.

solstice one of two times each year when the position of the noon sun is overhead at its farthest distance from the equator; this occurs when the sun is overhead at the Tropic of Cancer (about June 21) and the Tropic of Capricorn (about December 21).

solution dissolving in a fluid such as water, or the liquid containing dissolved material.

solution sinkhole a topographic depression formed mainly by the solution and removal of soluble rock, such as limestone.

sonar a system that uses sound waves for location and mapping underwater.

source region nearly homogeneous surface of land or ocean over which an air mass acquires its temperature and humidity characteristics.

southeast trades *see* trade winds.

Southern Oscillation the systematic variation in atmospheric pressure between the eastern and western Pacific Ocean.

South pole maximum south latitude (90° S), at the point marking the axis of rotation.

spatial distribution location and extent of an area or areas where a feature exists.

spatial interaction process whereby different phenomena are linked or interconnected, and, as a result, impact one another through Earth space.

spatial pattern arrangement of a feature as it is distributed through Earth space.

spatial science term used when defining geography as the science that examines phenomena as it is located, distributed, and interacting with other phenomena throughout Earth space.

specific humidity mass of water vapor present per unit mass of air, expressed as grams per kilogram of moist air.

speleology the science of studying caves.

speleothem a cave feature made from the deposition of minerals by groundwater.

spit beach feature attached to the mainland and built partially across a bay or inlet by the depositional action of longshore currents.

spring any surface outflow of groundwater, generally where the water table intersects the ground surface.

spring tide tide that occurs about the time of a new or full moon, the tide that rises highest and falls lowest from mean sea level

squall line narrow line of rapidly advancing storm clouds, strong winds, and heavy precipitation; usually develops in front of a fast-moving cold front.

stability condition of air when it is cooler than the surrounding atmosphere and resists the tendency to rise; the lapse rate of the surrounding atmosphere is less than that of *stable* air.

stalactite An icicle-like deposit of dripstone that hangs from a cave ceiling.

stalagmite a vertical deposit of dripstone that sticks up from a cave floor.

star dune a star-shaped sand dune formed by rising air currents.

stationary front frontal system between air masses of nearly equal strength; produces stagnation over one location for an extended period of time.

steppe climate characterized by middle-latitude semiarid vegetation, treeless and dominated by short bunch grasses.

stock individual deep-seated igneous mass of limited size; it is often associated with other igneous masses known generally as plutons.

storm local atmospheric disturbance often associated with rain, hail, snow, sleet, lightning, or strong winds.

storm surge rise in sea level due to wind and reduced air pressure during a hurricane or other severe storm.

storm track path frequently traveled by a cyclonic storm as it moves in a generally eastward direction from its point of origin.

strata (stratification) distinct layers or beds of sedimentary rock.

strato signifies a low-level cloud (i.e., from the surface to 2000 meters in elevation).

stratopause upper limit of stratosphere, separating it from the mesosphere.

stratosphere layer of atmosphere lying above the troposphere and below the mesosphere, characterized by fairly constant temperatures and ozone concentration.

stratus uniform layer of low sheetlike clouds, frequently grayish in appearance.

stratovolcano *see* composite cone.

stream a body of water that is flowing in a channel (a river is a stream, but generally larger); also a specific reference to a small river.

stream capacity how much sediment and dissolved load a stream can carry.

stream competence the largest particle size that a stream can carry.

stream gradient the drop of a streambed in a given distance, generally given in feet per mile or meters per kilometer.

stream hydrograph a graph of changes in stream flow over time.

stream terraces former banks of a stream or river that today are above the level of the stream.

stream load amount of material transported by a stream at a given instant; includes bed load, suspended load, and dissolved load.

striations gouges, grooves, and scratches produced in bedrock by rock fragments and boulders imbedded in a glacier.

strike the compass direction taken by a rock stratum or fault plane, which is at right angles to their dip.

strike-slip fault a fault with horizontal motion, where movement takes place along the strike of the fault.

structure the descriptive physical characteristics and arrangement of bedrock, such as folded, faulted, layered, fractured, massive.

subduction process associated with plate tectonic theory whereby an oceanic crustal plate is forced downward into the mantle beneath a lighter continental plate when the two converge.

submarine canyon steep-sided erosional valley cut into the continental shelf or continental slope.

subpolar lows east/west trending belts or cells of low atmospheric pressure located in the upper middle latitudes.

subsurface horizon buried soil layer that possesses specific characteristics essential to the identification of soils in the National Resources Conservation Service System. (Example of subsurface horizons may be found in Table 12.1.)

subsystem separate system operating within the boundaries of a larger Earth system.

subtropical highs cells of high atmospheric pressure centered over the eastern portions of the oceans in the vicinity of 30°N and 30°S latitude; source of the westerlies poleward and the trades equatorward.

subtropical jet stream high-velocity air current flowing above the sinking air of the subtropical high pressure cells; most prominent in the winter season.

succession progression of natural vegetation from one plant community to the next until a final stage of equilibrium has been reached with the natural environment.

surface of discontinuity three-dimensional surface with length, width, and height separating two different air masses; also referred to as a *front*.

sunspots temporary, dark, cool, spots on the solar surface. Their numbers follow an approximate 11-year cycle.

surge (glacial) sudden shift downslope of glacial ice, possibly caused by a reduction of basal friction with underlying bedrock.

suspended load solid particles, held in suspension, that are transported by flowing water.

swallow hole a depression where a river or stream disappears, flowing downward into the groundwater system.

swash onshore flow of water by waves.

swell regular longer-period sea wave traveling a significant distance from the area where it was generated by the wind.

syncline trough or downfold in a wave of crustal folding.

system group of interacting and interdependent units that together form an organized whole.

taiga term used to describe the northern coniferous forest of subarctic regions on the Eurasian landmass.

taku cold downslope wind in Alaska. (*See also* katabatic wind.)

talus (talus cone) rock debris in a cone-shaped deposit at the base of a steep slope or escarpment; usually a result of frost wedging and individual rockfalls with debris accumulating at the angle of rest.

tarn mountain lake in a glacial cirque.

tectonic processes processes that derive their energy from within Earth's interior and serve to create landforms by elevating, disrupting, and roughening Earth's surface.

temperature degree of heat or cold and its measurement.

temperature gradient rate of change of temperature with distance in any direction from a given point; refers to rate of change horizontally; a vertical temperature gradient is referred to as the *lapse rate*.

temperature inversion reverse of the normal pattern of vertical distribution of air temperature; in the case of inversion, temperature *increases* rather than decreases with increasing altitude.

tension stress resulting from divergent forces, pulling apart.

terminal moraine end moraine that marks the farthest advance of an alpine glacier or ice sheet.

terminus (snout) the lower end of a glacier.

terra rossa characteristic calcium-rich (developed over limestone bedrock) red-brown soils of the climatic regions surrounding the Mediterranean Sea.

terrestrial planets the four closest planets to the sun—Mercury, Venus, Earth, and Mars.

thematic map a map designed to present information or data about a specific theme, as in a population distribution map, a map of climate or vegetation.

thematic mapper (TM) a family of imaging systems that return images of Earth from Landsat satellites.

thermal infrared (TIR) scanning images made with scanning equipment that produces an image of heat differences.

thermocline vertical zone of ocean water where there is a sharp change in temperature with depth.

thermosphere highest layer of atmosphere extending from the mesopause to outer space.

Thornthwaite system climate classification based on moisture availability and of greatest use at the local level; climate types are distinguished by examining and comparing potential and actual evapotranspiration.

threshold condition within a system that causes dramatic and often irreversible change for long periods of time to all variables in the system.

thrust fault a reverse fault where compression has pushed the hanging wall on one side of the fault over the rocks on the other side.

thunder sound produced by the rapidly expanding, heated air along the channel of a lightning discharge.

thunderstorm intense convectional storm characterized by thunder and lightning, short in duration and often accompanied by heavy rain, hail, and strong winds.

tide periodic rise and fall of sea level in response to the gravitational interaction of the moon, sun, and Earth.

tidal current onshore or offshore flow of water that results from tidal fluctuation.

tidal range elevation distance between water levels at high tide and low tide.

till unsorted glacial drift, characterized by variation in size of deposit from clay particles to boulders.

till plain a broad area of low relief covered by glacial deposits.

tolerance ability of a species to survive under specific environmental conditions.

tombolo wave depositional beach feature connecting an island to the mainland.

topographic contour line line on a map connecting points that are the same elevation above mean sea level.

tornado small, intense, funnel-shaped cyclonic storm of very low pressure, violent updrafts, and converging winds of enormous velocity.

trace less than a measurable amount of rain or snow (i.e., less than 1 mm or 0.01 in.

trade winds consistent surface winds blowing in low latitudes from the subtropical highs toward the intertropical convergence zone; labeled northeast trades in the Northern Hemisphere and southeast trades in the Southern Hemisphere.

transform movement horizontal sliding of tectonic plates, alongside each other.

transpiration transfer of moisture from living plants to the atmosphere by the emission of water vapor, primarily from leaf pores.

transportation movement of Earth materials from one site to another as a result of the transporting power of water, wind, or glacial ice.

transverse dune a linear ridge-like sand dune that is oriented at right angles to the prevailing wind direction.

travertine calcium carbonate (limestone) deposits resulting from the evaporation in caves or caverns and near surface openings of groundwater saturated with lime.

tree line elevation in mountain regions above which cold temperatures and wind stress prohibit tree growth.

tributary a stream or river that flows into another stream or river.

trophic level number of feeding steps that a given organism is removed from the autotrophs (e.g., green plant—first level, herbivore—second level, carnivore—third level, etc.).

trophic structure organization of an ecosystem based on the feeding patterns of the organisms that comprise the ecosystem.

tropical region on Earth lying between the Tropic of Cancer (23½ degrees N latitude), and the Tropic of Capricorn (23½ degrees S latitude).

Tropic of Cancer parallel of latitude at 23½°N; the northern limit to the migration of the sun's vertical rays throughout the year.

Tropic of Capricorn parallel of latitude at 23½°S; the southern limit to the migration of the sun's vertical rays throughout the year.

tropopause boundary between the troposphere and stratosphere.

troposphere lowest layer of the atmosphere, exhibiting a steady decrease in temperature with increasing altitude and containing virtually all atmospheric dust and water vapor.

trough elongated area or "belt" of low atmospheric pressure; also glacial trough, a U-shaped valley carved by a glacier.

tsunami ocean wave produced by submarine earthquake, volcanic eruption, or landslide; not noticeable in deep ocean waters, but building to dangerous heights in shallow waters.

tundra climate characterized by treeless vegetation of polar regions and very high mountains, consisting of mosses, lichens, and low-growing shrubs and flowering plants.

turbidity current submarine flow of sediment-laden water.

typhoon a tropical cyclone found in the western Pacific, the same as a hurricane.

unconformity an interruption in the sequence of deposition mainly due to erosion.

uniformitarianism widely accepted theory that Earth's landforms have developed over exceedingly long periods of time as a result of processes that may be observed in the present landscape.

uplift mechanisms methods of lifting surface air aloft, they are: orographic, frontal, convergence (cyclonic), and convectional.

upper air westerlies system of westerly winds in the upper atmosphere, flowing in latitudes poleward of 20°.

upwelling upward movement of colder, nutrient-rich, subsurface ocean water, replacing surface water that is pushed away from shore by winds.

urban heat island *see* heat island.

U.S. Public Lands Survey System a method for locating and dividing land, used in much of the Midwest and western United States. This system divides land into six by six mile square *townships* consisting of thirty-six *sections* of land (each one square mile). Sections can also be subdivided into halves, quarter-sections, and quarter-quarter-sections.

valley breeze air flow upslope from the valleys toward the mountains during the day.

valley glacier *see* alpine glacier.

valley sink (uvala) a karst depression where a stream flows from the surface into the underground.

valley train outwash deposit from glacial meltwater, resembling an alluvial fan confined by valley walls.

variable one of a set of objects and/or characteristics of objects, which are interrelated in such a way that they function together as a system.

varve a pairing of organic-rich summer sediments and organic-poor winter sediments found in exposed lake beds; because each pair represents one year of time, counting varves is useful as a dating technique for recent Earth history.

veering wind shift the change in wind direction clockwise around the compass; e.g., east to southeast to south, to southwest, to west, and northwest.

ventifact wind-fashioned rock produced by wind abrasion (sandblasting).

verbal scale stating the scale of a map using words such as "one inch represents one mile."

vertical exaggeration a technique that stretches the height representation of terrain in order to emphasize topographic detail.

vertical rays sun's rays that strike Earth's surface at a 90 degree angle.

visualization a wide array of computer techniques used to vividly illustrate a place or concept, or the illustration produced by one of these techniques.

volcanic ash small fragments of lava (sand to dust sized) thrown into the atmosphere by a volcano.

volcanic neck the throat or conduit forming the passageway for molten rock in a volcano.

volcanism the upward movement of molten material (magma) and its cooling above Earth's surface.

v-shaped valley the typical shape of a stream valley where the gradient is steep.

warm front leading edge of a relatively warmer, less dense air mass advancing upon a cooler, denser air mass.

warping broad and general uplift or settling of Earth's crust with little or no local distortion.

wash (arroyo, wadi, barranca) generally steep-walled channel of an ephemeral stream in an arid region; the streambed is characteristically choked with coarse alluvium.

water budget relationship between evaporation, condensation, and storage of water within the Earth system.

water mining taking more groundwater out of an aquifer through pumping than is being replaced by natural processes in the same period of time.

water table upper limit of the zone of saturation below which all pore spaces are filled with water.

water vapor water in its gaseous form.

wave-cut bench gently sloping surface produced by wave erosion at the base of a sea cliff.

wave refraction bending of waves as they approach a shore, aligning themselves with the bottom contours of the surf zone.

weather atmospheric conditions, at a given time, in a specific location.

weathering physical (mechanical) fragmentation and chemical decomposition of rocks and minerals in Earth's crust.

westerlies surface winds flowing from the polar portions of the subtropical highs, carrying fronts, storms, and variable weather conditions from west to east through the middle latitudes.

wet adiabatic lapse rate rate at which a rising mass of air is cooled by expansion when condensation is taking place. The rate varies but averages 5°C/1000 m (3.2°F/1000 ft).

white frost a heavy coating of white crystalline frost.

wind air in motion from areas of higher pressure to areas of lower pressure; movement is generally horizontal, relative to the ground surface.

windward location on the side that faces toward the wind and is therefore exposed or unprotected; usually refers to mountain and island locations.

xerophytic vegetation type that has genetically evolved to withstand the extended periods of drought common to arid regions.

yazoo stream a stream tributary that flows parallel to the main stream for a considerable distance before joining it.

zone of ablation lower portion of a glacier, below the firn line, where melting, evaporation, and sublimation exceed net snow accumulation.

zone of accumulation subsoil or *B* horizon of a soil, characterized by deposition or illuviation of soil components by gravitational water; also the upper portion of a glacier, above the firn line, where net snow accumulation exceeds the melting, evaporation, and sublimation of snowfall.

zone of aeration upper groundwater zone above the water table where pore spaces may be alternately filled with air or water.

zone of depletion top layer, or *A* horizon, of a soil, characterized by the removal of soluble and insoluble soil components through leaching and eluviation by gravitational water.

zone of saturation zone immediately below the water table, where all pore spaces in soil and rock are filled with groundwater.

zone of transition an area of gradual change from one region to another.

World Map of Population Density